→	1B	2B	3A	4A	5A	6A	7A	0
							1 **H** 1.008	2 **He** 4.003
			5 **B** 10.81	6 **C** 12.01	7 **N** 14.01	8 **O** 16.00	9 **F** 19.00	10 **Ne** 20.18
			13 **Al** 26.98	14 **Si** 28.09	15 **P** 30.97	16 **S** 32.06	17 **Cl** 35.45	18 **Ar** 39.95
28 **Ni** 58.70	29 **Cu** 63.55	30 **Zn** 65.38	31 **Ga** 69.72	32 **Ge** 72.59	33 **As** 74.92	34 **Se** 78.96	35 **Br** 79.90	36 **Kr** 83.80
46 **Pd** 106.4	47 **Ag** 107.9	48 **Cd** 112.4	49 **In** 114.8	50 **Sn** 118.7	51 **Sb** 121.8	52 **Te** 127.6	53 **I** 126.9	54 **Xe** 131.3
78 **Pt** 195.1	79 **Au** 197.0	80 **Hg** 200.6	81 **Tl** 204.4	82 **Pb** 207.2	83 **Bi** 209.0	84 **Po** (209)	85 **At** (210)	86 **Rn** (222)

63 **Eu** 152.0	64 **Gd** 157.3	65 **Tb** 158.9	66 **Dy** 162.5	67 **Ho** 164.9	68 **Er** 167.3	69 **Tm** 168.9	70 **Yb** 173.0	71 **Lu** 175.0
95 **Am** (243)	96 **Cm** (247)	97 **Bk** (247)	98 **Cf** (251)	99 **Es** (252)	100 **Fm** (257)	101 **Md** (258)	102 **No** (259)	103 **Lr** (260)

§The International Union for Pure and Applied Chemistry has not adopted official names or symbols for these elements. All atomic weights have been rounded off to four significant figures.

INTRODUCTION TO Chemical Principles

FOURTH EDITION

Edward I. Peters

Department of Chemistry, West Valley College, Saratoga, California

SAUNDERS GOLDEN SUNBURST SERIES
SAUNDERS COLLEGE PUBLISHING
Philadelphia New York Chicago
San Francisco Montreal Toronto
London Sydney Tokyo Mexico City
Rio de Janeiro Madrid

Address orders to:
383 Madison Avenue
New York, NY 10017

Address editorial correspondence to:
West Washington Square
Philadelphia, PA 19105

Text Typeface: Times Roman
Compositor: General Graphic Services
Acquisitions Editor: John Vondeling
Project Editor: Carol Field
Copyeditor: Mark Hobbs
Art Director: Carol Bleistine
Text Design: Emily Harste
Cover Design: Lawrence R. Didona
Text Artwork: Larry Ward and Linda Maugeri
Production Manager: Tim Frelick
Assistant Production Manager: Jo Ann Melody

Cover credit: © Gary Gladstone/THE IMAGE BANK

Library of Congress Cataloging in Publication Data

Peters, Edward I.
Introduction to chemical principles.

(Saunders golden sunburst series)
Includes index.
1. Chemistry. I. Title.
QD33.P44 1985 540 85-8294
ISBN 0-03-002948-1

INTRODUCTION TO CHEMICAL PRINCIPLES ISBN 0-03-002948-1

Library of Congress catalog card number 85-8294.

456 032 98765432

CBS COLLEGE PUBLISHING
Saunders College Publishing
Holt, Rinehart and Winston
The Dryden Press

to my friend and colleague,
the late Dr. Peter P. Berlow,
Professor of Chemistry at Dawson College,
La Fontaine Campus, in Montreal, Quebec.

Preface

This book is addressed to the student who will take a one-semester or one-quarter introductory chemistry course before undertaking a college-level general chemistry program. The user of this book will learn to (a) read, write, and talk about chemistry, using a basic chemical vocabulary; (b) write routine chemical formulas, and write names when formulas are given; (c) write and balance ordinary chemical equations; (d) set up and solve elementary chemical problems, using dimensional analysis where applicable; and (e) "think" chemistry in some of the simpler theoretical areas, to visualize what is happening on the atomic or molecular level.

In planning this edition, the criterion for deciding what to add, what to delete, and what to change has been whether or not the revision would improve learning. Learning is, after all, what really counts. The following are among the more important learning enhancements in this edition:

- Most chapters open with a "Looking Back" section that identifies concepts or skills developed in earlier chapters that are required in the chapter about to be studied. Specific references enable the student to review the earlier material, if necessary.
- Performance goals (learning objectives) appear where and when they are most needed—at the beginning of the section where a topic is introduced, and again in the chapter summary, where they can be reviewed as a group when preparing for a test.
- "Quick Check" questions give the student immediate feedback on whether or not a concept has been grasped in studying a section.
- Semi-programmed problem examples invoke and develop a student's problem-solving skills by having him or her participate actively in solving the example, rather than just seeing how someone else solved it.
- Open space in semi-programmed examples is provided so that the student may solve the problem right in the book.
- Answered end-of-chapter problems include calculation setups.
- A "Terms and Concepts" list appears at the end of each chapter. It includes references to where each concept may be found within the chapter.
- A glossary at the end of the book is a dictionary-in-the-hand that is better than two-in-the-other-room (or library) for scientific terms.
- Readability has been carefully controlled. Vocabulary, sentence structure, and concept presentation have been held at a level that accommodates generally acknowledged problems in reading, while still crediting the readers with the maturity of young adults.
- An extensive review of basic mathematics appears in the appendix, including specific instructions on how to use a calculator to solve chemistry problems.
- Dimensional analysis is presented as a *thinking* process in which each step is a meaningful product of reasoning, not a random juggling of units until an acceptable answer is reached.
- Tear-out periodic tables are provided for simultaneous use as a reference and as a shield when solving programmed examples.
- Wherever possible, the student is encouraged to learn a small number of systems and procedures that can be applied broadly to both familiar and unfamiliar situations, rather than memorize a large number of isolated facts that too often appear to stand alone.

- Many tables and summaries are used throughout the text to encourage the student to pause, consider what has been presented, organize it in his or her thought, and get the "big picture."

End-of-chapter questions and problems are presented in two columns that are matched in the sense that problems that are next to each other involve the same concepts. Answers to most of the questions and all of the problems in the left column are given in the back of the book. Left-column answers that are absent are for questions that can be answered with direct quotes from the book. The numbers of those questions are printed in italics. All questions and problems in the right column are unanswered in the book, and are available for assignments. Complete answers appear in the teacher's guide.

Except for one or two problems on finding atomic mass from the abundance of isotopes, all end-of-chapter problems are new in this edition. A set of "Miscellaneous Questions" has been added to almost all chapters. In each case, the first question is a multiple-part concept drill of the "distinguish between" type, followed by another multiple-part true-false question. Other miscellaneous questions fall into all categories, including relevant and particularly challenging questions.

The total number of end-of-chapter questions has been increased substantially. If you count only the numbered questions, the increase is more than 20%. If the multiple-part true-false and "distinguish between" questions and quick checks are counted individually, the increase is above 60%.

As in the Third Edition, a conscious attempt has been made to furnish an instructor with the greatest possible flexibility in choosing an order of topics. If you wish to follow atomic structure (Chapters 4 and 5) with bonding, molecular structure, and a full treatment of nomenclature (Chapters 10 through 12), your students will never feel they missed something by skipping Chapters 6 through 9. Chapter 6, "Introduction to Chemical Formulas," is used in formula calculations, equations, and stoichiometry (Chapters 7 through 9), but everything in those chapters has been written so that any formula that is needed is readily available without the student's having studied Chapter 6. Intermolecular attractions, discussed in Chapter 14, are referred to in subsequent chapters, but never in a way that understanding depends on a comprehension of those attractions. These are among the more obvious options; provisions have been made for others as well.

This book was not written with the Keller Plan in mind, but for many years it has been used successfully in that format at West Valley College. A STUDENT GUIDE FOR INTRODUCTION TO CHEMICAL PRINCIPLES, written by Dr. Robert Kowerski of the College of San Mateo and myself, has all the elements of a PSI study guide without the Keller Plan jargon. It has proved to be equally useful in the conventional course. If the text and the study guide enable the PSI student to learn independently, they surely do the same for the student who attends formal lectures and takes typical midterm exams.

Many of the ideas for changes that appear in this edition of INTRODUCTION TO CHEMICAL PRINCIPLES came from reviewers and users of the prior edition. Valuable detailed reviews came from Professors Fred Redmore of Highland Community College, Freeport, Illinois, and Edward Genser, California State University, Hayward, California. Especially helpful user suggestions were received from Judith LoFaro, St. Louis Community College–Forest Park, St. Louis, Missouri; Doris Mourad, San Jose State University, San Jose, California; Barbara Perez, El Camino College, Torrance, California; and James R. Peter, Cerritos College, Norwalk, California. Professors William T. Scroggins of El Camino College, Torrance, California, Stan Ashbaugh of Orange Coast College, Costa Mesa, California, and Robert Kowerski of College of San Mateo, San Mateo, California, reviewed the manuscript and made recommendations ranging from the approach to major topic areas down to words and phrases that would be more easily understood. I am most thankful to all of these people—and to the students who are the real users of the book, and who offer very special insights.

Edward I. Peters

Contents Overview

Table of Contents

About Bubbles in a Fish Tank—A Prologue on the Methods of a Scientist

Have you ever noticed that bubbles rising from the bottom of a fish tank become larger as they approach the surface? At least I think they do. I must confess that I've never consciously observed and measured bubbles in a fish tank, but I am quite certain that they grow larger as they rise. To explain why I am so sure, let's consider a story—a bit of fiction that tells us something about chemistry and, for that matter, science in general.

Once there was a man who personally *observed* that bubbles increase in size as they rise from the bottom of a fish tank. Being a Curious Man, he *wondered why*. His *curiosity* drove him to find out. So he *thought* about it. Finally, he developed a *hypothesis*. He guessed that the volume of the bubble was smaller at the bottom because there was more water on top of the bubble to "push" it into a smaller size. As the bubble rises there is less pressure, or "push," over its surface, so it gets bigger.

Having figured out a possible explanation for his observation, the Curious Man wanted to check it. Being a practical man, as well as curious, he built a bubble. His bubble had the form of a sealed cylinder (Fig. P.1). It had a snug-fitting piston that was both air- and watertight, but was free to move up and down so the pressure of the air inside would equal the pressure on the outside. The volume of the bubble was therefore governed by the pressure on top of the piston.

Gleefully our Curious Man took his bubble to the nearest lake, where he *experimented* with it. He lowered it into the water to different depths, measuring the volume at each depth. His efforts were rewarded. He found that, like real bubbles, his artificial bubble became larger as it approached the surface. His

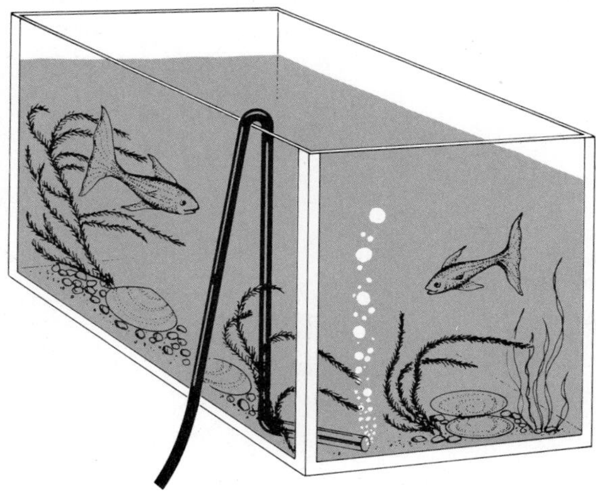

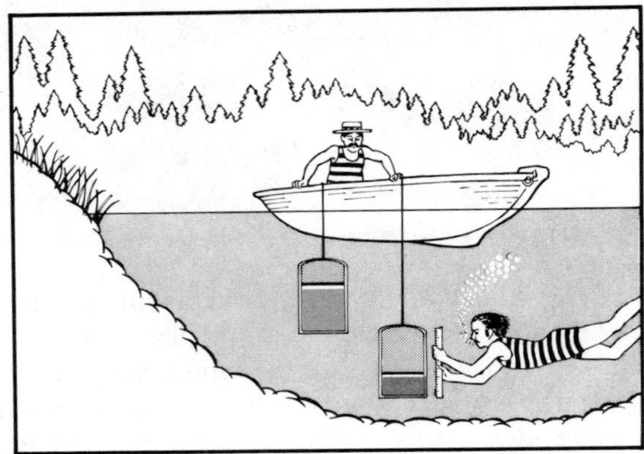

Figure P.1
The "bubble" experiment. The size of the bubble (colored) depends on the depth of the water in which it is submerged.

hypothesis was correct: the volume of a bubble does depend on the pressure on the outside. For that matter, he could say the pressure exerted by the gas inside the bubble was related to the volume, inasmuch as the inside and outside pressures were equal.

Excitedly, our Curious Man *communicated* his findings to anyone who would listen. Not many did; not many men are curious about bubbles. But one was. He was also *skeptical*. He didn't believe the reported results. Therefore he *repeated the experiments*. Curious Man No. 2 found that the experiments of Curious Man No. 1 were correct and gave *reproducible results*. Being an *intellectually honest* person, Curious Man No. 2 freely admitted his error in doubting Curious Man No. 1. The next time they gathered with their Curious Friends—curious in the same sense we have been using the term so far!—Curious Man No. 2 reported to all that he had checked the results himself.

One of the Curious Friends, Curious Woman, was not entirely satisfied. She thought there should be more data. She suggested the *hypothesis* that if you *measured* the pressure of the gas and its volume, you would find a quantitative relationship between them—a relationship between the amount of pressure and the amount of volume. She designed an experiment to test her hypothesis. Her experiment didn't require her to go down to the lake and get all wet, incidentally. She reasoned that you could measure the pressure more easily simply by putting weights on top of the piston (Fig. P.2).

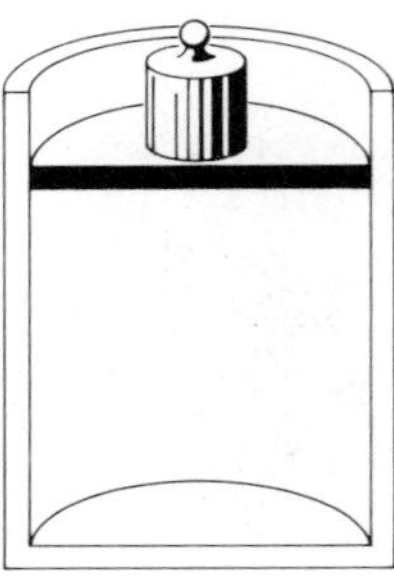

Figure P.2
The relationship between pressure and volume of a confined gas. At low pressure the volume is larger.

One morning Curious Woman conducted her experiment and *drew a graph* of her results (Fig. P.3). She found that if you multiply gas pressure by gas volume, you always get the same answer. Wisely she *checked her results* before telling them to her Curious Friends. She checked them many times, in fact, until she was quite sure of her findings.

At the next gathering of the Curious Friends, Curious Woman reported that the product of pressure times volume of a gas is a constant. Other Curious Persons picked up the idea and tried it in their laboratories. They got the same results, but only if the temperature and amount of gas were the same as when Curious Woman did her experiment. They found new relationships at different temperatures and amounts. Finally, one Curious Person put them all together and proposed an ''explanation'' for these *experimental facts*. His *theory* pictured a gas as made up of many tiny particles moving wildly about, causing all the experimental results recorded by the other Curious Persons.

It's been over a hundred years since Curious Woman first proposed that pressure × volume is constant (at fixed quantity and temperature, of course). Recently, the Society of Curious Persons has honored Curious Woman's relationship by elevating it to the status of a *law* of Science. It is now called Curious Woman's Law.

Now you know why I am so sure that bubbles become larger as they rise in a fish tank, even though I have not personally made this observation. It's the law. But even here, I must be cautious. Scientific laws are rarely found to be in error, but it has happened. To be absolutely certain, I'm going to conduct my own experiment. Where can I find a fish tank?

In this little story we have tried to provide a small glimpse of the character of chemistry. The observing, hypothesizing, experimenting, testing and retesting, theorizing, and finally, reaching conclusions have been going on for centuries. And they continue today more actively than ever before. Collectively they are often called the *scientific method*.

There is really no rigid order to the scientific method. Looking back over the history of science, though, the above features always seem to be there. They are the outcome of the day-to-day thinking of the scientifically curious

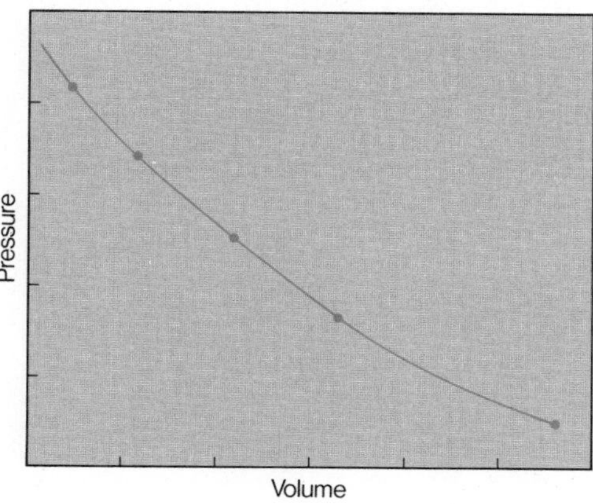

Figure P.3
Graph of pressure vs. volume of a gas.

person as he continually asks himself, "What do I already know that can be applied here? What is the next logical step I can take?" Out of such questions come new hypotheses, new experiments, new theories. Ultimately new laws are discovered.

Our story also lists in italicized words some of the qualities and actions of the scientist. He surely is an *observer,* and he is *curious* about and *thinks* about what he sees. He *develops a hypothesis* as a tentative explanation of his observations. He *conducts experiments* to test the hypothesis. If he finds something new, he *communicates* with his fellow scientists, usually through scientific journals. He combines the qualities of *skepticism* with *intellectual honesty,* both of which leave him free to receive and evaluate new information that reaches him through many sources.

Our story furnishes one more important insight into the nature of chemistry. Notice that the first two curious men concerned themselves with *What* was taking place and *How*. Answers to these questions are considered to be the *qualitative* part of chemistry. It was not until the Curious Woman entered the picture that measurements appeared. She recognized that *What* and *How* furnish only some of the answers, but they cannot be used to make reliable predictions about the extent of chemical activity. She added the vital question, *How much*? We see, then, that the study of chemistry is both qualitative and *quantitative*. In this book we will consider both of these essential areas.

While the characters and events in this story are obviously fictitious, the law around which it was built is very real. It is the product of one person, not three, and was first proposed in the 17th century by Robert Boyle (1627–1691). Boyle was one of the first scientists to devote himself to orderly and careful experimental investigation. His book, *The Sceptical Chymist* (1661), is sometimes regarded as the beginning of modern chemistry. You will study Boyle's law in Chapter 13 of this text.

1 Learning Chemistry

The next few pages may be among the most important pages you will read in this book. Nothing in them will be on a test question. They will teach you no chemistry. You will receive no grade for having read them. But they will have a lot to do with the chemistry you learn, your performance on tests, and the grades you will receive. Therefore, you are urged to read these pages carefully and consider what they suggest about the way you learn chemistry.

You and *learn* are the two most important words in the above paragraph. Your teacher may teach, but only you can learn. That responsibility belongs to you alone. This is why the first chapter is about learning chemistry. If you learn *how to learn* chemistry, the rest is much easier.

No matter what you may have heard, it is not hard to learn chemistry at the beginning level. Time-consuming, perhaps, but not hard. If you *decide* to succeed in chemistry, and do what must be done, you will succeed.

What must be done? Three things. First, you must spend the time required. Learning chemistry takes time, often more time than most other subjects. Furthermore, the time must be spent regularly, doing each assignment each day. You cannot learn chemistry all at once, just before a test. It must be learned a bit at a time, because what you learn in today's lecture depends on what you learned doing yesterday's assignment. If you haven't finished yesterday's assignment, today's lecture will make no sense. You must keep up; falling behind is the biggest problem of all when it comes to learning chemistry.

The second thing you must do to succeed in chemistry is concentrate. This means studying without distractions—without sounds, sights, people, or thoughts that take your attention away from chemistry. Every minute your mind wanders while you are studying must be added to your total study time, making that time one minute longer. If you have a limited amount of time for study, that minute is lost forever. What you learn has been reduced, and your grade will suffer from that lost learning.

The third requirement for success is that you follow the study procedures described in the rest of this chapter. They have been tried and tested. They work. You may later develop procedures that are better for you, but it is unlikely that you will begin with better procedures. It is best to start with something good and then to improve it as you go along.

Study Tools

This textbook is one of several tools you will use while learning chemistry. It must be supported by other tools. You do not just read when you study chemistry. You do things. You answer questions, not only in your mind but also by writing down the answers. This means you need a pencil. Some paper helps, too, although there is enough open space in this book for you to make notes, to answer questions, and to solve problems. There are many calculation problems in chemistry. To solve them in any reasonable amount of time you will need a calculator. In Appendix I, Part B, you will find a list of the mathematical operations your calculator must be able to perform to solve the problems in this book.

You simply cannot honestly claim to be studying chemistry without these tools. If you try, you will spend more time studying while you learn less.

Performance Goals

PG 1A Read the performance goal or goals at the beginning of each section; study the section; reread the performance goals; satisfy yourself that the goals have been reached; if so, go on; if not, go back—restudy the text until the goal has been met.

As you approach most sections in this text, you will encounter one or more "performance goals," as you did here. In many cases the performance goals will contain language that is unfamiliar but about to be introduced in the section. They give you some idea of what to expect and what to look for as you study the section. They tell you exactly what you should learn. You can then direct your attention to learning those things. This is the key to efficient study.

Notice that the performance goal above is written as an action. It describes *doing* something—performing some act. Each performance goal in this book should be thought of as the close of a sentence that begins, "After studying this section, you will be able to. . . ." The goal describes something you should learn to do in your study. You should find out whether or not it has been learned immediately after completing the section. Return to the performance goals, asking yourself the question, "Am I now able to do what is expected?" Demand of yourself that the honest answer to this question be "Yes!" If it is not, go over the material again until you can do what is expected.

Performance goals are designated in the text by PG, followed by a number and a letter. The number is the chapter number, and the letter identifies the performance goal within the chapter.

All performance goals in each chapter are repeated as a group at the end of the chapter. They serve as a chapter summary, and they are very helpful in reviewing for a test. The section number in which the performance goal first appears is also given for a handy reference.

The performance goals in the book are those the author has in mind for each section. They may or may not correspond to those of your instructor. If not, you should be alert to changes he or she may make, and adjust your study plan accordingly.

Looking Back

Learning chemistry is cumulative. You first learn some basic principle or skill, and later use that skill or principle to learn another that is more advanced. Sometimes there is a time gap between the first learning and its later application—a gap in which what was learned earlier might be forgotten. To help you bridge this gap, many chapters in this book begin with a section called "Looking Back."

A Looking Back section is a list of concepts and skills that were developed in earlier chapters and that will be used again in the present chapter. It tells you what you are expected to know and be able to do before you begin. It includes the section numbers where the topics were introduced. If you feel uncertain about any item, you can refer quickly to its earlier presentation and refresh your memory.

Quick Checks and Examples

Looking Back reminds you of skills you are presumed to have as you approach each section. Performance Goals tell you what you will learn in studying that section. The section develops the new material. The final step in the learning process—the clincher, if you wish—is an immediate check to determine whether or not you have caught on to the new ideas. This textbook does this in two ways.

Quick Check Questions. Chemical principles and theories are presented with words and illustrations. Hopefully you learn and understand these ideas as you study the text and figures. A "Quick Check" appears at the end of text sections that deal with concepts. The quick check is a few questions you will be able to answer instantly if you have caught the main ideas of the section. True–false questions are common. So are "Classify (something) as A or B" questions. Rarely should you have to reread the text to find an answer; but if you do have to read something again, *that instant is precisely the best possible moment to do it*. Once you have decided on the answers to the quick check questions, compare your answers with those that are given in the "Quick Check Answers" section at the back of the book.

Examples. "Skills" chemistry includes the things a student must be able to do as a result of his or her study. This includes writing chemical formulas and equations and solving problems. These skills are developed and checked by means of examples.

Nearly all examples are presented in a self-teaching style, a series of questions and answers that guide you to an understanding of a concept. You will reach that understanding only if you "play the game by the rules." The rules call for you to answer each question *before* you look at the textbook answer you know is printed less than an inch farther down the page. This way you force yourself to reason through the question and gain an understanding of the principle it illustrates.

Most examples include suggestions on how to begin. These are followed by a question that leads you to the first step in solving the problem. Space is provided for you to calculate and write your answer to the question. At that point short broken lines appear on each side of the page:

- - - - - - - -

Right below these lines is the answer to the question, including the mathematical setup if one is required. An explanation of the answer also may appear. More suggestions lead to a second question, space for your answer, and another set of broken lines. The pattern repeats until the example is completed.

To learn by these examples, you will have to cover the printed answers. Elsewhere in this book you will find shields that can easily be torn out of the book. One side of each shield is a periodic table from which you will obtain information needed for solving problems. The other side lists the steps you should follow when working on examples. These steps are:

1. When you come to an example, use the shield to cover the part of the page below the first set of dashed lines in the example.
2. Read the example question. Write in the open space above the shield, or on separate paper, any answers or calculations needed.
3. Move the shield down to the next broken lines.
4. Compare your answer with the one you can now read in the book. Be sure you understand the example up to that point before going on.
5. Repeat the procedure until you finish the example.

End-of-Chapter Questions and Problems

At the end of most chapters you will find a two-column set of questions and/or problems. They are matched side-by-side; they involve similar reasoning and, in the case of problems, similar calculations. Some questions are easy, requiring only the basic information presented in the chapter. Others are more demanding and require that you analyze a situation, apply chemical principles, and then explain or predict some event or calculate some result. The more difficult problems are identified by an asterisk (*).

Answers appear in the back of the book for most questions and all problems in the left-hand columns. Problem answers include calculation setups. The left-column answers that are not given are those that are direct quotations from the text. Numbers of left-column questions that are not answered are printed in italics. All questions in the right-hand column are without answers; they may be used for assignments by your instructor.

If you have difficulty working any problem from either column, you will have reached a critical point. You will be tempted to return to the examples in the chapter, find one that matches your problem, and then solve the assigned problem step-by-step as in the example. Resist that temptation! If you get stuck on a problem, it means you did not understand the corresponding example in the text. Leave the problem. Turn back to the example. Study it again, by itself, until you understand it thoroughly. Then return to the assigned problem with a fresh start and work it to the end without further reference to the example.

You should approach your problem work in chemistry with a clear sense of purpose. This purpose is not simply to get the right answer. It is, rather, to learn how to solve the problem and others like it. This is why you should not copy examples from the book. Copying examples does not produce learning. Demand learning of yourself before you consider any problem completed. More than anything else you do, this will assure your success in solving chemistry problems.

Terms and Concepts

As with any specialized field of study, chemistry has its own language. One of the main purposes of a first course in chemistry is to become familiar with this language so you will understand and use it properly.

A large part of any language is its vocabulary. At the end of each chapter you will find a list of the scientific terms introduced or used in the chapter. They are grouped according to the section in which they appeared so you may find them quickly. The presence of a word in this list does not mean its definition must be memorized. It means you should understand the term so that it will make sense to you if you read or hear it and so that you can use it properly in your own speaking and writing. Any definitions that must be memorized or explained are identified in the performance goals.

Glossary

You will no doubt run across technical words in this book whose meanings are not quite clear to you. A word may have been introduced earlier, and you need to refresh yourself on it, or it may be some other scientific word that you have not seen before. The Glossary is provided so you can look it up easily. The Glossary begins on page 551. You should think of the Glossary as a dictionary in your hand and *use it regularly*.

A Choice

In this chapter you have read about the procedures by which you will learn chemistry. You now have a choice to make. For the study of chemistry you may choose to schedule the time it requires on a regular basis, to establish good study habits, and to use the text as it should be used. If you make this choice, you will learn chemistry. You won't find it particularly difficult. If, on the other hand, you choose to study chemistry only now and then, mainly just before a test, if you do not solve problems right after they are assigned, if you "study" in the middle of distractions, if you take short cuts, or if you get behind, you will make chemistry harder than it really is.

If you ever begin to feel that chemistry is a difficult subject, remember these lines. Read this chapter again, and see if your study habits are at fault. Perhaps improving those habits can solve your problem.

At all stages of our lives we make choices. We then live with the consequences of those choices. Choose wisely—and enjoy chemistry.

2 Matter and Energy

In a broad sense, chemistry is the study of matter and the energy associated with chemical change. It is appropriate, then, that we begin the study of chemistry by examining these well-known but not widely understood parts of the physical universe.

2.1 PHYSICAL AND CHEMICAL PROPERTIES AND CHANGES

PG 2A Distinguish between physical and chemical properties.

2B Distinguish between physical and chemical changes.

Matter has mass. Matter also occupies space. These two characteristics combine to define matter. In order to examine matter, to understand what it does, or appears to do, we must be able to describe it clearly. Normally a sample of matter is described by listing its **physical properties.** Certain physical properties can be detected directly by our senses. They tell us how a material *looks* (the blackness of charcoal compared with the yellow of sulfur), *feels* (the hardness of glass compared with the softness of putty), *smells* (the odor of sour milk or the scent of a rose), or *tastes* (salt vs. sugar). Other physical properties can be measured in the laboratory. Among them are the temperatures at which materials boil or melt, called the *boiling* and *melting points,* or the *density,* or relative "heaviness," of the material.

Changes that alter the physical form of matter *without changing its chemical identity* are called **physical changes.** The melting of ice is a physical change. The substance is water both before and after the change. Dissolving sugar in water is another example of a physical change. The form of the sugar changes, but it is still sugar. The dissolved sugar may be recovered by simply evaporating the water, another physical change.

The chemist is interested in more than the physical properties of matter. He or she wants to know what sort of chemical reactions it can have. Paper burns. Iron rusts. Milk becomes sour. Eggs become rotten. Each of these is a **chemical change.** A chemical change can be recognized when one or more substances are chemically destroyed, and one or more new substances are formed. The **chemical properties** of a substance are simply a list of the chemical changes possible for that substance.

Chemical and physical properties lead to several ways in which matter may be classified, some of which will be considered now.

Quick Check 2.1: Are the following statements true or false?

1. Baking bread is a chemical change.
2. The flammability of gasoline is a physical property.
3. Ethyl alcohol boils at 78°C. This is a chemical property.
4. Grinding sugar into a powder is a physical change.

2.2 STATES OF MATTER: GASES, LIQUIDS, SOLIDS

PG 2C Identify and explain the differences between gases, liquids, and solids in terms of (a) visible properties and (b) particle movement.

The air we breathe, the water we drink, and the food we eat are examples of the gaseous, liquid, and solid **states of matter.** Water is the only substance we normally meet in all three states, as suggested in Figure 2.1. The differences among gases, liquids, and solids can be explained in terms of what is called the **kinetic molecular theory.** According to this theory, all matter consists of extremely tiny bits or particles that are in constant motion. (*Kinetic* refers to motion.) The "amount" of motion, or the speed at which the particles move, is related to temperature. At high temperatures the movement is fast, and at low temperatures it is relatively slow.

Because particles of matter are constantly moving, they tend to separate, to "fly apart" from each other. But particles of matter are also attracted to each other. This causes them to "stick together" if the attractions are strong enough to overcome the "fly apart" tendencies of their movement. At low temperatures, when the particles are moving slowly, they not only remain

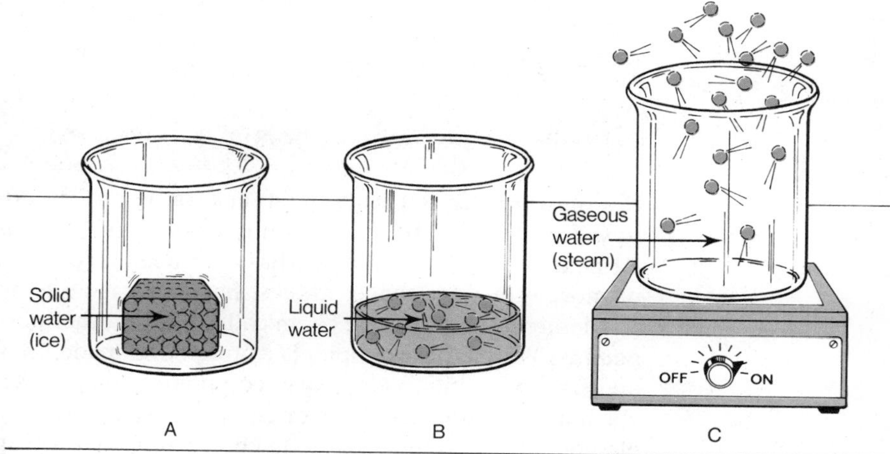

Figure 2.1
The three states of matter, illustrated by water. *A.* Solid water (ice) has definite volume and shape. Particles vibrate in place. *B.* Liquid water has definite volume, but shape of bottom of container. Particles move freely within that volume. *C.* Gaseous water (steam) has volume and shape of container if closed, escaping to atmosphere if open. Particle movement is completely independent.

together, but they also arrange themselves in fixed positions relative to each other. Movement is limited to vibrating, or shaking, in place. This is the nature of the **solid** state of matter. A solid has a definite volume, and the fixed positions of its particles give it a definite shape as well.

Particle movement becomes more vigorous when the temperature of a solid is increased. The particles are able to break the attractions that hold them in fixed positions, but they may not yet be moving fast enough to escape from each other. So they stay together but tumble and slip about among each other almost at will. They continue to occupy a definite volume, but now they settle to the bottom of the container that holds them, assuming the shape of that container. These are the features of the **liquid** state.

At still higher temperatures the particle movement is enough to overcome the attractive forces entirely. The particles are widely separated from each other, which causes the attractions between them to almost disappear. The particles move independently, filling the vessel that holds them. The sample acquires both the shape and volume of its container. These are the features of the **gaseous** state of matter; matter in this state is called a **gas.**

In Summary:

a. A solid has definite shape and volume. Particle movement consists of vibration in place within a rigid structure.
b. A liquid has definite volume, but takes the shape of its container up to that volume. Particles are free to move among themselves so long as they remain together.
c. A gas fills its container, taking both its shape and volume. Movement of particles is completely random.

Quick Check 2.2: Are the following statements true or false?

1. Particles move more freely in a gas than in a solid.
2. The volume of a liquid may change, but its shape cannot.

2.3 ELEMENTS AND COMPOUNDS

PG 2D Distinguish between elements and compounds.

If a pure substance (Section 2.4) cannot be broken down chemically—*decomposed* is the scientific term—into other pure substances, it is said to be an **element.** The fact that nobody has ever been able to decompose sulfur into two or more other substances shows that sulfur is an element.

Elements can combine chemically to form other pure substances called **compounds.** Water, for example, is a compound made up of the elements hydrogen and oxygen. Unlike elements, compounds can be separated into other pure substances, either elements or compounds. Water may be decomposed into its elements by passing an electric current through it (Fig. 2.2).

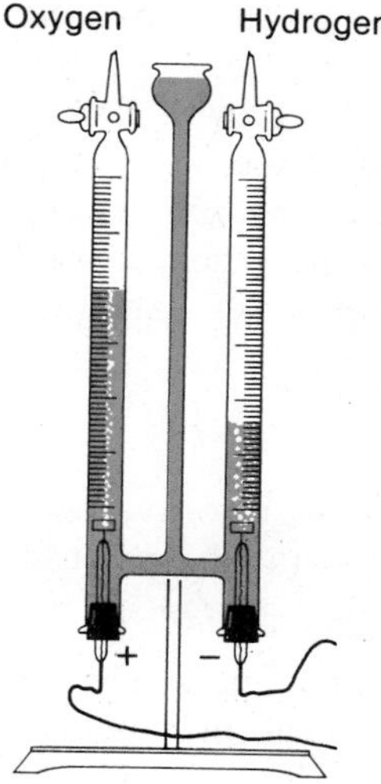

Figure 2.2
Decomposition of water into hydrogen and oxygen by means of an electric current.

Nature provides us with 88 elements. Most of our environment is made up of compounds containing a relatively small number of these elements (Table 2.1). It is natural that elements such as oxygen, silicon, and aluminum, which are so abundant in nature, are familiar to us. Some elements that make up only a very small percentage of the crust of the earth are readily recognized and have become vital to man. For example, such well-known elements as chromium, copper, nickel, tin, silver, and gold are absent from Table 2.1. Other elements that occur in lower percentages but are essential to modern technology include cobalt, vanadium, cadmium, and molybdenum.

Only a few elements occur uncombined in nature. Most of these are gases (nitrogen, oxygen, argon) that make up over 99% of the earth's atmosphere. Gold, silver, sulfur, and copper are among the few solid elements that exist in nature in elemental form. To obtain pure samples of nearly all other solid elements, it is necessary to obtain them from naturally occurring compounds or ores.

Most elements are solids at normal temperatures and pressures. At 25°C and one atmosphere, only two are liquids (mercury and bromine), while 11 elements, including hydrogen, fluorine, chlorine, oxygen, and nitrogen, are gases. The elements may also be divided into metals (iron, copper, magnesium, aluminum) and nonmetals (carbon, iodine). A few (boron, silicon) are often referred to as "metalloids" because they have properties between those of metals and nonmetals.

The properties of compounds are different from the properties of the elements from which they are formed. To illustrate, powdered iron is black in appearance and powdered sulfur is yellow. A mixture of these two elements, which appears gray, is readily separated into the elements. The iron in the mixture may be removed with a magnet. Sulfur can be taken from the mixture by shaking it in a liquid called carbon disulfide. The sulfur dissolves; the iron does not. If the iron–sulfur mixture is heated strongly, a chemical reaction occurs to produce a compound, iron sulfide. The compound is not attracted by a magnet, nor does it dissolve in carbon disulfide. In other words, iron sulfide has none of the properties of iron or sulfur.

Sodium chloride is another example. Neither sodium, a metal that reacts with both water and oxygen, nor chlorine, a poisonous gas with a suffocating

Table 2.1 Composition of Earth's Crust*

Element	Percent by Weight	Element	Percent by Weight
Oxygen	49.2	Sodium	2.6
Silicon	25.7	Potassium	2.4
Aluminum	7.5	Magnesium	1.9
Iron	4.7	Hydrogen	0.9
Calcium	3.4	All others	1.7

*The earth's "crust" includes the atmosphere and surface waters.

odor, is pleasant to work with. Yet the compound formed by these two elements, sodium chloride (table salt), is used to flavor food.

An important fact about compounds is summarized in the **Law of Definite Composition.** This law states that the percentage of weight of the elements in a compound is always the same, regardless of the source or method of preparation of the compound. Water, for example, whether it comes from a pond, river, or lake, from Europe, America, or Asia, or is the product of a chemical reaction, always contains 11.1% hydrogen and 88.9% oxygen.

2.4 PURE SUBSTANCES AND MIXTURES

PG 2E Distinguish between a pure substance and a mixture.

A **pure substance** is a single chemical, one kind of matter. It may be an element or a compound. A pure substance has its own set of physical and chemical properties, not exactly the same as any other pure substance. These properties may be used to identify the substance.

A **mixture** is a sample of matter that contains two or more pure substances, either elements or compounds. The properties of a mixture are determined by the substances in it, and they vary as the percentages of the different parts change.

Pure water and salt water offer a good example of the difference between the properties of pure substances and mixtures. At normal pressure, pure water boils at 100°C. Salt water boils at a higher temperature; how much higher depends on the amount of salt in the mixture. The salt and water can be separated by a process called **distillation** (Fig. 2.3). When the mixture is heated, the water boils off, is cooled and condensed (changed back to a liquid), and collected as pure water. The boiling temperature of the remaining salt water gradually increases because removal of water from the mixture increases the percentage of salt (Fig. 2.4). If boiling is continued long enough, the salt and water will be separated completely. This is another property of mixtures: they may be separated into pure components by physical changes.

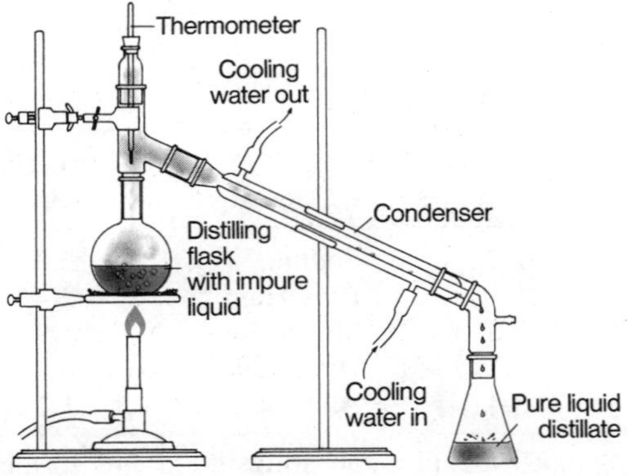

Figure 2.3
Laboratory distillation apparatus.

Quick Check 2.4: Are the following statements true or false?

1. Stannous fluoride, a compound formed by the elements tin and fluorine, is a mixture.
2. The compound silver chloride has a definite, fixed melting point.

2.5 HOMOGENEOUS AND HETEROGENEOUS MATTER

PG 2F Distinguish between homogeneous and heterogeneous matter.

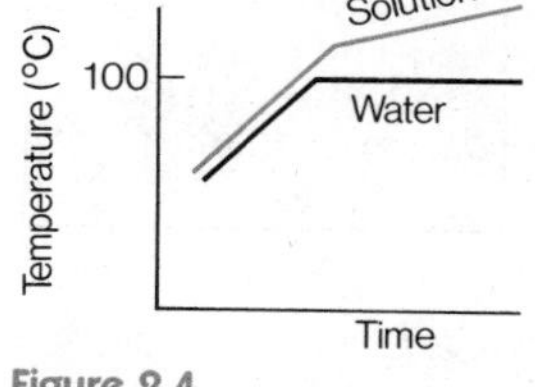

Figure 2.4
Comparison of boiling temperatures of a pure substance and an impure substance (solution). As water boils off, the solution that remains becomes more concentrated. This change in composition causes the boiling temperature to increase.

The prefixes *homo-* for ''same'' and *hetero-* for ''different'' are used with many words in the English language. **Homogeneous** matter has the same appearance, composition or make-up, and physical and chemical properties throughout. A small portion taken from any place in a larger sample of homogeneous matter is identical to a small portion taken from any other place in the larger sample. **Heterogeneous** matter is made up of different parts, often called **phases.** The phases are usually visibly different. More importantly, composition varies from one place in the sample to another. Each phase has its own unique properties and appearance. Therefore, a sample of matter generally may be classified as homogeneous or heterogeneous by its appearance alone.

Oil and water are both homogeneous liquids. When mixed in the same container, they quickly separate into two distinct liquid phases, forming a heterogeneous mixture (Fig. 2.5). Sugar and water, on the other hand, mix completely with each other to form a single liquid phase, a homogeneous mixture called a **solution.** Even a pure substance can be heterogeneous if it is present in two visibly distinct states. An ice cube floating in water is an example.

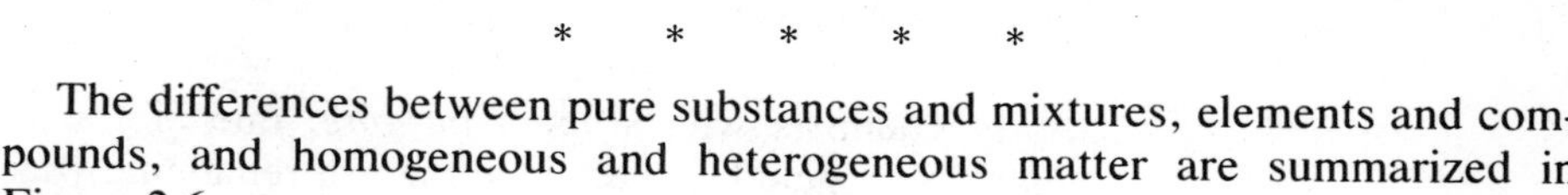

* * * * *

The differences between pure substances and mixtures, elements and compounds, and homogeneous and heterogeneous matter are summarized in Figure 2.6.

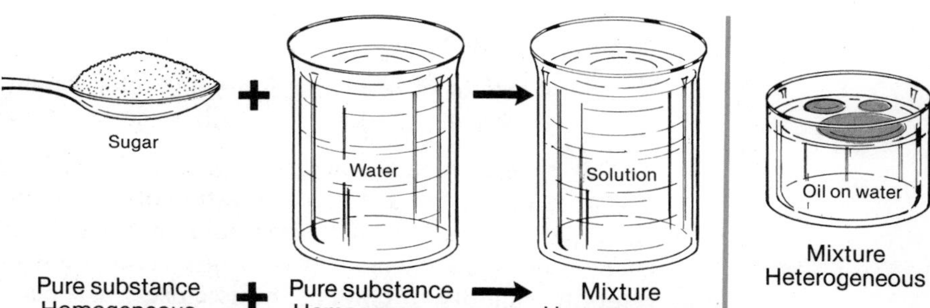

Figure 2.5
Examples of homogeneous and heterogeneous samples of matter.

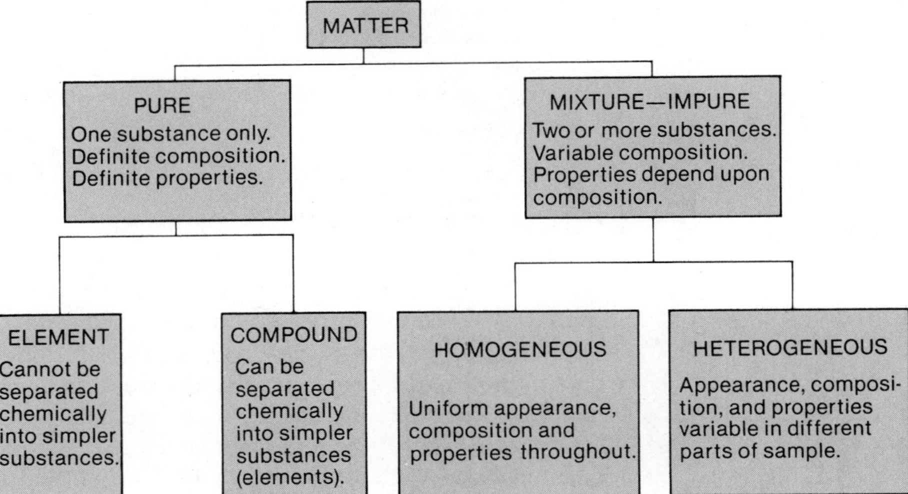

Figure 2.6
Summary of classificatior systems for matter.

Quick Check 2.5: Classify the following as heterogeneous or homogeneous:
(a) Oil and vinegar dressing (b) Freshly squeezed orange juice (c) Coney Island beach sand (d) Gasoline

2.6 THE LAW OF CONSERVATION OF MASS

PG 2G State the meaning of, or draw conclusions based on, the Law of Conservation of Mass.

Early chemists who studied *burning,* one of the most familiar of chemical changes, concluded that since the ash remaining was so much lighter than the object burned, something was "lost" in the reaction. Their reasoning was faulty. They did not realize that gases, which they could not see, took part in the reaction, both as reactants, or things used up, and as products, or things made. When a piece of wood burns, gaseous oxygen is a reactant that is lost in the reaction. The products, in addition to ash, include two gases, carbon dioxide and water vapor. If we take careful account of all the reactants and products, we find that

$$\begin{array}{c}\text{total weight of reactants}\\ \text{(wood + oxygen)}\end{array} = \begin{array}{c}\text{total weight of products}\\ \text{(ash + carbon dioxide + water vapor)}\end{array}$$

This type of weight balance applies to all ordinary chemical reactions. It was once called the law of conservation of matter, but now is referred to more accurately as the **Law of Conservation of Mass.** In other words, **in a chemical change, mass is conserved; it is neither created nor destroyed.** The word *mass* refers to quantity of matter, and is closely associated with the more familiar term *weight*. The difference between them will be pointed out in Chapter 3.

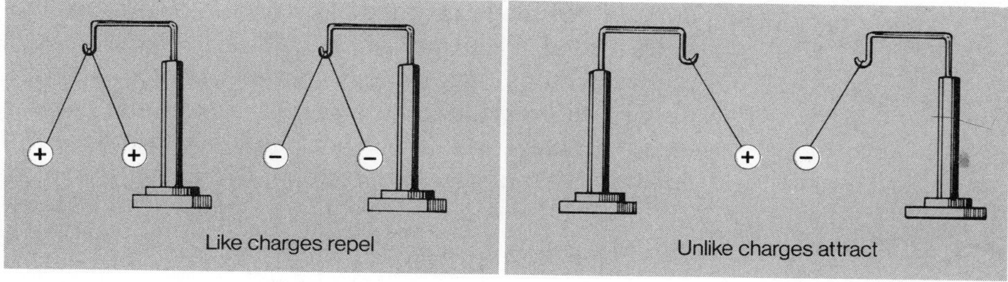
Like charges repel
Unlike charges attract

Figure 2.7
Electrostatic attraction and repulsion.

2.7 ELECTRICAL CHARACTER OF MATTER

PG 2H Match electrostatic forces of attraction and repulsion with combinations of positive and negative charges.

If you release an object held above the floor, it falls to the floor. This is caused by gravity, an invisible attractive force between the object and the earth. Although the force is invisible, its effect is very evident. There are two other invisible forces, both capable not only of attraction but also of repulsion. They are magnetic and **electrostatic forces.**

One of the most common electrostatic events occurs when you scrape your feet across a carpet on a dry day and then turn on a light switch, or perhaps "shock" another person by touching him. In the foot-scraping operation you become "charged with electricity." In the laboratory, we can charge a rubber rod by rubbing it with fur. If a pith ball* hung on a string is touched with the rod, the pith ball gains the same charge as the rod (Fig. 2.7). Two pith balls charged in the same manner push each other apart, indicating a *repulsion force between objects carrying the same kind of electrical charge*. Similarly, if two pith balls are touched with glass rods rubbed by silk, they repel each other, again illustrating repulsion between like charges. If one of the balls touched by the rubber rod is now brought near a ball touched by the glass rod, the balls are attracted toward each other. Obviously, the balls do not have like charges, which repel each other; instead they must have opposite charges. *Objects which are oppositely charged are attracted toward each other.*

These and numerous other examples lead to the conclusion that there are two, and only two, kinds of **electrical charges.** One is called **positive,** the other **negative**—adjectives assigned long before the nature of the charges was known. The attractive and repulsive forces between charged objects are called **electrostatic forces.** The electrostatic force between two like charges (+ and +, or − and −) is one of repulsion; the force between a positive and a negative charge is one of attraction. These relationships are illustrated in Figure 2.7.

Electrostatic forces show that matter has electrical properties. We now know that all matter consists of tiny particles called atoms and that these atoms

*Pith is a soft, spongy, substance made from plant fibers. Pith balls used in this experiment are typically ¼ inch in diameter.

contain positively charged particles called protons and negatively charged particles called electrons. In the rubbing operations described above, electrons (never protons) are moved from one object to another. The object that ends up with more electrons than protons has a net negative charge. The positively charged object has fewer electrons than protons. If the number of protons is equal to the number of electrons, the object is said to be electrically **neutral.**

2.8 ENERGY IN CHEMICAL CHANGE

PG 2I Distinguish between exothermic and endothermic changes.

2J Distinguish between potential energy and kinetic energy.

Changes in matter don't "just happen." Something causes each change. Frequently change occurs because matter absorbs or releases energy. For example, if you supply heat energy to a pan of water, it boils. Figure 2.2 showed that water may be chemically decomposed into its elements by means of electrical energy. **Energy** is absorbed—taken in—in each of these examples. Chemical or physical changes in which energy is *absorbed* are called **endothermic changes.**

By contrast, chemical or physical changes in which energy is *released* to the surroundings are called **exothermic changes.** If you strike a match and put your finger in the flame you learn very quickly that the chemical change in burning wood is releasing heat energy. It also releases light energy. Chemical changes in flashlight and automobile batteries release electrical energy. The physical change when steam condenses to water releases exactly the same energy that was required to boil the water in the first place. All these are examples of exothermic changes.

Most of the energy changes we will be concerned with will be in the form of heat. Other forms of energy will be left to more advanced courses. To understand the source of chemical energy, however, we will consider briefly a physicist's definition of energy and its closely associated concept, **work.**

Work, according to the physicist, is the application of a force over a distance. If you raise a book from the floor to a desk, you do work (Fig. 2.8 *A* and *B*). You exert the force required to lift the book against the gravitational attraction of the earth. You exert this force over the distance from the floor to the desk. **Energy may be defined as the ability to do work.** You had the energy required to do the work of raising the book. In fact, you transferred this energy to the book, which has more energy on the desk than it had on the floor. The book now has **potential energy** with respect to the floor (Fig. 2.8*B*). **Potential energy**

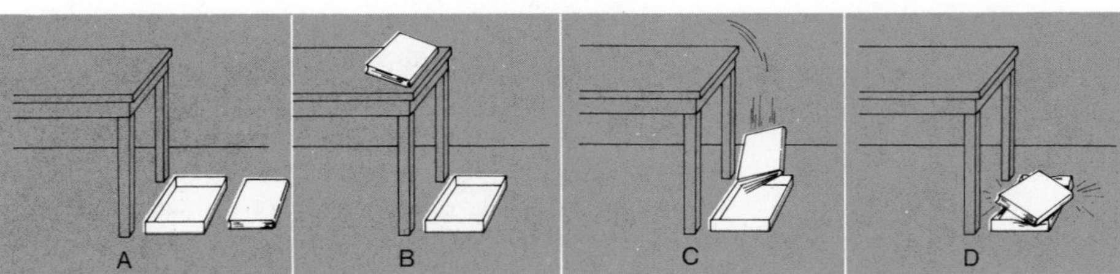

Figure 2.8
Potential and kinetic energy.

is energy possessed by an object because of its position. An object having potential energy is capable of doing work. This may be shown by pushing the book off the desk and allowing it to fall toward the box still on the floor. In the act of falling (Fig. 2.8*C*) the potential energy is converted into **kinetic energy, the energy possessed by an object because of its motion.** When the book hits the box (Fig. 2.8*D*), it does work on the box by crushing it.

Potential and kinetic energy are two forms of mechanical energy. While it is not necessary for you to know about these in detail in an introductory chemistry course, you should recognize that *potential energy is related to position* and *kinetic energy is energy of motion.* What is called chemical energy comes mainly from the positions of positively and negatively charged objects. The energy associated with chemical changes comes from changes in these positions, just as there is a change in potential energy of a book as it is moved between the floor and a desk.

At the beginning of this section, it was suggested that changes are produced by causes. Energy is one of the driving forces that produce change. When a book drops to the floor, there is a spontaneous change from a high potential energy to a low potential energy. This is an example of a general rule that changes tend to occur until a system is at the lowest energy available to it. In other words, there is a tendency toward a "minimization of energy."

The other driving force is called *entropy* and is sometimes described as "randomness" and "disorder." Entropy tends to increase spontaneously. Whenever a change occurs, it is the result of the net effect of changes in energy and entropy. There will be little occasion to refer to entropy in this book; but from time to time we will indicate that minimization of energy is responsible for a change.

Quick Check 2.8

A charged object is moved closer to another object that has the same charge.

1. Is this a change in kinetic energy or potential energy?
2. Is the energy change an increase or a decrease?

2.9 THE LAW OF CONSERVATION OF ENERGY

PG 2K State the meaning of, or draw conclusions based on, the Law of Conservation of Energy.

The various processes that take place when we drive an automobile offer a good illustration of energy conversions, changing energy from one form to another. Starting uses the chemical energy of the storage battery to produce electrical energy. As we drive along, chemical energy of the fuel is being converted into both heat energy and "mechanical" energy of the engine parts. This mechanical energy is transferred to the car as kinetic energy, or energy of motion. Some of this energy is converted to heat through friction between the tires and the road. The extent of this conversion increases dramatically if we slam on the brakes at an intersection. Potential energy, the energy associated with position, is gained as the car climbs a hill. It is changed back into kinetic energy if we coast down the hill.

Careful investigation of energy conversions such as these show that the energy "lost" in one form is always exactly equal to the energy "gained" in another form. This leads to another conservation law, the **Law of Conservation of Energy,** which states that, **in any ordinary change, energy is neither created nor destroyed.**

2.10 THE MODIFIED CONSERVATION LAW

Early in this century, Albert Einstein recognized a "sameness" between mass and energy. He suggested that it should be possible to convert one of these into the other. Such conversions are indeed going on in the world around us. In a few cases, enough mass is converted to energy to become dramatically apparent. This happens, for example, when a hydrogen bomb explodes or a nuclear reactor is used to produce electrical energy. Einstein's famous equation relates the amounts of mass and energy:

$$\Delta E = \Delta mc^2,$$

where ΔE is the energy change, Δm is the mass change, and c is the speed of light.

The amount of energy that could be produced from the conversion of mass is enormous. If it were possible to convert all of a given mass of coal to energy, that energy would be about 2.5 billion times as great as the energy derived from burning that same amount of coal! This is why nuclear energy is such an attractive alternative to the traditional sources of energy, an attraction clouded by serious questions of safety.

The suggestion that matter can be converted into energy, or energy into matter, might lead one to believe that the laws of conservation of mass and energy are not true. In ordinary (non-nuclear) chemical reactions the matter–energy conversion is so small it cannot be measured. For all practical purposes, the laws are as good today as when they were first proposed. For *all* chemical changes, both nuclear and non-nuclear, the laws may be combined into a single conservation law that states that the total amount of mass and energy in the universe is a constant. Expressed as an equation this becomes

$$(\text{mass} + \text{energy})_{\text{reactants}} = (\text{mass} + \text{energy})_{\text{products}}$$

CHAPTER 2 IN REVIEW

2.1 Physical and Chemical Properties and Changes
- **2A** Distinguish between physical and chemical properties.
- **2B** Distinguish between physical and chemical changes.

2.2 States of Matter: Gases, Liquids, Solids
- **2C** Identify and explain the differences between gases, liquids, and solids in terms of (a) visible properties and (b) particle movement.

2.3 Elements and Compounds
- **2D** Distinguish between elements and compounds.

2.4 Pure Substances and Mixtures
- **2E** Distinguish between a pure substance and a mixture.

2.5 Homogeneous and Heterogeneous Matter
- **2F** Distinguish between homogeneous and heterogeneous samples of matter.

2.6 The Law of Conservation of Mass

2G State the meaning of, or draw conclusions based on, the Law of Conservation of Mass.

2.7 Electrical Character of Matter

2H Match electrostatic forces of attraction and repulsion with combinations of positive and negative charges.

2.8 Energy in Chemical Change

2I Distinguish between exothermic and endothermic changes.

2J Distinguish between potential energy and kinetic energy.

2.9 The Law of Conservation of Energy

2K State the meaning of, or draw conclusions based on, the Law of Conservation of Energy.

2.10 The Modified Conservation Law

TERMS AND CONCEPTS

2.1 Matter
Physical property
Physical change
Chemical property
Chemical change

2.2 States of matter
Kinetic molecular theory

2.3 Element
Compound
Law of Definite Composition
Chemical formula

2.4 Pure substance
Mixture

2.5 Homogeneous
Heterogeneous
Phase
Solution

2.6 Law of Conservation of Mass

2.7 Electrostatic forces
Electrical charge

2.8 Endothermic change
Exothermic change
Energy
Potential energy
Kinetic energy

2.9 Law of Conservation of Energy

Most of these terms and many others appear in the Glossary. Use your Glossary regularly.

QUESTIONS AND PROBLEMS

Section 2.1

(1) Which of the following properties are physical, and which are chemical: (a) color of paper; (b) decay of bananas; (c) odor of perfume; (d) electrical conductivity of a metal; (e) does not react with oxygen.

(2) Describe each of the following changes as chemical or physical: (a) souring of milk; (b) dissolving alcohol in water; (c) making a wood carving; (d) rusting of iron; (e) burning a candle.

(3) What kind of a change, physical or chemical, is needed to make toast out of a piece of bread? On what do you base your answer?

(4) Suggest at least two ways to separate ball bearings from table tennis balls. On what property is each method based?

(22) Classify each of the following properties as chemical or physical: (a) hardness of a diamond; (b) combustibility of gasoline; (c) corrosive character of an acid; (d) elasticity of a rubber band; (e) taste of chocolate.

(23) Which among the following are physical changes: (a) blowing glass; (b) fermenting grapes; (c) forming a snowflake; (d) evaporation of dry ice; (e) decomposing a substance by heating it.

(24) Would you use a physical change or a chemical change to extract mercury from cinnabar (mecury sulfide), a dark red ore?

(25) Is separating the hydrogen from oxygen in water a physical change or a chemical change? Justify your answer.

Section 2.2

(5) Compare the volumes occupied by the same sample of matter when in the solid, liquid or gaseous states.

(26) Which of the three states of matter is most easily compressed? Suggest a reason for this.

(6) Explain in terms of the properties of gases, liquids, and solids why a snowman disappears when the temperature rises.

(7) The phrase, "When it rains, it pours," has been associated with a brand of table salt for decades. How can salt, a solid, be poured?

(27) Which is the most rigid state of matter? Suggest a reason for this.

(28) The word *pour* is commonly used in reference to liquids, but not solids or gases. Can either a solid or a gas be poured? Why, or why not?

Section 2.3

(8) Which of the following are elements, and which are compounds: (a) sulfur dioxide; (b) aluminum; (c) baking soda (sodium hydrogen carbonate); (d) carbon tetrachloride; (e) iodine; (f) chromium.

(9) How can you tell if a substance is an element or a compound?

(10) A white, crystalline material that looks like table salt gives off a gas when heated under certain conditions. There is no change in the appearance of the solid that remains, but it does not taste the same as it did originally. Was the beginning material an element or a compound? Explain your answer.

(29) Classify the following as compounds or elements: (a) silver bromide (used in photography); (b) calcium carbonate (limestone); (c) sodium hydroxide (lye); (d) uranium; (e) tin.

(30) Can a compound be decomposed into two other compounds?

(31) Metal A dissolves in nitric acid. The original metal can be recovered if Metal B is placed in the solution. Metal A becomes heavier after prolonged exposure to air. The procedure is faster if the metal is heated. From the evidence given, can you tell if Metal A definitely is or could be an element or a compound? If you cannot, what other information is necessary to make that classification?

Section 2.4

(11) If two elements are placed in the same container and there is no chemical change, is the result a compound or a mixture? If it is a mixture, what are the components of the mixture?

(12)* A clear, colorless liquid is distilled in an apparatus similar to that shown in Figure 2.3. The temperature remains constant throughout the distillation process. The liquid leaving the condenser is also clear and colorless. Both liquids are odorless, and they have the same freezing point. Is the starting liquid a pure substance or a mixture? What single bit of evidence in the above description is the most convincing reason for your answer?

(13) Aspirin is a pure substance. If you had the choice of buying a widely advertised brand of aspirin whose effectiveness is well known or the unproven product of a new manufacturer at half the price, which would you buy? Explain.

(32) If two elements are placed in the same container and there is a chemical change, is the result a compound or a mixture? If it is a mixture, what are the components of the mixture?

(33)* The density of a liquid is determined in the laboratory. The liquid is left in an open container overnight. The next morning the density is measured again and found to be greater than it was the day before. Is the liquid a pure substance or a mixture? Explain your answer.

(34) Graphite and diamonds are two forms of carbon, an element. If chunks of graphite are sprinkled among the diamonds on a jeweler's display tray, is the material on the tray a mixture or a pure substance?

Section 2.5

(14) Identify the homogeneous substances among the following: (a) a log; (b) an ordinary glass of water; (c) hamburger; (d) flour; (e) paint.

(15) Explain why milk is described as *homogenized.* What is unhomogenized milk? What are the advantages of homogenized milk over that which is not homogenized?

(35) Which are the heterogeneous items in the list that follows: (a) Sterling silver; (b) freshly opened root beer; (c) an ice cube; (d) scrambled eggs; (e) motor oil.

(36) Is the combination of graphite and diamonds described in Question 34 homogeneous or heterogeneous? Explain.

Section 2.6

(16) How would you expect the weight of a flash bulb before use to compare with the weight after use? Explain.

(37) If solid limestone is heated, the rock that remains weighs less than the original limestone. What do you conclude has happened?

Section 2.7

(17) Identify the net electrostatic force (attraction, repulsion, or none) between the following pairs of substances: (a) two positively charged table tennis balls; (b) a negatively charged piece of dust and a positively charged piece of dust; (c) a positively charged sodium ion and a positively charged potassium ion.

(38) What is the main difference between electrostatic forces and gravitational forces? Which is more similar to magnetic force? Can two or all three of these forces be exerted between two objects at the same time? Explain.

Section 2.8

(18) Are the following changes exothermic or endothermic: (a) cooking; (b) being burned by a match; (c) a match burning; (d) a light that is on; (e) digesting food?

(19) An automobile accelerates from 20 miles per hour to 55 miles per hour. Does its energy increase or decrease? Is the change in potential energy or kinetic energy?

(20)* There is always an increase in potential energy when an object is raised higher above the surface of the earth, i.e., when the distance between the earth and the object is increased. Increasing the distance between two electrically charged objects, however, may raise or lower potential energy. How can this be?

(39) Which of the following processes is exothermic: (a) being cooled by standing in front of a fan after perspiring; (b) freezing water; (c) listening; (d) running; (e) a falling rock striking the ground?

(40) As a child plays on a swing, at what point in her movement is her kinetic energy the greatest? At what point is potential energy at its maximum?

(41) In the gravitational field of the earth, an object always falls until it is stopped by some physical object that prevents it from falling farther. Two electrically charged objects, each of which is made up of unequal numbers of both positive and negative charges, will reach a certain separation distance and stay there without physical support. Can you suggest an explanation for this?

Section 2.9

(21) List the energy conversions that occur between water about to enter a hydroelectric dam to the burning of an electric light bulb in your home.

(42) Identify several energy conversions that occur regularly in your home. State whether each is good (useful), bad (wasteful), or sometimes good, sometimes bad.

General Questions

(*43*) Distinguish precisely and in scientific terms the differences between items in the following groups:
(a) Physical change, physical property, chemical change, chemical property.
(b) Gases, liquids, solids.
(c) Element, compound.
(d) Pure substance, mixture.
(e) Homogeneous matter, heterogeneous matter.
(f) Exothermic change, endothermic change.
(g) Potential energy, kinetic energy.

(44) Determine whether each statement that follows is true or false:
(a) The fact that paper burns is a physical property.
(b) Particles of matter are moving in gases and liquids, but not in solids.
(c) A heterogeneous substance has a uniform appearance throughout.
(d) Compounds are impure substances.
(e) If one sample of sulfur dioxide is 50% sulfur and 50% oxygen, then all samples of sulfur dioxide are 50% sulfur and 50% oxygen.
(f) A solution is a homogeneous mixture.
(g) Two positively charged objects attract each other, but two negatively charged objects repel each other.
(h) Mass is conserved in an ordinary (non-nuclear) endothermic chemical change, but not in an ordinary exothermic chemical change.
(i) Potential energy can be related to positions in an electric force field.
(j) Chemical energy can be converted to kinetic energy.
(k) Potential energy is more powerful than kinetic energy.
(l) A chemical change always destroys something, and always makes something.

(45) A "natural food" store advertises that "no chemicals" are present in any food sold in the store. If their ad is true, what do you expect to find in the store?

(*46*) Name some things you have used today that are not the result of a man-made chemical change.

(*47*) Name some pure substances you have used today.

(*48*) How many homogeneous substances can you reach without moving?

(49) Which among the following can be pure substances: mercury, milk, water, a tree, ink, iced tea, ice, carbon?

(50) Can you have a mixture of two elements and a compound of the same two elements?

(51) Can you have more than one compound made of the same two elements? If yes, try to give an example.

(52) Rain water comes from the oceans. Is rain water more pure, less pure, or of the same purity as ocean water? Explain.

(53) A large box contains a white powder of uniform appearance. A sample is taken from the top of the box and another sample is taken from the bottom. By analysis it is found that the percentage of oxygen in the sample from the top is 48.2%, whereas the sample from the bottom is 45.3% oxygen. Answer each question below independently and give a reason that supports your answer.
(a) Is the powder an element or a compound?
(b) Are the contents of the box homogeneous or heterogeneous?
(c) Can you be certain that the contents of the box are either a pure substance or a mixture?

(54) If energy can neither be created nor destroyed, as the Law of Conservation of Energy states, why are we so concerned about wasting our energy resources?

3 Measurements in Chemistry

Part I Calculations with Measured Quantities

3.1 INTRODUCTION TO MEASUREMENT

Chemistry is both qualitative and quantitative. In its qualitative role it explains *how* and *why* chemical changes occur. Quantitatively it measures *how much* of a substance is used or produced.

To find how much of a chemical is involved in a reaction, a chemist must make measurements. Making measurements is nothing new to you; you measure your height and weight, or the distance to school or work. If you shop in the United States, you buy milk by the quart or gallon, butter by the pound, and fabric by the yard. If you live in Canada–or just about any other place in the world, for that matter—the metric units of liters, grams, and meters are more familiar to you.*

Most scientific measurements are expressed in **SI units.** SI is an abbreviation for the French name for the International System of Units. (A summary of SI units appears in Appendix II.) Metric units for measurements of mass and length are part of the SI system. In this text, metric will be used almost exclusively. The only exceptions will be in the present chapter, which will include some of the important conversions between English and metric units. This is done to give you some "feeling" for the metric system in case you are not familiar with it. Several of the more important relationships between English and metric units are included in the conversion factor table in Appendix II, Table AP-3.

Two of the ways in which laboratory measurements are made are suggested in Figure 3.1. Liquid volume is measured in a number of ways, depending on the precision required; a simple graduated cylinder is often satisfactory. Mass is measured on a balance; again the accuracy requirement determines which instrument is used. Length is ordinarily measured with a meter stick or "ruler." Most of the common one-foot rulers sold in bookstores of American colleges have a centimeter scale on one edge. Thermometers used to measure temperature in the laboratory are graduated in Celsius (formerly centigrade) degrees, rather than in the Fahrenheit degrees familiar to Americans.

*If you live outside the United States you probably spell the volume unit *litre* and the length unit *metre*. These spellings correspond with the French pronunciations of the words, and it was the French who first used them. In this book we use the American spellings, which match the English pronunciations.

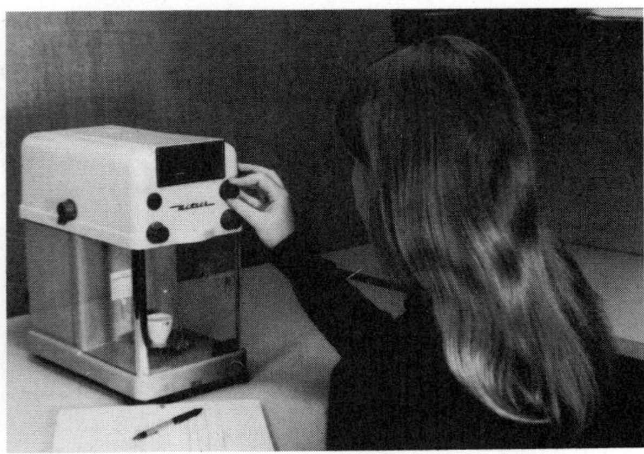

Figure 3.1
A student measures volume in a graduated cylinder and mass on an analytical balance.

3.2 DIMENSIONAL ANALYSIS

PG 3A Identify the "given quantity" in the statement of a problem. Beginning with the given quantity, set up and solve the problem by the dimensional analysis method.

How many inches are in 3 feet?

Already you have probably figured out the answer: 36 inches. How did you get it? You probably reasoned that there are 12 inches in 1 foot, so there must be 3 × 12 inches in 3 feet: 3 × 12 = 36. In doing this you used, without thinking about it, the problem-solving method called **dimensional analysis.*** Dimensional analysis is widely used in solving chemistry problems, and other problems too. It will be applied in about 80% of the problems in this book. We therefore introduce it early and in some detail so that you can begin building your skill in this important calculation method.

Dimensional analysis can be used whenever two quantities are directly proportional to each other and you must convert one quantity to the other. (See Appendix I, Part D for a discussion of proportionalities.) The number of inches in a given distance is proportional to the number of feet. This leads to the **conversion factor** or **conversion relationship**

$$12 \text{ inches} = 1 \text{ foot} \tag{3.1}$$

If both sides of Equation 3.1 are divided by 1 foot we get

$$\frac{12 \text{ inches}}{1 \text{ foot}} = \frac{1 \text{ foot}}{1 \text{ foot}} = 1 \tag{3.2}$$

When the numerator and denominator of a fraction are the same, as in a/a or, in this case, 1 foot/1 foot, the value of the fraction is 1. This is also true if the numerator and denominator are equal, as 12 inches/1 foot. Dividing both sides

*Other names for the dimensional analysis method are conversion factor, factor–label, unit cancellation, unity factor, factor–unit, and unary rate methods.

of Equation 3.1 by 12 inches, and extending it to include Equation 3.2, gives a second ratio that is equal to 1:

$$\frac{12 \text{ inches}}{12 \text{ inches}} = \frac{1 \text{ foot}}{12 \text{ inches}} = 1 = \frac{12 \text{ inches}}{1 \text{ foot}} \tag{3.3}$$

Turning briefly to algebra, it is a well known fact that if you multiply any quantity by 1, the product is equal to the original quantity:

$$a \times 1 = a \tag{3.4}$$

Sometimes the 1 takes the form of a fraction. For example,

$$\frac{by}{c} \times \frac{1}{y} = \frac{b}{c} \times \frac{y}{y} = \frac{b}{c} \times 1 = \frac{b}{c} \tag{3.5}$$

The procedure is usually shortened by "cancelling" any factor that is in both the numerator and the denominator:

$$\frac{by}{c} \times \frac{1}{y} = \frac{b\cancel{y}}{c} \times \frac{1}{\cancel{y}} = \frac{b}{c} \tag{3.6}$$

Now let's apply these ideas to find the number of inches in 3 feet. If the **given quantity,** 3 feet, is multiplied by 1 in the form of 12 inches/1 foot, the unit *feet (foot)* can be cancelled, just as the y was cancelled in Equation 3.6:

$$3 \text{ feet} \times \frac{12 \text{ inches}}{1 \text{ foot}} = \frac{3 \cancel{\text{feet}}}{1} \times \frac{12 \text{ inches}}{1 \cancel{\text{foot}}} = 36 \text{ inches} \tag{3.7}$$

This is a typical dimensional analysis setup for solving a problem. It begins with the given quantity, which is an *amount* of something. In particular, note that *the given quantity is not a relationship between two quantities*, as conversion relationships are. The given quantity always includes the unit in which the quantity is counted or measured, as 4 hours, 5 hot dogs, or 9 tons. The given quantity is multiplied by one or more conversion ratios that are equal to 1, arranged so the unwanted starting unit may be cancelled and replaced by a desired unit. Arithmetic treatment of the numbers in the setup gives the size of the final answer; algebraic treatment (cancellation) of the units in the setup gives the units of the final answer.

Much of your success in solving chemistry problems will depend on your ability to identify the given quantity. It tells you where to begin. Only in this chapter will the given quantity be mentioned specifically, but it will be used in the calculation setup of nearly all problems. To remind you of its importance, the given quantity will be enclosed in a lightly colored box, $\boxed{}$, in all calculation setups in this book.

Another term that is helpful in learning dimensional analysis is **unit path.** The unit path is the series of unit conversions used in a dimensional analysis setup. In the problem above, where a number of feet was converted to inches, the unit path was "feet to inches," which can be written as feet → inches, or ft → in. The unit path maps out the route to be travelled in getting from the given quantity to the final answer. To solve a problem, you must know the conversion relationship between each pair of units in the unit path.

Let's try another simple problem: How many feet are in 48 inches?

Again, you probably have the answer already. You didn't think about a unit path, but in terms of dimensional analysis you used inches → feet from the same conversion relationship, 12 inches = 1 foot. In effect, you reasoned, "If there are 12 inches in 1 foot, how many 12-inch 'packages' are in 48 inches?" In other words, how many times does 12 go into 48? The answer, of course, is 4, so 48 inches = 4 feet.

In the second problem the given quantity, 48 inches, must be *divided* by the conversion factor 12 inches/1 foot. Do you remember how to divide by a fraction? You invert the fraction, and then multiply. In more mathematical terms, *to divide by a fraction you multiply by its inverse, or reciprocal.* Thus

$$\boxed{48 \text{ in.}} \div \frac{12 \text{ in.}}{1 \text{ ft}} = \boxed{48 \not{\text{in.}}} \times \frac{1 \text{ ft}}{12 \not{\text{in.}}} = 4 \text{ ft}$$

The units again cancel so the only unit left, feet, is the unit in which the answer is to be expressed.

One of the advantages of dimensional anaylsis is the way it tells you if you have multiplied when you should have divided, or divided when you should have multiplied. Suppose in the second problem you had multiplied by mistake. Your setup would be

$$\boxed{48 \text{ in.}} \times \frac{12 \text{ in.}}{1 \text{ ft}} = 576 \text{ in.}^2/\text{ft}$$

Because of your familiarity with the numbers, you would have known 576 was wrong. But even if the numbers were not familiar to you, as they might not be later in this book, the "nonsense units" in.^{2}/ft would have alerted you to something being wrong. If the units of an answer are not correct, you can be almost positive that the numerical result is wrong. You should check your setup of the problem and correct the error.

Some conversion factors are established by definition. 12 inches = 1 foot is an example. The relationship is permanent, never changing. The units of measurement are for the same thing—distance, in this case. Other conversion relationships are temporary. They may have one value in one problem, and another value in another problem. Speed, measured in miles per hour, is an example. (The term *per* implies *divide*. See a discussion of this word in Appendix I, Part E.) If we drive a car at an average of 50 miles per hour, we can calculate the distance traveled in two, three, or any number of hours. As long as that speed is held, the number of miles driven is proportional to the number of hours, and 1 hour of driving is said to be "equivalent" to 50 miles. In this book these temporary equivalent relationships will be written with the symbol ≏, saving the equal sign, =, for true equalities. Thus the conversion relationship between miles and hours is 50 miles ≏ 1 hour *at 50 miles per hour.*

We will now use the 50-miles-per-hour relationship for the first example to be solved by the question-and-answer method described on page 7. You may wish to review the procedure at this time. It is also summarized on the opaque shields found elsewhere in this book. We suggest you tear out one of these shields and use it to cover parts of the example as you work through it. The shield may also be used as a bookmark and as a reference periodic table when you work examples in later chapters.

EXAMPLE 3.1 How long will it take to drive 250 miles at an average speed of 50 miles per hour?

The first step in solving this problem is to identify the given quantity. Remember that a given quantity is always a measurement or count of a single "thing." It is not a relationship between two measurable quantities. Write the given quantity here.

$$250 \text{ miles} \times \underline{\hspace{3cm}} =$$

The number of miles, a measurable distance, is the given quantity. The speed, 50 miles per hour, is a relationship between two measurable quantities, distance and time. From that relationship, 50 miles ≏ 1 hour within this problem.

Before solving the problem by dimensional analysis, think through to the answer. Decide in your mind whether you must multiply or divide by 50 to find the number of hours for the trip. Express your decision as a fraction equal to 1, to be used as a multiplier for 250 miles. The line for that fraction is already printed above. Write the numerator and denominator in their proper places, or repeat the setup and complete the calculation on a separate sheet of paper. Include units. Be sure the units cancel correctly. Then calculate your answer—please don't reach for a calculator for such a simple problem—and write it after the equal sign.

$$250 \text{ \sout{miles}} \times \frac{1 \text{ hour}}{50 \text{ \sout{miles}}} = 5 \text{ hours}$$

If you had multiplied by 50 miles/1 hour the units would not have canceled and your answer would have been 12 500 $\text{miles}^2/\text{hour}$. Both the number and the nonsense units would have told you that the answer was incorrect. (You may wish to check Appendix I, Part A to see why there is no comma in 12 500.)

We didn't mention it in Example 3.1, but notice that your unit path was miles → hours. The unit path idea is obvious in the one-step conversions in the problems so far. Its value becomes apparent in problems that must be solved by a series of steps.

The dimensional analysis method of problem solving may be summarized as follows:

1. Begin with the given quantity.
2. Establish the unit path from the given quantity to the wanted quantity. You must be given or otherwise know each conversion factor in the unit path.
3. Write the setup for the problem, multiplying or dividing in a logical sequence through each step of the unit path. Be sure to write all units.

$$\begin{pmatrix}\text{Given}\\ \text{quantity}\end{pmatrix} \times \begin{pmatrix}\text{One or more}\\ \text{conversion factors}\end{pmatrix} = \begin{matrix}\text{Wanted}\\ \text{quantity}\end{matrix}$$

4. Cancel units to be sure that the setup gives an answer expressed in the correct unit(s).
5. Multiply and/or divide to get the numerical value of the answer.

EXAMPLE 3.2 How many times does the hour hand of a clock go around in two weeks?

Write the given quantity first.

-----

2 weeks

Two weeks is the only quantity in the problem; it must be the given quantity.

The unit in which the answer is to be stated is revolutions of the hour hand. The unit path begins with weeks and ends with revolutions. Unless you happen to know how many times an hour hand goes around in one week, you cannot go from the beginning of the unit path to the end in one step. Additional information is required—information drawn from your knowledge about clocks, hours, weeks, and days. Can you suggest a unit path from weeks to revolutions that has a series of steps for which you know the conversion relationship between each pair of units in the path? Try it. Also write the required conversion relationship beneath each arrow.

weeks → → revolutions

-----

weeks → days → hours → revolutions
1 week = 7 days   1 day = 24 hours   12 hours = 1 revolution

Weeks are changed to days with the first relationship, then days may be changed to hours, and finally hours are changed to revolutions.

Now you are ready to take the first step. Below we have begun the setup. Decide whether you must multiply or divide by 7 days per week to change weeks to days. Write the necessary numerator and denominator in their proper places. Include units, and show the cancellation that is possible. Do not solve to a numerical answer.

$$\text{2 weeks} \times \underline{\qquad\qquad}$$

$$\text{2 }\cancel{\text{weeks}} \times \frac{\text{7 days}}{\text{1 }\cancel{\text{week}}} \times \underline{\qquad\qquad}$$

Normally, intermediate answers are not calculated in a dimensional analysis setup. If they were, and if the thought process was reasonable, the intermediate answer would be meaningful. In the setup this far, the "weeks" cancel and the calculation yields an intermediate answer of 14 days. An incorrect setup, 2 weeks $\times \frac{\text{1 week}}{\text{7 days}}$, would result in a nonsense unit, $\text{weeks}^2/\text{day}$, which would have signaled an error at this point.

The setup through the first multiplication indicates the number of days. Now *extend* the setup by multiplying a conversion ratio that changes *days* into *hours*. As before, do not solve the setup.

$$\boxed{2\text{ weeks}} \times \frac{7\ \cancel{\text{days}}}{1\ \cancel{\text{week}}} \times \frac{24\text{ hours}}{1\ \cancel{\text{day}}} \times \underline{\hspace{3cm}} =$$

Again notice that calculation of the setup this far would yield a meaningful answer in hours. This would not be true if the incorrect ratio $\frac{1\text{ day}}{24\text{ hours}}$ had been used.

The final step converts hours to revolutions by the third equivalence relationship listed in the unit path. Complete the setup, and this time solve for the numerical answer as well.

$$\boxed{2\ \cancel{\text{weeks}}} \times \frac{7\ \cancel{\text{days}}}{1\ \cancel{\text{week}}} \times \frac{24\ \cancel{\text{hours}}}{1\ \cancel{\text{day}}} \times \frac{1\text{ revolution}}{12\ \cancel{\text{hours}}} = 28\text{ revolutions}$$

Separating the arithmetic from the setup, we have

$$\frac{2 \times 7 \times 24}{12} = 28$$

This may be done mentally by first seeing that 24/12 is 2. This is another advantage you should seek in a dimensional analysis setup: setting up the entire problem frequently leads to simplification in calculations, and therefore fewer errors.

A word of caution: Used without thinking, dimensional analysis can become a very mechanical process. Do not permit this. You will gain the most benefit from dimensional analysis if you think your way through each problem. Do not simply juggle units until they "come out right." Chart the unit path from the given quantity to the desired quantity, and then thoughtfully consider each conversion. This will make each step of the problem meaningful, and you will *understand* the overall solution. Understanding, rather than an answer, should be your goal for every problem.

Dimensional analysis setups will be used in unit conversions throughout this book. Additional practice problems may be found at the end of the chapter.

Quick Check 3.2

Which among the following might be given quantities and, therefore, starting points for a dimensional analysis setup; and which are conversion relationships, or conversion factors?

(a) 10 days; (b) 2000 pounds = 1 ton; (c) 2.54 centimeters per inch; (d) 4 quarts; (e) 1 gallon; (f) 4 quarts per gallon.

Part II: SI (Metric) Units

3.3 LENGTH AND VOLUME UNITS IN THE METRIC SYSTEM

PG 3B Given an English–metric conversion table, convert a length measurement expressed in (a) English units, (b) meters, (c) kilometers, (d) centimeters, or (e) millimeters to each of the other units.

The standard unit of length is the **meter.** The meter was first defined as 1/10 000 000 of the distance from the North Pole to the equator. It is now identified more precisely as 1 650 763.73 times the wavelength of a certain line in the emission spectrum (see Section 5.1) of an elemental gas, krypton. The advantage of the modern definition over the original may not be obvious, but precision measurements of modern technology require such a definition. The meter is 39.37 inches long—about three inches longer than a yard.

As the English system uses both larger and smaller units than the yard for different purposes, so the meter must be modified for larger and smaller units. The metric system accomplishes this with multiples and submultiples of 10, each of which is identified by a prefix. The prefix for 1⁄10 is *deci-*. Therefore, 1⁄10 of a meter, or 0.1 meter, is a decimeter. (In scientific writing it is customary to place a zero before the decimal point of a number smaller than 1.) The prefix for 1⁄100 (or 0.01) metric unit is *centi-*; 0.01 meter is a centimeter. Similarly, using *milli-* for 1⁄1000 (or 0.001) unit, 0.001 meter is a millimeter. Figure 3.2 compares the smaller metric units with the English inch. The most familiar larger unit is 1000 times larger than the basic unit. Its prefix is *kilo-*; 1000 meters is a kilometer.

Table 3.1 shows the metric prefixes. *Centi-*, *milli-*, and *kilo-* are the most important; they should become a part of your vocabulary. You may recognize common uses of some of these prefixes. *Megaton* (1 million tons) is used to describe the size of hydrogen bombs, and *kilobucks* and *megabucks* have become slang expressions for one thousand and one million dollars.

The metric symbols in Table 3.1 are used as abbreviations for metric units. The letter m identifies the meter. The symbol for kilometer is formed by placing a k for *kilo-* in front of the m for meter: 1 km = 1 kilometer. Similarly, 1 cm is 1 centimeter, and 1 mm is 1 millimeter. The relationships among these distances are summarized in the three equations that follow on the next page:*

*See Appendix I, Part F, for information on exponential numbers.

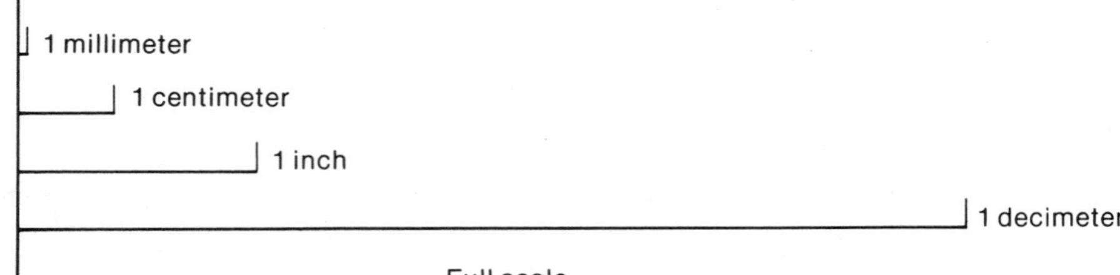

Figure 3.2
Comparison between English and metric units of length.

Table 3.1 Metric Prefixes*

Large Units			Small Units		
METRIC PREFIX	**METRIC SYMBOL**	**MULTIPLE**	**METRIC PREFIX**	**METRIC SYMBOL**	**SUBMULTIPLE**
tera-	T	10^{12}	deci-	d	$0.1 = 10^{-1}$
giga-	G	10^{9}	**centi-**	**c**	**0.01** $= 10^{-2}$
mega-	M	$1\ 000\ 000 = 10^{6}$	**milli-**	**m**	**0.001** $= 10^{-3}$
kilo-	**k**	**1 000** $= 10^{3}$	micro-	μ	$0.000\ 001 = 10^{-4}$
hecto-	h	$100 = 10^{2}$	nano-	n	10^{-9}
deka-	da	$10 = 10^{1}$	pico-	p	10^{-12}
Unit (gram, meter, liter) $1 = 10^{0}$			femto-	f	10^{-15}
			atto-	a	10^{-18}

*The most important prefixes are printed in boldface.

$$1\text{ km} = 1000\text{ m} = 10^{3}\text{m} \quad or \quad 1\text{ m} = 0.001\text{ km} = 10^{-3}\text{ km} \tag{3.8}$$

$$1\text{ cm} = 0.01\text{ m} = 10^{-2}\text{ m} \quad or \quad 1\text{ m} = 100\text{ cm} = 10^{2}\text{ cm} \tag{3.9}$$

$$1\text{ mm} = 0.001\text{ m} = 10^{-3}\text{ m} \quad or \quad 1\text{ m} = 1000\text{ mm} = 10^{3}\text{ mm} \tag{3.10}$$

The outstanding advantage of the metric system over the English system is the ease with which values may be converted from one unit to another. Conversions are all made by multiplying or dividing by multiples of 10—accomplished simply by moving the decimal point a certain number of places to the right or to the left. Until you become familiar with these conversions, you may be uncertain which direction the decimal point should move. A dimensional analysis setup of the problem helps remove this uncertainty.

EXAMPLE 3.3 How many meters are there in 28.6 cm?

First step: Identify the given quantity and unit path,

----- -----

28.6 cm; cm → m

The given quantity is now to be multiplied by a conversion factor that will change centimeters to meters. Equation 3.9 shows that there are 100 cm in one meter. From this, two factors are possible. Select the correct one and complete the problem.

----- -----

$$28.6 \text{ cm} \times \frac{1 \text{ m}}{100 \text{ cm}} = 0.286 \text{ m}$$

If you had mistakenly multiplied by 100 instead of dividing, your units would have been meaningless (cm^2/m), which would have alerted you to the error.

EXAMPLE 3.4 How many millimeters are in 3.04 cm?

Although the relationship between millimeters and centimeters has not been stated in an equation, you should be able to see it by comparing Equations 3.9 and 3.10. If 1000 mm and 100 cm are both equal to 1 m, then . . . Supply the numbers to complete the following relationship:

__________ mm = __________ cm

1000 mm = 100 cm

From Equations 3.9 and 3.10, 100 cm and 1000 mm are both equal to one meter, and therefore equal to each other. The relationship reduces to 10 mm = 1 cm, an equally acceptable answer.

Now complete the problem: 3.04 cm × __________ =

$$3.04 \text{ cm} \times \frac{1000 \text{ mm}}{100 \text{ cm}} = 30.4 \text{ mm} \quad or \quad 3.04 \text{ cm} \times \frac{10 \text{ mm}}{1 \text{ cm}} = 30.4 \text{ mm}$$

To convert between the English and metric systems, one of several conversion factors is required. (Many of these appear in Appendix II, Table AP-3.) Perhaps the most common length conversion is the relationship between centimeters and inches, which now defines the inch

$$1 \text{ in} = 2.54 \text{ cm} \qquad (3.11)$$

EXAMPLE 3.5 How many meters are in 7.60 ft? (Use Equation 3.11 to get from English to metric units.)

Because you have been limited to Equation 3.11 for the conversion from English to metric units, your unit path must have several steps. Write that unit path.

-----

ft → in → cm → m

Now start with the given quantity, complete the entire dimensional analysis setup, and calculate the answer.

$$7.60\ \cancel{ft} \times \frac{12\ \cancel{in}}{1\ \cancel{ft}} \times \frac{2.54\ \cancel{cm}}{1\ \cancel{in}} \times \frac{1\ m}{100\ \cancel{cm}} = 2.32^*\ m$$

Volume

PG 3C Given an English–metric conversion table, convert a volume measurement expressed in (a) English units, (b) liters, (c) milliliters, or (d) cubic centimeters to each of the other units.

The most common volume unit in the chemistry laboratory is the **cubic centimeter, or cm^3.** It is the volume of a cube one centimeter on an edge. (Compare with the English cubic inch, which represents the volume of a cube one inch on an edge.) Figure 3.3 illustrates metric volume measurements.

To convert a volume from one set of cubic units to another you must apply a length conversion three times. This is shown in the following example.

EXAMPLE 3.6 How many cubic inches are in a box 9.00 cm long by 6.00 cm wide by 4.00 cm high?

SOLUTION: To find the volume of a rectangular box you multiply the length by the width by the height. The volume of the box in cubic centimeters is

$$9.00\ cm \times 6.00\ cm \times 4.00\ cm = 216\ cm^3$$

To find the volume in cubic inches, each centimeter measurement should be converted to inches.

$$\underbrace{9.00\ \cancel{cm} \times \frac{1\ in}{2.54\ \cancel{cm}}}_{\text{length}} \times \underbrace{6.00\ \cancel{cm} \times \frac{1\ in}{2.54\ \cancel{cm}}}_{\text{width}} \times \underbrace{4.00\ \cancel{cm} \times \frac{1\ in}{2.54\ \cancel{cm}}}_{\text{height}} = \underbrace{13.2\ in^3}_{\text{volume}}$$

This expression may be simplified as

$$9.00\ \cancel{cm} \times 6.00\ \cancel{cm} \times 4.00\ \cancel{cm} \times \left(\frac{1\ in}{2.54\ \cancel{cm}}\right)^3 = 13.2\ in^3$$

or,

$$9.00\ \cancel{cm} \times 6.00\ \cancel{cm} \times 4.00\ \cancel{cm} \times \frac{1^3\ in^3}{2.54^3\ \cancel{cm^3}} = 13.2\ in^3$$

Notice that both the numbers and the units in both numerator and denominator are raised to a power when a fraction is to be raised to that power. (The algebra is explained in greater detail in Appendix I, Part C, if you wish to review it.)

*Calculated answers are rounded off according to the rules of significant figures, which are considered in Section 3.7.

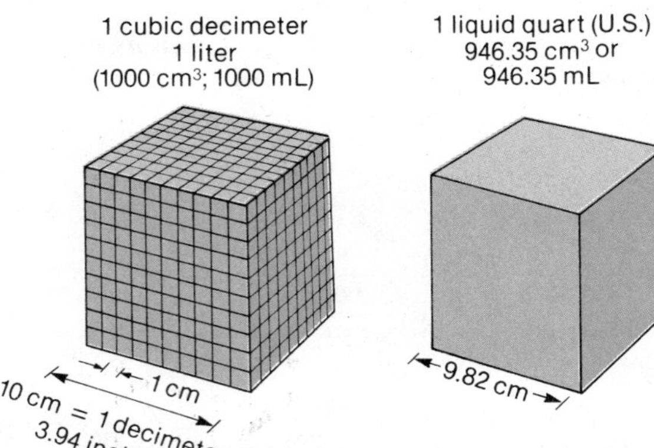

Figure 3.3
Comparison of common measures of volume.

EXAMPLE 3.7 Calculate the number of cubic meters of earth in 15.0 cubic yards of fill.

Appendix II, Table AP-3, gives you the necessary length conversion. Identify the given quantity, set up the problem, and calculate the answer.

$$150\ \text{yd}^3 \times \left(\frac{1\ \text{m}}{1.09\ \text{yd}}\right)^3 = 15.0\ \text{yd}^3 \times \frac{1^3\ \text{m}^3}{1.09^3\ \text{yd}^3} = 11.6\ \text{m}^3$$

This setup represents the unit path $\text{yd}^3 \rightarrow \text{m}^3$, using the conversion factor 1 m = 1.09 yd three times, or $1^3\ \text{m}^3 = 1.09^3\ \text{yd}^3$.

Volumes of liquids and gases are usually expressed in "capacity" units rather than cubic length units. The common metric capacity unit is the **liter,** which is exactly 1 cubic decimeter (1 dm^3) or 1000 cubic centimeters (1000 cm^3). This volume is very close to one U.S. quart—1.06 quarts, to be more exact. The smaller unit commonly encountered in the laboratory is the **milliliter**, mL. The prefix *milli-* again means 1/1000th part, so 1 milliliter = 0.001 liter, or 1000 milliliters = 1 liter. By definition there are also 1000 cubic centimeters in a liter. *This makes 1 cubic centimeter exactly equal to 1 milliliter.*

EXAMPLE 3.8 Calculate the number of liters in 0.500 cubic foot.

The necessary conversion factors are given in Appendix II, Table AP-3. Be careful about the final step in the setup.

$$0.500\ \cancel{ft^3} \times \frac{30.5^3\ \cancel{cm^3}}{1^3\ \cancel{ft^3}} \times \frac{1\ L}{1000\ \cancel{cm^3}} = 14.2\ L$$

Notice that the 1000 and the 1 in the final step are *not* cubed. The liter is already a volume unit, not the cube of a length unit.

QUICK CHECK 3.3

1. How many square centimeters are in a square meter?
2. How many cubic centimeters are in (a) 1 milliliter; (b) 1 liter?

3.4 MASS

PG 3D Distinguish between mass and weight.

3E Given an English–metric conversion table, convert a mass measurement, expressed in (a) English units, (b) grams, (c) kilograms, (d) centigrams, or (e) milligrams, to each of the other units.

Consider a tool carried by astronauts on a trip to the moon. Suppose that tool, on earth, weighs six ounces. If it were on the surface of the moon, it would weigh about one ounce. But halfway between the earth and moon it would be essentially weightless. Released in "midair" it would remain there, floating, until moved by one of the astronauts to some other location. Yet in all three locations it would be the same tool, having a constant quantity of matter.

Mass is a measure of quantity of matter. Weight is a measure of the force of gravitational attraction. Weight is proportional to mass, and the ratio between them depends upon where in the universe you happen to be. Fortunately this proportionality is essentially constant over the surface of the earth. When you "weigh" something—measure the force of gravity on that object—you can express this weight in terms of mass. By common usage, *weighing* an object is the same as measuring its mass. In the laboratory mass is measured on a *balance* (Fig. 3.1).

The basic unit of mass is the **kilogram** (kg). A kilogram weighs about 2.2 pounds and, therefore, is too large a unit for small scale work in chemistry. The more common laboratory unit of mass is the gram (g), which is $\frac{1}{1000}$ kilogram. The balances commonly used in college chemistry laboratories are capable of measuring centigrams ($\frac{1}{100}$ g) or milligrams ($\frac{1}{1000}$ g).

Conversion between the metric and English units may be made with either of the following factors:

$$454\ \text{grams} = 1\ \text{pound} = 16\ \text{ounces} \qquad (3.12)$$

$$1\ \text{kilogram} = 2.20\ \text{pounds} \qquad (3.13)$$

EXAMPLE 3.9 Find the mass in grams of an object that weighs 13.4 ounces.

Solve the problem completely.

$$13.4\ \cancel{oz} \times \frac{1\ \cancel{lb}}{16\ \cancel{oz}} \times \frac{454\ \cancel{g}}{1\ \cancel{lb}} = 380\ g$$

The setup might be shortened slightly by moving directly from ounces to grams using the extension of Equation 3.12:

$$13.4\ \cancel{oz} \times \frac{454\ g}{16\ \cancel{oz}} = 380\ g$$

EXAMPLE 3.10 (a) How many grams are in 0.285 kg? (b) How many kilograms are in 73.4 g?

These are simple metric conversions, using metric prefixes for mass measurements just as you used them for length measurements earlier.

(a) $0.285\ \cancel{kg} \times \frac{1000\ g}{1\ \cancel{kg}} = 285\ kg$ (b) $73.4\ \cancel{g} \times \frac{1\ kg}{1000\ \cancel{g}} = 0.0734\ kg$

Example 3.10 involves the change between grams and kilograms in both directions, using the conversion factor 1000 g = 1 kg. When you multiply or divide by 1000, you simply move the decimal point three places to the right for multiplication or three places to the left for division. This can be done mentally. To help you remember which way to go, note that there are 1000 little units in one big unit. In (a), you start with big units (kilograms), so the equivalent number of little units (grams) must be larger. You multiply by 1000, moving the decimal three places right. In (b), you begin with little units (grams), so there will be a smaller number of big units (kilograms). This means dividing by 1000—moving the decimal three places left.

The unit ↔ kilounit conversion will be made many times in this text. So will its equivalent conversion between units and milliunits, also different by a factor of 1000. Hereafter we will make these conversions without comment. You should do this too, reasoning out the direction to move the decimal as explained above. If you are ever uncertain, however, write the dimensional analysis setup and remove all doubts.

QUICK CHECK 3.4
In a given length, which is the larger number, the number of centimeters or the number of meters?

3.5 TEMPERATURE MEASUREMENT

PG 3F Given a temperature in °F or °C, convert from one scale to the other.

The familiar temperature scale in the United States is the Fahrenheit scale. The temperature scale normally used for scientific purposes is the Celsius scale. The Celsius scale is based on two "fixed" points, the normal freezing and boiling points of water. The freezing point is assigned a value of 0°C, and the boiling point 100°C. The difference between the fixed points is divided into 100 identical degrees. The corresponding temperatures on the Fahrenheit scale are 32°F for the freezing point of water and 212°F for the boiling point (Fig. 3.4)—a difference of 180 identical degrees.

Because of the way the temperature scales have been defined, the Celsius degree is 180/100 = 1.8 times as large as the Fahrenheit degree. They do not have a common zero, however, and therefore the temperatures are not directly proportional to each other. This rules out dimensional analysis as a way of converting from one temperature to the other.* Instead the two scales are related by the equation

$$°F - 32 = (1.8)(°C) \tag{3.14}$$

To solve a temperature conversion problem you solve Equation 3.14 for the unknown, substitute the given temperature and calculate the wanted temperature.

EXAMPLE 3.11 What is the Celsius temperature on a comfortable 72°F day? Solve to the nearest °C.

$$°C = \frac{°F - 32}{1.8} = \frac{72 - 32}{1.8} = 22$$

EXAMPLE 3.12 It's a cold day, about − 25°C. Calculate the corresponding Fahrenheit temperature to the nearest degree.

$$°F = (1.8)(°C) + 32 = (1.8)(-25) + 32 = -13$$

SI units include a third temperature scale known as the Kelvin or absolute scale. Kelvin degrees are identical to Celsius degrees, but zero on the Kelvin

*Dimensional analysis may be used to change between a temperature *difference* on one scale to a *difference* on the other:

1.8°F difference = 1.0°C difference

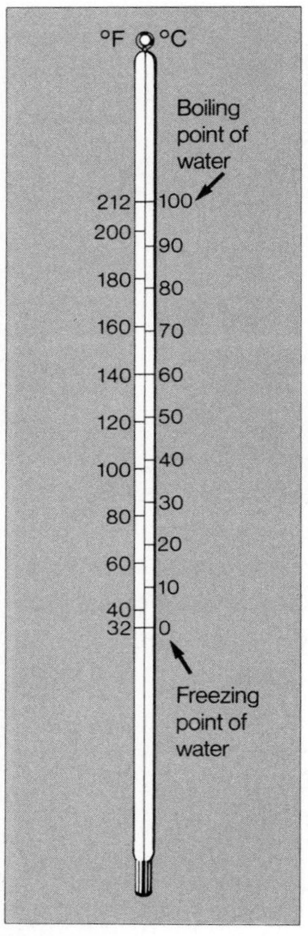

Figure 3.4
Comparison between Fahrenheit and Celsius temperature scales.

scale is 273.15°C below zero on the Celsius scale. The two scales are therefore related by the equation

$$K = °C + 273.15 \tag{3.15}$$

Notice that the degree sign, °, is not used with Kelvin temperatures. The Kelvin scale is based on an absolute zero, theoretically the lowest temperature possible. We will not be concerned with Kelvin temperatures until we study gases.

3.6 DENSITY

PG 3G Given two of the following, calculate the third: mass of a sample; volume occupied by the sample; density.

Density is an example of the combination of base units to define a physical property. Units produced by such combinations are called **derived units.** The formal definition of density is **mass per unit volume.** Expressed mathematically, this becomes

$$\text{density} = \frac{\text{mass}}{\text{volume}} \tag{3.16}$$

We can think of density as a measure of the "heaviness" of a substance, in the sense that a block of iron is heavier than a block of aluminum of the same size. Table 3.2 lists the densities of some common materials.

The definition of density establishes the units in which it is expressed. Mass is measured in grams; volume is measured in cubic centimeters. Therefore, according to Equation 3.16, the units of density are grams/cubic centimeter. Liquid densities may be expressed as grams/milliliter. (Recall that one milliliter is exactly equal to one cubic centimeter by definition.) There are, of course, other units in which density can be expressed, but they all must reflect the definition in terms of mass/volume. Examples include grams/liter and pounds/cubic foot.

In order to find the density of a substance, it is necessary to know both the mass and volume of the same quantity of the substance. Dividing the mass by the volume yields the density.

EXAMPLE 3.13 A 12.0-cm^3 piece of magnesium weighs 20.9 g. Find the density of magnesium.

----- -----

$$\text{density} = \frac{\text{mass}}{\text{volume}} = \frac{20.9\ \text{g}}{12.0\ \text{cm}^3} = 1.74\ \text{g/cm}^3$$

Table 3.2 Approximate Densities of Some Common Substances*

Helium (used in balloons)	0.000 17	Aluminum	2.7
Air	0.0012	Iron	7.8
Pine lumber	0.5	Copper	9.0
Maple lumber	0.6	Silver	10.5
Oak lumber	0.8	Lead	11.4
Water	1.0	Mercury	13.6
Glass	2.5	Gold	19.3

*Densities are given in grams per cubic centimeter at sea level and 20°C.

Specific gravity tells us how dense a substance is compared to some standard, usually water. More precisely, **specific gravity is the ratio of the density of a substance to the density of water at 4°C.** Expressed as an equation,

$$\text{specific gravity} = \frac{\text{density of substance (g/cm}^3\text{)}}{\text{density of water at 4°C}} \tag{3.17}$$

In Equation 3.17 the same units for density appear in both numerator and denominator. Therefore all units cancel, and specific gravity is a unitless or dimensionless term, a pure number.

In metric units the density of water at 4°C is 1.000 g/cm^3. Dividing the density of a substance by 1.000 g/cm^3 causes the specific gravity of a substance to be numerically equal to its density in grams per cubic centimeter. Accordingly, if the specific gravity of a substance is 6.3, we may conclude that its density is 6.3 g/cm^3.

Density furnishes a useful equivalence relationship for making conversions between mass and volume. Since the density of copper, for example, is 8.9 g/cm^3, this means that 8.9 g of copper represent the same quantity as 1.0 cm^3 of copper. Therefore we may say 8.9 g copper $\simeq$ 1.0 cm^3 copper. Use of density as a connecting link between mass and volume is shown in the following example.

EXAMPLE 3.14 The density of a certain kind of wood is 0.860 g/cm^3. Find the mass of a block of that wood measuring 16.0 cm × 5.00 cm × 3.00 cm.

SOLUTION: The volume of the wood is not given, but it can be found simply by multiplying the length–width–height dimensions. The product is the given quantity. The density may be used as a conversion factor in a dimensional analysis setup:

$$16.0\ \cancel{\text{cm}} \times 5.00\ \cancel{\text{cm}} \times 3.00\ \cancel{\text{cm}} \times \frac{0.860\ \text{g}}{1\ \cancel{\text{cm}^3}} = 206\ \text{g}$$

Many students like to solve density problems algebraically by rearranging the defining equation and substituting. For instance, in Example 3.14,

$$\text{density} = \frac{\text{mass}}{\text{volume}}$$

$$\text{mass} = \text{volume} \times \text{density}$$

$$\text{mass in grams} = 16.0\ \cancel{\text{cm}} \times 5.00\ \cancel{\text{cm}} \times 3.00\ \cancel{\text{cm}} \times \frac{0.860\ \text{g}}{1\ \cancel{\text{cm}^3}} = 206\ \text{g}$$

The resulting arithmetic is identical, and the algebraic approach is entirely correct. Part of the reason for introducing density at this point, however, is to give you practice in the use of dimensional analysis. We will continue to use it whenever possible, and we recommend that you do too.

EXAMPLE 3.15 The density of a certain copper-plating solution is 1.18 g/cm³. How many milliliters will be occupied by 150 g of the solution?

Recall that a cubic centimeter and a milliliter are identical volumes. Starting with the given quantity and interpreting density as an equivalence between mass and volume, the setup is straightforward. Complete the problem.

$$150\ \cancel{\text{g}} \times \frac{1\ \text{mL}}{1.18\ \cancel{\text{g}}} = 127\ \text{mL}$$

Part III Precision in Measurement

3.7 SIGNIFICANT FIGURES

Uncertainty in Measurement

No physical measurement is exact. Every measurement has some uncertainty associated with it. This is evident as you attempt to measure the length of a board with a series of meter sticks that are graduated in smaller and smaller subdivisions, as shown in Figure 3.5. In Figure 3.5A there are no graduations. Looking at the top illustration alone, the best that can be done is to estimate the length of the board as 0.6 or 0.7 m. The number of tenths of a meter is uncertain. Uncertainty is often added to a measurement as a "plus or minus." In this case the length of the board might be recorded as 0.6 ± 0.1 m, or 0.7 ± 0.1 m.

In Figure 3.5B the meter stick is graduated in tenths of a meter (decimeters), but numbered in centimeters. It is clear that the board is between 0.6 and 0.7 m, or between 60 and 70 cm. Can you estimate the next digit? The end of the

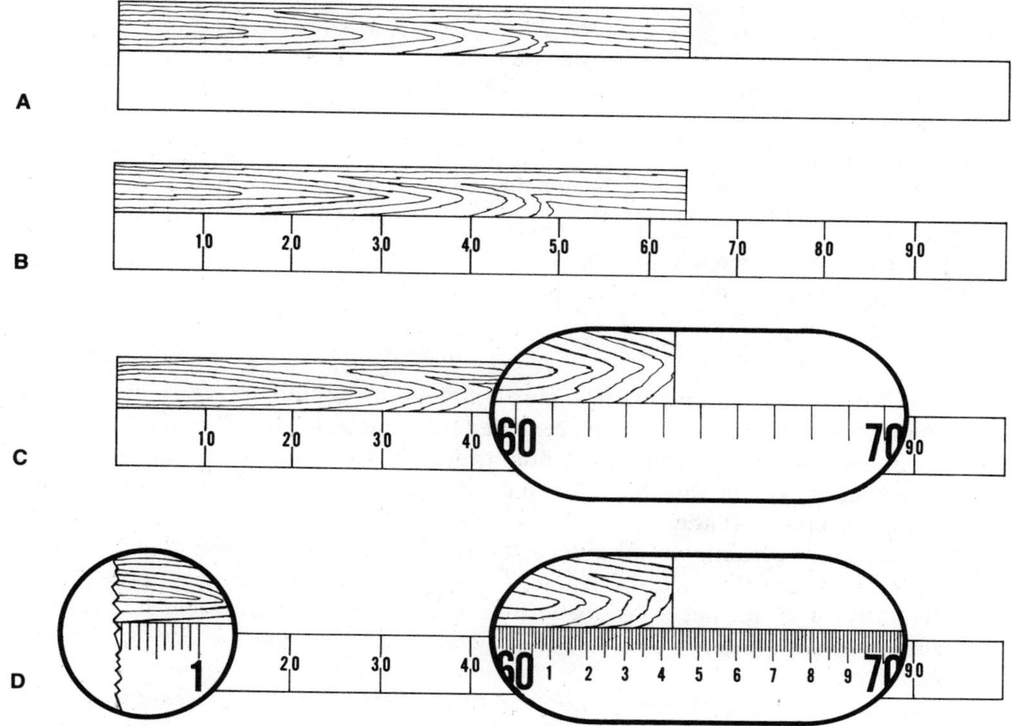

Figure 3.5
Significant figures in a measurement.

board is almost halfway between 60 and 70 cm, but it is hard to tell if it is closer to 64 or 65. Either number is reasonable, so we recognize that the number of centimeters is uncertain. The reading might be recorded as 64 ± 1 cm or 65 ± 1 cm. If the length is recorded in meters, the numerals are the same, but the decimal is two places to the left: 0.64 ± 0.01 m or 0.65 ± 0.01 m. The uncertain digit, the 4 or 5 in either expression of the length, is sometimes called the **doubtful digit**.

If the measuring stick is graduated in centimeters, as in Figure 3.5C, the length of the board may be estimated to the next decimal place. 64.2 ± 0.1 or 64.3 ± 0.1 cm, or 0.642 ± 0.001 or 0.643 ± 0.001 m, are acceptable measurements. The doubtful digit is now the 2 or 3. Extending graduations to millimeters in Figure 3.5D lets us select 64.3 cm as the better reading, but roughness at the ends of the board and meter stick plus the thickness of the graduation lines make it impractical to estimate tenths of a millimeter. To record such a distance would be misleading, suggesting measuring precision that simply is not present.

It is important in scientific work to make accurate measurements and to record them correctly. It is also important to know how reliable a measurement is—that is, how large an uncertainty is present. One way to satisfy these goals is to use **significant figures**. The number of significant figures in a quantity is **the number of digits that are known accurately plus the first uncertain digit—the doubtful digit**. It follows from this definition that, if a quantity is recorded correctly in terms of significant figures, *the doubtful digit is the last digit written*.

Table 3.3

Figure AP-3	Reading	Doubtful Digit Value	Significant Figures
A	0.6 ± 0.1 m	Tenths	1
B	0.64 ± 0.01 m 64 ± 1 cm	Hundredths Units	2 2
C & D	0.643 ± 0.001 m 64.3 ± 0.1 cm	Thousandths Tenths	3 3

All of the measurements in Figure 3.5 were presented correctly in terms of significant figures. These measurements are summarized in Table 3.3. In each case, the last digit shown is the doubtful digit. In C and D, for example, the uncertainty was in the third decimal place at 0.001 m. The recorded value, 0.643 m, ended at the third decimal place. When the same measurement is expressed in centimeters, the uncertainty is in the first decimal place, 0.1 cm, which is where the recorded value, 64.3 cm, ends. If this rule is applied consistently, it is not necessary to write the uncertainty when recording a measurement. It is simply understood that *all measurements are doubtful in the last digit shown.*

Another important observation can be made from the measurements in Figure 3.5: Only one measurement was made with the fully calibrated meter stick. Yet two values, 0.643 m and 64.3 cm, are given for that single measurement. The values and their numerical representation are equally precise. Hence the important conclusion: *The measurement process, not the units in which the result is expressed, determines the number of significant figures in a quantity. Therefore, the location of the decimal point has nothing to do with significant figures.*

Counting Significant Figures

PG 3H Given a measured quantity, state the number of significant figures in which it is expressed.

The definition of significant figures and the italicized and colored statements in the last three paragraphs set the rule for counting the number of significant figures in a measurement:

> *Begin with the first nonzero digit and end with the doubtful digit—the last digit shown.*

Application of this rule to Figure 3.5A shows that the first nonzero digit is 6 in 0.6 m. It is also the doubtful digit. The measurement begins and ends with the same digit. It is a one-significant-figure number. In Figure 3.5B the 6 is again the first nonzero digit and the 4 is doubtful. 0.64 and 64 are two-significant-figure numbers. In Figures 3.5C and 3.5D the 6 continues to be the first nonzero digit and the 3 is doubtful. 0.643 and 64.3 are three-significant-figure numbers.

Note again that the location of the decimal point has nothing to do with the number of significant figures.

The decimal point location requires special attention when zeros appear. Suppose the measurement in Figure 3.5D were to be expressed in kilometers. It would be 0.000 643 ± 0.000 001 km. This is still a three-significant-figure number; that was established by the measuring process. Remember: "The location of the decimal point has nothing to do with significant figures." Also always begin counting at the first nonzero digit. Specifically, *do not begin counting at the decimal point.*

What if the 0.643 m were to be expressed in nanometers: 643 000 000 nm? This must still be a three-significant-figure number, but the uncertainty is 1 000 000. How do we end the recorded value with the doubtful digit, 3? The answer is to write the number in exponential notation:* 6.43×10^{8} nm. In fact, exponential notation solves the very-small-number problem too: 6.43×10^{-4} km.

Sometimes—one time in ten, on the average—the doubtful digit is a zero. If so, it must be the last digit recorded. Suppose, for example, the length of a board is seventy five centimeters, plus or minus 0.1 cm. To record this as 75 cm is incorrect; it implies that uncertainty is ±1 cm. The correct way to write this number is 75.0 cm. If the measurement had been uncertain to 0.01 cm, it would be recorded as 75.00 cm. Always, *the doubtful digit is the last digit written, even if it is a zero to the right of the decimal point.*

Before you try counting significant figures, let's summarize everything so far. For any quantity,

1. The number of significant figures is the numbers that are known accurately plus the doubtful digit.
2. The doubtful digit is always the last digit written.
3. Begin counting with the first nonzero digit.
4. End counting with the doubtful digit.
5. Exponential notation must be used for very large numbers to show which zeros, if any, are significant.

The full meaning of Item 5 will appear in the discussion of the example below.

EXAMPLE 3.16 For each of the following quantities, write the correct number of significant figures.

45.26 ft ____	0.109 in ____	0.000 25 kg ____	2.3659×10^{-8} cm ____
163 mL ____	0.60 ft ____	62 700 cm ____	5.890×10^{5} L ____

45.26 ft 4	0.109 in 3	0.000 25 kg 2	2.3659×10^{-8} cm 5
163 mL 3	0.60 ft 2	62 700 cm ?	5.890×10^{5} L 4

Begin counting significant figures with the first nonzero digit. This is the 2 in 0.000 25 kg. The tail-end zeros in 0.60 ft and 5.890×10^{5} L identify them as the doubtful digit, and therefore significant. For 62 700 cm, see below.

*See Appendix I, Part F, for a discussion of exponential notation.

The 62 700 cm in Example 3.16 shows why exponential notation is required for very large numbers. The value has three significant figures if the 7 is doubtful, four significant figures if the first zero is doubtful, and five if the second zero is doubtful:

6.27×10^4	doubtful digit in hundreds	(three significant figures)
6.270×10^4	doubtful digit in tens	(four significant figures)
6.2700×10^4	doubtful digit in ones	(five significant figures)

Exact Numbers

Notice that everything that has been said about significant figures is related to *measured* quantities. Uncertainty in numbers appears only in measurements. Numbers found by counting are exact. A bicycle has exactly two wheels, and at any instant there may be exactly five apples in a fruit basket. Numbers in definitions are also exact. There are exactly 16 ounces in 1 pound, 1000 meters in 1 kilometer, and 12 eggs in 1 dozen eggs. Exact numbers are infinitely significant.

Rounding Off

PG 31 Round off given numbers to a stated number of significant figures.

Sometimes, when experimentally measured quantities are added, subtracted, multiplied, or divided, the answer contains figures that are not significant. When this happens the result must be rounded off. Rules for rounding off are as follows:

1. If the first digit to be dropped is less than 5, leave the digit before it unchanged.
2. If the first digit to be dropped is 5 or more, increase the digit before it by 1.

Other rules for rounding off vary if exactly 5 is the first digit to be dropped. Among other things, the proposed method is more useful in working with computers. For any individual round off by any method, every rule has a 50% chance of being "more correct." Even when "wrong," the rounded off result is acceptable because the only digit affected is the doubtful one.

EXAMPLE 3.17 Round off each of the following quantities to three significant figures.

a) 1.427 52 g/cm^3 __________ e) 45 853 cm __________

b) 643.349 cm^2 __________ f) 0.0394 449 8 m __________

c) 0.007 456 2 kg __________ g) 3.605×10^{-7} cm __________

d) 2.103×10^4 cm __________ h) 3.5000 sec __________

-----

a) 1.43 g/cm^3

b) 643 cm^2

c) 0.007 46 kg or 7.46×10^{-3} kg

d) 2.10×10^4 cm

e) 4.59×10^4 cm

f) 0.0394 m or 3.94×10^{-2} m.

g) 3.61×10^{-7} cm

h) 3.50 sec

Addition and Subtraction

PG 3U Add or subtract given quantities and express the result in the proper number of significant figures.

The **significant figure rule for addition and subtraction** can be stated as follows:

Round off the answer to the first column that has a doubtful digit.

Example 3.18 shows how this rule is applied:

EXAMPLE 3.18 A student weighs four different chemicals into a pre-weighed beaker. The individual weights and their sum are as follows:

Beaker	319.542 g
Chemical A	20.460 g
Chemical B	0.0639 g
Chemical C	38.2 g
Chemical D	4.173 g
Total	382.4389 g

Express the sum to the proper number of significant figures.

This sum is to be rounded off to the first column that has a doubtful digit. What column is this: hundreds, tens, units, tenths, hundredths, thousandths, or ten thousandths?

Tenths. The doubtful digit in 38.2 is in the tenths column. In all other numbers the doubtful digit is in the hundredths column or smaller.

According to the rule, the answer must now be rounded off to the nearest number of tenths. What answer do you report?

382.4 g

Another way to solve the above problem is to round off all numbers before adding. Rounding off everything to tenths gives 382.5 as the answer:

```
319.5
 20.5
  0.1
 38.2
  4.2
-----
382.5
```

Both methods yield uncertainty in the tenths digit, so both are technically acceptable. The recommended procedure is to use all the data in the calculation and then round off the final answer.

This example may be used to justify the rule for addition and subtraction. A sum or difference digit must be doubtful if any number entering into that sum or difference is doubtful or unknown. In the left addition below, all doubtful digits are shown in color, and all digits to the right of a colored digit are simply unknown.

```
319.542                       319.5¦42
 20.4609                       20.4¦609
  0.063                         0.0¦63
 38.2                          38.2¦
  4.173                         4.1¦73
--------                      --------
382.4389 = 382.4              382.4¦389 = 382.4
```

In the left addition the 4 in the tenths column is clearly the first doubtful digit.

The addition at the right above shows a mechanical way to locate the first doubtful digit in a sum. Draw a vertical line after the last column that has every space occupied and the doubtful digit in the sum will be just left of that line. The result must be rounded off to the line.

The same rule, procedure, and rationalization hold for subtraction.

EXAMPLE 3.19 In an experiment in which oxygen is produced by heating potassium chlorate in the presence of a catalyst, a student assembled and weighed a test tube, test tube holder and catalyst. He then added potassium chlorate and weighed the assembly again. The data were as follows:

Test tube, test tube holder, catalyst, and potassium chlorate	26.255 g
Test tube, test tube holder, and catalyst	24.05 g

The weight of potassium chlorate is the difference between these numbers. Express this weight in the proper number of significant figures.

$$\begin{array}{r} 26.255 \text{ g} \\ \underline{24.05\ \ \text{g}} \\ 2.205 \text{ g} = 2.21 \text{ g} \end{array}$$

The 24.05 is doubtful in the hundredths column, so the difference is rounded off to hundredths.

Multiplication and Division

PG 3K Multiply or divide given measurements and express the result in the proper number of significant figures.

The **significant figure rule for multiplication and division is as follows:**

Round off the answer to the same number of significant figures as the smallest number of significant figures in any factor.

Again, application will be illustrated by example:

EXAMPLE 3.20 The density of a certain gas is 1.436 grams per liter. Find the mass of 0.0573 liter of the gas. Express the answer in the proper number of significant figures.

The mass is found by multiplying the volume by the density. The setup of the problem is

$$0.0573 \text{ \sout{liter}} \times \frac{1.436 \text{ g}}{1 \text{ \sout{liter}}} =$$

According to the rule, the product may have no more significant figures than the least number of significant figures in any factor. What is the smallest number of significant figures in any factor?

Three. 0.0573 is a three-significant-figure number: 5.73×10^{-2}.

Now calculate the answer and round off to three significant figures.

0.0823 g. If you showed 0.082, you forgot that counting significant figures begins at the first nonzero digit.

Example 3.20 may be used to justify the rule for multiplication and division. If both quantities have final digits that are one number higher than their true values, the true answer would be $0.0572 \times 1.435 = 0.0821$. If both are too low by 1, the problem is $0.0574 \times 1.437 = 0.0825$. Uncertainty appears in the third significant figure, just as predicted. Alternatively, each product number into which a doubtful multiplier enters must itself be doubtful. Colored numbers indicate the doubtful digits in the detailed multiplication:

$$\begin{array}{r} 1.436 \\ \times\ 0.0573 \\ \hline 4308 \\ 10052 \\ 7180 \\ \hline 0.0822828 \end{array}$$

EXAMPLE 3.21 Assuming the numbers are derived from experimental measurements, solve

$$\frac{(2.86 \times 10^4)\,(3.163 \times 10^{-2})}{1.8} =$$

and express the answer in the correct number of significant figures.

5.0×10^2. The answer should not be shown as 500, as the number of significant figures could be read as one, two, or three. Two significant figures are set by the 1.8.

EXAMPLE 3.22 How many millimeters are in 0.6294 cm? Express the answer in the proper number of significant figures.

With 10 millimeters per centimeter, the arithmetic is simple enough. But how many significant figures? How would you state the answer?

6.294 mm

By definition, there are *exactly* 10 mm in 1 cm. Exact numbers are infinitely significant. They never limit the number of significant figures in a product or quotient.

Significant Figures and This Book

Most modern calculators report answers in all the digits they are able to display, usually eight or more. Such answers are unrealistic; they should never be used. Calculations in this book have generally been made using all digits given, and the final answer has been rounded off according to the rules given in this section.

Part IV: Energy

3.8 UNITS OF ENERGY

PG 3L State the relationship between the two principal scientific units of energy. Given one, calculate the other.

The *meter*, the *kilogram*, and the *kelvin* are three of the "base units" on which the SI system of units is built. Another is the unit of time, the *second* (*s*). Energy is measured in a derived unit that is a combination of the units of mass, distance, and time. It is called the **joule** (pronounced *jool*), and it is defined as 1 $kg \cdot m^2/sec^2$. While the joule is mechanical in nature, it is also suitable for any form of energy. This includes heat energy, the only kind of energy that is considered in this text.

The more familiar unit of heat energy is the **calorie,** the amount of heat required to raise the temperature of one gram of water one degree Celsius. The calorie has now been redefined in terms of the joule:

$$1 \text{ calorie} = 4.184 \text{ joules} \tag{3.18}$$

Both the calorie and the joule are small quantities of energy, so energy is often expressed in terms of the *kilo-* unit, which is 1000 times larger—the kilocalorie (kcal) and kilojoule (kJ):

$$1 \text{ kcal} = 1000 \text{ cal}; \quad 1 \text{ kJ} = 1000 \text{ J} \tag{3.19}$$

It follows that

$$1 \text{ kilocalorie} = 4.184 \text{ kilojoules} \tag{3.20}$$

The term *Calorie* used in nutrition is actually the kilocalorie. Capitalization of the first letter is used to distinguish the larger unit from the smaller. Caloric food requirements vary considerably among individuals. A small adult doing average physical work needs as little as 2000 to 4000 Calories per day. This requirement may increase to 5000 to 6000 Calories per day for a large man engaged in strenuous physical labor.

In this text we will use the SI energy units, the joule and the kilojoule. Each example or problem answer, however, will be followed by its equivalent in calories or kilocalories, enclosed in square brackets.

EXAMPLE 3.23 How many kilocalories are equivalent to 42.5 kJ?

$$42.5 \cancel{\text{kJ}} \times \frac{1 \text{ kcal}}{4.184 \cancel{\text{kJ}}} = 10.2 \text{ kcal}$$

3.9 ENERGY AND CHANGE OF TEMPERATURE: SPECIFIC HEAT

PG 3M Given (a) the mass of a pure substance, (b) its specific heat, and (c) its temperature change, or initial and final temperatures, calculate the heat flow.

3N Given the heat flow to or from a known mass of a substance, and its temperature change, or initial and final temperatures, calculate the specific heat of the substance.

In order to raise the temperature of a substance you must heat it; heat must flow into the substance. If the substance is to be cooled, there must be heat flow from the substance to its surroundings. Experiments indicate that heat flow is proportional to both the mass of the sample and its temperature change. Combining these proportionalities and introducing a proportionality constant* yields the equation

$$Q = m \times c \times \Delta T \qquad (3.21)$$

where Q is the heat flow in joules, m is the mass in grams, and c is the proportionality constant. The Greek letter delta, Δ, is used to indicate "change" from one value to another. Δ always means the final value of some quantity minus the initial value. Thus, ΔT means final temperature minus initial temperature, $T_f - T_i$. Depending on the sizes of the final and initial values, a delta quantity may be positive or negative.

The proportionality constant, c, is a property of a pure substance called its **specific heat. Specific heat is the heat flow required to change the temperature of one gram of a substance one degree Celsius.** The units of specific heat may be determined by solving Equation 3.21 for the proportionality constant:

$$\text{specific heat} = c = \frac{Q}{m \times \Delta T} = \frac{\text{joules}}{\text{grams}\cdot{}^\circ\text{C}} = \frac{\text{J}}{\text{g}\cdot{}^\circ\text{C}} \qquad (3.22)$$

These units are read "joules per gram degree." Specific heats of selected substances are given in Table 3.4.

Specific heat is a measure of the relative ease with which a given quantity of a substance may be heated or cooled. A substance with a low specific heat absorbs little energy in warming through a given temperature range, and loses little as it cools. This is why metals, which generally have specific heats less than 0.8 J/(g·°C), are well suited for pots and pans used for cooking.

To store heat energy, a substance with a high specific heat is best. In solar heating systems, rocks are used to store heat from the sun in daytime for use at night. Their specific heats are relatively high. They are therefore able to absorb a large amount of heat per unit of mass, which is later given back when needed. At 4.184 J/g·°C, water has one of the highest specific heats of all substances. This is why air temperatures near large bodies of water are usually warmer in winter and cooler in summer than nearby inland temperatures.

*See Appendix I, Part D, for a discussion of proportionalities and proportionality constants.

Table 3.4 Selected Specific Heats

Substance	$\frac{J}{g \cdot °C}$	$\frac{cal}{g \cdot °C}$	Substance	$\frac{J}{g \cdot °C}$	$\frac{cal}{g \cdot °C}$
ELEMENTS			Silver (s)	0.24	0.057
Aluminum	0.88	0.21	Silver (l)	0.32	0.076
Cadmium (s)	0.232	0.0554	Sulfur	0.732	0.175
Cadmium (l)	0.267	0.0637	Zinc (s)	0.38	0.092
Carbon			Zinc (l)	0.512	0.122
Diamond	0.50	0.12			
Graphite	0.71	0.17	COMPOUNDS		
Cobalt	0.46	0.11	Acetone	2.1	0.51
Copper	0.38	0.092	Benzene	1.8	0.42
Gold (s)	0.13	0.031	Carbon		
Gold (l)	0.148	0.0355	tetrachloride	0.84	0.20
Iron (s)	0.444	0.106	Ethanol	2.5	0.59
Iron (l)	0.452	0.108	Methanol	2.6	0.61
Lead	0.16	0.038	Ice [$H_2O(s)$]	2.1	0.49
Magnesium	1.0	0.24	Water [$H_2O(l)$]	4.18	1.00
Silicon	0.71	0.17	Steam [$H_2O(g)$]	2.0	0.48

Specific heat problems are always such that three of the four factors in Equation 3.21 are given or available, and you must solve for the fourth. The problem is best solved algebraically rather than by dimensional analysis, as we have done so far. Dimensional analysis is still used, however, to check the correctness of the algebraic operations. The next two examples illustrate the procedure.

EXAMPLE 3.24 How much heat is required to raise the temperature of 475 grams of water for a pot of tea from 14°C to 90°C? Answer in both joules and kilojoules.

Direct substitution into Equation 3.21 yields the heat flow in joules. For ΔT, substitute $T_{final} - T_{initial}$. Calculate Q in joules. Include units in your calculation setup.

----- -----

$$Q = m \times c \times \Delta T = 475 \not{g} \times \frac{4.184 \text{ J}}{\not{g} \cdot °\not{C}} \times (90 - 14)°\not{C} = 1.5 \times 10^5 \text{ J } [3.6 \times 10^4 \text{ cal}]$$

$$1.5 \times 10^5 \text{ J} = 1.5 \times 10^2 \text{ kJ } [36 \text{ kcal}]$$

ΔT = 76°C, a two-significant-figure number. The answer is therefore rounded to two significant figures. The conversion between joules and kilojoules, from small units to large units, involves dividing by 1000. (See discussion following Example 3.10.) To divide by 1000 (or 10^3), move the decimal three places left, or reduce the exponent of 10 by 3: $10^5/10^3 = 10^2$.

Notice how dimensional analysis checks the correctness of the substitutions into the algebraic solution of Example 3.24. You were asked to include units in your calculation setup of the problem, as we did in the answer. Notice that the units cancel properly, leaving only joules, the units of the answer. If the substitutions are performed correctly, the units *must be correct*. If the units are not correct, look for an error.

EXAMPLE 3.25 A student in a college laboratory observes a loss of 803 joules (192 cal) as 141 grams of aluminum cool from 31.7°C to 25.0°C. Calculate the specific heat of aluminum from these data.

Dimensional analysis will help you catch more algebra mistakes if you first solve the equation (Equation 3.21 in this case) for the unknown, and then substitute the known values, including units, and calculate the answer. This time your starting equation is already solved; it is Equation 3.22. A *loss* of heat means that Q is negative; Q = −803 J. Substitute away; don't forget the units. Think about the significant figures in your answer too.

$$c = \frac{-803\ \text{J}}{(141\ \text{g})(25.0 - 31.7)°\text{C}} = 0.85\ \text{J/g·°C} \qquad [0.20\ \text{cal/g·°C}]$$

ΔT is 25.0 − 31.7 = −6.7°C. This negative value combines with the negative numerator to yield a positive answer. Also, even though every number in the data is a three-significant-figure number, the *change* in temperature has only two significant figures. This reduces the answer to two significant figures.

CHAPTER 3 IN REVIEW

3.1 Introduction to Measurement

3.2 Dimensional Analysis

3A Identify the "given quantity" in the statement of a problem. Beginning with the given quantity, set up and solve the problem by the dimensional analysis method.

3.3 Length and Volume Units in the Metric System

3B Given an English–metric conversion table, convert a length measurement expressed in (a) English units, (b) meters, (c) kilometers, (d) centimeters, or (e) millimeters to each of the other units.

3C Given an English–metric conversion table, convert a volume measurement expressed in (a) English units, (b) liters, (c) milliliters, or (d) cubic centimeters to each of the other units.

3.4 Mass

3D Distinguish between mass and weight.

3E Given an English–metric conversion table, convert a mass measurement expressed in (a) English units, (b) grams, (c) kilograms, (d) centigrams, or (e) milligrams to each of the other units.

3.5 Temperature Measurement

3F Given a temperature in °F or °C, convert from one scale to the other.

3.6 Density

3G Given two of the following, calculate the third: mass of a sample; volume occupied by the sample; density.

3.7 Significant Figures

3H Given a measured quantity, state the number of significant figures in which it is expressed.

3I Round off given numbers to a stated number of significant figures.

3J Add or subtract given quantities and express the result in the proper number of significant figures.

3K Multiply or divide given measurements and express the result in the proper number of significant figures.

3.8 Units of Energy

3L State the relationship between the two principal scientific units of energy. Given one, calculate the other.

3.9 Energy and Change of Temperature: Specific Heat

3M Given (a) the mass of a pure substance, (b) its specific heat, and (c) its temperature change, or initial and final temperatures, calculate the heat flow.

3N Given the heat flow to or from a known mass of a substance, and its temperature change, or initial and final temperatures, calculate the specific heat of the substance.

TERMS AND CONCEPTS

3.1 SI units
3.2 Dimensional analysis
Conversion factor
Conversion relationship
Given quantity
Unit path
3.3 Meter
Metric prefixes: *milli-, centi-, kilo-*
Liter
3.4 Mass
Weight
Kilogram
3.5 Fahrenheit
Celsius
Kelvin
3.6 Density
Derived unit
Specific gravity
3.7 Uncertainty in measurement
Significant figures
Doubtful digit
Exact number
3.8 Joule
Calorie
3.9 Specific heat

Most of these terms and many others appear in the Glossary. Use your Glossary regularly.

QUESTIONS AND PROBLEMS

Section 3.2

(1) What is the distance in feet between two points 3 miles apart? There are 5280 feet in a mile.

(49) How long will it take to travel the 406 miles between Los Angeles and San Francisco at an average of 53 miles per hour?

(2) If light travels 186 000 miles per second, how many minutes are required for light to cover the 93 million miles between the sun and Earth?

(50) How many minutes does it take a car traveling 55 miles per hour to cover 2.5 miles?

(3) Bamboo can grow as much as 3.0 feet in 24 hours. At that rate, how many feet will it grow in 63 hours?

(51) In 1960 the meter was defined to be equal to 1 650 763.73 wavelengths of a certain line in the emission spectrum of Krypton-86. How many wavelengths are in 5.78 meters?

(4) The rate of exchange on a certain day was 1.29 Canadian dollars per American dollar. An American tourist in Vancouver bought a sweater that had a price tag of $39.95. What did she pay for the sweater in American funds?

(52) An American tourist in Ensenada, Mexico, was startled to see $1800 on a menu as the price for a steak dinner. He then realized that the price represented pesos, and the exchange rate that day was 195 pesos to the dollar. How much did he have to pay for his steak in American funds?

(5) A college chemistry laboratory served 825 students in one semester, during which 55 beakers were broken. The storeroom manager is about to order beakers to replace those that are expected to be lost during the full year ahead (2 semesters per year). The anticipated enrollment for the coming year is 900 students each semester. Assuming breakage for the coming year will be at the same rate as during the semester just ended, how many beakers should be ordered?

(6) At an annual chili cook-off, 125 contestants have produced an average of 5.0 pounds of chili each. If the average amount of chili powder per pound of chili was 0.054 ounces, how many pounds of chili powder were used by all the contestants?

(53) What would be the cost in dollars of nails for a fence 88 feet long if you needed 9 nails per foot of fence, there were 36 nails in a pound, and they sold for 69 cents per pound?

(54)* A student drives a total of 515 miles to visit her sister. Her auto gets an average of 25 miles per gallon in the city and 31 miles per gallon on the highway. If she spends 1.5 hours in the city at an average speed of 26 miles per hour and another 8.7 hours on the highway, what is her average speed on the highway? How much gasoline does she consume?

Section 3.3

Make each conversion indicated. Use exponential notation when appropriate.

(7) 25.9 km = ________________ m
4.27 mm = ________________ m
9.46 cm = ________________ mm
937 nm = ________________ m

(8) The height of Angel Falls in Venezuela is 399.9 meters. How high is this in (a) yards; (b) feet?

(9) What is the length of the Mississippi River in kilometers if it is 2350 miles long?

(10) The summit of Mount Everest is 29 002 ft above sea level. Express this height in kilometers.

(55) 98.0 m = ________________ cm
16.9 m = ________________ km
6.13 km = ________________ cm
0.384 gm = ________________ m

(56) The average height of Mbuti pygmies is 135 cm. How high is this in inches?

(57) The Sears Tower in Chicago is 1454 ft tall. How high is this in meters?

(58) The smallest brilliant cut diamond, exhibited in Paris in 1979, had a diameter of about 1/50 in. How many millimeters is this?

Make each conversion indicated. Use exponential notation when appropriate.

(11) 46.2 m^2 = ________________ cm^2
751 mm^3 = ________________ m^3
231 mL = ________________ L

(12) The actual dimensions of a 2-by-4 piece of lumber are 1.5 in × 3.5 in. Calculate the cross sectional area in square centimeters.

(13) A new sidewalk 20 ft long by 36 in wide will be 4 in thick. Calculate the volume of concrete that must be ordered in cubic meters.

(14) A German automobile has a 50.0-liter fuel tank. Express this volume in U.S. gallons.

(59) 0.593 m^2 = ________________ mm^2
346 cm^3 = ________________ m^3
122 L = ________________ dm^3

(60) The area of North Carolina is 136 308 km^2. Calculate this area in square miles.

(61)* In Europe, spring water for drinking fountains is sold in 19-cubic-decimeter bottles. What is this volume in cubic inches? in gallons?

(62) An office building is heated by oil-fired burners that draw fuel from a 496-gallon storage tank. What is the volume of the tank in liters and cubic meters?

(15) The displacement of an automobile engine is specified as 3.0 L. Calculate this volume in cubic inches, the dimensions formerly used in the United States.

(16) A weather balloon displaces 113 ft^3 of air. Express this in liters.

(17) The strongest wind ever recorded was measured at 231 miles per hour on the top of Mt. Washington in New Hampshire. Convert this to kilometers per minute.

(18) A cost-conscious driver who needs gas for her car has the choice of two filling stations. The sign in front of one shows $0.317/L, and the other $1.23 per gallon. At which station should she stop?

(63) A shoe box measures 12 in × 5.5 in × 3.5 in. Calculate its volume in cubic centimeters.

(64) A back yard swimming pool measures 26 ft by 15 ft, and its average depth is 5.9 ft. Calculate the volume of the pool in cubic meters.

(65) A driver averaged 209 miles per hour in a qualifying run for the Memorial Day race at the Indianapolis Speedway. How fast was he traveling in feet per second? (5280 feet = 1 mile.)

(66)* A 750-milliliter bottle of a soft drink costs 33 cents. A quart of the same beverage is priced at 41 cents. Which is the better buy?

Section 3.4

(19) A person can pick up a large rock from the water near the shore of a lake, but may not be able to pick up the same rock from the beach. Compare the mass and the weight of the rock when in the lake and when on the beach.

(20) 0.595 kg = ______________ g
14.9 g = ______________ cg
0.871 kg = ______________ mg
0.0913 pg = ______________ g

(21) The mass of a new silver dollar is 26.4 g. Convert that mass to ounces.

(22)* Olga Svensen has just given birth to a bouncing 6-lb, 7-oz baby boy. How should she describe the weight of her child to her sister, who still lives in Sweden—in metric units, of course?

(23) There are 115 mg of calcium in a 100-g portion of whole cow's milk. How many grams of calcium is this? How many pounds?

(24) The largest recorded difference of weight between spouses is 922 pounds. The husband weighed 1020 pounds, and his dear wife, 98 pounds. Express this difference in kilograms.

(67) A person stands on a scale in the elevator in a tall building. The elevator starts going up, rises rapidly at constant speed for half a minute, and then slows to a stop. Compare the person's weight as recorded by the scale and his mass while the elevator was standing still, during the starting period, during the constant rate period, and during the slowing period.

(68) 435 mg = ______________ g
26.0 kg = ______________ g
86.7 mg = ______________ cg
70.4 μg = ______________ g

(69) A popular breakfast cereal comes in a box containing 510 g. How many pounds of cereal is this?

(70) The payload of a small pick-up truck is 1450 pounds. What is this in kilograms?

(71) An Austrian boxer reads 64.5 kg when he steps on a balance (scale) in his gymnasium. Should he be classified as a welterweight (136–147 pounds) or a middleweight (148–160 pounds)?

(72) The Hope diamond is the world's largest blue diamond. It weighs 44.4 carats. If 1 carat is 200 mg, calculate the mass of the diamond in grams and ounces.

Section 3.5

Fill in the spaces in the tables below so that each temperature is expressed in all three scales.

(25)

Celsius	Fahrenheit	Kelvin
69		
	−29	
		111
	36	
		358
−141		

(73)

Celsius	Fahrenheit	Kelvin
	40	
		590
−13		
		229
440		
	−314	

(26) "Normal" body temperature is 98.6°F. What is this temperature in Celsius degrees?

(27) Energy conservationists suggest that air conditioners should be set so they do not turn on until the temperature tops 78°F. What is the Celsius equivalent of this temperature?

(28) The world's highest shade temperature was recorded in Libya at 58.0°C. What is its Fahrenheit equivalent?

(29)* At high noon on the Lunar equator the temperature may reach 243°F. At night the temperature may sink to −261°F. Express the difference in temperature in degrees Celsius.

(74) The boiling point of acetone, a component of nail polish remover, is 56°C. What is this temperature on the Fahrenheit scale?

(75) To save heating fuel, it is recommended that home thermostats be set at no more than 68°F. Express this on the Celsius scale.

(76) A French manufacturer of hot water heaters marks 60°C as "normal temperature" on its thermostats. What is this temperature in Fahrenheit degrees?

(77)* A welcome rainfall caused the temperature to drop by 33°F after a sweltering day in Chicago. What is this temperature drop in degrees Celsius?

Section 3.6

(30) Oil and water do not mix. When both are in the same container, the oil is in a layer on top of the water. Why?

(31) Magnesium is among the least dense of our common metals. Calculate its density and specific gravity if an 865-cm^3 block has a mass of 1.51×10^3 g.

(32) A rectangular block of iron 4.60 cm × 10.3 cm × 13.2 cm has a mass of 4.92 kg. Find its density.

(33) Find the mass of 35.3 mL of alcohol if its density is 0.790 g/mL.

(78)* Does all wood float on water? Do all metals sink in water? How can a ship made of steel float? What determines whether a solid will float on or sink in a liquid?

(79) 182 g benzene, an important organic solvent, fills a graduated cylinder to the 207-mL mark. Calculate the density and specific gravity of benzene.

(80) Densities of gases are usually measured in grams per liter. Calculate the density of air if the mass of 24.5 liters is 29.0 grams.

(81) Ether, a well-known anesthetic, has a density of 0.736 g/cm^3. What is the volume of 225 grams of ether?

(34) What is the volume of a half pound (227 grams) of olive oil if its density is 0.92 g/mL?

(35)* The fuel tank in an automobile has a capacity of 11.0 gallons. If the density of gasoline is 42.0 lb/ft^3, what is the mass of fuel in kilograms when the tank is full?

(82)* A recipe calls for a half cup of butter (125 cm^3). Calculate its mass in grams if its density is 0.86 g/cm^3. (1 cup = 0.25 quart.)

(83)* How many grams of milk are in a 12.0-fl-oz tumbler? The density of milk is 64.4 lb/ft^3, and there are 32 fluid ounces in a quart.

Section 3.7

In how many significant figures is each of the following expressed?

(36) 4.5609 g salt
0.10-in-diameter wire
12.3×10^{-3} kg fat
5310 cm^3 copper
0.0231 ft licorice
6.1240×10^6 L salt brine
328 mL ginger ale
1200.0 mg dye

(84) 21.45 g sugar
248 mL weed killer
0.528-in line
85 300 cm wire
0.50 ft spaghetti
0.00084 kg silver
5.4096×10^6 cm thread
2.140×10^{-3} L acid

Round off each of the following to three significant figures:

(37) 52.20 mL helium
17.963 g nitrogen
78.45 mg MSG
23 642 000 μm wavelength
0.0041962 kg lead

(85) 6.101×10^{-3} km rope
0.0680 g silver nitrate
2500 m cable
794 500 tons fertilizer
$199.90

(38) A solution is prepared by dissolving 2.86 grams of sodium chloride, 3.9 grams of ammonium sulfate, and 0.896 grams of potassium iodide in 246 grams of water. Calculate the total mass of the solution and express the sum in the proper number of significant figures.

(39) An empty beaker has a mass of 94.33 grams. After some chemical has been added, the mass is 101.209 grams. What is the mass of chemical in the beaker?

(40) One of the tourist attractions of Acapulco, Mexico, is the divers who plunge from a cliff 118 feet above the ocean. Convert this height to meters and express it in the proper number of significant figures.

(86) A moving van crew picks up the following items: a couch that weighs 126 pounds; a chair that weighs 45.8 pounds; a piano at 2.4×10^2 pounds; and several boxes having a total weight of 374.82 pounds. Calculate and express in the correct number of significant figures the total weight of the load.

(87) A buret contains 33.25 mL sodium hydroxide solution. A few minutes later, however, the volume is down to 32.9 mL because of a small leak. How many milliliters of solution have drained from the buret?

(88) The mass of a box was measured on a balance and found to be 114.63 kg. Calculate the mass of the box in pounds, rounded off to the correct number of significant figures.

Section 3.8

(41) Complete the following: (a) 60.5 cal = ______ J; (b) 8.32 kJ = ____________ cal; (c) 0.753 kJ = ____________ kcal.

(89) Complete the following: (a) 0.448 kcal = ______ kJ; (b) 703 J = ____________ cal; (c) 1480 J = ____________ kcal

(42) 912 cal are transferred in a reaction between ammonia and hydrochloric acid. Express this in joules and kilojoules.

(43) The burning of sucrose releases 3.94×10^3 cal/g. Calculate the kilojoules released when 56.7 grams of sucrose are burned.

(90) 394 kilojoules of energy are released when 12 grams of carbon are burned. Calculate the number of calories and kilocalories this represents.

(91) 5.8×10^2 kcal of heat are removed from the body by evaporation each day. How many kilojoules are removed in a week?

Section 3.9 *(See Table 3.4 for specific heat values.)*

(44) If you are going to heat some water to boiling to prepare tea, will it take more time or less time if you start with hot water than if you start with cold, or will the times be the same? Explain.

(45) How much heat is required to raise the temperature of 204 grams of lead from 22.8°C to 64.9°C?

(46) The 2.55 kg blade of an iron sword has been forged in a fire and is being cooled from 350°C to 25°C. Calculate the heat flow from the blade.

(47) To what temperature will 545 grams of cobalt be raised if, beginning at 25.0°C, it absorbs 3.14 kJ of heat?

(48)* A calorimeter contains 72.0 grams of water at 19.2°C. A 141-gram piece of tin is heated to 89.0°C and dropped into the water. The entire system eventually reaches 25.5°C. Assuming all of the heat gained by the water comes from the cooling of the tin—no heat loss to the calorimeter or surroundings—calculate the specific heat of the tin.

(92) Samples of two different metals, A and B, have the same mass. Both samples absorb the same amount of heat. The temperature of A increases by 10°C, and the sample of B increases by 12°C. Which metal has the higher specific heat?

(93) Find the number of joules released as 186 grams of zinc cool from 68°C to 31°C.

(94) How many kilojoules are required to cool 1.87 kilograms of gold from 88°C to 22°C?

(95) The mass of one dollar's worth of copper pennies is 318 grams. If 100 pennies at an outside temperature of 23°C lose 1.47 kJ when they are tossed into a fountain and drop to the fountain's water temperature, what is that temperature?

(96)* A certain kind of rock is being checked for its ability to store heat in a solar heating system. A 3.62-kg piece is heated in an oven until it is at a uniform temperature of 92°C. It is then placed in a calorimeter that contains 8.86 kg of water at 17.1°C. The final temperature of the system is 28.0°C. Assuming no heat loss to the surroundings, find the specific heat of the rock.

General Questions

(97) Distinguish precisely and in scientific terms the differences between items in each of the following groups:
(a) Given quantity, conversion relationship
(b) Milliliter, cubic centimeter
(c) Mass and weight
(d) Fahrenheit, Celsius, Kelvin (temperature scales)
(e) Density, specific gravity
(f) Joule, calorie

(98) Determine whether each statement that follows is true or false:
(a) The SI system includes metric units.
(b) Dimensional analysis can be used to change from one unit to another only when the quantities are directly proportional.
(c) A given quantity expresses the proportionality between two quantities.
(d) A unit path begins with the given quantity and ends with the units in which the answer is to be expressed.
(e) Per means multiply.
(f) There are 1000 kilounits in a unit.
(g) There are 10 milliunits in a centiunit.
(h) There are 100 cm^3 in a cubic meter.
(i) There are 1000 mL in a cubic centimeter.
(j) The mass of an object is independent of its location in the universe.

(k) To change kilounits to units, the decimal must be moved three places to the left.
(l) Celsius degrees are smaller than Fahrenheit degrees.
(m) Fahrenheit temperature is proportional to Celsius temperature.
(n) The doubtful digit is the last digit written when a number is expressed properly in significant figures.
(o) 76.2 g has the same meaning as 76.200 g.
(p) The number of significant figures in a sum may be more than the number of significant figures in any of the quantities added.
(q) The number of significant figures in a difference may be fewer than the number of significant figures in any of the quantities subtracted.
(r) The number of significant figures in a product may be more than the number of significant figures in any of the quantities multiplied.
(s) Specific gravity and density have the same units.
(t) In the SI system, heat is measured in Celsius degrees.
(u) ΔT is positive when a substance is heated.

(99) There is an open pit iron mine in Minnesota that averages 6.5 km long, 3.1 km wide, and 150 m deep. How many cubic miles of earth have been removed from this pit?

(100) How tall are you in (a) meters; (b) decimeters; (c) centimeters; (d) millimeters? Which of the four metric units do you think would be most useful in expressing people's height without resorting to decimal fractions?

(101) What do you weigh in (a) milligrams; (b) grams; (c) kilograms. Which of these units do you think is best for expressing a person's weight? Why?

(102) Using only the English–metric conversion factors in Table AP-4 in the Appendix, calculate the number of cubic inches in a gallon.

(103) One of the areas of disagreement in the conversion from English measure to metrics in the United States is the size of a cup, as the term is used in American recipes. One side argues in favor of a 250-mL cup, and the other side favors a 240-mL cup. Which is closer to the real cup, which is 8 fluid ounces? (There are 128 fluid ounces in a gallon. You should be able to take it from there.) Can you suggest reasons each side would put forth to support its position? Which would you prefer if you wanted to convert your favorite recipes to metrics?

(104) Standard typewriter paper in the United States is 8½ inches by 11 inches. What are these dimensions in centimeters?

(105) The element osmium is the most dense substance known at 22.5 g/cm^3. Calculate the mass in pounds of one cubic inch of osmium.

(106) What is the weight in ounces of a cubic foot of air if its density is 0.001 29 g/cm^3?

(107) The specific gravity of aluminum is 2.70. An ecology-minded student has gathered 126 empty aluminum cans for recycling. If there are 20 cans per pound, how many grams of aluminum does the student have, and what is their volume in cubic centimeters?

4 Atomic Theory and the Periodic Table: The Beginning

The cumulative character of chemistry—the way the study of chemistry keeps expanding on ideas that were introduced earlier in the course—first appears in this chapter. We therefore introduce the "Looking Back" feature of this text that was described in Chapter 1. In the left column you will find a section number followed by an idea that was considered in that section. Next to that in the right column you will find how the earlier concept is to be used in the new chapter. If you don't thoroughly understand the earlier material, you will probably have difficulty understanding how it is applied. In that case, it is suggested that you review the reference section.

LOOKING BACK	LOOKING AHEAD IN CHAPTER 4
2.3 Compounds can be separated by chemical change into simpler substances called elements, which cannot be separated further.	The simplest particle of an element is an atom. Atoms of different elements combine to form compounds.
2.7 Matter has electrical properties; it can be given a positive or negative charge.	Atoms contain positively charged protons and negatively charged electrons that are responsible for the electrical properties of matter.

Early in the 19th century, John Dalton, an English chemist and school teacher, revived the concept of the **atom** as the ultimate, indivisible, smallest-possible piece of matter. The idea had first been proposed by the Greeks as early as 400 BC. The history of atomic theory from Dalton to today offers example after example of the "scientific method" of thinking described in the Prologue. Some thought trends were sequential, each one leading to the next. Others, primarily those in the areas of physics, developed side-by-side with those in chemistry. During the period 1895 to 1932 there was a virtual explosion of progress as the different threads merged into atomic theory as it applies to chemistry. This is the atomic theory we will study in this chapter and the next.

Though we will focus on a theory developed nearly 60 years ago, this by no means suggests that new theories about the atom have not continued to appear. Nuclear research is conducted in all the scientifically advanced nations of the world. Some of its accomplishments and some of its problems are considered in Chapter 20.

4.1 DALTON'S ATOMIC THEORY

PG 4A List the main features of Dalton's atomic theory.

Dalton did a lot of thinking about two laws discussed in Chapter 2. The Law of Definite Composition (Section 2.3) states that the percentage by weight of each element in a compound is always the same. The Law of Conservation of Mass (Section 2.6) tells us that mass is conserved in a chemical change. In an attempt to explain these laws Dalton suggested his **atomic theory** in 1808. The main features of that theory are:

1. Each element is made up of tiny, individual particles called atoms.
2. Atoms are indivisible; they cannot be created or destroyed.
3. All atoms of each element are identical in every respect.
4. Atoms of one element are different from atoms of any other element.
5. Atoms of one element may combine with atoms of another element, usually in the ratio of small, whole numbers, to form chemical compounds.

As with many new theories, Dalton's suggestions were not readily received. From them, however, came a prediction that *must* be true if atomic theory is correct. Now known as the **Law of Multiple Proportions,** it states that when two elements combine to form more than one compound, the different weights of one element that combine with a fixed weight of the other are in a simple ratio of whole numbers (Fig. 4.1). When experiments confirmed the prediction, most scientists accepted Dalton's theory.

4.2 SYMBOLS OF THE ELEMENTS

As communication between chemists increased in the early 19th century, there was a need for an international shorthand to identify elements and compounds. John Dalton filled that need by using simple symbols to represent the elements. Some of these reflected the character of the element. Carbon, for example, was a solid black circle, suggesting charcoal. Compounds were represented by drawing the symbols touching each other.

Dalton's symbols were reasonably satisfactory when the number of known elements was small. As more elements were discovered, however, more was needed than little sketches. The drawings were replaced by the first letters of the names of the elements. The symbol for hydrogen is H, for oxygen, O, and for nitrogen, N. When more than one element begins with the same letter, two letters are used for the additional elements. C is the symbol for carbon, Ca represents calcium, Cr is chromium, and Cl is chlorine.

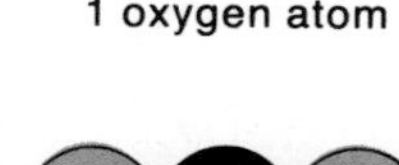

carbon monoxide
1 carbon atom
1 oxygen atom

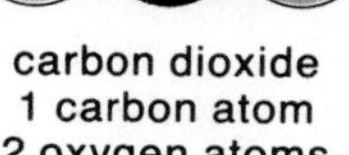

carbon dioxide
1 carbon atom
2 oxygen atoms

Figure 4.1

Example of Law of Multiple Proportions. Carbon and oxygen combine to form two compounds, carbon monoxide and carbon dioxide. A carbon monoxide molecule consists of one carbon atom and one oxygen atom. A carbon dioxide molecule has one carbon atom and two oxygen atoms. Considering both molecules, for a fixed number of carbon atoms—one in each molecule—the ratio of oxygen atoms is 1 to 2, or 1/2. If all oxygen atoms have the same mass, M, the mass ratio is also 1/2:

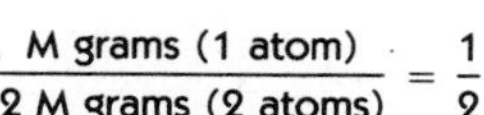

$$\frac{\text{M grams (1 atom)}}{\text{2 M grams (2 atoms)}} = \frac{1}{2}$$

The association of an element's symbol with one or two letters of its name makes it easy to learn most symbols, but not all. Some symbols come from Latin names. The symbol for iron, for example, comes from the Latin *ferrum*—Fe. The symbol for gold is Au, from *aurum;* for sodium, Na, from *natrium;* and for silver, Ag, from *argentum*.

A list of the elements and their symbols is inside the back cover of this book. A more useful reference is the **periodic table** inside the front cover. We will refer to these symbols and the periodic table throughout this chapter, and in the last section you will use the periodic table as an aid in learning some of the symbols.

4.3 SUBATOMIC PARTICLES

PG 4B Identify the features of Dalton's atomic theory that are no longer considered valid, and explain why.

4C Identify the three major subatomic particles by charge and approximate atomic mass, expressed in atomic mass units.

Despite the general acceptance of the atomic theory, it was soon to be challenged in some of its details. As early as the 1830's, laboratory experiments suggested that the atom contains even smaller parts, or **subatomic particles.** The brilliant works of Michael Faraday and William Crookes, among others, led to discovering within the atom a negatively charged particle that we now call the **electron.** The electron was identified in 1897 by J. J. Thomson. Its charge is apparently the smallest charge possible; it has been assigned a value of -1. The mass of an electron is extremely small, 9.107×10^{-28} g.

The second subatomic particle, the **proton,** was identified in 1919 by Ernest Rutherford. Its mass is about 1837 times as great as the mass of an electron, and it carries a $+1$ charge, equal in size but opposite in sign to the negative charge of the electron. The third particle, the **neutron** was discovered by James Chadwick in 1932. As its name suggests, it is electrically neutral. The mass of a neutron is slightly greater than the mass of a proton. Masses of atoms and parts of atoms are often expressed in **atomic mass units** (Section 4.6).

Today we know that atoms of all elements are made up of different combinations of many kinds of subatomic particles. Only the three described above are of interest in an introductory chemistry course. The properties of the electron, proton, and neutron are summarized in Table 4.1.

Quick Check 4.3

Are the following statements true or false?

1. The electron, proton, and neutron are the only three parts of the atom.
2. The electron has less mass than a proton or neutron.
3. The mass of a proton is about 1 g.

Table 4.1 Subatomic Particles

Subatomic Particle	Symbol	Fundamental Charge	Mass		Location	Discovered
			GRAMS	amu* (C^{12} = 12.000 00)		
Electron	e^-	−1	9.107×10^{-28}	0.000 549 ≈ 0	Outside nucleus	1897 Thomson
Proton	p or p^+	+1	1.672×10^{-24}	1.007 28 ≈ 1	Inside nucleus	1919 Rutherford
Neutron	n or n^0	0	1.675×10^{-24}	1.008 67 ≈ 1	Inside nucleus	1932 Chadwick

*An *amu* is a very small unit of mass used for atomic-sized particles. It is defined in Section 4.6.

4.4 THE NUCLEAR ATOM

PG 4D Describe the nuclear model of the atom, based on the Rutherford scattering experiment.

In 1911 Ernest Rutherford and his students performed a series of experiments that are described in Figures 4.2 and 4.3. The results of these experiments led to the following conclusions:

1. Every atom contains an extremely small, extremely dense nucleus.
2. All of the positive charge and nearly all of the mass of the atom are concentrated in the nucleus.
3. The nucleus is surrounded by a much larger volume of nearly empty space, which makes up the rest of the atom.

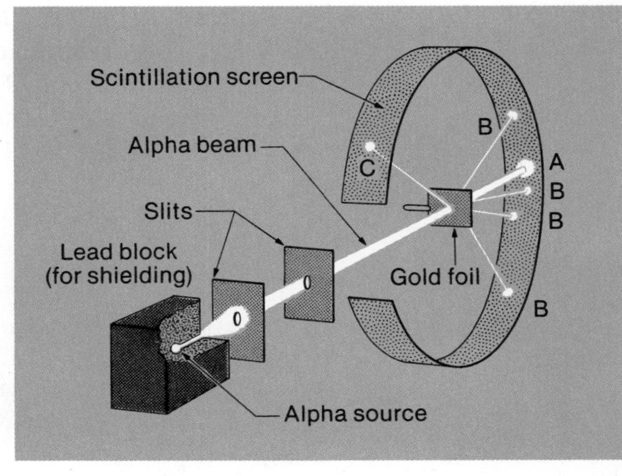

Figure 4.2
Rutherford scattering experiment. Using a natural radioactive source, a narrow beam of alpha particles (helium atoms stripped of their electrons) was directed at a very thin gold foil. Most of the particles passed right through the foil, striking a fluorescent screen at A, causing it to glow. Some of the particles were deflected, striking the screen at points such as those labeled B. The larger deflections were surprises, but the 0.001% of the total that were reflected at acute angles, C, were totally unexpected. Similar results were observed using foils of other metals.

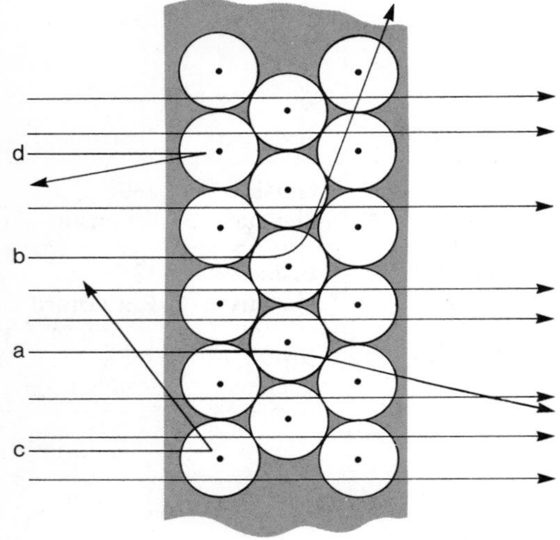

Figure 4.3
Interpretation of Rutherford scattering. The atom is pictured as consisting mostly of open space. At the center is a tiny and extremely dense nucleus that contains all of the positive charge of the atom and nearly all of its mass. The electrons are thinly distributed throughout the open space. Most of the positively charged alpha particles pass through the open space undeflected, not coming near any gold nuclei. Those that would pass close to a nucleus (a and b) are repelled by electrostatic force and thereby deflected. The few particles that are on a "collision course" with gold nuclei are repelled backward at acute angles (c and d). Calculations based on the results of the experiment indicate that the diameter of the open-space atom is from 10 000 to 100 000 times greater than the diameter of the nucleus.

4. The space outside the nucleus is very thinly populated by electrons, the total charge of which exactly balances the positive charge of the nucleus.

This description of the atom is called the **nuclear model of the atom.**

The "emptiness" of the atom is difficult to visualize. If the nucleus of the atom were the size of a pea, the distance to its closest neighbor would be about 0.6 mile, or 1 km. Between them would be almost nothing—only a small number of electrons of negligible size and mass. If it were possible to eliminate all of this nearly empty space and pack nothing but nuclei into a sphere the size of a period on this page, that sphere could, for some elements, weigh as much as a million tons!

When protons and neutrons were later discovered, it was concluded that these relatively massive particles make up the nucleus of the atom. But electrons were already known in 1911, and it was natural to wonder what they did in the vast open space they occupied. The most wisely held opinion was that they traveled in circular orbits around the nucleus, much as planets move in orbits around the sun. The atom would then have the character of a miniature solar system. This is called the **planetary model of the atom.** In Chapter 5 we will examine this model more closely—and find out why it is wrong.

Quick Check 4.4
Are the following statements true or false?
1. Atoms are like small, hard spheres.
2. An atom is electrically neutral.
3. An atom consists mostly of empty space.

4.5 ISOTOPES

PG 4E Explain what isotopes of an element are and how they differ from each other.

4F For an isotope of any element whose chemical symbol is known, given one of the following, state the other two: (a) nuclear symbol; (b) number of protons and neutrons in the nucleus; (c) atomic number and mass number.

After more than a hundred years, another feature of Dalton's atomic theory was shown to be incorrect. All atoms of an element are not identical; some have more mass than others. We now know that every atom of an element has the same number of protons. This number is called the **atomic number** and is represented by the symbol **Z**. Atoms are electrically neutral, so the number of electrons must be the same as the number of protons. But electron masses are so small that their total contribution to atomic mass may usually be disregarded. We therefore conclude that differences in mass between atoms of an element must be caused by different numbers of neutrons in their nuclei. **Atoms of the same element that have different masses are called isotopes.**

An isotope of an element is identified by what we will call its **nuclear symbol.** The nuclear symbol of an isotope is written

$$^{A}_{Z}\text{Sy} \qquad \textit{or} \qquad {}^{A}\text{Sy}$$

where Sy is the symbol of the element and A is the **mass number, the total number of protons and neutrons in the nucleus.** As the symbol alone is sufficient to identify the element, the atomic number, Z, may be omitted; but it is useful in writing nuclear equations (Chapter 20). Mass number may be calculated from the equation

$$\begin{aligned} \text{mass number} &= \text{number of protons} + \text{number of neutrons} \\ A &= Z + \text{number of neutrons} \end{aligned} \tag{4.1}$$

The name of an isotope is the elemental name followed by the mass number. $^{16}_{8}O$, for example, is called "oxygen sixteen," and is written "oxygen-16."

Two natural isotopes of carbon are $^{12}_{6}C$ and $^{13}_{6}C$, or carbon-12 and carbon-13. From the name and symbol of each isotope and from Equation 4.1 it is possible to find the number of neutrons in each nucleus. In carbon-12, if you subtract the atomic number (protons) from the mass number (protons + neutrons), you get the number of neutrons: $12 - 6 = 6$. In carbon-13 there are 7 neutrons: $13 - 6 = 7$.

If you know the number of protons and neutrons in an atom, you can readily determine its mass number and nuclear symbol. A nucleus having 12 protons and 14 neutrons has an atomic number of 12, the same as the number of protons. The mass number, according to Equation 4.1, is $12 + 14 = 26$. You can find the symbol of the element in the periodic table inside the front cover of this book. The number at the top of each box in the table is the atomic number. The elemental symbol corresponding to $Z = 12$ is Mg, for magnesium. The nuclear symbol is $^{26}_{12}Mg$ or ^{26}Mg.

EXAMPLE 4.1 Write the name and number of protons, neutrons, and electrons in a calcium isotope if its nuclear symbol is $^{42}_{20}Ca$.

$^{42}_{20}Ca$: Calcium-42; 20 protons; 22 neutrons; 20 electrons

The name of the isotope is the elemental name followed by the mass number (superscript). The number of protons is the atomic number (subscript). The number of neutrons is, from Equation 4.1, the difference between the mass number and the atomic number. The number of electrons is equal to the number of protons.

EXAMPLE 4.2 Write the nuclear symbol for the sulfur isotope that has 18 neutrons. $Z = 16$ for sulfur, and its elemental symbol is S.

$^{34}_{16}S$ or ^{34}S

The mass number is the sum of the number of protons (16) and neutrons (18).

EXAMPLE 4.3 Write the symbol and name of the lead isotope having 82 protons and 122 neutrons in its nucleus.

The mass number is found from Equation 4.1, as in Example 4.2. Find the elemental symbol from the periodic table inside the front cover. Remember that the number at the top of each box is the atomic number of the element.

$^{204}_{82}Pb$ or ^{204}Pb; lead-204

The atomic number is the same as the number of protons, 82. In the periodic table the box for $Z = 82$ has the symbol Pb. This is from the Latin *plumbum,* meaning lead.

EXAMPLE 4.4 Write the number of protons, neutrons, and electrons in an atom of zinc-70.

Use the table inside the back cover of the book for the atomic number of zinc.

-----

30 protons, 40 neutrons, and 30 electrons

The table inside the back cover gives 30 as the atomic number for zinc. This is the number of protons. As in Example 4.1, the number of neutrons is the mass number minus the atomic number, and the number of electrons is equal to the number of protons.

Quick Check 4.5

Are the following statements true or false?

1. The atomic number of an element is the number of protons in the nucleus.
2. All atoms of a specific element have the same number of protons.
3. The difference between isotopes of an element is a difference in the number of neutrons in the nucleus.
4. The mass number of an atom is always equal to or larger than the atomic number.

4.6 ATOMIC MASS

PG 4G Define the atomic mass unit, amu.

4H Given the relative abundances of the natural isotopes of an element, calculate the atomic mass.

The mass of an atom is very small—much too small to be measured on a balance. Nevertheless, early chemists did find ways to isolate samples of elements that contained the same number of atoms. These samples were weighed and compared. Because the ratio of the masses of equal numbers of atoms of different elements is the same as the ratio of the masses of individual atoms,* it was possible to develop a scale of relative atomic weights. In its modern version, this scale compares the mass of atoms of different elements to the mass of an atom of carbon-12, to which is assigned the value of exactly 12 atomic mass units. Consequently, **one atomic mass unit is defined as exactly 1/12 of the mass of one atom of carbon-12.** This mass is 1.66×10^{-24} g.

As a practical matter, almost every sample of any element consists of a mixture of the natural isotopes of that element.† Fortunately, the percentage, or fraction, of each isotope is generally the same, regardless of the source of the sample. Table 4.2 lists the natural isotopes of some of the more common elements and their relative abundance in nature. From these it is possible to calculate the **atomic mass** of an element, which is defined as the **average mass of the atoms of an element compared to an atom of carbon-12 at exactly 12 atomic mass units.**

The following example shows how the atomic mass of an element is calculated.

*If a is the mass of one atom of A, b is the mass of one atom of B, and N is any number of atoms, then $N \times a$ = mass of N atoms of A and $N \times b$ = mass of N atoms of B.

$$\frac{N \times a}{N \times b} = \frac{N}{N} \times \frac{a}{b} = \frac{a}{b}$$

†Isotopes of some elements do not occur in nature but have been prepared in laboratories.

Table 4.2 Percent Abundance of Some Natural Isotopes

Symbol	Mass (amu)	Percent	Symbol	Mass (amu)	Percent
${}^{1}_{1}H$	1.007 825	99.985	${}^{19}_{9}F$	18.998 40	100
${}^{2}_{1}H$	2.014 0	0.015	${}^{32}_{16}S$	31.972 07	95.0
${}^{3}_{2}He$	3.016 03	0.000 13	${}^{33}_{16}S$	32.971 46	0.76
${}^{4}_{2}He$	4.002 60	100	${}^{34}_{16}S$	33.967 86	4.22
${}^{12}_{6}C$	12.000 00	98.89	${}^{36}_{16}S$	35.967 09	0.014
${}^{13}_{6}C$	13.003 35	1.11	${}^{35}_{17}Cl$	34.968 85	75.53
${}^{14}_{7}N$	14.003 07	99.63	${}^{37}_{17}Cl$	36.965 90	24.47
${}^{15}_{7}N$	15.000 11	0.37	${}^{39}_{19}K$	38.963 71	93.1
${}^{16}_{8}O$	15.994 91	99.759	${}^{40}_{19}K$	39.974	0.001 18
${}^{17}_{8}O$	16.994 74	0.037	${}^{41}_{19}K$	40.961 84	6.88
${}^{18}_{8}O$	17.994 77	0.204			

EXAMPLE 4.5 The natural distribution of isotopes of magnesium is 78.70% ${}^{24}_{12}Mg$ at a mass of 23.985 04 amu, 10.13% ${}^{25}_{12}Mg$ at 24.985 84 amu, and 11.17% ${}^{26}_{12}Mg$ at 25.982 59 amu. Calculate the atomic mass of magnesium.

SOLUTION: The "average" magnesium atom consists of 78.70% of an atom that has a mass of 23.985 04 amu. It therefore contributes 0.7870 × 23.985 04 = 18.88 amu to the mass of the "average" atom. A similar calculation for the other isotopes is added:

$$
\begin{array}{lcl}
0.7870 \times 23.985\ 04 \text{ amu} & = & 18.88 \text{ amu} \\
0.1013 \times 23.985\ 84 \text{ amu} & = & 2.531 \text{ amu} \\
\underline{0.1117} \times 25.982\ 59 \text{ amu} & = & \underline{2.902 \text{ amu}} \\
1.0000 \quad \text{average atom} & = & 24.31 \text{ amu}
\end{array}
$$

The presently accepted value of the atomic weight of magnesium is 24.305. This matches the calculated value to four significant figures.

Now you try an atomic mass calculation.

EXAMPLE 4.6 Calculate the atomic mass of potassium (symbol K), using data from Table 4.2.

Proceed as in Example 4.5.

-----　　　　　-----

$$
\begin{array}{llll}
0.931 & \times\ 38.963\ 71\ \text{amu} & = & 36.3\ \text{amu} \\
0.000\ 011\ 8 & \times\ 39.974\ \text{amu} & = & 0.000\ 472\ \text{amu} \\
0.0688 & \times\ 40.961\ 84\ \text{amu} & = & \underline{2.82\ \text{amu}} \\
 & & & 39.1\ \text{amu}
\end{array}
$$

The accepted value of the atomic mass of potassium is 39.098 amu.

4.7 THE PERIODIC TABLE

PG 4I Distinguish between groups and periods in a periodic table and identify them by number.

4J Given the atomic number of an element, use a periodic table to find the symbol and atomic mass of that element, and identify the period and group in which it is found.

During the period of research on the atom, even before any subatomic particles were identified, other chemists searched for an order among the elements. In 1869 it was found independently by two men at the same time. Dmitri Mendeleev and Lothar Meyer observed that when elements are arranged according to what was then called their atomic weights, certain properties repeat at regular intervals. Mendeleev reached this conclusion by studying chemical properties, while Meyer examined physical properties.

Mendeleev and Meyer arranged the elements in tables so that the elements with similar properties were in the same column or row. These were the first **periodic tables** of the elements. The arrangements were not perfect; in order for all elements to fall into the proper groups, it was necessary to switch a few, interrupting the orderly increase in atomic weight. This was partly because of errors in atomic weights as they were known in 1869. More important, nearly 50 years later it was found that the correct ordering property is atomic number (Z) rather than atomic weight.

As Dalton used atomic theory to predict the Law of Multiple Proportions correctly, so Mendeleev used his periodic table to predict the existence and properties of elements unknown in 1869. Noting certain "blanks" in the table, he reasoned that the blank spaces were there simply because nobody had yet discovered the elements that belonged there. By averaging the properties of the elements above and below or on each side of the unknown elements, he forecast the properties the elements would have when discovered. One element about which he made these predictions is germanium. The predictions and the presently accepted values of the properties of germanium are summarized in Table 4.3.

Figure 4.4 is a modern periodic table; another copy is inside the front cover of this book. A partial periodic table is printed on the opaque shields provided for working examples. You will find yourself referring to periodic tables throughout your study of chemistry.

Each box in the periodic tables in this book contains three items of information about the element it represents. The atomic number is on top, the

Table 4.3 The Predicted and Observed Properties of Germanium

Property	Predicted by Mendeleev	Observed
Atomic weight	72	72.60
Density of metal	5.5 g/cm^3	5.36 g/cm^3
Color of metal	Dark gray	Gray
Formula of oxide	GeO_2	GeO_2
Density of oxide	4.7 g/cm^3	4.703 g/cm^3
Formula of chloride	$GeCl_4$	$GeCl_4$
Density of chloride	1.9 g/cm^3	1.887 g/cm^3
Boiling point of chloride	Below 100°C	86°C
Formula of ethyl compound	$Ge(C_2H_5)_4$	$Ge(C_2H_5)_4$
Boiling point of ethyl compound	160°C	160°C
Density of ethyl compound	0.96 g/cm^3	Slightly less than 1.0 g/cm^3

chemical symbol is in the middle, and the atomic mass is on the bottom (Fig. 4.5). The boxes are arranged according to atomic number in horizontal rows called **periods.** Periods are identified by number from top to bottom, but the numbers are not usually printed. Periods vary in length. The first period has two elements, the second and third have eight elements each, and the fourth

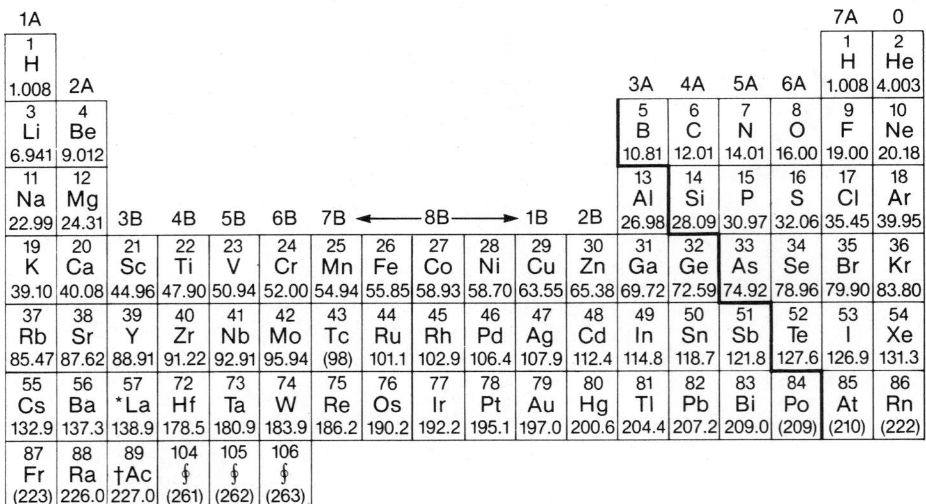

*Lanthanide series	58 Ce 140.1	59 Pr 140.9	60 Nd 144.2	61 Pm (145)	62 Sm 150.4	63 Eu 152.0	64 Gd 157.3	65 Tb 158.9	66 Dy 162.5	67 Ho 164.9	68 Er 167.3	69 Tm 168.9	70 Yb 173.0	71 Lu 175.0
†Actinide series	90 Th 232.0	91 Pa 231.0	92 U 238.0	93 Np 237.0	94 Pu (244)	95 Am (243)	96 Cm (247)	97 Bk (247)	98 Cf (251)	99 Es (252)	100 Fm (257)	101 Md (258)	102 No (259)	103 Lr (260)

∮ The International Union for Pure and Applied Chemistry has not adopted official names or symbols for these elements.

All atomic weights have been rounded off to four significant figures.

Figure 4.4
Periodic Table of the Elements.

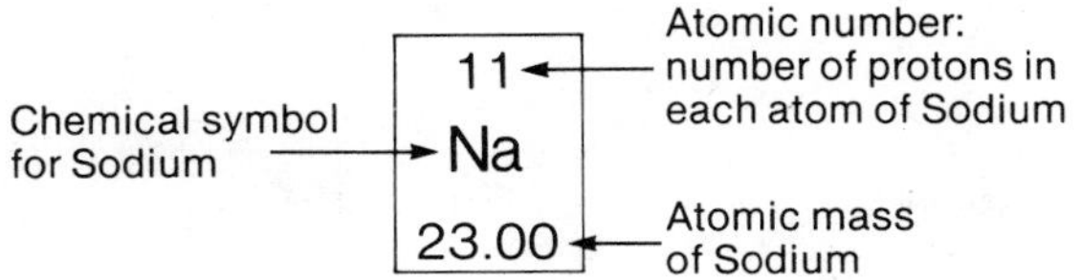

Figure 4.5
Sample box from the periodic table, showing the atomic number, symbol, and atomic mass of the element sodium.

and fifth have eighteen. Period 6 has 32 elements, including atomic numbers 58 through 71, which are printed separately at the bottom to keep the table from becoming too wide. Period 7 also has 32 elements, but the discovery and naming of elements beyond Z = 106 has not yet been confirmed.

The periodic table places elements with similar chemical properties in vertical columns called **groups,** or **chemical families.** Groups are numbered across the top of the table. If you locate the element whose atomic number is 30, you will find it in the fourth period in Group 2B. Its symbol is Zn (zinc), and its atomic mass is 65.38 amu. Elements in the A groups of the periodic table are called **representative elements,** and those in the B groups are **transition elements.**

EXAMPLE 4.7 List the atomic number, chemical symbol, and atomic mass of the fourth period element in Group 7A.

Z = 35; symbol, Br; atomic mass, 79.90 amu

In Group 7A, the second column from the right side of the table, you find Z = 1 in period 1, Z = 9 in period 2, Z = 17 in period 3, and Z = 35 in period 4. The element is bromine.

EXAMPLE 4.8 List the group, period, symbol and atomic mass for tin (Z = 50).

Group 4A; period, 5; symbol, Sn; atomic mass, 118.7 amu.

The first element in Group 4A, Z = 6, is the second period. Counting down from there, Z = 50 is in Period 5.

You may wonder why the periodic table has such an unusual shape. Remember that the arrangement is based primarily on the periodic appearance of physical and chemical properties. Placing elements with similar properties in the same column dictates the shape. The *reason* the properties produce the shape they do has to do with the arrangement of electrons in the atom. You will study this in Chapter 5.

4.8 ELEMENTAL SYMBOLS AND THE PERIODIC TABLE

PG 4K Given the name (or symbol) of an element in Figure 4.6, write the symbol (or name).

It has been indicated that you will use the periodic table as a reference in different ways. Its first application will be as an aid in learning the symbols of the elements you will be using in this course. Memorizing symbols is much easier if the location of each element in the table is learned at the same time. Here's how to do it:

Table 4.4 lists the name, symbol, and atomic number of the elements whose symbols are to be learned. Figure 4.6 is a periodic table showing the atomic numbers and symbols of the same elements. Their names are listed in alphabetical order beneath the table. Study Table 4.4 briefly. Try to learn the symbol that goes with each element, but don't spend more than a few minutes in this effort. Then do not look at Table 4.4, but turn to Figure 4.6. Run through the symbols mentally and see how many elements you can name. If you get stuck on one, glance at the alphabetical list below the periodic table and see if that will jog your memory. If you still can't get it, note the atomic number and check back to Table 4.4 for the elemental name. Do this a few times until you become fairly quick in naming most of the elements from the symbols.

Then reverse the process. Look at the alphabetical list beneath the periodic table in Figure 4.6. For each name, mentally "write"—in other words, *think*—the symbol. Whether you can think it or not, glance up to the periodic table

Table 4.4 Table of Common Elements

Atomic Number	Symbol	Element	Atomic Number	Symbol	Element	Atomic Number	Symbol	Element
1	H	Hydrogen	13	Al	Aluminum	28	Ni	Nickel
2	He	Helium	14	Si	Silicon	29	Cu	Copper
3	Li	Lithium	15	P	Phosphorus	30	Zn	Zinc
4	Be	Beryllium	16	S	Sulfur	35	Br	Bromine
5	B	Boron	17	Cl	Chlorine	36	Kr	Krypton
6	C	Carbon	18	Ar	Argon	47	Ag	Silver
7	N	Nitrogen	19	K	Potassium	50	Sn	Tin
8	O	Oxygen	20	Ca	Calcium	53	I	Iodine
9	F	Fluorine	24	Cr	Chromium	56	Ba	Barium
10	Ne	Neon	25	Mn	Manganese	80	Hg	Mercury
11	Na	Sodium	26	Fe	Iron	82	Pb	Lead
12	Mg	Magnesium	27	Co	Cobalt			

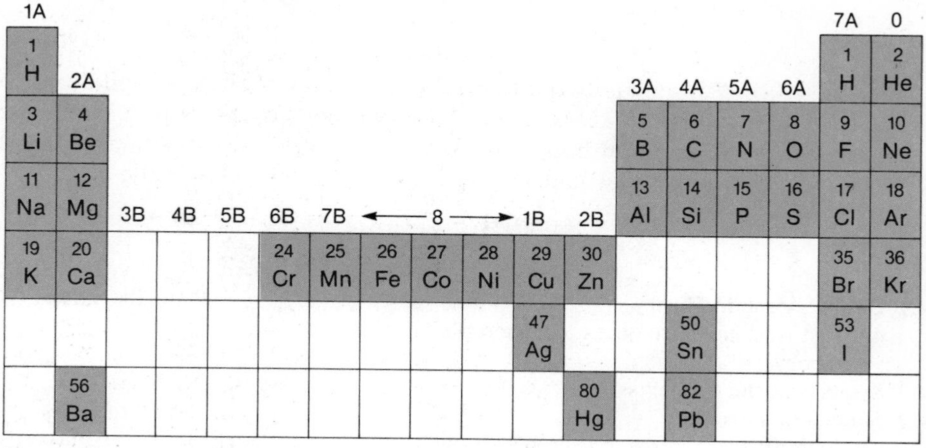

1A	2A	3B	4B	5B	6B	7B	8	8	8	1B	2B	3A	4A	5A	6A	7A	0
1 H																1 H	2 He
3 Li	4 Be											5 B	6 C	7 N	8 O	9 F	10 Ne
11 Na	12 Mg											13 Al	14 Si	15 P	16 S	17 Cl	18 Ar
19 K	20 Ca				24 Cr	25 Mn	26 Fe	27 Co	28 Ni	29 Cu	30 Zn					35 Br	36 Kr
										47 Ag			50 Sn			53 I	
	56 Ba										80 Hg		82 Pb				

Figure 4.6
Partial periodic table showing the symbols and locations of the more common elements. The symbols above and the list that follows identify the elements you should be able to recognize or write, referring only to a complete periodic table. Associating the names and symbols with the table makes learning them much easier. The elemental names are:

aluminum	bromine	chromium	iodine	magnesium	nitrogen	silver
argon	calcium	copper	iron	manganese	oxygen	sodium
barium	carbon	fluorine	krypton	mercury	phosphorus	sulfur
beryllium	chlorine	helium	lead	neon	potassium	tin
boron	cobalt	hydrogen	lithium	nickel	silicon	zinc

and find the element. Again, if you have trouble, use Table 4.4 as a temporary help. Repeat the procedure several times, taking the elements in random order. Move in both directions, from name to symbol in the periodic table and from symbol in the periodic table to name.

When you feel reasonably sure of yourself, try the following example. Do not refer to either Table 4.4 or Figure 4.6. Instead, use the more complete periodic table that is on one side of the tear-out shield that is provided for that purpose.

EXAMPLE 4.9 For each elemental symbol below, write the name; for each name, write the symbol.

N ______________ P ______________ Carbon __________ Potassium __________

F ______________ Cl ______________ Aluminum __________ Zinc __________

I ______________ Fe ______________ Copper __________ Bromine __________

N, nitrogen P, phosphorus Carbon, C Potassium, K
F, fluorine Cl, chlorine Aluminum, Al Zinc, Zn
I, iodine Fe, iron Copper, Cu Bromine, Br

Look closely at your symbols for aluminum, zinc, copper, and bromine in the above example. If you wrote AL, ZN, CU, or BR, the symbol is wrong. The letters are right, but the symbol is not. Whenever writing a symbol that has two letters, the first letter is always capitalized, but THE SECOND LETTER IS ALWAYS WRITTEN IN LOWER CASE, or as a small letter. You can enjoy a long and happy life with a pile of Co in your house, but the day will be your last that you stay in a garage that has CO in the air!*

CHAPTER 4 IN REVIEW

4.1 Dalton's Atomic Theory

4A List the main features of Dalton's atomic theory.

4.2 Symbols of the Elements

4.3 Subatomic Particles

4B Identify the features of Dalton's atomic theory that are no longer considered valid, and explain why.

4C Identify the three major subatomic particles by charge and approximate atomic mass, expressed in atomic mass units.

4.4 The Nuclear Atom

4D Describe the nuclear model of the atom, based on the Rutherford scattering experiment.

4.5 Isotopes

4E Explain what isotopes of an element are and how they differ from each other.

4F For an isotope of any element whose chemical symbol is known, given one of the following, state the other two: (a) nuclear symbol; (b) number of protons and neutrons in the nucleus; (c) atomic number and mass number.

4.6 Atomic Mass

4G Define the atomic mass unit, amu.

4H Given the masses and relative abundances of the natural isotopes of an element, calculate the atomic mass.

4.7 The Periodic Table

4I Distinguish between groups and periods in a periodic table and identify them by number.

4J Given the atomic number of an element, use a periodic table to find the symbol and atomic mass of that element, and identify the period and group in which it is found.

4.8 Elemental Symbols and the Periodic Table

4K Given the name (or symbol) of an element in Figure 4.6, write the symbol (or name).

TERMS AND CONCEPTS

4.1 Dalton's atomic theory
Law of Multiple Proportions

4.2 Elemental symbol
Periodic table

4.3 Subatomic particle
Electron
Proton
Neutron
Atomic mass unit

4.4 Nuclear model of the atom
Planetary model of the atom

4.5 Atomic number
Isotopes
Nuclear symbol
Mass number

4.6 Atomic mass unit (defined)
Atomic mass

4.7 Period (in periodic table)
Group (in periodic table)
Chemical family
Representative element
Transition element

Most of these terms and many others appear in the Glossary. Use your Glossary regularly.

QUESTIONS AND PROBLEMS

Section 4.1

(*1*) List the major points in Dalton's atomic theory.

(25) According to Dalton's atomic theory, can more than one compound be made from atoms of the same two elements?

*Co is the metal, cobalt. CO is the deadly gas, carbon monoxide, that is present in automobile exhaust.

(2) How does Dalton's atomic theory account for the Law of Conservation of Mass?

(3) The brilliance with which magnesium burns makes it ideal for use in flares and flash bulbs. Compare the mass of magnesium that burns with the mass of magnesium in the magnesium oxide ash that forms. Explain this in terms of the atomic theory.

(26) Show that the Law of Definite Composition is explained by Dalton's atomic theory.

(27) The chemical name for limestone, a compound of calcium, carbon, and oxygen, is calcium carbonate. When heated, limestone decomposes into solid calcium oxide and gaseous carbon dioxide. From the names of the products, tell where the atoms of each element may be found after the reaction. How does the atomic theory explain this?

Section 4.2

(4)* Two compounds of mercury and chlorine are mercury(I) chloride and mercury(II) chloride. The amount of mercury(I) chloride that contains 71 g of chlorine has 402 g of mercury; the amount of mercury(II) chloride having 71 g of chlorine has 201 g of mercury. Show how the Law of Multiple Proportions is illustrated by these figures.

(28)* Sodium oxide and sodium peroxide are two compounds made up of the elements sodium and oxygen. 62 g of sodium oxide consist of 46 g of sodium and 16 g of oxygen; 78 g of sodium peroxide are made up of 46 g of sodium and 32 g of oxygen. Show how these figures confirm the Law of Multiple Proportions.

Section 4.3

(5) Identify and explain that part of Dalton's atomic theory that was first proved to be incorrect.

(29) Compare the three major parts of an atom in charge and mass.

Section 4.4

(6) How can we account for the fact that most of the alpha particles in the Rutherford scattering experiment passed directly through a solid sheet of gold?

(7) What major conclusions were drawn from the Rutherford scattering experiment?

(8) The Rutherford experiment was performed and its conclusions reached before protons and neutrons were discovered. Why, when they were found, was it believed that they were in the nucleus of the atom?

(30) How can we account for the fact that, in the Rutherford scattering experiment, some of the alpha particles were deflected from their paths through the gold foil, and some even bounced back at various angles?

(31) What name is given to the central part of an atom?

(32) Describe the activity of electrons according to the planetary model of the atom that appeared after the Rutherford scattering experiment.

Section 4.5

(9) Compare the number of protons and electrons in an atom; the number of protons and neutrons; the number of electrons and neutrons.

(10) Can two different isotopes of an element have the same mass number? Explain.

(33) Can two different elements have the same atomic number? Explain.

(34) Can atoms of two different elements have the same mass number? Explain.

(11) From the information given on each line of the following table, complete as many blanks as you can without looking at any reference. If there are any unfilled spaces, continue while referring to your periodic table. As a last resort, check the table inside the back cover of your book.

NAME OF ELEMENT	NUCLEAR SYMBOL	ATOMIC NUMBER	MASS NUMBER	NUMBER OF PROTONS	NUMBER OF NEUTRONS	NUMBER OF ELECTRONS
	$^{70}_{33}As$					
			19	9		
		24			28	
	$^{197}_{79}Au$					
Iron			57			
					40	34

Section 4.6

While this set of questions is based on material in Section 4.6, some parts of some of the problems assume that you have also studied Section 4.7 and recognize the periodic table as a source of atomic masses.

(*12*) What is an atomic mass unit?

(13) The average mass of boron atoms is 10.81 amu. How would you explain what this means to a friend who had never taken chemistry?

(14) Lithium exists in two natural isotopes. The atomic mass of one is 6.01512 amu, and the other is 7.01600 amu. Why is not the atomic mass of lithium on the periodic table 6.51556, the average of these two numbers?

(15) Isotopic data for boron permit calculation of its atomic mass to more than the four significant figures in 10.81 from the periodic table. If 19.78% of boron atoms have an atomic mass of 10.0129 amu and the mass of the remainder is 11.00931 amu, find the average mass in as many significant figures as those data will allow.

(35) What advantage does the atomic mass unit have over grams when speaking of the mass of an atom or subatomic particle?

(36) The mass of an "average atom" of a certain element is 3.34 times as great as the mass of an atom of carbon-12. Using either the periodic table or the table of atoms inside the back cover, identify the element.

(37)* Roughly three quarters of the atoms of an element have a mass number of 63, and the mass number of the remainder is 65. Without calculating, estimate the atomic mass of the element to the first decimal. Locate the element on the periodic table and write its symbol.

(38) 68.9257 amu is the mass of 60.4% of the atoms of an element. There is only one other natural isotope of that element, and its atomic mass is 70.9249 amu. Calculate the average atomic mass of the element and, using the periodic table and/or the table inside the back cover of this book, write its symbol and name.

In the next three problems in each column, you are given the atomic masses (amu) and percentage abundances of the natural isotopes of different elements. (1) Calculate the atomic mass of each element from these data. (2) Using the tables inside the front and back covers, identify the element.*

	ATOMIC MASS (amu)	PERCENTAGE ABUNDANCE
(16)	106.9041	51.82
	108.9047	48.18
(17)	120.9038	57.25
	122.9041	42.75
(18)	135.907	0.193
	137.9057	0.250
	139.9053	88.48
	141.9090	11.07

(19)* The atomic mass of lithium on the periodic table is 6.941 amu. From this number and the values in Example 14, calculate the percentage distribution between the two isotopes.

	ATOMIC MASS (amu)	PERCENTAGE ABUNDANCE
(39)	62.9298	69.09
	64.9278	30.91
(40)	184.9530	37.07
	186.9560	62.93
(41)	57.9353	67.88
	59.9332	26.23
	60.9310	1.19
	61.9283	3.66
	63.9280	1.08

(42)* The CRC Handbook, a large reference book of chemical and physical data from which many of the values in this book are taken, lists two isotopes of rubidium (Z = 37). The atomic mass of 72.15% of rubidium atoms is 84.9117 amu. Through a typographical oversight, the atomic mass of the second isotope is not printed. Calculate that atomic mass.

Section 4.7

(20) Write the symbols of the elements in Group 2A of the periodic table. Write the atomic numbers of the elements in Period 3.

(21) Locate on the periodic table each pair of elements whose atomic numbers are given below. Does each pair belong to the same period or the same chemical family? (a) 12 and 16; (b) 7 and 33; (c) 2 and 10; (d) 42 and 51.

(22) Referring only to a periodic table, list the atomic masses of the elements having atomic numbers 24, 50, and 77.

(23) What are the atomic masses of potassium and sulfur?

(43) How many elements are in Period 5 of the periodic table? Write the atomic numbers of the elements in Group 1B.

(44) Locate on a periodic table each element whose atomic number is given and identify first the number of the period it is in, and then the number of the group: (a) 20; (b) 14; (c) 43.

(45) Using only a periodic table for reference, list the atomic masses of the elements whose atomic numbers are 29, 82, and 55.

(46) Write the atomic masses of helium and aluminum.

Section 4.8

(24) The names, atomic numbers, or symbols of the elements in Figure 4.6 are entered into Table 4.5. Fill in the open spaces, referring only to a periodic table for any information you may require.

Table 4.5 Table of Elements

Name of Element	Atomic Number	Symbol of Element	Name of Element	Atomic Number	Symbol of Element
Sodium					Mg
		Pb		8	
Aluminum			Phosphorus		
	26				Ca
		F	Zinc		
Boron					Li
	18		Nitrogen		
Silver				16	
	6			53	
Copper			Barium		
		Be			K
Krypton				10	
Chlorine			Helium		
	1				Br
		Mn			Ni
	24		Tin		
Cobalt				14	
	80				

Miscellaneous Questions

(47) Distinguish precisely and in scientific terms the differences between items in each of the following groups:
(a) Atom, subatomic particle
(b) Electron, proton, neutron
(c) Nuclear model of the atom, planetary model of the atom
(d) Atomic number, mass number
(e) Chemical symbol of an element, nuclear symbol
(f) Atom, isotope
(g) Atomic mass, atomic mass unit
(h) Atomic mass of an element, atomic mass of an isotope
(i) Period, group or family (in the periodic table)
(j) Representative element, transition element

(48) Determine whether each statement that follows is true or false:
(a) Dalton proposed that atoms of different elements always combine on a one-to-one basis.
(b) According to Dalton, all oxygen atoms have the same diameter.
(c) The mass of an electron is about the same as the mass of a proton.
(d) There are subatomic particles in addition to the electron, proton, and neutron.
(e) The mass of an atom is uniformly distributed throughout the atom.

(f) Most of the particles fired into the gold foil in the Rutherford experiment were not deflected.
(g) The masses of the proton and electron are equal but opposite in sign.
(h) Isotopes of an element have different electrical charges.
(i) The atomic number of an element is proportional to its atomic mass.
(j) An oxygen-16 atom has the same number of protons as an oxygen-17 atom.
(k) The mass of a carbon-12 atom is exactly 12 g.
(l) Periods are arranged vertically in the periodic table.
(m) The atomic mass of the second element in the right column of the periodic table is 10 amu.
(n) Nb is the symbol of the element for which Z = 41.
(o) Elements in the same column of the periodic table have similar properties.
(p) The element for which Z = 38 is in both Group 2A and the fifth period.

(49) The first experiments to suggest that an atom consisted of smaller particles showed that one particle had a negative charge. From that fact, what could be said about the charge of other particles that might be present?

(50)* When Thomson identified the electron, he found that the ratio of its charge to its mass (the e/m ratio) was the same regardless of the element from which the electron came. Positively charged particles found at about the same time did not all have the same e/m ratio. What does that suggest about the mass, particle charge, and minimum number of particles present in the positive particles from different elements?

(51)* Why were scientists inclined to think of an atom as a miniature solar system in the planetary model of the atom? What are the similarities and differences between electrons in orbit around a nucleus and planets in orbit around the sun?

(52)* The existence of isotopes did not appear until nearly a century after Dalton proposed the atomic theory, and then it appeared in experiments more closely associated with physics than with chemistry. What does this suggest about the chemical properties of isotopes?

(53)* A carbon-12 atom contains six electrons, six protons, and six neutrons. Assuming the mass of the atom is the sum of the masses of those parts as given in Table 4.1, calculate the mass of the atom. Why is it not exactly 12 amu, as the definition of atomic mass unit would suggest?

(54) Using the figures in Question 53 calculate the percentage each kind of subatomic particle contributes to the mass of a carbon-12 atom.

(55)* The element carbon occurs in two crystal forms, diamond and graphite. The density of the diamond form is 3.51 g/cm^3, and of graphite, 2.25 g/cm^3. The volume of a carbon atom is 1.9×10^{-24} cm^3. As stated in Section 4.5, one atomic mass unit is 1.66×10^{-24} g.
(a) Calculate the density of a carbon atom.
(b) Suggest a reason for the density of the atom being so much larger than the density of either form of carbon.
(c) The radius of a carbon atom is roughly 1×10^5 times larger than the radius of the nucleus. What is the volume of that nucleus? (Hint: Volume is proportional to the cube of the radius.)
(d) Calculate the density of the nucleus.
(e) The radius of a period on this page is about 0.02 cm. The volume of a sphere that size is 4×10^{-5} cm^3. Calculate the mass of that period if it were completely filled with carbon nuclei. Express the mass in tons.

5 Atomic Theory and the Periodic Table: A Modern View

LOOKING BACK

2.8 Minimization of energy is a driving force in physical and chemical change.

4.3 The atom has a dense nucleus surrounded by space that is thinly populated by electrons.

4.7 and 4.8 The properties of elements recur in a periodic fashion. A periodic table has elements with similar properties (families) in columns called Groups.

LOOKING AHEAD IN CHAPTER 5

Electrons in atoms move spontaneously to the lowest possible energy level.

The distribution of electrons is quantized in energy levels and orbitals.

The shape of the periodic table depends on the arrangement of electrons in the atom. The similar chemical properties of a family are partly the result of a noble gas electron configuration of valence electrons.

5.1 THE BOHR MODEL OF THE HYDROGEN ATOM

PG 5A Describe the Bohr model of the hydrogen atom.

5B Explain the meaning of quantized energy levels in an atom and show how these levels are related to the discrete lines in the spectrum of that atom.

5C Distinguish between ground state and excited state.

In 1913, Niels Bohr, a Danish physicist, proposed a model of the hydrogen atom that opened the way to the understanding of atomic structure. His starting point was the planetary model of the atom. According to this model, an atom has an extremely dense nucleus that is responsible for all of the positive charge in the atom, and nearly all of its mass. Negatively charged electrons of very small mass travel in orbits around the nucleus. The orbits are huge compared to the nucleus, which means that most of the atom is empty space.

Bohr also drew on other scientific observations made late in the 19th century and the first decade of the 20th. Among them were:

1. When white light is passed through a prism, it produces a **continuous spectrum** (Fig. 5.1). This is a rainbow-like spreading out of light in an orderly arrangement of colors. When light from a particular element passes through a prism, it gives a **line spectrum** consisting of separate, or *discrete,* lines of color. Plate 1, a color photograph that faces page 84, shows both a continuous spectrum and the line spectra of several elements.

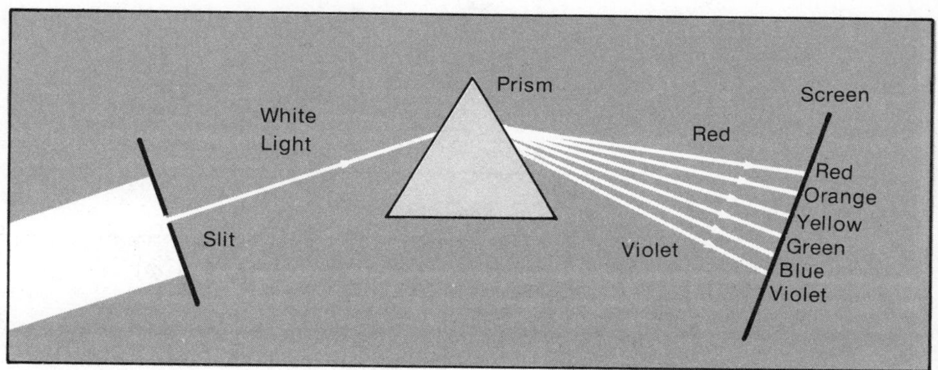

Figure 5.1
Spectrum of white light. White light is a combination of light waves having all the colors of the rainbow. When the light passes through a prism, the colors are separated into a spectrum, which may be projected onto a screen.

2. Light can be described by wave properties such as speed or velocity (c), wavelength (λ), and frequency (ν)* (Figs. 5.2 and 5.3).
3. Light has properties that suggest it is made up of individual bundles, or **quanta,** of energy. The energy of each quantum is calculated by the equation $E = h\nu$, where E is the energy, h is a fixed number known as *Planck's constant,* and ν is the frequency.
4. Physics describes mathematical relationships among radii, speed, energy, and forces when one object moves in an orbit around another, as the planets move around the sun. Physics also furnishes information about electrostatic attraction between plus and minus charges.

From these beginnings, Bohr made a bold assumption. He assumed that the energy possessed by the electron in a hydrogen atom and the radius of its orbit are **quantized.** The amount of something that is quantized is limited to specific values; it may never be between two of those values. By contrast, an amount is **continuous** if it may have any value; between any two values there is an infinite number of other acceptable values. **Quantized energy levels mean that the electron in the atom may have, at any instant, any one of several possible energies, but at no time may it have an energy between them.** An example of something that is quantized is shown in Figure 5.4.

*λ is the Greek letter *lambda*, and ν is the Greek letter *nu*. Do not confuse ν with lower case V.

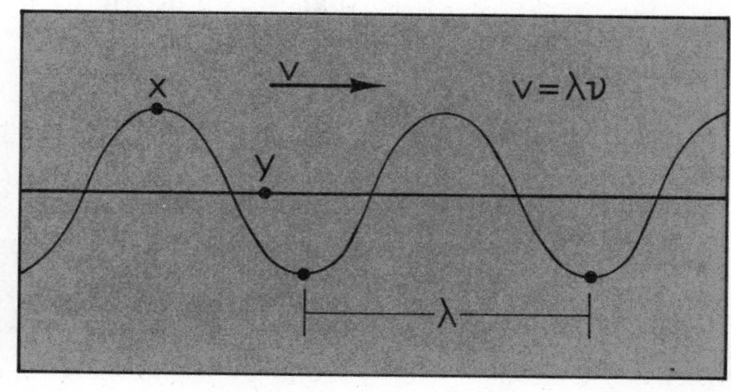

Figure 5.2
Wave properties. Waves may be represented mathematically by the curve shown. All waves have measurements such as wavelength, λ, the distance between corresponding points on consecutive waves; velocity, v, or, in the case of light, c, the linear speed of a point x on a wave; and frequency, ν, the number of wave cycles that pass a point y each second. Reflection, refraction, and diffraction are other wave properties.

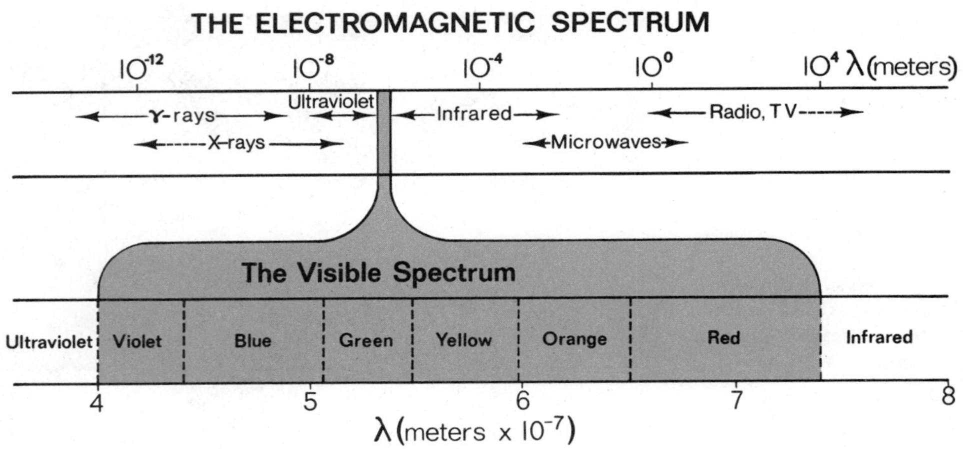

THE ELECTROMAGNETIC SPECTRUM

10^{-12} 10^{-8} 10^{-4} 10^{0} 10^{4} λ(meters)

γ-rays
X-rays
Ultraviolet
Infrared
Microwaves
Radio, TV

The Visible Spectrum

Ultraviolet | Violet | Blue | Green | Yellow | Orange | Red | Infrared

4 5 6 7 8

λ(meters x 10^{-7})

Figure 5.3
The electromagnetic spectrum. The electromagnetic spectrum covers a span from the very high energy, shortwave gamma rays of 10^{-10} meters or less to the low energy long radio waves at 10^4 meters or more. The visible spectrum is but a small portion of the electromagnetic spectrum, covering a wavelength range of about 4×10^{-4} to 7×10^{-4} meters.

Bohr calculated the values of his quantized energy levels using an equation that contains an integer, n—1, 2, 3 . . . and so forth. The results for the integers 1 to 4 are shown in Figure 5.5. The lowest energy level, when n = 1, is called the **ground state.** Higher energy levels, when n = 2 or more, are called **excited states.** The electron is normally found in the ground state. If the atom absorbs energy, the electron can be "excited," or raised to one of the higher energy levels. It cannot stay at that level, but falls back to the ground state. In doing this it releases light energy, $h\nu$, that is equal to the energy difference between the two levels. If this energy is in the visible part of the spectrum, it appears as one of the lines in the spectrum of hydrogen.

Using different values of n in his equation, Bohr was able to calculate the energies of all known lines in the spectrum of the hydrogen atom. He also predicted additional lines and their energies. When the lines were found, his predictions were proved to be correct. In fact, all of Bohr's calculations correspond with measured values to within one part per thousand. It certainly seemed that Bohr had found the answer to the structure of the atom.

There were problems, however. First, hydrogen is the *only* atom that fits the Bohr model. The model fails for any atom with more than one electron. Second, it is a fact that a charged body moving in a circle radiates energy. This means the electron itself should lose energy and promptly—in about 0.000 000 000 01 second!—crash into the nucleus. To do otherwise is to violate the Law of Conservation of Energy (Section 2.9). So, for 13 years, science did what it does so often when faced with contradictions. It accepted a theory that was known to be only partly correct. It worked on the faulty parts,

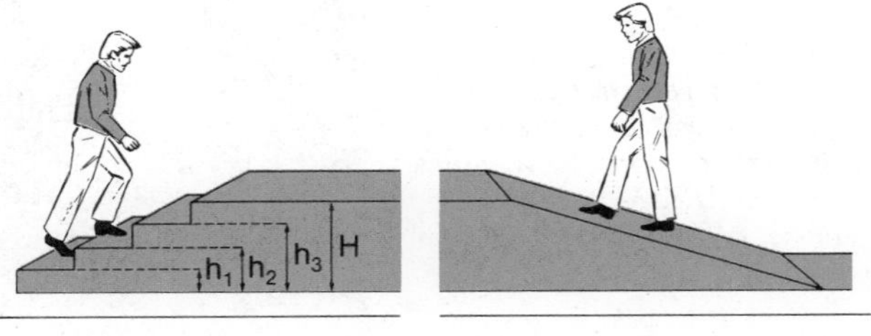

Figure 5.4
The quantization, or quantum, concept. Man on ramp can stop at any level above ground; his elevation above ground is not quantized. Man on stairs can stop only on a step; his elevation is quantized at h_1, h_2, h_3, or H.

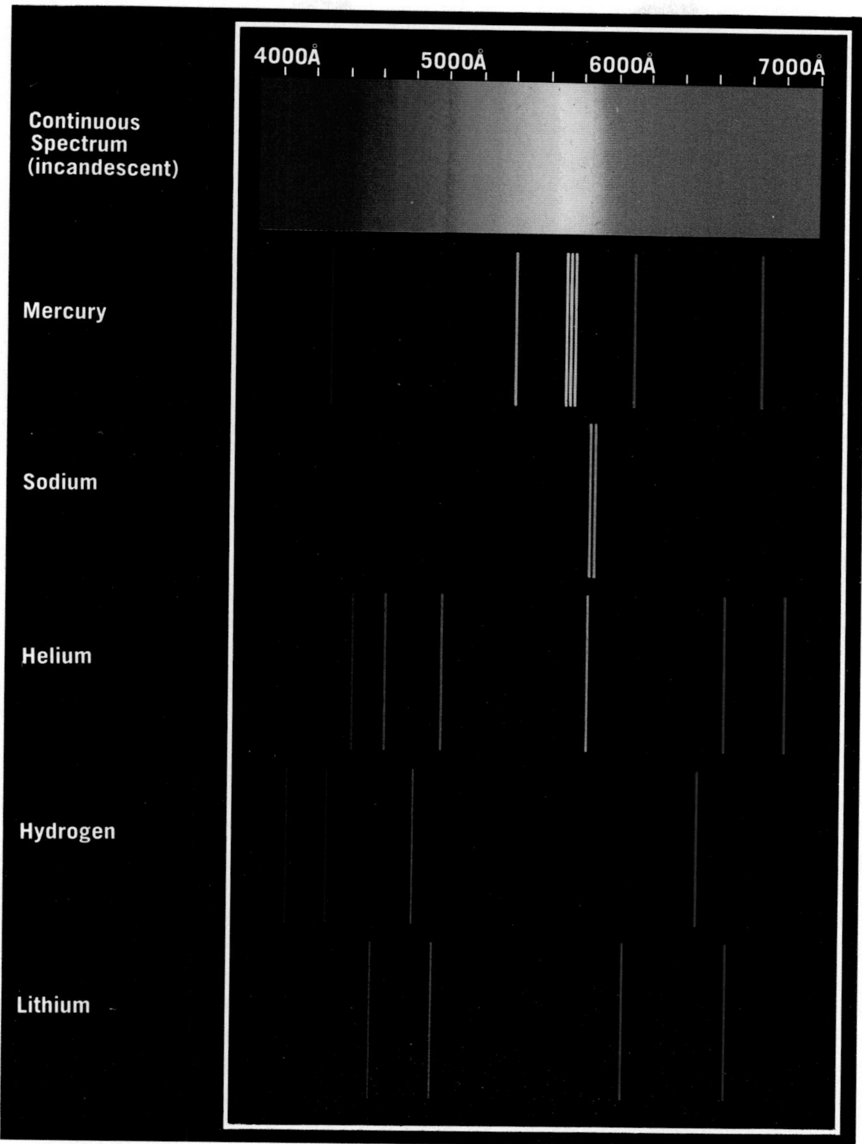

Color Plate 1.
Spectrum chart. Continuous "rainbow" spectrum at top is from "white" light of an incandescent lamp. Beneath it are the bright line spectra that characterize certain elements.

Color Plate 2.
Physical properties of liquid oxygen. Oxygen is attracted into a magnetic field; one consequence is that liquid oxygen can be suspended between the poles of an electromagnet. Both the paramagnetism and the blue color are due to the unpaired electrons in the O_2 molecule.

Color Plate 3.
When a strip of zinc is placed in a solution containing Cu^{2+} ions (left), an oxidation-reduction reaction occurs. The final result is shown at the right. Copper metal plates out and the blue color due to Cu^{2+} fades.

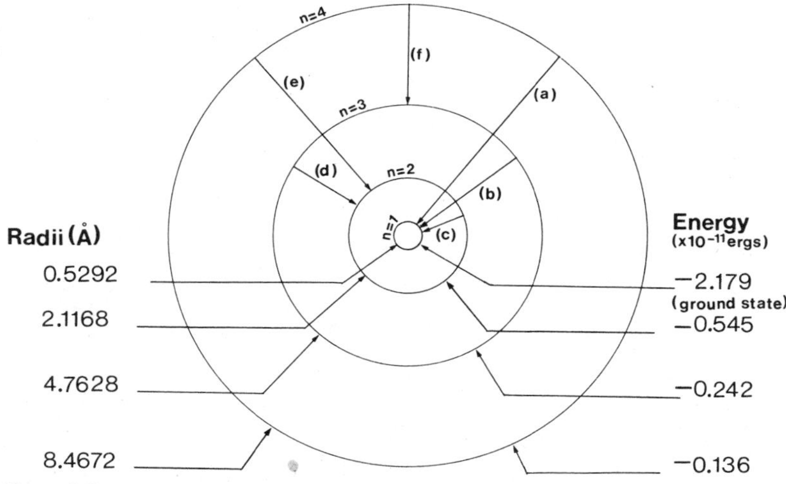

Figure 5.5
The Bohr model of the hydrogen atom. The electron is "allowed" to circle the nucleus only at certain radii and with certain energies, the first four of which are shown. An electron in the "ground state" n = 1 level can absorb the exact amount of energy to raise it to any other level, e.g., n = 2, 3, or 4. An electron at such an "excited state" is unstable and drops back to the n = 1 level in one or more steps. Electromagnetic energy is radiated with each step. Jumps a to c are in the ultraviolet portion of the spectrum, d and e are in the visible range, and f is in the infrared region. (Å is the symbol for angstrom, a length unit: 1 nm = 10 Å. The *erg* is an energy unit: 1 J = 1×10^7 ergs.)

modified them, improved them, and ultimately replaced them with more accurate concepts.

Niels Bohr made two huge contributions to the development of modern atomic theory. First, he suggested a reasonable explanation for atomic spectra in terms of electron energies. Second, he introduced the idea of quantized electron energy levels in the atom. These levels appear in modern theory as **principal energy levels;** they are identified by the **principal quantum number,** n.

Quick Check 5.1

Are the following true or false:

1. Two of the following are quantized: (a) the speed of automobiles on a freeway; (b) paper money in the U.S.; (c) soup on a grocery store shelf; (d) water coming from a faucet; (e) a person's weight.
2. Bohr described mathematically the orbits of electrons in a sodium (Z = 11) atom.

5.2 THE QUANTUM MECHANICAL MODEL OF THE ATOM

In 1924, Louis de Broglie, a French physicist, suggested that matter in motion has properties normally associated with waves, and that these properties are important in subatomic particles. In the period 1925 to 1928, Erwin Schrödinger

Table 5.1 Principal Energy Levels, Sublevels, and Orbitals in the Quantum Mechanical Model of the Atom*

Principal Energy Levels	Number of Sublevels	Identification of Sublevels						
n = 1	1 Orbital per sublevel	1s 1						
n = 2	2 Orbitals per sublevel	2s 1	2p 3					
n = 3	3 Orbitals per sublevel	3s 1	3p 3	3d 5				
n = 4	4 Orbitals per sublevel	4s 1	4p 3	4d 5	4f 7			
n = 5	5 Orbitals per sublevel	5s 1	5p 3	5d 5	5f 7	5g 9		
n = 6	6 Orbitals per sublevel	6s 1	6p 3	6d 5	6f 7	6g 9	6h 11	
n = 7	7 Orbitals per sublevel	7s 1	7p 3	7d 5	7f 7	7g 9	7h 11	7i 13

*Principal energy levels are identified by numbers, 1, 2, 3, . . . 7. Each principal energy level has a number of sublevels equal to the identifying number of the principal energy level. Only four sublevels, s, p, d, and f, are required for the elements now known, but if more are discovered the electrons will be assigned to the g, h, and i sublevels. Sublevels are divided into orbitals. Each s sublevel has 1 orbital, each p sublevel has 3 orbitals, a d sublevel has 5 orbitals, and an f sublevel has 7 orbitals. The sublevels shown in color are not required to accommodate ground state electrons of elements now known.

applied the principles of wave mechanics and developed the **quantum mechanical model** of the atom, the heart of which is the wave equation. This model has been tested for half a century. It explains more satisfactorily than any other theory all experimental observations to date, and no exceptions to the theory have appeared. It is the theory that is generally accepted today.

Unfortunately, the quantum mechanical model does not give a very good idea of the appearance of an atom. One physicist has noted, ". . . that the question, 'What does an atom look like?' has no meaning, much less an answer."* The quantum concept is abstract and mathematical, rather than physical. You might keep this in mind as you read the following sections.

As its name suggests, the quantum mechanical model of the atom keeps the quantized energy levels introduced by Bohr. It also states that more than just the principal quantum number is needed to describe the electron energy. It requires three quantum numbers that specify (1) the principal energy level, (2)

*H. E. White: Modern College Physics. D. Van Nostrand, 1962.

the sublevel, and (3) the orbital. These are summarized in Table 5.1, which you may follow as they are examined separately.

Principal Energy Levels

PG 5D Identify the principal energy levels within an atom and state the energy trend among them.

Following the Bohr model, principal energy levels are identified by the principal quantum number, n. The first principal energy level is n = 1, the second is n = 2, and so on. (An older system uses the letters, K, L, M . . . to designate principal energy levels.) Mathematically, there is no end to the number of principal energy levels, but the seventh level is the highest occupied by ground state electrons among the elements now known.

The energy possessed by an electron depends on the principal energy level in which it is found. An electron in the n = 2 level has more energy than an n = 1 electron; an n = 3 electron is at higher energy than an electron in the n = 2 level. This trend holds generally, but the principal energy levels for elements other than hydrogen represent a *range* of energies. Beginning with n = 3 and n = 4, the ranges overlap. We therefore find that a high-energy n = 3 electron is at a higher level than a low-energy n = 4 electron.

Sublevels

PG 5E For each principal energy level, state the number of sublevels, identify them by letter, and state the energy trend among them.

According to the quantum mechanical model of the atom, principal energy levels consist of one or more **sublevels** (Table 5.1). These are identified by the letters **s**, **p**, **d**, and **f**. (The letters come from terms formerly used in spectroscopy.) *The total number of sublevels within a given principal energy level is equal to n, the principal quantum number*. For n = 1 there is one sublevel, designated 1s. At n = 2 there are two sublevels, 2s and 2p. When n = 3 there are three sublevels, 3s, 3p, and 3d; n = 4 has four sublevels, 4s, 4p, 4d, and 4f. Quantum theory describes sublevels beyond f when n = 5 or more, but these are not needed by elements known today. The energy of an electron is described by the sublevel in which it is found. An electron in the 2s sublevel is called a "2s electron," and an electron in the 3p sublevel is a "3p electron."

Just as electron energies increase through the principal energy levels, so do they increase through the sublevels. Within a given principal energy level, an electron in the s sublevel is at lower energy than one in the p sublevel, a p electron is lower in energy than a d electron, and a d is lower than an f. This is what makes it possible for principal energy levels to overlap. The highest n = 3 electron is a 3d electron, while the lowest n = 4 electron is a 4s electron. A 3d electron is at higher energy than a 4s electron. But, among s electrons, energy always increases from 1s to 2s to 3s, etc. Similarly, among p electrons, $2p < 3p < 4p$, etc., in terms of electron energies. The same trends hold for d and f electrons.

Electron Orbitals

PG 5F Sketch the shapes of s and p orbitals.

5G State the number of orbitals in each sublevel.

According to modern atomic theory, it is not possible to know at the same time both the position of an electron in an atom and the path it travels. It is possible, however, to describe mathematically a region in space around a nucleus where there is a "high probability" of finding an electron. These regions are called **orbitals.** Notice the uncertainty of the *orbital* description, stated in terms of probability, compared to a Bohr description of *orbit* that states exactly where the electron is, where it was, and where it is going. The mathematically described shapes of the s, p, and d orbitals are shown in Figure 5.6. The f orbitals are more complex and difficult to illustrate in two dimensions.

Table 5.1 summarizes the sublevels and orbitals in the first seven principal energy levels. Each principal energy level has one s orbital. Figure 5.6 shows all s orbitals to be *spherical* in shape. Both the size of an s orbital and the electron energy increase as the principal quantum number increases from 1s to 2s to 3s . . . up to 7s, the largest s orbital required by any known element.

Table 5.1 and Figure 5.6 indicate three p orbitals when n is 2 or more. Each p orbital consists of a pair of lobes oriented at right angles to the other two

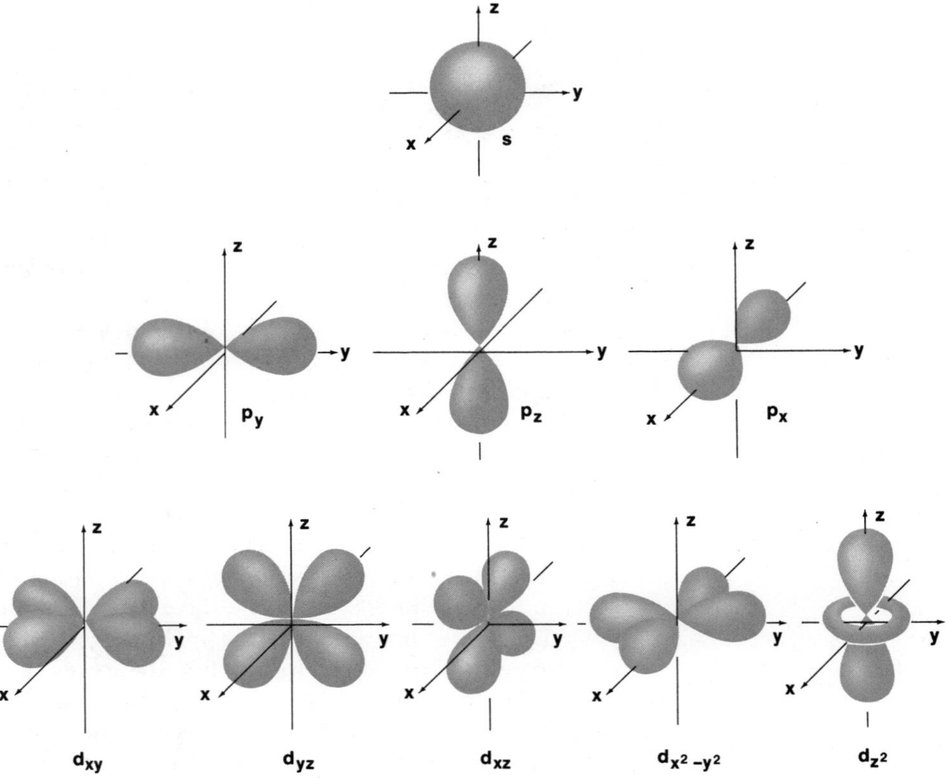

Figure 5.6
Shapes of electron orbitals according to the quantum mechanical model of the atom.

pairs at their common center, the nucleus of the atom. These are described as lying along the x, y, and z axes. As with the s orbitals, both size and energy increase with increasing principal quantum number, n.

The d orbitals shown in Figure 5.6 are more complex, and we will not be concerned with their shapes. It should be noted, however, that the first d orbitals appear at n = 3, and there are five of them (Table 5.1). The table also shows that there are seven f orbitals for each quantum number beginning at n = 4. The sizes and energies of both d and f orbitals increase as n becomes larger.

The Pauli Exclusion Principle

PG 5H State the restrictions on the electron population of an orbital.

In the beginning of this section we said that the quantum mechanical model of the atom provides three quantum numbers by which the energy of an electron is specified. The principal quantum number, n, was identified. We did not name the other two quantum numbers, but we described them in the sublevels and orbitals. In addition to these, what is known as the **Pauli exclusion principle** requires a fourth quantum number, not coming directly from quantum mechanics, but added to make the theory match experimental observations. Its effect is to restrict the population of any orbital to two electrons. At any time, an orbital may be (a) unoccupied, (b) occupied by one electron, or (c) occupied by two electrons. No other occupancy condition is possible.

Summary of the Quantum Mechanical Model of the Atom

1. Generally speaking, energy increases with increasing principal quantum number: $n = 1 < n = 2 < n = 3 \ldots$ Specifically, for any given sublevel, s, p, d, or f, energy increases with increasing principal quantum number: 1s < 2s < 3s . . . ; 2p < 3p < 4p . . . ; etc.
2. For any given value of n, energy increases through the sublevels in the order s p d f: 2s < 2p; 3s < 3p < 3d; 4s < 4p < 4d < 4f; etc.
3. For any value of n there are n sublevels:

n	*Sublevels*	
1	s	(one sublevel)
2	s, p	(two sublevels)
3	s, p, d	(three sublevels)
4	s, p, d, f	(four sublevels)

4. The number of orbitals for each sublevel is

Sublevel	*Orbitals*
s	1
p	3
d	5
f	7

5. An orbital may be occupied by 0, 1, or 2 electrons, but never more than 2.

Quick Check 5.2

Without looking at the above summary, are the following true or false?

1. There is one s orbital when n = 2, two s orbitals when n = 2, three s orbitals when n = 3, and so on.
2. All n = 3 orbitals are at lower energy than all n = 4 orbitals.
3. There is no d sublevel when n = 2.
4. There are five d orbitals at both the fourth and sixth principal energy levels.

5.3 ELECTRON CONFIGURATION

Development of Electron Configurations

Figure 5.7 is an electron energy level diagram that illustrates the above summary. Relative energies of sublevels are plotted vertically, and principal quan-

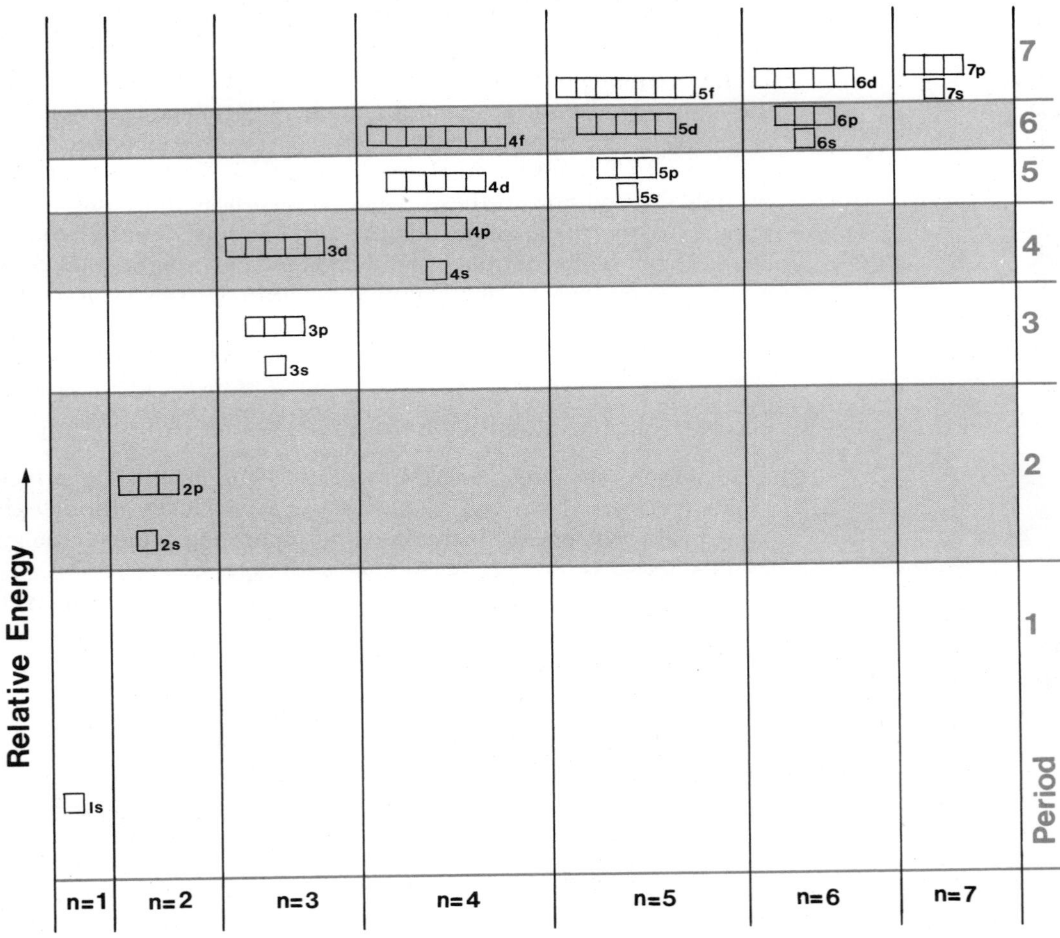

Figure 5.7

Electron energy level diagram. Each column contains one box for each orbital in the n = 1, n = 2, n = 3 . . . n = 7 principal energy levels. The boxes are grouped by sublevels s, p, d and f, as far as required within each principal energy level. All sublevels are positioned vertically on the page according to a general energy level scale shown at the left. Any orbital that is higher on the page than a second orbital is therefore higher in energy than the second orbital. In filling orbitals from the lowest energy level, each orbital will generally hold two electrons before any orbital higher in energy accepts an electron. The colored lines and the scale at the right correlate energies of the sublevels with the periods of the periodic table.

1 H $1s^1$																	2 He $1s^2$
3 Li $1s^2$ $2s^1$	4 Be $1s^2$ $2s^2$											5 B $1s^2$ $2s^2 2p^1$	6 C $1s^2$ $2s^2 2p^2$	7 N $1s^2$ $2s^2 2p^3$	8 O $1s^2$ $2s^2 2p^4$	9 F $1s^2$ $2s^2 2p^5$	10 Ne $1s^2$ $2s^2 2p^6$
11 Na [Ne] $3s^1$	12 Mg [Ne] $3s^2$											13 Al [Ne] $3s^2 3p^1$	14 Si [Ne] $3s^2 3p^2$	15 P [Ne] $3s^2 3p^3$	16 S [Ne] $3s^2 3p^4$	17 Cl [Ne] $3s^2 3p^5$	18 Ar [Ne] $3s^2 3p^6$
19 K [Ar] $4s^1$	20 Ca [Ar] $4s^2$	21 Sc [Ar]$4s^2$ $3d^1$	22 Ti [Ar]$4s^2$ $3d^2$	23 V [Ar]$4s^2$ $3d^3$	24 Cr [Ar]$4s^1$ $3d^5$	25 Mn [Ar]$4s^2$ $3d^5$	26 Fe [Ar]$4s^2$ $3d^6$	27 Co [Ar]$4s^2$ $3d^7$	28 Ni [Ar]$4s^2$ $3d^8$	29 Cu [Ar]$4s^1$ $3d^{10}$	30 Zn [Ar]$4s^2$ $3d^{10}$	31 Ga [Ar]$4s^2$ $3d^{10}4p^1$	32 Ge [Ar]$4s^2$ $3d^{10}4p^2$	33 As [Ar]$4s^2$ $3d^{10}4p^3$	34 Se [Ar]$4s^2$ $3d^{10}4p^4$	35 Br [Ar]$4s^2$ $3d^{10}4p^5$	36 Kr [Ar]$4s^2$ $3d^{10}4p^6$

Figure 5.8
Ground state electron configurations of neutral gaseous atoms.

tum numbers are plotted horizontally. Each box represents one orbital, which may contain 0, 1, or 2 electrons. Each s sublevel has one orbital, each p sublevel has three orbitals, each d has five, and each f has seven. The different backgrounds separate the sublevels by periods in the periodic table.

Many chemical properties of an element depend on its **electron configuration,** the ground state distribution of electrons among the orbitals of a gaseous atom. Two rules guide the assignment of electrons to orbitals:

1. At ground state, the electrons will fill the *lowest* energy orbitals available, as shown in Figure 5.7.
2. No orbital can have more than two electrons.

1s. An atom of hydrogen (H, Z = 1) has only one electron, and it occupies the lowest energy orbital in any atom. Figure 5.7 shows this is the 1s orbital. We indicate the total number of electrons in any sublevel by a superscript number. Therefore, the electron configuration of hydrogen is written $1s^1$, showing one electron in the 1s orbital. Helium (He, Z = 2) has two electrons in the neutral atom, and both fit into the 1s orbital. The electron configuration of helium is therefore $1s^2$.

These and other electron configurations to be developed are shown in the first four periods of the periodic table in Figure 5.8.

2s. Lithium (Li, Z = 3) has three electrons. The first two fill the 1s orbital as before. The third electron goes to the next orbital up the energy scale with a vacancy, which Figure 5.7 shows to be the 2s orbital. The electron configuration for lithium is therefore $1s^2 2s^1$. In a similar manner beryllium (Be, Z = 4) divides its four electrons between the two lowest orbitals, filling them both: $1s^2 2s^2$. These two configurations are shown in Figure 5.8 also.

2p. The first four electrons of boron (B, Z = 5) fill 1s and 2s orbitals. The fifth electron goes to the next highest level, the 2p, according to Figure 5.7. The configuration for boron is $1s^2 2s^2 2p^1$. Similarly carbon (C, Z = 6) has a configuration $1s^2 2s^2 2p^2$. Though we will not emphasize the point, the two 2p electrons occupy different p orbitals. Electrons half-fill each orbital in a sublevel before completely filling any of them. The next four elements increase the number of electrons in the three 2p orbitals until they are filled with six electrons for neon (Ne, Z = 10): $1s^2 2s^2 2p^6$. All these configurations appear in Figure 5.8.

3s and 3p. The first ten electrons of sodium ($Z = 11$) are distributed in the same way as the the electrons in neon. The eleventh sodium electron is added as a 3s electron: $1s^22s^22p^63s^1$. The electron configurations for all elements whose atomic numbers are greater than 10 begin with the neon configuration, $1s^22s^22p^6$. That part of those configurations is frequently shortened to the **neon core,** represented by [Ne]. For sodium this becomes $[Ne]3s^1$; for magnesium, $[Ne]3s^2$; for aluminum, $[Ne]3s^23p^1$; and so on to argon, $[Ne]3s^23p^6$. The sequence is exactly as it was in Period 2. The neon core is used for Period 3 in Figure 5.8.

4s. Potassium (K, $Z = 19$) repeats at the 4s level the development of sodium at the 3s level. Its complete configuration is $1s^22s^22p^63s^23p^64s^1$. All configurations for atomic numbers greater than 18 distribute their first 18 electrons in the configuration of argon (Ar, $Z = 18$), $1s^22s^22p^63s^23p^6$, which may be shortened to the **argon core,** [Ar]. Accordingly, the configuration for potassium may be written $[Ar]4s^1$; and calcium (Ca, $Z = 20$) is $[Ar]4s^2$.

3d. Figure 5.7 predicts that five 3d orbitals are next available for electron occupancy. At two electrons per orbital they should accommodate the next ten elements. The first three of these, scandium, Sc, titanium, Ti, and vanadium, V, atomic numbers 21 to 23, fill in order as predicted. The configuration for vanadium is $1s^22s^22p^63s^23p^64s^23d^3$, or $[Ar]4s^23d^3$.* Chromium (Cr, $Z = 24$) is the first element to break the orderly sequence in which lowest energy orbitals are filled. Its configuration is $[Ar]4s^13d^5$, rather than the expected $[Ar]4s^23d^4$. This is generally attributed to an extra stability found when the d sublevel is half-filled or completely filled. Manganese (Mn, $Z = 25$) puts us back on the track, only to be derailed again at copper (Cu, $Z = 29$): $[Ar]4s^13d^{10}$. Zinc (Zn, $Z = 30$) has the expected configuration of $[Ar]4s^23d^{10}$. Examine the sequence for atomic numbers 21 to 30 in Figure 5.8 and note the two exceptions.

4p. By now the pattern should be clear; you should expect atomic numbers 31 to 36 to fill in sequence the next orbitals available, which are the 4p orbitals. They do, as shown in Figure 5.8.

Our consideration of electron configurations ends with atomic number 36, krypton. Were we to continue, we would find the higher s and p orbitals fill in perfect accord with the procedures used thus far. The 4d, 4f, 5d, and 5f orbitals have several variations like those with chromium and copper, so configurations involving those orbitals must be looked up. But you should be able to reproduce easily the configurations for the first 36 elements—not from memory, but from their correlation with the periodic table.

Electron Configurations and the Periodic Table

Figure 5.8 shows that specific sublevels are being filled in different regions of the periodic table. This is indicated by color in Figure 5.9. In Groups 1A and 2A the s sublevels are the highest occupied energy levels. The p orbitals are filled in order across Groups 3A to 0. The d electrons appear in the B groups

*Some chemists prefer to write this configuration $1s^22s^22p^63s^23p^63d^34s^2$, or $[Ar]3d^34s^2$, putting the 3d before the 4s. This is an equally acceptable alternative. There is perhaps some advantage at this point in listing the sublevels in the order in which they fill, so we will continue to use this order in this text.

and Group 8. Finally, the f electrons show up in the lanthanide and actinide series beneath the table.

Notice that, when you read the periodic table from left to right across the periods in Figure 5.9, you get the order of increasing sublevel energy. The first period gives only the 1s sublevel. Period 2 takes you through 2s and 2p. Similarly, the third period covers the 3s and 3p sublevels. Period 4 starts with 4s, follows with 3d, and ends up with 4p, and so forth. Ignoring the minor variations in Periods 6 and 7, the complete list is

1s 2s 2p 3s 3p 4s 3d 4p 5s 4d 5p 6s 4f 5d 6p 7s 5f 6d 7p

Now compare this list with the order of increasing sublevel energy from Figure 5.7. They are exactly the same! *The periodic table therefore replaces the electron energy diagram as a guide to the order of increasing sublevel energy.*

Think, for a moment, how remarkable this is. Mendeleev and Meyer constructed their periodic tables on the bases of physical and chemical properties of the elements. They suggested no reason for the periodic recurrence of these properties. They knew nothing of electrons, protons, nuclei, wave equations, nor quantized energy levels. Yet, when these things were found some 60 years later, the match between those first periodic tables and the quantum mechanical model of the atom was nearly perfect!

If you are ever required to list the sublevels in order of increasing energy without reference to a periodic table, the diagram produced by the sublevels in Table 5.1 is helpful and is easily recalled. Beginning at the upper left, the diagonal lines pass through the sublevels in the sequence required.

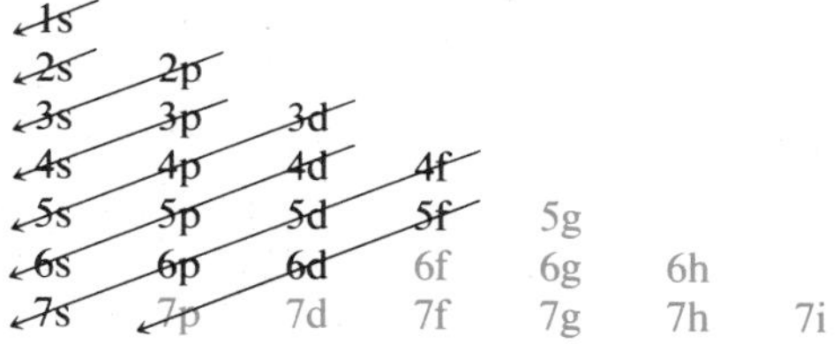

The sublevels shown in color are not needed for the elements known today, but the mathematics of the quantum mechanical model predict the order of increasing energy indefinitely.

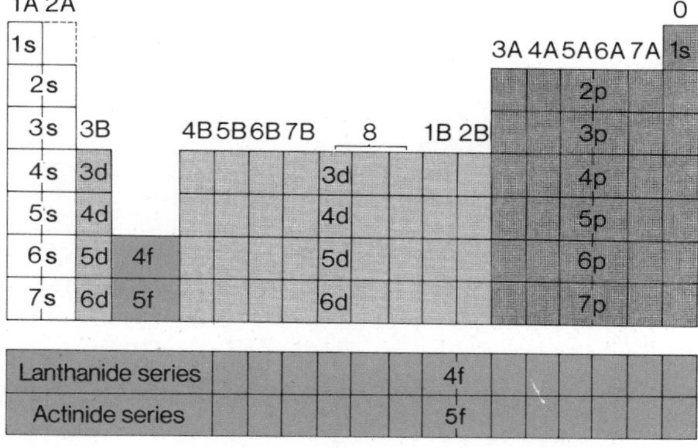

Figure 5.9
Arrangement of periodic table according to atomic sublevels. Highest energy sublevels occupied at ground state are s sublevels in Groups 1A and 2A; p sublevels in Groups 3A to 0, except for helium; d sublevels in Groups 1B to 7B and 8; and f sublevels in the lanthanide and actinide series.

Writing Electron Configurations

PG 5I Referring only to a periodic table, write the ground state electron configuration of a gaseous atom of any element up to atomic number 36.

If, with help from a periodic table, you can (1) establish the number of electrons in the highest occupied energy sublevel of an atom and (2) list the sublevels in order of increasing energy, you can write the electron configuration of that atom.

There is one electron in the highest occupied energy sublevel for each element in Group 1A: ns^1. There are two electrons in the highest occupied energy sublevel of a Group 2A element: ns^2. In Groups 3A to 0, the number of p electrons is found by counting from the left; it increases in order from one to six:

Group:	3A	4A	5A	6A	7A	0
p Electrons:	1	2	3	4	5	6
Electron configuration:	np^1	np^2	np^3	np^4	np^5	np^6

A similar count-from-the-left order appears among the 3d electrons, except that it interrupted for chromium (Z = 24) and copper (Z = 29):

Group:	3B	4B	5B	6B	7B	←	8B	→	1B	2B
d Electrons:	1	2	3	5	5	6	7	8	10	10
Electron configuration:	$3d^1$	$3d^2$	$3d^3$	$3d^5$	$3d^5$	$3d^6$	$3d^7$	$3d^8$	$3d^{10}$	$3d^{10}$

Fix these electron populations and their positions in the periodic table firmly in your thought now. Then cover both the above summaries and the answers below the dashes as you respond to the next example. You may refer to a full periodic table, of course.

EXAMPLE 5.1 Write the electron configuration of the highest occupied energy sublevel for each of the following elements:

beryllium (Z = 4) _____ phosphorus (Z = 15) _____ manganese (Z = 25) _____

-----

beryllium, $2s^2$; phosphorus, $3p^3$; manganese, $3d^5$

Counting from the left, (a) beryllium is the second box among the 2s sublevel elements, so its electron configuration is $2s^2$; (b) phosphorus is in the third box among the 3p sublevel elements, so its configuration is $3p^3$; and (c) manganese is in the fifth box among 3d sublevel elements, so it is $3d^5$.

You are now ready to write electron configurations. The procedure is as follows:

1. Locate the element in the periodic table. From its position in the table, identify and write the electron configuration of its highest occupied energy sublevel. (Leave room for writing lower energy sublevels to its left.)
2. To the left of what has already been written, list in order of increasing energy all lower energy sublevels.
3. For each filled lower energy sublevel, write as a superscript the number of electrons that fill that sublevel. (There are two s electrons, ns^2; six p electrons, np^6; and ten d electrons, nd^{10}. Exceptions: for chromium and copper the 4s sublevel has only one electron, $4s^1$.)
4. Confirm that the total number of electrons is the same as the atomic number.

The last step checks the correctness of your final result. The atomic number is the number of protons in the nucleus of an atom, which is equal to the number of electrons. Therefore, the sum of the superscripts in an electron configuration, which is the total number of electrons in the atom, must be the same as the atomic number. For example, the electron configuration of oxygen ($Z = 8$) is $1s^22s^22p^4$. The sum of the superscripts is $2 + 2 + 4 = 8$, the same as the atomic number.

EXAMPLE 5.2 Write the complete electron configuration for chlorine (Cl, $Z = 17$).

SOLUTION: By steps from the above procedure:

1) From its position in the periodic table (Group 7A, Period 3), the electron configuration of the highest occupied energy sublevel of chlorine is $3p^5$.
2) The sublevels having lower energies than 3p can be "read" across the periods from left to right in the periodic table, as in Figure 5.10: 1s 2s 2p 3s $3p^5$. If the neon core were to be used, these would be represented by [Ne]3s $3p^5$.
3) A filled s orbital has two electrons, and a filled p orbital has six. Filling in these numbers yields $1s^22s^22p^63s^23p^5$, or $[Ne]3s^23p^5$.
4) $2 + 2 + 6 + 2 + 5 = 17 = Z$

EXAMPLE 5.3 Write the complete electron configuration (no Group 0 core) for potassium (K, $Z = 19$).

First, what is the electron configuration of the highest occupied energy sublevel? (When you write the answer, leave space for the lower energy sublevels.)

_____ _____

$4s^1$ (Group 1A elements have one s electron.)

Now list to the left of $4s^1$ above all lower energy sublevels in order of increasing energy.

_____ _____

1s 2s 2p 3s 3p $4s^1$

Finally, add the superscripts that show how many electrons fill the lower energy sublevels. Check the final result.

----- -----

$1s^22s^22p^63s^23p^64s^1$. $2 + 2 + 6 + 2 + 6 + 1 = 19 = Z$

Rewrite the configuration with a core from the closest Group 0 element with a smaller atomic number.

----- -----

$[Ar]4s^1$

EXAMPLE 5.4 Develop the electron configuration for cobalt, (Co, Z = 27).

This is your first example with d electrons. The procedure is the same. Write both a complete configuration and one with a Group 0 core.

----- -----

$1s^22s^22p^63s^23p^64s^23d^7$, or $[Ar]4s^23d^7$

By steps, (1) $3d^7$; (2) 1s 2s 2p 3s 3p 4s $3d^7$ or [Ar]4s $3d^7$; (3) $1s^22s^22p^63s^23p^64s^23d^7$ or $[Ar]4s^23d^7$; (4) $2 + 2 + 6 + 2 + 6 + 2 + 7 = 27 = Z$.

5.4 VALENCE ELECTRONS

PG 5J Using n for the highest occupied energy level, write the configuration of the valence electrons of any Group A element in the periodic table.

5K Write the Lewis (electron dot) symbol for an atom of any representative element.

It is now known that many of the similar chemical properties of elements in the same column of the periodic table are related to the arrangement of electrons in the highest occupied s and p sublevels. These are, for example, the single 3s electron in sodium, $1s^22s^22p^63s^1$, or $[Ne]3s^1$, and the six electrons in the 3s and 3p orbitals of sulfur, $1s^22s^22p^63s^23p^4$, or $[Ne]3s^23p^4$. These highest energy s and p electrons are called **valence electrons.**

Look again at Figure 5.8. Notice the valence electrons of the elements in Group 1A. They are $1s^1$ for hydrogen, Z = 1; $2s^1$ for lithium, Z = 3; $3s^1$ for sodium, Z = 11; and $4s^1$ for potassium, Z = 19. There is one electron in the s orbital in each case. The trend continues for atomic numbers 37, 55, and 87 as $5s^1$, $6s^1$, and $7s^1$, respectively. This regularity for Group 1A elements is sometimes summarized by saying their valence electron configuration is ns^1, where n is any principal energy level.

Similar trends exist for other groups of representative elements. The valence electron configurations for Group 2A elements is ns^2. In Group 3A the p electrons appear: ns^2np^1. They increase in order to Group 0, which has the general configuration ns^2np^6. These configurations are given for all representative element groups in the second row of Table 5.2. Notice that, except for Group 0, the number of valence electrons, or the total of the electrons in the s and p orbitals, is the same as the group number. For example, in Group 3A, $2 + 1 = 3$.

EXAMPLE 5.5 If necessary, refer only to a periodic table while answering this question. (a) Write the electron configuration for the highest occupied energy level for the Group 6A elements. (b) Identify the group whose electron configuration for the highest occupied energy level is ns^2np^2.

(a) ns^2np^4; (b) Group 4A

(a) A Group 6A element has six valence electrons. The first two must be in the ns sublevel, and the remaining four must be in the np sublevel. (b) The total number of valence electrons is $2 + 2 = 4$; therefore Group 4A.

Another way to show valence electrons uses **Lewis symbols,** which are also called **electron dot symbols.** The symbol of the element is written, and then surrounded by that number of dots that matches the number of valence electrons. The dot symbols for the representative elements in Period 3 are given in Table 5.2. Paired electrons, those that occupy the same orbital, are usually placed on the same side of the symbol, and single occupants of an orbital are by themselves. This is not a hard and fast rule; exceptions are common if other positions better serve a particular purpose. Group 0 atoms have a full set of 8 valence electrons, two in the s orbital and six in the p orbitals. This is sometimes called an **octet** of electrons.

Table 5.2 Lewis Symbols of the Elements

Group	**1A**	**2A**	**3A**	**4A**	**5A**	**6A**	**7A**	**0**
Highest energy electron configuration	ns^1	ns^2	ns^2np^1	ns^2np^2	ns^2np^3	ns^2np^4	ns^2np^5	ns^2np^6
Number of valence electrons	1	2	3	4	5	6	7	8
Lewis symbol third period element	Na·	Mg:	Al:	·Si:	·P:	·S:	:Cl:	:Ar:

EXAMPLE 5.6 Write the electron dot symbols for the elements whose atomic numbers are 38 and 52.

Locate the element whose atomic numbers are 38 and 52 in the periodic table. Write their symbols, and then surround them with dots corresponding to the number of valence electrons.

$\text{Sr}:$ $\quad \cdot\dot{\underset{\cdot\cdot}{\text{Te}}}:$

The periodic table gives the symbol Sr for atomic number 38. (The element is strontium, one isotope of which is present in radioactive fallout.) The element is in Group 2A, indicating two valence electrons. Te (tellurium) is the symbol for atomic number 52. It is in Group 6A, so six electron dots.

5.5 TRENDS IN THE PERIODIC TABLE

When we introduced the periodic table in Section 4.8, it was noted that Mendeleev and Meyer had been trying to "organize" some of the recurring chemical and physical properties of the elements. Some of these properties will be examined in this section.

Ionization Energy

A sodium atom, Na, Z = 11, has eleven protons in its nucleus and eleven electrons outside the nucleus. (We'll not be concerned with the twelve neutrons also present in most sodium atoms.) One of the electrons is a valence electron. Mentally separate the valence electron from the other ten. This is pictured in the larger block of Figure 5.10. The valence electron, with its −1 charge, is still part of the neutral atom. The rest of the atom has 11 protons and 10 electrons (11 plus charges and 10 minus charges), giving it a net charge of +1. Work must be done to pull the electron away from that +1 charge. After the valence electron has been removed from the atom, the positively charged particle that remains is called a sodium ion, Na^+.

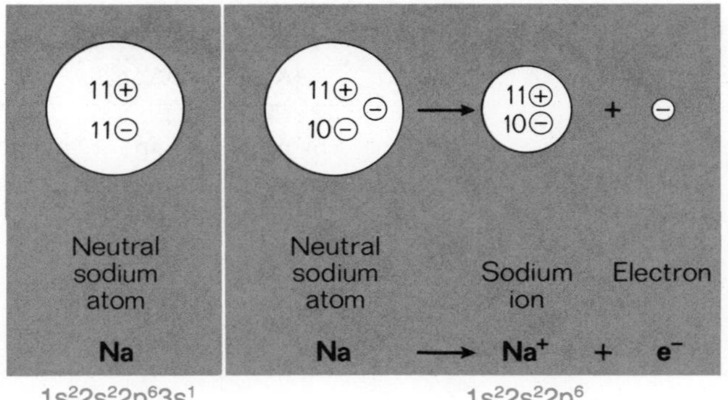

Figure 5.10
The formation of a sodium ion from a sodium atom.

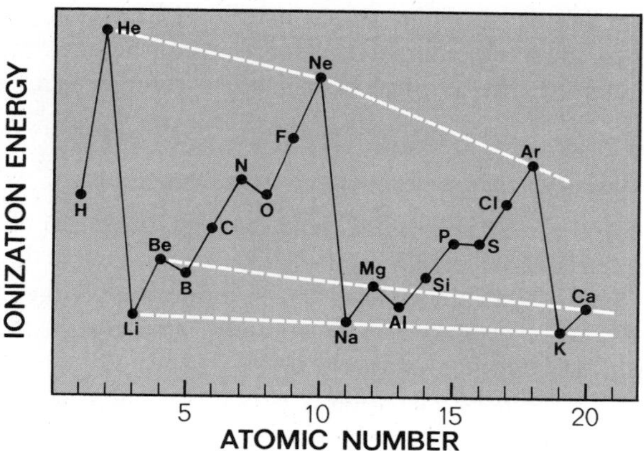

Figure 5.11
Ionization energy, plotted as a function of atomic number, to show periodic properties of elements. Dotted white lines show that ionization energies of elements in the same family decrease as atomic number increases.

In general, **an ion is an atom or group of atoms that has an electric charge because it has more or fewer electrons than protons.** An ion with a positive charge is a **cation** (cat′-ion, not ca′-shun); an ion with a negative charge is an **anion.** If the ion is formed from a single atom, it is a **monatomic ion.** (*Mono-* is a prefix that means one.)

The energy required to remove one electron from a neutral gaseous atom of an element is the ionization energy of that element. Ionization energy is one of the more striking examples of a periodic property, particularly when graphed (Fig. 5.11). Notice the similarity of the shape of the graph between atomic numbers 3 and 10 (Period 2 in the periodic table) and atomic numbers 11 and 18 (Period 3). Notice also that the three peaks are elements in Group 0 at the right end of the table, and the three low points are from Group 1A. Observe the trends; without exception, if you draw lines connecting elements in the same group in the periodic table, the lines slant down to the right. This indicates lower ionization energies with increasing atomic numbers within the group. If the graph is extended to include higher atomic numbers, the same general shapes and trends are found through the s and p sublevel portions of all periods.

The energy required to remove a second electron from an atom is the **second ionization energy** of that element. The third electron requires the third ionization energy, and so forth. In all cases, ionization energy is very high when an electron must be removed from a full octet—a noble gas configuration. This adds to our belief that valence electrons are largely responsible for the chemical properties of an element.

Chemical Families

PG 5L Explain, from the standpoint of electron configuration, why certain groups of elements make up chemical families.

5M Identify in the periodic table the following chemical families: noble gases; alkali metals; alkaline earths; halogens.

Elements with similar chemical properties appear in the same group, or vertical column, in the periodic table. Several of these groups of elements make up a

chemical family. Family trends are most apparent among the representative elements, those elements in the A groups of the periodic table. As these families are discussed, the symbol X is used to refer generally to any member of the family.

Noble Gases. Valence electrons: Z = 2, ns^2 He: ; others, ns^2np^6 :Ẍ: The elements of Group 0 are known as **noble gases** (Fig. 5.12). In chemistry, the word *noble* means unreactive. Until the 1960's it was believed that these elements formed no compounds at all; then a few compounds of krypton (Z = 36) and xenon (Z = 54) were prepared. Nevertheless, as a group, their inactivity is still their outstanding chemical property.

The inactivity of the noble gases is believed to be the result of their filled valence electron sublevels. The two electrons of helium fill the 1s orbital, the only valence orbital helium has. All other elements in the group have a complete octet of electrons. This configuration apparently represents a "minimization of energy" (Section 2.8) arrangement of electrons that is very stable. The high ionization energies of a full octet have already been noted, so the noble gases resist forming positively charged ions. Nor do the atoms tend to gain electrons to form negatively charged ions.

The noble gases are an excellent example of the periodic trends in physical properties. Without exception, the density, melting point, and boiling point all increase as you move down the column in the periodic table.

Alkali Metals. Valence electrons: ns^1 X· With the exception of hydrogen, Z = 1, the elements in Group 1A are known as **alkali metals** (Fig. 5.12). (Francium, Z = 87, is a radioactive element about which we know little but which we presume behaves as an alkali metal.) It has already been noted that members of this group have low ionization energies. The single valence electron is easily lost, forming an ion with a +1 charge. All ions with a +1 charge tend to combine with negatively charged ions in the same way. This is why the chemical properties of the family are similar.

Figure 5.11 shows that ionization energy decreases among the alkali metals as atomic number increases. The higher the energy level in which the ns^1 electron is located, the farther it is from the nucleus and, therefore, the more easily it is removed. As a direct result of this, the **reactivity** of the element—

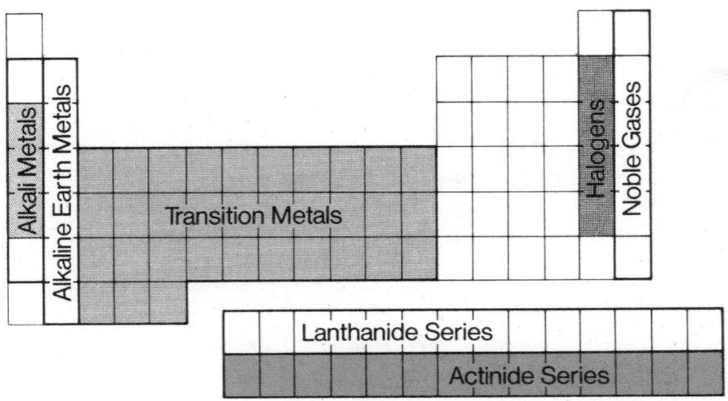

Figure 5.12
Chemical families and regions in the periodic table.

that is, its tendency to react with other elements and form compounds—increases as you go down the column.

What stops an alkali metal atom from losing two electrons to form an ion with a +2 charge? Consider what is left after one electron is gone. Using sodium as an example, the sodium ion has ten electrons (Fig. 5.10). They are distributed among the orbitals in exactly the same way the electrons of the noble gas neon are distributed, $2s^22p^6$. This is described by saying a Na^+ ion is **isoelectronic** with a neon atom. (The prefix *iso-* means *same,* as in isotope.) To remove the second electron would be to take it from a full octet. Thus the second ionization energy of sodium is very high, so there is no tendency to form a Na^{2+} ion. The same reasoning applies to all the alkali metals.

Group 1A elements are metals, although they don't normally look like metals. This is because exposure to air leads to the formation of an oxide coating that hides the bright metallic luster that can be seen in a freshly cut sample. These elements possess other common metallic properties too, such as being good conductors of heat and electricity and being able to form wires and thin foils.

The alkali metals show distinct trends in their physical properties. Among these, the density increases as the atomic number increases. Boiling points and melting points generally decrease as you go down the periodic table, with the single exception of cesium, which boils at a temperature slightly higher than the boiling point of rubidium. (Exceptions are always challenging and frequently instructive when one seeks an explanation for them.)

Alkaline Earths. Valence electrons: ns^2 X: Group 2A elements are called **alkaline earths** or **alkaline earth metals.** Both first and second ionization energies are relatively low, so the two valence electrons are given up readily to form ions with a +2 charge. Again, the ions have the configuration of a noble gas. If magnesium, $[Ne]3s^2$ loses two electrons, only the [Ne] core is left.

Trends like those noted with the alkali metals are seen also with the alkaline earths. Reactivity again increases as you go down the column in the periodic table. Physical property trends are less evident among the alkali earths.

Halogens. Valence electrons: ns^2np^5 :Ẋ: Four elements in Group 7A, fluorine ($Z = 9$), chlorine ($Z = 17$), bromine ($Z = 35$), and iodine ($Z = 53$) make up the family known as the **halogens,** or *salt-formers*. (Astatine ($Z = 85$), a radioactive element about which little is known, probably would fit into the group if there was enough information to classify it.) The easiest way for a halogen to reach a full octet of electrons is to gain one. This gives it the configuration of a noble gas and forms a stable ion with a −1 charge. The *tendency* to gain the electron is greater the closer the added electron is to the nucleus. Consequently, reactivity is greatest for fluorine at the top of the group, and least for iodine at the bottom. Density, melting point, and boiling point all increase steadily with increasing atomic number among the halogens.

Hydrogen, $Z = 1$. Valence electron: $1s^1$ H· You have probably wondered why hydrogen appears twice in the periodic table, at the top of Groups 1A and 7A. Hydrogen is neither an alkali metal nor a halogen. In forming compounds, hydrogen atoms combine with atoms of other elements in the same ratio as alkali metal atoms, which justifies its placement in Group 1A. The way the compounds are formed, however, is different. Hydrogen can also gain an electron to reach the noble gas configuration of helium, forming an ion with a −1 charge. In this way it behaves like a halogen. The way the periodic table is used makes it handy to have hydrogen in both positions, though it really stands alone as an element.

Summary While family properties exist among elements in Groups 3A–6A of the periodic table, they are less important, so we will not go into them. At least some elements in all these groups except 4A form monatomic ions that are isoelectronic with noble gas atoms. To the extent that they do this, they have similar chemical properties. When you study Section 10.4 you will see how the valence electrons contribute to the formation of covalent bonds, again yielding similar chemical properties among elements whose atoms have the same number of valence electrons.

The Transition Metals

PG 5N Identify the transition elements and the representative elements in the periodic table.

Periodic table trends are most pronounced among the **representative elements,** those elements that are found in the A groups of the table. Elements in the B groups and Group 8 are called **transition elements,** or, because they are all metals, **transition metals**. These are the elements whose 3d, 4d, and 5d sublevels are "filling" as you work through the electron configuration scheme. The d electrons are responsible for some of the properties of transition elements, including some trends that are less apparent than those in the representative elements.

The transition elements include some of our better known metals, such as iron, gold, copper, silver, chromium, zinc, nickel, and mercury.

Although the elements within a given transition series differ from each other chemically, they are, as a group, distinguished from the 1A and 2A metals by being less reactive.

Atomic Size

PG 5O Predict how and explain why atomic size varies with position in the periodic table.

Figure 5.13 shows the sizes of Group A atoms. Moving across the table from left to right, the atoms become smaller. This trend is perhaps most obvious in the second period, where the element at the far left, lithium, has a radius more than twice that of fluorine at the far right. As we go down the table, within a given group, atoms ordinarily increase in size. We see, for example, that cesium, the last element listed in Group 1A, has a radius nearly twice that of lithium.

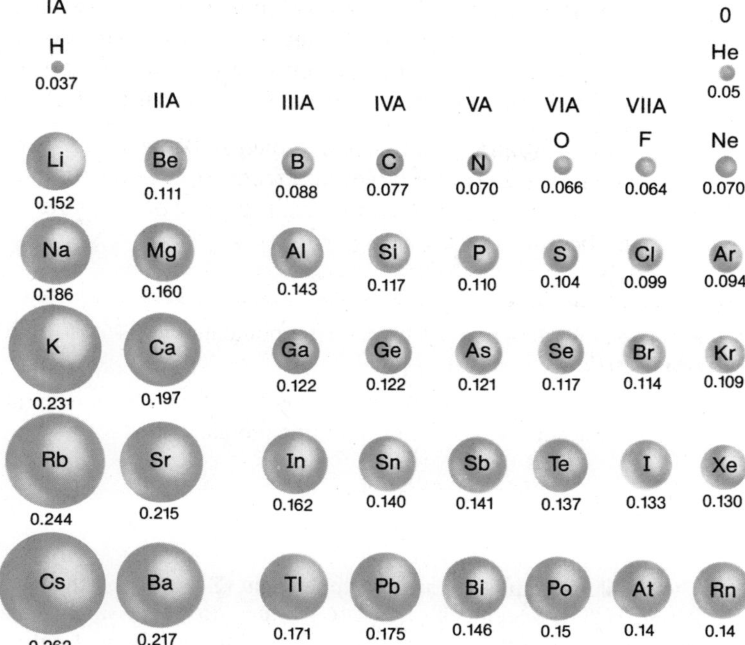

Figure 5.13
Atomic radii of the representative elements. Atomic radii increase as one goes down a group in the periodic table and, in general, decrease in going across a row from left to right.

These observations about atomic size are believed to be the result of three influences:

1. *Number of occupied principal energy levels.* As the number of occupied principal energy levels increases, the atom becomes larger. Electrons in higher energy levels are generally farther from the nucleus. This influence alone accounts for the increase in size as you move down a column in the periodic table. For example, compare sodium (Z = 11) and lithium (Z = 3) atoms. Sodium atoms are larger. The highest electron of a sodium atom is a 3s electron, but for lithium it is a 2s electron.
2. *Nuclear charge.* Atoms generally become smaller from left to right across each row of the table. Within a period, the highest energy electrons are all in the same principal energy level. As the number of protons in the nucleus increases across the row, the positive nuclear charge increases. This causes the nucleus to exert a stronger attraction for the outermost electrons, thus pulling them closer to the nucleus. Compare sodium (Z = 11) and magnesium (Z = 12) atoms. Both have 3s electrons in their highest energy levels. The 12 protons in the nucleus of a magnesium atom attract those 3s electrons more strongly than the 11 protons in the nucleus of a sodium atom. The magnesium electrons are therefore closer to the nucleus, making the magnesium atom smaller.
3. *Shielding effect.* The attraction of the positively charged nucleus for the negatively charged outermost electrons is partially canceled, or "shielded," by the repulsion of electrons in lower energy levels. This is one effect that

makes nuclear charge less important than the number of occupied energy levels in determining atomic size. Compare sodium (Z = 11) and lithium (Z = 3) atoms again. Even though a sodium atom has nearly four times as much nuclear charge, it is larger than a lithium atom.

In summary, atomic size generally increases from right to left across any row of the periodic table, and from top to bottom in any column. The smallest atoms are toward the upper right corner of the table, and the largest are toward the bottom left corner.

EXAMPLE 5.7 Referring only to a periodic table, list atomic numbers 15, 16, and 33 in order of increasing atomic size.

The summary in the paragraph just before this example should tell you how to identify the smallest and the largest of the three atoms. . . .

16 – 15 – 33

The smallest atom is toward the upper right (Z = 16) and the largest is toward the bottom left (Z = 33). Specifically, Z = 16 (sulfur) atoms are smaller than Z = 15 (phosphorus) atoms because sulfur atoms have a higher nuclear charge to attract the highest energy 3s and 3p electrons in both atoms. The phosphorus atom is smaller than the Z = 33 (arsenic) atom because the 3s and 3p electrons of phosphorus are at a lower energy than the 4s and 4p electrons of arsenic.

Metals and Nonmetals

PG 5P Identify metals and nonmetals in the periodic table.

Both physically and chemically the alkali metals, alkaline earths, and transition elements are metals. Chemically, an element is known as a metal if it can lose one or more electrons and become a positively charged ion. The larger the atom, the more easily the outermost electron is removed. Therefore, the metallic character of elements in a group increases as you go down a column in the periodic table.

It has also been noted that the size of an atom becomes smaller as the nuclear charge increases across a period in the table. The larger number of protons also holds the outermost electrons more strongly, making it more difficult for them to be lost. This makes the metallic character of elements *decrease* as you go from left to right across the period.

The distinction between metals, elements that lose electrons in chemical reactions, and nonmetals, those that do not, is not a sharp one. It can be drawn roughly as a step-like line beginning between atomic numbers 4 and 5 in Period 2, and ending between 84 and 85 in Period 6 (Fig. 5.14). Most elements next to the line on either side appear in some compounds as metals and in others as nonmetals. They are often loosely classified as **metalloids**.

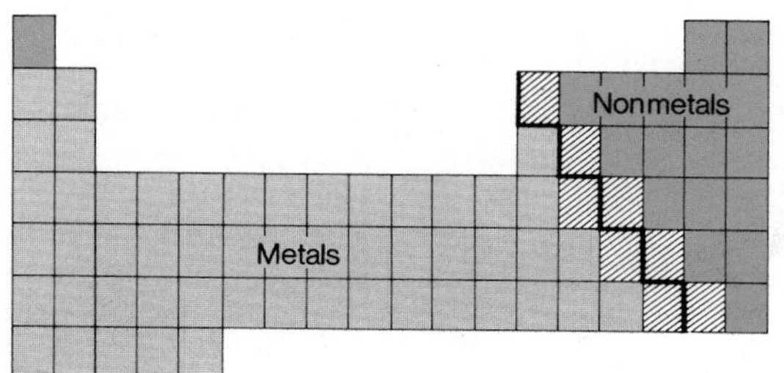

Figure 5.14
Metals and nonmetals. Shaded elements are metalloids, having properties intermediate between those of metals and nonmetals.

Quick Check 5.5

Are the following true or false:

1. The number of valence electrons frequently determines the chemical properties of an element.
2. The alkaline earths are in Group 2A of the periodic table.
3. Elements whose atomic numbers are 22, 41, and 74 are all transition elements.
4. Bromine (Z = 35) atoms are larger than calcium (Z = 20) atoms.
5. Elements whose atomic numbers are 7, 33, and 83 are all nonmetals.

CHAPTER 5 IN REVIEW

5.1 The Bohr Model of the Hydrogen Atom

- **5A** Describe the Bohr model of the hydrogen atom.
- **5B** Explain the meaning of quantized energy levels in an atom and show how these levels are related to the discrete lines in the spectrum of that atom.
- **5C** Distinguish between ground state and excited state.

5.2 The Quantum Mechanical Model of the Atom

- **5D** Identify the principal energy levels within an atom and state the energy trend among them.
- **5E** For each principal energy level, state the number of sublevels, identify them by letter, and state the energy trend among them.
- **5F** Sketch the shapes of s and p orbitals.
- **5G** State the number of orbitals in each sublevel.
- **5H** State the restrictions on the electron population of an orbital.

5.3 Electron Configuration

- **5I** Referring only to a periodic table, write the ground state electron configuration of a gaseous atom of any element up to atomic number 36.

5.4 Valence Electrons

- **5J** Using n for the highest occupied energy level, write the configuration of the valence electrons of any Group A element in the periodic table.
- **5K** Write the Lewis (electron dot) symbol for an atom of any representative element.

5.5 Trends in the Periodic Table

- **5L** Explain, from the standpoint of electron configuration, why certain groups of elements make up chemical families.
- **5M** Identify in the periodic table the following chemical families: noble gases; alkali metals; alkaline earths; halogens.
- **5N** Identify the transition elements and the representative elements in the periodic table.
- **5O** Predict how and explain why atomic size varies with position in the periodic table.
- **5P** Identify metals and nonmetals in the periodic table.

TERMS AND CONCEPTS

5.1 Spectrum (continuous, line)
Discrete
Quantum, quanta
Quantized
Continuous
Ground state
Excited state
Principal energy level
Principal quantum number

5.2 Quantum mechanical model of the atom
Sublevels, s, p, d, f
Orbital
Pauli exclusion principle
5.3 Electron configuration
Neon core
Argon core
5.4 Valence electrons
Lewis (electron dot) symbols
Octet of electrons
5.5 Ion; monatomic ion; cation; anion
Ionization energy
Second ionization energy
Chemical family
Noble gases (family)
Alkali metal (family)
Reactivity
Isoelectronic
Alkaline earths; alkaline earth metal (family)
Halogens (family)
Representative elements
Transition metals (elements)
Shielding effect
Metal, nonmetal (in terms of electron behavior)
Metalloid

Most of these terms and many others appear in the Glossary. Use the Glossary regularly.

QUESTIONS AND PROBLEMS

Section 5.1

(*1*) What is meant by a "discrete line spectrum"? What kind of spectra do not have discrete lines? Have you ever seen a spectrum that does not have discrete lines? Have you ever seen one with discrete lines?

(2)* What is meant by the statement that something behaves "like a wave"?

(3) Which of the following are quantized: (a) milk from a cow; (b) milk from a store; (c) wheels in an automobile factory; (d) temperature; (e) amount of rainfall.

(*4*) "Electron energies are quantized within the atom." What is the meaning of this statement?

(*5*) What experimental evidence leads us to believe that electron energy levels are quantized? Explain.

(6) What is meant when an atom is "in its ground state"?

(*7*) Draw a sketch of an atom according to the Bohr model. Describe the atom with reference to the sketch.

(*8*) Identify the shortcomings of the Bohr theory of the atom.

(47) The visible spectrum is a small part of the whole electromagnetic spectrum. Name several other parts of the whole spectrum that are included in our everyday vocabulary.

(48) Identify measurable wave properties that are used in describing light.

(49) Which among the following are *not* quantized? (a) shoelaces; (b) cars passing through a toll plaza in a day; (c) birds in an aviary; (d) flow of a river in m^3/hr; (e) percentage of salt in a solution.

(50) What kind of light would be emitted by atoms if the electron energy were not quantized? Explain.

(51) What must be done to an atom, or what must happen to an atom, before it can emit light? Explain.

(52) Which atom is most apt to emit light, one in the ground state or one in an excited state? Why?

(53) Using a sketch of the Bohr model of an atom, explain the source of the observed lines in the spectrum of hydrogen.

(54) Identify the major advances that came from the Bohr model of the atom.

Section 5.2

(*9*) What is the meaning of the "principal energy levels" of an atom?

(55) Compare the relative energies of the principal energy levels within the same atom.

(*10*) What are sublevels and how are they identified?

(*11*) What is an orbital? Describe in words and by sketch the shapes of s and p orbitals.

(12) At n = 2 there are two p sublevels; at n = 3 there are three p sublevels. Comment on these statements.

(13) Although we may draw the 4s orbital with the shape of a ball, there is some probability of finding the electron outside the ball we draw. Comment on this statement.

(*14*) What is the significance of the Pauli exclusion principle?

(15) An electron in an orbital of a p sublevel follows a path similar to a Figure 8. Comment on this statement.

(16) What is the maximum number of d orbitals when n = 6?

(17) What general statement may be made about the energies of the principal energy levels? About the energies of the sublevels?

(18) Is the quantum mechanical model of the atom consistent with the Bohr model?

(56) How many sublevels are present in each principal energy level?

(57) How many orbitals are in the s sublevel? the p sublevel? the d sublevel? the f sublevel?

(58) Each p sublevel contains six orbitals. Comment on this statement.

(59) The principal energy level with n = 6 contains six sublevels, although not all six are occupied for any element now known. Comment on this statement.

(60) An orbital may hold one electron or two electrons, but no other number. Comment on this statement.

(61) What is your opinion of the common picture showing one or more electrons whirling around the nucleus of an atom?

(62) What is the largest number of electrons that can occupy the 4p orbitals? the 3d orbitals? the 5f orbitals?

(63) "Energies of the principal energy levels of an atom sometimes overlap." Explain how this is possible.

(64) Which of the following statements is true? The quantum mechanical model of the atom includes (a) all of the Bohr model; (b) part of the Bohr model; (c) no part of the Bohr model. Justify your answer.

Section 5.3

(19) What is the meaning of the symbol $3p^4$?

(20) The electron configuration of an element is $1s^22s^22p^5$. What element is this? What period is it in, and what group is it in?

(*21*) What is meant by [Ne] in [Ne] $3s^23p^1$?

(22)* The vertical axis of Figure 5.7 represents electron energy. It is not to scale, but the spacing is used to suggest relative energy values. From this illustration, suggest a possible explanation for the irregularities in filling 4s and 3d orbitals. Why are similar irregularities even more frequent at higher energy levels?

(65) What do you conclude about the symbol $2d^5$? (Careful. . . .)

(66) $1s^22s^22p^63s^23p^4$ is the electron configuration of an element. What element is it? In which period and group of the periodic table is it located?

(67) What is the argon core? What is its symbol, and what do you use it for?

(68)* What relationship, if any, is there between periods and principal energy levels? Are both identified on the periodic table? If so, how? If not, which one is clearly indentified? Can the other be inferred from the table? Figure 5.7 may help in answering all of these questions.

For the next two questions in each column, write a complete electron configuration for an atom of the element shown. Do not use a noble gas core.

(23) Nitrogen (Z = 7) and titanium (Z = 22).

(24) Calcium (Z = 20) and copper (Z = 29).

(25) If it can be done, rewrite the electron configurations in Questions 26 and 27 with a neon or argon core.

(26) Use a noble gas core to write the electron configuration of germanium (Z = 32).

(27)* Write the electron configurations you would expect for barium (Z = 56) and technetium (Z = 43).

(69) Magnesium (Z = 12) and nickel (Z = 28).

(70) Chromium (Z = 24) and selenium (Z = 34).

(71) If it can be done, rewrite the electron configurations in Questions 69 and 70 with a neon or argon core.

(72) Use a noble gas core to write the electron configuration of aluminum (Z = 13).

(73) Write the electron configurations you would expect for iodine (Z = 53) and tungsten (Z = 74).

Section 5.4

(28) What are valence electrons? What is the maximum number of valence electrons an atom can have?

(29) What two ways are used to represent valence electrons? Give an example of each.

(30) For which group of the periodic table does ns^2np^2 represent the valence electrons?

(74) Why are valence electrons important? Are they the only electrons in an atom that have this importance?

(75) What are the valence electrons of aluminum, Z = 13? Write both of the ways they may be represented.

(76) Using n for the principal quantum number, write the electron configuration for the valence electrons of Group 6A atoms.

Section 5.5

(31) Explain the meaning of *ionization energy*. What is meant by *first* ionization energy; *second* ionization energy?

(32) Suggest an explanation for the down-to-the-right slant of the dashed lines in Figure 5.11.

(33) What does it mean when two elements are in the same chemical family?

(34) What is the highest energy electron configuration associated with the alkali metals?

(35) Explain the meaning of *isoelectronic*. Expressed as ns^xnp^y, what electron configuration is isoelectronic with a noble gas?

(36) Identify the chemical family of each of the following elements: iodine (Z = 53); rubidium (Z = 37).

(77)* Why do you suppose the second ionization energy of an element is always greater than the first ionization energy?

(78)* Why is the definition of atomic number based on the number of protons in an atom rather than the number of electrons?

(79) What is it about strontium (Z = 38) and barium (Z = 56) that makes them members of the same chemical family?

(80) To what family does the electron configuration ns^2np^5 belong?

(81)* Why is an "electron configuration that is isoelectronic with a noble gas" important in explaining the similar properties of members of a chemical family?

(82) Identify the chemical family of each of the following elements: krypton (Z = 36); beryllium (Z = 4).

(37) Explain how the electron configurations of magnesium and calcium are responsible for the many chemical similarities between these elements.

(38) Where in the periodic table do you find the representative elements? Are they metals or nonmetals?

(39) Identify an atom in Group 5A that is (a) larger than an atom of arsenic (Z = 33) and (b) smaller than an atom of arsenic.

(*40*) Why does atomic size increase as you go down a column in the periodic table?

(41) Even though an atom of germanium (Z = 32) has more than twice the nuclear charge of an atom of silicon (Z = 14), the germanium atom is larger. Explain.

(42)* Compare the ionization energies of aluminum (Z = 13) and chlorine (Z = 17). Why are these values as they are?

(*43*) What property of atoms of an element determines that the element is a metal rather than a nonmetal?

(83) Account for the chemical similarities between chlorine and iodine in terms of their electron configurations.

(84) Where in the periodic table do you find the transition elements? Are they metals or nonmetals?

(85) Identify an atom in Period 4 that is (a) larger than an atom of arsenic (Z = 33) and (b) smaller than an atom of arsenic.

(86) Why does atomic size decrease as you go left to right across a row in the periodic table?

(87) Describe the shielding effect. Explain its influence on atomic size.

(88)* Suggest reasons why the ionization energy of magnesium (Z = 12) is greater than the ionization energies of sodium (Z = 11), aluminum (Z = 13), and calcium (Z = 20).

(89) Do elements become more metallic or less metallic as you (a) go down a group in the periodic table, (b) move left to right across a period in the periodic table? Support your answers with examples.

Use the "periodic table," Figure 5.15, to answer Questions 44–46 and 90–92. Answer with letters (D, E, X, Z, etc.) from that table.

(44) Give the letters that are in the positions of (a) halogens and (b) alkali metals.

(45) Give the letters that correspond with transition elements.

(46) List the elements D, E, and G in order of decreasing atomic size (largest atom first).

(90) Give the letters that are in the positions of (a) alkaline earth metals and (b) noble gases.

(91) List the letters that correspond with nonmetals.

(92) List the elements Q, X, and Z in order of increasing atomic size (smallest atom first).

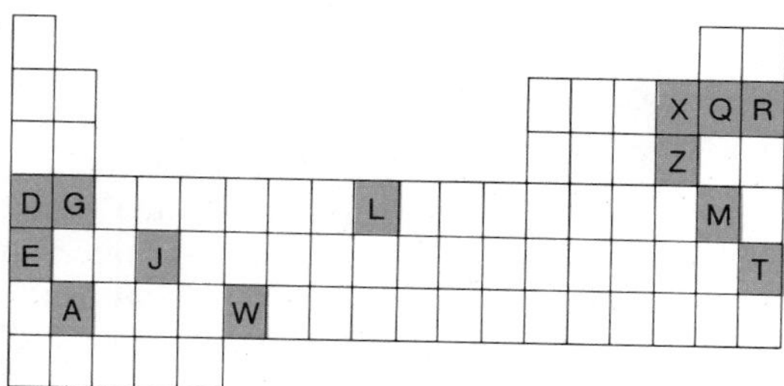

Figure 5.15
Chart for Questions 44 to 46 and 90 to 92.

Miscellaneous Questions

(93) Distinguish precisely, and in scientific terms, the differences between items in each of the following groups:
(a) Continuous spectrum, discrete line spectrum
(b) Quantized, continuous
(c) Ground state, excited state
(d) Principal energy level, principal quantum number
(e) Bohr model of the atom, quantum mechanical model of the atom
(f) Principal energy level, sublevel, orbital
(g) s, p, d, f (sublevels)
(h) Orbit, orbital
(i) First, second, third, . . . ionization energies
(j) Representative elements, transition elements
(k) Metal, nonmetal, metalloid

(94) Determine whether each statement that follows is true or false:
(a) Electron energies are quantized in excited states but not in the ground state.
(b) Line spectra of the elements are experimental evidence of the quantization of electron energies.
(c) Energy is released as an electron passes from ground state to an excited state.
(d) The energy of an electron may be between two quantized energy levels.
(e) The Bohr model explanation of line spectra is still thought to be correct.
(f) The quantum mechanical model of the atom describes orbits in which electrons travel around the nucleus.
(g) Orbitals are regions in which there is a high probability of finding an electron.
(h) All energy sublevels have the same number of orbitals.
(i) The 3p orbitals of an atom are larger than its 2p orbitals, but smaller than its 4p orbitals.
(j) At a given sublevel, the maximum number of d electrons is 5.
(k) The halogens are found in Group 7A of the periodic table.
(l) The dot structure of the alkaline earths is X· , where X is the symbol of any element in the family.
(m) Stable ions formed by alkaline earth metals are isoelectronic with noble gas atoms.
(n) Atomic numbers 23 and 45 both belong to transition elements.
(o) Atomic numbers 52, 35, and 18 are arranged in order of increasing atomic size.
(p) Atomic numbers 7, 16, and 35 are all nonmetals.

(95)* One of the successes of the Bohr model of the atom was its explanation of the lines in atomic spectra. Does the quantum mechanical model also have a satisfactory explanation for these lines? Justify your answer.

(96)* Though the quantum model of the atom makes predictions for atoms of all elements, most of the quantitative confirmation of these predictions is limited to substances whose formulas are H, He^+, and H_2^+. From your knowledge of electron configurations and the limited information about chemical formulas given in this chapter, can you identify a single feature that all three of these substances have that makes them unique and probably the easiest substances to investigate?

(97) Color Plate 1 is a color illustration of the spectra of several elements as they are seen in a spectroscope. Hydrogen has four lines, and one of those is sometimes difficult to see. Yet we say there are many more lines in the hydrogen spectrum, and, for that matter, all the other spectra in the illustration. Why do we not see them?

(98)* What do you suppose are the electron configuration and the formula of the monatomic ion formed by scandium ($Z = 21$)?

(99)* Carbon does not form a stable monatomic ion. Suggest a reason for this.

(100)* Consider the block of elements in Periods 2 to 6 and Groups 5A to 7A. In Group 7A are the halogens, a distinct chemical family with many properties shared by the different elements, except for radioactive astatine ($Z = 85$). Polonium ($Z = 84$) is also radioactive, but the other elements in Group 6A have enough similarity to be considered a family; they are called the chalcogen family. In Group 5A, however, "family" similarities are weak, and those that exist belong largely to nitrogen and phosphorus, and to some extent, arsenic ($Z = 33$). Why do you suppose family similarities break down in Group 5A?

(101)* Xenon (Z = 54) was the first noble gas to be chemically combined with another element, and it and krypton (Z = 36) make up nearly all of the small number of noble gas compounds known today. Note the ionization energy trend begun with the other noble gases in Figure 5.11. What do these facts suggest about the relative reactivities of the noble gases and the character of noble gas compounds?

(102)* Iron (Z = 26) forms two monatomic ions, Fe^{2+} and Fe^{3+}. From which orbitals do you expect electrons are lost in forming these ions?

(103)* Color Plate 1 indicates a general increase in first ionization energy from left to right across a row of the periodic table, with two sharp breaks in both periods shown. (a) Suggest a reason why ionization energy should increase as atomic number increases within a period. (b) Can you correlate features of electron configurations with the locations of the breaks?

6 Introduction to Chemical Formulas

LOOKING BACK

2.7 and 2.8 Particles with like charges repel and those with opposite charges attract to find a minimum potential energy.

4.1 Atoms of different elements combine to form compounds that are represented by chemical formulas.

4.3 Protons and electrons are charged subatomic particles.

4.7 Groups (columns) in the periodic table identify families of elements that have similar chemical properties.

4.8 The symbols of the common elements shown in Figure 4.6 are to be written while referring to a periodic table.

5.4 and **5.5** The number of valence electrons in an atom can be determined from the atom's position in the periodic table. An atom of a representative element forms a monatomic ion by gaining or losing electrons to become isoelectronic with a noble gas atom.

If you have studied Sections 5.4 and 5.5, you will recognize that the ions described in Chapter 6 are produced as the valence electrons adjust themselves to the configuration of a noble gas.

LOOKING AHEAD IN CHAPTER 6

These forces are responsible for the formation of chemical bonds between atoms.

Chemical formulas give the ratios in which the atoms combine to form compounds.

When an atom gains or loses electrons, it acquires a charge and becomes an ion.

The periodic table is used to write the formulas of nonatomic ions.

The symbols of the common elements are used to write chemical formulas.

The formula of a monatomic ion formed by a representative element can be written from the position of the element in the periodic table.

In this chapter you will be introduced to a small part of chemical nomenclature, the system of names and formulas of chemical substances. The idea is that a small sample of the system can be learned and mastered more readily than a large one. Once you learn the system and practice it with a limited number of compounds, you will be able to expand your skill easily when the topic is presented in more detail in Chapter 12.

The important thing now is that you learn the *system*. Some names, formulas, and rules must be memorized. The names and formulas are examples of the rules, and the rules make up the system. If you just memorize some names and formulas, they are all you will know and they will soon be forgotten. But, if you learn the system, you will be able to apply it to names and formulas of substances you have never heard of before. Furthermore you will remember how to do this throughout your study of chemistry. Keep this in mind as you study this chapter. We cannot overemphasize this advice: *learn the system.*

6.1 THE MEANING OF A CHEMICAL FORMULA

PG 6A Given a chemical formula, state the number of atoms of each element in the formula unit.

6B Given the number of atoms of each element in a formula unit of a pure substance, write the chemical formula.

A **chemical formula** uses elemental symbols to identify the elements in a pure substance, either element or compound. A subscript number following the symbol of an element tells how many atoms of that element are in a "formula unit" of the substance. *Formula unit* is a general term by which all substances may be considered as independent particles, although in fact they may be parts of a group of particles.

The formula of water is H_2O. H is the elemental symbol for hydrogen, and O is the symbol for oxygen. The subscript 2 after H indicates that the formula unit of water contains two hydrogen atoms. The absence of a subscript after a symbol, as with oxygen, indicates that there is only one atom of that element in the formula unit. Thus, the formula unit of water contains two hydrogen atoms and one oxygen atom.

EXAMPLE 6.1 State the number of atoms of each element in the formula unit of the following common substances:

CO_2, a gas that you exhale: ______

CH_4, the fuel for homes heated with natural gas: ______

$NaHCO_3$, baking soda: ______

$C_{12}H_{22}O_{11}$, table sugar: ______

CO_2: one carbon atom, two oxygen atoms
CH_4: one carbon atom, four hydrogen atoms
$NaHCO_3$: one sodium atom, one hydrogen atom, one carbon atom, three oxygen atoms
$C_{12}H_{22}O_{11}$: 12 carbon atoms, 22 hydrogen atoms, 11 oxygen atoms

EXAMPLE 6.2 Write the formula of a compound whose formula unit contains two sodium atoms, two sulfur atoms, and three oxygen atoms.

$Na_2S_2O_3$

6.2 ATOMS AND COMBINATIONS OF ATOMS: CHEMICAL BONDS

PG 6C Distinguish between monatomic, diatomic, and polyatomic particles.

6D Distinguish between ions, anions, and cations.

6E Distinguish between molecular substances and ionic compounds.

6F Distinguish between covalent bonds and ionic bonds.

If a sample of a pure substance were to be cut in half, and one half cut in half again, and the process repeated over and over with one half being divided each time, where would it end? What is the ultimate particle that could not be further divided without destroying the substance in a chemical change?

The answer to the above question depends on the structure of the ultimate particle. Some elemental particles consist of a single atom. These are called **monatomic,** the prefix *mono-* meaning *one*. Some elements and some compounds have two or more atoms in their ultimate particles. If the number is two, they are **diatomic;** three, **triatomic;** and so forth. Any particle may be described as **polyatomic** if it has more than one atom. (*Poly-* means *many*).

When two or more atoms combine in definite number combinations to form particles that behave as independent units, those units are called **molecules.** Some elements exist as molecules, and are said to be **molecular.** Nitrogen, for example, exists as diatomic molecules. Carbon monoxide is a **molecular compound** that is diatomic. Figure 6.1 illustrates molecules of other elements and compounds.

The electrostatic forces that hold atoms of the same or different elements together are **chemical bonds.** The forces yield a minimization of energy when

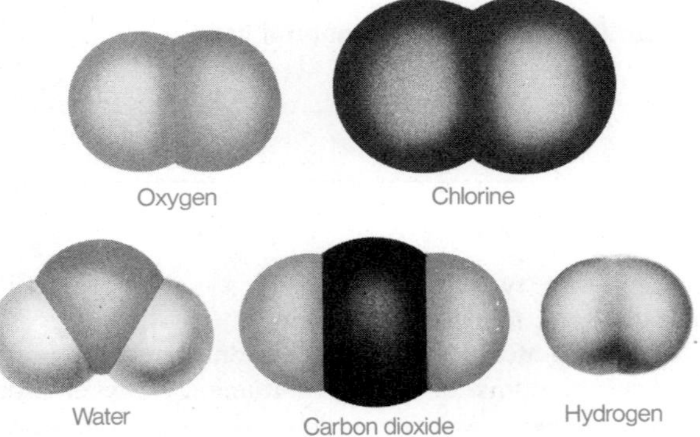

Figure 6.1
Molecules of elements and compounds. Drawings are all to the same scale.

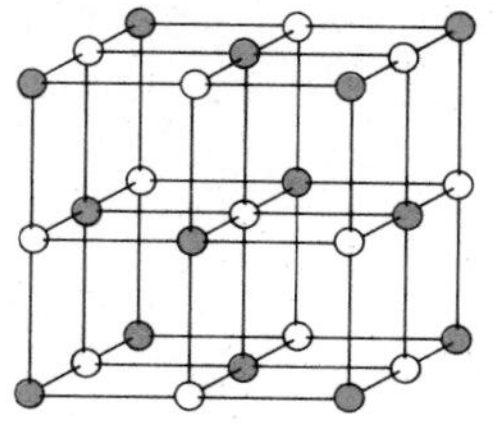

Figure 6.2
Crystal structure of sodium chloride.

electrons are arranged in a way that is like the arrangement in the noble gases, the elements in Group 0 in the periodic table. Two atoms can reach this arrangement by sharing electrons to form **covalent bonds.** This kind of bonding is found in molecular compounds.

Another kind of bond results when electrons are transferred from one species to another, producing positively and negatively charged particles. The particles may be monatomic or polyatomic, and they are called **ions.** Many monatomic ions—all of those that appear in this chapter—have the electron configuration of a noble gas atom. An ion with a positive charge is a **cation** (cat′-ion, not ca-shun), and an ion with a negative charge is an **anion.** A minimization of energy results when anions and cations arrange themselves in a crystal that has a distinct geometrical pattern (see Fig. 6.2). The result is an **ionic compound** in which the ions are held in place by **ionic bonds.**

Ionic and covalent bonding will be studied in greater detail in Chapter 10. Here we are concerned only with the names and formulas of the substances they produce.

Quick Check 6.2

1. For what minimum number of atoms is the word *polyatomic* used?
2. What kind of forces hold bonded atoms together?
3. How do two atoms form a covalent bond?
4. Is the charge on a cation positive or negative? What name is given to the ion of the opposite charge?

6.3 THE FORMULAS OF ELEMENTS

PG 6G Given the name (or formula) of an element listed in Figure 4.6, write its formula (or name).

The formula unit of most elements is the single atom. The chemical formula is the symbol of the element. You learned to write these symbols in Section 4.8, using only the periodic table as a guide. Examples are iron, Fe, aluminum, Al, and copper, Cu.

The formulas of the elements that exist as diatomic molecules are the elemental symbols followed by a subscript 2. *These elements and their symbols must be memorized.* They are the most abundant element in air, nitrogen, N_2; the two elements in water, hydrogen and oxygen, H_2 and O_2; and the halogens in Group 7A of the periodic table, fluorine, F_2, chlorine, Cl_2, bromine, Br_2, and iodine, I_2.

CAUTION: Forgetting to write the formulas of these seven elements as diatomic molecules is probably the most common mistake made by beginning chemistry students. Be alert to it; avoid it. Also note that the subscript 2 for these elements does not necessarily extend to compounds. In a compound the subscript may or may not be 2. Water, H_2O, is an obvious example.

EXAMPLE 6.3 Write the formulas of the following elements:

calcium nickel bromine argon

\-\-\-\-\- \-\-\-\-\-

calcium, Ca; nickel, Ni; bromine, Br_2; argon, Ar

Bromine is one of the halogens that forms diatomic molecules. The other formulas are simply the symbols of the elements.

6.4 COMMON BINARY MOLECULAR COMPOUNDS

PG 6H Given the name (or formula) of water, ammonia, carbon monoxide, or carbon dioxide, write its formula (or name).

A **binary compound** is a compound that contains two elements. (The prefix, *bi-*, means *two*.) The general nomenclature system for binary molecular compounds is presented in Chapter 12. Here we will include only four familiar substances, two of which are known by their common names and two by their chemical names. Their names and formulas are:

Water	H_2O	The most familiar of all compounds
Ammonia	NH_3	A common household cleanser
Carbon monoxide	CO	A deadly gas in automobile exhaust
Carbon dioxide	CO_2	A product of burning

The names and formulas of these compounds should be memorized. Notice that the prefixes before "oxide" in carbon monoxide and carbon dioxide tell the number of atoms in the molecule. Water and carbon dioxide molecules are shown in Figure 6.1.

6.5 MONATOMIC IONS

PG 6I Given a periodic table and the name and atomic number of a representative element, or a source from which that name and number may be found, write the name and formula of the monatomic ion formed by that element.

Three rules govern the names and formulas of monatomic ions:

The formula of a monatomic ion is the symbol of the element from which it came, followed by the ionic charge written as a superscript.

If the charge on the ion is +1, a + sign is used without a number: Na^+. If the charge is +2, −2, +3 or −3, the charge's size is written before the sign, as in S^{2-} and Al^{3+}. IMPORTANT: Notice the difference between the formula of an element and the formula of an ion: ions are *always* charged; elements are not. *It is not the formula of an ion if the charge is left off.*

The name of a monatomic cation is the name of the element followed by the word ion.

The name of a monatomic anion is the name of the element changed to end in -ide, *followed by the word* ion.

The formulas of monatomic ions produced by representative elements can be predicted by the location of the element in the periodic table. Recalling that elements with similar chemical properties appear in the same group in the table, it follows that ions formed by all members of the group have the same charge.

The formation of an ion can be shown by a chemical equation. An equation begins with the formulas of the starting substances. This is followed by an arrow that points to the final substances, or products. Using sodium to illustrate the formation, formula, and name of Group 1A ions,

Group 1A, sodium ion: $Na \rightarrow Na^+ + e^-$ (6.1)

Group 2A elements release two electrons per atom, yielding ions with a +2 charge, and Group 3A ions have a +3 charge because their atoms lose three electrons. Magnesium and aluminum are examples:

Group 2A, magnesium ion: $Mg \rightarrow Mg^{2+} + 2\ e^-$ (6.2)

Group 3A, aluminum ion: $Al \rightarrow Al^{3+} + 3\ e^-$ (6.3)

Carbon, silicon, and germanium, the top three elements in Group 4A, do not form monatomic ions. Compounds of these elements are molecular rather than ionic. Tin (Sn) and lead (Pb), appear in both ionic and molecular compounds, but their monatomic ions do not fit into the pattern being developed here. These ions are included in Chapter 12.

The heavy stair-step line that divides the periodic table into two sections separates the metals in the larger section on the left from the nonmetals in the smaller section on the right. Nonmetal atoms of Groups 5A, 6A and 7A form monatomic ions by *gaining* 3, 2, and 1 electrons, respectively, rather than by losing them. These ions are therefore anions; they have negative charges:

Group 5A, phosphide ion: $P + 3\ e^- \rightarrow P^{3-}$ (6.4)

Group 6A, sulfide ion: $S + 2\ e^- \rightarrow S^{2-}$ (6.5)

Group 7A, chloride ion: $Cl + e^- \rightarrow Cl^-$ (6.6)

Notice how the names of the anions are formed by changing the ending of the elemental name to -ide. Phosph*orus* becomes phosph*ide*, sul*fur* becomes sul*fide*, and chlor*ine* becomes chlor*ide*.*

Figure 6.3 summarizes the monatomic ions you should know, and relates them to a periodic table which will be your guide in writing formulas. You already know the elements and their symbols from Figure 4.6. If you tie these ideas together, you will find that your mastery of Group A monatomic ions is complete.

*As elements, phosphorus, sulfur, and chlorine exist as polyatomic molecules that must be separated into atoms before ions can form.

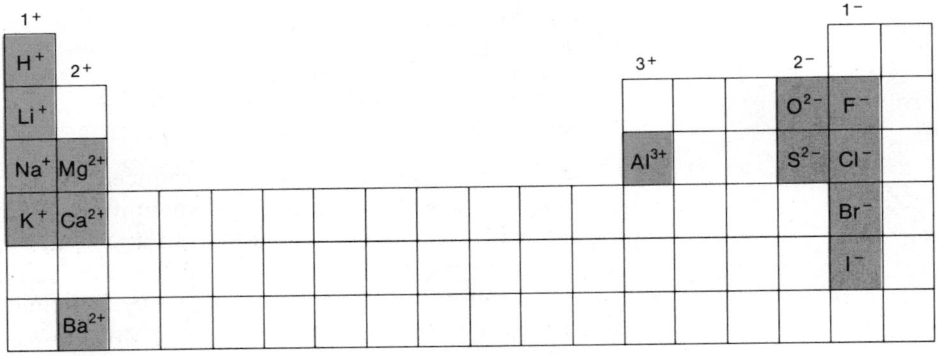

Figure 6.3
Partial periodic table showing the symbols and locations of some of the more important monatomic ions. This is the minimum list of monatomic ions you should be able to use in writing formulas of ionic compounds, referring only to a complete periodic table. Associating the charge on an ion with its location in the periodic table aids in memorizing ionic charges.

EXAMPLE 6.4 Referring only to a periodic table (*not* Fig. 6.3), write the chemical formulas of lithium atoms, lithium ions, oxygen atoms, and oxide ions.

Lithium atoms, Li; lithium ions, Li^+; oxygen atoms, O; oxide ions, O^{2-}

All ions from Group 1A have a +1 charge and therefore carry a + superscript. All Group 6A ions have a −2 charge, and therefore a 2− superscript. The question asks specifically for the formula of oxygen *atoms*, O, rather than oxygen or oxygen molecules, O_2.

EXAMPLE 6.5 Referring only to a periodic table, write the name and formula of the monatomic ion formed by each of the following elements and classify the ion as a cation or an anion:

barium	nitrogen
bromine	potassium

-----

barium: barium ion, Ba^{2+}, cation; nitrogen: nitride ion, N^{3-}, anion

bromine: bromide ion, Br^-, anion; potassium; potassium ion, K^+, cation

Quick Check 6.5
How do you tell the charge on a monatomic ion formed by a representative element?

6.6 THE IONS OF HYDROGEN

Hydrogen appears twice in the periodic table, at the head of Group 1A and at the top of Group 7A. Its second position suggests that it forms hydride ions, H^-. It does, but in only a few compounds.

The term "hydrogen ion" refers to H^+, as would be expected by the position of hydrogen in Group 1A of the periodic table. This species does not exist as such in pure compounds, but it is said to be present in **acids. An acid may be considered as a solution containing the hydrogen ion formed when certain hydrogen-bearing compounds react with water.** Hydrogen chloride, a gaseous compound having the formula HCl, is typical. When bubbled through water its reaction may be represented by

$$HCl + H_2O \rightarrow H_3O^+ + Cl^- \tag{6.7}$$

H_3O^+ is the **hydronium ion,** which may be thought of as a hydrogen ion connected to a water molecule: $H \cdot H_2O^+$. Removing a water molecule from both sides of Equation 6.6 leaves

$$HCl \rightarrow H^+ + Cl^- \tag{6.8}$$

In this form hydrogen chloride appears to *ionize,* producing the hydrogen chloride ion. When referring to acids it is understood that the hydrogen ion present is "hydrated," or attached to one or more water molecules, even though it may be written simply as H^+.*

The water solution of hydrogen chloride is hydrochloric acid. Both the compound and the acid have the formula HCl. The way the formula is used generally shows if it refers to the gas or the water solution. If more precise identification is needed, HCl(g) is used for the gaseous compound, and HCl(aq) is its **aqueous** solution, hydrochloric acid. (*Aqueous* is from the Latin *aqua,* meaning *water.*)

Quick Check 6.6
What is the difference between a hydrogen ion and a hydride ion?

6.7 ACIDS AND THE IONS DERIVED FROM THEM

PG 6J Given the name (or formula) of an acid or ion in Table 6.1, write its formula (or name).

Table 6.1 gives the names of five acids, including hydrochloric acid, their total ionization equations, similar to Equation 6.8, and the names of the anions resulting from their total ionization. *Total ionization* is the removal of all the hydrogens that can separate from an acid molecule. A large part of the nomenclature system in this book is based on this table. *The names and formulas of the acids in the table should be memorized.*

*Most chemists and chemistry textbooks use the hydrogen ion as H^+. A substantial number of chemists object to this simplification, however, and prefer to discuss only the hydronium ion. If you take chemistry courses from several teachers you are apt to meet champions of both causes.

Table 6.1 Acids and Anions

Acid	Ionization Equation	Ion Name
Hydrochloric acid	$HCl \rightarrow H^+ + Cl^-$	
Nitric acid	$HNO_3 \rightarrow H^+ + NO_3^-$	Nitrate
Sulfuric acid	$H_2SO_4 \rightarrow 2\ H^+ + SO_4^{2-}$	Sulfate
Carbonic acid*	$H_2CO_3 \rightarrow 2\ H^+ + CO_3^{2-}$	Carbonate
Phosphoric acid*	$H_3PO_4 \rightarrow 3\ H^+ + PO_4^{3-}$	Phosphate

*The carbonic and phosphoric acid ionizations occur only slightly in water solutions. They are used here to illustrate the derivation of the formulas and names of the carbonate and phosphate ions, both of which are quite abundant from sources other than their parent acids.

Table 6.1 contains important features that illustrate the nomenclature system. Notice the following:

1. Each acid contains at least one hydrogen, written first in the formula, that can separate as H^+ in ionization. If a chemical formula begins with H, it is usually an acid.
2. The first acid, HCl, contains only two elements—hydrogen and a nonmetal. Acids having only two elements are called **binary acids.** *The name of a binary acid begins with hydro-, followed by the name of the nonmetal, changed so it ends with* -ic. For HCl, chlorine → *hydro*chlor*ic*.
3. The anion from the total ionization of a binary acid is a monatomic anion, named by the rule for all monatomic anions. Example: chlorine → chlor*ide*.
4. Each of the other four acids have three elements—hydrogen, a nonmetal, and oxygen. Acids that contain oxygen are called **oxyacids.** *These particular oxyacids*—the ones to be memorized—*are named by changing the name of the nonmetal so it ends in* -ic. Examples: nitrogen → nitr*ic*; sulfur → sulfur*ic*.
5. The anion of an oxyacid is made up of the central nonmetal and oxygen. Because it contains oxygen it is called an ***oxyanion.*** *The name of an anion derived from an oxyacid whose name ends in* -ic *is the name of the central element changed to end with* -ate. You might say the *-ic* of the acid name is replaced by *-ate*. Examples: nitr*ic* → nitr*ate;* sulfur*ic* → sulf*ate*.
6. The formula of the anion resulting from the total ionization of an acid is the formula of the acid without the hydrogen(s). The amount of negative charge, −1, −2, or −3, is equal to the number of hydrogen ions removed from the neutral molecule. Examples: *one* H^+ from $HNO_3 \rightarrow NO_3^-$, with a minus *one* charge; *two* H^+ from $H_2SO_4 \rightarrow SO_4^{2-}$, with a minus *two* charge.

Before leaving Table 6.1, we should point out that its only purpose is to show the relationship between the names and formulas of acids and ions. It is not intended to describe the chemical properties of the substances listed, although all of the ionizations shown do take place, some to a large extent, and some only slightly. The acids are not the usual sources of the anions, all of which are present in abundance in the crust of the earth.

6.8 OTHER POLYATOMIC IONS

PG 6K Given the name (or formula) of the ammonium or hydroxide ion, write the formula (or name).

The **ammonium ion, NH_4^+,** is the only polyatomic cation we will consider in this text. It forms when a hydrogen ion is added to an ammonia molecule.

$$\underset{\text{hydrogen ion}}{H^+} + \underset{\text{ammonia}}{NH_3} \rightarrow \underset{\text{ammonium ion}}{NH_4^+} \quad (6.9)$$

Be sure you know and use the small but important difference between the names and formulas of the compound ammonia, NH_3, and the ammonium ion, NH_4^+.

The **hydroxide ion, OH^-,** is responsible for the properties of some members of a large class of compounds known as **bases.** While available from many sources, it can be thought of as coming from the ionization of water into a hydrogen and hydroxide ion. Writing the formula of water as HOH, this takes the form of a typical acid ionization:

$$HOH \rightarrow H^+ + OH^- \quad (6.10)$$

This ionization does, in fact, occur in water, but to a very, very small extent—about one molecule in ten million.

6.9 SUMMARY OF NAMES AND FORMULAS OF IONS

1. The name of a monatomic cation is the name of the element, followed by the word *ion*.
2. The name of a monatomic anion is the name of the element, changed to end with *-ide,* followed by the word *ion*.
3. The charge on a monatomic ion from an A group of the periodic table can be determined from the group it is in, as follows: 1A, +1; 2A, +2; 3A, +3; 5A, −3; 6A, −2; 7A, −1.
4. The hydrogen ion, H^+, comes from the ionization of an acid, a molecular compound having a formula beginning with H.
5. The formula for an anion coming from the total ionization of an acid is the formula of the acid minus the hydrogen(s), with a superscript charge that is the negative of the number of hydrogen(s) lost by the original acid.
6. The name of an oxyanion that comes from an oxyacid whose name ends in *-ic* is the name of the central element in the acid changed to end in *-ate*. In effect, the *-ic* of the acid name is replaced by *-ate*.
7. Ammonium ion is NH_4^+. Hydroxide ion is OH^-.

Once these nomenclature rules are firmly fixed in your mind, you are ready to write the names and formulas of ionic compounds.

6.10 NAMES AND FORMULAS OF IONIC COMPOUNDS

PG 6L Give a periodic table and the name (or formula) of any ionic compound whose ions are from the following, write its formula (or name):

Monatomic cation from Group 1A, 2A, or 3A of the periodic table, or the ammonium ion;

Monatomic anion from Group 6A or 7A of the periodic table, or the nitrate, sulfate, carbonate, phosphate, or hydroxide ion.

An ionic compound consists of a large group of positive and negative ions, usually arranged in a precise geometric pattern called a crystal. Sodium chloride, ordinary table salt, is a typical ionic compound. Its crystal form is that of the cube shown in Figure 6.2. An ionic compound has no independent molecules. Instead, its formula unit is the simplest ratio in which the ions appear in the crystal. In sodium chloride the number of sodium ions is equal to the number of chloride ions, so the formula is NaCl.

Names of Ionic Compounds

The rule for naming ionic compounds is:

The name of an ionic compound is the name of the cation, followed by the name of the anion.

All you have to do is to recognize the ions that make up the compound.

EXAMPLE 6.6 For each of the following ionic compounds, identify the cation and the anion, by name and by symbol, and write the name of the compound:

FORMULA	CATION	ANION	NAME
KCl			
$AlPO_4$			
$BaSO_4$			
NaF			

FORMULA	CATION	ANION	NAME
KCl	potassium ion, K^+	chloride ion, Cl^-	potassium chloride
$AlPO_4$	aluminum ion, Al^{3+}	phosphate ion, PO_4^{3-}	aluminum phosphate
$BaSO_4$	barium ion, Ba^{2+}	sulfate ion, SO_4^{2-}	barium sulfate
NaF	sodium ion, Na^+	fluoride ion, F^-	sodium fluoride

Formulas of Ionic Compounds

The rule for writing formulas of ionic compound is:

> *The formula of an ionic compound is the formula of the cation followed by the formula of the anion, each taken as many times as is necessary to yield a total charge of zero.* (Note: *Individual ionic charges are not normally written in the formula of a compound.) Whenever a polyatomic ion appears more than once in a chemical formula, that ion is enclosed in parentheses.*

Chemical compounds are electrically neutral. For ionic compounds this means that the formula unit must have an equal number of positive and negative charges. A net zero charge is achieved by taking the cations and anions in such numbers that the positive and negative charges are balanced.

Cations generally have charges of +1, +2, or +3; anion charges are −1, −2, and −3. This leads to four possible charge ratios that must be combined to produce a formula that is neutral. Table 6.2 includes every possible combination. In the 1:1 ratio, there is one of both ions in the formula of the compound. In the 2:1 ratio, there are two ions with the charge of 1 and one ion with the opposite charge of 2. The 3:1 ratio is similar. The 3:2 ratio is brought to a net 0 by taking two of the 3's and three of the 2's. Study Table 6.2 carefully until you understand how each formula is produced.

It is neither necessary nor desirable to write equations when writing formulas, as in Table 6.2; but at the beginning they may help you visualize the process. Another temporary aid some students like to use is to include the ionic charges in the formula. For example, they would write $Ba^{2+}S^{2-}$ for BaS. If this helps at the beginning, do it—but don't forget to remove the charges! They are not part of the formula. If you use either aid, break your dependence on it as soon as possible. Better still, learn how to combine the ions mentally with neither aid, and write the formulas directly.

Table 6.2 Formulas of Ionic Compounds

Charge Ratio	Cation	Anion	$\frac{\text{Cation}}{\text{Anion}}$ Ratio	Charge Combination	Example
1:1	+1	−1	1/1	$+1 + (-1) = 0$	$Na^{+} + Cl^{-} \rightarrow NaCl$
	+2	−2	1/1	$+2 + (-2) = 0$	$Ba^{2+} + S^{2-} \rightarrow BaS$
	+3	−3	1/1	$+3 + (-3) = 0$	$Al^{3+} + N^{3-} \rightarrow AlN$
2:1	+1	−2	2/1	$2(+1) + (-2) = 0$	$2\ Na^{+} + S^{2-} \rightarrow Na_2S$
	+2	−1	1/2	$+2 + 2(-1) = 0$	$Ba^{2+} + 2\ Cl^{-} \rightarrow BaCl_2$
3:1	+1	−3	3/1	$3(+1) + (-3) = 0$	$3\ Na^{+} + N^{3-} \rightarrow Na_3N$
	+3	−1	1/3	$+3 + 3(-1) = 0$	$Al^{3+} + 3\ Cl^{-} \rightarrow AlCl_3$
3:2	+2	−3	3/2	$3(+2) + 2(-3) = 0$	$3\ Ba^{2+} + 2\ N^{3-} \rightarrow Ba_3N_2$
	+3	−2	2/3	$2(+3) + 3(-2) = 0$	$2\ Al^{3+} + 3\ S^{2-} \rightarrow Al_2S_3$

When you thoroughly understand the process, try these examples:

EXAMPLE 6.7 Write the formulas of potassium chloride and potassium hydroxide.

First be sure you know the formulas of the ions, and then combine them.

----- -----

potassium chloride, KCl; potassium hydroxide, KOH

The ions are K^+, Cl^-, and OH^-. Both compounds are 1:1 combinations like NaCl in Table 6.2.

EXAMPLE 6.8 Write the formulas of calcium chloride and calcium hydroxide.

The anions are the same as before, but the cation now has a +2 charge: Ca^{2+}. Beyond that, there is a small catch to the question—something you haven't met before. Nevertheless, try to write the formulas without further instructions.

----- -----

calcium chloride, $CaCl_2$; calcium hydroxide, $Ca(OH)_2$. Comments follow.

Did you write $CaOH_2$ for calcium hydroxide? If so, be assured you have lots of company! That formula indicates an atomic ratio of one atom of calcium, one atom of oxygen, and two atoms of hydrogen. The correct ratio is one calcium ion for every two *hydroxide ions*: Ca^{2+} + OH^- + OH^-. In two hydroxide ions there are two atoms of oxygen, so there must be two atoms of oxygen in the formula unit of calcium hydroxide. The atom count could be satisfied with CaO_2H_2; but that formula is also incorrect, as it fails to show that the hydroxide ion, OH^-, is a structural unit in the compound.

The rule is

> *To show more than one polyatomic ion in a formula, enclose that ion in parenthesees and show the number of these ions with a subscript after the parentheses.*

The subscript applies to everything in the parentheses immediately before it. Note that parentheses are used only to enclose polyatomic ions, never monatomic ions. The formula for calcium chloride is $CaCl_2$, not $Ca(Cl)_2$.

EXAMPLE 6.9 Write the formulas of potassium nitrate and calcium nitrate.

Think carefully about the formulas you have written already and what has just been said about parentheses.

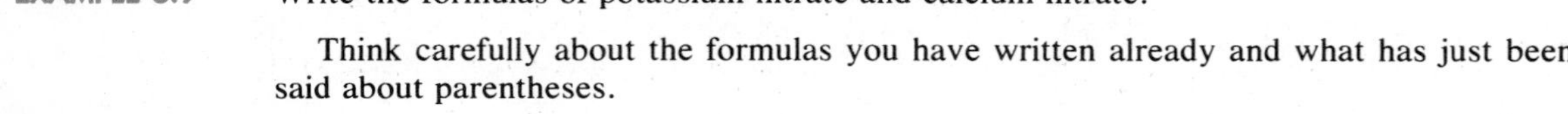

----- -----

potassium nitrate, KNO_3; calcium nitrate, $Ca(NO_3)_2$

Potassium nitrate is a +1 and −1 combination, just like potassium hydroxide in Example 6.7. Calcium nitrate is a +2 and −1 combination, just like calcium hydroxide in Example 6.8. The subscript 3 at the end of NO_3^- does not make the handling of nitrate different from how you treat hydroxide. The entire symbol is enclosed in parentheses.

EXAMPLE 6.10 Write the formulas of ammonium nitrate and ammonium carbonate. Again think carefully as you apply the rules you have learned.

----- -----

ammonium nitrate, NH_4NO_3; ammonium carbonate, $(NH_4)_2CO_3$

Like all ionic compounds, the formula for ammonium nitrate has the formula of the cation followed by the formula of the anion. This is not changed just because both ions happen to contain the same element, nitrogen. In $(NH_4)_2CO_3$ it is a polyatomic cation that is taken twice and must be enclosed in parentheses.

EXAMPLE 6.11 Write the formula of barium phosphate.

----- -----

barium phosphate, $Ba_3(PO_4)_2$

This +2 and −3 combination corresponds to Ba_3N_2 in Table 6.2.

EXAMPLE 6.12 Write the formula for each compound listed below:

calcium oxide | potassium sulfide

magnesium bromide | aluminum sulfate

----- -----

calcium oxide, CaO | potassium sulfide, K_2S

magnesium bromide, $MgBr_2$ | aluminum sulfate, $Al_2(SO_4)_3$

Practice is the only way to be sure you will master the skill of writing chemical formulas. Tables 6.3 and 6.4 at the end of the chapter are two formula-writing exercises covering all ionic compounds included in this chapter. We strongly recommend that you complete both of these exercises.

6.11 HYDRATES

PG 6M Given the formula of a hydrate, state the number of water molecules associated with each formula unit of the anhydrous compound.

6N Given the name (or formula) of a hydrate, write its formula (or name). (This performance goal is limited to hydrates of ionic compounds discussed in this chapter.)

Some compounds, when crystallized from water solutions, form solids that include water molecules as part of the crystal structure. Such water is referred to as **water of crystallization** or **water of hydration.** The compound is said to be **hydrated,** and is called a **hydrate.** Hydration water can usually be driven from a compound by heating, leaving the **anhydrous** compound.

Copper sulfate is an example of a hydrate. The anhydrous compound, $CuSO_4$, is a nearly white powder. Each formula unit of $CuSO_4$ combines with five water molecules in the hydrate, which is a dark blue crystal. The formula of the hydrate, $CuSO_4 \cdot 5\ H_2O$, and its name, copper sulfate 5-hydrate, illustrate the nomenclature system for hydrates. The equation for the dehydration of this compound is

$$CuSO_4 \cdot 5\ H_2O \rightarrow CuSO_4 + 5\ H_2O \tag{6.11}$$

EXAMPLE 6.13 (a) How many water molecules are associated with each formula unit of anhydrous sodium carbonate in $Na_2CO_3 \cdot 10\ H_2O$? Name the hydrate.

(b) Write the formula of nickel chloride 6-hydrate if the formula of the anhydrous compound is $NiCl_2$.

----- -----

(a) Ten; sodium carbonate 10-hydrate
(b) $NiCl_2 \cdot 6\ H_2O$

6.12 SUMMARY

Figure 6.4 summarizes the pure substances whose names and formulas you have learned in this chapter. Study it and the caption beneath it. Learn to recognize the different kinds of substances whose names and formulas you must know and how they are combined and related. As urged at the beginning of the chapter, *learn the system*.

Skill in chemical nomenclature is gained only by much practice. The end-of-chapter questions, and particularly Tables 6.3 and 6.4, provide ample opportunity. Occasions to apply your skill appear constantly in the chapters ahead. If your formula-writing skill is perfected now, you will find that the substances added in Chapter 12 fit neatly into the system you learned here.

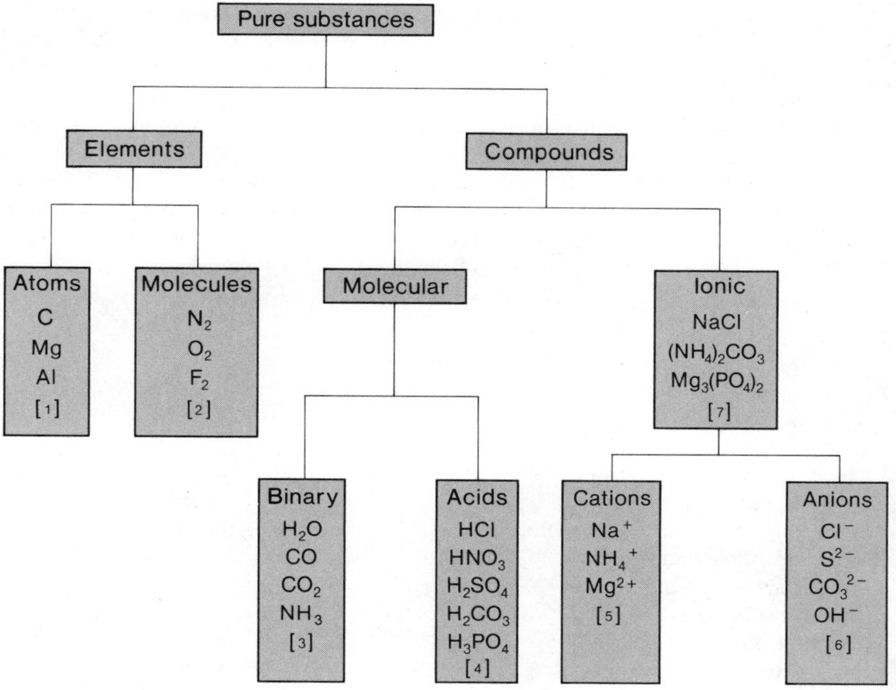

Figure 6.4
Classifications and examples of substances studied in this chapter. The formula unit of most elements is the individual atom, and the elemental symbol is its formula [1]. Seven elements form diatomic molecules; three are shown [2]. Only four binary molecular compounds are included in this chapter [3]. The water solutions of five molecular compounds are common acids [4]. The ammonium ion is the only polyatomic cation considered [5]. The formulas of monatomic ions [5] and [6] can be determined from the periodic table. Polyatomic anion formulas [6] can be derived from the ionization of oxyacids. The formula of an ionic compound [7] is found by combining a cation and an anion in such numbers that the net charge is zero.

CHAPTER 6 IN REVIEW

6.1 The Meaning of a Chemical Formula

- **6A** Given a chemical formula, state the number of atoms of each element in the formula unit.
- **6B** Given the number of atoms of each element in a formula unit of a pure substance, write the chemical formula.

6.2 Atoms and Combinations of Atoms: Chemical Bonds

- **6C** Distinguish between monatomic, diatomic, and polyatomic particles.
- **6D** Distinguish between ions, anions, and cations.
- **6E** Distinguish between molecular substances and ionic compounds.
- **6F** Distinguish between covalent bonds and ionic bonds.

6.3 The Formulas of the Elements

- **6G** Given the name (or formula) of an element listed in Figure 4.6, write its formula (or name).

6.4 Common Binary Molecular Compounds

- **6H** Given the name (or formula) of water, ammonia, carbon monoxide, or carbon dioxide, write its formula (or name).

6.5 Monatomic Ions

- **6I** Given a periodic table and the name and atomic number of a representative element or a source from which that name and number may be found, write the name and formula of the monatomic ion formed by that element.

6.6 The Ions of Hydrogen

6.7 Acids and the Ions Derived from Them

- **6J** Given the name (or formula) of an acid or ion in Table 6.1, write its formula (or name).

6.8 Other Polyatomic Ions

- **6K** Given the name (or formula) of the ammonium or hydroxide ions, write the formula (or name).

6.9 Summary of Names and Formulas of Ions

6.10 Names and Formulas of Ionic Compounds

6L Given a periodic table and the name (or formula) of any ionic compound whose ions are from the following, write its formula (or name):

Monatomic cation from Group 1A, 2A, or 3A of the periodic table, or the ammonium ion;

Monatomic anion from Group 6A or 7A of the periodic table, or the nitrate, sulfate, carbonate, phosphate, or hydroxide ion.

6.11 Hydrates

6M Given the formula of a hydrate, state the number of water molecules associated with each formula unit of the anhydrous compound.

6N Given the name (or formula) of a hydrate, write its formula (or name). (This performance goal is limited to hydrates of ionic compounds covered in this chapter.)

6.12 Summary

TERMS AND CONCEPTS

6.1 Chemical formula
Formula unit

6.2 Monatomic, diatomic, polyatomic
Molecule, molecular
Molecular compound
Chemical bond
Covalent bond
Ion, cation, anion
Ionic compound
Ionic bond

6.4 Binary

6.6 Acid
Hydronium ion
Ionize
Hydrated
Aqueous

6.7 Total ionization
Oxyacid
Oxyanion

6.11 Water of crystallization (hydration)
Hydrate
Anhydrous compound

Most of these terms and many others appear in the Glossary. Use the Glossary regularly.

QUESTIONS AND PROBLEMS

Section 6.1

For Questions 1 to 4 and 34 to 37, for each formula, state the identity of each element in the compound and the number of atoms of that element in the formula unit; or, from the given number of atoms of each element in the formula unit, write the formula.

(1) N_2O.

(2) SO_2Cl_2.

(3) One silicon, four hydrogen atoms.

(4) Three carbon, eight hydrogen, one oxygen atoms.

(34) PCl_3.

(35) CH_3CHO.

(36) Two hydrogen, one sulfur, three oxygen atoms.

(37) Four hydrogen, two phosphorus, seven oxygen atoms.

Section 6.2

(5) If something is *diatomic*, what does that mean?

(38) If *poly-* means *many*, why can we use the term when speaking of a two-atom species?

(*6*) Which term *monatomic, diatomic,* or *polyatomic,* best fits the definition of a molecule?

(*7*) How do two atoms form a covalent bond?

(8)* What is a crystal? Which kinds of compounds, molecular, ionic, or both, form crystals? Do elements exist as crystals?

(39) It is common to speak of a molecular compound, but not so common to speak of a molecular element. Is there such a thing as a molecular element? If yes, can you given an example?

(40)* Can two atoms form an ionic bond? Explain.

(41)* The word *cation* was given to the ions that move toward the cathode in an electrolytic cell. Similarly, anions move toward the anode. What are the charges on the cathode and anode of a cell?

Section 6.3

(*9*) The stable form of seven elements is the diatomic molecule. Write the names and formulas of those elements.

(10) Write the formulas of the elements iron, chlorine, chromium, and beryllium.

(42) The elements of Group 0 of the periodic table are stable as monatomic atoms. Write their formulas.

(43) Write the formulas of calcium, silver, barium, and hydrogen.

Section 6.4

(11) What characteristic must a compound have if it is to be classified as molecular?

(*12*) Write the names of H_2O and CO_2.

(44) What is the difference between a binary compound and any other compound?

(45) Write the formulas of carbon monoxide and ammonia.

Section 6.5

(*13*) Distinguish betweeen monatomic ions and polyatomic ions.

(14) How does a monatomic anion differ from a monatomic cation?

(*15*) Explain how monatomic cations are formed from atoms.

(46) Identify the polyatomic ions among the following: K^+, S^{2-}, Br^-, NH_4^+, CO_3^{2-}, Al^{3+}.

(47) Identify the monatomic cations in Question 37.

(48) Explain how monatomic anions are formed from atoms.

(16) The table below gives the atomic numbers of several elements from the A groups of the periodic table. Fill in the blanks with the ion name and symbol for the monatomic ion formed by each element, and classify that ion as a cation or anion. You may refer to a periodic table.

ATOMIC NUMBER	ION NAME	ION FORMULA	CATION OR ANION
11			
53			
3			
20			
16			

(17) You are not expected to be familiar with the elements selenium (Z = 34) and rubidium (Z = 37), but from their positions in the periodic table, you should be able to make intelligent guesses as to the names and formulas of the monatomic ions they will form. Complete the following table.

ELEMENT	ION NAME	ION FORMULA	CATION OR ANION
Selenium			
Rubidium			

(18) Write the names of F^-, K^+, Al^{3+}, O^{2-}.

(49) What are the names of Mg^{2+}, Cl^-, Ba^{2+}, and S^{2-}?

Section 6.6

(*19*) What property identifies a substance as an acid?

(50) How do you recognize the formula of an acid?

(*20*) What is the difference between a hydrogen ion and a hydronium ion?

(51)* Write the equations for the reaction between gaseous hydrogen iodide and water to produce the hydronium ion. Follow with the equation for the ionization of hydrogen iodide.

(21) Name a property of hydrogen that justifies its position at the top of Group 1A in the periodic table.

(52) What reason is there for placing hydrogen at the head of Group 7A in the periodic table?

Section 6.7

(*22*) Write the formula of hydrochloric acid.

(53) Show how the hydrogen and chloride ions are derived from the ionization of hydrochloric acid.

(23) Write the formula of nitric acid, the formula of the anion derived from its ionization, and the name of the anion.

(54) Write the formula of phosphoric acid, the formula of the anion derived from its ionization, and the name of the anion.

(*24*) Write the name of H_2CO_3, the formula of the anion derived from its ionization, and the name of the anion.

(55) Write the name of H_2SO_4, the formula of the anion derived from its ionization, and the name of the anion.

(25)* The formula of chloric acid is $HClO_3$. Write the ionization equation for chloric acid and the name of the oxyanion it forms.

(56)* The formula of acetic acid is $HC_2H_3O_2$. Write the ionization equation for acetic acid and the name of the oxyanion it forms. [Hint: Don't be misled by hydrogen appearing twice in the formula of the acid.]

Section 6.8

(*26*) What is the name of the OH^- ion?

(57) Write the formula of the ammonium ion.

Section 6.10

The formula-writing exercises of Tables 6.3 and 6.4 are the principal questions for Section 6.11. For Questions 27 to 30 and 58 to 61, given the name of an ionic compound, write the formula; or, given the formula, write the name. Refer only to a periodic table while answering these questions.

(27) Lithium chloride; CaS.

(58) KF; magnesium oxide.

(28) Ammonium nitrate; $BaCO_3$.

(59) NaOH; aluminum phosphate.

(29) Barium bromide, K_3PO_4.

(60) CaI_2; sodium sulfate.

(30) Magnesium phosphate; $(NH_4)_2SO_4$.

(61) $Al_2(CO_3)_3$; lithium sulfide.

Table 6.3 Formula-Writing and Nomenclature Exercise Number 1

Instructions: For each box write the chemical formula and name of the compound formed by the cation at the head of the column and the anion at the left of the row. Correct formulas and names are listed on page ANS-10.

IONS	Li^+	Mg^{2+}	NH_4^+	Al^{3+}	Na^+	Ba^{2+}	K^+	Ca^{2+}
Br^-								
SO_4^{2-}								
OH^-								
F^-								
O^{2-}								
NO_3^-								
PO_4^{3-}								
Cl^-								
S^{2-}								
I^-								
CO_3^{2-}								

Table 6.4 Formula-Writing Exercise Number 2

Instructions: For each box write the chemical formula of the compound formed by the cation at the head of the column and the anion at the left of the row. Refer only to the periodic table in completing this exercise. Correct formulas are listed on page ANS-10.

IONS	Aluminum	Ammonium	Barium	Calcium	Lithium	Magnesium	Potassium	Sodium
Bromide								
Carbonate								
Chloride								
Fluoride								
Hydroxide								
Iodide								
Nitrate								
Oxide								
Phosphate								
Sulfate								
Sulfide								

Section 6.11

(*31*) Distinguish between a hydrate and an anhydrous compound.

(**32**) How many water molecules are associated with one anhydrous formula unit of calcium chloride in $CaCl_2 \cdot 2\ H_2O$? Write the name of the compound.

(**33**) One hydrate of barium hydroxide contains eight molecules of water per formula unit of the anhydrous compound. Write the formula and name of the hydrate.

(**62**) Among the following, identify all hydrates and anhydrous compounds: $NiSO_4 \cdot 6\ H_2O$; KCl; $Na_3PO_4 \cdot 12\ H_2O$.

(**63**) How many water moelcules are associated with one anhydrous formula unit of magnesium sulfate in $MgSO_4 \cdot 7\ H_2O$? Write the name of the compound.

(**64**) Write the formula of ammonium phosphate 3-hydrate and potassium sulfide 5-hydrate.

Miscellaneous Questions

(*65*) Distinguish precisely and in scientific terms the differences between items in each of the following groups:
(a) Monatomic, diatomic, and polyatomic particles
(b) Ions, anions, and cations
(c) Molecular compounds and ionic compounds
(d) Covalent bonds and ionic bonds
(e) Atoms and diatomic molecules
(f) Symbol of an element and formula of an element
(g) Molecular compound and binary molecular compound
(h) Binary acid, oxyacid
(i) Hydrogen ion, hydronium ion, hydride ion
(j) Hydrates and anhydrous compounds

(**66**) Determine whether each statement that follows is true or false:
(a) A polyatomic particle has at least three atoms in it.
(b) Anions and cations are both charged particles.
(c) Anions have a positive charge.
(d) Ionic compounds exist as tiny independent units.
(e) Ionic compounds are held together by covalent bonds.
(f) Covalent bonds are formed when two atoms share one or more pairs of electrons.
(g) All elements are stable as diatomic molecules.
(h) The formula of most elements is the same as the symbol of the element.
(i) Binary compounds contain two elements.
(j) Hydrogen is a part of all acids discussed in this chapter.
(k) An oxyacid contains hydrogen, oxygen, and a nonmetal.
(l) The hydrogen ion and the hydronium ion are identical.
(m) A chemical formula tells what elements are in a pure substance and the number of atoms of each element in the formula unit.
(n) Water of crystallization is the amount of water needed to crystallize a certain amount of a compound.
(o) When an oxyacid ionizes, it always produces a polyatomic anion.
(p) The ocean is an aqueous solution.

(**67**)* It was stated in the introduction to the chapter that, if you learned the rules of nomenclature, you would be able to apply them to things you have never heard of before. It is not expected that you will be able to do this until after you study Chapter 12. But you may have caught on already. Let's try just a few. As you answer these questions, remember that elements in the same family form similar compounds.
(a) Write the formula of strontium sulfide (Z = 38 for strontium).
(b) Write the formula of indium selenide (Z = 49 for indium; Z = 34 for selenium).
(c) What is the name of $GaCl_3$? (Ga is gallium, Z = 31).
(d) What is the name of Na_2Te? (Te is tellurium, Z = 52.)
(e) IO_3^- is the formula of the iodate ion. Write the name and formula of the acid whose total ionization produces it.

7 Chemical Formula Calculations

LOOKING BACK

3.2 Dimensional analysis is a method for solving chemistry problems.

3.7 Addition, subtraction, multiplication, and division are governed by rules of significant figures.

4.6 The atomic mass of an element is the average mass of its atoms compared to an atom of carbon-12, expressed in atomic mass units.

4.7 and 4.8 The symbols of the common elements were located in the periodic table, which shows the atomic masses of all elements.

LOOKING AHEAD IN CHAPTER 7

Dimensional analysis is applied in chemical formula calculations.

Answers in this chapter are rounded off according to the rules of significant figures.

Molecular mass, formula mass, and molar mass are calculated from the atomic masses of the elements in the species.

The periodic table is your source of molar masses for chemical formula calculations.

Several examples in this chapter require you to write the formula of a compound from its name. This assumes you have studied Chapter 6, "Introduction to Nomenclature." If so, we encourage you to write those formulas while using a periodic table as your only reference. If you will not study nomenclature until Chapter 12—an option available to your instructor—skip the "write the formula" step of the examples and move on to the second question. Or, if you wish, you may refer to the answers to Formula Writing Exercise No. 2 on page 132. This is a complete list of the ionic compounds whose formulas were developed in Chapter 6.

7.1 MOLECULAR MASS; FORMULA MASS

PG 7A Distinguish between atomic mass, molecular mass, and formula mass.

In Section 4.5 you learned that the atomic mass of an element is the average mass of its atoms, expressed in atomic mass units, amu. But what about compounds? Is there such a thing as a "compound mass"? The answer is yes. It is based on the chemical formula of the compound. This formula tells how many atoms of each element are present in the smallest structural unit of the compound. Its **formula mass** is simply the sum of the atomic masses in its "formula unit."

The term *formula mass* is not widely used, although it is the most accurate term for ionic compounds. An ionic compound is made up of an assembly of positively and negatively charged ions, and its formula represents the simplest ratio in which these ions combine. A molecular substance, on the other hand, actually exists as independent particles—molecules—having the number of atoms of each element shown in its formula. The mass of a molecule is its **molecular mass.** Defined exactly as atomic mass is defined, **molecular mass (or formula mass) is the average mass of molecules (or formula units) compared to the mass of an atom of carbon-12 at exactly 12 atomic mass units.**

7.2 THE MOLE CONCEPT

PG 7B Define *mole.* Identify the number that corresponds to one mole.

Dalton's atomic theory says that compounds consist of atoms of different elements combined in a ratio of small whole numbers. To prepare a compound from its elements we must therefore "count out" the proper number of atoms of the different elements. But, if we wish to count out equal numbers of oxygen and carbon atoms so that they can react to produce carbon monoxide, CO, it will be a time-consuming process. This is because of the tremendous number of atoms of each element in the tiniest of samples. There must be a better way—and chemists have found it.

In Section 4.5 it was shown that the ratio of masses of individual atoms of two elements is the same as the ratio of masses of equal numbers of atoms of those elements. If the mass of an oxygen atom is 16.0 amu, and the mass of a carbon atom is 12.0 amu, then any quantities of oxygen and carbon that have a mass ratio of 16.0/12.0 contain the same number of atoms. Therefore, if you wish to count out an equal number of atoms of any two elements, all you must do is weigh out samples having a mass ratio that is the same as the ratio of their atomic masses.

Chemists use the SI unit for amount of substance, the **mole** (abbreviated **mol**), to organize the "counting by weighing" idea. A mole is defined as **that amount of any species that contains the same number of units as the number of atoms in exactly 12 g of carbon-12.** This number is called **Avogadro's number, N,** in honor of the man whose interpretation of experiments with gases led to an early method of estimating atomic weights.

The definition of the mole does not identify the number of atoms in exactly 12 g of carbon-12. That number must be found by experiment. Since a carbon atom is extremely small we would expect the number of atoms in exactly 12 g of carbon-12 to be huge. This is the case; the number has been estimated to be 602 000 000 000 000 000 000 000. In exponential notation this is 6.02×10^{23}. The following equivalence results from the definition of a mole:

$$\text{One mole of any species} = 6.02 \times 10^{23} \text{ units of that species} \qquad (7.1)$$

To get a better idea of what is meant by a mole, let us consider a more familiar counting unit, the dozen. Suppose a dozen were defined as the number of eggs in the carton in which they are sold at the local supermarket. By experiment (walking down to the store, opening a box of eggs, and counting them) you can determine that this number is 12. In buying eggs, you rarely use the number 12 directly; rather, you think in terms of dozens. When is the last time, for example, you saw a shopping list with "24 eggs" written on it? Would it not be written simply as "two dozen eggs"? Just as two dozen eggs expresses a quantity of eggs, two moles of carbon expresses a quantity of carbon, a quantity that can be thought of as containing a certain number of carbon atoms. If you ever have difficulty understanding *mole* in a sentence, substitute the word *dozen* and you will see what it means.

Let's think for a moment about this number—Avogadro's number. Its size is tremendous; it staggers the imagination. Many attempts have been made to create a word picture of it, but they can only hint at its vast size by leaving the reader with another huge number that likewise defies the imagination. For example, if you were to try to count the atoms in exactly 12 g of carbon-12, and proceeded at the rate of 100 atoms per minute without interruption, 24 hours per day, 365 days per year, the task would require 11 500 000 000 000 000 years. This is about two million times as long as the earth has been in existence! If we enlisted the aid of the entire population of the earth, about 4.5 billion people, all counting at the same rate, the time required could be brought down to a somewhat more reasonable 2 540 000 years.

Is it any wonder the chemist prefers to think in terms of moles rather than dealing with actual numbers of molecules?

Quick Check 7.2
What does 1 g of hydrogen have in common with 4 g of helium? (Note: The atomic mass of hydrogen is 1 amu; of helium, 4 amu.)

7.3 MOLAR MASS

PG 7C Define molar mass.

Molar mass is the mass in grams of one mole of any species. As defined, the units of molar mass are grams per mole. The concept is particularly useful in finding the number of grams in a specified number of moles, or vice versa, as we will see in Section 7.6. *Molar mass* is a general term that may be applied to any species, atoms, molecules, or formula units.*

There is an important relationship you should catch at this point. Because the definitions of atomic mass, mole, and molar mass are all tied to "exactly 12 g of carbon-12," *the atomic mass of an element is numerically equal to its molar mass*. The atomic mass of sodium is 23.0 amu; the molar mass of sodium atoms is 23.0 g/mol. The periodic table is therefore a source of molar masses of atoms in grams per mole, as well as atomic masses in atomic mass units.

*Before SI units came into general use, *gram atomic weight, gram molecular weight, and gram formula weight* were widely used for the molar mass of elements, molecular substances, and ionic compounds, respectively.

Similarly, the molecular mass of a molecular compound and the formula mass of an ionic compound are numerically equal to the molar masses of those compounds.

Quick Check 7.3
The molecular mass of water is 18.0 amu. What is the molar mass of water? State the principle on which you base your answer.

7.4 CALCULATION OF MOLAR MASS

PG 7D Calculate the molar mass of any substance whose chemical formula is known.

What is the molar mass of CO? Based on the time-honored truth that the whole is equal to the sum of its parts, the mass of 1 mol of CO is equal to the sum of the masses of all the atoms it contains. Each molecule of CO contains one carbon atom and one oxygen atom. Therefore 1 mol of CO molecules contains 1 mol of carbon atoms (12.0 g) and 1 mol of oxygen atoms (16.0 g). The molar mass of CO is 12.0 g C + 16.0 g O = 28.0 g/mol.

A word about significant figures: The periodic tables in this book give atomic masses to four significant figures. If the data of a problem are such that a four-significant-figure answer is required, all four digits in the table must be used. Otherwise, our procedure will be to read and use atomic masses to three significant figures. If problem data limit an answer to one or two significant figures, the final result will be rounded off accordingly.

EXAMPLE 7.1 Calculate the molar mass of sodium fluoride, NaF.

The procedure is straightforward. Solve the problem, remembering that the periodic table is your best source for atomic masses (molar masses of atoms).

----- -----

23.0 g Na + 19.0 g F = 42.0 g NaF/mol

EXAMPLE 7.2 Calculate the molar mass of calcium fluoride.

No formula this time. This you must furnish. Refer to page 72 for help, if necessary.

----- -----

CaF_2

Now ask yourself, how many moles of calcium and how many moles of fluorine are in 1 mol of CaF_2 units. Answer, please.

----- -----

There are 1 mol Ca and 2 mol F in 1 mol CaF_2.

Now find the grams of calcium fluoride per mole.

----- -----

40.1 g Ca + 2(19.0) g F = 78.1 g CaF_2/mol

EXAMPLE 7.3 Find the molar mass of magnesium hydroxide.

First you need the formula of magnesium hydroxide. It is. . . .

----- -----

$Mg(OH)_2$

You know the procedure. Although this problem does introduce a variation not met previously, you probably will be able to think it through. All the way to the answer, please.

----- -----

24.3 g Mg + 2(16.0) g O + 2(1.0) g H = 58.3 g $Mg(OH)_2$/mol

This problem requires you to recognize that 1 mol contains 1 mol of magnesium ions and 2 mol of hydroxide ions. Each hydroxide ion contains one oxygen atom and one hydrogen atom. Hence, there are 2 mol of oxygen and 2 mol of hydrogen in the calculations.

EXAMPLE 7.4 Calculate the molar mass of aluminum sulfate.

The formula, please. . . .

----- -----

$Al_2(SO_4)_3$

Now to be sure the formula is correctly interpreted, how many moles of aluminum, sulfur, and oxygen atoms are in 1 mol of $Al_2(SO_4)_3$?

----- -----

2 mol Al, 3 mol S, 12 mol O

Each sulfate ion contains one sulfur atom and four oxygen atoms. There are three sulfate ions in the formula unit. This makes three sulfur atoms and 3 × 4 = 12 oxygen atoms in the formula unit.

Now calculate the molar mass.

----- -----

2(27.0) g Al + 3(32.1) g S + 12(16.0) g O = 342 g $Al_2(SO_4)_3$/mol

EXAMPLE 7.5 Find the molar mass of $CuSO_4 \cdot 5\ H_2O$. (Cu is copper, Z = 29.)

The formula of a hydrate shows the number of water molecules locked into a crystal with each formula unit of the compound. Another way of writing the hydrate formula that may make its interpretation easier is $CuSO_4(H_2O)_5$. Now list the number of moles of atoms of each element in 1 mol of formula units: copper, sulfur, oxygen (careful!) and hydrogen.

----- -----

1 mol Cu, 1 mol S, 9 mol O (4 from $CuSO_4$ + 5 from H_2O), 10 mol H.

Complete the calculation of molar mass.

----- -----

63.5 g Cu + 32.1 g S + 9(16.0) g O + 10(1.0) g H = 250 g $CuSO_4 \cdot 5\ H_2O$/mol

EXAMPLE 7.6 Find the molar mass of bromine vapor, Br_2.

This time you are looking for the molar mass of an element. The problem is easy, but you may wonder about something as you proceed. Write the answer.

----- -----

2(79.9) g Br = 160 g Br_2/mol

Did you wonder about whether or not you should multiply the atomic mass of bromine by 2? Bromine is not stable as an individual atom but forms diatomic molecules, molecules consisting of two atoms. This is why the formula is Br_2. In 1 mol of Br_2 there are 2 mol of Br atoms.

Example 7.6 teaches an important lesson: always perform your molar mass calculation for the formula unit exactly as that formula is written. This also emphasizes the importance of having the correct formula. This is critical when working with the seven elements that form diatomic molecules as elements, but must be thought of as atoms when they are in a compound. Keep this in mind when you meet a slightly more complicated example a few pages ahead.

7.5 MASS RELATIONSHIPS BETWEEN ELEMENTS IN A COMPOUND; PERCENTAGE COMPOSITION

PG 7E For any compound whose formula is known, given the mass of a sample, calculate the mass of any element in the sample. Or, given the mass of any element in the sample, calculate the mass of the sample or the mass of any other element in the sample.

7F Calculate the percentage composition of any compound whose formula is known.

The numbers used in calculating the molar mass of a compound furnish a one-step conversion between the mass of any sample of the compound and the mass of any element in the sample. In Example 7.4 you found that 1 mol of aluminum sulfate, $Al_2(SO_4)_3$, is made up of $2 \times 27.0 = 54.0$ g Al, plus $3 \times 32.1 = 96.3$ g S, plus $12 \times 16.0 = 192$ g O, a total of 342 g/mol. For aluminum sulfate these numbers give us equivalent relationships between the masses of the elements and the compound:

$$54.0 \text{ g Al} \mathrel{\hat{=}} 96.3 \text{ g S} \mathrel{\hat{=}} 192 \text{ g O} \mathrel{\hat{=}} 342 \text{ g } Al_2(SO_4)_3 \qquad (7.2)$$

If the number of grams of any species in Equation 7.2 is known, the mass of any other species may be calculated.

EXAMPLE 7.7 How many grams of aluminum are in 138 g of $Al_2(SO_4)_3$?

The unit path is grams of given quantity → grams of wanted quantity. Set up the problem and solve.

$$138 \text{ g } \cancel{Al_2(SO_4)_3} \times \frac{54.0 \text{ g Al}}{342 \text{ g } \cancel{Al_2(SO_4)_3}} = 21.8 \text{ g Al}$$

The mass relationships between elements in a compound are often expressed in its percentage composition. (If percentage problems are troublesome to you, check Appendix I, Part E.) The percentage of one element in a compound is found by dividing the mass of that element in a sample by the mass of the whole sample, and then multiplying by 100:

$$\% \text{ of element} = \frac{\text{grams of emenet}}{\text{grams of sample}} \times 100 \qquad (7.3)$$

The most convenient "sample" to use is 1 mol. Again, using the numbers for aluminum sulfate, its percentage composition is

$$\text{Al: } \frac{54.0 \text{ g Al}}{342 \text{ g } Al_2(SO_4)_3} \times 100 = 15.8\% \text{ Al}$$

$$\text{S: } \frac{96.3 \text{ g S}}{342 \text{ g } Al_2(SO_4)_3} \times 100 = 28.2\% \text{ S}$$

$$\text{O: } \frac{192 \text{ g O}}{342 \text{ g } Al_2(SO_4)_3} \times 100 = \underline{56.1\% \text{ O}}$$

$$\text{total percent} = 100.1\%$$

The sum of the percentages must be 100.0. A difference of 0.1 or 0.2 may be due to rounding off.

The same procedure may be used for any compound.

EXAMPLE 7.8 Calculate the percentage composition of calcium nitrate.

First you need the formula of calcium nitrate. It is. . . .

$Ca(NO_3)_2$

Now calculate the molar mass of calcium nitrate. You will have to find the grams of each element to determine percentage. Carry it all the way, and check to see if your percentages total 100.

Element	*Grams*	*Percent*
Ca	1 × 40.1 = 40.1 g Ca	$\frac{40.1}{164} \times 100 =$ 24.4% Ca
N	2 × 14.0 = 28.0 g N	$\frac{28.0}{164} \times 100 =$ 17.1% N
O	6 × 16.0 = 96.0 g O	$\frac{96.0}{164} \times 100 =$ 58.5% O
	164.1 g $Ca(NO_3)_2$	100.0%

7.6 CONVERSION BETWEEN MASS AND NUMBER OF MOLES

PG 7G Given the number of grams (or moles) of a chemical species of known or calculable molar mass, find the number of moles (or grams).

The "counting by weighing" value of the mole concept is based on the ready conversion of the mass of a chemical species into a number of moles, and vice versa. Using dimensional analysis, this is a simple operation. It is a part of many chemical problems, and is one of the most important skills to be learned in the introductory course. The operation is based on the equivalence between grams and moles of a chemical that comes from the mole concept. If MM is the number of grams of X in 1 mol (the molar mass of X) then

$$\text{MM g X} \simeq 1 \text{ mol X} \tag{7.4}$$

The following examples illustrate the two possible conversions.

EXAMPLE 7.9 How many moles of aluminum sulfate, $Al_2(SO_4)_3$ are in 150 g? The molar mass, from Example 7.4, is 342 g/mol.

Applying Equation 7.4 to the problem at hand yields the equivalence 342 g $Al_2(SO_4)_3$ $\simeq$ 1 mol $Al_2(SO_4)_3$. Your unit path is from the given quantity, grams, to the desired quantity, moles: g $\rightarrow$ mol. If 1 mol is 342 g, then the given quantity of 150 g is what number of moles? Set up and solve.

$$150 \text{ g } \cancel{Al_2(SO_4)_3} \times \frac{1 \text{ mol } Al_2(SO_4)_3}{342 \text{ g } \cancel{Al_2(SO_4)_3}} = 0.439 \text{ mol } Al_2(SO_4)_3$$

EXAMPLE 7.10 You are carrying out a laboratory reaction that requires 0.0250 mol of $NiCl_2$. How many grams of the compound do you weigh out?

First, find the molar mass of $NiCl_2$. (Ni is nickel, Z = 28.)

58.7 g Ni + 2(35.5) g Cl = 130 g $NiCl_2$/mol

Use molar mass to convert from the given quantity of moles to grams. If 1 mol is 238 g, how many grams is 0.0250 mol?

$$0.0250 \cancel{\text{ mol } NiCl_2} \times \frac{130 \text{ g } NiCl_2}{1 \cancel{\text{ mol } NiCl_2}} = 3.25 \text{ g } NiCl_2$$

7.7 NUMBER OF ATOMS, MOLECULES, OR FORMULA UNITS IN A SAMPLE

PG 7H Given the mass or number of moles of a pure substance whose formula is known, find the number of atoms, molecules, or formula units. Or, given the number of atoms, molecules, or formula units, find the number of moles or mass.

Equation 7.1 (1 mol = 6.02×10^{23} units) gives us a simple one-step conversion between moles and units. In principle the following example is no more complicated than telling how many eggs are in four dozen.

EXAMPLE 7.11 How many sodium atoms are in 4.00 mol of sodium?

$$4.00 \text{ mol Na} \times \frac{6.02 \times 10^{23} \text{ Na atoms}}{1 \text{ mol Na}} = 24.1 \times 10^{23} \text{ Na atoms} = 2.41 \times 10^{24} \text{ Na atoms}$$

To find the number of units of a substance in a given mass, you must begin the setup one step earlier with a grams → moles conversion.

EXAMPLE 7.12 How many molecules are in 500 grams of water (about one pint)?

SOLUTION: Using Equation 7.4 first, and then Equation 7.1, the unit path begins with the given quantity: grams → moles → molecules.

$$500 \text{ g } H_2O \times \frac{1 \text{ mol } H_2O}{18.0 \text{ g } H_2O} \times \frac{6.02 \times 10^{23} \text{ } H_2O \text{ molecules}}{1 \text{ mol } H_2O} = 1.67 \times 10^{25} \text{ } H_2O \text{ molecules}$$

In the reverse problem, finding the mass of a given number of units, the procedure is logically reversed. . . .

EXAMPLE 7.13 What is the mass of one billion billion (1.00×10^{18}) molecules of ammonia, NH_3?

At one point the molar mass of ammonia will be required. Find it first.

17.0 g NH_3/mol

Now solve the problem.

$$1.00 \times 10^{18} \text{ } NH_3 \text{ molecules} \times \frac{1 \text{ mol } NH_3}{6.02 \times 10^{23} \text{ } NH_3 \text{ molecules}} \times \frac{17.0 \text{ g } NH_3}{1 \text{ mol } NH_3} = 2.82 \times 10^{-5} \text{ g } NH_3$$

This very small mass, about $\frac{6}{100\,000\,000}$ of a pound, suggests again the enormous number of molecules in a mole.

The next example shows how important it is to identify chemical substances by formula.

EXAMPLE 7.14 (a) How many atoms of chlorine are in 48.5 grams of chlorine atoms, Cl? (b) How many atoms of chlorine are in 48.5 grams of chlorine gas, Cl_2?

Part (a) is straightforward. Set up and solve.

$$48.5\ \cancel{g\ Cl} \times \frac{1\ \cancel{mol\ Cl}}{35.5\ \cancel{g\ Cl}} \times \frac{6.02 \times 10^{23}\ Cl\ atoms}{1\ \cancel{mol\ Cl}} = 8.22 \times 10^{23}\ Cl\ atoms$$

In Part (b), the idea is the same, but there is an important difference. You might find it helpful to write the unit path. Then set up and solve the problem.

unit path: g Cl_2 → mol Cl_2 → Cl_2 molecules → Cl atoms

$$48.5\ \cancel{g\ Cl_2} \times \frac{1\ \cancel{mol\ Cl_2}}{71.0\ \cancel{g\ Cl_2}} \times \frac{6.02 \times 10^{23}\ \cancel{Cl_2\ molecules}}{1\ \cancel{mol\ Cl_2}} \times \frac{2\ Cl\ atoms}{1\ \cancel{Cl_2\ molecule}} = 8.22 \times 10^{23}\ Cl\ atoms$$

Answers (a) and (b) must be the same. 48.5 grams of chlorine must have the same number of atoms no matter how they are "packaged." Notice in (a), where you work with chlorine atoms, formula Cl, you use the mass of one mole of chlorine atoms—the atomic mass, 35.5 grams. In (b) you work with chlorine molecules, formula Cl_2. You therefore use the molar mass of chlorine molecules—the molecular mass, 71.0 grams. In each case the mass used matches the formula of the substance.

If you write the chemical formula for the species whose molar mass you are using—and calculate the molar mass straight from the chemical formula—your setups will be correct. *Always include the chemical formula in the dimensional analysis setup.*

7.8 SIMPLEST (EMPIRICAL) FORMULA OF A COMPOUND

Simplest Formulas and Molecular Formulas

PG 7I Distinguish between a simplest formula and a molecular formula.

The percentage composition of the compound ethene (also called ethylene) is 85.7% carbon and 14.3% hydrogen. Its chemical formula is C_2H_4. The percentage composition of the compound propene (also called propylene) is likewise 85.7% carbon and 14.3% hydrogen. Its formula is C_3H_6. These are, in fact, two of a whole series of compounds having the general formula C_nH_{2n}, where n is an integer. In ethene and propene, n = 2 and 3, respectively. All compounds in this series have the same percentage composition.

C_2H_4 and C_3H_6 are typical **molecular formulas.** They show the number of carbon and hydrogen atoms actually present in a molecule of ethene and propene. They are formulas of real chemical substances. If, in the general formula C_nH_{2n}, we let n be 1, the resulting formula is CH_2. This formula is the **simplest formula** for all compounds having the general formula C_nH_{2n}. The **simplest formula shows the simplest ratio of atoms of the elements in the compound.** All subscripts are reduced to their lowest terms: they have no common divisors. Simplest formulas can be determined from percentage composition data—as we will shortly see—which are found by chemical analysis. Hence their other name, **empirical formula,** which means a formula based on experience or experiment.

Simplest formulas may or may not be molecular formulas of real chemical compounds. There happens to be no known stable compound that has the formula CH_2—and there is good reason to believe that no such compound can exist. On the other hand, the molecular formula of dinitrogen tetroxide is N_2O_4. The subscript numbers have a common divisor, 2. Dividing by 2 we reach the simplest formula, NO_2. This simplest formula is also the molecular formula of a real chemical compound, nitrogen dioxide. In other words, NO_2 is *both* the simplest formula and the molecular formula for nitrogen dioxide, as well as the simplest formula for dinitrogen tetroxide.

EXAMPLE 7.15 For each formula below that could be a simplest formula, write EF; for each formula that could not be a simplest formula, write the simplest formula corresponding to the formula shown: C_4H_{10}; C_2H_6O; Hg_2Cl_2; $(CH)_6$.

C_4H_{10}: C_2H_5; C_2H_6O: EF; Hg_2Cl_2: HgCl; $(CH)_6$: CH

Determination of a Simplest Formula

PG 7J Given data from which the ratio of relative masses of the elements in a compound can be determined, write the simplest formula of the compound.

To determine the simplest formula of a compound from its percentage composition we must find the whole number ratio of atoms of the elements in the compound. The numbers in this ratio become the subscripts in the simplest formula of the compound. The procedure by which this is done is as follows:

1. Determine the relative masses of different elements in the compound.
2. Convert these relative masses to the relative numbers of moles of atoms of the elements.
3. Express the number of moles of atoms as the smallest possible ratio of integers.

The integers found in the last step are the subscripts in the simplest formula.

It is usually helpful in a simplest formula problem to organize the calculations in a table. The headings in the table will be

ELEMENT	GRAMS	MOLES	MOLE RATIO	FORMULA RATIO	SIMPLEST FORMULA

We will illustrate the procedure for finding simplest formulas with the compound ethene. As noted, ethene is 85.7% carbon and 14.3% hydrogen. The first step in the above procedure calls for converting the percentage figures into relative masses of the elements. Thinking of percent as the number of parts of one element per 100 parts of the compound—or number of *grams* of one element per 100 *grams* of the compound—it follows that a 100-g sample of the unknown contains 85.7 g of carbon and 14.3 g of hydrogen. From this we see that *percentage composition figures may always be interpreted directly as the relative masses of different elements* in satisfying the first step of the procedure. The elemental symbol and the relative grams of each element are entered into the table:

ELEMENT	GRAMS	MOLES	MOLE RATIO	FORMULA RATIO	SIMPLEST FORMULA
C	85.7				
H	14.3				

We are now ready to find the number of moles of atoms of each element, Step 2 in the procedure. This is a direct one-step conversion from grams to moles, as in Section 7.6.

ELEMENT	GRAMS	MOLES	MOLE RATIO	FORMULA RATIO	SIMPLEST FORMULA
C	85.7	$\frac{85.7 \text{ g C}}{12.0 \text{ g/mol}} = 7.14$			
H	14.3	$\frac{14.3 \text{ g H}}{1.0 \text{ g/mol}} = 14.3$			

Students sometimes question the conversion of grams of hydrogen, or any elemental gas that forms diatomic molecules, to moles of atoms. They tend to divide by the molar mass of *molecules*—the *molecular* mass—rather than the molar mass of *atoms*—the *atomic* mass. But a chemical formula expresses the ratio of moles of *atoms* of the different elements, so the molar mass of atoms, or atomic mass, must be used.

It is the ratio of these moles of atoms that must now be expressed in the smallest possible ratio of integers, Step 3 in the procedure. This is most easily done by *dividing each number of moles by the smallest number of moles*. In this problem the smallest number of moles is 7.14. Thus,

ELEMENT	GRAMS	MOLES	MOLE RATIO	FORMULA RATIO	SIMPLEST FORMULA
C	85.7	7.14	$\frac{7.14}{7.14} = 1.00$		
H	14.3	14.2	$\frac{14.2}{7.14} = 1.99$		

The ratio of *atoms* of the elements in a compound is the same as the ratio of *moles* of atoms in the compound. To see this in a more familiar setting, the ratio of handlebars to wheels in bicycles is ½. In four dozen bicycles there are four dozen handlebars and eight dozen wheels. The ratio of dozens is 4/8, which is the same as ½. Thus the numbers in the Mole Ratio column are in the same ratio as the subscripts in the simplest formula.

The simplest formula subscripts must be whole numbers. The numbers in the Mole Ratio column must therefore be expressed as a ratio of integers. In this case the ratio 1.00/1.99 becomes ½, and the empirical formula is CH_2.

ELEMENT	GRAMS	MOLES	MOLE RATIO	FORMULA RATIO	SIMPLEST FORMULA
C	85.7	7.14	1.00	1	
H	14.2	14.2	1.99	2	CH_2

It is often necessary to adjust a number in the Mole Ratio column by a few hundredths to get an integer, as 1.99 is adjusted to 2 in this example. These minor changes correct for experimental errors or significant figure roundoffs.

If either quotient in the Mole Ratio column is not very close to a whole number, the Formula Ratio may be found by multiplying both quotients by a small integer. This is shown in the next example.

EXAMPLE 7.16 The mass of a piece of iron (Z = 26) is 1.34 g. Exposed to oxygen under conditions in which oxygen combines with all of the iron to form a pure oxide of iron, the final mass increases to 1.92 g. Find the simplest formula of the compound.

As before, relative weights of the elements contained in the compound are required—but this time they are not obtained from percentage composition values. The number of grams of iron in the final compound is given in the problem. How many grams of oxygen combine with 1.34 g of iron if the iron oxide produced weighs 1.92 g?

-----

grams oxygen = grams iron oxide − grams iron

= 1.92 grams − 1.34 grams = 0.58 gram

The table is started below, and the symbols and masses of elements are entered. Step 2 is to compute the number of moles of atoms of each element. Do so, and put the results into the table.

ELEMENT	GRAMS	MOLES	MOLE RATIO	FORMULA RATIO	SIMPLEST FORMULA
Fe	1.34				
O	0.58				

ELEMENT	GRAMS	MOLES	MOLE RATIO	FORMULA RATIO	SIMPLEST FORMULA
Fe	1.34	$\dfrac{1.34 \text{ g Fe}}{55.8 \text{ g/mol}} = 0.0240$			
O	0.58	$\dfrac{0.58 \text{ g O}}{16.0 \text{ g/mol}} = 0.036$			

Recalling that the mole ratio figures are obtained by dividing each number of moles by the smallest, find those numbers and place them in the table.

ELEMENT	GRAMS	MOLES	MOLE RATIO	FORMULA RATIO	SIMPLEST FORMULA
Fe	1.34	0.0240	$\dfrac{0.0240}{0.0240} = 1.00$		
O	0.58	0.036	$\dfrac{0.036}{0.0240} = 1.5$		

This time the numbers in the mole ratio column are not both integers or very close to integers. But they can be changed to integers and kept in the same ratio by multiplying both of them by the same small integer. Find the smallest whole number that will yield integers when used as a multiplier for 1.00 and 1.5, use it to obtain the formula ratio figures, complete the table, and write the simplest formula of the compound.

ELEMENT	GRAMS	MOLES	MOLE RATIO	FORMULA RATIO	SIMPLEST FORMULA
Fe	1.34	0.0240	1.00	2	Fe_2O_3
O	0.58	0.036	1.5	3	

Multiplication of the mole ratio numbers by 2 yields $1.00 \times 2 = 2$ and $1.5 \times 2 = 3$, both whole numbers.

The decimal part of a mole ratio determines what multiplier to use. If the mole ratio is about 1.33 or 1.67, multiplication by 3 yields 4 or 5:

$$1.33 \times 3 = 4 \quad or \quad 1.67 \times 3 = 5$$

A mole ratio of 1.25 or 1.75 should be multiplied by 4 to get 5 or 7:

$$1.25 \times 4 = 5 \quad or \quad 1.75 \times 4 = 7$$

You are not likely to find more complicated ratios in the beginning course.

The procedure is the same for compounds containing more than two elements.

EXAMPLE 7.17 An organic compound is found to contain 20.0% carbon, 2.2% hydrogen, and 77.8% chlorine. Determine the simplest formula of the compound.

ELEMENT	GRAMS	MOLES	MOLE RATIO	FORMULA RATIO	SIMPLEST FORMULA

ELEMENT	GRAMS	MOLES	MOLE RATIO	FORMULA RATIO	SIMPLEST FORMULA
C	20.0	1.67	1.00	3	
H	2.2	2.2	1.3	4	$C_3H_4Cl_4$
Cl	77.8	2.19	1.31	4	

Notice that the experimental results in this problem yield mole ratio figures 1.3 and 1.31, both close to .33 in the decimal part of the number. Multiplication of the mole ratio figures by 3 yields integers for the formula ratio column: $1.00 \times 3 = 3$; $1.3 \times 3 = 3.9$ or 4; $1.31 \times 3 = 3.93$ or 4.

Determination of a Molecular Formula

PG 7K Given the molar mass of a compound and data from which its simplest formula can be determined, write the molecular formula of the compound.

The determination of a simplest formula is an important step in finding the molecular formula of an unknown compound. Suppose, for example, the simplest formula of a compound is CH_2, and a different experiment has shown that its molar mass is 70.0 g/mol. The molecular formula is made up of a certain number of multiples of simplest formula units. If the multiple is one, the molecular formula is CH_2; if two, C_2H_4; if three, C_3H_6; and so forth, with the number of H atoms always being twice the number of C atoms. If n is the number of multiples, the molecular formula is C_nH_{2n}.

If you calculate the molar mass of the simplest formula and divide it into the molar mass of the compound, the result will be the number of simplest formula units in the molecular formula—the n that is needed. The molar mass of CH_2 is $12.0 + 2.02 = 14.0$ g/mol. $70.0 \div 14.0 = 5$ simplest formula units in one molecule. The molecular formula is therefore C_5H_{10}.

This determination can be expressed in an equation. If SF is the simplest formula unit, MF is the molecular formula, and MM is molar mass,

$$\text{simplest formula units in 1 molecule} = \frac{\text{MM of MF}}{\text{MM of SF}} \tag{7.5}$$

EXAMPLE 7.18 An unknown compound is found in the laboratory to be 91.8% silicon and 8.2% hydrogen. Another experiment indicates that the molar mass of the compound is 122 g/mol. Find the simplest and molecular formulas of the compound.

Start by finding the simplest formula.

ELEMENT	GRAMS	MOLES	MOLE RATIO	FORMULA RATIO	SIMPLEST FORMULA

ELEMENT	GRAMS	MOLES	MOLE RATIO	FORMULA RATIO	SIMPLEST FORMULA
Si	91.8	3.27	1.00	2	Si_2H_5
H	8.2	8.2	2.5	5	

To use Equation 7.5, you must have the molar mass of the simplest formula unit. Find the molar mass of Si_2H_5.

61.2 g Si_2H_5/mol

Calculate the number of simplest formula units in the molecule and write the molecular formula.

$n = 122/61.2 = 2 \qquad (Si_2H_5)_2 = Si_4H_{10}$

In Chapter 13 you will learn one way to find the molar mass of an unknown compound from experimental data. You will then have a complete picture of how a chemical formula may be determined.

CHAPTER 7 IN REVIEW

7.1 Molecular Mass; Formula Mass

7A Distinguish between atomic mass, molecular mass, and formula mass.

7.2 The Mole Concept

7B Define *mole*. Identify the number that corresponds to one mole.

7.3 Molar Mass

7C Define molar mass.

7.4 Calculation of Molar Mass

7D Calculate the molar mass of any substance whose chemical formula is known.

7.5 Mass Relationship Between Elements in a Compound; Percentage Composition

7E For any compound whose formula is known, given the mass of a sample, calculate the mass of any element in the sample. Or, given the mass of any element in the sample, calculate the mass of the sample or the mass of any other element in the sample.

7F Calculate the percentage composition of any compound whose formula is known.

7.6 Conversion Between Mass and Number of Moles

7G Given the number of grams (or moles) of a chemical species of known or calculable molar mass, find the number of moles (or grams).

7.7 Number of Atoms, Molecules or Formula Units in a Sample

7H Given the mass or number of moles of a pure substance whose formula is known, find the number of atoms, molecules, or formula units. Or, given the number of atoms, molecules, or formula units, find the number of moles or mass.

7.9 Simplest Empirical Formula of a Compound

7I Distinguish between a simplest formula and a molecular formula.

7J Given data from which the ratio of relative masses of elements in a compound can be determined, write the simplest formula of the compound.

7K Given the molar mass of a compound and data from which the simplest formula can be determined, write the molecular formula of the compound.

TERMS AND CONCEPTS

7.1 Formula mass
Formula unit
Molecular mass
7.2 Mole
Avogadro's number
7.3 Molar mass
7.5 Percentage composition
7.8 Molecular formula
Simplest formula
Empirical formula

Most of these terms and many more are defined in the Glossary. Use your Glossary regularly.

QUESTIONS AND PROBLEMS

Only names are given for those compounds whose formulas were introduced in Chapter 6. For references, you may find all of these formulas on page ANS-10. You are encouraged to write formulas with no assistance other than a periodic table, using the reference page only when necessary.

Section 7.1

(*1*) It may be said that because atomic, molecular, and formula masses are all comparative masses, they are conceptually alike. What, then, is their difference?

(32) Why is it proper to speak of the molecular mass of water, but not the molecular mass of sodium nitrate?

(*2*) In what units are atomic, molecular, and formula mass expressed? Define those units.

(33) Which of the three terms, *atomic mass, molecular mass,* or *formula mass,* is most appropriate for each of the following: NH_3; CaO; Ba; N_2; Na_2CO_3?

(3)* Before 1961, when the present scale of atomic mass was adopted, the atomic *weight* of an element was defined as the average weight of an atom compared to the average weights of oxygen atoms at exactly 16. Compare the numerical values of atomic weights by that definition with those of atomic mass as we use the term today.

(34)* Before 1961, physicists based their values of atomic weight (see Question 3) on the weight of an atom of oxygen-16 as exactly 16. Chemists based their values on the average weight of natural oxygen isotopes as exactly 16. Which scale had the higher values, and how much higher were they (by what ratio or percentage were they higher)? (Table 4.2 will help.)

Section 7.2

(4) Explain the meaning of the term *mole*. Why is it used in chemistry?

(5) Give the name and value of the number associated with the mole.

(6) How many grams of carbon have the same number of atoms as 64.12 g of sulfur?

(35) What do quantities representing 1 mol of iron, 1 mol of ammonia, and 1 mol of calcium carbonate (limestone) have in common?

(36) Is the mole a number? Explain.

(37)* Define Avogadro's number. Why has its value changed over the years?

Section 7.3

(7) How does molar mass differ from molecular mass? What about them is the same?

(8)* What is the molar mass of oxygen? You should be uncertain about which of two answers to give for that question. Give both, and state the conditions under which each is correct.

(38) What is the molar mass of phosphorus?

(39) What is the molar mass of nitrogen, which makes up nearly 80% of the atmosphere as an elemental gas?

Section 7.4

For Problems 9 and 40, calculate to three significant figures the molar mass of each substance listed.

(9) (a) Potassium iodide
(b) Hydrogen
(c) Sodium nitrate
(d) Magnesium phosphate
(e) Barium hydroxide 8-hydrate
(f) Propanol, C_3H_7OH
(g) Copper(II) sulfate, $CuSO_4$
(h) Chromium(III) chloride, $CrCl_3$
(i) Aluminum oxalate 4-hydrate, $Al_2(C_2O_4)_3 \cdot 4\ H_2O$
(j) Sodium acetate, $NaC_2H_3O_2$

(40) (a) Lithium chloride
(b) Bromine
(c) Aluminum carbonate
(d) Ammonium sulfate
(e) Calcium chloride 2-hydrate
(f) Butane, C_4H_{10}
(g) Silver nitrate, $AgNO_3$
(h) Manganese dioxide, MnO_2
(i) Cobalt(II) chloride 6-hydrate, $CoCl_2 \cdot 6\ H_2O$
(j) Zinc phosphate, $Zn_3(PO_4)_2$

Section 7.5

(10) Lithium fluoride is used as a flux when welding or soldering aluminum. How many grams of lithium are in 1 lb (454 g) of lithium fluoride?

(11) Potassium sulfate is found in some fertilizers as a source of potassium. How many grams of potassium can be obtained from 16.3 g of the chemical?

(12) Zinc cyanide, $Zn(CN)_2$, is an important compound in zinc electroplating. How many grams of compound must be dissolved in a test bath in a laboratory to introduce 158 g of zinc into the solution?

(13) Wulfenite, $PbMoO_4$, is an ore from which molybdenum, an important element used in making

(41) Ammonium bromide is a raw material in the manufacture of photographic film. What mass of bromine is found in 7.50 g of the compound?

(42) Magnesium oxide is used in making bricks to line very high temperature furnaces. If a brick contains 1.82 kg of the oxide, what is the mass of magnesium in the brick?

(43) Calculate the grams of oxygen in 445 g of table sugar, $C_{12}H_{22}O_{11}$.

(44) Acetone, CH_3COCH_3, is a solvent that is widely used in manufacturing many organic chemicals.

steel alloys, is extracted. Calculate the mass of molybdenum (Z = 42) that may be obtained from 78.0 kg of wulfenite.

(14) How large a sample of the insecticide calcium chlorate, $Ca(ClO_3)_2$, must be set aside if the sample is to contain 2.29 g of combined chlorine?

How many grams of hydrogen are in 87.1 g CH_3COCH_3?

(45) Strontium nitrate, $Sr(NO_3)_2$, is responsible for the red colors in firework displays. What is the largest mass of strontium nitrate that can be used for a purpose that can contain no more than 7.86 g of nitrogen?

For Problems 15 and 46, calculate the percentage composition of each of the following compounds, selected from Problems 9 and 40:

(15) (a) sodium nitrate
(b) magnesium phosphate
(c) chromium(III) chloride, $CrCl_3$
(d) copper(II) sulfate, $CuSO_4$

(46) (a) silver nitrate, $AgNO_3$
(b) ammonium sulfate
(c) aluminum carbonate
(d) manganese(IV) oxide, MnO_2

(16) Find the percentage of water in aluminum oxalate 4-hydrate? (See Problem 9i.)

(47) Find the percentage of cobalt(II) chloride 6-hydrate that is the anhydrous compound. (See Problem 40 i.)

(17)* A chemical process requires nickel ion in solution, which is to be added in the form of a soluble nickel compound. The manager has a choice of using nickel sulfate 6-hydrate at $1.48/lb or nickel chloride 6-hydrate at $2.13/lb. For each compound, calculate the price per kilogram of nickel to determine which is the better buy.

(48)* Sodium acetate solutions are used in equilibrium experiments in college chemistry laboratories. The anhydrous compound is priced at $25.70/kg, but the trihydrate $NaC_2H_3O_2 \cdot 3\ H_2O$ is much less expensive at $17.45/kg. If cost per unit quantity of sodium acetate is the only consideration, should the stockroom manager buy the anhydrous compound or the trihydrate?

Section 7.6

For Problems 18 and 49, find the number of moles for each mass of substance selected from Problems 9 and 40.

(18) (a) 8.59 g H_2
(b) 9.85 g KI
(c) 427 g $CuSO_4$
(d) 22.8 g $Al_2(C_2O_4)_3 \cdot 4\ H_2O$

(49) (a) 70.3 g C_4H_{10}
(b) 25.9 g $Zn_3(PO_4)_2$
(c) 0.405 g MnO_2
(d) 827 g $CoCl_2 \cdot 6\ H_2O$

For Problems 19 and 50, calculate the mass for each given number of moles of substances selected from Problems 9 and 40.

(19) (a) 0.581 mol C_3H_7OH
(b) 4.28 mol $NaC_2H_3O_2$
(c) 0.0913 mol $Mg_3(PO_4)_2$
(d) 0.148 mol $Ba(OH)_2 \cdot 8\ H_2O$

(50) (a) 0.0551 mol Br_2
(b) 0.0898 mol $AgNO_3$
(c) 3.40 mol $(NH_4)_2SO_4$
(d) 0.0876 mol $CaCl_2 \cdot 2\ H_2O$

Section 7.7

For Problems 20 and 51, determine how many atoms, molecules, or formula units are in each of the following:

(20) (a) 0.818 mol K
(b) 70.3 g C_2H_6
(c) 3.78 g NH_4Cl
(d) 0.629 mol Al
(e) 6.57 g $Ca(NO_3)_2$
(f) 0.184 mol CS_2

(51) (a) 3.84 mol CH_4
(b) 7.39 g Ne
(c) 0.469 g LiBr
(d) 0.0342 mol NaOH
(e) 24.3 g CH_3OH
(f) 309 g $Mg(OH)_2$

For Problems 21 and 52, calculate the number of moles in each of the following:

(21) (a) 1.84×10^{22} Ar atoms
(b) 9.24×10^{24} formula units of KOH

(52) (a) 3.71×10^{21} C_2H_2 molecules
(b) 5.01×10^{23} formula units of $NaNO_3$

For Problems 22 and 53, calculate the mass of

(22) (a) 4.11×10^{22} molecules of N_2O
(b) 1.03×10^{24} K atoms

(53) (a) 6.24×10^{21} Pb atoms
(b) 7.01×10^{23} formula units of $CaCl_2$

(23) How many atoms of carbon has a young man given his bride-to-be if the engagement ring has a 0.500-carat diamond? There are 200 mg in a carat. (The price of diamonds doesn't seem so high when figured at dollars per atom.)

(24) 495 g is a typical mass for a glass of water. Calculate the number of water molecules you drink when quenching your thirst with this quantity of our most common liquid.

(25) (a) How many molecules are in 3.61 g F_2?
(b) How many atoms are in 3.61 g F_2?
(c) How many atoms are in 3.61 g F?
(d) What is the mass of 3.61×10^{23} atoms of F?
(e) What is the mass of 3.61×10^{23} molecules of F_2?

(54) On June 29, 1984, the financial pages quoted the price of gold at \$373/troy oz (1 troy oz = 31.1 g). What is the price of a single atom of gold (Z = 79)?

(55) One who sweetens coffee with two teaspoons of sugar, $C_{12}H_{22}O_{11}$, uses about 0.600 g. How many sugar molecules is this?

(56) (a) What is the mass of 5.24×10^{24} N atoms?
(b) What is the mass of 5.24×10^{24} N_2 molecules?
(c) How many atoms are in 52.4 g N?
(d) How many molecules are in 52.4 g N_2?
(e) How many atoms are in 52.4 g N_2?

Section 7.8

(26) Explain why C_6H_{10} must be a molecular formula, while C_7H_{10} could be a molecular formula, a simplest formula, or both.

(27) A certain compound is 52.2% carbon, 13.0% hydrogen, and 34.8% oxygen. Find the simplest formula of the compound.

(28) 11.89 g of iron are exposed to a stream of oxygen until they react to product 16.99 g of a pure oxide of iron. What is the simplest formula of the product?

(29) A compound is 17.2% C, 1.44% H, and 81.4% F. Find its simplest formula.

(30) A coolant widely used in automobile engines is 38.7% carbon, 9.7% hydrogen, and 51.6% oxygen. Its molar mass is 62.0 g/mol. What is the molecular formula of the compound?

(31) 73.1% of a compound is chlorine, 24.8% is carbon, and the balance is hydrogen. If the molar mass of the compound is 97 g/mol, find the molecular formula.

(57) From the following list, identify each formula that could be a simplest formula. Write the simplest formulas of the other compounds. C_2H_6O; Na_2O_2; $C_2H_4O_2$; N_2O_5.

(58) A compound analyzes 29.1% sodium, 40.5% sulfur, and 30.4% oxygen. Calculate the simplest formula of the compound.

(59) A 6.49-g sample of a compound of nitrogen and oxygen is found to contain 4.80 g of oxygen. Find the simplest formula of the compound.

(60) What is the simplest formula of a compound that is 19.2% sodium, 1.7% hydrogen, 25.8% phosphorus, and 53.3% oxygen.

(61) 88 g/mol is the molar mass of a compound that is 54.6% C, 9.0% H, and 36.4% oxygen. Find the molecular formula of the compound.

(62) A compound that is 23.1% aluminum, 15.4% carbon, and 61.5% oxygen has a molar mass of 234 g/mol. What is the "molecular formula" of the compound, which is actually an ionic substance.

Miscellaneous Questions

(63) Distinguish precisely and in scientific terms the differences between items in each of the following groups:
(a) Atomic mass, molecular mass, formula mass, molar mass
(b) Mole, molecule
(c) Mole, Avogadro's number
(d) Molecular formula, simplest formula

(64) Classify each of the following statements as true or false:
(a) The term *molecular mass* applies mostly to ionic compounds.
(b) Molar mass is measured in atomic mass units.
(c) In its practical application, a *mole* represents Avogadro's number.
(d) Grams are larger than atomic mass units; therefore molar mass is numerically larger than atomic mass.
(e) The molar mass of hydrogen is read directly from the periodic table, whether it be monatomic hydrogen, H, or hydrogen gas, H_2.
(f) A simplest formula is always a molecular formula, although a molecular formula may or may not be simplest formula.

(65) Would you need a truck to transport 10^{25} atoms of copper?

(66) The stable form of elemental phosphorus is a tetratomic molecule. Calculate the number of molecules and atoms in 85.0 g P_4.

(67)* The quantitative significance of "take a deep breath" varies, of course, with the individual. When one person did so, he found that he inhaled 2.95×10^{22} molecules of the mixture of nitrogen and oxygen we call air. Assuming this mixture has an average molar mass of 29 g/mol, what is his apparent lung capacity in grams of air?

(68)* Assuming gasoline to be pure octane, C_8H_{18}—actually it is a mixture of iso-octane and other hydrocarbons—an automobile getting 25.0 miles/gal would consume 5.62×10^{23} molecules per mile. Calculate the mass of this amount of fuel.

(69)* 27.65 g of a certain compound containing only carbon and hydrogen are burned completely in oxygen. All the carbon is changed to 86.9 g of CO_2 and all the hydrogen is changed to 35.5 g of H_2O. What is the simplest formula of the original compound? (Hint: Find the grams of carbon and hydrogen in the original compound.)

(70)* A certain hydrate has the general formula $Co_aS_bO_c \cdot X\ H_2O$. 43.0 g of the compound are heated to drive off the water, leaving 26.1 g of anhydrous compound. Further analysis shows that the percentage composition of the anhydride is 42.4% Co, 23.0% S, and 34.6% O. Find the simplest formula of (a) the anhydrous compound and (b) the formula of hydrate.

(71) The use of marijuana forms a compound that is 76.4% carbon, 9.2% hydrogen, and 14.5% oxygen. Its molar mass is 330 g/mol. What is the molecular formula of the compound?

8 Chemical Reactions and Equations

LOOKING BACK

2.3 and 6.1 Elements and compounds are represented by formulas that tell how many atoms of each element are present in the smallest unit of the substance.

2.6 The Law of Conservation of Mass is that the total mass of the products of a reaction is equal to the total mass of the reactants.

LOOKING AHEAD IN CHAPTER 8

Chemical changes are represented by chemical equations that show the formulas of the reactants and the products.

The number of atoms of each element must be balanced on the two sides of a chemical equation.

Some examples in Chapter 8 continue to encourage the development of formula-writing skills by furnishing only the names of compounds whose formulas were introduced in Chapter 6. If you did not study Chapter 6, skip the opening question of these examples and start with the formulas given. Recall that the names and formulas of ionic compounds from Chapter 6 are listed on page ANS-10.

8.1 EVOLUTION OF A CHEMICAL EQUATION

Sodium reacts vigorously with water, producing hydrogen gas and a solution of sodium hydroxide, and releasing heat (Fig. 8.1). A chemist will reduce that 17-word description of a chemical change to "sodium plus water yields hydrogen plus sodium hydroxide solution plus heat." In writing, the chemist substitutes symbols for words and produces a **chemical equation:**

$$Na + H_2O \rightarrow H_2 + NaOH(aq) + heat \tag{8.1}$$

The substances on the left side of the arrow are called **reactants,** and those on the right are the **products.** The arrow is read "yields" or "produces."

The (aq) after the formula of sodium hydroxide indicates the substance is in aqueous (water) solution. This is an example of how the "state" of a species is indicated in an equation. The conventional symbols for true states of matter are (g) for gas, (ℓ) for liquid, and (s) for solid. Although they are often omitted unless important in describing a reaction, they will generally be included in equations in this book.

Figure 8.1
Sodium reacting with water. A. Small piece of sodium dropped into a beaker of water. B. Sodium forms a "ball" that dashes erratically over water surface, releasing hydrogen as it reacts. C. Solution of sodium hydroxide, NaOH(aq), which is hot because of heat released in the reaction. WARNING: Do not "try" this experiment, as it is dangerous, potentially splattering hot alkali into eyes and onto skin and clothing.

Nearly all chemical reactions involve some energy transfer, usually in the form of heat, as shown in Equation 8.1. Energy terms are generally omitted from equations unless there is a specific reason for including them. Equations that include the gain or loss of heat are considered in Chapter 9.

Eliminating the energy term from and adding state designations to Equation 8.1 gives

$$Na(s) + H_2O(\ell) \rightarrow H_2(g) + NaOH(aq) \qquad (8.2)$$

The Law of Conservation of Mass (Section 2.6) says the total mass of the products of a reaction is the same as the total mass of the reactants. Dalton provided for this in his theory by saying atoms were neither created nor destroyed in a chemical change, but that the same number of atoms of the different elements were simply rearranged. Equation 8.2 does not satisfy this condition. There are two atoms of hydrogen in H_2O on the left side of the equation, but three atoms of hydrogen on the right—two in H_2 and one in NaOH. The equation is not "balanced."

An equation is balanced by placing a coefficient in front of one or more of the formulas, indicating it is used more than once. Hydrogen is short on the left side of Equation 8.2, so let's try two water molecules:

$$Na(s) + 2\ H_2O(\ell) \rightarrow H_2(g) + NaOH(aq) \qquad (8.3)$$

At first glance, this hasn't helped; indeed it seems to have made matters worse. The hydrogen is still out of balance (four on the left, three on the right) and, furthermore, oxygen is now unbalanced (two on the left, one on the right). We are short one oxygen and one hydrogen atom on the right side. But look closely. Oxygen and hydrogen are part of the same compound on the right, and there is one atom of each in that compound. If we take two NaOH units

$$Na(s) + 2\ H_2O(\ell) \rightarrow H_2(g) + 2\ NaOH(aq) \qquad (8.4)$$

there are four hydrogens and two oxygens on both sides of the equation. These elements are now in balance. But, alas, the sodium has been *un*balanced.

Correction of this condition, however, should be obvious:

$$2\ Na(s) + 2\ H_2O(\ell) \rightarrow H_2(g) + 2\ NaOH(aq) \qquad (8.5)$$

The equation is now balanced. Note that, in the absence of a numerical coefficient, as with H_2, the coefficient is assumed to be 1.

Balancing an equation involves some important do's and don'ts that are apparent in this example:

DO: *Balance the equation entirely by use of coefficients placed before the different chemical formulas.*

DON'T: *Change a correct chemical formula in order to make an element balance.*

DON'T: *Add some real or imaginary chemical species to either side of the equation just to make an element balance.*

There is a great temptation toward either of the "don'ts." A moment's thought should show why they are improper. The original equation expresses the *correct* formula for each species present. If you change or add a formula, even if the new formula is for a real chemical, it is *not a species in the reaction* and does not belong in the equation. Coefficients alone must be used in balancing equations.

Quick Check 8.1

Are the following true or false?

1. The equation $C_2H_4O + 3\ O_2 \rightarrow 2\ CO_2 + 2\ H_2O$ is balanced.
2. The unbalanced equation $H_2 + O_2 \rightarrow H_2O$ may be balanced by changing it to $H_2 + O_2 \rightarrow H_2O_2$.

8.2 THE QUANTITATIVE MEANING OF A CHEMICAL EQUATION

PG 8A Given a chemical equation, interpret it in terms of (a) atoms, molecules, and/or formula units and (b) moles.

Quantitatively, Equation 8.5 can be interpreted in two ways. The "molecular" meaning of the equation is that two sodium atoms react with two water molecules to produce one hydrogen molecule and two sodium hydroxide units. Each formula in the equation represents one unit of that substance. It may be an atom, a molecule, or a formula unit. The coefficient in front of the formula tells how many such units are involved, relative to the indicated numbers of units of other species.

As both sides of an algebraic equation may be multiplied by the same number, so may both sides of a chemical equation. Multiplying by 2,

$$4\ Na(s) + 4\ H_2O(\ell) \rightarrow 2\ H_2(g) + 4\ NaOH(aq) \qquad (8.6)$$

The number in a dozen, 12, is also a good multiplier:

$$2(12)\ Na(s) + 2(12)\ H_2O(\ell) \rightarrow 1(12)\ H_2(g) + 2(12)\ NaOH(aq)$$

This could be read or written correctly as

$$2 \text{ dozen } Na(s) + 2 \text{ dozen } H_2O(\ell) \rightarrow 1 \text{ dozen } H_2(g) + 2 \text{ dozen } NaOH(aq)$$

Similarly,

$$2(6.02 \times 10^{23})\ Na(s) + 2(6.02 \times 10^{23})\ H_2O(\ell) \rightarrow 1(6.02 \times 10^{23})\ H_2(g) + 2(6.02 \times 10^{23})\ NaOH(aq)$$

may be thought of as

$$2 \text{ moles } Na(s) + 2 \text{ moles } H_2O(\ell) \rightarrow 1 \text{ mole } H_2(g) + 2 \text{ moles } NaOH(aq)$$

We may refer to this last equation as a "molar" interpretation of a chemical equation. With the understanding that the coefficients refer to *moles* rather than atoms, molecules, or formula units, the interpretation of the equation may be applied directly to Equation 8.5.

The molar interpretation provides another way to balance chemical equations. Returning to the unbalanced equation,

$$Na(s) + H_2O(\ell) \rightarrow H_2(g) + NaOH(aq) \qquad (8.2)$$

we find that the left side of the equation has two moles of hydrogen atoms, whereas the right side has three moles of hydrogen atoms. A single coefficient, ½, applied to hydrogen balances the equation:

$$Na(s) + H_2O(\ell) \rightarrow \tfrac{1}{2}\, H_2(g) + NaOH(aq) \qquad (8.7)$$

This says that one mole of sodium reacts with one mole of water to form half a mole of hydrogen plus one mole of sodium hydroxide. There are now two moles of hydrogen atoms on each side of the equation, which is balanced in other atoms as well. From a molecular standpoint you cannot properly think of half of a hydrogen molecule or any fractional part of any atom or molecule any more than you can think of half an egg—but half a *mole* of molecules is fine, just as acceptable as half a dozen eggs.

In chemistry, as in algebra, equations are *usually* written with the lowest set of whole number (integer) coefficients possible, and you should conform to this practice. There are occasions, however, when fractional coefficients or coefficients all divisible by the same integer are required.

8.3 WRITING CHEMICAL EQUATIONS

There are several kinds of chemical reactions. If you can identify the kind of reaction that may occur between given reactants, it is much easier to predict the products of the reaction and write its equation. We will now examine different types of reactions and learn how to write the equations that describe them. The procedure may be broken into three steps:

1. Determine what kind of reaction it is.
2. Write the correct chemical formula for each reactant on the left and each product on the right.
3. *Using coefficients only* (do NOT change a formula, or add another), balance the number of atoms of each element on each side of the equation.

8.4 COMBINATION REACTIONS

PG 8B Given the identity of a compound that is formed from two or more simpler substances, write the equation for the reaction.

Reactions in which two or more substances, often elements, combine to form a compound are called **combination,** or **synthesis, reactions.** A general equation for a combination reaction is

$$A + B \rightarrow C \quad (8.8)$$

EXAMPLE 8.1 When charcoal (carbon) is burned completely in air, carbon dioxide, CO_2, is formed. Write the equation for the reaction.

A word description of a reaction often assumes prior knowledge of some fact. This is an example. It assumes that you are already aware that something that burns in air is reacting chemically with the oxygen in the air. It is **oxidized.*** In other words, oxygen is an unidentified reactant in this example. Another reactant, carbon, is identified; so is the product, carbon dioxide, CO_2. This gives two reactants and one product, just what is necessary to identify the change as a combination reaction (Step 1). Next you must write the formulas of the reactants on the left, and the products on the right (Step 2). Don't forget that oxygen forms diatomic molecules. (As noted earlier, we will include state symbols in writing equations. You may omit them, unless your instructor requires otherwise.)

----- -----

$C(s) + \quad O_2(g) \rightarrow \quad CO_2(g)$

The final step is to balance the number of atoms of each element *by use of coefficients only*. Complete the equation.

----- -----

$C(s) + O_2(g) \rightarrow CO_2(g)$

Sometimes balancing is easy, as when all coefficients are 1!

Limiting the amount of air (oxygen) in burning carbon yields a combination reaction with a different result.

*In the next section and in Chapter 18 you will learn that "oxidize" and "oxidation" have meanings beyond reacting with oxygen.

EXAMPLE 8.2 When carbon is burned in a limited quantity of air, poisonous carbon monoxide, $CO(g)$, is produced. Write the equation.

First, write the unbalanced equation.

$C(s) + O_2(g) \rightarrow CO(g)$ (unbalanced)

Now count up the atoms of each element on each side of the equation and balance by inserting coefficients as needed.

$2\ C(s) + O_2(g) \rightarrow 2\ CO(g)$ *or* $C(s) + \frac{1}{2}\ O_2(g) \rightarrow CO(g)$

As noted previously, unless there is some specific reason for doing otherwise, the smallest whole number coefficients are usually used. The first equation is preferred.

If, in balancing an equation, you end with fractional coefficients, the easiest way to remove them is to multiply by the lowest common denominator of the fractions. Thus, for $C(s) + \frac{1}{2}\ O_2(g) \rightarrow CO(g)$, multiplication by 2 yields $2 \times \frac{1}{2} = 1$ for the coefficient of oxygen, and $2 \times 1 = 2$ for the coefficients of carbon and carbon monoxide.

EXAMPLE 8.3 Write the equation for the formation of sodium chloride by direct combination of its elements.

Begin with the formulas of the reactants and products.

$Na(s) + Cl_2(g) \rightarrow NaCl(s)$ (unbalanced)

You did recall, did you not, that chlorine is one of those elements that forms diatomic molecules?

Complete the equation.

$2\ Na(s) + Cl_2(g) \rightarrow 2\ NaCl(s)$

8.5 DECOMPOSITION REACTIONS

PG 8C Given the identity of a compound that is decomposed into simpler substances, either compounds or elements, write the equation for the reaction.

Decomposition reactions are the opposite of combination reactions, in that chemical compounds break down into simpler substances, usually elements. A general decomposition equation is

$$D \rightarrow E + F \tag{8.9}$$

EXAMPLE 8.4 When water is electrolyzed it decomposes into its elements (Fig. 2.2). Write the equation.

"Electrolyzing" a substance involves running an electric current through it, a process used in many chemical reactions. You know the formulas for water and the elements in it, so you should be able to produce the unbalanced equation readily. Balancing is straightforward. Go all the way.

----- -----

$2\ H_2O(\ell) \rightarrow 2\ H_2(g) + O_2(g)$

Decomposition reactions do not always go all the way to the elements, as the following example illustrates.

EXAMPLE 8.5 Heating potassium chlorate, $KClO_3(s)$, releases oxygen, leaving solid potassium chloride. Write the equation.

From the description of the reaction you must identify reactants and products. Write their formulas in the unbalanced equation.

----- -----

$KClO_3(s) \rightarrow \quad KCl(s) + \quad O_2(g)$ (unbalanced)

Inspection balancing is a bit more complicated this time. A quick glance shows the potassium and chlorine already balanced; only the oxygen remains. Oxygen comes three to a "package" in $KClO_3$ on the left, and two to a "package" in O_2 on the right. What is the smallest number of packages of three that can be repackaged two at a time and have none left over? How many packages of two will there be? It's like tennis balls, packed three in a can. What is the smallest number of cans that must be opened to furnish two balls to each of what number of courts? With that hint, see if you can balance the oxygen. Disregard potassium and chlorine.

----- -----

$2\ KClO_3(s) \rightarrow \quad KCl(s) + 3\ O_2(g)$ (unbalanced)

Two potassium chlorates, each with 3 oxygens, give six atoms of oxygen; three oxygen molecules, with atoms two at a time, also give six atoms of oxygen. The oxygen is in balance.

Complete the equation.

----- -----

$$2\ KClO_3(s) \rightarrow 2\ KCl(s) + 3\ O_2(g)$$

Now use a slightly different approach. Going back to the unbalanced equation,

$$KClO_3(s) \rightarrow \quad KCl(s) + \quad O_2(g) \qquad \text{(unbalanced)}$$

and noting again that oxygen is the only unbalanced element, observe also that oxygen is in elemental form on the right side of the equation. Any time all elements are in balance except one that is in elemental form, the only necessary step is to apply the proper coefficient to that element to complete the balancing. With three oxygen atoms on the left, how many oxygen *molecules*, must be used to obtain three oxygen *atoms* on the right?

----- -----

$1\frac{1}{2}$, or $\frac{3}{2}$

The improper fraction form, $\frac{3}{2}$, is more useful than the mixed form, $1\frac{1}{2}$, in seeing the implied multiplication, $\frac{3}{2} \times 2 = 3$:

$$\frac{3}{2}\ \cancel{\text{molecules}} \times \frac{2\ \text{atoms}}{1\ \cancel{\text{molecule}}} = 3\ \text{atoms}$$

Insert the fractional coefficient in the proper space to produce a balanced equation.

----- -----

$$KClO_3(s) \rightarrow KCl(s) + \tfrac{3}{2}\ O_2(g)$$

Now multiplication of the entire equation by 2 will reproduce the balanced equation with whole-number coefficients:

$$2\ KClO_3(s) \rightarrow 2\ KCl(s) + 3\ O_2(g)$$

EXAMPLE 8.6 Lime, CaO(s), and carbon dioxide gas, CO_2, are the products of the thermal decomposition of limestone, $CaCO_3(s)$. Write the equation.

----- -----

$$CaCO_3(s) \rightarrow CaO(s) + CO_2(g)$$

8.6 COMPLETE OXIDATION OR BURNING OF ORGANIC COMPOUNDS

PG 8D Write the equation for the complete oxidation or burning of any compound consisting only of carbon, hydrogen, and possibly oxygen.

A large number of organic compounds consist of two or three elements: carbon and hydrogen, or carbon, hydrogen, and oxygen. When such compounds are burned in an excess of air, or otherwise oxidized completely, the end products are always the same: carbon dioxide, $CO_2(g)$, and steam, $H_2O(g)$.* When the

*In some cases you may wish to consider *liquid* water, $H_2O(\ell)$, as the product. We will use either in the coming pages, while recognizing that the distinction is important when writing thermochemical equations (Section 9.6).

steam condenses it becomes the "white smoke" so commonly seen rising from chimneys.

In writing the following equations, you will be given only the identity of the compound that "burns" or is "oxidized." Understand that these terms mean reacting with oxygen. Remember that oxygen is diatomic. And finally, keep in mind that carbon dioxide and water are always the only products. Thus, the general equation for a complete oxidation reaction is always

$$C_xH_yO_z + O_2 \rightarrow CO_2 + H_2O \qquad (8.10)$$

plus the coefficients required for balancing.

EXAMPLE 8.7 Write the equation for the complete burning of the liquid hydrocarbon pentane, C_5H_{12}, in air.

Reactants and products are both known, so the unbalanced equation should be written readily.

----- -----

$C_5H_{12}(\ell) + \quad O_2(g) \rightarrow \quad CO_2(g) + \quad H_2O(g)$ (unbalanced)

Quite often in balancing equations it is advisable to balance first the elements other than hydrogen and oxygen, then balance hydrogen, and finally oxygen. This is particularly true when elemental oxygen is in the equation, as it is here. Working in the order carbon, hydrogen, oxygen, you should be able to balance this equation.

----- -----

$C_5H_{12}(\ell) + 8\ O_2(g) \rightarrow 5\ CO_2(g) + 6\ H_2O(g)$

Five carbons in C_5H_{12} require 5 carbon dioxides. Twelve hydrogens in C_5H_{12} call for 6 waters. This adds up to a total of 10 oxygens in CO_2 + 6 in H_2O, or 16 atoms of oxygen, all to come from the oxygen of the air. Thus 8 O_2 molecules are required.

EXAMPLE 8.8 Write the equation for the complete burning of ethane, $C_2H_6(g)$.

----- -----

$2\ C_2H_6(g) + 7\ O_2(g) \rightarrow 4\ CO_2(g) + 6\ H_2O(g)$

Working with one molecule of ethane produces

$$C_2H_6(g) + O_2(g) \rightarrow 2\ CO_2(g) + 3\ H_2O(g) \qquad \text{(unbalanced)}$$

with the oxygen yet to balance. There are 4 + 3 or 7 oxygen atoms on the right, which takes $\frac{7}{2}$ O_2 molecules on the left ($\frac{7}{2} \times 2 = 7$). Therefore

$$C_2H_6(g) + \tfrac{7}{2}\ O_2(g) \rightarrow 2\ CO_2(g) + 3\ H_2O(g)$$

Multiplying by 2 to clear the fractional coefficient yields the answer given.

Notice how the coefficient of oxygen was selected in Examples 8.7 and 8.8. In both cases, the number of oxygen molecules was one half the number of oxygen atoms required. The 16 atoms needed in Example 8.7 came from $\frac{1}{2} \times 16 = \frac{16}{2} = 8\ O_2$ molecules; the seven atoms needed in Example 8.8 came from $\frac{1}{2} \times 7 = \frac{7}{2}\ O_2$ molecules. If n oxygen atoms are needed, an oxygen coefficient of n/2 will balance the equation. If n is an even number, all coefficients are integers. If n is odd, you must multiply the equation by 2 to get integers.

Be careful on this next one.

EXAMPLE 8.9 Write the equation for the complete burning of butanol, $C_4H_9OH(\ell)$, in air.

----- -----

$$C_4H_9OH(\ell) + 6\ O_2(g) \rightarrow 4\ CO_2(g) + 5\ H_2O(g)$$

If you did not get this answer, you probably counted only 9 hydrogen atoms in C_4H_9OH (there are 10); or overlooked the fact that the one oxygen in C_4H_9OH takes care of one of the 13 oxygens on the right, leaving only 12 to come from O_2.

Quick Check 8.6
What unnamed reactants and products are always present in an equation for the complete burning of a compound containing carbon, hydrogen, and oxygen?

8.7 REACTIONS OCCURRING IN WATER SOLUTION

The reactions we have considered in Examples 8.1 through 8.9 have all dealt with pure substances in the gas, liquid, or solid state. Many reactions in the laboratory and in nature occur in aqueous solutions. The equations for three such reactions are considered in this section.

Just because two potential reactants are put together, there is no guarantee that they will react with each other. Many laboratory experiments have been performed to find the rules by which we may predict with reasonable certainty whether or not a reaction will occur. In this chapter the purpose is to learn how to write the equations for reactions that do take place. Later, in Chapters 16 and 18, aqueous solution reactions will be examined more closely and from the standpoint of *net ionic equations*. You will learn then how to predict whether or not a reaction will occur.

Redox Reactions—"Single Replacement" Type

PG 8E Given the reactants of a redox reaction ("single replacement" type only), write the equation.

Many elements are capable of replacing other elements from aqueous solutions. This is one kind of **oxidation–reduction reaction,** or **"redox" reaction,** that will be studied in greater detail in Chapter 18. The equation for such a reaction makes it appear as if one element is replacing another in a compound. It is a **"single replacement" equation;** in fact, the reactions are sometimes called **single replacement reactions.** The general equation is

$$M(s, g, \text{or } \ell) + NX(aq) \rightarrow MX(aq) + N(s, g, \text{or } \ell) \qquad (8.11)$$

The reactants in a single replacement equation are always an element and a compound. If the element is a metal, located to the left of the stair-step line on the periodic table, it will replace the metal or hydrogen in the compound. If the element is a nonmetal, located to the right of the stair-step line, it will replace the nonmetal in the compound. All three possibilities are in the next three examples.

EXAMPLE 8.10 Write the single replacement equation for the reaction between elemental calcium and hydrochloric acid.

Begin by writing the formulas of the reactants to the left of the arrow.

----- -----

$Ca(s) + HCl(aq) \rightarrow$

Now decide which element in the compound, hydrogen or chlorine, will be replaced by the calcium. Reread the paragraph before this example if you need help. Then write the formulas of the products to the right of the arrow above.

----- -----

$Ca(s) + HCl(aq) \rightarrow H_2(g) + CaCl_2(aq)$

A metal will replace the positive ion in a solution, which is another metal or hydrogen in the compound. Hydrogen molecules are diatomic and must be written as H_2 in an equation.

Now balance the equation.

----- -----

$Ca(s) + 2\ HCl(aq) \rightarrow H_2(g) + CaCl_2(aq)$

EXAMPLE 8.11 Copper reacts with a solution of silver nitrate, $AgNO_3$. Write the equation for the reaction.

In this case an elemental metal is replacing the metal in a compound. The unbalanced equation, please.

----- -----

$Cu(s)$ + $AgNO_3(aq)$ → $Ag(s)$ + $Cu(NO_3)_2(aq)$ (unbalanced)

An important technique appears in this equation. Notice that the nitrate ion, NO_3^-, appears on both sides of the equation: as a part of $AgNO_3$, on the left, and as a part of $Cu(NO_3)_2$, on the right. When a polyatomic ion is unchanged in a reaction, it may be balanced as a distinct unit, just as atoms of an element are balanced. The equation is balanced for that matter, except for the nitrate ion; there is one nitrate on the left, and there are two on the right. Balance them as the next step.

----- -----

$Cu(s)$ + 2 $AgNO_3(aq)$ → $Ag(s)$ + $Cu(NO_3)_2(aq)$ (unbalanced)

The equation now has the nitrates balanced, but it is unbalanced in another respect. Complete the equation.

----- -----

$Cu(s)$ + 2 $AgNO_3(aq)$ → 2 $Ag(s)$ + $Cu(NO_3)_2(aq)$

EXAMPLE 8.12 If chlorine gas is bubbled through a solution of sodium bromide, an aqueous solution of bromine and sodium chloride results.

This time the elemental reactant is a nonmetal, so it will replace the nonmetal in the compound. Write the unbalanced equation.

----- -----

$Cl_2(g)$ + $NaBr(aq)$ → $Br_2(aq)$ + $NaCl(aq)$

Now balance the equation.

----- -----

$Cl_2(g)$ + 2 $NaBr(aq)$ → $Br_2(aq)$ + 2 $NaCl(aq)$

Precipitation Reactions

PG 8F Given the reactants in a precipitation reaction, write the equation.

When solutions of two compounds are mixed, a positive ion from one compound may combine with a negative ion from the other to form a solid compound that settles to the bottom. The solid is a **precipitate;** the reaction is a **precipitation reaction.** In the equation for a precipitation reaction, ions of the two reactants appear to change partners. The equation, and sometimes the reaction itself, is called a **double replacement equation** or **double replacement reaction.** The general equation has the form

$$MY(aq) + NX(aq) \rightarrow MX(aq) + NY(s) \qquad (8.12)$$

The next two examples illustrate chemical precipitations.

EXAMPLE 8.13 If solutions of sodium chloride and silver nitrate, $AgNO_3$, are combined, silver chloride, AgCl, precipitates. Write the equation.

Begin by writing the formulas of the reactants to the left of the arrow.

$NaCl(aq) + AgNO_3(aq) \rightarrow$

Now write the formulas of the products on the right side of the equation. Notice that silver chloride, AgCl, formed by the silver ion from silver nitrate and the chloride ion from sodium chloride, has been identified as the precipitate. The other two ions make up the second product, which remains in solution.

$NaCl(aq) + AgNO_3(aq) \rightarrow AgCl(s) + NaNO_3(aq)$

Now for the balancing.

$NaCl(aq) + AgNO_3(aq) \rightarrow AgCl(s) + NaNO_3(aq)$

The equation is balanced, with all coefficients equal to 1. In checking this balance the nitrate ion may again be regarded as a distinct unit. There is one nitrate on each side of the equation.

EXAMPLE 8.14 A precipitate forms when solutions of potassium hydroxide and aluminum nitrate are combined. Write the equation.

Recalling the form of a double replacement equation, write the unbalanced equation. Do not attempt state designations.

$KOH(aq) + Al(NO_3)_3(aq) \rightarrow Al(OH)_3(s) + KNO_3(aq)$ (unbalanced)

It is the aluminum hydroxide that precipitates. At this point you have no reason to know this; you will learn how to recognize a precipitate in Chapter 16.

In this equation there are two polyatomic ions, OH^- and NO_3^-, which appear unchanged in the reaction. They may be treated as units in balancing the equation. Doing so makes the remainder of the balancing routine.

$3\ KOH(aq) + Al(NO_3)_3(aq) \rightarrow Al(OH)_3(s) + 3\ KNO_3(aq)$

The three hydroxides in $Al(OH)_3$ on the right side of the unbalanced equation require three KOH on the left, and the three nitrates in $Al(NO_3)_3$ on the left require three KNO_3 on the right. Balancing the polyatomic ions leaves the potassium and aluminum in balance.

Neutralization

PG 8G Given the reactants in a neutralization reaction, write the equation.

An acid is as a solution that contains hydrogen ions, H^+ (see Section 6.7). A solution that contains hydroxide ions, OH^-, is called a **base.** When an acid is added to an equal amount of a base, each hydrogen ion forms a chemical bond with a hydroxide ion to produce a molecule of water. The acid and base **neutralize** each other; the process is called **neutralization.*** An ionic compound called a **salt** is also formed; it usually remains in solution. Neutralization reactions are described by double replacement equations. In general,

$$\underset{\text{acid}}{HX(aq)} + \underset{\text{base}}{MOH(aq)} \rightarrow \underset{\text{water}}{HOH(\ell)} + \underset{\text{salt}}{MX(aq)} \tag{8.13}$$

The formula of water has been written as HOH in Equation 8.13, rather than the usual H_2O. The HOH formula makes it a little easier to treat the hydroxide ion as a polyatomic unit in balancing neutralization equations. Whichever water formula you use, remember that the compound is *molecular,* not ionic. It is *not correct* to think of water as "hydrogen hydroxide."

EXAMPLE 8.15 Write the equation for the reaction between hydrochloric acid and sodium hydroxide solutions.

Begin with the formulas of the reactants to the left of the arrow.

$HCl(aq) + NaOH(aq) \rightarrow$

Next use the double replacement form of an equation to write the formulas of the products.

$HCl(aq) + NaOH(aq) \rightarrow HOH(\ell) + NaCl(aq)$

Now balance the equation.

$HCl(aq) + NaOH(aq) \rightarrow HOH(\ell) + NaCl(aq)$

The equation is balanced, with all coefficients equal to 1.

*There are other kinds of acids that do not contain hydrogen ions, and bases that do not contain hydroxide ions, and the reactions between them are also neutralization reactions. The H^+ and OH^- neutralization is the most common, and the only one we will consider here.

EXAMPLE 8.16 Write the equation for the reaction between sulfuric acid and solid aluminum hydroxide.

The products here are similar to the earlier ones. Write the unbalanced equation.

$H_2SO_4(aq)$ + $Al(OH)_3(s)$ → $HOH(\ell)$ + $Al_2(SO_4)_3(aq)$ (unbalanced)

Balancing this time is a bit trickier than before, but everything falls into place if you treat the polyatomic ions, SO_4^{2-} and OH^-, as distinct units in balancing.

$3\ H_2SO_4(aq) + 2\ Al(OH)_3(s) \rightarrow 6\ HOH(\ell) + Al_2(SO_4)_3(aq)$

Starting from the unbalanced equation and balancing first the sulfate ions, three on the right in $Al_2(SO_4)_3$ require three H_2SO_4:

$3\ H_2SO_4(aq) + Al(OH)_3(s) \rightarrow Al_2(SO_4)_3(aq) + HOH(\ell)$ (unbalanced)

Next, two aluminums in $Al_2(SO_4)_3$ on the right require two $Al(OH)_3$ on the left:

$3\ H_2SO_4(aq) + 2\ Al(OH)_3(s) \rightarrow Al_2(SO_4)_3(aq) + HOH(\ell)$ (unbalanced)

Two aluminum hydroxides make six hydroxide ions available for the formation of six water molecules:

$3\ H_2SO_4(aq) + 2\ Al(OH)_3(s) \rightarrow Al_2(SO_4)_3(aq) + 6\ HOH(\ell)$

The hydroxide step also balances the hydrogens other than those in the hydroxide ions, six from three molecules of H_2SO_4 on the left, and six from six water molecules on the right. All oxygen is polyatomic ions, so it too should be in balance. It is, with 18 atoms on each side.

Your sequence of steps in balancing this equation may have been different, but the final result should be the same.

Quick Check 8.7

1. What kind(s) of reactants are present in a single replacement equation?
2. What kind(s) of reactants are present in a double replacement equation?
3. Name two kinds of reactions that are described by double replacement equations.

8.8 OTHER REACTIONS

PG 8H Given a word description of a chemical reaction in which all reactants and products are identified, write the chemical equation for the reaction.

In addition to the reactions already considered, there are many others that do not fit into any of the classifications above. We will use a few of these to provide further practice in balancing equations.

EXAMPLE 8.17 When rust, Fe_2O_3, is treated with hydrogen gas, it is reduced to metallic iron, with water vapor as a second product. Write and balance the equation.

$Fe_2O_3(s) + 3\ H_2(g) \rightarrow 2\ Fe(s) + 3\ H_2O(g)$

From the unbalanced equation, $Fe_2O_3(s) + H_2(g) \rightarrow Fe(s) + H_2O(g)$, 3 waters balance the oxygen. Six hydrogens in the 3 H_2O molecules require a coefficient of 3 for the H_2. Fe is balanced by a coefficient of 2. Whenever an equation can be balanced completely except for one or more elemental substances, balancing those elements last is readily accomplished by whatever coefficients are necessary. Note that this is a single replacement redox reaction that does not occur in water solution.

EXAMPLE 8.18 Carbon disulfide, $CS_2(\ell)$, reacts with oxygen gas to produce two gaseous products, carbon dioxide, CO_2, and sulfur dioxide, SO_2. Write the equation.

$CS_2(\ell) + 3\ O_2(g) \rightarrow CO_2(g) + 2\ SO_2(g)$

There is one carbon on each side of the unbalanced equation. Sulfur is readily balanced by a coefficient of 2 for the sulfur dioxide. Elemental oxygen is balanced last.

EXAMPLE 8.19 Balance the following equation:

$PCl_5(s) + \quad H_2O(\ell) \rightarrow \quad H_3PO_4(aq) + \quad HCl(aq)$

$PCl_5(s) + 4\ H_2O(\ell) \rightarrow H_3PO_4(aq) + 5\ HCl(aq)$

Taking chlorine first, one PCl_5 on the left requires 5 HCl on the right. Next, phosphorus is already balanced. Then hydrogen: three from H_3PO_4 and five from HCl is a total of eight on the right, requiring 4 H_2O on the left. Oxygen, as frequently happens when not present in elemental form, is balanced by the time we get to it.

8.9 SUMMARY OF TYPES OF REACTIONS AND EQUATIONS

In Section 8.3, the first step in writing chemical equations is ''Determine what kind of a reaction it is.'' Because the chapter is organized according to types of reactions, it was not necessary for you to make this decision in working out the examples. In the miscellaneous questions at the end of this chapter (and in any tests you happen to take!), where the reaction type is not specified, you will find writing the equation much easier if you can classify it. Table 8.1 has been prepared to help you develop this skill. It summarizes all of the reaction types in this chapter, the kinds of reactants that engage in each type, and the products.

Table 8.1 Summary of Types of Reactions and Equations

Reactants	Reaction Type	Products	Reference
Element + element Element + compound Compound + compound	Combination	One compound	Equation 8.8
One compound	Decomposition	Element + element Element + compound Compound + compound	Equation 8.9
O_2* + compound of C, H, and O	Complete oxidation or burning	CO_2* + H_2O*	Equation 8.10
Element + ionic compound or acid	Oxidation–Reduction†	Element + ionic compound	Equation 8.11
Two ionic compounds	Precipitation§	Two ionic compounds	Equation 8.12
Acic + base	Neutralization§	Ionic compound (salt) + H_2O	Equation 8.13

*The reactant oxygen and the products carbon dioxide and water are usually not mentioned in the description of a reaction of this kind.

†There are several kinds of oxidation–reduction reactions. The one described in this chapter occurs in water solution. Its equation is called a single replacement equation.

§Precipitation and neutralization reactions are usually described by a double replacement equation.

8.10 MISCELLANEOUS SYMBOLS USED IN CHEMICAL EQUATIONS

When it is necessary to show more information about a chemical reaction than the basic equation provides, other symbols are sometimes used. State symbols (s), (ℓ), and (g), and for solutions, (aq), which are used throughout this book, are examples. An alternate way of indicating the formation of a precipitate is to place an arrow pointing down next to its formula, as in the equation of Example 8.13:

$$NaCl + AgNO_3 \rightarrow AgCl\downarrow + NaNO_3$$

Formation of a gas may be shown by an arrow pointing up, as in the equation of Example 8.10:

$$Ca + 2\,HCl \rightarrow H_2\uparrow + CaCl_2$$

If a reaction is brought about by the application of heat, as in Example 8.6, it may be indicated by placing a Δ above the arrow:

$$CaCO_3(s) \xrightarrow{\Delta} CaO(s) + CO_2(g)$$

The use of other reaction conditions is also shown above, and sometimes above and below, the arrow in the equation. The electrolysis of water (Example 8.4) may be designated as

$$2\,H_2O(\ell) \xrightarrow{\text{electrolysis}} 2\,H_2(g) + O_2(g)$$

Many reactions are carried out in the presence of a catalyst, a substance that speeds up the rate of a reaction without being consumed itself. Manganese dioxide, MnO_2, is a catalyst for the decomposition of potassium chlorate (Example 8.5):

$$2\ KClO_3(s) \xrightarrow{MnO_2} 2\ KCl(s) + 3\ O_2(g)$$

Both heat and the catalyst may be shown:

$$2\ KClO_3(s) \xrightarrow[MnO_2]{\Delta} 2\ KCl(s) + 3\ O_2(g)$$

CHAPTER 8 IN REVIEW

8.1 Evolution of a Chemical Equation

8.2 Meaning of a Chemical Equation

8A Given a chemical equation, interpret it in terms of (a) atoms, molecules, and/or formula units, and (b) moles.

8.3 Writing Chemical Equations

8.4 Combination Reactions

8B Given the identity of a compound that is formed from two or more simpler substances, write the equation for the reaction.

8.5 Decomposition Reactions

8C Given the identity of a compound that is decomposed into simpler substances, either compounds or elements, write the equation for the reaction.

8.6 Complete Oxidation or Burning of Organic Compounds

8D Write the equation for the complete oxidation or burning of any compound consisting only of carbon, hydrogen, and possibly oxygen.

8.7 Reactions Occurring in Water Solutions

8E Given the reactants of a redox reaction ("single replacement" type only), write the equation.

8F Given the reactants in a precipitation reaction, write the equation.

8G Given the reactants in a neutralization reaction, write the equation.

8.8 Other Reactions

8H Given a word description of a chemical reaction in which all reactants and products are identified, write the chemical equation for the reaction.

8.9 Summary of Types of Reactions and Equations

8.10 Miscellaneous Symbols Used in Chemical Equations

TERMS AND CONCEPTS

8.1 Chemical equations
Reactants
Products
Aqueous
"State" symbol: (g), (ℓ), (s), (aq)
Balanced equation

8.4 Combination (synthesis) reaction
Burn, burning
Oxidize, oxidation

8.5 Decomposition reaction

8.7 Oxidation–reduction (redox) reaction
"Single replacement" equation, reaction
Precipitate; precipitation reaction
"Double replacement" equation, reaction
Acid
Base
Neutralize, neutralization
Salt

8.10 Catalyst

Most of these terms and many more are defined in the Glossary. Use your Glossary regularly.

EQUATION BALANCING EXERCISE

Instructions: Balance the following equations, for which correct chemical formulas are already written. Balanced equations are in the answer section.

(1) $Na + \quad O_2 \rightarrow \quad Na_2O$

(2) $H_2 + \quad Cl_2 \rightarrow \quad HCl$

(3) P + O_2 → P_2O_3
(4) $KClO_4$ → KCl + O_2
(5) Sb_2S_3 + HCl → $SbCl_3$ + H_2S
(6) NH_3 + H_2SO_4 → $(NH_4)_2SO_4$
(7) CuO + HCl → $CuCl_2$ + H_2O
(8) Zn + $Pb(NO_3)_2$ → $Zn(NO_3)_2$ + Pb
(9) $AgNO_3$ + H_2S → Ag_2S + HNO_3
(10) Cu + S → Cu_2S
(11) Al + H_3PO_4 → H_2 + $AlPO_4$
(12) $NaNO_3$ → $NaNO_2$ + O_2
(13) $Mg(ClO_3)_2$ → $MgCl_2$ + O_2
(14) H_2O_2 → H_2O + O_2
(15) BaO_2 → BaO + O_2
(16) H_2CO_3 → H_2O + CO_2
(17) $Pb(NO_3)_2$ + KCl → $PbCl_2$ + KNO_3
(18) Al + Cl_2 → $AlCl_3$
(19) P + O_2 → P_2O_5
(20) NH_4NO_2 → N_2 + H_2O
(21) H_2 + N_2 → NH_3
(22) Cl_2 + KBr → Br_2 + KCl
(23) $BaCl_2$ + $(NH_4)_2CO_3$ → $BaCO_3$ + NH_4Cl
(24) $MgCO_3$ + HCl → $MgCl_2$ + CO_2 + H_2O
(25) P + I_2 → PI_3
(26) PbO_2 → PbO + O_2
(27) Al + HCl → $AlCl_3$ + H_2
(28) $Fe_2(SO_4)_3$ + $Ba(OH)_2$ → $BaSO_4$ + $Fe(OH)_3$
(29) Al + $CuSO_4$ → $Al_2(SO_4)_3$ + Cu
(30) $KClO_3$ → KCl + O_2
(31) Mg + N_2 → Mg_3N_2
(32) C_6H_{14} + O_2 → CO_2 + H_2O
(33) $FeCl_2$ + Na_3PO_4 → $Fe_3(PO_4)_2$ + NaCl
(34) Li_2O + HOH → LiOH
(35) HgO → Hg + O_2
(36) $CaSO_4 \cdot 2\,H_2O$ → $CaSO_4$ + H_2O
(37) C_3H_7CHO + O_2 → CO_2 + H_2O
(38) $NaHCO_3$ + HCl → NaCl + H_2O + CO_2
(39) $Bi(NO_3)_3$ + NaOH → $Bi(OH)_3$ + $NaNO_3$
(40) FeS + HBr → $FeBr_2$ + H_2S
(41) $Zn(OH)_2$ + H_2SO_4 → $ZnSO_4$ + HOH
(42) P_4O_{10} + H_2O → H_3PO_4
(43) C_4H_9OH + O_2 → CO_2 + H_2O
(44) CaC_2 + H_2O → C_2H_2 + $Ca(OH)_2$
(45) $CaCO_3$ + H_3PO_4 → $Ca_3(PO_4)_2$ + CO_2 + H_2O
(46) PCl_5 + H_2O → H_3PO_4 + HCl
(47) CaI_2 + H_2SO_4 → HI + $CaSO_4$
(48) C_3H_7COOH + O_2 → CO_2 + H_2O
(49) $Mg(CN)_2$ + HCl → HCN + $MgCl_2$
(50) $(NH_4)_2S$ + $HgBr_2$ → NH_4Br + HgS

QUESTIONS AND PROBLEMS

Some questions among the following give only the names of compounds whose formulas were introduced in Chapter 6. For reference, you may find all of these formulas on page ANS-10. You are encouraged to write formulas with no assistance other than a periodic table, using the reference page only if necessary. This is the way you will build and improve your formula writing skills.

Section 8.2

(1) Write a sentence giving the "molecular" meaning of the equation for the complete oxidation of benzene:

$$2\ C_6H_6 + 15\ O_2 \rightarrow 12\ CO_2 + 6\ H_2O$$

(35) State the "molar" interpretation of the equation in 8.1. Explain the difference in the two interpretations.

Section 8.4

Write balanced chemical equations for the combination reactions described in Questions 2 to 7 and 36 to 41.

(2) Calcium forms calcium oxide by reaction with the oxygen in the air.

(3) P_4O_{10} is the product of the direct combination of phosphorus and oxygen.

(4) When potassium contacts fluorine gas—two highly reactive elements—potassium fluoride is produced.

(5) Silicon reacts with chlorine to form silicon tetrachloride, $SiCl_4$.

(6) Fluorine is an extremely reactive element, one that will react with all other elements except the lighter noble gases. Write the equation for its reaction with oxygen, in which oxygen difluoride, OF_2, is the product.

(7)* Magnesium nitride is one of the few metallic nitrides. Its formula is what you would expect from the positions of the elements in the periodic table, and it is formed by direct combination of the elements.

(36) Lithium combines with oxygen to form lithium oxide.

(37) Boron ($Z = 5$) combines with oxygen to form B_2O_3.

(38) Calcium combines with bromine to make calcium bromide.

(39) Phosphorus tribromide, PBr_3, is produced by direct combination of phosphorus and bromine.

(40) Powdered antimony ($Z = 51$) ignites when sprinkled into chlorine gas, producing antimony trichloride, $SbCl_3$.

(41)* There are only a few carbides—compounds made up of a metal and carbon. Although aluminum carbide is not ionic, its formula is as would be expected from the periodic table positions of the elements, counting carbon at -4 to balance aluminum's $+3$. Write the equation for its formation by direct combination of the elements.

Section 8.5

Write the equations for the following decomposition reactions in Questions 8 to 11 and 42 to 45.

(8) Melted crystals of table salt, sodium chloride, may be decomposed to its elements by electrolysis.

(9) Oxygen was discovered by the thermal decomposition of mercury(II) oxide, HgO, into its elements.

(42) Pure hydrogen iodide, HI, decomposes spontaneously to its elements.

(43) Silver oxide, Ag_2O, may be decomposed to its elements by heating.

(10) Carbonated beverages contain carbonic acid, H_2CO_3, an unstable compound that decomposes into carbon dioxide and water.

(11) Hydrogen peroxide, H_2O_2, the familiar bleaching compound, decomposes slowly into water and oxygen.

(44) Sulfurous acid, H_2SO_3, decomposes spontaneously to water and sulfur dioxide.

(45) When calcium hydroxide—sometimes called slaked lime—is heated, it decomposes to calcium oxide—or lime—and water vapor.

Section 8.6

Write equations for the complete oxidation of the organic compounds listed in Questions 12 to 15 and 46 to 49:

(12) Propane, C_3H_8, a component of "bottled gas" used in heating mobile homes.

(13) Acetylene, C_2H_2, the fuel used in welding.

(14) Acetaldehyde, CH_3CHO, a raw material used in the manufacture of vinegar, perfumes, dyes, plastics, and other organic products.

(15) Table sugar, $C_{12}H_{22}O_{11}$.

(46) Acetic acid, CH_3COOH, the acid in vinegar.

(47) Ethyl alcohol, C_2H_5OH, the alcohol in alcoholic beverages.

(48) Butane, C_4H_{10}, a raw material used in the production of fuels and in the manufacture of synthetic rubber.

(49) Glycerine, $C_3H_8O_3$, used in making soap, cosmetics, and explosives.

Section 8.7

For each pair of reactants given in Questions 16 to 30 and 50 to 64, write the equation for the water solution reaction that is most apt to occur. If necessary, use page ANS-10 as a source of formulas of salts and hydroxides. Common acids in these questions are hydrochloric, HCl; nitric, HNO_3; sulfuric H_2SO_4; and phosphoric, H_3PO_4. Classify the reaction as redox, precipitation, or neutralization. It is not necessary to predict which product will be a solid in a precipitation reaction.

(16) Magnesium is placed into sulfuric acid.

(17) Bromine is added to a solution of sodium iodide.

(18) Magnesium chloride and sodium fluoride solutions are combined.

(19) Potassium hydroxide solution is treated with nitric acid.

(20) Solid magnesium hydroxide reacts with hydrochloric acid.

(21) Barium chloride and sodium sulfate solutions are combined.

(22) When barium metal is placed into water, hydrogen gas bubbles off and a solution of barium hydroxide remains.

(50) Calcium reacts with hydrobromic acid.

(51) Chlorine gas is bubbled through aqueous potassium iodide.

(52) Calcium nitrate and potassium fluoride solutions are combined.

(53) Sulfuric acid reacts with barium hydroxide solution.

(54) Sodium hydroxide is added to phosphoric acid.

(55) Sodium hydroxide solution is added to magnesium bromide solution.

(56) Potassium and water react vigorously to yield hydrogen gas and a solution of potassium hydroxide.

Questions 23 to 29 and 57 to 63 include compounds not mentioned in Chapter 6. Their formulas are given whether they are reactants or products. You may have to supply the formula of an unnamed product if it is listed on page ANS-10.

(23) When silver nitrate, $AgNO_3$, and potassium bromide solutions are combined, silver bromide, $AgBr$, is produced.

(24) Lithium is added to an acidified solution of manganese(II) chloride, $MnCl_2$.

(25) Silver sulfide, Ag_2S, precipitates when sodium sulfide solution is added to a solution of silver nitrate, $AgNO_3$.

(26) The concentration of a sodium hydroxide solution can be found by titration with oxalic acid, $H_2C_2O_4$.

(27) Copper(II) iodate, $Cu(IO_3)_2$, precipitates when sodium iodate solution, $NaIO_3(aq)$, is added to a solution of copper(II) sulfate, $CuSO_4(aq)$.

(28) Magnesium reacts with a solution of nickel chloride, $NiCl_2(aq)$.

(57) Lead nitrate solution, $Pb(NO_3)_2(aq)$, reacts with a solution of sodium iodide to form lead iodide, PbI_2.

(58) A solution of chromium(III) nitrate, $Cr(NO_3)_3$, is one product of the reaction between metallic chromium and tin(II) nitrate, $Sn(NO_3)_2$.

(59) Ammonium sulfide solution added to a solution of copper(II) nitrate, $Cu(NO_3)_2$, yields a precipitate of copper(II) sulfide, CuS.

(60)* Only the first hydrogen comes off a sulfamic acid molecule, HNH_2SO_3, when it is titrated with a potassium hydroxide solution.

(61) The appearance of solid silver chromate, Ag_2CrO_4, indicates the end of a reaction when silver nitrate solution, $AgNO_3(aq)$, comes into contact with a solution of sodium chromate, $Na_2CrO_4(aq)$.

(62)* If manganese is placed in a solution of chromium(III) chloride, $CrCl_3(aq)$, one of the reaction products is manganese(II) chloride, $MnCl_2(aq)$.

Questions 29, 30, 63, and 64 involve ionic compounds with familiar anions, whose formulas were given in Chapter 6, but cations that were not mentioned earlier. The formulas of the reactants are given. Knowing that the total charge on a compound is zero, it is possible to figure out the charge on the cation. The formula of an unnamed ionic product can then be written. Test your skill in the remaining questions.

(29)* Silver iodide precipitates when solutions of silver nitrate, $AgNO_3$, and potassium iodide are combined.

(30)* Solutions of lead nitrate, $Pb(NO_3)_2(aq)$, and copper(II) sulfate, $CuSO_4(aq)$, are combined.

(63)* When aqueous nickel chloride, $NiCl_2$, reacts with a solution of sodium carbonate, a green precipitate forms.

(64)* A solution of potassium hydroxide reacts with a solution of zinc chloride, $ZnCl_2(aq)$.

Section 8.8

For each general reaction described in Questions 31 to 34 and 65 to 68, write the chemical equation.

(31) Metallic zinc ($Z = 30$) reacts with steam at high temperatures to produce solid zinc oxide, ZnO, and hydrogen gas.

(32) When solid barium oxide is placed into water, a solution of barium hydroxide is produced.

(33) Igniting a mixture of powdered iron(II) oxide, FeO, and aluminum produces a vigorous reaction in which aluminum oxide and iron are the products.

(34) Solid iron(III) oxide, Fe_2O_3, reacts with gaseous carbon monoxide to produce iron and carbon dioxide.

(65) A solid oxide of iron, $Fe_3O_4(s)$, and hydrogen are the products of the reaction between iron ($Z = 26$) and steam.

(66) Sulfuric acid is produced when gaseous sulfur trioxide, $SO_3(g)$, reacts with water.

(67) At high temperatures carbon monoxide and steam react to produce carbon dioxide and hydrogen.

(68) Magnesium nitride, Mg_3N_2, and hydrogen are the products of the reaction between magnesium and ammonia.

MISCELLANEOUS QUESTIONS

(*69*) Distinguish precisely and in scientific terms the differences between items in each of the following groups:
(a) Reactant, product
(b) (g), (ℓ), (s), (aq)
(c) Burn, oxidize
(d) Combination reaction, decomposition reaction
(e) Single replacement, double replacement
(f) Acid, base, salt
(g) Precipitation, neutralization

(70) Classify each of the following statements as true or false:
(a) In a chemical reaction, reacting substances are destroyed and new substances are formed.
(b) A chemical equation expresses the quantity relationships between reactants and products in terms of moles.
(c) Elements combine to form compounds in combination reactions.
(d) Compounds decompose into elements in decomposition reactions.
(e) All carbon is changed to carbon dioxide when a carbon-bearing compound burns.
(f) A nonmetal cannot replace a nonmetal in a single replacement redox reaction.
(g) A redox reaction is described by a double replacement equation.
(h) A catalyst is always a reactant, never a product, in a chemical reaction.

(71) Each reactant or pair of reactants listed below is *potentially* able to participate in one type of reaction described in this chapter. In each case, name the type of reaction and complete the equation.
(a) $Pb + Cu(NO_3)_2 \rightarrow$
(b) $Mg(OH)_2 + HBr \rightarrow$
(c) $C_5H_{10}O + O_2 \rightarrow$
(d) $Na_2CO_3 + CaSO_4 \rightarrow$
(e) $LiBr \rightarrow$
(f) $NH_4Cl + AgNO_3 \rightarrow$
(g) $Ca + Cl_2 \rightarrow$
(h) $F_2 + NaI \rightarrow$
(i) $Zn(NO_3)_2 + Ba(OH)_2$
(j) $Cu + NiCl_2 \rightarrow$

(72) Hydrogen, nitrogen, oxygen, fluorine, chlorine, bromine, and iodine exist as diatomic molecules at the temperatures and pressures at which most reactions occur. Under these normal conditions, when may the formulas of these elements be written without a subscript 2 in a chemical equation?

(73) ''Acid rain'' is rainfall that contains sulfuric acid originating in organic fuels that contain sulfur. The process occurs in three steps. The sulfur is first burned to sulfur dioxide, SO_2. In sunlight the SO_2 reacts with oxygen in the air to produce sulfur trioxide, SO_3. When water falls through the SO_3, they react to produce sulfuric acid, H_2SO_4. Write the equation for each step in the process and tell what kind of reaction it is.

(74) One of the harmful effects of acid rain is its attack on structures made of limestone, which includes marble structures, ancient ruins, and many famous statues. Write the equation you would expect for the reaction between acid rain (see Question 73) and limestone, $CaCO_3$. Write a second equation with the same reactants, showing that the expected but unstable carbonic acid is decomposed to carbon dioxide and water.

(75) The tarnish that appears on silverware is silver sulfide, Ag_2S, that is formed when silver is exposed to sulfur-bearing compounds in the presence of oxygen in the air. When hydrogen sulfide, H_2S, is the compound, water is the second product. Write the equation for the reaction.

(76) Sulfur combines directly with three of the halogens, but not with iodine. The products, however, do not have similar chemical formulas. When sulfur reacts with fluorine, SF_6 is the most common product; with chlorine, the product is usually SCl_2; and bromine generally yields S_2Br_2. Write the equation for each reaction.

(77) One source of the tungsten ($Z = 74$) filament used in light bulbs is tungsten(VI) oxide, WO_3. It is heated with hydrogen at high temperatures. The hydrogen removes the oxygen from the oxide, forming steam. Write the equation for the reaction and classify it as one of the reaction types discussed in this chapter.

9 Quantity Relationships in Chemical Reactions

LOOKING BACK	LOOKING AHEAD IN CHAPTER 9
2.8 Energy is released in an exothermic change and absorbed in an endothermic change.	Heat of reaction, ΔH is negative for an exothermic reaction and positive for an endothermic reaction.
3.2 Dimensional analysis was introduced as a problem-solving method.	Dimensional analysis is applied in solving stoichiometry problems.
3.8 Energy is measured in joules.	ΔH, measured in joules, is related to mass of reactants and products.
3.9 $\Delta X = X(\text{final}) - X(\text{initial})$	$\Delta H = H\ (\text{products}) - H\ (\text{reactants})$
4.8 The symbols of the common elements were located in the periodic table, which also shows the atomic masses of all elements.	The periodic table is your source of atomic masses for molar mass calculations.
7.2 The mole is the unit of quantity in chemistry. It represents Avogadro's number of atoms, molecules, or formula units.	The connecting link between quantities of substances in a chemical reaction is the numbers of moles.
7.3, 7.6 Molar mass is the mass in grams of one mole of a pure substance. It may be used to convert between mass and moles of that substance.	Converting between moles and mass is the first and third steps in a stoichiometry problem.
8.1 All chemical reactions involve a change in energy.	Energy changes in reactions are included in chemical equations and are calculated.
8.2 The coefficients in a chemical equation express the mole relationship between substances in the reaction.	Converting from moles of one substance to moles of another, using coefficients from the chemical equation, is the second step in a stoichiometry problem.
8.3 to 8.9 Chemical equations were written for different types of chemical reactions.	A chemical equation is the basis for solving any stoichiometry problem.

9.1 CONVERSION RELATIONSHIPS FROM A CHEMICAL EQUATION

PG 9A Given a chemical equation, or a reaction for which the equation can be written, and the number of moles of one species in the reaction, calculate the number of moles of any other species.

A balanced chemical equation tells how many moles of individual reactants are required for the reaction, as well as the number of moles of diifferent species that will be produced. Using the equation

$$PCl_5(s) + 4\ H_2O(\ell) \rightarrow H_3PO_4(aq) + 5\ HCl(aq),$$

we see that four moles of water are required to react with each mole of PCl_5. *For this reaction* there is, in effect, an equivalence between PCl_5 and H_2O: 1 mole $PCl_5 \simeq 4$ moles H_2O. Furthermore, for this reaction, each mole of PCl_5 that reacts will produce one mole of H_3PO_4 and five moles of HCl. The equivalence may thus be extended to the products:

$$1\ \text{mol}\ PCl_5 \simeq 4\ \text{mol}\ H_2O \simeq 1\ \text{mol}\ H_3PO_4 \simeq 5\ \text{mol HCl} \qquad (9.1)$$

Notice that *the numbers in the series of mole relationships derived from an equation are the same as the coefficients in the equation.*

If the number of moles of one species in a reaction, either reactant or product, is known, the equivalences establish a unit path to the moles of any other species. A single-step conversion is involved. If, for example, 3.20 moles of PCl_5 react according to the above equation, how many moles of HCl will be produced? If five moles of HCl are produced by one mole of PCl_5 (1 mol $PCl_5 \simeq 5$ mol HCl), then the given quantity of PCl_5 (3.20 mol) will yield

$$3.20\ \cancel{\text{mol}\ PCl_5} \times \frac{5\ \text{mol HCl}}{1\ \cancel{\text{mol}\ PCl_5}} = 16.0\ \text{mol HCl}$$

The procedure may be applied to any equation.

EXAMPLE 9.1 How many moles of oxygen are required to burn 2.40 moles of ethane, C_2H_6, according to the equation

$$2\ C_2H_6(g) + 7\ O_2(g) \rightarrow 4\ CO_2(g) + 6\ H_2O(\ell)$$

From the equivalences available in the equation, what unit path and what specific equivalence are required for this problem?

-----

mol $C_2H_6 \rightarrow$ mol O_2; 2 mol $C_2H_6 \simeq 7$ mol O_2

The problem requires conversion from moles of ethane to moles of oxygen, so only those two species are needed. The numbers are the coefficients in the equation. For every two moles of C_2H_6, seven moles of O_2 are required.

Complete the problem.

$$2.40\ \cancel{\text{mol}\ C_2H_6} \times \frac{7\ \text{mol}\ O_2}{2\ \cancel{\text{mol}\ C_2H_6}} = 8.40\ \text{mol}\ O_2$$

EXAMPLE 9.2 In the reaction $N_2(g) + 3\ H_2(g) \rightarrow 2\ NH_3(g)$, how many moles of hydrogen are required to produce 4.20 moles of ammonia (NH_3)?

$$4.20\ \cancel{mol\ NH_3} \times \frac{3\ mol\ H_2}{2\ \cancel{mol\ NH_3}} = 6.30\ mol\ H_2$$

The conversion of moles of one substance to moles of another, moles given → moles wanted, is a step that appears in every chemical reaction problem having to do with amounts of reactants or products.

9.2 MASS CALCULATIONS: STOICHIOMETRY

PG 9B Given a chemical equation, or a reaction for which the equation can be written, and the number of grams or moles of one species in the reaction, find the number of grams or moles of any other species.

Section 7.6 showed how to change the mass of a chemical of known formula to moles, or the number of moles to grams. Unit paths are *grams → moles* or *moles → grams*. The conversion ratio is molar mass, grams/mole, in both directions. In Section 9.1 we have just seen how to follow a unit path of *moles given species → moles wanted species,* using coefficients from the equation for the conversion ratio. By combining these techniques we have a unit path from grams of one substance to grams of another:

$$\begin{matrix}\text{grams of} \\ \text{given species}\end{matrix} \rightarrow \begin{matrix}\text{moles of} \\ \text{given species}\end{matrix} \rightarrow \begin{matrix}\text{moles of} \\ \text{wanted species}\end{matrix} \rightarrow \begin{matrix}\text{grams of} \\ \text{wanted species}\end{matrix} \qquad (9.2)$$

As an equation, using G for the given species and W for what is wanted,

$$g\ G \times \frac{mol\ G}{g\ G} \times \frac{mol\ W}{mol\ G} \times \frac{g\ W}{mol\ W} = g\ W \qquad (9.3)$$

Source of conversion ratio	molar mass of G	from equation	molar mass of W

In words, these three steps are

1. Change the quantity of given species to moles:
   g G → mol G, Section 7.6
2. Change the moles of given species to moles of wanted species:
   mol G → mol W, Section 9.1
3. Change the moles of wanted species to units required:
   mol W → g W, or other units, Section 7.6

We will refer to this three-step procedure as a ''stoichiometry pattern.''

Stoichiometry is the name given to finding the amount of one substance in a reaction from a known quantity of another. "Quantity" may be expressed in a number of ways. Officially, according to the SI system of units, quantity is the number of moles. We more commonly think in terms of measurable units, such as grams for mass and liters for volume. For the moment we will concentrate on grams; at the end of the chapter we will solve an example involving liters of a gas.

Occasionally you may be given the quantity of one substance in grams and asked to find the number of moles of a second species. In this case the first two steps of the stoichiometry pattern complete the problem. Or, you may be given the moles of one substance and asked to find the grams of another. Steps 2 and 3 accomplish this objective.

The burning of ethane may be used to illustrate the method of stoichiometry.

EXAMPLE 9.3 Calculate the number of grams of oxygen that are required to burn 150 grams of ethane, C_2H_6, in the reaction $2\ C_2H_6(g) + 7\ O_2(g) \rightarrow 4\ CO_2(g) + 6\ H_2O(\ell)$.

SOLUTION: Step 1 calls for the conversion of the given quantity to moles—in this case, the conversion of 150 g of ethane to moles of ethane as in Section 7.6. The setup begins

$$150\ \cancel{g\ C_2H_6} \times \frac{1\ mol\ C_2H_6}{30.0\ \cancel{g\ C_2H_6}} \times \underline{\qquad\qquad} \times \underline{\qquad\qquad} =$$

In Step 2 the setup is extended to convert moles of C_2H_6 to moles of O_2, as in Section 9.1. The conversion equivalence is 7 moles $O_2 \mathrel{\hat{=}}$ 2 moles C_2H_6, from the equation:

$$150\ \cancel{g\ C_2H_6} \times \frac{1\ \cancel{mol\ C_2H_6}}{30.0\ \cancel{g\ C_2H_6}} \times \frac{7\ mol\ O_2}{2\ \cancel{mol\ C_2H_6}} \times \underline{\qquad\qquad} =$$

Finally, in Step 3, the moles of oxygen are converted to grams, as in Section 7.6, p. 138.

$$150\ \cancel{g\ C_2H_6} \times \frac{1\ \cancel{mol\ C_2H_6}}{30.0\ \cancel{g\ C_2H_6}} \times \frac{7\ \cancel{mol\ O_2}}{2\ \cancel{mol\ C_2H_6}} \times \frac{32.0\ g\ O_2}{1\ \cancel{mol\ O_2}} = 560\ g\ O_2$$

EXAMPLE 9.4 The equation for burning heptane, C_7H_{16}, is

$$C_7H_{16}(\ell) + 11\ O_2(g) \rightarrow 7\ CO_2(g) + 8\ H_2O(\ell)$$

How many grams of oxygen are required to burn 3.50 moles of C_7H_{16}?

The given quantity is *moles* of heptane. In other words, we already have moles of the given substance, so the first step in the stoichiometry pattern is completed. The equation gives us the molar equivalence between heptane and oxygen. Start with the given moles of heptane, and set up, but do not calculate, the number of moles of the wanted substance, oxygen.

$$3.50 \text{ mol } C_7H_{16} \times \frac{11 \text{ mol } O_2}{1 \text{ mol } C_7H_{16}} \times \frac{\quad}{\quad}$$

From the equation, 11 mol O_2 ≏ 1 mol C_7H_{16}. This establishes the conversion factor.

Changing from moles of O_2 to grams of O_2 via molar weight in Step 3 in the stoichiometric sequence and finishes the problem. Complete the setup by inserting the required numerator and denominator. Calculate the answer.

$$3.50 \text{ mol } C_7H_{16} \times \frac{11 \text{ mol } O_2}{1 \text{ mol } C_7H_{16}} \times \frac{32.0 \text{ g } O_2}{1 \text{ mol } O_2} = 1.23 \times 10^3 \text{ g } O_2$$

EXAMPLE 9.5 How many moles of H_2O will be produced in the heptane burning reaction that also yields 115 g of CO_2? $C_7H_{16}(\ell) + 11\ O_2(g) \rightarrow 7\ CO_2(g) + 8\ H_2O(\ell)$

This time we are given grams of one species and must find moles of another. The first two steps in the stoichiometric pattern are all that is necessary. Set up, but do not solve, the first step, beginning with the given quantity.

$$115 \text{ g } CO_2 \times \frac{1 \text{ mol } CO_2}{44.0 \text{ g } CO_2} \times \frac{\quad}{\quad}$$

By Step 2, and by drawing the proper molar equivalence from the equation, extend the setup and solve the problem.

$$115 \text{ g } CO_2 \times \frac{1 \text{ mol } CO_2}{44.0 \text{ g } CO_2} \times \frac{8 \text{ mol } H_2O}{7 \text{ mol } CO_2} = 2.99 \text{ mol } H_2O$$

Now we'll go all the way from grams to grams.

EXAMPLE 9.6 How many grams of CO_2 will be produced by burning 66.0 g of heptane? The equation is

$$C_7H_{16}(\ell) + 11\ O_2(g) \rightarrow 7\ CO_2(g) + 8\ H_2O(\ell)$$

Beginning with the given quantity, set up, but do not solve, the first step, changing 66.0 g of heptane to moles.

$$66.0\ \cancel{g\ C_7H_{16}} \times \frac{1\ mol\ C_7H_{16}}{100\ \cancel{g\ C_7H_{16}}} \times \underline{\qquad\qquad}$$

Step 2 converts the moles of given species (heptane) to moles of wanted species (carbon dioxide). Extend the setup that far.

$$C_7H_{16}(\ell) + 11\ O_2(g) \rightarrow 7\ CO_2(g) + 8\ H_2O(\ell)$$

$$66.0\ \cancel{g\ C_7H_{16}} \times \frac{1\ \cancel{mol\ C_7H_{16}}}{100\ \cancel{g\ C_7H_{16}}} \times \frac{7\ mol\ CO_2}{1\ \cancel{mol\ C_7H_{16}}} \times \underline{\qquad\qquad} =$$

Finally, convert moles of CO_2 to grams, Step 3. Calculate the numerical answer.

$$66.0\ \cancel{g\ C_7H_{16}} \times \frac{1\ \cancel{mol\ C_7H_{16}}}{100\ \cancel{g\ C_7H_{16}}} \times \frac{7\ \cancel{mol\ CO_2}}{1\ \cancel{mol\ C_7H_{16}}} \times \frac{44.0\ g\ CO_2}{1\ \cancel{mol\ CO_2}} = 203\ g\ CO_2$$

Sometimes it is more convenient to measure mass in larger or smaller units. An analytical chemist, for example, works in milligrams and millimoles of substances, whereas a chemical engineer in industry is more apt to think in kilograms and kilomoles. The mass $\rightleftharpoons$ mole conversions in Steps 1 and 3 are performed in exactly the same way. Just as 1 mol $C_7H_{16} \triangleq$ 100 g C_7H_{16} in Example 9.6, so 1 kmol $C_7H_{16} \triangleq$ 100 kg C_7H_{16}:

$$\frac{1\ \cancel{mol}\ C_7H_{16}}{100\ \cancel{g}\ C_7H_{16}} \times \frac{1\ kmol}{1000\ \cancel{mol}} \times \frac{1000\ \cancel{g}}{1\ kg} = \frac{1\ kmol\ C_7H_{16}}{100\ kg\ C_7H_{16}}$$

The 1000 factors cancel in changing from units to kilounits. The same is true if milliunits are used.

EXAMPLE 9.7 How many milligrams of nickel chloride, $NiCl_2$, are in a solution if 503 mg AgCl are precipitated in the reaction $2\ AgNO_3(aq) + NiCl_2(aq) \rightarrow 2\ AgCl(s) + Ni(NO_3)_2(aq)$?

Set up and solve the problem completely. The setup is exactly the same as it would be with grams and moles, except that the unit g becomes mg and mol becomes mmol.

$$503\ \cancel{mg\ AgCl} \times \frac{1\ \cancel{mmol\ AgCl}}{144\ \cancel{mg\ AgCl}} \times \frac{1\ \cancel{mmol\ NiCl_2}}{2\ \cancel{mmol\ AgCl}} \times \frac{130\ mg\ NiCl_2}{1\ \cancel{mmol\ NiCl_2}} = 227\ mg\ NiCl_2$$

9.3 PERCENT YIELD

PG 9C Given two of the following, or information from which two of the following may be determined, calculate the third: theoretical yield, actual yield, and percent yield.

Example 9.6 indicated that the burning of 66.0 g of heptane will produce 203 g of CO_2. This is called a **theoretical yield,** where *yield* is the amount produced. In actual practice, factors such as impure reactants, incomplete reactions, and side reactions cause the **actual yield** to be less than the calculated yield. Knowing the actual yield found in the laboratory and the theoretical yield calculated from the equation for the reaction, we can find the **percent yield:**

$$\frac{\text{actual yield}}{\text{theoretical yield}} \times 100 = \text{percent yield} \qquad (9.4)$$

If only 180 g of carbon dioxide resulted in Example 9.6 instead of the theoretical 203 g, the percent yield would be

$$\frac{\text{actual yield}}{\text{theoretical yield}} \times 100 = \frac{180\ \cancel{g}}{203\ \cancel{g}} \times 100 = 88.7\%$$

EXAMPLE 9.8 A solution containing excess* sodium sulfate is added to a second solution containing 3.18 g of barium nitrate. Barium sulfate precipitates. (a) Calculate the theoretical yield of barium sulfate. (b) If the actual yield is 2.69 g, calculate the percent yield. The equation is

$$Ba(NO_3)_2(aq) + Na_2SO_4(aq) \rightarrow BaSO_4(s) + 2\ NaNO_3(aq)$$

Calculation of the theoretical yield is a typical stoichiometry problem. Solve part (a).

$$3.18\ \cancel{g\ Ba(NO_3)_2} \times \frac{1\ \cancel{mol\ Ba(NO_3)_2}}{261\ \cancel{g\ Ba(NO_3)_2}} \times \frac{1\ \cancel{mol\ BaSO_4}}{1\ \cancel{mol\ Ba(NO_3)_2}} \times \frac{233\ g\ BaSO_4}{1\ \cancel{mol\ BaSO_4}} = 2.84\ g\ BaSO_4$$

Now, if the actual yield is 2.69 g, find the percent yield by substituting into Equation 9.3.

$$\frac{2.69\ g\ BaSO_4\ \text{(actual)}}{2.84\ g\ BaSO_4\ \text{(theoretical)}} \times 100 = 94.7\%$$

The basic meaning of percent is *parts per hundred*—the number of parts of one kind of thing in 100 parts of all kinds of things (see Appendix I, Part E). Applied to percent yield, it means grams of actual product per 100 grams of theoretical product. This leads to an equivalence that can be used in a dimensional analysis approach to percent yield problems. An X% yield may be interpreted as

$$\text{X grams actual product} \mathrel{\hat=} \text{100 grams theoretical product} \qquad (9.5)$$

*The word "excess" as used here means "more than enough," i.e., more than enough sodium sulfate than is required to precipitate all the barium in 3.18 g of barium nitrate.

Notice that percent yield is concerned with actual and theoretical grams of *product,* not reactant.

EXAMPLE 9.9 A procedure for preparing sodium sulfate is summarized in the equation

$$2\ S(s) + 3\ O_2(g) + 4\ NaOH(aq) \rightarrow 2\ Na_2SO_4(aq) + 2\ H_2O(\ell)$$

The percent yield of the process is 82.2%. Find the number of grams of sodium sulfate that will be recovered from the reaction of 36.9 g of sodium hydroxide.

Using the stoichiometry procedures already explained, you can calculate the grams of Na_2SO_4 that can be produced by 36.9 g of NaOH. That quantity is the theoretical yield. But you will not recover that much; you will recover only 82.2% of that amount. This is the strategy for solving the problem. Begin with the given quantity and set up the three steps to the theoretical grams of Na_2SO_4. Do not calculate the answer.

$$36.9\ \text{g NaOH} \times \frac{1\ \text{mol NaOH}}{40.0\ \text{g NaOH}} \times \frac{2\ \text{mol}\ Na_2SO_4}{4\ \text{mol NaOH}} \times \frac{142\ \text{g}\ Na_2SO_4\ \text{(theoretical)}}{1\ \text{mol}\ Na_2SO_4} \times \underline{\qquad\qquad}$$

We asked you not to calculate the answer to this point, but we did. The theoretical yield from the above setup is 65.5 g Na_2SO_4. Now forget that number for a moment; we'll return to it shortly.

The above setup represents the unit path g NaOH → mol NaOH → mol Na_2SO_4 → g Na_2SO_4 (theoretical). Can you extend that path one more step to the actual yield: g Na_2SO_4 (theoretical) → g Na_2SO_4 (actual)? For every 100 g of theoretical product, there will be only 82.2 g of actual product. This is the conversion factor that is expressed in Equation 9.5. Insert above the necessary numbers and units to complete the setup. Then calculate the answer.

$$36.9\ \text{g NaOH} \times \frac{1\ \text{mol NaOH}}{40.0\ \text{g NaOH}} \times \frac{2\ \text{mol}\ Na_2SO_4}{4\ \text{mol NaOH}} \times \frac{142\ \text{g}\ Na_2SO_4\ \text{(theoretical)}}{1\ \text{mol}\ Na_2SO_4}$$

$$\times \frac{82.2\ \text{g}\ Na_2SO_4\ \text{(actual)}}{100\ \text{g}\ Na_2SO_4\ \text{(theoretical)}} = 53.8\ \text{g}\ Na_2SO_4\ \text{(actual)}$$

The above setup represents the overall unit path g NaOH → mol NaOH → mol Na_2SO_4 → g Na_2SO_4 (theoretical) → g Na_2SO_4 (actual). It also illustrates the dimensional analysis "parts per 100" approach to solving percentage problems.

Before you started to solve the problem, the strategy was laid out: find the theoretical yield, and then take 82.2% of that quantity. As noted above, the theoretical yield is 65.5 g Na_2SO_4. Thus,

$$82.2\%\ \text{of}\ 65.5\ \text{g}\ Na_2SO_4 = 0.822 \times 65.5\ \text{g}\ Na_2SO_4 = 53.8\ \text{g}\ Na_2SO_4$$

The decimal factor 0.822, coming from 82.2%, is the same as the dimensional analysis factor, $\dfrac{82.2\ \text{g}\ Na_2SO_4\ \text{(actual)}}{100\ \text{g}\ Na_2SO_4\ \text{(theoretical)}}$.

In Example 9.9 a percent yield conversion appears at the end of the stoichiometric pattern, Equation 9.3. Sometimes it must be used at the beginning. Suppose a chemical manufacturer wants to make X grams of a product. Suppose also that he knows his actual yield is only 50% of the theoretical yield. If he is going to *get* X grams of actual product, he must use enough reactant to produce 2X grams of theoretical product because he will recover only half that amount. The next example shows how this is figured.

EXAMPLE 9.10 Sodium nitrite, $NaNO_2$, is produced from sodium nitrate, $NaNO_3$, by the reaction $2\ NaNO_3 \rightarrow 2\ NaNO_2 + O_2$. How many grams of $NaNO_3$ must be used to produce an actual yield of 60.0 g $NaNO_2$ if the percent yield is 76.3%?

In this problem, 60.0 g $NaNO_2$ is only 76.3% of the theoretical yield from which the amount of reactant must be calculated. 76.3% of what theoretical yield is 60.0 g $NaNO_2$ (actual)? The easiest way to solve this is by dimensional analysis. The unit path is g $NaNO_2$ (actual) → $NaNO_2$ (theoretical), and Equation 9.4 furnishes the conversion relationship. Set up the conversion, but do not calculate the answer.

$$60.0\ \text{g NaNO}_2\ \text{(actual)} \times \frac{100\ \text{g NaNO}_2\ \text{(theoretical)}}{76.3\ \text{g NaNO}_2\ \text{(actual)}} \times \underline{\qquad\qquad}$$

The conventional arithmetic approach to this percentage problem is

$$76.3\%\ \text{of X} = 0.763\text{X} = 60.0$$

$$\text{X} = \frac{60.0}{0.763} = \cdots$$

The setups are essentially the same.

The setup so far represents the theoretical yield. Extend it with the usual three steps of the stoichiometry pattern and calculate the answer.

$$60.0\ \text{g NaNO}_2\ \text{(actual)} \times \frac{100\ \text{g NaNO}_2\ \text{(theoretical)}}{76.3\ \text{g NaNO}_2\ \text{(actual)}} \times \underline{\qquad\qquad}$$

$$60.0\ \text{g NaNO}_2\ \text{(actual)} \times \frac{100\ \text{g NaNO}_2\ \text{(theoretical)}}{76.3\ \text{g NaNO}_2\ \text{(actual)}} \times \frac{1\ \text{mol NaNO}_2}{69.0\ \text{g NaNO}_2}$$

$$\times \frac{2\ \text{mol NaNO}_3}{2\ \text{mol NaNO}_2} \times \frac{85.0\ \text{g NaNO}_3}{1\ \text{mol NaNO}_3} = 96.9\ \text{g NaNO}_3$$

In Example 9.10 the percentage conversion appears before the stoichiometry pattern, and in Example 9.9 it is after. How do you know where to put it? It is *always* applied to the product, never to a reactant. *Yield* always refers to a product, so that is where percent yield conversions belong. You can get some right answers by using it on reactants, but the thought process is more complicated and less logical—and errors are much more likely.

9.4 LIMITING REAGENT PROBLEMS

PG 9D Given a chemical equation or information from which it may be determined, and initial quantities of two or more reactants, (a) identify the limiting reagent; (b) calculate the theoretical yield of a specified product, assuming complete use of the limiting reagent; and (c) calculate the quantity of the reactant initially in excess that remains unreacted.

Zinc reacts with sulfur to form zinc sulfide: $Zn(s) + S(s) \rightarrow ZnS(s)$. Suppose you put three moles of zinc and two moles of sulfur into a reaction vessel and cause them to react until one is totally used up. How many moles of zinc sulfide will result? Also, how many moles of which element will remain unreacted?

This question is something like asking: "How many pairs of gloves can you assemble out of 20 left gloves and 30 right gloves, and how many unmatched gloves, and for which hand, will be left over?" The answer, of course, is 20 pairs of gloves. After you have assembled 20 pairs you run out of left gloves, even though you have 10 right gloves remaining. Similarly, in the zinc sulfide question where the elements combine on a 1-to-1 mole ratio, you will run out of sulfur after two moles of zinc sulfide form. One mole of zinc, the excess reactant, will remain unreacted. Sulfur, **the reactant that is completely used up by the reaction, is called the limiting reagent.** This analysis is summarized below:

Equation	Zn +	S →	ZnS
Moles at start	3	2	0
Moles used (−) or produced (+)	−2	−2	+2
Moles left	1	0	2

There is another way to think about the zinc sulfide reaction. Start again with three moles of zinc and two moles of sulfur. How many moles of ZnS can be produced by 3 moles of zinc, assuming there is an excess of sulfur? Three, of course: 1 mol Zn $\hat{=}$ 1 mol ZnS, so 3 moles of zinc can produce 3 moles of zinc sulfide. Similarly, how many moles of zinc sulfide can be formed from 2 moles of sulfur if there is an excess of zinc? By the same reasoning, 2 can be formed. Now combine these. The available zinc can produce 3 moles of zinc sulfide if there is enough sulfur, but the sulfur available can produce only 2 moles of zinc sulfide. The smaller product quantity, 2 moles of zinc sulfide, is the most the reaction can yield before running out of sulfur, the limiting reagent.

Now for the moles of zinc that remain unchanged: How many moles of zinc are used by all of the limiting reagent, sulfur? From the equation, 1 mol Zn $\hat{=}$

1 mol S. Therefore, 2 moles of sulfur will use 2 moles of zinc. If you start with 3 moles of zinc and use 2, then 1 mole of zinc (3 − 2) will be left.

You now have the idea behind solving limiting reagent problems. The reasoning is not difficult in terms of moles, but calculations become a bit more complicated when grams are introduced. To illustrate, consider the next example.

EXAMPLE 9.11 How many grams of ZnS (97.5 g/mol) can be produced by 8.50 g Zn and 6.84 g S in $Zn + S \rightarrow ZnS$? How many grams of which element will remain unreacted?

Begin by finding the number of grams of ZnS that can be formed by 8.50 g Zn, assuming that excess sulfur is present.

$$8.50 \text{ g Zn} \times \frac{1 \text{ mol Zn}}{65.4 \text{ g Zn}} \times \frac{1 \text{ mol ZnS}}{1 \text{ mol Zn}} \times \frac{97.5 \text{ g ZnS}}{1 \text{ mol ZnS}} = 12.7 \text{ g ZnS}$$

Now find the grams of ZnS that can come from 6.84 g S, assuming excess zinc.

$$6.84 \text{ g S} \times \frac{1 \text{ mol S}}{32.1 \text{ g S}} \times \frac{1 \text{ mol ZnS}}{1 \text{ mol S}} \times \frac{97.5 \text{ g ZnS}}{1 \text{ mol ZnS}} = 20.8 \text{ g ZnS}$$

These two results show that there is enough zinc to produce 12.7 grams of zinc sulfide, and enough sulfur to form 20.8 grams of zinc sulfide. Which element is the limiting reagent, and how many grams of zinc sulfide are possible?

Zinc is the limiting reagent. It will yield 12.7 grams of zinc sulfide.

Notice that, in this problem, the substance present in greater mass was the limiting reagent. You cannot draw any conclusions about the limiting reagent by the relative masses of reactants.

Now you must find the number of grams of sulfur that are left over. How many grams of sulfur will react with all of the limiting reagent, 8.50 g Zn?

$$8.50\ \cancel{g\ Zn} \times \frac{1\ \cancel{mol\ Zn}}{65.4\ \cancel{g\ Zn}} \times \frac{1\ \cancel{mol\ S}}{1\ \cancel{mol\ Zn}} \times \frac{32.1\ g\ S}{1\ \cancel{mol\ S}} = 4.17\ g\ S$$

If you start with 6.84 grams of sulfur and use 4.17 grams, how many will be left?

6.84 g S at start − 4.17 g S used = 2.67 g S unreacted

The procedure for solving limiting reagent problems may be summarized as follows:

1. For each reactant, calculate the amount of product that can result from the given quantity of that reactant. The *smallest* amount calculated in this way is the quantity of product that will form. The reactant that produces the smallest amount of product is the limiting reagent.
2. For each reactant other than the limiting reagent, calculate the amount that will react with all of the limiting reagent.
3. For each reactant other than the limiting reagent, subtract the amount that reacts (Step 2, above) from the quantity initially present. This is the amount of that substance which remains unreacted.

EXAMPLE 9.12 A solution containing 29.0 grams of calcium nitrate is added to a solution containing 33.0 grams of sodium fluoride. Calcium fluoride precipitates, as shown by the equation

$$Ca(NO_3)_2(aq) + 2\ NaF(aq) \rightarrow CaF_2(s) + 2\ NaNO_3(aq)$$

How many grams of calcium fluoride will precipitate? How many grams of which reactant are in excess?

Begin by calculating the grams of CaF_2 that can be precipitated by 29.0 g $Ca(NO_3)_2$, assuming that calcium nitrate is the limiting reagent.

$$29.0\ \cancel{g\ Ca(NO_3)_2} \times \frac{1\ \cancel{mol\ Ca(NO_3)_2}}{164\ \cancel{g\ Ca(NO_3)_2}} \times \frac{1\ \cancel{mol\ CaF_2}}{1\ \cancel{mol\ Ca(NO_3)_2}} \times \frac{78.1\ g\ CaF_2}{1\ \cancel{mol\ CaF_2}} = 13.8\ g\ CaF_2$$

Now do the same for 33.0 g NaF.

$$33.0\ \cancel{g\ NaF} \times \frac{1\ \cancel{mol\ NaF}}{42.0\ \cancel{g\ NaF}} \times \frac{1\ \cancel{mol\ CaF_2}}{2\ \cancel{mol\ NaF}} \times \frac{78.1\ g\ CaF_2}{1\ \cancel{mol\ CaF_2}} = 30.7\ g\ CaF_2$$

Which reactant, $Ca(NO_3)_2$ or NaF, is the limiting reagent? How many grams of CaF_2 will it form?

$Ca(NO_3)_2$ is the limiting reagent. It will form 13.8 g CaF_2.

The next step (Step 2, above) is to calculate the grams of NaF that will react with the limiting quantity of $Ca(NO_3)_2$.

$$20.9\ \cancel{g\ Ca(NO_3)_2} \times \frac{1\ \cancel{mol\ Ca(NO_3)_2}}{164\ \cancel{g\ Ca(NO_3)_2}} \times \frac{2\ \cancel{mol\ NaF}}{1\ \cancel{mol\ Ca(NO_3)_2}} \times \frac{42.0\ g\ NaF}{1\ \cancel{mol\ NaF}} = 14.9\ g\ NaF$$

Finally, how many grams of NaF will remain unreacted (Step 3, above)?

33.0 g NaF at start − 14.9 g NaF used = 18.1 g NaF unreacted.

9.5 GAS STOICHIOMETRY AT STANDARD TEMPERATURE AND PRESSURE

PG 9E Given a chemical equation, or a reaction for which the equation can be written, and the number of grams of one species or the STP volume of one species, find the number of grams or the STP volume of any other species.

It was mentioned in Section 9.2 that measured quantities in chemical reactions could be in volume units as well as mass units. Gases are usually measured in terms of volume. It is a fact that the **molar volume—the volume occupied by one mole—of a gas at a given temperature and pressure** is the same for all gases. Molar volume is like molar mass in being a "measurable quantity per mole." While the units of molar mass are grams per mole, the units of molar volume are liters per mole.

Unlike molar mass, molar volume depends on temperature and pressure. A widely used reference point for measuring gas volumes is 0°C and 1 atmosphere of pressure. This particular combination is referred to as **standard temperature and pressure,** often abbreviated **STP.** At STP the molar volume of all gases is 22.4 L/mol:

$$22.4\ L\ X(g) \simeq 1\ mol\ X \qquad (9.6)$$

Equation 9.6 furnishes a second way to change between quantity and moles in the first and third steps of the stoichiometry pattern. Equation 9.2 may now be expanded to

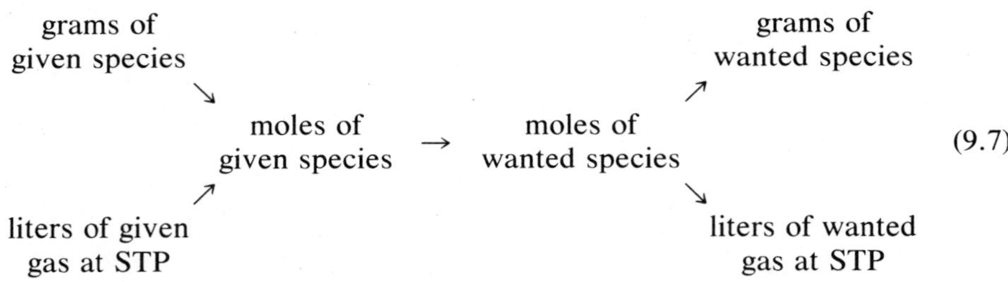

EXAMPLE 9.13 What volume of hydrogen, measured at STP, can be released by 42.7 g of zinc as it reacts with hydrochloric acid. The equation is $Zn + 2\ HCl \rightarrow H_2 + ZnCl_2$.

The first two steps in the stoichiometry pattern are as before. Change the given quantity mass to moles, and then change moles of given to moles of wanted. Set up the problem that far, but do not solve.

----- -----

$$42.7\ \cancel{g\ Zn} \times \frac{1\ \cancel{mol\ Zn}}{65.4\ \cancel{g\ Zn}} \times \frac{1\ mol\ H_2}{1\ \cancel{mol\ Zn}} \times \underline{\hspace{3cm}}$$

To this point in the problem we have moles of hydrogen. Call it X. If each mole of hydrogen occupies 22.4 L, how many liters are occupied by X moles? The unit path for this step is mol → L, based on Equation 9.6. Will you multiply by 22.4 or divide to get the answer? Complete the problem.

----- -----

$$42.7\ \cancel{g\ Zn} \times \frac{1\ \cancel{mol\ Zn}}{65.4\ \cancel{g\ Zn}} \times \frac{1\ \cancel{mol\ H_2}}{1\ \cancel{mol\ Zn}} \times \frac{22.4\ L\ H_2}{1\ \cancel{mol\ H_2}} = 14.6\ L\ H_2$$

EXAMPLE 9.14 How many grams of oxygen are needed to react with 18.6 L of sulfur dioxide, measured at STP, according to the equation

$$2\ SO_2(g) + O_2(g) \rightarrow 2\ SO_3(g)$$

Each 22.4 L of gas is one mole. You have 18.6 L. Will you multiply by 22.4 or divide to get moles? Set up, but do not solve, the conversion of the given quantity to moles.

----- -----

$$18.6\ \cancel{L\ SO_2} \times \frac{1\ mol\ SO_2}{22.4\ \cancel{L\ SO_2}} \times \underline{\hspace{3cm}} \times \underline{\hspace{3cm}} =$$

Complete the setup and calculate the answer in the usual way.

$$18.6\ \cancel{L\ SO_2} \times \frac{1\ \cancel{mol\ SO_2}}{22.4\ \cancel{L\ SO_2}} \times \frac{1\ \cancel{mol\ O_2}}{2\ \cancel{mol\ SO_2}} \times \frac{32.0\ g\ O_2}{1\ \cancel{mol\ O_2}} = 13.3\ g\ O_2$$

At temperatures and pressures other than standard, the molar volumes of gases are not 22.4 L/mol. The approach to gas stoichiometry problems must take this into account. You will learn how this is done in Chapter 13.

9.6 THERMOCHEMICAL EQUATIONS

PG 9F Given a chemical equation, or information from which it may be written, and the heat (enthalpy) of reaction, write the thermochemical equation in two forms.

At the beginning of Chapter 8 it was pointed out that nearly all chemical changes involve an energy transfer, usually in the form of heat. This heat flow is referred to as **heat of reaction** or **enthalpy of reaction. Enthalpy, designated H, may be thought of as "heat content," the amount of heat possessed by a chemical system—the chemicals involved in a reaction—at a given temperature and pressure.**

The absolute enthalpy of a substance or of a system cannot be measured directly. However, a change in enthalpy can be measured in the laboratory. "Heat of reaction" is therefore equivalent to "change of enthalpy," or ΔH.

If you burn two moles of ethane, C_2H_6, 3080 kJ (kilojoules) are given off. The reaction is exothermic (see Section 2.8). The products of the reaction have a smaller heat content than the reactants had; the final value of H is smaller than the initial value. ΔH is therefore a negative quantity; $\Delta H = -3080$ kJ.* ΔH is always negative for an exothermic reaction and, by similar reasoning, positive for an endothermic reaction.

An equation that includes the change in energy is a **thermochemical equation.** There are two kinds of thermochemical equations. One simply writes the ΔH of the reaction to the right of the conventional equation. For the burning of ethane, this is

$$2\ C_2H_6(g) + 7\ O_2(g) \rightarrow 4\ CO_2(g) + 6\ H_2O(\ell) \qquad \Delta H = -3080\ kJ \tag{9.8}$$

The second form includes energy in the equation as if it were a reactant or product. In an exothermic reaction, heat is "produced," so it appears on the product side of the equation:

$$2\ C_2H_6(g) + 7\ O_2(g) \rightarrow 4\ CO_2(g) + 6\ H_2O(\ell) + 3080\ kJ \tag{9.9}$$

*Recall that Δ always means final value minus initial value: $\Delta X = X(\text{final}) - X(\text{initial})$. If X(final) is smaller than X(initial), ΔX is less than zero, a negative quantity. See Section 3.9.

In an endothermic reaction, energy must be added to the reactants to make the reaction happen. Heat is a "reactant." The thermal decomposition of potassium chlorate is an example:

$$2\ KClO_3(s)\ +\ 89.5\ kJ \rightarrow 2\ KCl(s)\ +\ O_2(g)$$

$$2\ KClO_3(s) \rightarrow 2\ KCl(s)\ +\ 3\ O_2(g) \qquad \Delta H = +89.5\ kJ$$

When writing thermochemical equations, the state designations (g), (ℓ), (s), and (aq) *must* be used. The equation is meaningless without them because the size of the enthalpy change depends on the state of the reactants and products. If Equation 9.8 is written with water in the gaseous state, for example, $2\ C_2H_6(g) + 7\ O_2(g) \rightarrow 4\ CO_2(g) + 6\ H_2O(g)$, the value of ΔH is -2820 kJ.

EXAMPLE 9.15 The thermal decomposition of $CaCO_3(s)$ to $CaO(s)$ and $CO_2(g)$ is an endothermic reaction requiring 176 kJ per mole of $CaCO_3(s)$ decomposed. Write the thermochemical equation in two forms.

$$CaCO_3(s) \rightarrow CaO(s)\ +\ CO_2(g) \qquad \Delta H = 176\ kJ$$

$$CaCO_3(s)\ +\ 176\ kJ \rightarrow CaO(s)\ +\ CO_2(g)$$

9.7 THERMOCHEMICAL STOICHIOMETRY

PG 9G Given a thermochemical equation, or information from which it may be written, calculate the amount of heat evolved or absorbed for a given amount of reactant or product; alternately, calculate how many grams of a reactant are required to produce a given amount of heat.

The molar equivalence relationships that may be drawn from a chemical equation include energy in thermochemical equations. From

$$2\ C_2H_6(g)\ +\ 7\ O_2(g) \rightarrow 4\ CO_2(g)\ +\ 6\ H_2O(\ell)\ +\ 3080\ kJ \qquad (9.10)$$

the following equivalences may be established:

$$2\ mol\ C_2H_6 \triangleq 7\ mol\ O_2 \triangleq 4\ mol\ CO_2 \triangleq 6\ mol\ H_2O \triangleq 3080\ kJ \qquad (9.11)$$

Heats of reaction may be related quantitatively to amounts of chemical species by changing the stoichiometry pattern to convert between moles of one species and kilojoules of energy.

EXAMPLE 9.16 How many kilojoules of heat are evolved by the burning of 84.0 g of ethane, C_2H_6, according to Equation 9.10?

This problem is a two-step conversion. The unit path is from given quantity to energy: grams ethane → moles ethane → kilojoules. The second conversion is based on the equivalence that may be drawn from Equation 9.11. Go all the way to the answer.

$$84.0\ \cancel{g\ C_2H_6} \times \frac{1\ \cancel{mole\ C_2H_6}}{30.0\ \cancel{g\ C_2H_6}} \times \frac{3080\ kJ}{2\ \cancel{moles\ C_2H_6}} = 4.31 \times 10^3\ kJ$$

The final conversion comes from the fact that 2 mol C_2H_6 are equivalent to 3080 kJ, according to Equations 9.10 and 9.11.

EXAMPLE 9.17 What mass of hexane, $C_6H_{14}(\ell)$, must be burned in order to provide 8.00×10^3 kJ of heat? The thermochemical equation is

$$2\ C_6H_{14}(\ell) + 19\ O_2(g) \rightarrow 12\ CO_2(g) + 14\ H_2O(\ell) \quad \Delta H = -8280\ kJ;\ or$$

$$2\ C_6H_{14}(\ell) + 19\ O_2(g) \rightarrow 12\ CO_2(g) + 14\ H_2O(\ell) + 8280\ kJ$$

In this problem the given quantity is kilojoules. The path is the reverse of that in the previous example. Carry it all the way to an answer.

$$8000\ \cancel{kJ} \times \frac{2\ \cancel{moles\ C_6H_{14}}}{8280\ \cancel{kJ}} \times \frac{86.0\ g\ C_6H_{14}}{1\ \cancel{mole\ C_6H_{14}}} = 166\ g\ C_6H_{14}$$

A word about algebraic signs: In these examples we have been able to disregard the sign of ΔH because of the way the questions were worded. The question, "How much heat . . .?" is answered simply with a number of kilojoules. The wording of the question tells whether the heat is gained or lost. In Example 19.12, the question was "What mass of . . .?" Obviously that answer must be positive regardless of the sign of ΔH. However, whenever a question is of the form "What is the value of ΔH?" the algebraic sign is an essential part of the answer.

CHAPTER 9 IN REVIEW

9.1 Conversion Relationships from a Chemical Equation

9A Given a chemical equation, or a reaction for which the equation can be written, and the number of moles of one species in the reaction, calculate the number of moles of any other species.

9.2 Mass Calculations: Stoichiometry

9B Given a chemical equation, or a reaction for which the equation can be written, and the number of grams or moles of one species in the reaction, find the number of grams or moles of any other species.

9.3 Percent Yield

9C Given two of the following, or information from which two of the following may be determined, calculate the third: theoretical yield, actual yield, and percent yield.

9.4 Limiting Reagent Problems

9D Given a chemical equation, or information from which it may be determined, and initial quantities of two or more reactants, (a) identify the limiting reagent; (b) calculate the theoretical yield of a specified product, assuming complete use of the limiting reagent; and (c) calculate the quantity of the reactant initially in excess that remains unreacted.

9.5 Gas Stoichiometry at Standard Temperature and Pressure

9E Given a chemical equation, or a reaction for which the equation can be written, and the number of grams of one species or the STP volume of one species, find the number of grams or the STP volume of any other species.

9.6 Thermochemical Equations

9F Given a chemical equation, or information from which it may be written, and the heat (enthalpy) of reaction, write the thermochemical equation in two forms.

9.7 Thermochemical Stoichiometry

9G Given a thermochemical equation, or information from which it may be written, calculate the amount of heat evolved or absorbed for a given amount of reactant or product; alternately, calculate how many grams of a reactant are required to produce a given amount of heat.

TERMS AND CONCEPTS

9.2 Stoichiometry; "stoichiometry pattern"
9.3 Theoretical yield
Actual yield
Percent yield
9.4 Limiting reagent
9.5 Molar volume
Standard temperature and pressure; STP
9.6 Heat, or enthalpy, of reaction, H
Heat content
System (chemical)
Change in enthalpy, ΔH
Thermochemical equation
9.7 Thermochemical stoichiometry

Most of these terms, and many others, are defined in the Glossary. Use your Glossary regularly.

QUESTIONS AND PROBLEMS

Sections 9.1 and 9.2

Problems 1–7: Butane, C_4H_{10}, is a common fuel for heating homes in areas not serviced by natural gas. The equation for its combustion is to be used for the first seven problems. It is

$$2\ C_4H_{10} + 13\ O_2 \rightarrow 8\ CO_2 + 10\ H_2O.$$

(1) How many moles of oxygen are required to burn 3.40 mol of butane?

(2) How many moles of carbon dioxide will be produced when 4.68 mol of butane are burned?

(3) How many moles of water will be produced along with 0.568 mol of carbon dioxide?

(4) How many grams of butane can be burned by 1.42 mol of oxygen?

(5) 9.43 g of oxygen are used in burning butane. How many moles of water result?

Problems 38–44: The first step in the Ostwald process for manufacturing nitric acid is the reaction between ammonia and oxygen described by the equation

$$4\ NH_3 + 5\ O_2 \rightarrow 4\ NO + 6\ H_2O.$$

This equation is to be used for the next seven problems.

(38) How many moles of ammonia will react with 49.6 mol of oxygen?

(39) How many moles of NO will result from the reaction of 9.45 mol of ammonia?

(40) If 10.3 mol of water are produced, how many moles of NO will also be produced?

(41) How many moles of ammonia can be oxidized by 303 g of oxygen?

(42) If the reaction consumes 37.8 mol of ammonia, how many grams of water will be produced?

(6) Calculate the number of grams of carbon dioxide that will be produced by burning 78.4 g of butane.

(7) How many grams of oxygen are used in a reaction that produces 43.8 g of water?

(8) A reaction that produces great heat for welding and incendiary bombs is the "thermit" reaction: $Fe_2O_3 + 2\ Al \rightarrow Al_2O_3 + 2\ Fe$. How many grams of Fe_2O_3 can be converted to aluminum oxide by the reaction of 47.1 g of aluminum?

(9) A biological process whereby large starch molecules are converted to a sugar may be represented by the equation $C_{600}H_{1000}O_{500} + 50\ H_2O \rightarrow 50\ C_{12}H_{22}O_{11}$. Calculate the number of grams of sugar that will result from the reaction of 100 g of starch by this reaction.

(10) A common type of fire extinguisher depends on the reaction of sodium hydrogen carbonate, $NaHCO_3$, with sulfuric acid to produce carbon dioxide that develops a pressure to squirt water or foam onto a fire. The equation is $2\ NaHCO_3 + H_2SO_4 \rightarrow Na_2SO_4 + 2\ H_2O + 2\ CO_2$. If a fire extinguisher were designed to hold 600 g of sodium hydrogen carbonate, how many grams of sulfuric would be required to react with all of it?

(11) One of the methods for manufacturing sodium sulfate, widely used in making the kraft paper for grocery bags, involves the reaction $4\ NaCl + 2\ SO_2 + 2\ H_2O + O_2 \rightarrow 2\ Na_2SO_4 + 4\ HCl$. Calculate the number of kilograms of NaCl required to produce 5.00 kg of sodium sulfate.

(12) Trinitrotoluene is the chemical name for the explosive commonly known as TNT. Its formula is $C_7H_5N_3O_6$. TNT is manufactured by the reaction of toluene, C_7H_8, with nitric acid: $C_7H_8 + 3\ HNO_3 \rightarrow C_7H_5N_3O_6 + 3\ H_2O$. (a) How much nitric acid is needed to react completely with 1.90 kg C_7H_8? (b) How many kilograms of $C_7H_5N_3O_6$ can be produced in the reaction?

(13) Iron is extracted from its most abundant ore, *hematite*, Fe_2O_3, in a blast furnace. The essential chemical reaction is $Fe_2O_3 + 3\ CO \rightarrow 2\ Fe + 3\ CO_2$. Use the stoichiometry pattern to calculate the yield of iron from 778 kg of Fe_2O_3. There is another way you might get the same result. If you can think of it, perform the calculation to check your answer.

(43) How many grams of ammonia are required to produce 307 g of NO?

(44) If 7.05 g of water result from the reaction, what will be the yield of NO?

(45) The reaction of a dry cell may be represented by $Zn + 2\ NH_4Cl \rightarrow ZnCl_2 + 2\ NH_3 + H_2$. Calculate the number of grams of zinc consumed during the release of 9.32 g of ammonia in such a cell.

(46) The explosion of nitroglycerine is described by the equation $4\ C_3H_5(NO_3)_3 \rightarrow 12\ CO_2 + 10\ H_2O + 6\ N_2 + O_2$. How many grams of carbon dioxide are produced by the explosion of 69.7 g of nitroglycerine?

(47) Soaps are produced by the reaction of sodium hydroxide with naturally occurring fats. The equation for one such reaction is $C_3H_5(C_{17}H_{35}COO)_3 + 3\ NaOH \rightarrow C_3H_5(OH)_3 + 3\ C_{17}H_{35}COONa$, the last compound being the soap. Calculate the number of grams of NaOH required to produce 165 g of soap by this method.

(48) One way of making sodium thiosulfate, the "hypo" in photographic developing, is described by the equation $Na_2CO_3 + 2\ Na_2S + 4\ SO_2 \rightarrow 3\ Na_2S_2O_3 + CO_2$. How many grams of sodium carbonate are required to produce 494 g of sodium thiosulfate?

(49) The hard water "scum" that forms a ring around the bathtub is an insoluble soap, $Ca(C_{18}H_{35}O_2)_2$. It is formed when a soluble soap, $NaC_{18}H_{35}O_2$, reacts with the calcium ion that is responsible for the hardness in the water: $2\ NaC_{18}H_{35}O_2 + Ca^{2+} \rightarrow Ca(C_{18}H_{35}O_2)_2 + 2\ Na^+$. How many milligrams of scum can be formed from 825 mg of soap?

(50) Pig iron from a blast furnace contains several impurities, one of which is phosphorus. Additional iron ore, Fe_2O_3, is included in the charge with pig iron in making steel. The oxygen in the ore oxidizes the phosphorus by the reaction $12\ P + 10\ Fe_2O_3 \rightarrow 3\ P_4O_{10} + 20\ Fe$. If a sample of slag from the furnace contains 397 mg P_4O_{10}, how many grams of Fe_2O_3 were used in making it?

Sodium carbonate, Na_2CO_3, also known as washing soda, *and sodium hydrogen carbonate, $NaHCO_3$, better known as* baking soda, *are two very important industrial chemicals that can be made by the Solvay process. The next two questions in both columns are based on reactions in the Solvay process.*

(14) How many grams of ammonium hydrogen carbonate, NH_4HCO_3, will be formed by the reaction of 81.2 g of ammonia in the reaction $NH_3 + H_2O + CO_2 \rightarrow NH_4HCO_3$?

(15) What mass of $NaHCO_3$ must decompose to produce 448 g of Na_2CO_3 in $2\ NaHCO_3 \rightarrow Na_2CO_3 + H_2O + CO_2$?

(16)* A phosphate rock quarry yields rock that is 79.4% calcium phosphate, which is the raw material used in preparing the fertilizer *superphosphate*, $Ca(H_2PO_4)_2$. The rock is reacted with sulfuric acid according to the equation $Ca_3(PO_4)_2 + 2\ H_2SO_4 \rightarrow Ca(H_2PO_4)_2 + 2\ CaSO_4$. What is the smallest number of kilograms of rock that must be processed to yield 0.500 ton of fertilizer?

(17)* Silver chloride is a waste product from silver spray operations in making mirrors and phonograph records. The silver may be recovered by dissolving the silver chloride in a sodium cyanide (NaCN) solution, followed by reducing the silver with zinc. The overall equation is $2\ AgCl + 4\ NaCN + Zn \rightarrow 2\ NaCl + Na_2Zn(CN)_4 + 2\ Ag$. What minimum amount of sodium cyanide is needed to dissolve all the silver chloride from 40.1 kg of a sludge that is 23.1% silver chloride?

(51) How much NaCl is needed to react completely with 61.7 g NH_4HCO_3 in $NaCl + NH_4HCO_3 \rightarrow NaHCO_3 + NH_4Cl$?

(52) By-product ammonia is recovered from the process by the reaction $Ca(OH)_2 + 2\ NH_4Cl \rightarrow CaCl_2 + 2\ H_2O + 2\ NH_3$. How many grams of $CaCl_2$ can be produced along with 57.8 g of NH_3?

(53)* Carborundum, SiC, is widely used as an abrasive in industrial grinding wheels. It is prepared by the reaction of sand, SiO_2, with the carbon in coke, which is primarily carbon: $SiO_2 + 3\ C \rightarrow SiC + 2\ CO$. How many kilograms of carborundum can be prepared from 624 kg of coke that is 88.9% carbon?

(54)* $PbO_2 + Pb + 2\ H_2SO_4 \rightarrow 2\ PbSO_4 + 2\ H_2O$ is the chemical equation that describes what happens in an automobile storage battery as it generates electrical energy. (a) What fraction of the lead in the $PbSO_4$ comes from PbO_2 and what fraction comes from elemental lead? (b) If the operation of the process produces 44.7 g of $PbSO_4$, how many grams of PbO_2 must have been consumed?

Section 9.3

(18) Copper is extracted from chalcocite, a copper(I) sulfide ore, by a reaction with oxygen that may be represented by the equation $Cu_2S + O_2 \rightarrow 2\ Cu + SO_2$. If the treatment of 41.9 g of pure Cu_2S by the process yields 29.2 g of copper, calculate the percent yield.

(19) A commercial method for preparing hydrogen chloride is the treatment of sodium chloride with sulfuric acid: $2\ NaCl + H_2SO_4 \rightarrow Na_2SO_4 + 2\ HCl$. How much HCl will be made from 557 kg NaCl if the percent yield is 82.6%?

(20) Commercial preparation of the once-popular dry cleaning solvent carbon tetrachloride, CCl_4, is by the reaction $CS_2 + 2\ S_2Cl_2 \rightarrow CCl_4 + 6\ S$. How many grams of S_2Cl_2 are needed to prepare 38.5 g of CCl_4 if the percent yield is expected to be 85%?

(55) The function of "hypo" (see Problem 48) in photographic developing is to remove excess silver bromide by the reaction $2\ Na_2S_2O_3 + AgBr \rightarrow Na_3Ag(S_2O_3)_2 + NaBr$. What is the percent yield if the reaction of 8.71 g $Na_2S_2O_3$ produces 2.61 g NaBr?

(56) Calcium cyanamide, $CaCN_2$, is a common fertilizer. When mixed with water in the soil it reacts to produce calcium carbonate and ammonia: $CaCN_2 + 3\ H_2O \rightarrow CaCO_3 + 2\ NH_3$. How much ammonia can be obtained from 7.25 g $CaCN_2$ in a laboratory experiment in which the percent yield will be 94.3%?

(57) The Haber process for making ammonia from the nitrogen of the air is given by the equation $N_2 + 3\ H_2 \rightarrow 2\ NH_3$. Calculate the mass of hydrogen that must be supplied to make 5.00×10^2 kg NH_3 in a system that has an 85.5% yield.

(21) The laboratory preparation of chlorine is by the reaction $MnO_2 + 4\ HCl \rightarrow MnCl_2 + 2\ H_2O + Cl_2$. What is the percent yield if 5.95 g MnO_2 produces 4.22 g Cl_2?

(22) Formaldehyde, HCHO, widely used as a preservative in biology laboratories, is prepared commercially by the reaction $2\ CH_3OH + O_2 \rightarrow 2\ HCHO + 2\ H_2O$. The percent yield is 84.9% in a manufacturing plant. How much formaldehyde can be expected from the processing of 397 kg CH_3OH?

(23)* Hydrogen and chlorine are produced simultaneously in commercial quantities by the electrolysis of salt water: $2\ NaCl + 2\ H_2O \rightarrow 2\ NaOH + H_2 + Cl_2$. Yield on the process is 61%, based on the NaCl in the salt water. What mass of water that is 9.6% NaCl must be processed to obtain 105 kg of chlorine?

(58) Calculate the percent yield in the photosynthesis reaction by which carbon dioxide is converted to sugar if 7.03 g CO_2 yields 3.71 g $C_6H_{12}O_6$. The equation is $6\ CO_2 + 6\ H_2O \rightarrow C_6H_{12}O_6 + 6\ O_2$.

(59) Ethyl acetate, $CH_3COOC_2H_5$, is manufactured by the reaction between acetic acid, CH_3COOH, and ethanol, C_2H_5OH, by the reaction $CH_3COOH + C_2H_5OH \rightarrow CH_3COOC_2H_5 + H_2O$. How much acetic acid must be used to get 62.5 kg of product if the percent yield is 71.4%?

(60)* A dry mixture of hydrogen chloride and air is passed over a heated catalyst in the Deacon process for manufacturing chlorine. Oxidation occurs by the following reaction: $4\ HCl + O_2 \rightarrow 2\ Cl_2 + 2\ H_2O$. If the conversion is 58% complete, how many tons of chlorine can be recovered from 1.4 tons of HCl? (Hint: Whatever you can do with kilomoles and millimoles you can also do with ton moles.)

Section 9.4

(24) The fertilizer ammonium nitrate can be made by direct combination of ammonia with nitric acid: $NH_3 + HNO_3 \rightarrow NH_4NO_3$. If 74.4 g NH_3 are reacted with 159 g of HNO_3, how many grams of NH_4NO_3 can be produced? Also, calculate the mass of unreacted reagent that is in excess.

(25) The well-known insecticide, DDT, is made by the reaction $CCl_3CHO + 2\ C_6H_5Cl \rightarrow (ClC_6H_4)_2CHCCl_3$ (DDT) $+ H_2O$. In a laboratory test to determine percent yield, 3.19 g CCl_3CHO were reacted with 4.54 g C_6H_5Cl. Calculate the theoretical yield of DDT, identify the reagent that is in excess, and find the amount that is unreacted.

(26) Sodium carbonate can neutralize nitric acid by the reaction $2\ HNO_3 + Na_2CO_3 \rightarrow 2\ NaNO_3 + H_2O + CO_2$. Is 135 g Na_2CO_3 enough to neutralize a solution that contains 188 g HNO_3? How many grams of CO_2 will be released in the reaction?

(27)* The chemistry by which fluorides retard tooth decay is the hard, acid-resisting calcium fluoride layer that forms by the reaction $SnF_2 + Ca(OH)_2 \rightarrow CaF_2 + Sn(OH)_2$. If, at the time of a treatment, there are 239 mg $Ca(OH)_2$ on the teeth and the dentist uses a SnF_2 mixture that contains 305

(61) A solution containing 1.46 g of $BaCl_2$ is added to a solution made up of 2.14 g of Na_2CrO_4. Find the number of grams of $BaCrO_4$ that can precipitate. Also determine which reactant was in excess, as well as the number of grams over the amount required by the limiting reagent.

(62) $2\ NaIO_3 + 5\ NaHSO_3 \rightarrow I_2 + 3\ NaHSO_4 + 2\ Na_2SO_4 + H_2O$ is the equation for one method of preparing iodine. If 6.38 kg $NaIO_3$ are reacted with 7.80 kg $NaHSO_3$, how many kilograms of iodine can be produced? Which reagent will be left over? How many kilograms will be left?

(63) A mixture of tetraphosphorus trisulfide, P_4S_3, and powdered glass is in the white tip of strike-anywhere matches. The compound is made by the direct combination of the elements, $P_4 + 3\ S \rightarrow P_4S_3$. 145 g of phosphorus are mixed with the full contents of a 4-oz (126 g) bottle of sulfur. How many grams of compound can be formed? How much of which element will be left over?

(64)* In the recovery of silver from silver chloride waste (see Problem 17) a certain quantity of waste material is estimated to contain 175 g of silver chloride. The treatment tanks are charged with 45 g of zinc and 145 g of sodium cyanide, NaCN. Is there enough of the two reactants to recover all

mg SnF_2, has enough of the mixture been used to convert all of the $Ca(OH)_2$? If no, what minimum additional amount should have been used? If yes, by what number of milligrams was the amount in excess? (Sn is tin, Z = 50.)

of the silver from the AgCl? If no, how many grams of silver chloride will remain? If yes, how many more grams of silver chloride could have been treated by the available Zn and NaCN?

Section 9.5

(28) The discovery of oxygen occurred from the decomposition of mercury(II) oxide: $2\ HgO \rightarrow 2\ Hg + O_2$. What STP volume of oxygen can be produced by the reaction of 28.9 g of the oxide?

(29) How many liters of oxygen, measured at STP, are required for the complete combustion of 72.0 g C_7H_{16} (heptane, a component of gasoline)? (You will have to write the equation for the reaction.) If air is effectively 21% oxygen by volume (more accurately, 21 mole percent oxygen, which is proportional to "volume percent"), how many liters of air are needed by this amount of fuel?

(30) When nitric acid, NO, is released in air, it immediately combines with oxygen to produce nitrogen dioxide, NO_2: $2\ NO + O_2 \rightarrow 2\ NO_2$. If 0.345 L NO, measured at STP, react, how many liters of NO_2, also measured at STP, will be formed? (A pleasant surprise awaits you in this problem.)

(65) Calculate the number of grams of HCl that must react with limestone, $CaCO_3$, to liberate 650 mL CO_2, measured at STP, in the reaction $2\ HCl + CaCO_3 \rightarrow CaCl_2 + CO_2 + H_2O$.

(66) The conventional laboratory method for preparing oxygen is to heat potassium chlorate, $KClO_3$, in the presence of a catalyst. The oxygen is driven off, leaving a residue of potassium chloride. Write the equation for the reaction and calculate the number of milliliters of gas, measured at STP, that will result from the decomposition of 2.06 g $KClO_3$.

(67) Much of the destructive capability of nitroglycerine is the result of the rapid production of large volumes of gaseous products. (See Problem 46.) Assuming all gases are measured at STP, what total volume of $CO_2(g)$, $H_2O(g)$, and $O_2(g)$ will accompany the formation of 2.4 L $N_2(g)$?

Section 9.6

Thermochemical equations may be written in two ways, one with a heat term as a part of the equation, and alternately with ΔH set apart from the regular equation. In the examples that follow, write the equations for the reactions described in both forms. Recall that state designations of all substances are essential in thermochemical equations.

(31) When propane, $C_3H_8(g)$—a primary constituent in bottled gas used to heat trailers and rural homes—is burned to form gaseous carbon dioxide and liquid water, 2220 kJ of heat energy are released for every mole of butane consumed.

(32) In "slaking" lime, CaO(s), by converting it to solid calcium hydroxide through reaction with water, 65.3 kJ of heat are released for each mole of calcium hydroxide formed.

(33) The extraction of elemental aluminum from aluminum oxide is a highly endothermic electrolytic process. The reaction may be summarized by an equation in which the oxide reacts with carbon to yield aluminum and carbon dioxide gas. The energy requirement is 540 kJ/mol of aluminum metal produced. (Caution on the value of ΔH.)

(68) 2820 kJ of energy are absorbed from sunlight in the photosynthesis reaction in which carbon dioxide and water combine to produce sugar, $C_6H_{12}O_6$, and release oxygen.

(69) The electrolysis of water is an endothermic reaction, absorbing 286 kJ for each mole of liquid water decomposed to its elements.

(70) The reaction in an oxyacetylene torch is highly exothermic, releasing 1310 kJ of heat for every mole of acetylene, $C_2H_2(g)$, burned. The end products are gaseous carbon dioxide and liquid water.

Section 9.7

(34) How many milliliters of water can be decomposed into its elements by 356 kJ of energy? The thermochemical equation is 2 $H_2O(\ell)$ + 572 kJ → 2 $H_2(g)$ + $O_2(g)$.

(35) ΔH = 2.82×10^3 kJ for the photosynthesis reaction, whereby plants use energy from the sun to form carbohydrates from carbon dioxide and water. The equation is 6 CO_2 + 6 H_2O → $C_6H_{12}O_6$ + 6 O_2. How much energy is required to form 1 lb (454 g) of simple sugar, $C_6H_{12}O_6$?

(36) One of the fuels sold as "bottled gas" is butane, C_4H_{10} (see Questions 1 to 7). Calculate the energy that may be obtained by burning 1.50 kg C_4H_{10} if ΔH = -5.77×10^3 kJ for the reaction 2 $C_4H_{10}(g)$ + 13 $O_2(g)$ → 8 $CO_2(g)$ + 10 $H_2O(\ell)$.

(37)* In 1866 a young chemistry student conceived the electrolytic method of obtaining aluminum from its oxide. This method is still used. ΔH = 1.97×10^3 kJ for the reaction 2 $Al_2O_3(s)$ + 3 C(s) → 4 Al(s) + 3 $CO_2(g)$. The large amount of electrical energy required limits the process to areas of cheap power. How many kilowatt-hours of energy are needed to produce one pound (454 g) of aluminum by this process if 1 kw-hr = 3.60×10^3 kJ?

(71) Quicklime, the common name for calcium oxide, CaO, is made by heating limestone, $CaCO_3$, in slowly rotating kilns about 8 ft in diameter and nearly 200-ft long. The reaction is $CaCO_3(s)$ + 178 kJ → CaO(s) + $CO_2(g)$. How many kilojoules are required to decompose 5.00 kg of limestone?

(72) The quicklime produced as described in the foregoing problem is frequently converted to calcium hydroxide, or, as it is sometimes called, slaked lime, by an exothermic reaction with water: CaO(s) + $H_2O(\ell)$ → $Ca(OH)_2(s)$ + 66.5 kJ. How many grams of quicklime were processed in a reaction that produced 230 kJ of energy?

(73) How many grams of octane, a component of gasoline, would you have to use in your car to convert it to 1.26×10^5 kJ of energy? ΔH = 1.09×10^4 kJ for the reaction 2 $C_8H_{18}(\ell)$ + 25 $O_2(g)$ → 16 $CO_2(g)$ + 18 $H_2O(\ell)$.

(74)* Nitroglycerine is the explosive ingredient in many industrial dynamites. In Problem 67 it was indicated that much of its destructive force came from the sudden creation of large volumes of gaseous products. A great deal of energy is released too! ΔH = -6.17×10^3 kJ for the equation as written in Problem 46. Calculate the number of pounds of nitroglycerine that must be used in a blasting operation that requires 2.09×10^4 kJ of energy.

Miscellaneous Questions

(75) Distinguish precisely and in scientific terms the differences between items in each of the following groups:
(a) Theoretical, actual, and percent yield
(b) Limiting reagent, excess reagent
(c) Molar mass, molar volume
(d) Heat of reaction, enthalpy of reaction
(e) H, ΔH
(f) Chemical equation, thermochemical equation
(g) Stoichiometry, thermochemical stoichiometry

(76) Classify each of the following statements as true or false:
(a) Coefficients in a chemical equation express the molar proportions among both reactants and products.
(b) A stoichiometry problem can be solved with an unbalanced equation.
(c) In solving a stoichiometry problem, the change from quantity of given substance to quantity of wanted substance is based on masses.
(d) Percent yield is actual yield expressed as a percent of theoretical yield.
(e) The quantity of product of any reaction can be calculated only through the moles of the limiting reagent.
(f) All gases have the same molar volume at STP.
(g) ΔH is positive for an endothermic reaction, negative for an exothermic reaction.

(77) One of the few ways of "fixing" nitrogen, meaning to make a nitrogen compound from the elemental nitrogen in the atmosphere, is by the reaction $Na_2CO_3 + 4\,C + N_2 \rightarrow 2\,NaCN + 3\,CO$. Calculate (a) the grams of Na_2CO_3 required to react with 35 g N_2, and (b) the STP volume of the CO released.

(78) Ammonia has been considered as a fuel that would yield no pollutants. It reacts with oxygen according to the equation $4\,NH_3 + 3\,O_2 \rightarrow 2\,N_2 + 3\,H_2O$. If ammonia that occupies 68.7 L at STP reacts, (a) what volume of oxygen, also measured at STP, would be required, and (b) what mass of nitrogen would be produced?

(79) Potassium superoxide, KO_2, is converted to oxygen by reaction with exhaled CO_2 in oxygen masks used in emergency breathing situations. If the percent yield is 76.2%, what volume of oxygen, measured at STP, is produced by 42.2 g of KO_2? The equation is $4\,KO_2 + 2\,CO_2 \rightarrow 2\,K_2CO_3 + 3\,O_2$.

(80) Methyl alcohol (methanol), a substance known to have caused death or blindness when drunk under the mistaken impression that it is the alcohol used in intoxicating beverages, is produced by reacting hydrogen and carbon monoxide: $2\,H_2 + CO \rightarrow CH_3OH$. What volume of carbon monoxide, measured at STP, must be used to produce 578 g CH_3OH if the percent yield is 78.6%?

(81)* A student was given a 1.6240-g sample of a mixture of sodium nitrate and sodium chloride and asked to find the percentage of each compound in the mixture. She dissolved the sample and added a solution that contained an excess of silver nitrate, $AgNO_3$. The silver ion precipitated all of the chloride ion in the mixture as AgCl. It was filtered, dried, and weighed. Its mass was 2.056 g. What was the percentage of each compound in the mixture?

(82)* A crude iron ore was submitted to a laboratory for analysis. The iron was present as FeS_2, and the ore contained no other sulfides. 19.86 g of the ore was roasted in air at high temperature until all of the sulfur was changed to SO_2. The gas was collected, adjusted to STP, and its volume measured at 0.65 L. What is the percentage of iron in the ore?

(83)* 1.382 g of impure copper were dissolved in nitric acid to produce a solution of $Cu(NO_3)_2$. The solution went through a series of steps in which $Cu(NO_3)_2$ was changed to $Cu(OH)_2$, then to CuO, and then to a solution $CuCl_2$. This was treated with an excess of a soluble phosphate, precipitating all the copper in the original sample as pure $Cu_3(PO_4)_2$. The precipitate was dried and weighed. Its mass was 2.637 g. Find the percent copper in the original sample.

(84)* How many grams of magnesium nitrate, $Mg(NO_3)_2$, must be used to precipitate as magnesium hydroxide all the hydroxide ion in 50.0 mL 17.0% NaOH, the density of which is 1.19 g/mL? The precipitation reaction is $2\,NaOH + Mg(NO_3)_2 \rightarrow Mg(OH)_2 + 2\,NaNO_3$.

10 Chemical Bonding

LOOKING BACK	LOOKING AHEAD IN CHAPTER 10
2.7 Particles with unlike charges attract; those with like charges repel.	Chemical bonding is the result of electrostatic attractions and repulsions.
5.3 The number of valence electrons, the ground state electrons in the highest occupied s and p orbitals, can be found from an element's position in the periodic table. These are shown as dots in a Lewis diagram. An octet is complete when all eight valence electrons are present.	The number of valence electrons determines how monatomic ions and covalent bonds are formed.
5.5 Monatomic anions and cations formed by representative elements are generally isoelectronic with noble gas atoms.	Anions and cations form ionic bonds by which ions and cations are arranged in crystals, the usual form of ionic compounds.
5.5 Atomic size is determined by (a) the number of occupied energy levels, (b) the nuclear charge, and (c) the shielding effect.	Electronegativity is the effect of the same three causes, while the number of occupied energy levels and nuclear charge determine ionic size.

10.1 MONATOMIC IONS WITH NOBLE GAS ELECTRON CONFIGURATIONS

PG 10A Identify the monatomic ions that are isoelectronic with a given noble gas atom, and write the electron configurations of those ions.

In Chapter 5 you learned that representative elements in the same chemical family form monatomic ions having the same charge. They do this by gaining or losing valence electrons until they become isoelectronic with a noble gas atom. The ion has a full octet of valence electrons. There are six monatomic ions that are isoelectronic with a neon atom, which has a total of ten electrons. The manner in which the ions are formed is reviewed in Tables 10.1 and 10.2.

The fact that all Group 1A monatomic ions have a +1 charge, all Group 2A monatomic ions have a +2 charge, and all Group 7A monatomic ions have a −1 charge, to name just three groups, suggests that the pattern built around the neon configuration is duplicated around other noble gas atoms. With modifications for hydrogen, lithium, beryllium, and boron (see below), this is the case.

Table 10.1 Formation of Monatomic Cations

Element	Atom		Monatomic Ion		Electron(s)
Sodium	(11 p^+, 11 e^-)	→	(11 p^+, 10 e^-)	+	e^-
	$Na\cdot$	→	Na^+	+	$\cdot(e^-)$
	$1s^22s^22p^63s^1$	→	$1s^22s^22p^6$	+	e^-
Magnesium	(12 p^+, 12 e^-)	→	(12 p^+, 10 e^-)	+	$e^- + e^-$
	$Mg\cdot$ (with second dot)	→	Mg^{2+}	+	$\cdot(e^-) + \cdot(e^-)$
	$1s^22s^22p^63s^2$	→	$1s^22s^22p^6$	+	2 e^-
Aluminum	(13 p^+, 13 e^-)	→	(13 p^+, 10 e^-)	+	$e^- + e^- + e^-$
	$\cdot Al\cdot$ (with third dot)	→	Al^{3+}	+	$\cdot(e^-) + \cdot(e^-) + \cdot(e^-)$
	$1s^22s^22p^63s^23p^1$	→	$1s^22s^22p^6$	+	3 e^-

Table 10.1
A neon atom has ten protons and ten electrons. The electron configuration is $1s^22s^22p^6$. The eight 2s and 2p electrons constitute the stable octet of valence electrons. A neutral sodium atom has 11 electrons and 11 protons (see above). This is one electron more than a neon atom. If the sodium atom loses its single 3s electron, the ten remaining electrons have the same configuration as a neon atom. There are then 10 electrons and 11 protons, yielding a net charge of +1 for the sodium ion. The Lewis diagram of the sodium atom in the table shows the valence electron that is lost. Also shown are the configurations and electron and proton counts for both the atom and the ion.

There are two valence electrons in magnesium ($3s^2$) and three valence electrons in aluminum ($3s^23p^1$). These are two and three electrons more than a neon atom. As the valence electrons are lost, dropping the total electron count to the ten of neon, the cations formed are Mg^{2+} and Al^{3+}.

EXAMPLE 10.1 Write the electron configurations for the calcium and chloride ions, Ca^{2+} and Cl^-. With what noble gases are these ions isoelectronic?

To begin, note the locations of calcium and chlorine in the periodic table. Write their Lewis symbols and from them state the number of electrons that must be gained or lost to achieve complete octets at the highest energy level.

----- -----

Ca: must lose two electrons to achieve an octet, and :C̣̈l: must gain one, to achieve an octet.

You are now ready to answer the main questions: the electron configuration for Ca^{2+} and Cl^-, please, and the noble gases with which they are isoelectronic.

----- -----

Both ions are isoelectronic with argon, $1s^22s^22p^63s^23p^6$

The calcium atom starts with the configuration $1s^22s^22p^63s^23p^64s^2$. In losing two electrons to yield Ca^{2+}, it reaches the electron configuration of argon. Chlorine, with configuration $1s^22s^22p^63s^23p^5$, must gain one electron to become Cl^-, which is isoelectronic with argon.

EXAMPLE 10.2 Identify at least one more cation and one more anion that are isoelectronic with an atom of argon.

K^+, Sc^{3+}, S^{2-}, and P^{3-}

The formations of K^+, S^{2-}, and P^{3-} are identical to the formations of Na^+, O^{2-}, and N^{3-}, respectively, the ions immediately above them in the periodic table. You have not yet had information that would lead you to include the scandium ion, Sc^{3+}, in your answer to the question. If given the electron configuration of scandium ($Z = 21$) $1s^22s^22p^63s^23p^64s^23d^1$, you might have guessed the charge on the ion would be $+3$ because removal of the three highest energy electrons leaves the configuration of the noble gas, argon. Your guess would have been correct.

Table 10.2 Formation of Monatomic Anions

Element	Atom		Electron(s)		Monatomic Ion	
Fluorine	(9 p⁺ 9 e⁻)	+	e^-	→	(9 p⁺ 10 e⁻)	
	$:\ddot{F}:$	+	$\cdot(e^-)$	→	$[:\ddot{F}:]^-$	F^-
	$1s^22s^22p^5$	+	e^-	→	$1s^22s^22p^6$	
Oxygen	(8 p⁺ 8 e⁻)	+	$e^- + e^-$	→	(8 p⁺ 10 e⁻)	
	$\cdot\dot{O}:$	+	$\cdot(e^-) + \cdot(e^-)$	→	$[:\ddot{O}:]^{2-}$	O^{2-}
	$1s^22s^22p^4$	+	$2\ e^-$	→	$1s^22s^22p^6$	
Nitrogen	(7 p⁺ 7 e⁻)	+	$e^- + e^- + e^-$	→	(7 p⁺ 10 e⁻)	
	$\cdot\dot{N}:$	+	$\cdot(e^-) + \cdot(e^-) + \cdot(e^-)$	→	$[:\ddot{N}:]^{3-}$	N^{3-}
	$1s^22s^22p^3$	+	$3\ e^-$	→	$1s^22s^22p^6$	

*Nitride ion, N^{3-}, is not common, but does exist in a small number of compounds.

Table 10.2

Anions are formed by gaining enough electrons to complete the octet of valence electrons. This is the opposite of losing electrons to form cations (Table 10.1). Neutral fluorine, oxygen, and nitrogen atoms are one, two, and three electrons short of the ten of neon. As they gain these electrons to match neon, the ion charges are -1, -2, and -3: F^-, O^{2-}, and N^{3-}.

It was noted earlier that these ideas must be modified when applied to hydrogen, lithium, beryllium, and boron. Beryllium and boron tend to form covalent bonds by sharing electrons, rather than ionic bonds by losing electrons. Covalent bonds are discussed later in this chapter. Some bonding features of beryllium and boron are covered specifically in Section 11.4.

Hydrogen and lithium form ions that are isoelectronic with a noble gas atom, but they do not have a complete octet of electrons. The noble gas they duplicate is helium, which has the electron configuration of $1s^2$. Lithium, Li, with electron configuration $1s^2 2s^1$, loses its 2s electron to form the lithium ion, Li^+. Hydrogen, H, with configuration $1s^1$, gains an electron to reach the helium configuration and form the hydride ion, H^-.

You may wonder about the hydrogen ion, H^+. This ion does not normally exist by itself, but rather as a "hydrated" hydrogen ion, $H \cdot H_2O^+$, commonly called the hydronium ion and written H_3O^+. This polyatomic ion exists in aqueous acid solutions, and is not properly a part of a consideration of monatomic ions.

Figure 10.1 is a periodic table showing most of the monatomic ions that are isoelectronic with noble gases, as well as the noble gases themselves. Referring to the table, this section may be summarized in the following generalization: *Metal atoms with 1, 2, or 3 electrons more than the preceding noble gas tend to acquire noble gas configurations by losing these electrons, thereby forming cations with charges of +1, +2, and +3, respectively. Nonmetal atoms with 1, 2, or 3 electrons fewer than the following noble gas reach noble gas configurations by gaining these electrons, thereby forming anions with charges of −1, −2, and −3, respectively.*

Not all monatomic ions are isoelectronic with atoms of noble gases. Cations of the elements to the right of Group 3B in the periodic table generally contain

Figure 10.1
Monatomic ions having noble gas electron configurations. Each color group includes one noble gas atom and the monatomic ions that are isoelectronic with it. The number show that ionic radius in angstroms. (An angstrom is a unit of length equal to 0.1 nanometer.) For all isoelectronic species, size decreases as nuclear charge increases.

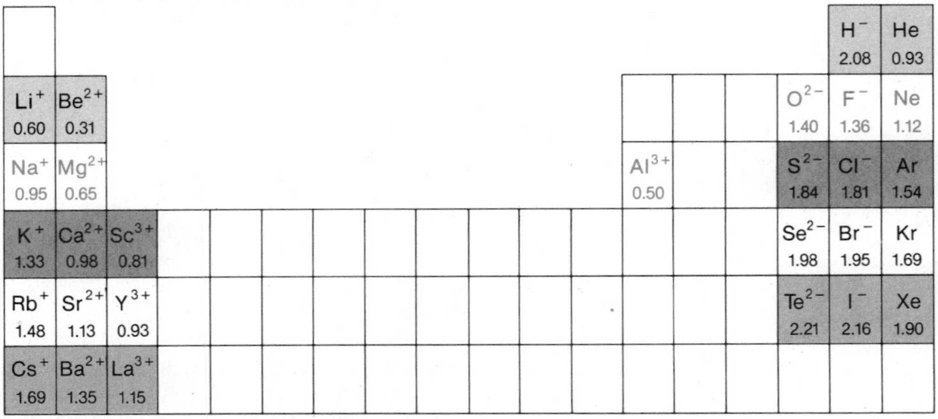

																H^- 2.08	He 0.93
Li^+ 0.60	Be^{2+} 0.31														O^{2-} 1.40	F^- 1.36	Ne 1.12
Na^+ 0.95	Mg^{2+} 0.65											Al^{3+} 0.50			S^{2-} 1.84	Cl^- 1.81	Ar 1.54
K^+ 1.33	Ca^{2+} 0.98	Sc^{3+} 0.81													Se^{2-} 1.98	Br^- 1.95	Kr 1.69
Rb^+ 1.48	Sr^{2+} 1.13	Y^{3+} 0.93													Te^{2-} 2.21	I^- 2.16	Xe 1.90
Cs^+ 1.69	Ba^{2+} 1.35	La^{3+} 1.15															

d electrons that are not present in the preceding noble gas atom. We will not attempt to describe these ions in terms of electron configuration, except to note that they usually differ from the parent atom by the absence of the highest energy s electrons and sometimes a d electron.

Quick Check 10.1
Which ions among the following are isoelectronic with noble gas atoms:

Cu^{2+} S^{2-} Fe^{3+} Ag^{+} Ba^{2+}

10.2 SIZES OF IONS

PG 10B Compare and explain the sizes of given isoelectronic monatomic ions.

10C Compare and explain the sizes of given monatomic ions formed by elements in the same group in the periodic table.

The numbers in the boxes in Figure 10.1 are ionic radii in angstroms.* Like the sizes of atoms, the sizes of ions increase as you go down a column in the periodic table. Each step down represents an additional principal energy level that is occupied with electrons, yielding a larger radius. This corresponds with the first of the three influences that determine atomic size, as discussed in Section 5.5.

Unlike the sizes of atoms, the sizes of ions do not decrease across each period in the table. This is not an irregularity in periodic table trends, as it first appears, but a significant regularity in the sizes of isoelectronic species. Notice that for each isoelectronic group of ions and noble gas atom, indicated by a color group in the table, size decreases as atomic number increases. Atomic number is the number of protons in the nucleus, or the nuclear charge. In other words, for a noble gas atom and the ions that have the identical electron configuration, size decreases as nuclear charge increases. The larger the nuclear charge, the more strongly it pulls the electrons to the nucleus. This matches the second of the three influences that determine atomic size, as discussed in Section 5.5.

The third atomic size influence, the shielding effect of inner electrons, has no part in fixing ionic size. Atoms and ions that are isoelectronic all have the same shielding.

The sizes of ions apparently are important in establishing some physical and chemical properties. For example, in some respects lithium is "out of step" with other alkali metals in its chemical properties. The beryllium ion is strongly hydrated, unlike other alkaline earth ions, and, as already noted, beryllium tends to form covalent bonds rather than ionic bonds. These and other features are partly the result of the smallness of the ions.

*An angstrom is a length unit that is equal to 10^{-10} meter, or 0.1 nm.

Quick Check 10.2: Which statements *and explanations* are true and which are false?

1. Ca^{2+} is larger than S^{2-} because the valence electrons of Ca^{2+} are in a higher energy level.
2. S^{2-} is smaller than Na^+ because S^{2-} has a higher nuclear charge.

10.3 IONIC BONDS

In Section 10.1 we considered the formation of monatomic ions as neutral atoms that "gain" or "lose" electrons. For most elements this is not a common event, but rather an accomplished fact. The natural occurrence of many elements is in ionic compounds or solutions of ionic compounds. Nowhere, for example, are sodium or chlorine atoms to be found; but there are large natural deposits of solid sodium chloride that are made up of sodium and chloride ions. The compound may also be obtained by evaporating sea water, which contains the ions in solution. In solid form, the ions are arranged in a definite geometric pattern called a **crystal** (see Fig. 10.2). They are held in fixed position by strong electrostatic forces called **ionic bonds.** This is an example of an **ionic compound.**

Sodium chloride can be produced by direct combination of its elements. Diatomic chlorine molecules are broken into chlorine atoms in the presence of sodium. It is then that a sodium atom can lose an electron to become a sodium ion, and a chlorine atom gains that electron to become a chloride ion. Using Lewis diagrams,

$$\text{Na}\cdot + \cdot\ddot{\underset{\cdot\cdot}{\text{Cl}}}: \rightarrow \text{Na}^+ + \left[:\ddot{\underset{\cdot\cdot}{\text{Cl}}}:\right]^- \rightarrow \text{NaCl crystal}$$

The ionic crystal (Fig. 10.2) forms from the ions produced in this way.

There are many kinds of ionic crystals. The ions present do not have to be in a 1:1 ratio, as with sodium chloride. If the compound is calcium chloride, in which each calcium atom has two valence electrons to lose, there must be two chlorine atoms to receive them:

$$\text{Ca}: + \begin{matrix}\cdot\ddot{\underset{\cdot\cdot}{\text{Cl}}}: \\ \\ \cdot\ddot{\underset{\cdot\cdot}{\text{Cl}}}:\end{matrix} \rightarrow \text{Ca}^{2+} + 2\left[:\ddot{\underset{\cdot\cdot}{\text{Cl}}}:\right]^- \rightarrow \text{CaCl}_2 \text{ crystal}$$

The 1:2 ratio of calcium ions to chloride ions in calcium chloride is reflected in the formula of the compound, $CaCl_2$. Several combinations of numbers and charges enter into ionic crystals, but always in such proportions as to yield a compound that is electrically neutral.

Ionic crystals are not limited to monatomic ions; polyatomic ions also form crystal structures. Atoms in a polyatomic ion are held together by covalent bonds. Figure 10.3 shows a model of a calcium carbonate crystal. Carbonate ions, one of which is circled, have a carbon atom in the middle that is covalently bonded to three oxygen atoms. Not only is the carbonate ion a distinct unit in the structure of the crystal, it also behaves as a unit in chemical changes.

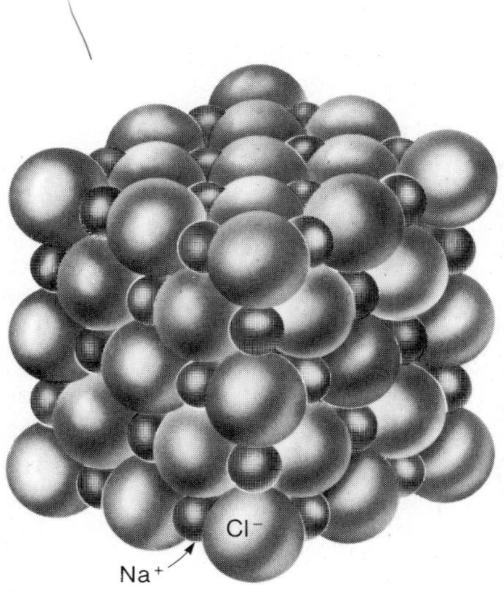

Figure 10.2
Arrangement of ions in sodium chloride. The small spheres represent sodium ions, and the large spheres are chloride ions. The number of positive charges is the same as the number of negative charges, making the crystal electrically neutral.

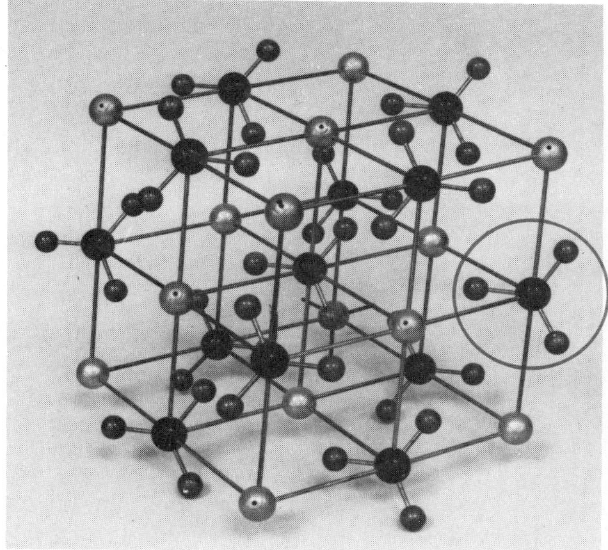

Figure 10.3
Model of calcium carbonate. Each carbon atom is covalently bonded to three oxygen atoms to form a carbonate ion (circled). There are equal numbers of calcium ions, Ca^{2+}, and carbonate ions, CO_3^{2-}, yielding a compound that is electrically neutral.

Ionic bonds are very strong. Because of this, nearly all ionic compounds are solids at room temperature. Solid ionic compounds are poor conductors of electricity because the ions are locked in place in the crystal. To melt an ionic compound there must be enough kinetic energy to break the ionic bonds and free the ions from each other. This takes a high temperature, often 800°C or more. When the substance is melted or dissolved, the crystal is destroyed and the ions are free to move. The ions are then able to carry electric current, so liquid ionic compounds and their solutions are good conductors.

10.4 COVALENT BONDS

PG 10D Distinguish between ionic and covalent bonds.

Ionic bonding explains quite well the properties of many compounds, but not all. All compounds known to be ionic conduct electricity when melted. But other compounds, notably most of those made up of nonmetals, such as water, are nonconductors in the liquid state. What is responsible for this difference between ionic compounds and the "other kind?"

Hydrogen fluoride, HF, and methane CH_4, are compounds of the other kind. Both are gases at room temperature and pressure. When condensed to liquids, they are nonconductors, like water. These and other facts lead us to believe the ultimate particle of these compounds is an electrically neutral unit composed of a small number of atoms of the different elements in the compound. These units are called **molecules;** the compounds are **molecular compounds.** The chemical formula of a molecular compound expresses the number of atoms of each element in the molecule.

In 1916, G. N. Lewis proposed that two atoms in a molecule are held together by a **covalent bond** in which they share one or more pairs of electrons. The idea is that when the bonding electrons can spend most of their time between two atoms, they attract *both* positively charged nuclei and "couple" the atoms to each other, much as two railroad cars are held together by the coupler between them. The result is a bond that is permanent until broken by a chemical change.

The simplest molecule and the simplest covalent bond appear in hydrogen, H_2. The formation of a molecule of H_2 can be represented as

$$\mathrm{H\cdot} + \mathrm{\cdot H} \rightarrow \mathrm{H:H} \quad \text{or} \quad \mathrm{H{-}H}$$

The two dots or the straight line drawn between the two atoms represents the covalent bond that holds the atoms together. In modern terms we say that the *electron cloud* or *charge density* formed by the two electrons is concentrated in the region between the two nuclei. This is where there is the greatest probability of locating the bonding electrons. The atomic orbitals of the separated atoms are said to *overlap* (Fig. 10.4).

A similar approach shows the formation of the covalent bond between two fluorine atoms to form a molecule of F_2, and between one hydrogen atom and one fluorine atom to form an HF molecule:

$$:\ddot{\underset{\cdot\cdot}{\mathrm{F}}}\cdot + \cdot\ddot{\underset{\cdot\cdot}{\mathrm{F}}}: \rightarrow :\ddot{\underset{\cdot\cdot}{\mathrm{F}}}:\ddot{\underset{\cdot\cdot}{\mathrm{F}}}: \quad \text{or} \quad :\ddot{\underset{\cdot\cdot}{\mathrm{F}}}{-}\ddot{\underset{\cdot\cdot}{\mathrm{F}}}: \quad \text{or} \quad \mathrm{F{-}F}$$

$$\mathrm{H}\cdot + \cdot\ddot{\underset{\cdot\cdot}{\mathrm{F}}}: \rightarrow \mathrm{H}:\ddot{\underset{\cdot\cdot}{\mathrm{F}}}: \quad \text{or} \quad \mathrm{H}{-}\ddot{\underset{\cdot\cdot}{\mathrm{F}}}: \quad \text{or} \quad \mathrm{H{-}F}$$

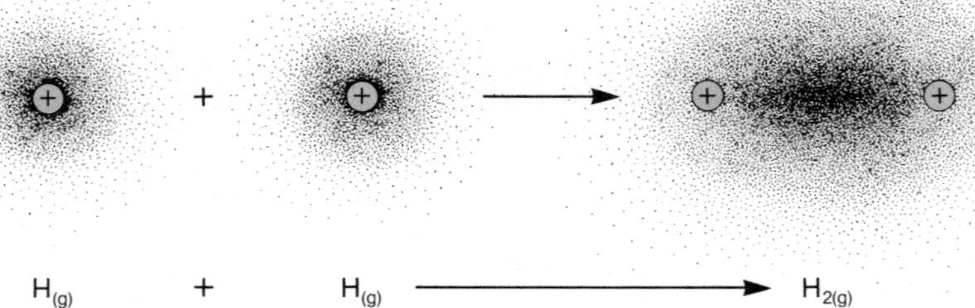

Figure 10.4
The formation of a hydrogen molecule from two hydrogen atoms. Each dot represents the instantaneous position of an electron. The "charge clouds" of the 1s orbitals of two hydrogen atoms are said to overlap and form a covalent bond. The bonding electrons are believed to spend most of their time between the two nuclei, as suggested by the heavier density of electron position dots in that area.

Fluorine has seven valence electrons. The 2s orbital and two of the 2p orbitals are filled, but the remaining 2p orbital has only one electron. The F_2 bond is considered to be formed by the overlap of the half-filled 2p orbitals of two fluorine atoms. In the HF molecule, the bond forms from the overlap of the half-filled 1s orbital of a hydrogen atom with the half-filled 2p orbital of a fluorine atom.

When used to show the bonding arrangement between atoms in a molecule, electron dot diagrams are commonly called **Lewis diagrams, Lewis formulas,** or **Lewis structures.** Notice that the unshared electron pairs of fluorine are shown for two of the three Lewis diagrams for F_2 and HF above, but not for the third. Technically they should always be shown, but they are frequently omitted when not required by the context in which the diagrams appear. Unshared electron pairs are often called **lone pairs.**

When two bonding electrons are shared by two atoms, the electrons effectively "belong" to both atoms. They are therefore counted as valence electrons for each bonded atom. Thus each hydrogen atom in H_2 and the hydrogen atom in HF have two electrons, the same number as an atom of the noble gas helium. Each fluorine atom in F_2 and the fluorine atom in HF have eight valence electrons, matching neon and the other noble gas atoms. These and many similar observations lead us to believe that the stability of a noble gas electron configuration contributes to the formation of covalent bonds, just as it contributes to the formation of ions. This generalization is known as the **octet rule,** or **rule of eight** because each atom has "completed its octet."

The tendency toward a complete octet of electrons in a bonded atom reflects the natural tendency for a system to move to the lowest energy state possible. The thermochemical equation for the formation of one mole of hydrogen molecules from two moles of hydrogen atoms is

$$H(g) + H(g) \rightarrow H_2(g) + 435 \text{ kJ}$$

This equation shows that the energy of one mole of H_2 molecules is 435 kJ lower than the energy of two moles of separate H atoms.

Quick Check 10.4: Are the following true or false?

1. Atoms in molecular compounds are held together by covalent bonds.
2. A lone pair of electrons is not shared between two atoms.
3. Covalent bonds are common between atoms of two metals.
4. An octet of valence electrons usually represents a low energy state.

10.5 POLAR AND NONPOLAR COVALENT BONDS

PG 10E Distinguish between polar and nonpolar bonds.

10F Given the electronegativities of two elements, classify the bond between them as nonpolar covalent, polar covalent, or primarily ionic. Identify the positive and negative ends of polar bonds.

As we might expect, the two electrons joining the atoms in the H_2 molecule are shared equally by the two nuclei. Another way of saying this is that the charge density is centered in the overlap region between the bonded atoms, as shown in Figure 10.4. **A bond in which the distribution of bonding electron charge is symmetrical, or centered, is said to be nonpolar.** A bond between identical atoms, as in H_2 or F_2, is always nonpolar.

In an HF molecule, the charge density of the bonding electrons is shifted toward the fluorine atom and away from the hydrogen atom (Fig. 10.5). **A bond with an unsymmetrical distribution of bonding electron charge is a polar bond.** The fluorine atom in an HF molecule acts as a negative pole, and the hydrogen atom is a positive pole.

Bond polarity may be described in terms of the **electronegativities** of the bonded atoms. **Electronegativity is a measure of the relative ability of two atoms to attract the pair of electrons forming a single covalent bond between them.** High electronegativity identifies an element with a strong attraction for bonding electrons.

Electronegativity values of Group A elements are shown in Figure 10.6. Notice that electronegativities tend to be greater at the top of any column. This is because the bonding electrons are closer to the nucleus in a smaller atom, and are therefore attracted by it more strongly. Electronegativities also increase from left to right across any row of the periodic table. This matches the increase in nuclear charge among atoms whose bonding electrons are in the same principal energy level. Perhaps you recognize these two explanations. They are identical with those given for atomic and ionic sizes (Sections 5.5 and 10.2). In general, electronegativities are highest at the upper right region of the periodic table, and lowest in the lower left region.

You can estimate the polarity of a bond by calculating the difference between the electronegativity values for the two elements: the greater the difference, the more polar the bond. In nonpolar H_2 and F_2 molecules, where two atoms of the same element are bonded, the electronegativity difference is zero. In the polar HF molecule the electronegativity difference is 4.0 for fluorine minus 2.1 for hydrogen, or 1.9. A bond between carbon and chlorine, for example, with an electronegativity difference of $3.0 - 2.5 = 0.5$, is more polar than an H—H bond, but less polar than an H—F bond.

The more electronegative element toward which the bonding electrons are displaced acts as the "negative pole" in a polar bond. This is sometimes indicated by using an arrow rather than a simple dash, with the arrow pointing to the negative pole. In a bond between hydrogen and fluorine this is $H \rightarrow F$.

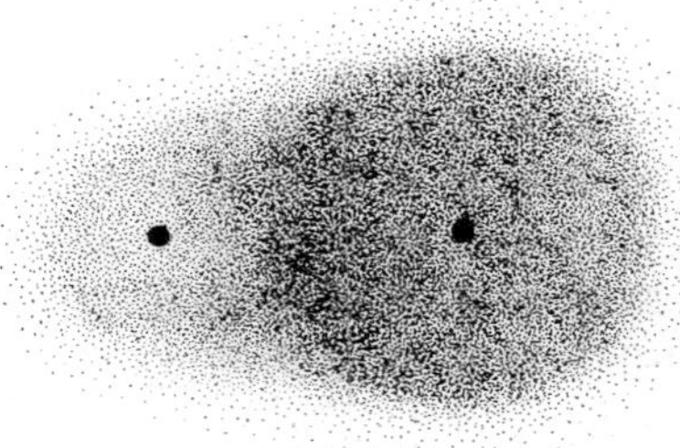

Figure 10.5
A polar bond. Fluorine in a molecule of HF has a higher electronegativity than hydrogen. The bonding electron pair is therefore shifted toward fluorine. The nonsymmetrical distribution of charge yields a polar bond.

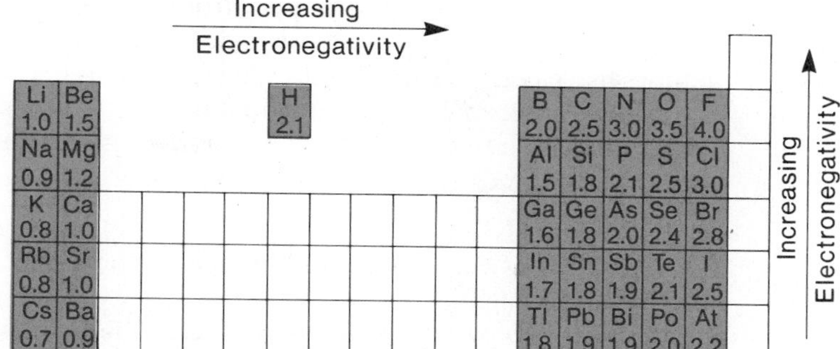

Increasing
Electronegativity
Li 1.0
Be 1.5
Na 0.9
Mg 1.2
K 0.8
Ca 1.0
Rb 0.8
Sr 1.0
Cs 0.7
Ba 0.9
H 2.1
B 2.0
C 2.5
N 3.0
O 3.5
F 4.0
Al 1.5
Si 1.8
P 2.1
S 2.5
Cl 3.0
Ga 1.6
Ge 1.8
As 2.0
Se 2.4
Br 2.8
In 1.7
Sn 1.8
Sb 1.9
Te 2.1
I 2.5
Tl 1.8
Pb 1.9
Bi 1.9
Po 2.0
At 2.2
Increasing
Electronegativity

Figure 10.6 Electronegativities of the representative elements.

EXAMPLE 10.3 Using data from Figure 10.7, arrange the following bonds in order of increasing polarity, and circle the element that will act as the negative pole.

H—O H—S

H—P H—C

Locate the elements in the table, calculate the differences in electronegativity, and enter those differences.

----- -----

H—O: $3.5 - 2.1 = 1.4$ H—S: $2.5 - 2.1 = 0.4$

H—P: $2.1 - 2.1 = 0.0$ H—C: $2.5 - 2.1 = 0.4$

Now arrange the bonds in order from the least polar to the most polar.

----- -----

H—P, H—S, H—C, H—O or H—P, H—C, H—S, H—O

The H—P bond is nonpolar (electronegativity difference = 0). H—O is the most polar bond because it has the largest electronegativity difference (1.4). The other bonds, with equal electronegativity differences, have about the same polarity.

Now draw a circle around the atom in each bond that will act as the negative pole.

----- -----

H—P H—(S) H—(C) H—(O)

Since sulfur, carbon, and oxygen are all more electronegative than hydrogen, the electron density in these three bonds is shifted away from hydrogen toward the other element. There is no electronegativity difference in the H—P bond, so neither atom will act as a negative pole.

Electronegativity numbers can also be used to predict whether a bond will be nonpolar covalent, polar covalent, or ionic. We must be cautious, however, not to give these predictions more credit than they deserve. In the first place, there is no sharp difference between the three bond classifications. The whole range of polarities passes *gradually* from a pure nonpolar covalent bond between identical atoms to the most ionic bond in cesium fluoride, CsF. Various authors use electronegativity differences anywhere from 1.6 to 1.9 as the crossover point between covalent and ionic bonds, which shows how arbitrary the classifications are. Only laboratory measurements can really determine the polarity of a bond. It is probably safe to say that:

1. The only truly nonpolar bond is between identical atoms.
2. The bond between atoms with an electronegativity difference less than 1.6 is polar covalent.
3. An electronegativity difference greater than 1.9 identifies a bond that is primarily ionic.
4. If the electronegativity difference is from 1.6 to 1.9, the bond may be considered as very strongly polar or slightly ionic.

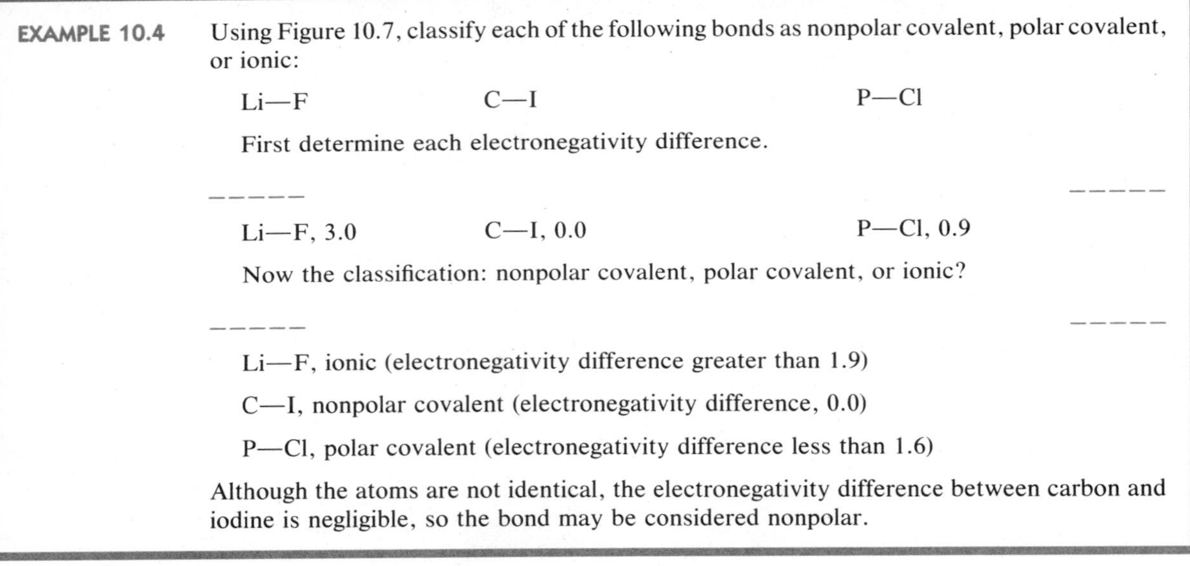

EXAMPLE 10.4 Using Figure 10.7, classify each of the following bonds as nonpolar covalent, polar covalent, or ionic:

Li—F C—I P—Cl

First determine each electronegativity difference.

----- -----

Li—F, 3.0 C—I, 0.0 P—Cl, 0.9

Now the classification: nonpolar covalent, polar covalent, or ionic?

----- -----

Li—F, ionic (electronegativity difference greater than 1.9)

C—I, nonpolar covalent (electronegativity difference, 0.0)

P—Cl, polar covalent (electronegativity difference less than 1.6)

Although the atoms are not identical, the electronegativity difference between carbon and iodine is negligible, so the bond may be considered nonpolar.

In Chapter 11 you will use bond polarities to predict the polarities of molecules. In Chapter 14 you will find that molecular polarity is largely responsible for the physical properties of many compounds.

10.6 MULTIPLE BONDS

So far our consideration of covalent bonds has been limited to the sharing of one pair of electrons by two bonded atoms. Such a bond is called a **single bond.** In many molecules we find two atoms bonded by two pairs of electrons; this is a **double bond.** When two atoms are bonded by three pairs of electrons it is called a **triple bond.** All four electrons in a double bond and all six electrons in a triple bond are counted as valence electrons for each of the bonded atoms.

Probably the most abundant substance containing a triple bond is in the nitrogen molecule, N_2. Its Lewis diagram may be thought of as the combination of two nitrogen atoms, each with three unpaired electrons:

$$:\dot{\underset{\cdot}{N}}\cdot + \cdot\dot{\underset{\cdot}{N}}: \rightarrow :N\vdots\vdots N: \quad \text{or} \quad :N\equiv N:$$

Counting the bonding electrons for both atoms, each nitrogen atom is satisfied with a full octet of electrons.

There is experimental evidence to support the idea of **multiple bonds,** a general term that includes both double and triple bonds. A triple bond is stronger and the distance between bonded atoms is shorter than the same measurements for a double bond between the same atoms, and a double bond is shorter and stronger than a single bond. Bond strength is measured as the energy required to break a bond. The triple bond in N_2 is among the strongest bonds known. This is why elemental nitrogen is so stable and unreactive in the earth's atmosphere.

In Chapter 11 you will examine multiple bonds that are found in more complex molecules.

CHAPTER 10 IN REVIEW

10.1 Monatomic Ions With Noble Gas Electron Configurations

10A Identify the monatomic ions that are isoelectronic with a given noble gas atom, and write the electron configurations of those ions.

10.2 Sizes of Ions

10B Compare and explain the sizes of given isoelectronic monatomic ions.

10C Compare and explain the sizes of given monatomic ions formed by elements in the same group in the periodic table.

10.3 Ionic Bonds

10.4 Covalent Bonds

10D Distinguish between ionic and covalent bonds.

10.5 Polar and Nonpolar Covalent Bonds

10E Distinguish between polar and nonpolar bonds.

10F Given the electronegativities of two elements, classify the bond between them as nonpolar covalent, polar covalent, or primarily ionic. Identify the positive and negative ends of polar bonds.

10.6 Multiple Bonds

TERMS AND CONCEPTS

10.3 Crystal
Ionic bond
Ionic compound

10.4 Molecule
Molecular compound
Covalent bond
Electron cloud
Charge density
Overlap
Lewis diagram, formula, structure
Lone pair
Octet rule, rule of eight

10.5 Nonpolar bond, polar bond
Electronegativity

10.6 Single, double, triple, multiple bond

Most of these terms, and many others, are defined in the Glossary. Use your Glossary regularly.

QUESTIONS AND PROBLEMS

Section 10.1

(1) Referring only to a periodic table, identify those third-period elements that form monatomic ions that are isoelectronic with a noble gas atom. Write the symbol for each such ion (example: Ca^{2+} in the fourth period).

(19) Write the electron configuration of each third period monatomic ion identified in Question 1. Also identify the noble gas atoms having the same configurations.

(2) Identify two negatively charged monatomic ions that are isoelectronic with neon.

(20) Identify by symbol two positively charged monatomic ions that are isoelectronic with krypton ($Z = 36$).

(3) Write the symbols of two ions that are isoelectronic with the chloride ion.

(21) Write the symbols of two ions that are isoelectronic with the barium ion.

(4)* A monatomic ion with a -2 charge has the electron configuration $1s^22s^22p^63s^23p^64s^23d^{10}4p^6$. (a) What neutral noble gas atom has the same electron configuration? (b) What is the monatomic ion with a -2 charge and this configuration? (c) Write the symbol of an ion with a $+1$ charge that is isoelectronic with the two above species.

(22)* If the monatomic ions in Question 10.4(b) and (c) combine to form an ionic compound, what will be the formula of that compound? Which atom will be at the negative end of each bond in the compound?

Section 10.2

(5) Identify the largest ion among the following: Li^+; F^-; S^{2-}.

(23) Which among the following is the smallest ion: Se^{2-} ($Z = 34$); Br^-; Te^{2-} ($Z = 52$)?

(6) Arrange the following ions in order of their increasing size (smallest ion first): Br^-; Ca^{2+}; Cl^-; K^+.

(24) List the following ions in order of their decreasing size (largest ion first): Ba^{2+}; Cs^+ ($Z = 55$); I^-; Rb^+ ($Z = 37$).

(7) State the size trend among ions and noble gas atoms that have the same electron configuration. Suggest an explanation for this trend.

(25) State the size trend among monatomic ions derived from elements in the same group in the periodic table. Suggest an explanation for this trend.

Section 10.3

(8) Using Lewis symbols, show how ionic bonds are formed by atoms of sulfur and potassium, leading to the correct formula of potassium sulfide.

(26) Aluminum oxide is an ionic compound. Sketch the transfer of electrons from aluminum atoms to oxygen atoms that accounts for the chemical formula of the compound.

Section 10.4

(9) Explain why ionic bonds are called electron *transfer* bonds, and covalent bonds are known as electron *sharing* bonds.

(27) Show how atoms achieve the stability of noble gas atoms in forming covalent bonds.

(10) Compare the bond between potassium and chlorine in potassium chloride with the bond between two chlorine atoms in chlorine gas. Which bond is ionic, and which is covalent? Describe how each bond is formed.

(28) Considering bonds between the following pairs of elements, which are most apt to be ionic and which are most apt to be covalent: sodium and sulfur; fluorine and chlorine; oxygen and sulfur? Explain your choice in each case.

(11) Show how a covalent bond forms between an atom of iodine and an atom of chlorine, yielding a molecule of ICl.

(12)* "The bond between a metal atom and a nonmetal atom is most apt to be ionic, whereas the bond between two nonmetal atoms is most apt to be covalent." Explain why this statement is true.

(*13*)* Does the energy of a system tend to increase, decrease, or remain unchanged as two atoms form a covalent bond?

(29)* Sketch the formation of two covalent bonds by an atom of sulfur in making a molecule of hydrogen sulfide, H_2S.

(30)* The bond between two metal atoms is neither ionic nor covalent. Explain, according to the octet rule, why this is so.

(31)* How do the energy and stability of bonded atoms and noble gas electron configurations appear to be related in forming covalent and ionic bonds?

Section 10.5

(*14*) What is meant by saying that a bond is *polar* or *nonpolar*? What bonds are completely nonpolar?

(15) Consider the following bonds: F—Cl; Cl—Cl; Br—Cl; I—Cl. Arrange these bonds in order of increasing polarity (lowest polarity first). Based on Figure 10.6, classify each bond as (a) nonpolar or (b) polar covalent.

(16) For each polar bond in Question 15, identify the atom that acts as the negative pole.

(17) What is electronegativity? Why are the noble gases not included in the electronegativity table?

(32) Compare the electron cloud formed by the bonding electron pair in a polar bond with that in a nonpolar bond.

(33) List the following bonds in order of decreasing polarity: K—Br; S—O; N—Cl; Li—F; C—C. Classify each bond as (a) nonpolar, (b) polar covalent, or (c) primarily ionic.

(34) For each bond in Question 33, identify the positive pole, if any.

(35) Identify the trends in electronegativities that may be observed in the periodic table.

Section 10.6

(*18*) What is a multiple bond? Distinguish between single, double, and triple bonds.

(36) Double bonds and triple bonds conform to the octet rule. Could a quadruple (4) bond obey that rule? a quintuple (5) bond?

Miscellaneous Questions

(*37*) Distinguish precisely and in scientific terms the differences between items in each of the following groups:
- (a) Ionic compound, molecular compound
- (b) Ionic bond, covalent bond
- (c) Lone pair, bonding pair (of electrons)
- (d) Nonpolar bond, polar bond
- (e) Single, double, triple, multiple bonds

(38) Classify each of the following statements as true or false:
- (a) K^+ is larger than Ca^{2+} because calcium has more neutrons in its nucleus than potassium.
- (b) It is reasonable to predict that Mg^{2+} is larger than Cl^- because monatomic metal ions are generally larger than monatomic nonmetal ions in the same period.
- (c) A single bond between carbon and nitrogen is polar covalent.
- (d) A bond between phosphorus and sulfur will be less polar than a bond between phosphorus and chlorine.
- (e) The electronegativity of calcium is less than that of aluminum.
- (f) Potassium ions are larger than aluminum ions.
- (g) Strontium ($Z = 38$) ions are isoelectronic with bromide ions.
- (h) The monatomic ion formed by selenium ($Z = 34$) is expected to be isoelectronic with a noble gas atom.

(i) It is reasonable to expect that one of the ions of cobalt is isoelectronic with a noble gas atom.

(j) Elements in Group 4A do not normally form monatomic ions.

(39) What is the electron configuration of the hydrogen ion, H^+? Explain your answer to this question.

(40) O^{2-}, F^-, and Mg^{2+}. These species are listed in order of size, either increasing or decreasing. Decide which is the order, and then explain it.

(41) Identify the pair among the following that are not isoelectronic: (a) Ne and Na^+; (b) S^{2-} and Cl^-; (c) Mg^{2+} and Ar; (d) K^+ and S^{2-}; (e) Ba^{2+} and Te^{2-} ($Z = 52$).

(42) Which bond formed between atoms of two elements whose atomic numbers are given would be expected to be the most ionic: (a) 8 and 16; (b) 11 and 35; (c) 17 and 20; (d) 3 and 53; (e) 9 and 55.

(43) List the following in order of increasing size: Al^{3+} Ar Cl^- K

(44) Which orbitals of each atom overlap in forming a bond between bromine and oxygen?

(45) Which are larger, potassium atoms or potassium ions? Why?

(46) Do ionic bonds appear in molecular compounds? Do covalent bonds appear in ionic compounds?

(47) If you did not have an electronegativity table, could you predict the relative electronegativities of elements whose positions are (B)—(A) [B lower left, A upper right] in the B periodic table? What about elements whose positions are (X)—(Y) [X upper left, Y lower right]? In both cases, explain why or why not.

(48) Is there any such thing as a completely nonpolar bond? If yes, give an example. Is there any such thing as a completely ionic bond? If yes, give an example. Explain both answers.

11 The Structure of Molecules

LOOKING BACK

10.4 Atoms form covalent bonds by sharing one or more pairs of electrons in such a way that both atoms become isoelectronic with a noble gas atom, usually in accord with the octet rule.

10.5 A bond is nonpolar when its charge distribution is symmetrical, and polar when its charge distribution is not symmetrical.

10.6 Double and triple bonds sometimes are needed to satisfy the octet rule.

LOOKING AHEAD IN CHAPTER 11

The octet rule will be used to draw Lewis diagrams of polyatomic molecules, including ions.

A molecule is nonpolar when its charge distribution is symmetrical, even if it has polar bonds, and polar if the charge distribution is not symmetrical.

When the number of electrons needed to bond atoms by single bonds is greater than the number of valence electrons available, a double or triple bond is formed.

So far in our consideration of covalent bonds, we have concentrated on the bonds between two atoms. If a molecule has only two atoms, such as H_2, F_2, and HF, all of which were discussed in Section 10.4, the description of the bond describes the molecule as well. But when an atom is bonded to two or more atoms, it is natural to wonder how the bonds are arranged in molecules and how the arrangements affect physical and chemical properties of the substance. The first of these questions is addressed in this chapter; in Chapter 14, you will study the second.

11.1 DRAWING LEWIS DIAGRAMS BY INSPECTION

PG 11A Draw the Lewis diagram for any molecule or polyatomic ion made up of representative elements (Sections 11.1 to 11.3).

Lewis diagrams can often be drawn by inspection if all atoms have the electron configuration of a noble gas. Each atom except hydrogen has an octet of electrons; hydrogen has two electrons, matching helium. In Section 10.4 you saw how the single 1s electron of hydrogen forms a bond with the single unpaired 2p electron of fluorine to form an HF molecule:

$$\mathrm{H}\cdot + \cdot\ddot{\underset{\cdot\cdot}{\mathrm{F}}}: \rightarrow \mathrm{H}:\ddot{\underset{\cdot\cdot}{\mathrm{F}}}: \quad or \quad \mathrm{H}—\ddot{\underset{\cdot\cdot}{\mathrm{F}}}:$$

In general, unpaired electrons from two different atoms are capable of pairing to form covalent bonds, and do so until all atoms reach the electron configuration of a noble gas.

The Lewis symbols for oxygen, nitrogen, and carbon are

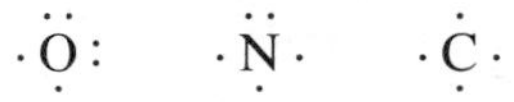

$$\cdot\ddot{\underset{\cdot}{O}}: \qquad \cdot\ddot{\underset{\cdot}{N}}\cdot \qquad \cdot\dot{\underset{\cdot}{C}}\cdot$$

Oxygen has six valence electrons, two of which are unpaired. It needs two more electrons to complete its octet. It can reach this point if each unpaired oxygen electron forms a bond with the electron from one hydrogen atom. In other words, it does twice what the fluorine atom does once in forming a molecule of HF. The result is a molecule of water, H_2O. With nitrogen there are three unpaired electrons, so it can form three bonds, one with each of three different hydrogen atoms. This yields a molecule of ammonia, NH_3. Carbon has four valence electrons, so it forms covalent bonds with four hydrogen atoms to produce a molecule of methane, CH_4.* The Lewis diagrams for these compounds are shown:

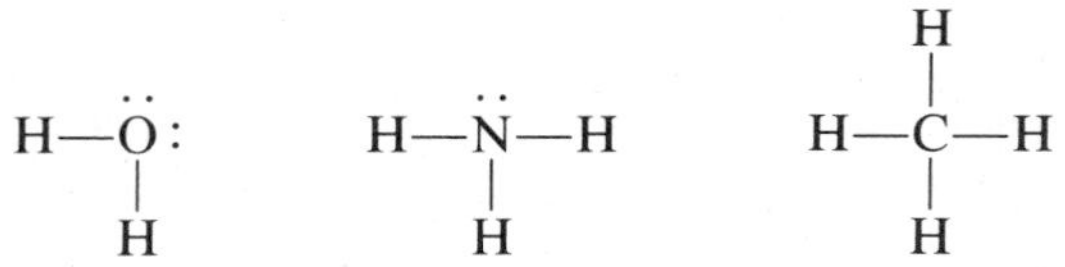

```
 H—Ö:        H—N̈—H          H
   |           |             |
   H           H          H—C—H
                             |
                             H
  H2O         NH3          CH4
```

EXAMPLE 11.1 Draw Lewis diagrams for carbon tetrachloride, CCl_4, and phosphorus tribromide, PBr_3.

A carbon atom has four valence electrons. It requires four more electrons to complete its octet and gets them by forming four bonds. Write the Lewis symbol for chlorine. From that symbol state the number of electrons each chlorine atom must gain to complete its octet.

$\cdot\ddot{\underset{\cdot\cdot}{Cl}}:$ The atom has seven valence electrons and needs one to reach eight.

Assuming that new bonds will be formed when each atom contributes one electron to the bond, draw the Lewis diagram of CCl_4 in the space at the right.

-----

*A carbon atom, with its electron configuration $1s^22s^22p^2$, actually has only two unpaired electrons, the two 2p electrons. It is a fact, however, that carbon atoms form covalent bonds with four hydrogen atoms. One explanation for this involves "hybridized" orbitals, a topic beyond the scope of this text.

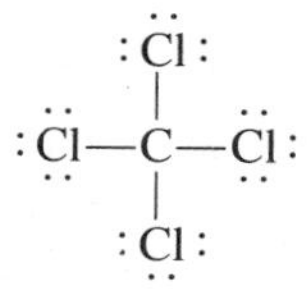

```
       ..
      :Cl:
   ..  |  ..
  :Cl—C—Cl:
   ..  |  ..
      :Cl:
       ..
```

Each of four electrons from a carbon atom forms a bond with the unpaired electron from a separate chlorine atom. This is just like the Lewis diagram for methane, CH_4.

Now draw the Lewis symbols of a phosphorus atom and a bromine atom. For each, state the number of electrons required to complete the octet.

– – – – –   – – – – –

```
  ..                               ..
 ·P·  Three electrons required.   ·Br:  One electron required.
  ·                                ..
```

Finally, the Lewis diagram for PBr_3.

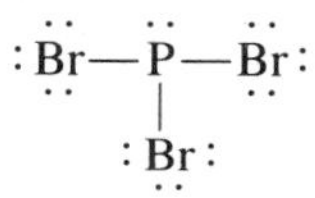

– – – – –   – – – – –

```
  ..   ..   ..
 :Br—P—Br:
  ..   |    ..
     :Br:
      ..
```

Notice the similarity between this diagram and that for NH_3. The central elements are both in Group 5A and have five valence electrons. The other element in each case requires one additional electron to reach a noble gas configuration. The inspection methods for drawing the diagrams are the same.

In all examples so far, covalent bonds have been formed when each atom contributes one electron to the bonding pair. This is not always the case. Many bonds are formed where one atom contributes both electrons and the other atom offers only an empty orbital. This is called a **coordinate covalent bond.** To illustrate, an ammonium ion is produced when a hydrogen ion is bonded to the unshared electron pair of the nitrogen atom in an ammonia molecule:

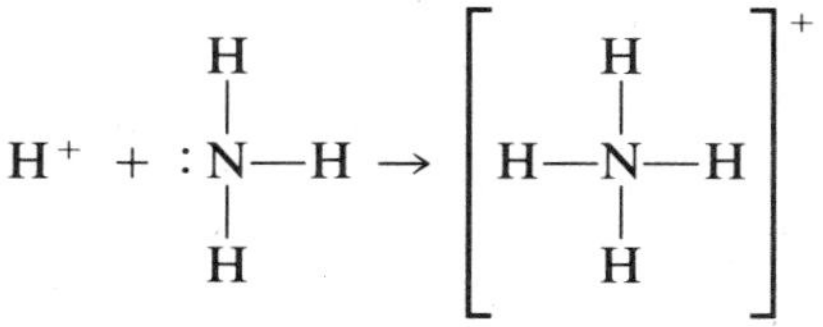

```
             H              ┌     H      ┐+
             |              │     |      │
   H+  +  :N—H   →   │ H—N—H │
             |              │     |      │
             H              └     H      ┘
```

The four bonds in an ammonium ion are identical. This shows that a coordinate covalent bond is the same as a bond formed by one electron from each atom.

Notice that in drawing the Lewis diagram of an ion, the diagram is enclosed in brackets and the charge is shown as a superscript.

11.2 MULTIPLE BONDS

Sometimes the number of electrons and the number of atoms in a molecule do not permit a Lewis diagram in which every bond is formed by a single pair of electrons, and each atom has the configuration of a noble gas. The development of the diagram for ethene, C_2H_4, illustrates both the problem and the solution. To emphasize the contrast, this will be shown side by side with the Lewis diagram of ethane, C_2H_6, a compound where the problem is not present. In both molecules the carbon atoms are bonded to each other, leaving six electrons—three on each carbon—still available for bonding:

$$\cdot\dot{\underset{\cdot}{C}}—\dot{\underset{\cdot}{C}}\cdot$$

With ethane there are six hydrogen atoms to form covalent bonds with the six available electrons, but with ethene there are only four hydrogen atoms:

$$\begin{array}{ccccc} & H & & H & \\ & | & & | & \\ H— & C & — & C & —H \\ & | & & | & \\ & H & & H & \end{array} \qquad\qquad \begin{array}{ccccc} & H & & H & \\ & | & & | & \\ H— & C & — & C & —H \\ & \cdot & & \cdot & \end{array}$$

Ethane, C_2H_6 Ethene, C_2H_4

At this point each carbon atom in ethane is surrounded by a full octet of electrons, but the carbon atoms in ethene have only seven electrons around them. The problem is resolved, however, if the unpaired electrons in ethene combine to form a *second* bond between the two carbon atoms.

$$\begin{array}{ccccc} & H & & H & \\ & | & & | & \\ H— & C & — & C & —H \\ & \cdot\nearrow & & \nwarrow\cdot & \end{array} \rightarrow \begin{array}{ccccc} & H & & H & \\ & | & & | & \\ H— & C & = & C & —H \end{array}$$

With two pairs of electrons between the carbon atoms, each carbon atom is surrounded by eight electrons, satisfying the octet rule. When two atoms are bonded by *two* pairs of electrons, they are held together by a **double bond.**

The idea of multiple bonds was introduced in Section 10.6. The N_2 molecule was used to illustrate a triple bond between two nitrogen atoms. Multiple bonding is not limited to bonds between atoms of the same element. For example, the Lewis diagrams of formaldehyde and the cyanide ion are, respectively,

$$\begin{array}{l} \qquad\ddot{O}: \\ \quad /\!\!/ \\ H—C \\ \quad \diagdown \\ \qquad H \end{array} \quad \text{and } [:C\equiv N:]^-.$$

11.3 DRAWING COMPLEX LEWIS DIAGRAMS

Lewis diagrams are not readily drawn by inspection for some of the larger or more complex molecules and polyatomic ions. The procedure that follows may be used to sketch the diagram for any species that obeys the octet rule. Each step will be illustrated by drawing the Lewis diagram for BrO_2^-.

1. Draw a tentative diagram for the molecule or ion, joining atoms by single bonds. Place electron dots around each symbol except hydrogen so the total number of electrons for each atom is eight. In some cases, only one arrangement of atoms is possible. In others, two or more structures may be drawn. Ultimately chemical or physical evidence must be used to decide which of the possible structures is correct. A few general rules will help you in making diagrams that are most likely to be correct:

a. A hydrogen atom always forms one bond; a carbon atom normally forms four bonds.
b. When several carbon atoms appear in the same molecule, they are often bonded to each other. In some compounds they are arranged in a closed loop; however, we will avoid such so-called cyclic compounds in this text.
c. In compounds or ions having two or more oxygen atoms and one atom of another nonmetal, the oxygen atoms are usually arranged *around* the central nonmetal atom.
d. In an oxyacid (hydrogen + oxygen + a nonmetal, such as H_2SO_4 or HNO_3), hydrogen is usually bonded to an oxygen atom, which is then bonded to the nonmetal: H—O—X, where X is a nonmetal.

For BrO_2^-, general rule 1c says the oxygen atoms should be distributed around the bromine atom: $:\ddot{\underset{\cdot\cdot}{O}}-\ddot{\underset{\cdot\cdot}{Br}}-\ddot{\underset{\cdot\cdot}{O}}:$

2. Count the electrons in your diagram.

In $:\ddot{\underset{\cdot\cdot}{O}}-\ddot{\underset{\cdot\cdot}{Br}}-\ddot{\underset{\cdot\cdot}{O}}:$ there are eight lone pairs (16 electrons) and two bonding pairs (4 electrons), a total of 20 electrons.

3. Find the total number of valence electrons available. For a molecule, this is the sum of the valence electrons contributed by each atom in the molecule. For a polyatomic ion, this total must be adjusted to account for the charge on the ion. An ion with a -1 charge will have one more electron than the number of valence electrons in the neutral atoms; a -2 ion, two more; a $+1$ ion, one less; and so forth.

Recall that the number of valence electrons of a Group A element is the same as the group number. In BrO_2^- there are seven valence electrons from bromine (Group 7A), six from each oxygen (Group 6A), and one from the -1 charge: $7 + 2(6) + 1 = 20$ electrons.

4. Compare the numbers in Steps 2 and 3. If they are the same, the diagram is complete. If they are different, modify the diagram with multiple bonds. If your diagram has two electrons more than the number available, there will be one double bond. If the difference is four, there will be a triple bond or two double bonds. Multiple bonds should be used only when necessary, and then use as few as possible.

For BrO_2^-, Steps 2 and 3 both yielded 20 electrons. Therefore, $\left[:\ddot{\underset{\cdot\cdot}{O}}-\ddot{\underset{\cdot\cdot}{Br}}-\ddot{\underset{\cdot\cdot}{O}}:\right]^-$ is the correct diagram. It has been enclosed in square brackets because it is a polyatomic ion.

EXAMPLE 11.2 Write Lewis diagrams for the ClF molecule and the ClO^- ion.

Step 1 is to draw the tentative diagrams, surrounding each atom by eight electrons. There is only one possibility for each species.

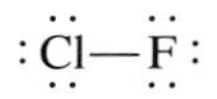

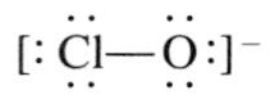

$:\ddot{\underset{..}{Cl}}—\ddot{\underset{..}{F}}:$ $[:\ddot{\underset{..}{Cl}}—\ddot{\underset{..}{O}}:]^-$

Now count the electrons in each diagram. Remember that the dash between symbols represents two electrons.

Both diagrams have 14 electrons.

Now add up the total number of valence electrons in each species. For the ion, increase the total by 1 to account for the −1 charge.

ClF, 14; ClO^-, 14

In ClF, simply add the valence electrons of the neutral atoms. Both Cl and F are in Group 7A, so they both have seven valence electrons. In ClO^-, there are seven valence electrons from chlorine (Group 7A) plus six from oxygen (Group 6A), plus one for the −1 charge on the ion, a total of 14.

The final step is to compare the number of valence electrons with the number in the tentative diagrams. If they are the same, the diagrams are complete. Such is the case with both species, so the tentative diagrams are the final diagrams.

The above example illustrates the fact that two species having the same number of electrons and the same number of atoms have similar Lewis diagrams, whether they are molecules or polyatomic ions.

EXAMPLE 11.3 Draw the Lewis diagram for the sulfite ion, SO_3^{2-}.

Begin with the tentative diagram. Recall Item c under Step 1 in the procedure.

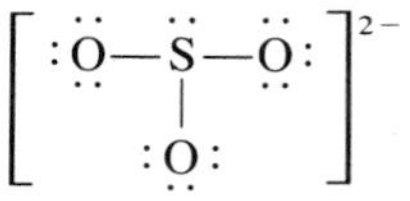

$$\left[\begin{array}{c}:\ddot{\underset{..}{O}}—\ddot{S}—\ddot{\underset{..}{O}}: \\ | \\ :\underset{..}{O}:\end{array}\right]^{2-}$$

How many electrons appear in the diagram?

26

Now count the valence electrons in SO_3^{2-}. Don't forget the ionic charge.

------ ------

6 (sulfur, Group 6A) + 3(6) (3 oxygens, Group 6A) + 2 (−2 charge) = 26

Anything else?

------ ------

No. The electrons in the diagram and the electrons available are the same, so the tentative diagram is the final diagram.

EXAMPLE 11.4 Draw the Lewis diagram for SO_2.

Begin with the tentative diagram.

------ ------

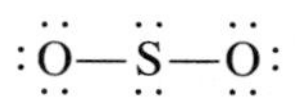

:Ö—S̈—Ö:

How many electrons are in the diagram?

------ ------

20

How many valence electrons are in an SO_2 molecule?

------ ------

6 (sulfur, Group 6A) + 2(6) (2 oxygens, Group 6A) = 18

Is the tentative diagram the final diagram?

------ ------

No.

The number of electrons in the tentative diagram (20) is two more than the total number of valence electrons available (18). A difference of 2 indicates one double bond. See if you can modify the structure so it has one double bond, and adjust the number of lone pairs so each atom continues to obey the octet rule. Check to be sure your new diagram has 18 electrons in it.

------ ------

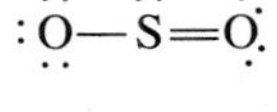

:Ö—S̈=Ȯ.

The mechanics of the double bond adjustment are discussed below.

Let's analyze the change from the tentative diagram to the final diagram in Example 11.4. Whenever you must reduce the number of electrons in your tentative diagram by 2, look for two atoms that are (1) bonded to each other and (2) each of which has a lone pair. Remove from the diagram the two lone pairs (takes away 4 electrons) and draw a second bond between the atoms (puts 2 electrons back). Taking away 4 electrons and putting 2 back reduces the total number of electrons by 2. For SO_2,

$$:\ddot{\underset{..}{O}}-\ddot{\underset{..}{S}}-\ddot{\underset{..}{O}}: \qquad :\ddot{\underset{..}{O}}-\ddot{S}=\ddot{O}:$$

Remove 4 Return 2

If a diagram requires a triple bond, each bonded atom in the tentative diagram will have two lone pairs. The above procedure is simply performed twice.

You may wonder if it makes any difference on which side of the sulfur atom the double bond is placed. It does not, for the limited purpose of learning how to draw Lewis diagrams. It is a fact, however, that experimentally the bonds are identical, and have strengths and lengths that are between those normally found with single bonds and double bonds connecting sulfur and oxygen atoms. This condition is known as *resonance,* and is frequently shown as

$$\dot{\underset{\cdot}{O}}=\ddot{S}-\ddot{\underset{..}{O}}: \leftrightarrow :\ddot{\underset{..}{O}}-\ddot{S}=\dot{\underset{\cdot}{O}}$$

In this text we will not be concerned with resonance structures beyond this point of information.

The rules we are following are readily applied to simple organic* molecules, which always contain carbon atoms, usually include hydrogen atoms, and may contain atoms of other elements, notably oxygen. If oxygen is present in an organic compound, it usually forms two bonds. If you remember that carbon forms four bonds and hydrogen forms one, and that two or more carbon atoms often bond to each other (Rules 1a and 1b at the beginning of this section), your tentative structures are most apt to be correct.

EXAMPLE 11.5 Write the Lewis diagram for propane, C_3H_8.

What is the tentative diagram?

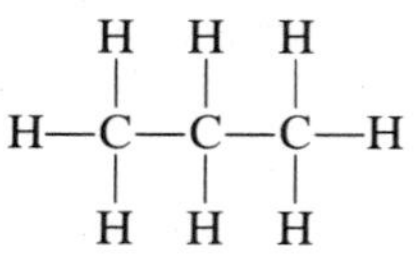

----- -----

```
    H   H   H
    |   |   |
H — C — C — C — H
    |   |   |
    H   H   H
```

How many electrons are in the diagram?

----- -----

*Organic chemistry is the chemistry of compounds containing carbon, other than certain "inorganic" carbon compounds such as carbonates, CO and CO_2. A knowledge of bonding, including Lewis diagrams, is important in organic chemistry.

20

How many valence electrons are in C_3H_8?

3(4) (3 carbons, Group 4A) + 8(1) (8 hydrogens, Group 1A) = 20

The electron counts are the same, so the tentative diagram is the final diagram.

The Lewis diagrams of methane (Section 11.1), ethane (Section 11.2), and propane are drawn side by side for comparison:

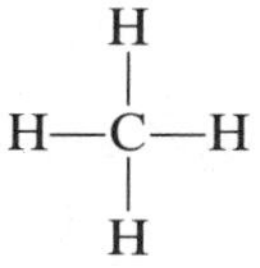

```
     H              H  H              H  H  H
     |              |  |              |  |  |
  H—C—H          H—C—C—H          H—C—C—C—H
     |              |  |              |  |  |
     H              H  H              H  H  H
```

Methane, CH_4 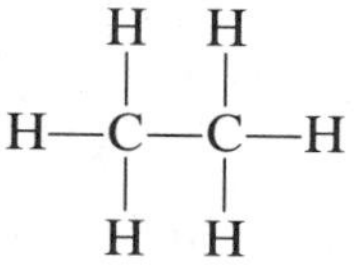   Ethane, C_2H_6 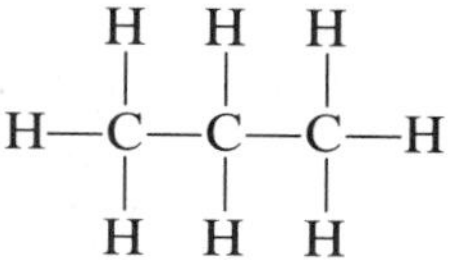   Propane, C_3H_8

Notice that each compound differs from the one before it by one

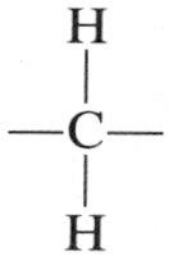

```
  H
  |
 —C—
  |
  H
```

unit. In fact, there is no end, theoretically, to the number of such units that might be inserted in the chain. The next member of the *alkane series,* as this is called, is C_4H_{10}, butane. It is possible to draw two structures of butane, and both are real but different compounds, each with its own set of physical properties. The diagrams are

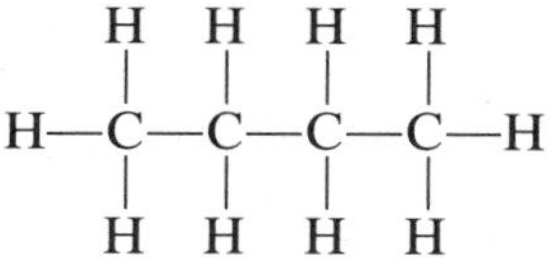

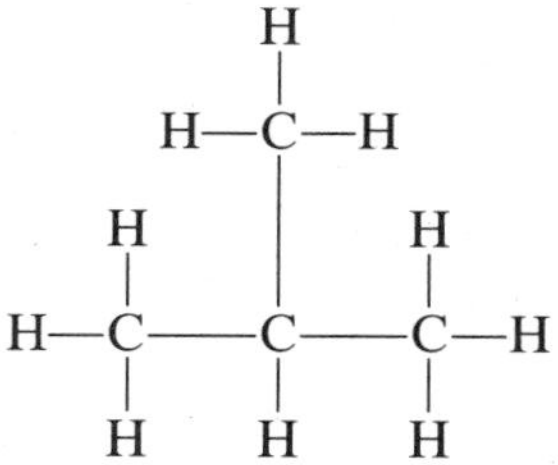

```
                               H
                               |
                             H—C—H
   H  H  H  H                  |
   |  |  |  |              H   |   H
H—C—C—C—C—H                |   |   |
   |  |  |  |            H—C———C———C—H
   H  H  H  H              |   |   |
                           H   H   H
```

Compounds having the same molecular formula but different structures are called **isomers** of each other. Among the alkanes the number of isomers increases dramatically as the number of carbon atoms increases. There are three pentanes, C_5H_{12}, five hexanes, C_6H_{14}, 35 nonanes, C_9H_{20}, over 300 000 with the formula $C_{20}H_{42}$, and over 100 000 000 having the formula $C_{30}H_{62}$! Needless to say, they have not all been studied. This gives you some idea why there are so many organic compounds—and we haven't even mentioned compounds that contain elements other than carbon and hydrogen. One kind of soap, for example, is $C_{18}H_{35}O_2Na$.

EXAMPLE 11.6 Draw the Lewis diagram of acetylene, C_2H_2

As usual, begin with the tentative diagram.

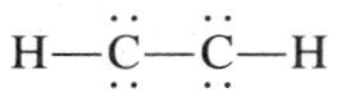

----- -----

```
   ..  ..
H—C—C—H
   ..  ..
```

How many electrons are in the tentative diagram?

----- -----

14

How many valence electrons are in C_2H_2?

----- -----

2(4) (2 carbons, Group 4A) + 2(1) (2 hydrogens, Group 1A) = 10

The difference in electron counts is 14 − 10 = 4. What do we do about that?

----- -----

A difference of 4 in electron count is adjusted by either a triple bond or two double bonds.

In this case there is only one choice. Complete the diagram, if you haven't already. Check the electron count.

----- -----

H—C≡C—H There are 10 electrons in the diagram.

EXAMPLE 11.7 Draw a Lewis diagram for C_2H_6O.

Start with a tentative diagram. Remember how many bonds carbon, hydrogen, and oxygen usually form. Also, bond the carbons to each other.

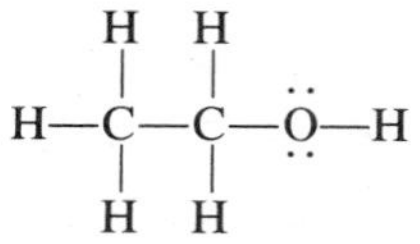

----- -----

```
   H  H
   |  |   ..
H—C—C—O—H
   |  |   ..
   H  H
```

Combine the remaining steps. Count the electrons in the diagram, count the valence electrons, and adjust the diagram if there is a difference between them.

----- -----

The number of electrons in the diagram is the same as the number of valence electrons. The tentative diagram is the final diagram.

The compound in Example 11.7 is ethyl alcohol. If we had not insisted that the carbon atoms be bonded together you might have produced the diagram for an isomer of alcohol, another well-known compound, dimethyl ether:

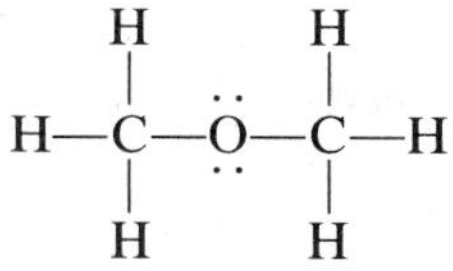

```
     H       H
     |  ..   |
H—C—O—C—H
     |  ..   |
     H       H
```

11.4 EXCEPTIONS TO THE OCTET RULE

Lewis diagrams for many substances may be drawn with the help of the octet rule, but some substances do not "obey" the rule. Two common oxides of nitrogen, NO and NO_2, have an odd number of electrons. It is therefore impossible to write Lewis diagrams for these compounds in which each atom is surrounded by eight electrons. Phosphorus pentafluoride, PF_5, places five electron-pair bonds around the phosphorus atom, and six pairs surround sulfur in SF_6.

Certain molecules whose Lewis diagrams obey the octet rule do not have the properties that would be predicted. Oxygen, O_2, was not used to introduce the double bond on page 214 for that reason. On paper, O_2 appears to have an ideal double bond:

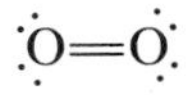

```
:O=O:
```

But liquid oxygen is *paramagnetic,* meaning that it is attracted by a magnetic field. (See Color Plate 2 opposite page 85. It shows liquid oxygen being held in place between the poles of a magnet.) This is characteristic of molecules that have unpaired electrons in their structures. This might suggest a Lewis diagram that has each oxygen surrounded by seven electrons:

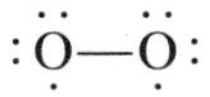

```
 ..   ..
:O—O:
 .    .
```

But this is in conflict with other evidence that the oxygen atoms are connected by something other than a single bond. In essence, it is impossible to write a single Lewis diagram that satisfactorily explains all the properties of molecular oxygen.

Two other substances for which satisfactory octet rule diagrams can be drawn, but are contradicted experimentally, are the fluorides of beryllium and boron. We might even expect BeF_2 and BF_3 to be ionic compounds, but laboratory evidence strongly supports covalent structures having the Lewis diagrams:

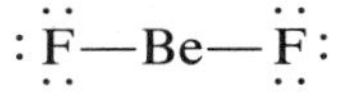

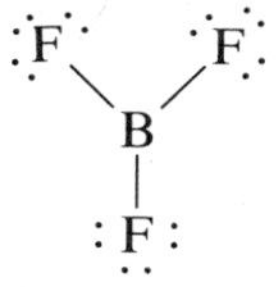

```
 ..        ..         ..     ..
:F—Be—F:         :F:     :F:
 ..        ..           \   /
                          B
                          |
                         :F:
                          ..
```

In these compounds the beryllium and boron atoms are surrounded by two and three pairs of electrons, respectively, rather than four.

11.5 MOLECULAR GEOMETRY

A Lewis diagram shows the order in which atoms are bonded to each other in a molecule, and by how many electron pairs they are bonded, but the diagram tells little about the geometry of the molecule. **Molecular geometry** identifies the **bond angles** in the molecule and its **shape.** A bond angle is the angle between any two bonds formed by the same atom (Fig. 11.1). It turns out that the shape of a molecule plays an important role in determining the physical and chemical properties of a substance, as you will see in Chapter 14.

To illustrate the shortcomings of a Lewis diagram, glance at the three shown for H_2O, NH_3, and CH_4 (Section 11.1). All three diagrams suggest that the bond angle between the central atom and any two hydrogen atoms is a right angle. This is not so. All of the bond angles are near 109°. In two dimensions it is impossible to draw four 109° angles around the central carbon atom in CH_4, or three such angles around a central nitrogen atom in NH_3. There is no way a two-dimensional Lewis diagram can give an accurate description of a three-dimensional shape.

Water is a two-dimensional molecule, since its three atoms lie in one plane. The Lewis diagram for water can be drawn with the proper bond angle. Even so, the diagram is not completely satisfactory because it provides no space for the unshared electron pairs, which should be placed above and below the plane of the paper.

If the atoms in a three-atom molecule are not in a straight line, as with water, the geometry of the molecule is said to be **bent** (Fig. 11.1). If the three atoms are in a straight line, as with carbon dioxide, it is a **linear** molecule, and the bond angle is 180°.

When a molecule has four or more atoms, they may or may not lie in the same plane. If they are in the same plane, the molecule is **planar.** Among molecules that are *not* planar, several geometric shapes are known. We will study only two, leaving the more complex forms to advanced courses. Each of these two looks like a pyramid with a triangular base. If the base and all three sides of a five-atom pyramid are identical equilateral triangles, the molecule has the shape of a **tetrahedron*** (Fig. 11.2A). The other pyramidal structure has four atoms and is called a **trigonal pyramid** to indicate a triangular base. It looks like a tetrahedron that has been "squashed down" (Fig. 11.2B).

*In solid geometry, a tetrahedron is the simplest regular solid. A *regular* solid is a solid figure with identical equilateral faces. A cube is a regular solid, having six identical squares for its faces.

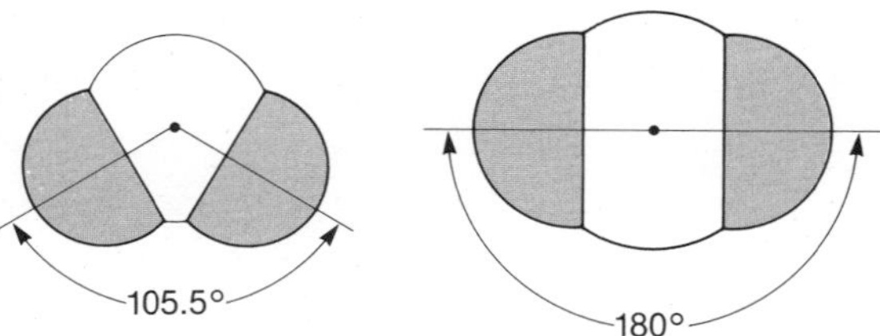

Figure 11.1
Bond angle. If an atom forms bonds with two other atoms, the angle between the bonds is the bond angle. In a water molecule the bonds form an angle of 105.5°. In a carbon dioxide molecule the bonds lie in a straight line. The bond angle is 180°.

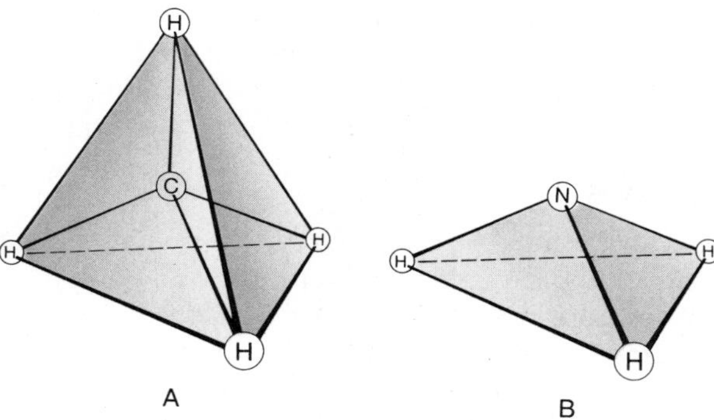

Figure 11.2
A. Methane, CH_4, is a typical five-atom molecule. The four hydrogen atoms are at the corners of a tetrahedron, and the carbon atom is at its center. The molecule is three dimensional. If the top hydrogen atom and the carbon atom are in the plane of the paper, the large hydrogen atom in the base is closer to you than the paper, and the other hydrogen atoms are behind the page.
B. Ammonia, NH_3, is a four-atom molecule having the shape of a pyramid with a triangular base. It is like the CH_4 molecule without the top hydrogen atom. The nitrogen atom is in the plane of the paper, the large hydrogen atom is in front of the paper, and the smaller hydrogen atoms are behind the page.

Quick Check 11.5

1. What is the only molecular geometry available to a two-atom molecule?
2. Name two possible molecular geometries for a three-atom molecule.
3. What two molecular geometries for four-atom molecules are described in this section?
4. Name the only five-atom molecular geometry described in this section.

11.6 ELECTRON-PAIR REPULSION: ELECTRON PAIR-GEOMETRY

No single theory or model yet developed succeeds in explaining all the molecular shapes that have been observed in the laboratory. A theory that explains one group of molecules fails when applied to another group. Each approach has advantages—and limitations. Chemists therefore use them all within the areas to which they apply, fully recognizing that there is still much to learn about how atoms are assembled in molecules.

In this text we will explore only one of the models used to explain molecular geometry. It is called the **valence-shell electron-pair repulsion principle, VSEPR.** According to this theory, the pairs of electrons that surround any atom in a molecule will, because of repulsion between similarly charged species, arrange themselves as far from each other as possible. The electron pairs may be shared in a covalent bond, or they may be unshared; it makes no difference, as far as electron-pair repulsion is concerned. The **electron-pair geometry** tells how electron pairs are arranged around a central atom.

Let us pause to distinguish clearly between two kinds of geometry. Electron-pair geometry describes the arrangement of all *electron pairs,* shared or unshared, around a central atom. Molecular geometry describes the arrangement of all *atoms* around the central atom to which they are bonded. The two are related: molecular geometry is the direct effect of electron-pair geometry.

Earlier in this chapter we drew Lewis diagrams for molecules in which central atoms were surrounded by an octet of electrons, four electron pairs. Some pairs make covalent bonds with other atoms, and some are lone pairs. In Section 11.4 you saw that not all molecules obey the octet rule. Sometimes a central atom is surrounded by only two or three pairs of electrons. We must therefore determine how two, three, or four electron pairs distribute themselves around a central atom if they are to be as far from each other as possible. It is a geometric fact that, when this condition is met, all angles formed between any two pairs of electrons and the central atom are equal. The three electron-pair geometries that result are shown in Figure 11.3. The reasoning behind those geometries is explained in the caption. The geometries are summarized below and in Table 11.1.

1. If the central atom is surrounded by two electron pairs, the atom is on the line between the pairs, the geometry is linear, and the "electron-pair" angle is 180°.
2. If the central atom is surrounded by three electron pairs, the atom is at the center of an equilateral triangle formed by the pairs, the geometry is **trigonal** (triangular) **planar,** and the "electron-pair" angle is 120°.
3. If the central atom is surrounded by four electron pairs, the atom is at the center of a tetrahedron formed by the pairs, the geometry is **tetrahedral,** and the "electron-pair" angle is 109.5°, often called the **tetrahedral angle.**

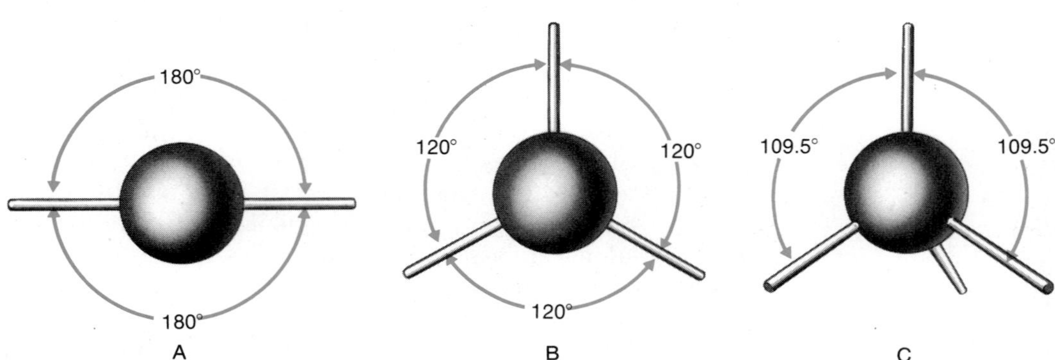

Figure 11.3
Electron-pair geometry. "Tinker-toy" models show the arrangement of two, three, and four electron pairs (sticks) around a central atom (ball). A. According to the electron-pair repulsion principle, two sticks are as far from each other as possible when they are diametrically opposite each other. The geometry is linear, and the angle formed is 180°. B. Three sticks are as far from each other as possible when equally spaced on a circumference of the ball. The sticks and the center of the ball are in the same plane, and the angles are 120°. The geometry is trigonal (triangular) planar. C. Four electron pairs are as far from each other as possible when arranged to form a tetrahedron. The angles are 109.5°, which are sometimes called "tetrahedral angles."

Table 11.1 Electron-Pair and Molecular Geometries

Line	Electron Pairs	Bonded Atoms	Electron-Pair Geometry	Ball and Stick Model	Ideal Bond Angle	Molecular Geometry	Lewis Diagram	Ball and Stick Model	Space Filling Model	Example	Actual Bond Angle
1	2	2	Linear		180°	Linear	A—B—A			BeF_2	180°
2	3	3	Trigonal (triangular) planar	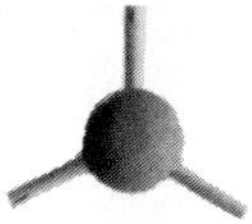	120°	Trigonal (triangular) planar	A A / B / A (B bonded to three A)	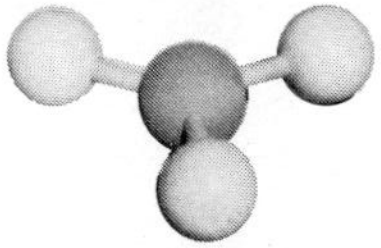		BF_3	120°
3	4	4	Tetrahedral		109.5°	Tetrahedral	A / A—B—A / A			CH_4	109.5°
4	4	3	Tetrahedral		109.5°	Trigonal (triangular) pyramid	A—B̈—A / A			NH_3	107.5°
5	4	2	Tetrahedral		109.5°	Bent	A—B—A (with two lone pairs on B) or A—B: / A	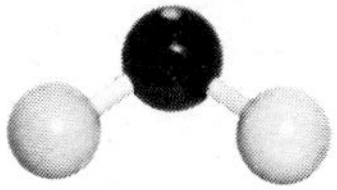		H_2O	105.5°

Quick Check 11.6

According to the VSEPR principle, what electron-pair geometry and bond angle are associated with

1. two electron pairs around a central atom?
2. three electron pairs around a central atom?
3. four electron pairs around a central atom?

11.7 MOLECULAR GEOMETRY BASED ON ELECTRON-PAIR REPULSION

PG 11B Given or having derived the Lewis diagram of a molecule in which a second period central atom is surrounded by two, three, or four pairs of electrons, predict the electron-pair geometry and the molecular geometry around that atom.

According to the electron-pair repulsion theory, bond angles between atoms are roughly the same as the "electron-pair" angles stated above. Minor changes appear if one or two electron pairs are unshared. We will give the molecular geometries for all combinations of electron pairs, including those in which some pairs are unshared. The descriptions that follow are summarized in Table 11.1. Line references below are to that table.

Two Electron Pairs, Two Bonded Atoms. Two electron pairs yield a linear electron-pair geometry (BeF_2, Line 1). Each electron pair bonds a fluorine atom to the central beryllium atom, creating a linear molecular geometry. The bond angle is 180°.

Three Electron Pairs, Three Bonded Atoms. Three electron pairs yield a trigonal planar electron-pair geometry (BF_3, Line 2). Each electron pair bonds a fluorine atom to the central boron atom, creating a trigonal planar molecular geometry. Each F—B—F angle is 120°.

Four Electron Pairs, Four Bonded Atoms. Four electron pairs produce a tetrahedral electron-pair geometry (CH_4, Line 3). Each electron pair bonds a hydrogen atom to the central carbon atom. The molecular geometry is the same as the electron-pair geometry, tetrahedral. Each H—C—H bond angle is the tetrahedral angle, 109.5°.

Four Electron Pairs, Three Bonded Atoms. As before, four electron pairs produce a tetrahedral electron-pair geometry (NH_3, Line 4). Only three of the electron pairs are bonded to hydrogen atoms, however. The four atoms make a molecule that has the shape of a low pyramid with the nitrogen atom at the top and the three hydrogen atoms forming an equilateral triangular base. This molecular geometry is a trigonal pyramid.

The bond angles in NH_3 do not conform to the "ideal" tetrahedral angle predicted by the four electron pairs. The absence of a hydrogen atom to compete with the nitrogen atom in attracting the unshared pair evidently allows the lone pair to be drawn closer to the nitrogen nucleus than the other three pairs. The lone pair appears to "push" the other three pairs closer to each other. This makes each H—N—H bond angle slightly less than the tetrahedral angle, 107.5° instead of 109.5°.

Four Electron Pairs, Two Bonded Atoms. Again, a tetrahedral electron-pair geometry is predicted (H_2O, Line 5). Only two hydrogen atoms are bonded to the central oxygen atom. The two lone pairs push the bonding pairs closer to each other than the tetrahedral angle, producing an actual bond angle of 105.5°. The three atoms, not in a straight line, have a bent molecular geometry.

EXAMPLE 11.8 Predict the electron-pair and molecular geometries of carbon tetrachloride, CCl_4.

The Lewis diagram, drawn in Example 11.1, is shown at the right. From this you should establish the number of electron pairs around the central atom and the number of atoms bonded to the central atom. Both geometries follow.

:C̈l:
:C̈l—C—C̈l:
:C̈l:

With four electron pairs around carbon, all bonded to other atoms, both geometries are tetrahedral.

EXAMPLE 11.9 Describe the shape of a molecule of boron trihydride, BH_3.

First draw the Lewis diagram. Remember that boron has only three valence electrons to contribute to covalent bonds. From the structure answer the question.

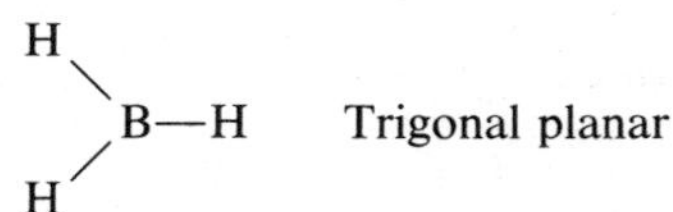

Three electron pairs yield both an electron-pair geometry and a molecular geometry that are trigonal planar with 120° bond angles.

EXAMPLE 11.10 Predict the electron pair geometry and shape of a molecule of dichlorine oxide, Cl_2O.

:C̈l—Ö:
:C̈l:

Electron pair geometry: tetrahedral; molecular geometry: bent

Oxygen has four electron pairs around it, yielding an electron pair geometry that is approximately tetrahedral. Only two of the electron pairs are bonded to other atoms, so the molecule is bent. The structure is similar to that of water.

11.8 THE GEOMETRY OF THE MULTIPLE BOND

Experimental evidence shows that the two or three electron pairs in a multiple bond behave as a single electron pair in establishing molecular geometry. This appears if we compare beryllium difluoride, carbon dioxide, and hydrogen cyanide, whose Lewis diagrams are

$$:\ddot{\underset{..}{F}}-Be-\ddot{\underset{..}{F}}: \qquad \ddot{\underset{..}{O}}=C=\ddot{\underset{..}{O}} \qquad H-C\equiv N:$$

All three molecules are linear; their bond angles are 180°. The two electron pairs in BeF_2 are as far from each other as possible, and there are no lone pairs. According to the VSEPR principle, this is responsible for the 180° bond angle in that compound. Similarly, there are no lone pairs on the carbon atoms of the other two molecules. In one case the carbon is flanked by two double bonds, and in the other one single bond and one triple. Evidently the second and third electron pairs in double and triple bonds don't count when it comes to establishing molecular geometry.

Further evidence supporting this conclusion comes from comparing the bond angles in boron trifluoride and formaldehyde, shown in Sections 11.2 and 11.4:

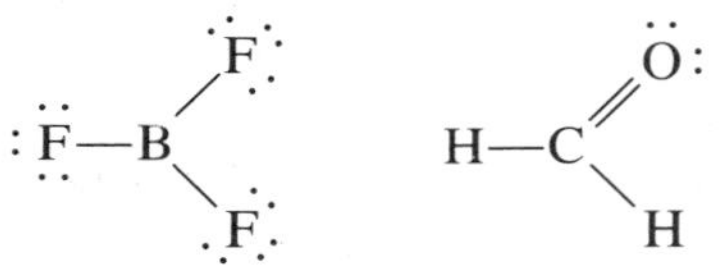

```
          .F:                  ..
          /  .                 O:
  :F—B                 H—C//
   ..     \                   \
          .F.                  H
```

There are no lone pairs, and the shapes are both planar triangular with 120° bond angles. This is the angle predicted for three electron pairs under the VSEPR principle. The second pair of electrons in the double bond in formaldehyde doesn't count when it comes to molecular geometry.

11.9 POLARITY OF MOLECULES

PG 11C Given or having determined the Lewis diagram of a molecule, predict whether the molecule is polar or nonpolar.

We previously considered the polarity of covalent bonds. Now that we have some idea about how atoms are arranged in molecules, we are ready to discuss the polarity of molecules themselves. **A polar molecule is one in which there is an unsymmetrical distribution of charge,** resulting in + and − poles. A simple example is the HF molecule. The fact that the bonding electrons are closer to the fluorine atom gives the fluorine end of the molecule a partial negative charge, while the hydrogen end acts as a positive pole. In general, any diatomic molecule in which the two atoms differ from each other will be at least slightly polar. Other examples are HCl and BrCl. In both of these molecules the chlorine atom acts as a negative pole.

When a molecule has more than two atoms we must know something about the bond angles in order to decide whether the molecule is polar or nonpolar. Consider, for example, the two triatomic molecules, BeF_2 and H_2O. Despite the presence of two strongly polar bonds, the linear BeF_2 *molecule* is nonpolar. Since the fluorine atoms are symmetrically arranged around Be, the two polar

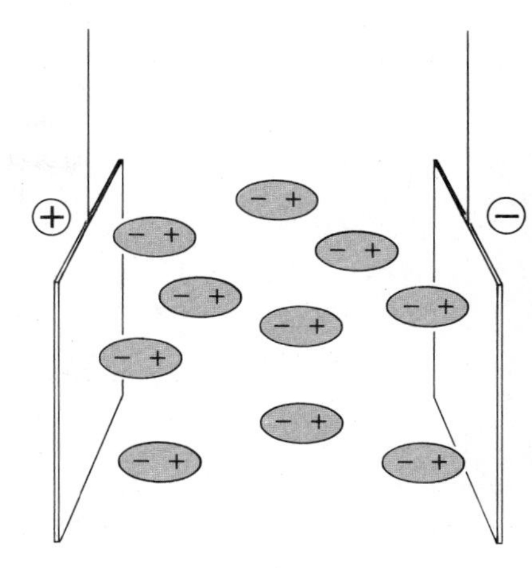

Field off

Field on

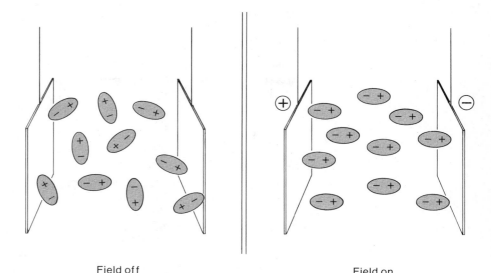

Figure 11.4
Orientation of polar molecules in an electric field. Two plates, immersed in a liquid whose molecules are polar, are connected through a switch to a source of an electric field. With the switch open the orientation of the molecules is random (left). When the switch is closed, the molecules tend to line up with the positive end toward the negative plate, and the negative pole toward the positive plate.

Cl Cl Cl C Cl

CCl_4

Nonpolar
(dipoles from polar bonds cancel due to symmetry)

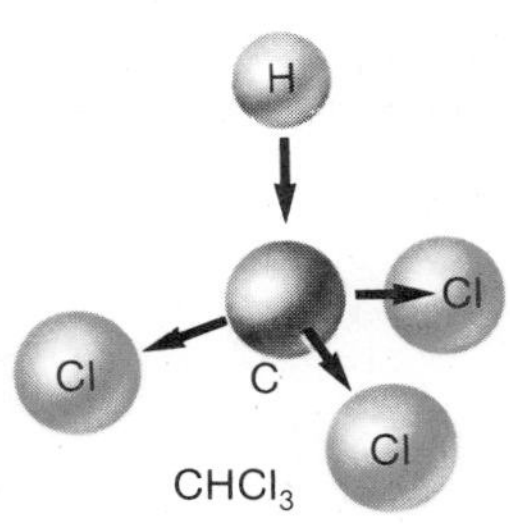

Polar
(dipoles from polar bonds do not cancel)

Figure 11.5
Polar and nonpolar molecules. Polar bonds in CCl_4 are arranged symmetrically, so they cancel and the molecule is nonpolar. Polar bonds in $CHCl_3$ lack symmetry, so they do not cancel and the molecule is polar.

Be—F bonds cancel each other. This may be shown as

$$:\ddot{\underset{\cdot\cdot}{F}} \leftarrow Be \rightarrow \ddot{\underset{\cdot\cdot}{F}}:$$

in which the arrows point to the more electronegative atoms.

In contrast, the bent water molecule is polar; the two polar bonds do not cancel each other because the molecule is not symmetrical around a horizontal axis.

$$\begin{array}{c} (-) \\ \dot{\ddot{O}}\cdot \\ \nearrow \quad \nwarrow \\ H\ (+)\ H \end{array}$$

The negative pole is located at the more electronegative oxygen atom; the positive pole is midway between the two hydrogen atoms. In an electric field, water molecules tend to line up with the hydrogen atoms pointing toward the plate with the negative charge and the oxygen atoms toward the plate with the positive charge (Fig. 11.4).

Another molecule that is nonpolar despite the presence of polar bonds is CCl_4. The four C—Cl bonds are themselves polar, with the bonding electrons displaced toward the chlorine atoms. But because the four chlorines are symmetrically distributed about the central carbon atom (Fig. 11.5), the polar bonds cancel each other. If one of the chlorine atoms in CCl_4 is replaced by hydrogen, the symmetry of the molecule is destroyed. The chloroform molecule, $CHCl_3$, is polar.

EXAMPLE 11.11 Is the BF_3 molecule polar? Is the NH_3 molecule polar?

The geometries of both of these molecules are described in Table 11.1. Consider BF_3 first. Sketch the Lewis diagram, with arrows pointing to the more electronegative element. Is the molecule polar?

----- -----

BF_3 is nonpolar.

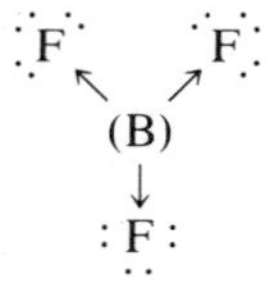

```
 :F:     :F:
    ↖   ↗
     (B)
      ↓
     :F:
```

NH_3:

Even though fluorine is more electronegative than boron, the three fluorine atoms are arranged symmetrically around the boron atom. The polar bonds cancel.

Now sketch the trigonal pyramidal structure of NH_3, with arrows pointing to the more electronegative element. Is it polar or nonpolar?

----- -----

NH_3 is polar.

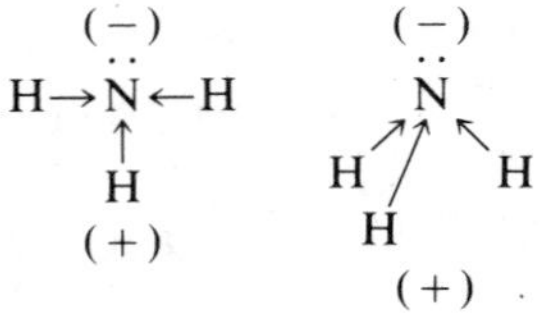

```
    (−)             (−)
     ..              ..
 H→N←H              N
     ↑            ↗↗  ↖
     H          H  /    H
    (+)           H
                 (+)
```

The bonding electrons in ammonia are displaced toward the more electronegative nitrogen atom. The bonds do not cancel in the unsymmetrical pyramidal shape, so the molecule is polar. The diagram at the right, which attempts to show the molecular shape, better suggests the charge displacement toward the nitrogen atom.

CHAPTER 11 IN REVIEW

11.1 Drawing Lewis Diagrams by Inspection

11A Draw the Lewis diagram for any molecule or polyatomic ion made up of representative elements.

11.2 Multiple Bonds

11.3 Drawing Complex Lewis Diagrams

11.4 Exceptions to the Octet Rule

11.5 Molecular Geometry

11.6 Electron-Pair Repulsion; Electron-Pair Geometry

11.7 Molecular Geometry Based on Electron-Pair Repulsion

11B Given or having derived the Lewis diagram of a molecule in which a second-period central atom is surrounded by two, three, or four pairs of electrons, predict the electron-pair geometry and the molecular geometry around that atom.

11.8 The Geometry of the Multiple Bond

11.9 Polarity of Molecules

11C Given or having determined the Lewis diagram of a molecule, predict whether the molecule is polar or nonpolar.

TERMS AND CONCEPTS

11.1 Coordinate covalent bond
11.2 Double bond; multiple bond
11.3 Isomer
11.5 Molecular geometry
Bond angle
Shape (molecular)
Bent (geometry)
Linear (geometry)
Planar (geometry)
Tetrahedron
Trigonal pyramid (geometry)
11.6 Valence-shell electron-pair repulsion, VSEPR
Electron-pair geometry
Trigonal planar (geometry)
Tetrahedral (geometry)
Tetrahedral angle
11.9 Polar, nonpolar molecule

Most of these terms and many others are defined in the Glossary. Use your Glossary regularly.

QUESTIONS AND PROBLEMS

Sections 11.1 to 11.3

Write Lewis diagrams for each of the following sets of molecules:

(1) HBr; H_2S; PH_3.

(2) OF_2; CO; SO_4^{2-}.

(3) IO^-; BrO_4^-; H_2SO_4.

(4) C_4H_{10}; C_4H_8; C_4H_6.

(5)* CH_3F; CH_2F_2; CF_4.

(6)* All possible isomers of C_5H_{12}.

(7)* Two isomers of C_5H_{10}.

(8)* Two isomers of C_3H_6O.

(9)* Formic acid, HCOOH, a compound produced by ants.

(10) HS^-; H_2S; H_2S_2; H_2S_3; H_2S_4; H_2S_5.

(23) BrF; SF_2; PF_3.

(24) OH^-; ClO_3^-; NO_3^-.

(25) IO_2^-; H_3PO_4; HSO_4^-.

(26)* CH_4O; C_2H_4O; $C_2H_6O_2$.

(27) CH_2ClF; CBr_2F_2; ClBrClF.

(28)* All possible isomers of C_6H_{14}.

(29)* Two isomers of C_3H_8O, one of which, an ether, does not have all of the carbon atoms bonded to each other.

(30)* Two isomers of $C_2H_2Cl_2$.

(31)* Propionic acid, C_2H_5COOH.

(32) Hydrogen peroxide, H_2O_2, and the peroxide ion, O_2^{2-}.

Section 11.4

(*11*) Why is it not possible to draw a Lewis diagram that obeys the octet rule if the species has an odd number of electrons?

(12)* Two iodides of arsenic (Z = 33) are AsI_3 and AsI_5. One of these iodides has a Lewis diagram that conforms to the octet rule, and one does not. Draw the diagram that is possible, and explain why the other cannot be drawn.

(33) Because of its five valence electrons, it is sometimes difficult to fit a nitrogen atom into a Lewis diagram that obeys the octet rule. Why is this so? Without actually drawing them, can you tell for which of the following species it is impossible to draw such a diagram? N_2O; NO_2; NO_2^-; NF_3; NO; N_2O_3; N_2O_4; N_2O_5; NOCl; NO_2Cl.

(34)* In Section 11.4 it says that five electron pairs surround the phosphorus atom in PF_5, and six surround sulfur in SF_6. How can this be when there is a total of four s and p orbitals?

Section 11.7

For each molecule, or for the atom specified in a molecule, describe (a) the electron-pair geometry, and (b) the molecular geometry predicted by the electron-pair repulsion principle.

(13) BeH_2; CF_4; OF_2.

(14) IO_4^-; ClO_2^-; CO_3^{2-}.

(15) Each carbon atom in C_2H_5OH.

(16) Nitrogen atom in CH_3NH_2.

(35) BH_3; NF_3; HF.

(36) ClO^-; IO_3^-; NO_3^-.

(37) Oxygen atom in C_2H_5OH.

(38) Silicon atom in SiF_4.

Section 11.8

For the atom specified in each of the following molecules, describe (a) the electron-pair geometry, and (b) the molecular geometry predicted by the electron-pair repulsion principle.

(17) Each carbon atom in C_2H_4.

(18) Carbon atom in HCHO.

(39) Each carbon atom in C_2H_2.

(40) Carbon atom in SCN^-.

Section 11.9

(*19*) Explain how the carbon tetrafluoride molecule, CF_4, which contains four polar bonds (electronegativity difference 1.5), can be nonpolar.

(20) Compare the polarities of the HCl and HI molecules. In each case identify the end of the molecule that is more negative.

(21) Compare the polarities of the H_2O and H_2S molecules. Which molecule is more polar? What would you predict about the polarity of the H_2Te molecule ($Z = 52$ for Te)?

(22)* Sketch the water molecule, paying particular attention to the bond angle and using arrows to indicate the polarity of the individual bonds. Then sketch the methanol molecule, $HOCH_3$, again using arrows to show bond polarity and predicting the approximate shape of the molecule around the oxygen atom. Estimate the relative polarities of the water and methanol molecules, and explain your prediction.

(41) The nitrogen-fluorine bond has an electronegativity difference of 1.0—less than the electronegativity difference between carbon and fluorine in CF_4. Yet the NF_3 molecule is polar, while CF_4 is nonpolar. How can this be?

(42) Compare the polarities of the following molecules: ClF; Cl_2; BrCl; ICl. In each case identify the end of the molecule that is more negative.

(43) Draw Lewis diagrams of the CF_4 and CH_2F_2 molecules, using arrows pointing to the more electronegative element in each bond. From these diagrams show that CH_2F_2 is polar and CF_4 is not.

(44)* As noted in Section 11.3, there are two possible Lewis structures for the compound with the molecular formula C_2H_6O, and both are real compounds. C_2H_5OH is ethanol, or ethyl alcohol, and CH_3OCH_3 is diethyl ether, the anesthetic. Sketch these molecules with arrows to indicate the direction of bond polarity around the oxygen atom. Predict the relative polarities of these molecules. Predict also the relative polarity between CH_3OH and C_2H_5OH. What would you expect of the polarity of $C_5H_{11}OH$?

Miscellaneous Questions

(*45*) Distinguish precisely and in scientific terms the differences between items in each of the following groups:
(a) Covalent bond, coordinate covalent bond
(b) Single bond, double bond, triple bond, multiple bond
(c) Molecular geometry, electron-pair geometry
(d) Bent geometry, trigonal planar geometry

(e) Trigonal planar geometry, trigonal pyramid geometry
(f) Trigonal pyramid geometry, tetrahedral geometry
(g) Polar molecule, nonpolar molecule

(46) Classify each of the following statements as true or false:
(a) If an atom is bonded to two or more other atoms, there is no difference between a coordinate covalent bond and any other bond.
(b) Multiple bonds can form only between two atoms of the same element.
(c) Hydrogen atoms never form double bonds.
(d) When hydrogen, oxygen, and a nonmetal are in the same molecule, oxygen is usually located between hydrogen and the nonmetal atom.
(e) Molecules cannot be made up of a ring of carbon atoms.
(f) Only valence electrons can participate in forming bonds.
(g) Molecular geometry around an atom may or may not be the same as the electron-pair geometry around the atom, but the molecular geometry is the effect of the electron-pair geometry.
(h) An atom can never be surrounded by more than four electron pairs.
(i) If the number of electron pairs around an atom is the same as the number of other atoms to which it is bonded, the molecular and electron-pair geometries are the same.
(j) The CO_2 molecule is linear, but the SO_2 molecule is bent.
(k) A molecule is polar if it contains polar bonds.

(47)* Draw Lewis diagrams for two isomers of C_3H_4.

(48)* $H_2C_2O_4$ is the formula of oxalic acid. The two carbon atoms are bonded to each other, and the molecule is symmetrical. Draw its Lewis diagram.

(49) One isomer of C_5H_{10} has its carbon atoms in a ring. Draw the Lewis diagram.

(50) Draw Lewis diagrams for these five acids of chlorine: HCl, HClO, $HClO_2$, $HClO_3$, and $HClO_4$.

(51)* Compare Lewis diagrams for CCl_4, SO_4^{2-}, ClO_4^-, and PO_4^{3-}. Identify two things about these diagrams and how they are drawn that they share. From these generalizations, can you predict the Lewis diagrams of the SiO_4^{4-} ion and CI_4?

12 Inorganic Nomenclature

LOOKING BACK

4.8 The names and symbols of the common elements, and their positions in the periodic table, were learned.

5.5 Charges and formulas of monatomic ions formed by representative elements are:

GROUP	CHARGE	EXAMPLE	GROUP	CHARGE	EXAMPLE
1A	+1	Na^+	5A	−3	P^{3-}
2A	+2	Mg^{2+}	6A	−2	S^{2-}
3A	+3	Al^{3+}	7A	−1	Cl^-

6.1 A chemical formula gives the symbol and number of atoms of each element in a formula unit of an element, compound, or ion.

6.2 A chemical unit may be described as monatomic, diatomic, triatomic, or polyatomic, based on the number of atoms in the unit. The unit may be an atom, a molecule, an anion or cation, or an ionic compound.

6.3 Hydrogen, nitrogen, oxygen, fluorine, chlorine, bromine, and iodine form stable diatomic molecules as uncombined elements. Their formulas are H_2, N_2, O_2, F_2, Cl_2, Br_2, and I_2, respectively. The formulas of other elements are generally written as the elemental symbol.

LOOKING AHEAD IN CHAPTER 12

The names of elements in a compound are used in the name of the compound, and the symbols are used in the formula of the compound.

Charges and formulas of monatomic ions of transition elements are added, and all are used in writing formulas and naming chemical substances.

The system is described by which formulas of elements, compounds, and ions are written.

New molecular compounds, ions, and ionic compounds are added to those learned earlier.

Knowledge of the diatomic molecules of the elements listed is assumed. Technically the formula of phosphorus is P_4, and of sulfur, S_8, although writing these formulas as P and S is generally acceptable.

The *Looking Back* references to Sections 6.1 to 6.3 summarize those sections briefly. They are appropriate beginnings to the more advanced treatment of chemical nomenclature in this chapter. If Chapter 6 was not a part of your earlier assignments, you may find it helpful to read those sections before beginning Chapter 12.

A very important point: The nomenclature presented in this chapter is part of a *system* that is based on different classes of chemical compounds. If you learn the system, you will be able to apply it to any chemical in that class, even though you may never have heard of the chemical before. This is much better, and also easier, than trying to memorize the names and formulas of substances without a point of reference. As in Chapter 6, we strongly advise that you *learn the system*.

The language of chemistry is like most languages in having many variations—many dialects, if you wish. Few people who work with chemicals speak a pure dialect. The standard for such a dialect, if one really exists, has been established by the International Union of Pure and Applied Chemistry (IUPAC), whose function it is to unify chemical terminology as it develops in laboratories throughout the world. The system they use is called the Stock system. At the other extreme we have common names, names given by craftsmen who identified a substance by its use or by some obvious physical or chemical property (Appendix III). In between are various levels of custom and formality.

In this chapter you will become familiar with the "professional" language of chemistry, the language used by chemists today. In areas where both old and new terms are employed, we will mention both, but lean toward the more modern terminology thereafter. This is, after all, the way a language grows.

12.1 OXIDATION STATE; OXIDATION NUMBER

The names of many chemical compounds include the **oxidation state** or **oxidation number** of an element in the compound. The terms are synonymous and are used interchangeably. Oxidation numbers are a sort of electron bookkeeping system that keeps track of electrons in oxidation–reduction reactions. These reactions will be considered in some detail in Chapter 18. Right now we are interested in the rules by which oxidation numbers are assigned. They are:

1. The oxidation number of any elemental substance is 0 (zero).
2. The oxidation number of a monatomic ion is the same as the charge on the ion.
3. The oxidation number of combined oxygen is -2, except in peroxides (-1) and superoxides ($-\frac{1}{2}$). (We will not emphasize peroxides or superoxides in this text.)
4. The oxidation number of combined hydrogen is $+1$, except in hydrides (-1).
5. In any molecular or ionic species, the sum of the oxidation numbers of all atoms in the formula unit is equal to the charge on the species.

12.2 NAMES AND FORMULAS OF CATIONS AND MONATOMIC ANIONS

PG 12A Given the name (or formula) of any ion in Figure 12.1, write the formula (or name) of that ion.

Oxidation number Rule 2 furnishes another way to look at the charge on a monatomic ion. Ions formed by Group 1A elements have a +1 charge. As ions, the elements are said to be in the +1 oxidation state. Similarly, ions formed by Group 2A elements have a +2 charge and are in the +2 oxidation state, and the oxidation number of Group 3A monatomic ions is +3. These are the only oxidation states known to Group 1A, 2A, and 3A elements as monatomic ions.

The name of a monatomic cation is the name of the element, followed by *ion*. Thus Na^+ is *sodium ion,* Ca^{2+} is *calcium ion,* and Zn^{2+} is *zinc ion*.

Some transition elements—elements in the B groups or Group 8 of the periodic table—are capable of forming two monatomic cations that differ in charge. Iron is one example. If a neutral iron atom loses two electrons, the ion has a +2 charge, Fe^{2+}. A monatomic ion of iron may also be in the +3 oxidation state, Fe^{3+}. In this case the neutral atom has lost three electrons. Copper is another common element that is capable of more than one oxidation state as a monatomic ion. Its oxidation state may be +1, as in Cu^+, or it may be +2, as in Cu^{2+}.

For the purpose of naming chemical compounds, it is necessary to distinguish between the two monatomic ions of iron. This is done in two ways. The older system applies either the *-ic* or the *-ous* suffix to the stem of the Latin name of the element. The Latin name for iron is *ferrum*. The *-ic* suffix is always applied to the higher oxidation state, and *-ous* to the lower. Hence the name of the Fe^{3+} ion is *ferric;* the *-ous* suffix applied to the Fe^{2+} ion gives us *ferrous*. By this system the name of $FeCl_2$, made up of an Fe^{2+} ion and two Cl^- ions, is ferrous chloride. When Fe^{3+} combines with three Cl^- ions the resulting compound is ferric chloride, $FeCl_3$.

In this text we emphasize the newer Stock system whereby an ion is identified by its English name, followed immediately by the oxidation state, written in Roman numerals and enclosed in parentheses. Thus, Fe^{2+} is written iron(II), and Fe^{3+} is written iron(III). In speaking, iron(II) becomes "iron two," and iron(III) is "iron three." Applied to the two chlorides, $FeCl_2$ is iron(II) chloride and $FeCl_3$ is iron(III) chloride.

As a general rule, if an element forms only one monatomic cation, its oxidation number is not included in the name. Na^+ is sodium ion, not sodium (I) ion. If an element forms two monatomic cations, the oxidation number is used in the name to distinguish between them.

Monatomic anions form when neutral atoms gain one or more electrons (Section 5.5). The name of a monatomic anion is formed by changing the name of the element so it ends in *-ide*. Thus, from chlorine we have *chloride* for Cl^-; from oxygen we have *oxide* for O^{2-}; and from sulfur we have *sulfide* for S^{2-}.

Figure 12.1 is a partial periodic table that shows most of the monatomic ions you are apt to find in an introductory course, as well as three polyatomic ions. The ammonium ion, NH_4^+, has been placed beneath the alkali metal ions it so closely resembles. It is the only polyatomic cation that is considered in this text. The hydroxide ion, OH^-, has been placed at the bottom of Group 7A. It is not like the halogens chemically, but the formulas of its compounds match the formulas of the corresponding halides.

The remaining diatomic ion is the mercury(I) ion, Hg_2^{2+}. It is made up of two mercury atoms that are covalently bonded, which have then lost two

1+	2+											3+		3−	2−	1−	
H^+																H^-	
Li^+	Be^{2+}													N^{3-}	O^{2-}	F^-	
Na^+	Mg^{2+}											Al^{3+}		P^{3-}	S^{2-}	Cl^-	
K^+	Ca^{2+}				Cr^{2+} Cr^{3+}	Mn^{2+} Mn^{3+}	Fe^{2+} Fe^{3+}	Co^{2+} Co^{3+}	Ni^{2+}	Cu^+ Cu^{2+}	Zn^{2+}					Br^-	
										Ag^+			Sn^{2+} Sn^{4+}			I^-	
	Ba^{2+}										Hg_2^{2+} Hg^{2+}		Pb^{2+} Pb^{4+}				
NH_4^+																OH^-	

Figure 12.1
Partial periodic table of common ions. Notes: (1) Tin (Sn) and lead (Pb) form monatomic ions in a +2 oxidation state. In their +4 oxidation states tin and lead compounds are more covalent than ionic, but they are frequently named as if they were ionic compounds. (2) Hg_2^{2+} is a diatomic elemental ion. Its name is mercury(1), indicating a +1 charge from each atom in the ion. (3) Ammonium ion, NH_4^+, and hydroxide ion, OH^-, are included as two important polyatomic ions. They form compounds that are similar to the compounds formed by the ions above them.

electrons, giving the ion a +2 charge. It is as if each atom loses one electron, giving it the name *mercury(I)*, or *mercurous* by the older system. The diatomic ion behaves as a single unit, as do all polyatomic ions.

EXAMPLE 12.1 Referring only to a periodic table (not Fig. 12.1), write the formula for each ion whose name is given, and the name where the formula is given.

barium ion	I^-
zinc ion	Ag^+
cobalt(II) ion	Cr^{3+}
rubidium ion (Z = 37)	Se^{2-} (Se is selenium, Z = 34)

The last item in each column involves an element that is not among those you are expected to recognize. The location of the element in the periodic table should give you all the information you need to write the formula or name requested.

-----

barium ion, Ba^{2+}	I^-, iodide ion
zinc ion, Zn^{2+}	Ag^+, silver ion
cobalt(II) ion, Co^{2+}	Cr^{3+}, chromium(III) ion
rubidium ion, Rb^+	Se^{2-}, selenide ion

The (II) in cobalt(II) identifies the oxidation state and charge as +2.

Chromium is capable of more than one oxidation state, so that oxidation state must appear in the name of the ion. It is the same as the ionic charge, so chromium(III).

There is only one ionic charge for each of the first two elements in each column, so those oxidation states do not appear in the names of the ions.

Atomic number 37 in the periodic table shows that Rb is the symbol for rubidium. Its appearance in Group 1A establishes the ionic charge as +1.

Given that selenium is the name of an element whose monatomic ion has a negative charge, the name of the ion is found by applying an *-ide* suffix to the elemental name: selenide.

12.3 ACIDS AND THE ANIONS DERIVED FROM THEIR TOTAL IONIZATION

PG 12B Given the name (or formula) of an acid of a Group 4A, 5A, 6A, or 7A element, or of an ion derived from the total ionization of such an acid, write its formula (or name).

The Hydrogen and Hydronium Ions

The most familiar form of an **acid** is a hydrogen-bearing compound that reacts with water to produce a **hydronium ion** and an anion. The ionizable hydrogen is written first in the conventional formula of an acid. If HX represents an acid, its "ionization equation" is

$$HX + H_2O(\ell) \rightarrow H_3O^+(aq) + X^-(aq) \tag{12.1}$$

Lewis diagrams show the bond breaking and bond forming:

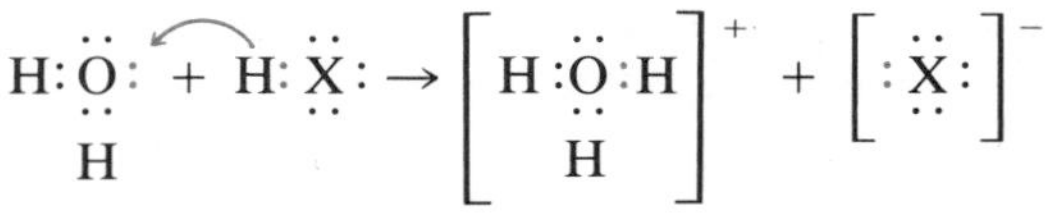

$$H{:}\ddot{\underset{\cdot\cdot}{O}}{:} \underset{H}{} + H{:}\ddot{\underset{\cdot\cdot}{X}}{:} \rightarrow \left[H{:}\ddot{\underset{\cdot\cdot}{O}}{:}H \atop H\right]^+ + \left[{:}\ddot{\underset{\cdot\cdot}{X}}{:}\right]^-$$

H_3O^+ is the hydronium ion. Most chemists and textbook authors prefer to simplify the ionization equation by subtracting a water molecule from each side. If a water molecule is removed from a hydronium ion, only a hydrogen ion, H^+, is left:

$$HX \xrightarrow{H_2O} H^+(aq) + X^-(aq) \tag{12.2}$$

The hydrogen ion does not exist by itself but is bonded covalently to a water molecule. Nevertheless, we will use this common simplification and refer to the hydrogen ion, except where the context requires the hydronium ion.*

A hydrogen atom consists of one proton and one electron. To form a hydrogen ion, H^+, the neutral atom must lose its electron. That leaves only the proton; a hydrogen ion is simply a proton. Acids are sometimes classified by the number of hydrogen ions, or protons, that can be released by a single

*There are strong arguments in favor of the hydronium ion. If your instructor prefers it over the hydrogen ion, by all means adopt his or her recommendation.

molecule. An acid that fits the general formula HX (HCl, HNO_3) is a **monoprotic acid.** If an acid has two ionizable hydrogens, H_2Y (H_2SO_4, H_2CO_3), it is **diprotic;** and a **triprotic acid,** H_3Z (H_3PO_4), has three ionizable hydrogens.

This section discusses only the "total ionization" of an acid, meaning that all of the ionizable hydrogen is removed from the acid molecule. The total ionization equations for diprotic and triprotic acids are therefore

$$H_2Y \xrightarrow{H_2O} 2\ H^+(aq) + Y^{2-}(aq) \tag{12.3}$$

$$H_3Z \xrightarrow{H_2O} 3\ H^+(aq) + Z^{3-}(aq) \tag{12.4}$$

The intermediate ions coming from the stepwise ionization of diprotic and triprotic acids are discussed in the next section.

Binary Acids

A **binary acid** has only two elements, hydrogen and another nonmetal. The best known binary acid is hydrochloric acid, HCl. The name, *hydrochloric,* shows how all binary acids are named. The name begins with the prefix, *hydro-*. This is followed by the name of the other nonmetal, chlorine, changed so it ends in *-ic*. Hence, *hydro*-chlor-*ic*.

One of the characteristics of chemical families—elements in the same group in the periodic table—is that their members usually form similar compounds. This is true of the binary acids of the halogens. Rather than simply supplying the names and formulas of these acids, we'll give you the opportunity to find them yourself, and thereby learn the system. We'll even include a binary acid that does not contain a halogen.

EXAMPLE 12.2 For each of the following names, write the formula; for each formula, write the name.

hydrofluoric acid	HI
hydrobromic acid	H_2S

Chlorine, fluorine, bromine, and iodine are all from the same family. If you know the formula of hydrochloric acid, you should be able to find the formulas of hydrofluoric and hydrobromic acids by substitution of elemental symbols. Then, if you reverse the thought process, you should be able to write the names of HI and H_2S.

-----

hydrofluoric acid, HF	HI, hydroiodic acid
hydrobromic acid, HBr	H_2S, hydrosulfuric acid

The names are established by the rule for naming binary acids: the prefix *hydro-* followed by the elemental name changed to end in *-ic*.

The anion that comes from the ionization of a binary acid is a monatomic anion. It is named by the system described in Section 12.2.

Oxyacids That End in *-ic* and Their Oxyanions

An acid that contains oxygen in addition to hydrogen and another nonmetal is an **oxyacid.** When a hydrogen ion is removed from an oxyacid, the oxygen remains covalently bonded to the other nonmetal as part of an **oxyanion.** The ionization of chloric acid, $HClO_3$, shows this:

$$HClO_3(aq) \rightarrow H^+(aq) + ClO_3^- \quad (12.5)$$

The name of the anion, ClO_3^-, is *chlorate.*

Chloric acid and the chlorate ion are a perfect example of the nomenclature system that follows.* Specifically, *for the total ionization of any oxyacid whose name ends in* -ic,

1. *the formula of the anion is the formula of the acid without the hydrogen(s), with*
2. *a negative charge equal to the number of hydrogens in the original acid; and*
3. *the name of the anion is the name of the acid with the* -ic *changed to* -ate.

Chloric acid, $HClO_3$, without the hydrogen is ClO_3. The difference is one hydrogen, so the charge on the ion is -1: ClO_3^-. To get the name, change the *-ic* in chlor*ic* to *-ate*: chlor*ate*.

The names and formulas of five *-ic* acids should be memorized. The names and formulas of their anions can be found by the above rules. These acids, their ionization equations, and the anion names are given in Table 12.1. Hydrochloric acid is included to make the table a complete summary of the acid/anion combinations that illustrate the nomenclature system.

*The oxidation number of chlorine in $HClO_3$ and ClO_3^- is $+5$. If hydrogen is $+1$ and each oxygen is -2, according to Rules 3 and 4 from Section 12.1, Rule 5 gives chlorine its $+5$ value for either species. The *-ic* suffix for the acid name and the *-ate* suffix for the anion are usually assigned to the substance in which the nonmetal is in its most common oxidation state.

Table 12.1 Acids and Anions

Acid	Ionization Equation	Ion Name
Hydrochloric acid	$HCl \rightarrow H^+ + Cl^-$	Chloride
Chloric acid	$HClO_3 \rightarrow H^+ + ClO_3^-$	Chlorate
Nitric acid	$HNO_3 \rightarrow H^+ + NO_3^-$	Nitrate
Sulfuric acid	$H_2SO_4 \rightarrow 2\ H^+ + SO_4^{2-}$	Sulfate
Carbonic acid*	$H_2CO_3 \rightarrow 2\ H^+ + CO_3^{2-}$	Carbonate
Phosphoric acid*	$H_3PO_4 \rightarrow 3\ H^+ + PO_4^{3-}$	Phosphate

*The carbonic and phosphoric acid ionizations occur only slightly in water solutions. They are used here to illustrate the derivation of the formulas and names of the carbonate and phosphate ions, both of which are quite abundant from sources other than their parent acids.

Bromine and iodine form oxyacids comparable to oxyacids of chlorine. In name and formula they may be substituted for chlorine in chloric acid and the anion derived from it. These ideas do not extend to fluorine because that element doesn't happen to form oxyacids.

EXAMPLE 12.3 Complete the name and formula blanks in the following table:

ACID NAME	ACID FORMULA	ANION FORMULA	ANION NAME
bromic acid			
		IO_3^-	

ACID NAME	ACID FORMULA	ANION FORMULA	ANION NAME
bromic acid	$HBrO_3$	BrO_3^-	bromate
iodic acid	HIO_3	IO_3^-	iodate

Thought process: Bromic acid corresponds with chloric acid, $HClO_3$. The acid and ion formulas come from substituting Br for Cl in the corresponding chlorine formulas. The anion name for an *-ic* acid is the name of the central element changed to end in *-ate*.

IO_3^- corresponds to ClO_3^-, the chlorate ion that comes from $HClO_3$, chloric acid. The acid and ion names and formulas are the same, except that iodine replaces chlorine. Hence, chlorate becomes iodate as the name of IO_3^-, chloric becomes iodic as the name of the acid, and $HClO_3$ becomes HIO_3 as the acid formula. The acid formula can also be derived directly from the anion formula. The ion has a -1 charge, so the acid must have one hydrogen: HIO_3.

Other Oxyacids and Their Oxyanions

Chlorine forms five acids that furnish quite a complete picture of the nomenclature of acids and the anions derived from their total ionization. Hydrochloric acid, HCl, and one oxyacid, chloric acid, $HClO_3$ have already been discussed. All five are assembled in Table 12.2. The Lewis diagrams show how the same three elements can form four different oxyacids. Notice that the hydrogen is always bonded to an oxygen atom that forms a link between it and the chlorine atom. This is characteristic of ionizable hydrogens in all oxyacids.

Table 12.3 isolates the *system* of nomenclature that, once learned, can be applied to many acids and their ions. It is a system of prefixes (beginnings) and suffixes (endings) based on the number of oxygen atoms in the *-ic* acid. Chlor*ic*

Table 12.2 Acids of Chlorine

Acid	Lewis Diagram	Ionization Equation	Anion Name
Perchloric	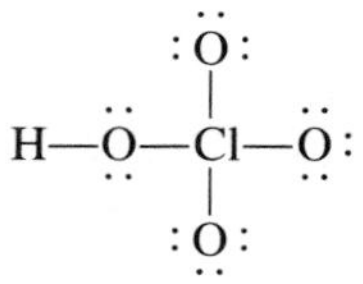 :O: │ H—O—Cl—O: │ :O:	$HClO_4 \rightarrow H^+ + ClO_4^-$	Perchlorate
Chloric	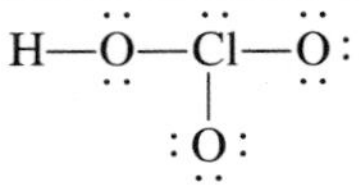 H—O—Cl—O: │ :O:	$HClO_3 \rightarrow H^+ + ClO_3^-$	Chlorate
Chlorous	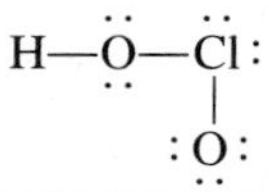 H—O—Cl: │ :O:	$HClO_2 \rightarrow H^+ + ClO_2^-$	Chlorite
Hypochlorous	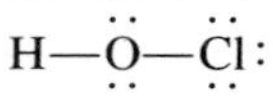 H—O—Cl:	$HClO \rightarrow H^+ + ClO^-$	Hypochlorite
Hydrochloric	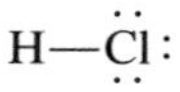 H—Cl:	$HCl \rightarrow H^+ + Cl^-$	Chloride

acid and the chlor*ate* ion have three oxygens. From that starting point, notice in both Tables 12.2 and 12.3:

1. If the number of oxygens is one larger than the number in the *-ic* acid, the prefix *per-* is placed before both the acid and anion names: $HClO_4$ is *per*chloric acid, and ClO_4^- is the *per*chlorate ion.

Table 12.3 Prefixes and Suffixes in Acid and Anion Nomenclature (Acids and Anions of Chlorine Given as Examples)

Line	Oxygen Atoms Compared to *-ic* Acid and *-ate* Anion	Acid Prefix and/or Suffix (Example)	Anion Prefix and/or Suffix (Example)
1	One more	*per-ic* (perchloric)	*per-ate* (perchlorate)
2	Same	*-ic* (chloric)	*-ate* (chlorate)
3	One fewer	*-ous* (chlorous)	*-ite* (chlorite)
4	Two fewer	*hypo-ous* (hypochlorous)	*hypo-ite* (hypochlorite)
5	No oxygen	*hydro-ic* (hydrochloric)	*-ide* (chloride)

2. If the number of oxygens is one smaller than the number in the *-ic* acid, the suffixes *-ic* and *-ate* are replaced with *-ous* and *-ite*. $HClO_2$ is chlor*ous* acid, and ClO_2^- is the chlor*ite* ion.
3. If the number of oxygens is one smaller than the number in the *-ous* acid (two smaller than the number in the *-ic* acid), the prefix *hypo-* is placed before both the acid and anion names, while keeping the *-ous* and *-ite* suffixes: HClO is *hypo*chlor*ous* acid and ClO^- is the *hypo*chlor*ite* ion.*

As we have seen before, bromine and iodine can be substituted for chlorine in $HClO_3$ and ClO_3^-. These substitutions can also be made in the other acids of chlorine. Try your skill on the following:

EXAMPLE 12.4 Fill in the name and formula blanks in the following table. Try to do it without referring to Tables 12.2 and 12.3, but use them if absolutely necessary.

ACID NAME	ACID FORMULA	ANION FORMULA	ANION NAME
periodic			
	HBrO		
		IO_2^-	

ACID NAME	ACID FORMULA	ANION FORMULA	ANION NAME
periodic	HIO_4	IO_4^-	periodate

Let's think about the top line only. The prefix *per-*, applied to the memorized formula of chloric acid, $HClO_3$, means one more oxygen atom. Therefore, perchloric acid is $HClO_4$. Substituting iodine for chlorine we have periodic acid as HIO_4. Remove one hydrogen ion to get the anion formula, IO_4^-, with a negative charge equal to the number of hydrogens removed from the neutral molecule. The prefix *per-* is applied to the anion name as it is to the acid name. As perchloric acid → perchlorate ion, so periodic acid → periodate ion.

If you now want to reconsider any of your entries on the other two lines, make any changes you wish. The answers follow.

-----

*According to the rules for oxidation numbers in Section 12.1, the oxidation state of chlorine is +7 in $HClO_4$ and ClO_4^-, +3 in $HClO_2$ and ClO_2^-, and +1 in HClO and ClO^-.

ACID NAME	ACID FORMULA	ANION FORMULA	ANION NAME
periodic	HIO_4	IO_4^-	periodate
hypobromous	HBrO	BrO^-	hypobromite
iodous	HIO_2	IO_2^-	iodite

Reasoning processes are similar for all lines in the table. In the second line there are two fewer oxygen atoms than in chloric acid, $HClO_3$, and in the third line one fewer. Prefixes and suffixes match those for the corresponding chlorine substances in Tables 12.2 and 12.3.

Nitric, sulfuric, and phosphoric acids have important variations with different numbers of oxygen atoms. We will limit ourselves to two of these. The acid and anion nomenclature system in Table 12.3 remains the same. See if you can apply it to this new situation.

EXAMPLE 12.5 Fill in the name and formula blanks in the following table.

ACID NAME	ACID FORMULA	ANION FORMULA	ANION NAME
	HNO_2		
			sulfite

ACID NAME	ACID FORMULA	ANION FORMULA	ANION NAME
nitrous	HNO_2	NO_2^-	nitrite
sulfurous	H_2SO_3	SO_3^{2-}	sulfite

HNO_2 has one fewer oxygen atoms than nitr*ic* acid, HNO_3, so its name must be nitr*ous* acid. One hydrogen ion must be removed from the acid formula to produce the ion, NO_2^-. The anion from an *-ous* acid has an *-ite* suffix: nitrous acid → nitrite ion.

From the memorized sulfuric acid, H_2SO_4, you have the sulfate ion, SO_4^{2-}. The sulfite ion has one fewer oxygen, SO_3^{2-}. If the anion has a -2 charge, the acid must have two hydrogens: H_2SO_3. The name of the acid with one fewer oxygens than sulfuric acid is sulfurous acid.

The following example gives you the opportunity to practice what you have learned about acid and anion nomenclature. If you have memorized what you must, and know how to apply the rules that have been given, you will be able to write the required names and formulas with reference to nothing other than a periodic table. If you have really mastered the system, you will be able to extend it to the last substance in each column. They have not been mentioned anywhere in this chapter.

EXAMPLE 12.6 For each of the following names, write the formula; for each formula, write the name.

phosphoric acid	CO_3^{2-}
sulfate ion	HF
bromous acid	NO_2^-
periodate ion	H_2SO_3
nitric acid	PO_4^{3-}
telluric acid (tellurium, Z = 52)	SeO_3^{2-} (Se, selenium Z = 34)

-----

phosphoric acid, H_3PO_4 (Table 12.1)

sulfate ion, SO_4^{2-} (Table 12.1)

bromous acid, $HBrO_2$ ($HClO_2$ and Tables 12.2 and 12.3)

periodate ion, IO_4^- (Example 12.4)

nitric acid, HNO_3 (Table 12.1)

telluric acid, H_2TeO_4

CO_3^{2-}, carbonate ion (Table 12.1)

HF, hydrofluoric acid, (Example 12.2)

NO_2^-, nitrite ion (Example 12.5)

H_2SO_3, sulfurous acid (Example 12.5)

PO_4^{3-}, phosphate ion (Table 12.1)

SeO_3^{2-}, selenite ion

References in parentheses tell where each name or formula may be found, or a starting point from which it may be figured out. The last item in each column includes elements from Group 6A, the same chemical family as sulfur. The formula of telluric acid matches that of sulfuric acid, H_2SO_4. From sulfuric acid the name of SO_4^{2-} is sulfate ion. One fewer oxygen atom makes it SO_3^{2-}, sulfite ion. Substitution of selenium for sulfur in name and formula gives SeO_3^{2-}, selenite ion.

Summary

Table 12.4 shows formulas of all the acids whose names and formulas can be figured out from the six acids whose names and formulas have been memorized. The six key acids are highlighted. The prefixes and suffixes are in the first column. Formulas and names of all anions can be derived from Table 12.3. If you have memorized the key acids and understand the *system* of nomenclature, you should be able to reason from any highlighted acid to any other acid or anion.

Table 12.4 Formulas of Acids Considered in Section 12.3

Prefixes, Suffixes	Group 7A	Group 6A	Group 5A	Group 4A
per- -ic per- -ate	$HClO_4$ $HBrO_4$ HIO_4			
-ic -ate	$HClO_3$ $HBrO_3$ HIO_3	H_2SO_4 H_2SeO_4 H_2TeO_4	H_3PO_4 and HNO_3	H_2CO_3
-ous -ite	$HClO_2$ $HBrO_2$ HIO_2	H_2SO_3 H_2SeO_3 H_2TeO_3	HNO_2	
hypo- -ous hypo- -ite	HClO HBrO HIO			
hydro- -ic -ide	HF HCl HBr HI	H_2S H_2Se H_2Te		

12.4 NAMES AND FORMULAS OF ANIONS DERIVED FROM THE STEPWISE IONIZATION OF POLYPROTIC ACIDS

PG 12C Given the name (or formula) of an ion formed by the stepwise ionization of a polyprotic acid from a Group 4A, 5A, or 6A element, write its formula (or name).

Polyprotic acids do not lose their hydrogens all at once, but rather one at a time. The intermediate anions produced are stable chemical species that are the negative ions in many ionic compounds. The ionization of carbonic acid, H_2CO_3, is an example:

$$H_2CO_3 \xrightarrow{-H^+} HCO_3^- \xrightarrow{-H^+} CO_3^{2-} \qquad (12.6)$$

The intermediate ion, HCO_3^-, is the hydrogen carbonate ion—a logical name, since the ion is literally a hydrogen ion bonded to a carbonate ion. The ion is also called the bicarbonate ion.

Phosphoric acid, H_3PO_4, has three steps in its ionization process:

$$H_3PO_4 \xrightarrow{-H^+} H_2PO_4^- \xrightarrow{-H^+} HPO_4^{2-} \xrightarrow{-H^+} PO_4^{3-} \qquad (12.7)$$

$H_2PO_4^-$ is the dihydrogen phosphate ion, signifying two hydrogen ions attached to a phosphate ion. HPO_4^{2-} is the monohydrogen phosphate ion, or simply hydrogen phosphate ion. It is essential that the prefix *di-* be used in naming the $H_2PO_4^-$ ion to distinguish it from HPO_4^{2-}. The prefix *mono-* in monohy-

Table 12.5 Names and Formulas of Anions Derived from the Stepwise Ionization of Acids

Acid	Ion	Names of Ions	
		PREFERRED	OTHER
H_2CO_3	HCO_3^-	Hydrogen carbonate	Bicarbonate Acid carbonate
H_2S	HS^-	Hydrogen sulfide	Bisulfide Acid sulfide
H_2SO_4	HSO_4^-	Hydrogen sulfate	Bisulfate Acid sulfate
H_2SO_3	HSO_3^-	Hydrogen sulfite	Bisulfite Acid sulfite
H_3PO_4	$H_2PO_4^-$	Dihydrogen phosphate	Monobasic phosphate
$H_2PO_4^-$	HPO_4^{2-}	Hydrogen phosphate	Dibasic phosphate

drogen phosphate is optional; absence of a prefix in a case like this is understood to mean *one*.

If you recognize the logic of this part of the nomenclature system you will be able to extend it to intermediate ions from the stepwise ionization of hydrosulfuric, sulfuric, and sulfurous acids. All of these are shown in Table 12.5.

12.5 NAMES AND FORMULAS OF OTHER ACIDS AND IONS

Organic Acids. Generally, organic acids ionize only slightly in water, but the anions produced are often abundant from other sources. Acetic acid, $HC_2H_3O_2$, the component of vinegar that is responsible for its odor and taste, is a good example. The ionization equation is

$$HC_2H_3O_2 \rightarrow H^+ + C_2H_3O_2^- \qquad (12.8)$$

Notice that only the hydrogen written first in the formula ionizes; the others do not (see below). $C_2H_3O_2^-$ is the acetate ion.

An organic chemist is more apt to write CH_3COOH for the formula of acetic acid. This "line formula," as it is called, shows the presence of the *carboxyl group*, —COOH, which is the source of ionizable hydrogen in almost all organic acids. The uniqueness of that hydrogen, compared to the other three, appears if Equation 12.3 is written with line formulas and Lewis diagrams:

$$CH_3COOH \rightarrow CH_3COO^- + H^+ \qquad (12.9)$$

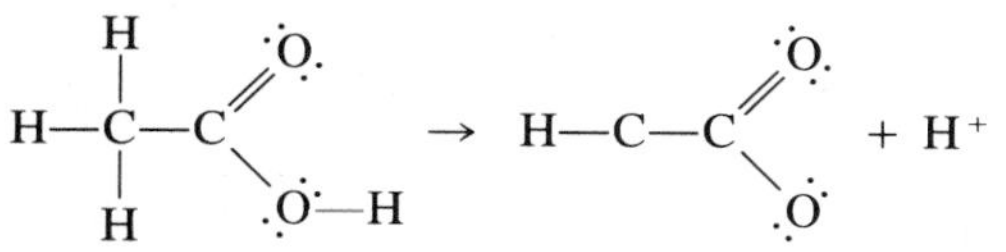

```
    H     :O.                      :O.
    |    //                        //
 H—C—C           →    H—C—C           + H⁺        (12.10)
    |    \                         \
    H     .O:—H                     .O:
```

In discussing the structure of the oxyacids of chlorine, it was mentioned that ionizable hydrogens are bonded to oxygen which, in turn, is bonded to a nonmetal. This is true for organic acids, too. The hydrogens linked to carbon through oxygen can ionize, whereas hydrogens bonded directly to carbon do not.

Hydrocyanic Acid, HCN. When hydrocyanic acid ionizes it produces the cyanide ion, CN^-. This ion and the hydroxide ion are the only two common polyatomic anions that have an *-ide* ending, a suffix otherwise reserved for monoatomic anions and binary molecular compounds.

Polyatomic Anions from Transition Elements. Chromium and manganese form some polyatomic anions that theoretically may be traced to acids, but only the ions are important. Their names and formulas are the chromate ion, CrO_4^{2-}, the dichromate ion, $Cr_2O_7^{2-}$, and the permanganate ion, MnO_4^-.

Other Ions. The performance goals have identified the important ions whose names and formulas you should recognize and be able to write. There are, of course, many others. Some of these, plus the ions already discussed, are listed in Tables 12.6 and 12.7. We recommend that you use these tables as a reference for the less common ions, and only as a last resort if you happen to forget one of the ions you should know.

Table 12.6 Cations

Ionic Charge: +1		Ionic Charge: +2		Ionic Charge: +3	
ALKALI METALS: GROUP 1A		ALKALINE EARTHS: GROUP 2A		GROUP 3A	
Li^+	Lithium	Be^{2+}	Beryllium	Al^{3+}	Aluminum
Na^+	Sodium	Mg^{2+}	Magnesium	Ga^{3+}	Gallium
K^+	Potassium	Ca^{2+}	Calcium		
Rb^+	Rubidium	Sr^{2+}	Strontium		
Cs^+	Cesium	Ba^{2+}	Barium		
TRANSITION ELEMENTS		TRANSITION ELEMENTS		TRANSITION ELEMENTS	
Cu^+	Copper(I)			Cr^{3+}	Chromium(III)
Ag^+	Silver	Cr^{2+}	Chromium(II)	Mn^{3+}	Manganese(III)
POLYATOMIC IONS		Mn^{2+}	Manganese(II)	Fe^{3+}	Iron(III)
		Fe^{2+}	Iron(II)	Co^{3+}	Cobalt(III)
NH_4^+	Ammonium	Co^{2+}	Cobalt(II)		
OTHERS		Ni^{2+}	Nickel		
H^+	Hydrogen	Cu^{2+}	Copper(II)		
	or	Zn^{2+}	Zinc		
H_3O^+	Hydronium	Cd^{2+}	Cadmium		
		Hg_2^{2+}	Mercury(I)		
		Hg^{2+}	Mercury(II)		
		OTHERS			
		Sn^{2+}	Tin(II)		
		Pb^{2+}	Lead(II)		

12.6 FORMULAS OF IONIC COMPOUNDS

PG 12D Given the name of any ionic compound made up of ions that are included in Performance Goals 12A, 12B, or 12C, write the formula of that compound.

It is easy to write the formula of an ionic compound when you have a firm grasp on the names and formulas of ions. It is done in two steps:

1. Write the formula of the cation, followed by the formula of the anion, omitting the charges.
2. Insert subscripts to show the number of each ion needed in the formula unit to make the sum of the charges equal to zero.
   a. If only one ion is needed, omit the subscript.
   b. If a polyatomic ion is needed more than once, enclose the formula of the ion in parentheses and place the subscript after the parentheses.

EXAMPLE 12.7 Write the formula for each compound listed below.

barium hydroxide

potassium chlorate

copper(II) chloride

magnesium hydrogen sulfate

sodium hydrogen telluride (tellurium, Z = 52)

ammonium nitrate

sodium hydrogen carbonate

calcium hypobromite

iron(III) sulfite

strontium dihydrogen phosphate (strontium, Z = 38)

-----

barium hydroxide, $Ba(OH)_2$
potassium chlorate, $KClO_3$
copper(II) chloride, $CuCl_2$
magnesium hydrogen sulfate, $Mg(HSO_4)_2$
sodium hydrogen telluride, NaHTe

ammonium nitrate, NH_4NO_3
sodium hydrogen carbonate, $NaHCO_3$
calcium hypobromite, $Ca(BrO)_2$
iron(III) sulfite, $Fe_2(SO_3)_3$
strontium dihydrogen phosphate, $Sr(H_2PO_4)_2$

The ions of all compounds except the last in each column are among those you should know and recognize. Tellurium is in the same family as sulfur, so the hydrogen telluride ion should correspond with the hydrogen sulfide ion, HS^-. It is the intermediate step in the ionization of hydrotelluric acid, H_2Te, giving HTe^-. Strontium ion has a charge of +2, according to its position in Group 2A. It therefore requires two dihydrogen phosphate ions, $H_2PO_4^-$, to reach a total charge of zero.

Table 12.7 Anions

Ionic Charge: -1			
HALOGENS: GROUP 7A		**OXYANIONS**	
		ClO_4^-	Perchlorate
F^-	Fluoride	ClO_3^-	Chlorate
Cl^-	Chloride	ClO_2^-	Chlorite
Br^-	Bromide	ClO^-	Hypochlorite
I^-	Iodide		
		BrO_3^-	Bromate
ACIDIC ANIONS		BrO_2^-	Bromite
HCO_3^-	Hydrogen carbonate	BrO^-	Hypobromite
HS^-	Hydrogen sulfide		
HSO_4^-	Hydrogen sulfate	IO_4^-	Periodate
HSO_3^-	Hydrogen sulfite	IO_3^-	Iodate
$H_2PO_4^-$	Dihydrogen phosphate		
		NO_3^-	Nitrate
OTHER ANIONS		NO_2^-	Nitrite
SCN^-	Thiocyanate		
CN^-	Cyanide	OH^-	Hydroxide
H^-	Hydride	$C_2H_3O_2^-$	Acetate
		MnO_4^-	Permanganate

Ionic Charge: -2	
GROUP 6A	
O^{2-}	Oxide
S^{2-}	Sulfide
OXYANIONS	
CO_3^{2-}	Carbonate
SO_4^{2-}	Sulfate
SO_3^{2-}	Sulfite
$C_2O_4^{2-}$	Oxalate
CrO_4^{2-}	Chromate
$Cr_2O_7^{2-}$	Dichromate
ACIDIC ANION	
HPO_4^{2-}	Hydrogen phosphate
DIATOMIC ELEMENTAL	
O_2^{2-}	Peroxide

Ionic Charge: -3	
GROUP 5A	
N^{3-}	Nitride
P^{3-}	Phosphide
OXYANION	
PO_4^{3-}	Phosphate

12.7 NAMES OF IONIC COMPOUNDS

PG 12E Given the formula of an ionic compound made up of identifiable ions, write the name of the compound.

The name of an ionic compound is the name of the cation followed by the name of the anion. If you recognize the two ions, you have the name of the compound.

There is one place where you are apt to be uncertain about the name of a "familiar" ion. For example, what is the name of Fe_2O_3? Iron oxide is not an adequate answer; it fails to distinguish between the two possible oxidation states of iron. Is it iron(II) oxide or iron(III) oxide? To decide, you must use a combination of oxidation number Rules 3 and 5 (Section 12.1). Rule 3 states that oxygen has an oxidation number of -2. Fe_2O_3 has three oxygen atoms in the formula unit, so oxygen contributes $3(-2) = -6$ to the total oxidation number for the formula unit.

Oxidation number Rule 5 requires that the total oxidation number of the formula unit be equal to the charge on the unit. In a compound this charge is zero. This means a total of $+6$ must come from the two atoms of iron in the formula, or $+3$ from each atom. The compound is therefore iron(III) oxide. FeO is iron(II) oxide. That conclusion would be reached by recognizing that the -2 of a single oxide ion is balanced by the $+2$ of a single iron(II) ion.

In writing or speaking the name of an ionic compound containing a metal that is capable of more than one possible oxidation state, it is essential that the compound name include the oxidation state of that metal.

EXAMPLE 12.8 Write the name of each compound below.

LiBr	$NaHSO_3$
$Mg(IO_4)_2$	K_2HPO_4
$AgNO_3$	$ZnCO_3$
$MnCl_3$	HgS
Hg_2Br_2	$(NH_4)_2SeO_4$ (selenium, Z = 34)

LiBr, lithium bromide
$Mg(IO_4)_2$, magnesium periodate
$AgNO_3$, silver nitrate
$MnCl_3$, manganese(III) chloride
Hg_2Br_2, mercury(I) bromide
$NaHSO_3$, sodium hydrogen sulfite
K_2HPO_4, potassium hydrogen phosphate
$ZnCO_3$, zinc carbonate
HgS, mercury(II) sulfide
$(NH_4)_2SeO_4$, ammonium selenate

In $MnCl_3$, three −1 charges from three Cl^- ions require +3 from the manganese ion, so it is the manganese(III) ion. Similarly, one −2 charge from the sulfide ion in HgS must be balanced by +2 from a mercury(II) ion. In Hg_2Br_2 the two −1 charges from two Br^- ions are balanced by the +2 charge from the diatomic mercury(I) ion. In $(NH_4)_2SeO_4$, selenium substitutes for its family member, sulfur, in sulfate ion, SO_4^{2-}, so SeO_4^{2-} is the selenate ion.

Before moving to the next section, notice that only once so far have you used a prefix that suggests a number. The $H_2PO_4^-$ ion has been called the dihydrogen phosphate ion, and HPO_4^{2-} may be called the monohydrogen phosphate ion. The dichromate ion, $Cr_2O_7^{2-}$, was mentioned, and it appears in Table 12.7. Number prefixes are technically acceptable in naming compounds containing other ions, but they are not commonly found. With the above exceptions, we recommend that you do not use them for ionic compounds. Number prefixes are required for molecular compounds. This is covered in the next section.

12.8 BINARY MOLECULAR COMPOUNDS

PG 12F Given the name (or formula) of a binary molecular compound, write its formula (or name).

Atoms in molecular compounds are held together by covalent bonds, in which two atoms share one or more pairs of electrons. Molecular compounds are sometimes called **covalent compounds.** The two elements in a binary molecular compound are generally *both nonmetals*. You may use this rule to distinguish between binary molecular compounds and binary ionic compounds, for which you already know the nomenclature rules.

Names of binary molecular compounds consist of two words:

1. The first word is the name of the element appearing first in the chemical formula, including a prefix to indicate the number of atoms of that element in the molecule.
2. The second word is the name of the element appearing second in the chemical formula, changed to end in *-ide*, and also including a prefix to indicate the number of atoms of that element in the molecule.

Elemental symbols usually appear in the order of their increasing electronegativities (Section 10.5). The electronegativity of oxygen is less than that of fluorine, but more than that of chlorine. Therefore, oxygen appears before fluorine in OF_2, but after chlorine in Cl_2O.

The same two nonmetals often form more than one binary compound. Their names are distinguished by the prefixes mentioned in the rules above. Silicon and chlorine form silicon tetrachloride, $SiCl_4$, and disilicon hexachloride, Si_2Cl_6. The prefix *tetra-* identifies four chlorine atoms in a molecule of $SiCl_4$. In Si_2Cl_6 *di-* indicates two silicon atoms and *hexa-* shows six chlorine atoms in the molecule. Technically, $SiCl_4$ should be monosilicon tetrachloride, but the prefix *mono-* for one is usually omitted. If an element has no prefix in the name of a binary molecular compound, you may assume that there is only one atom of that element in the molecule.

Table 12.8 Numerical Prefixes Used in Chemical Names

Number	Prefix
1	mono-
2	di-
3	tri-
4	tetra-
5	penta-
6	hex-
7	hepta-
8	octa-
9	nona-
10	deca-

Table 12.8 gives the first ten number prefixes. The letter "o" in *mono,* and the letter "a" in prefixes 4 to 10, are omitted if the resulting word "sounds better." This usually occurs when the next letter is a vowel. For example, a compound whose formula ends in O_5 is a *pentoxide* rather than a *pentaoxide*.

The oxides of nitrogen are ideal for practicing the nomenclature of binary molecular compounds.

EXAMPLE 12.9 For each name below, write the formula; for each formula, write the name.

nitrogen monoxide	NO_2
dinitrogen oxide	N_2O_3
dinitrogen pentoxide	N_2O_4

-----

nitrogen monoxide, NO	NO_2, nitrogen dioxide
dinitrogen oxide, N_2O	N_2O_3, dinitrogen trioxide
dinitrogen pentoxide, N_2O_5	N_2O_4, dinitrogen tetroxide

Nitrogen dioxide could be correctly identified as mononitrogen dioxide. Two of the above compounds continue to be called by their older names: N_2O is nitrous oxide and NO is nitric oxide.

Notice that hydrochloric acid is a water solution of hydrogen chloride, a gaseous binary molecular compound. Both the acid and the compound have the same formula, HCl. When necessary, state symbols may be used to distinguish between them. HCl(g) clearly identifies the compound, hydrogen chloride, while HCl(aq) refers specifically to the water solution, hydrochloric acid. The same distinction may be used for any binary acid and its parent compound. The procedure is not necessary for oxyacids, however. HNO_3 is always nitric acid, never "hydrogen nitrate."

CHAPTER 12 IN REVIEW

12.1 Oxidation State; Oxidation Number

12.2 Names and Formulas of Cations and Monatomic Anions

12A Given the name (or formula) of any ion in Figure 12.1, write the formula (or name) of that ion.

12.3 Acids and the Anions Derived from Their Total Ionization

12B Given the name (or formula) of an acid of a Group 4A, 5A, 6A, or 7A element, or of an ion derived from the total ionization of such an acid, write its formula (or name).

12.4 Names and Formulas of Anions Derived from the Stepwise Ionization of Polyprotic Acids

12C Given the name (or formula) of an ion formed by the stepwise ionization of a polyprotic acid from a Group 4A, 5A, or 6A element, write its formula (or name).

12.5 Names and Formulas of Other Acids and Ions

12.6 Formulas of Ionic Compounds

12D Given the name of any ionic compound made up of ions that are included in Performance Goals 12A, 12B, or 12C, write the formula of that compound.

12.7 Names of Ionic Compounds

12E Given the formula of an ionic compound made up of identifiable ions, write the name of the compound.

12.8 Binary Molecular Compounds

12F Given the name (or formula) of a binary molecular compound, write its formula (or name).

TERMS AND CONCEPTS

Introduction

International Union of Pure and Applied Chemistry (IUPAC)

Stock system

12.1 Oxidation state; oxidation number

12.3 Acid

Hydronium ion

Monoprotic, diprotic, triprotic

Binary acid

Oxyacid; oxyanion

12.4 Stepwise ionization

Most of these terms and many others are defined in the Glossary. Use your Glossary regularly.

QUESTIONS AND PROBLEMS

To answer all questions in this nomenclature chapter, write the name of the species for which the formula is given, or the formula if the name is given. In naming ions or ionic compounds, use oxidation numbers when necessary to avoid uncertainty, but not otherwise. For reference materials, try to limit yourself to a periodic table giving no more information than the one inside the front cover of this book. Atomic numbers are included in questions involving elements that are not in Figure 4.6.

Looking Back

(1) Cu; Kr; Mn; N_2.

(25) Cr; Cl_2; Be; S.

(2) Sodium; hydrogen; silicon; lead.

(26) Boron; silver; argon; iodine.

Section 12.2

(3) Ca^{2+}; Cr^{3+}; Zn^{2+}; P^{3-}; Br^{-}.

(27) Cu^{+}; I^{-}; K^{+}; Hg_2^{2+}; S^{2-}.

(4) Lithium ion; ammonium ion; nitride ion; fluoride ion; mercury(II) ion.

Section 12.3

(5) $HNO_3(aq)$; $H_2SO_3(aq)$; perchloric acid; selenic acid (selenium: Z = 34).

(6) Sulfate ion; chlorite ion; IO_3^-; BrO^-.

Section 12.4

(7) Hydrogen carbonate ion; dihydrogen phosphate ion; HSO_4^-.

Section 12.6

(8) Potassium sulfide; copper(II) nitrate; sodium hydrogen carbonate.

Section 12.7

(9) $MgSO_3$; AlF_3; $PbCO_3$.

Section 12.8

(10) SO_2; N_2O; phosphorus tribromide; hydrogen iodide.

(28) Iron(III) ion; hydride ion; oxide ion; aluminum ion; barium ion.

(29) $HClO(aq)$; $H_2TeO_3(aq)$ (Te is tellurium, Z = 52); bromic acid; phosphoric acid.

(30) Selenite ion (selenium: Z = 34); periodate ion; BrO_2^-; NO_2^-.

(31) HPO_4^{2+}; HS^-; hydrogen sulfite ion.

(32) Barium sulfate; chromium(III) oxide; calcium hydrogen phosphate.

(33) $CuSO_4$; $Ba(OH)_2$; Hg_2I_2.

(34) Dichlorine oxide; uranium hexafluoride (uranium: Z = 92); HBr (g); P_2O_3.

From this point items in the nomenclature exercise are selected at random from any section of the chapter. Unless marked with an asterisk (*), all names and formulas are included in the performance goals and should be found with reference to no more than the periodic table. Ions in compounds marked with an asterisk are included in Tables 12.6 and 12.7, or, if the unfamiliar ion is monatomic, the atomic number of the element is given.

(11) Hydrogen sulfite ion; potassium nitrate; $MnSO_4$; SO_3.

(12) BrO_3^-; $Ni(OH)_2$; silver chloride; silicon hexafluoride.

(13) Tellurate ion (tellurium: Z = 52); iron(III) phosphate; $NaC_2H_3O_2$*; H_2S (g).

(14) HPO_4^{2-}; CuO; sodium oxalate*; ammonia.

(15) Hypochlorous acid; chromium(II) bromide; $KHCO_3$; $Na_2Cr_2O_7$.*

(16) Co_2O_3; Na_2SO_3; mercury(II) iodide; aluminum hydroxide.

(17) Calcium dihydrogen phosphate; potassium permanganate*; NH_4IO_3; H_2SeO_4 (Se is selenium, Z = 34).

(35) Perchlorate ion; barium carbonate; NH_4I; PCl_3.

(36) HS^-; $MgSO_3$; aluminum nitrate; oxygen difluoride.

(37) Mercury(I) ion; cobalt(II) chloride; SiO_2; $LiNO_2$.

(38) N^{3-}; $Ca(ClO_3)_2$; iron(III) sulfate; phosphorus pentabromide.

(39) Tin(II) fluoride; potassium chromate*; LiH; $FeCO_3$.

(40) HNO_2; $Zn(HSO_4)_2$; potassium cyanide*; copper(I) fluoride.

(41) Magnesium nitride; lithium bromate; $NaHSO_3$; KSCN.*

(18) Hg_2Cl_2; $HIO_4(aq)$; cobalt(II) sulfate; lead(II) nitrate.

(19) Uranium trifluoride (uranium: Z = 92); barium peroxide*; $MnCl_2$; $NaClO_2$.

(20) K_2TeO_4 (Te is tellurium, Z = 52); $ZnCO_3$; chromium(II) chloride; acetic acid*.

(21) Barium chromate*; calcium sulfite; CuCl; $AgNO_3$.

(22) Na_2O_2*; $NiCO_3$; iron(II) oxide; hydrosulfuric acid.

(23) Zinc phosphide; cesium nitrate (cesium: Z = 55); NH_4CN*; S_2F_{10}.

(24) N_2O_3; $LiMnO_4$*; indium selenide (indium: Z = 49; selenium: Z = 34); mercury(I) thiocyanate.*

(42) $Ni(HCO_3)_2$; CuS; chromium(III) iodide; potassium hydrogen phosphate.

(43) Selenium dioxide (selenium: Z = 34); magnesium nitrite; $FeBr_2$; Ag_2O.

(44) SnO; $(NH_4)_2Cr_2O_7$*; sodium hydride; oxalic acid*.

(45) Cobalt(III) sulfate; iron(III) iodide; $Cu_3(PO_4)_2$; $Mn(OH)_2$.

(46) Al_2Se_3 (Se is selenium, Z = 34); $MgHPO_4$; potassium perchlorate; bromous acid.

(47) Strontium iodate (strontium: Z = 38); sodium hypochlorite; Rb_2SO_4 (Rb is rubidium, Z = 37); P_2O_5.

(48) ICl; $AgC_2H_3O_2$*; lead(II) dihydrogen phosphate; gallium fluoride (gallium: Z = 31).

13 The Gaseous State

LOOKING BACK	LOOKING AHEAD IN CHAPTER 13
Prologue. A fictional experiment was used to illustrate the scientific method as it appeared in the discovery of Boyle's Law.	A real experiment leading to Boyle's Law is described. Similar descriptions are used to show the experimental basis for all the gas laws developed in the chapter.
7.8 Given the molar mass of a substance and data from which its simplest formula could be determined, the molecular formula was written.	Given only data from which the molar mass and simplest formulas of a substance may be found, molecular formulas will be written.
Chapter 8. Chemical equations were written for various kinds of reactions.	Chemical equations will be written as a necessary part of solving gas stoichiometry problems.
9.2 and 9.5. A three-step stoichiometry pattern establishes the following unit path for solving stoichiometry problems: quantity given substance → moles given substance → moles wanted substance → quantity wanted substance. Quantity was measured in grams of any substance or STP volume of a gas. At STP, 22.4 L ≏ 1 mol.	The method of converting between moles and gas volume at any temperature and pressure is developed and then applied in the first and third steps of the stoichiometry pattern.
Appendix I, Parts C and D. Solving algebraic equations and proportionalities are discussed.	Proportional relationships derived from experiments are changed into equations which are solved algebraically for an unknown.

13.1 PROPERTIES OF GASES

The air that surrounds us is a sea of mixed gases, called the atmosphere. It is not necessary, then, to search very far to find a gas whose properties we may study. Some of the familiar characteristics of air—in fact, of all gases—are the following:

1. Gases May Be Compressed. A fixed quantity of air may be made to occupy a smaller volume by applying pressure. Figure 13.1A shows a quantity of air in a cylinder having a leak-proof piston that can be moved to change the volume occupied by the air. Push the piston down by applying more force, and the volume of air is reduced (Fig. 13.1B).

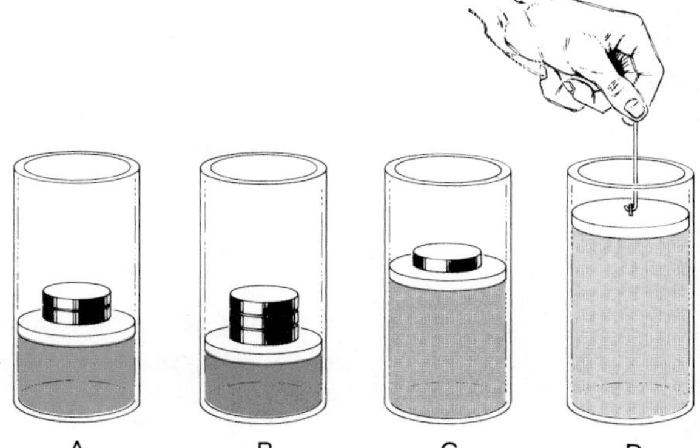

Figure 13.1
Properties of gases. The piston and cylinder show that gases may be compressed and that they expand to fill uniformly the volume available to them.

2. Gases Expand To Fill Their Containers Uniformly. If less force were applied to the piston, as shown in Figure 13.1C, air would respond immediately, pushing the piston upward, expanding to fill the larger volume uniformly. If the piston were pulled up (Fig. 13.1D), air would again expand to fill the additional space.

3. All Gases Have Low Density. The density of air is 0.0013 g/cm^3. The density of water is 770 times greater than the density of air; and iron is 6000 times more dense than air.

4. Gases May Be Mixed. "There's always room for more," is a phrase that may be applied to gases. You may add the same or a different gas to that gas already occupying a rigid container of fixed volume, provided there is no chemical reaction between them.

5. A Confined Gas Exerts Constant Pressure On The Walls Of Its Container Uniformly In All Directions. This pressure, illustrated in Figure 13.2, is a unique property of a gas, independent of external factors such as gravitational forces.

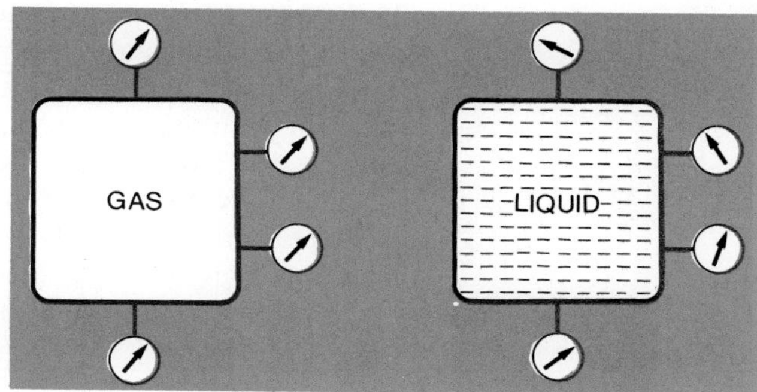

Figure 13.2
Gas pressures are exerted uniformly in all directions; liquid pressures depend on the depth of the liquid.

13.2 THE KINETIC THEORY OF GASES AND THE IDEAL GAS MODEL

PG 13A Explain physical properties of gases, or physical phenomena relating to gases, in terms of the ideal gas model.

In trying to account for the properties of gases, scientists have devised the **kinetic theory of gases.** This theory is actually the best understood portion of the **kinetic molecular theory** referred to in Section 2.2. The theory describes an **ideal gas model*** by which we can visualize the nature of a gas by comparing it with a physical system that can be seen, or at least readily imagined. The main features of the ideal gas model are:

1. Gases consist of molecular particles moving at any given instant in straight lines.
2. Molecules collide with each other and with the container walls without loss of energy.
3. Gas molecules behave as independent particles; attractive forces between them are negligible.
4. Gas molecules are very widely spaced.
5. The actual volume of molecules is negligible compared to the space they occupy.

Particle motion explains why gases fill their containers. It also suggests how they exert pressure. When an individual particle strikes a container wall it exerts a force at the point of collision. When this is added to billions upon billions of similar collisions occurring continuously, the total effect is the steady force that is responsible for gas pressure.

There can be no loss of energy as a result of these collisions. If the particles lost energy, or slowed down, the combined forces would become smaller and the pressure would gradually decrease. Furthermore, because of the relationship between temperature and average molecular speed (Section 13.5), temperature would drop if energy were lost in collisions. But these things do not happen, so we conclude that energy is not lost in molecular collisions, either with the walls or between molecules.

Gas molecules must be widely spaced; otherwise the density of a gas would not be so low. One gram of liquid water at the boiling point occupies 1.04 cm^3. When changed to steam at the same temperature, the same number of molecules fills 1670 cm^3, an expansion of 1600 times. If the molecules were touching each other in the liquid state, they must be widely separated in the vapor state. The compressibility and mixing ability of gases are also attributable to the open space between the molecules. Finally, it is because of the large intermolecular distance that attractions between molecules are negligible.

In summary, the ideal gas model pictures a gas as consisting of a large number of independent and widely spaced molecules, moving in random and chaotic fashion at high speed. Individual molecules move in straight lines until they collide with other molecules or the wall of the container without loss of energy.

*An "ideal" gas is one that conforms precisely to the equations that will be developed in this chapter. Most real gases at normal temperatures and pressures approach ideal behavior quite closely, but at certain conditions every real gas departs from this ideal.

13.3 GAS MEASUREMENTS

PG 13B List the measurable properties of a gas.

13C Given a gas pressure in atmospheres, torr, millimeters of mercury, centimeters of mercury, inches of mercury, pounds per square inch, pascals, or kilopascals, express that pressure in each of the other units.

Experiments with a gas usually involve measuring or controlling its quantity, temperature, volume, and pressure. Weighing a gas is different from weighing a liquid or solid, but readily accomplished. The amount of gas is most often expressed in moles. Unlike solids and liquids, a gas always fills its container, so its volume is equal to the volume of the container. The laboratory method of measuring pressure is not familiar, so we will examine it more closely.

By definition, pressure is the force exerted on a unit area:

$$\text{pressure} = \frac{\text{force}}{\text{area}} \quad \textit{or} \quad P = \frac{F}{A} \tag{13.1}$$

Units of pressure come from the definition. In the English system, if force is measured in pounds and area in square inches, the pressure unit is pounds per square inch (psi). The SI unit of pressure is the **pascal,** which is **one newton per square meter.** (The *newton* is the SI unit of force.) One pascal is a very small pressure; the kilopascal is a more practical unit. The **millimeter of mercury,** or its equivalent, the **torr,** and the **atmosphere** are the common units for expressing pressure.

Weather bureaus generally report *barometric* pressure, the pressure exerted by the atmosphere at a given weather station, in inches or centimeters of mercury. It is measured by a device known as a **barometer,** developed by Evangelista Torricelli in the 17th century (Fig. 13.3). On a day when the mercury column in a barometer is 752 mm high, we say that atmospheric pressure is 752 mm Hg.

Barometric pressure at sea level on an average day is 760.0 mm or 29.92 inches. This pressure is arbitrarily called **one standard atmosphere** of pressure.

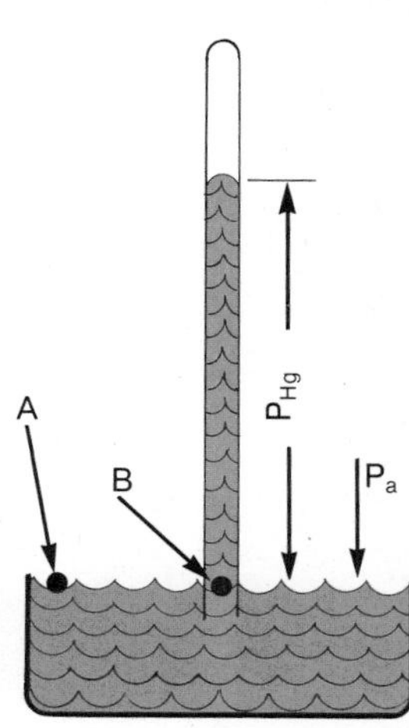

Figure 13.3
Mercury barometer. Two operational principles govern the mercury barometer. (1) The total pressure at any point in a liquid system is the sum of the pressures of each gas or liquid phase above that point. (2) The total pressures at any two points at the same level in a liquid system are always equal. Point A at the liquid surface outside the tube is at the same level as Point B inside the tube. The only thing exerting downward pressure at A is the atmosphere; P_a represents atmospheric pressure. The only thing exerting downward pressure at Point B is the mercury above that point, designated P_{Hg}. A and B being at the same level, the pressures at these points are equal: $P_a = P_{Hg}$.

The atmosphere unit is particularly useful in referring to very high pressures. The English unit that is equal to one atmosphere is 14.69 pounds per square inch. In summary, the different pressure units and their relationships to each other are:

$$1.000 \text{ atm} = 14.69 \text{ lb/in.}^2 = 29.92 \text{ in. Hg} = 76.00 \text{ cm Hg} = 760.0 \text{ mm Hg} = 760.0 \text{ torr} = 1.013 \times 10^5 \text{ Pa} = 101.3 \text{ kPa} \quad (13.2)$$

where Pa and kPa are the pascal and kilopascal, respectively.

The *torr* is the recently introduced substitute for the *millimeter of mercury,* honoring the work of Torricelli. Both terms are widely used, and the choice between them is one of personal preference. The advantage of *millimeter of mercury* is that it has physical meaning; it may be read by direct observation of an open-end **manometer,** the instrument by which pressure is most commonly measured in the laboratory (Fig. 13.4). *Torr,* however, is both easier to say and write. We will use *torr* hereafter in this text.

Outside of the laboratory, mechanical gauges are used to measure gas pressure. A typical tire gauge is probably the most familiar. Mechanical gauges show the pressure *above* atmospheric pressure, rather than the absolute pressure measured by a manometer. Even a flat tire contains air that exerts pressure. If it did not, the entire tire would collapse, not just the bottom. The pressure of the gas remaining in a flat tire is equal to atmospheric pressure. If a tire gauge shows 25 psi, that is the **gauge pressure** of the gas (air) in the tire. The absolute pressure is nearly 40 psi—the 25 psi shown by the gauge plus about 15 psi from the atmosphere.

A pressure given in one unit is changed to another unit by a one-step conversion, using the required relationship from Equation 13.2.

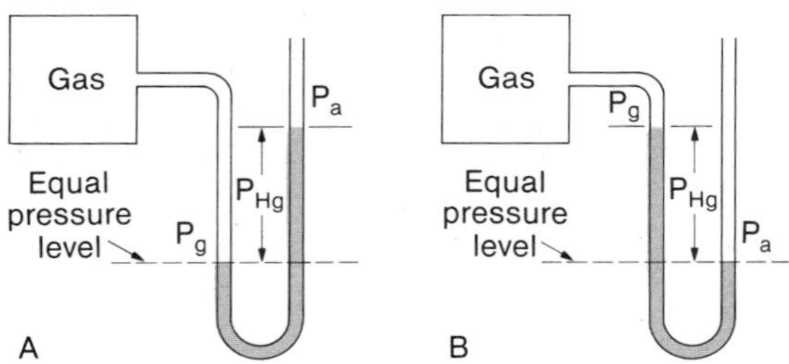

Figure 13.4
Open-end manometers. Open-end manometers are governed by the same principles as mercury barometers (Fig. 11.3). The pressure of the gas, P_g, is exerted on the mercury surface in the closed (left) leg of the manometer. Atmospheric pressure, P_a, is exerted on the mercury surface in the open (right) leg. Using a meter stick, the difference between these two pressures, P_{Hg}, may be measured directly in millimeters of mercury (torr). Gas pressure is determined by equating the total pressures at the lower liquid level. In A, the pressure in the left leg is the gas pressure, P_g. Total pressure at the same level in the right leg is the pressure of the atmosphere, P_a, plus the pressure difference, P_{Hg}. Equating the pressures, $P_g = P_a + P_{Hg}$. In B, total pressure in the closed leg is $P_g + P_{Hg}$, which is equal to the atmospheric pressure, P_a. Equating and solving for P_g yields $P_g = P_a - P_{Hg}$. In effect, the pressure of a gas, as measured by a manometer, may be found by adding the pressure difference to, or subtracting the pressure difference from, atmospheric pressure; $P_g = P_a \pm P_{Hg}$.

EXAMPLE 13.1 The pressure inside a steam boiler is 1050 psi. Express this pressure in atmospheres.

Set up and solve the problem.

$$1050 \cancel{\text{psi}} \times \frac{1 \text{ atm}}{14.7 \cancel{\text{psi}}} = 71.4 \text{ atm}$$

13.4 BOYLE'S LAW

PG 13D Given the initial volume (or pressure) and initial and final pressures (or volumes) of a fixed quantity of gas at constant temperature, calculate the final volume (or pressure).

Robert Boyle, in the 17th century, investigated the quantitative relationship between pressure and volume of a fixed amount of gas at constant temperature. A modern laboratory experiment finds this relationship with a mercury-filled manometer such as that shown in Figure 13.5. A graph (Fig. 13.5C) of pressure vs. volume measured in the experiment suggests an inverse proportionality between the variables. Expressed mathematically,

$$P \propto \frac{1}{V} \tag{13.3}$$

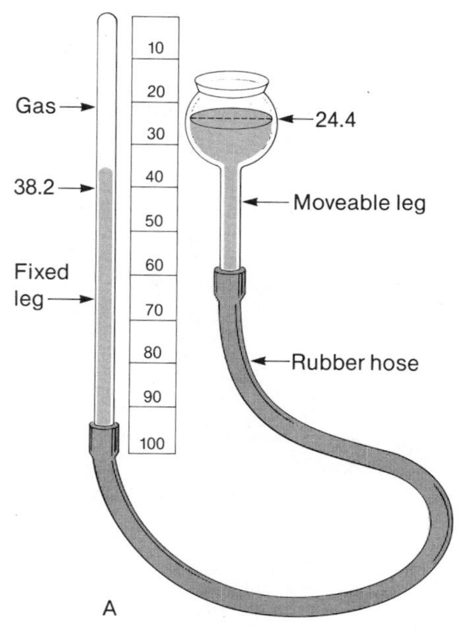

Pressure–Volume Data

Pressure (torr)	Volume (mL)	P x V (torr)(mL)
550	12.6	6930
668	10.3	6880
753	9.19	6920
842	8.17	6880
917	7.46	6840

B

Pressure (mm Hg): 500, 600, 700, 800, 900, 1000

Volume (mL): 7, 8, 9, 10, 11, 12, 13

C

Figure 13.5 Results of a student's experimental determination of the pressure–volume relationship of a fixed quantity of gas at constant temperature. By raising or lowering the moveable leg of the apparatus (A), the pressure and volume of the trapped gas may be determined. The first two columns of the table (B) are the student's data. A plot of pressure vs. volume (C) indicates that the variables are inversely proportional to each other. This is confirmed by the constant (within experimental error) product of pressure × volume, shown in the third column of the table.

Introducing a proportionality constant (see Appendix I, Part D), gives

$$P = k_1 \frac{1}{V} \tag{13.4}$$

Multiplying both sides of the equation by V yields

$$PV = k_1 \tag{13.5}$$

Within experimental error, PV is indeed a constant (Fig. 13.5B).

Boyle's Law, which this experiment illustrates, states that **for a fixed quantity of gas at constant temperature, pressure is inversely proportional to volume** (Fig. 13.6). Equation 13.5 represents the usual mathematical statement of Boyle's Law. Since the product of P and V is constant, when either factor increases the other must decrease, and vice versa. This is what is meant by an inverse proportionality.

From Equation 13.5, we see that $P_1V_1 = k_1 = P_2V_2$, or

$$P_1V_1 = P_2V_2 \tag{13.6}$$

where subscripts 1 and 2 refer to first and second measurements of pressure and volume of the gas sample at constant temperature. Solving for V_2,

$$V_2 = V_1 \frac{P_1}{P_2} \tag{13.7}$$

In other words, *if the pressure of a confined gas is changed, the final volume may be calculated by multiplying the initial volume by a ratio of pressures*—a pressure correction. The inverse proportionality between pressure and volume leads to one of two possibilities:

1. If final pressure is greater than initial pressure, then final volume must be less than initial volume. Therefore, the pressure correction must be a ratio less than 1—the numerator must be smaller than the denominator.
2. If final pressure is less than initial pressure, then final volume must be more than initial volume. Therefore, the pressure correction must be a ratio more than 1—the numerator must be larger than the denominator.

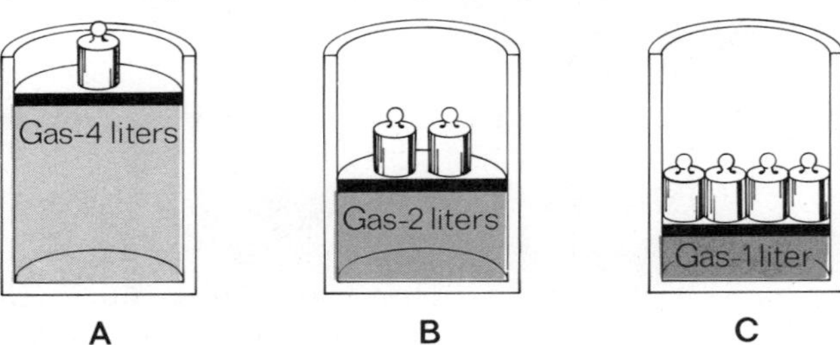

Figure 13.6
Boyle's Law. A fixed quantity of gas is confined to a cylinder, as in A, at a given pressure and constant temperature. If pressure is doubled, as in B, the volume is reduced to one half its original value. If pressure is doubled again, now four times the original pressure, the volume is reduced to one fourth of its original value (C).

The above statements are the keys to the "reasoning" method of solving pressure–volume problems. By knowing the inverse character of the pressure–volume relationship, you may *reason* whether final volume is more or less than initial volume and thereby choose a pressure correction greater or less than one.

Solve the following example by the reasoning method:

EXAMPLE 13.2 A certain gas sample occupies 3.25 liters at 740 torr. Find the volume of the gas sample if the pressure is changed to 790 torr. Temperature remains constant.

From the statement of the problem, does pressure increase (____) or decrease (____)?

It increases—from 740 to 790 torr.

Will this cause an increase (____) or decrease (____) in volume?

Decrease. Pressure and volume are inversely related: as one goes up, the other goes down.

The new volume will be found by applying a pressure correction to the initial volume—by multiplying the initial volume by a ratio of initial and final pressures. The ratio will be either $\frac{740 \text{ torr}}{790 \text{ torr}}$ or $\frac{790 \text{ torr}}{740 \text{ torr}}$. Bearing in mind that the final volume must be less than the initial volume, which ratio is correct?

$$\frac{740 \text{ torr}}{790 \text{ torr}}$$

If the final volume, the product of the multiplication, is to be less than the initial volume, the initial volume must be multiplied by a ratio less than 1.

Now complete the problem.

$$3.25 \text{ L} \times \frac{740 \cancel{\text{torr}}}{790 \cancel{\text{torr}}} = 3.04 \text{ L}$$

Example 13.2 may also be solved by substitution into Equation 13.6 or 13.7. This "formula" method is equally correct and preferred by some. These alternatives will be mentioned as they arise. Which method is "better" is, of course, a matter of opinion. It is recommended that you use the method presented in your chemistry class. In this text we will generally use the reasoning approach.

EXAMPLE 13.3 1.44 liters of gas at 0.935 atmosphere are compressed to a volume of 0.275 liter. Find the new pressure in atmospheres.

Will the reduction in volume cause an increase (____) or decrease (____) in pressure?

Increase. Pressure and volume are inversely related.

The correction factor this time is a ratio of volumes. Complete the problem.

$$0.935 \text{ atm} \times \frac{1.44 \not{L}}{0.275 \not{L}} = 4.90 \text{ atm}$$

This problem might also be solved by substitution into Equation 13.6.

13.5 ABSOLUTE TEMPERATURE

PG 13E Given a temperature in degrees Celsius (or kelvins), convert it to kelvins (or degrees Celsius).

We know that the temperature of things can be reduced until they become very cold. But how cold? Is there a bottom limit to temperature? Experiments suggest that there is. One such experiment performed by students in college laboratories shows how pressure varies with temperature when the volume of a gas sample is held constant. The experiment is described in Figure 13.7. Another experiment that measures the volume of a sample at different temperatures and constant pressure gives similar results. This includes a graph that is similar to Figure 13.7C, except that pressure is replaced by volume. These experiments, as well as many far more sophisticated investigations of low-temperature phenomena, predict an *absolute zero* temperature at −273°C.

The SI temperature scale is the **Kelvin temperature scale** (Section 3.5), which has its zero at −273°C. All temperatures on the Kelvin scale therefore have positive values. The size of the kelvin is the same as the size of the Celsius degree. This leads to the equation

$$\text{temperature (K)} = \text{temperature (°C)} + 273 \qquad (13.8)$$

by which temperatures expressed in either scale may be converted to the other. From Equation 13.8 the freezing point of water, 0°C, is 273 K. Under SI, the word "degree" and its symbol are not used. The freezing point of water is therefore "two-hundred seventy-three kelvins," or simply "two-hundred seventy-three K."

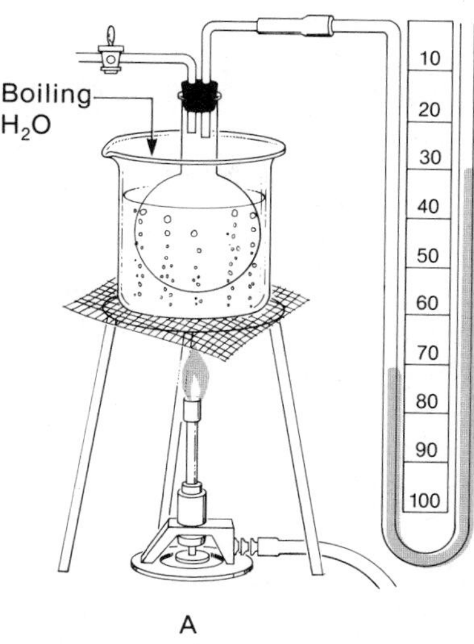

Pressure–Temperature Data

Temperature (°C)	Pressure (torr)
100	936
25	761
0	691
−79	497

B

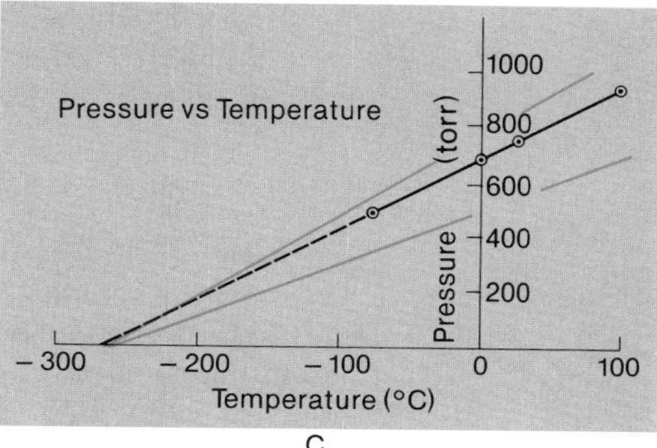

C

Figure 13.7 Results of the student's experimental determination of the pressure–temperature relationship of a fixed quantity of gas at constant volume. The flask is immersed in a liquid bath at different temperatures (A). The pressures are measured at each temperature and tabulated (B). A graph of the data (C) is plotted in black. The white lines in C show what the graph would be if the experiment were repeated with larger or smaller quantities of gas. Extrapolation to zero pressure suggests an "absolute zero" temperature at −273°C.

EXAMPLE 13.4 Tomorrow's weather forecast is for a high temperature of 77°F, or 25°C. Express this temperature in kelvins.

25°C + 273 = 298 K

To understand why there is an absolute zero in temperature it is necessary to recognize what is measured by temperature. Experiments indicate that temperature is a measure of the average translational kinetic energy of the particles. Translational kinetic energy is the energy of motion as a particle goes from one place to another. It is expressed mathematically as $\frac{1}{2}mv^2$, where m is the mass of the particle and v is its velocity. There is no reason to believe that the mass of a particle changes as temperature is reduced, so we conclude that particle velocity is less at lower temperatures. At absolute zero we imagine that all translational molecular movement stops.

Research with gases at absolute zero is impossible because the gases condense to liquids before absolute zero is reached. Other experiments support the concept of absolute zero, however, and fix the temperature somewhat more accurately at −273.15°C. No attempt to penetrate this bottom temperature barrier has ever been successful, although some researchers claim to have reached within 0.0014 K of the theoretical zero.

13.6 CHARLES' LAW

PG 13F Given the initial volume, and initial and final temperatures of a fixed quantity of gas at constant pressure, calculate the final volume.

A graph of volume or pressure vs. *absolute* temperature may be obtained by shifting the vertical axis of the graph in Figure 13.7 to −273°C. This is shown in Figure 13.8. A straight line passing through the origin is the graph of a direct proportionality. We conclude, therefore, that $V \propto T$ and $P \propto T$, where V is volume, T is absolute temperature, and P is pressure. Applying a proportionality constant to the volume–temperature relationship yields

$$V = k_2T \tag{13.9}$$

Equation 13.9 is the mathematical expression of what is known as **Charles' Law: the volume of a fixed quantity of gas at constant pressure is proportional to absolute temperature.** The physical significance of Charles' Law is illustrated in Figure 13.9.

Dividing both sides of Equation 13.9 by T yields

$$\frac{V}{T} = k_2 \tag{13.10}$$

From this it follows that $\frac{V_1}{T_1} = k_2 = \frac{V_2}{T_2}$, or

$$\frac{V_1}{T_1} = \frac{V_2}{T_2} \tag{13.11}$$

where the subscripts 1 and 2 again refer to first and second measurements of the two variables. Solving for V_2,

$$V_2 = V_1 \times \frac{T_2}{T_1} \tag{13.12}$$

Thus, if the temperature of a confined gas is changed at constant pressure, the final volume may be found by multiplying the initial volume by a *ratio of absolute temperatures*—a temperature correction. Again there are two possibilities, this time dictated by the *direct* proportionality between temperature and volume:

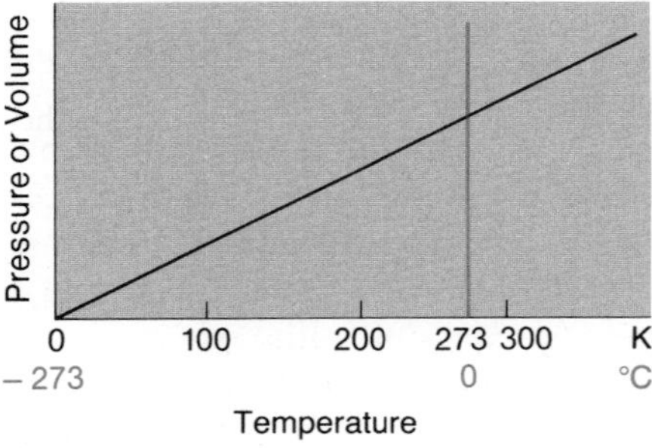

Figure 13.8
Graph of pressure or volume vs. absolute temperature, with other variables held constant. This graph is produced from Figure 13.7, with the vertical axis moved to absolute zero at −273 K. The straight line through the origin shows the pressure to be directly proportional to the absolute temperature at constant volume and quantity, and volume to be directly proportional to absolute temperature at constant pressure and quantity.

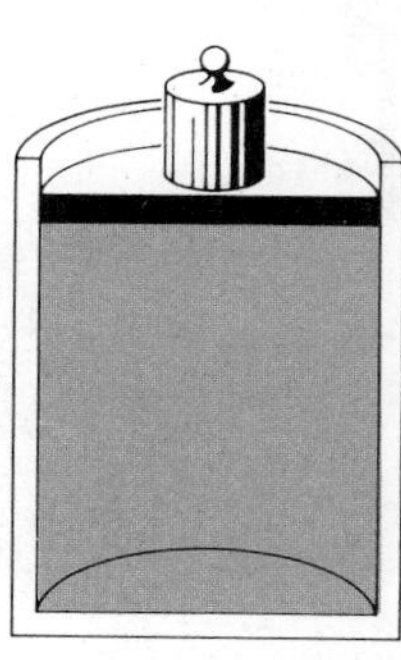
Weight on the pistons remains constant, so pressure on the gas remains constant.
27°C (300 K)
327°C (600 K)

Figure 13.9
Illustration of Charles' Law. If the absolute temperature is doubled, the volume is doubled at constant pressure and amount of gas.

1. If final temperature is greater than initial temperature, then final volume must be greater than initial volume. Therefore, the temperature correction must be a ratio more than 1—the numerator must be larger than the denominator.
2. If final temperature is less than initial temperature, then final volume must be less than initial volume. Therefore, the temperature correction must be a ratio less than 1—the numerator must be smaller than the denominator.

One extremely important fact must be remembered in working gas law problems involving temperature: the proportional relationships apply to *absolute* temperatures, not to Celsius temperatures. Before solving gas law problems, Celsius temperatures must be converted to kelvins.

EXAMPLE 13.5 2.42 liters of a gas, measured at 22°C, are heated to 45°C at constant pressure. What will be the new volume of the gas?

From the conditions of the problem, will the final volume be more or less than 2.42 liters?

More. Gas volume varies directly with temperature; if temperature increases, volume must also increase.

What temperature ratio will you use as a multiplier to find the new volume?

$$\frac{318\ K}{295\ K} \qquad \frac{(45 + 273)\ K}{(22 + 273)\ K} = \frac{318\ K}{295\ K}$$

Always be sure to convert to K for gas law problems.

Complete the problem: convert 2.42 L at 295 K to L at 318 K.

$$2.42\ L \times \frac{318\ \cancel{K}}{295\ \cancel{K}} = 2.61\ L$$

Direct substitution into Equation 13.12 would yield the same result.

13.7 THE IDEAL GAS EQUATION

PG 13G Write the equation that shows how the measurable properties of an ideal gas are related to each other (the ideal gas equation).

If we assume that the laws of physics can be applied to the kinetic theory of gases and to the ideal gas model, it is possible to derive an equation that includes all four measurable properties of gases: pressure (P), volume (V), absolute temperature (T), and quantity in moles (n). This equation is

$$PV = nRT \qquad (13.13)$$

R is a constant known as the **universal gas constant.** Its value is the same for any gas or mixture of gases that behaves like an ideal gas. The equation is called the **ideal gas equation** or the **ideal gas law.**

The ideal gas equation can also be derived from measurements of pressure, volume, absolute temperature, and quantity of gas. This derivation makes no assumption about the kinetic theory of gases or their ideal behavior. Boyle's Law and Charles' Law are specific examples of the ideal gas equation. The fact that theoretical calculations and experimental data yield the same equation makes us quite confident that the kinetic theory of gases and the ideal gas model provide a true picture of the nature of a gas.

13.8 PRESSURE, VOLUME, AND TEMPERATURE CHANGES WITH A FIXED SAMPLE OF GAS

PG 13H For a fixed quantity of a confined gas, given the initial volume, pressure, and temperature and the final pressure and temperature, calculate the final volume.

Quite often it is necessary to make calculations involving the pressure, volume, and temperature of a given sample of gas. In this case both n and R in Equation 13.13 are constant. Dividing both sides of the equation by T gives

$$\frac{PV}{T} = nR = K \tag{13.14}$$

where K is a constant for that gas sample. Using subscripts 1 and 2 for first and second measurements of the three variables, $\frac{P_1V_1}{T_1} = K = \frac{P_2V_2}{T_2}$, or

$$\frac{P_1V_1}{T_1} = \frac{P_2V_2}{T_2} \tag{13.15}$$

If the pressure, volume, and temperature of a fixed quantity of gas (P_1, V_1, and T_1) are known at any time, and any two of the variables (any two of P_2, V_2, and T_2) for the same gas sample are known at another time, the missing variable can be calculated. If the unknown is the final volume, V_2, Equation 13.16 may be solved for that unknown:

$$V_2 = V_1 \times \frac{P_1}{P_2} \times \frac{T_2}{T_1} \tag{13.16}$$

This shows that the final volume may be found by multiplying the initial volume by a pressure correction and a temperature correction:

$$\text{Final V} = \text{initial V} \times \text{pressure ratio} \times \text{temperature ratio} \tag{13.17}$$

Both of these ratios can be reasoned out just as they were in earlier examples.

EXAMPLE 13.6 A certain gas occupies 3.40 liters at 65°C and 680 torr. What volume will it occupy if it is cooled to room temperature, 21°C, and compressed to 800 torr?

We will approach this problem by first setting up for the volume change caused by a change in pressure, holding temperature constant. Then we will find the further volume change caused by a change in temperature, holding pressure constant. By steps, what will happen to the volume because of the change in pressure from 680 torr to 800 torr, considering temperature constant? Will volume increase (____) or decrease (____)?

----- -----

Volume will *decrease* if pressure increases; they are inversely proportional.

Now begin the setup of the problem with the 3.40-L volume multiplied by the proper ratio as pressure increases from 680 torr to 800 torr. Do not solve.

----- -----

$$3.40 \text{ L} \times \frac{680 \not{\text{torr}}}{800 \not{\text{torr}}} \times \underline{\hspace{3cm}}$$

3.40 liters is the volume at 680 torr and 65°C. Solving the setup as far as it is written would give the volume at 800 torr and 65°C. Now extend the setup by applying the proper temperature correction to get the volume at 800 torr and 21°C. Solve for the answer.

----- -----

$$3.40\text{ L} \times \frac{680\ \cancel{\text{torr}}}{800\ \cancel{\text{torr}}} \times \frac{294\ \cancel{\text{K}}}{338\ \cancel{\text{K}}} = 2.51\text{ L}$$

Reducing temperature reduces volume; they are directly proportional. The temperature correction is therefore less than 1.

Direct substitution into Equation 13.15 or Equation 13.16 would reproduce the above setup and yield the same result.

You now have two ways to solve volume problems by the gas laws: you may *reason* your way through temperature and pressure corrections, or you may solve the problems algebraically by Equation 13.15. If your instructor states a preference, by all means adopt it, at least for the present. If you must memorize an equation, either by your own choice or by teacher direction, Equation 13.15 is recommended. With it you may solve for any variable, given the other five.

13.9 STANDARD TEMPERATURE AND PRESSURE

PG 13I Given the volume of a gas at one temperature and pressure (or at STP), find the volume it would occupy at STP (or at a stated temperature and pressure).

The volume of a fixed quantity of gas depends on its temperature and pressure; if either changes, the volume changes. It is therefore not possible to state the amount of gas in volume units without also specifying the temperature and pressure. To meet this need, 0°C (273 K) and 1 atmosphere (760 torr) have been adopted as **standard temperature and pressure (STP).** Many gas laws problems require changing volume to or from STP. The problems are solved in the same manner as Example 13.5.

EXAMPLE 13.7 What would be the volume at STP of 4.06 liters of nitrogen, measured at 712 torr and 28°C?

Set up the problem in its entirety and solve.

$$4.06\text{ L} \times \frac{712\ \cancel{\text{torr}}}{760\ \cancel{\text{torr}}} \times \frac{273\ \cancel{\text{K}}}{301\ \cancel{\text{K}}} = 3.45\text{ L}$$

13.10 APPLICATIONS OF THE IDEAL GAS EQUATION

PG 13J Write the variation of the ideal gas equation that includes the mass of the gas sample and its molar mass.

13K Given values for all except one of the variables in either form of the ideal gas equation, calculate the value of that remaining variable.

Before the ideal gas equation, Equation 13.13, can be applied, you must have a value for the gas constant, R. This value has been determined in the laboratory. Solving Equation 13.13 for R gives

$$R = \frac{PV}{nT} \tag{13.18}$$

It is an experimental fact that one mole of any ideal gas occupies a volume of 22.4 liters at STP. Substituting these values of pressure, volume, moles, and absolute temperature into Equation 13.18 gives both the values and units of R:

$$R = \frac{1.00 \text{ atm} \times 22.4 \text{ L}}{1.00 \text{ mol} \times 273 \text{ K}} = 0.0821 \frac{\text{L·atm}}{\text{mol·K}} \tag{13.19}$$

$$R = \frac{760 \text{ torr} \times 22.4 \text{ L}}{1.00 \text{ mol} \times 273 \text{ K}} = 62.4 \frac{\text{L·torr}}{\text{mol·K}} \tag{13.20}$$

The choice between these values of R for a given problem is dictated by the units in the problem. *It is essential that the measurement units of pressure, volume, temperature, and quantity correspond with the units of R.*

A useful variation of the ideal gas equation may be found as follows: If the mass of any chemical species, g, is divided by the molar mass, MM, the quotient is the number of moles: $\frac{\text{grams}}{\text{grams/mole}}$ = moles. Therefore $\frac{g}{MM}$ may be substituted for its equivalent, n, in Equation 13.13:

$$PV = \frac{g}{MM} RT \tag{13.21}$$

In using the ideal gas equation, Equation 13.13 or 13.21, the recommended procedure is to solve the equation algebraically for the unknown, substitute known values of the other variables, and calculate the answer. Include and cancel units in the usual way; it is your best check on the correctness of your setup. While the equations may be used to determine the value of any unknown when the others are given, we will limit our examples to the calculation of moles, volume, and molar mass.

EXAMPLE 13.8 What volume will be occupied by 0.393 mole of nitrogen at 738 torr and 24°C?

SOLUTION: We begin by solving Equation 13.13 for the required volume:

$$V = \frac{nRT}{P}$$

Notice that pressure is given in torr. We therefore use the value of R in which torr is the pressure unit, Equation 13.20. Substituting this and other given data,

$$V = \frac{nRT}{P} = \frac{0.393 \text{ mol} \times \frac{62.4 \text{ L·torr}}{\text{mol·K}} \times (273 + 24) \text{ K}}{738 \text{ torr}}$$

$$= \frac{0.393 \cancel{\text{mol}}}{738 \cancel{\text{torr}}} \times \frac{62.4 \text{ L·}\cancel{\text{torr}}}{\cancel{\text{mol}}\text{·}\cancel{\text{K}}} \times (273 + 24) \cancel{\text{K}} = 9.87 \text{ L}$$

EXAMPLE 13.9 How many moles of ammonia are in a 5.00-L gas cylinder at 18°C if they exert a pressure of 8.65 atm?

Solve PV = nRT for n, substitute, and compute the answer.

$$n = \frac{PV}{RT} = \frac{8.65 \text{ atm} \times 5.00 \text{ L}}{\frac{0.0821 \text{ L·atm}}{\text{mol·K}} \times (273 + 18) \text{ K}} = 8.65 \cancel{\text{atm}} \times \frac{\text{mol·}\cancel{\text{K}}}{0.0821 \cancel{\text{L}}\text{·}\cancel{\text{atm}}} \times \frac{5.00 \cancel{\text{L}}}{291 \cancel{\text{K}}} = 1.81 \text{ mol}$$

Notice that, to divide by R, $\frac{0.0821 \text{ L·atm}}{\text{mol·K}}$, you multiply by its inverse, 1/R, $\frac{\text{mol·K}}{0.0821 \text{ L·atm}}$ (Appendix I, Part C-5).

One of the most useful applications of the ideal gas equation is determining the molar mass of an unknown substance in the vapor state. The following examples illustrate the method.

EXAMPLE 13.10 1.67 g of an unknown liquid are vaporized at a temperature of 125°C. Its volume is measured as 0.421 L at 749 torr. Calculate the molar mass.

Using Equation 13.21 we can solve for the molar mass, substitute, and calculate the answer. Complete the problem.

$$MM = \frac{gRT}{PV} = \frac{1.67 \text{ g} \times \frac{62.4 \text{ L·torr}}{\text{mol·K}} \times (273 + 125) \text{ K}}{749 \text{ torr} \times 0.421 \text{ L}}$$

$$= \frac{1.67 \text{ g}}{749 \cancel{\text{torr}}} \times \frac{398 \cancel{\text{K}}}{0.421 \cancel{\text{L}}} \times \frac{62.4 \cancel{\text{L}}\text{·}\cancel{\text{torr}}}{\text{mol·}\cancel{\text{K}}} = 132 \text{ g/mol}$$

EXAMPLE 13.11 Find the molar mass of an unknown gas if its density is 1.45 g/L at 25°C and 756 torr.

Density does not appear in the ideal gas equation as such, but its component units, grams/liter, do. The problem is solved just as the last one, interpreting density as the mass of 1.45 grams and the volume as 1.00 liter. Complete the problem.

$$MM = \frac{gRT}{PV} = \frac{1.45\ g}{1.00\ \cancel{L}} \times \frac{62.4\ \cancel{L}\cdot\cancel{torr}}{mol\cdot\cancel{K}} \times \frac{298\ \cancel{K}}{756\ \cancel{torr}} = 35.7\ g/mol$$

Real gases do not always obey the ideal gas equation exactly. Deviations are most likely to occur at relatively high pressure and/or low temperature, which cause a gas to condense to a liquid. Intermolecular attractions become strong, so the gas no longer behaves like an ideal gas. Fortunately, deviations from the equation are negligible for most gases over wide ranges of temperature and pressure, so the equation is generally satisfactory for quantitative work.

13.11 MOLAR VOLUME AT STANDARD TEMPERATURE AND PRESSURE

PG 13L Define molar volume. State the molar volume of any gas at STP.

13M Given the volume (or number of moles) of any gas at STP, find the number of moles (or volume).

13N Given two of the following for any gas at STP, find the third: grams, volume, molar mass.

13O Given gas density at STP (or molar mass), find molar mass (or gas density at STP).

Molar volume is similar to molar mass; as molar mass represents grams per mole, molar volume is liters per mole. In Section 13.10 it was noted that 22.4 liters is the volume of one mole of any ideal gas at standard temperature, 0°C (273 K), and one atmosphere. Thus, for any gas, X,

$$22.4\ L\ X\ (\text{gas at STP}) \simeq 1\ mol\ X \qquad (13.22)$$

Equation 13.22 furnishes a one-step unit path between the volume of any gas at STP and moles of that gas.

Because standard temperature and pressure are so frequently used as reference conditions for gases, the molar volume at these conditions is a useful quantity. You must be aware, however, of the restrictions placed on the value 22.4 liters per mole. It is the molar volume of a *gas*—never a solid or a liquid—*if that volume is measured at standard temperature and pressure,* not some other combination of temperature and pressure. Do not use 22.4 liters per mole unless these conditions are satisfied.

Three quantities are usually involved in molar volume problems. They are density, measured in grams per liter; molar mass, grams per mole; and molar volume, 22.4 liters per mole at STP. While density and molar mass are unique for each gas, molar volume at STP is 22.4 liters per mole for all gases. The use of molar volume is illustrated by the following examples.

EXAMPLE 13.12 Find the volume of 0.350 mole of helium at STP.

At 22.4 liters per mole, the calculation for this problem should be apparent. Solve completely.

----- -----

$$0.350 \cancel{\text{mol}} \times \frac{22.4 \text{ L}}{1 \cancel{\text{mol}}} = 7.84 \text{ L}$$

EXAMPLE 13.13 Find the volume of 12.0 grams of oxygen at STP.

The only difference between this example and the one before it is that the quantity is given in grams rather than moles. Conversion of 12.0 grams of oxygen to moles is straightforward, and converting moles to liters as in the previous example completes the problem.

----- -----

$$12.0 \cancel{\text{g } O_2} \times \frac{1 \cancel{\text{mol } O_2}}{32.0 \cancel{\text{g } O_2}} \times \frac{22.4 \text{ L } O_2}{1 \cancel{\text{mol } O_2}} = 8.40 \text{ L } O_2$$

EXAMPLE 13.14 Find the density of ammonia, NH_3, at STP.

This time there is no "given quantity." But we do know, or can find, two things about ammonia. Its molar volume at STP is 22.4 liters per mole, and its molar mass is 17.0 grams per mole. The mole is the connecting link. One mole weighs 17.0 grams, and one mole—the same quantity—occupies 22.4 liters. We seek the density, grams per liter. With that information you can complete the problem.

----- -----

$$\frac{17.0\ \text{g/}\cancel{\text{mol}}}{22.4\ \text{L/}\cancel{\text{mol}}} = \frac{17.0\ \text{g}}{22.4\ \text{L}} = 0.759\ \text{g/L}$$

Without the reasoning shown above, the problem may be solved strictly from the units. Knowing that grams are in the numerator of the answer, it is reasonable to assume that grams will be in the numerator of the setup of the problem. Starting with molar mass, grams per mole, by what must we multiply to get grams per liter? Apparently moles in the denominator of grams per mole must be replaced by liters—and we know the relationship between liters and moles of gas at STP. Therefore,

$$\frac{17.0\ \text{g}}{1\ \cancel{\text{mol}}} \times \frac{1\ \cancel{\text{mol}}}{22.4\ \text{L}} = 0.759\ \text{g/L}$$

EXAMPLE 13.15 The density of an unknown gas at STP is 1.25 grams per liter. Estimate the molar mass of the gas.

If one liter weighs 1.25 grams, and there are 22.4 liters in one mole, what is the mass of one mole? Set up and solve.

$$\frac{1.25\ \text{g}}{1\ \cancel{\text{L}}} \times \frac{22.4\ \cancel{\text{L}}}{1\ \text{mol}} = 28.0\ \text{g/mol}$$

13.12 GAS STOICHIOMETRY

PG 13P Given a chemical equation, or a reaction for which the equation can be written, and the number of grams of one species, or the volume of any gaseous species at specified temperature and pressure, find the number of grams of any other species, or volume of any other gaseous species at specified temperature and pressure.

In Section 9.2 you learned a three-step pattern for solving stoichiometry problems. It is repeated here for your ready reference:

1. Change the quantity of given species to moles.
2. Change the moles of given species to moles of wanted species.
3. Change the moles of wanted species to the units required.

Between Sections 9.2 and 9.5, the quantity was expressed either in grams of any species or STP volume of a gaseous substance. The volume ↔ mole conversions of Steps 1 and 3 were via molar volume at STP, 22.4 L/mol for any gas. If the change was from liters to moles in the first step, you divided by 22.4 L/mol; if the change was from moles to liters in the last step, you multiplied by 22.4 L/mol. In this section you will convert between moles and volume of a gaseous species at *any* temperature and pressure, not just STP. The unit path of the stoichiometry pattern is thereby expanded to

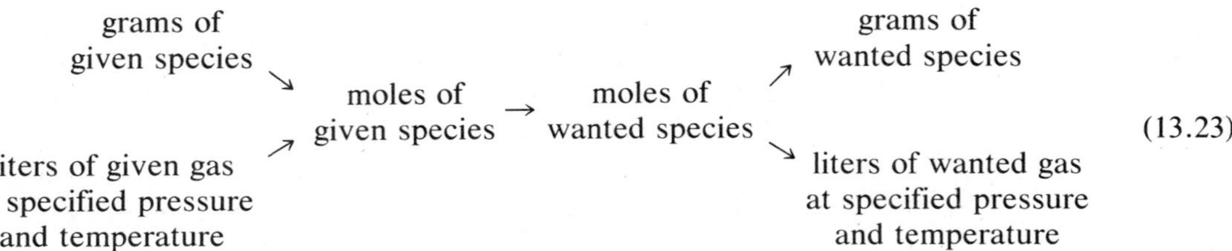

(13.23)

22.4 L/mol is the molar volume of a gas at STP. That number must not be used at any other combination of temperature and pressure. Then what *is* the number by which to change between liters and moles at nonstandard conditions? It can be found by solving the ideal gas equation for molar volume, V/n:

$$V/n = RT/P^* \tag{13.24}$$

Thus, if you wish to multiply by molar volume, multiply by RT/P; if you wish to divide by molar volume, multiply by its inverse, P/RT.

You can memorize Equation 13.24 and the sentence that follows it if you wish, but we don't recommend it. Much better to *think* multiplying or dividing by *molar volume,* liters per mole, which can be determined by reason rather than by memorization. Then think entirely in terms of R. R includes molar volume, V/n. Rearranging Equation 13.18,

$$R = \frac{PV}{nT} = \frac{V}{n} \times \frac{P}{T} = \text{molar volume} \times \frac{P}{T} \tag{13.18}$$

Therefore, if you want to multiply by molar volume, multiply by R; and if you want to divide by molar volume, divide by R. Include units. Then you can insert temperature and pressure where they must be to cancel the temperature and pressure units in R.

EXAMPLE 13.16 How many grams of ammonia can be produced by the reaction of 3.85 L of hydrogen, measured at 15.0 atm and 85°C. The equation is $N_2(g) + 3\ H_2(g) \rightarrow 2\ NH_3(g)$.

Step 1 of the stoichiometry pattern calls for converting the given quantity, a volume of hydrogen at stated temperature and pressure, to moles. To change volume to moles, do you multiply by molar volume (liters per mole) or divide by molar volume? Decide, and then do that with R. Write the setup that far; postpone temperature and pressure for the moment.

----- -----

*At STP, RT/P = 22.4: $\frac{0.0821\ \text{L}\cdot\cancel{\text{atm}}}{\text{mol}\cdot\cancel{\text{K}}} \times \frac{273\ \cancel{\text{K}}}{1\ \cancel{\text{atm}}} = 22.4\ \text{L/mol}$

$$3.85\ \cancel{L}\ H_2 \times \frac{mol \cdot K}{0.0821\ \cancel{L} \cdot atm} \times \underline{\qquad\qquad}$$

To change liters to moles, you divide by the number of liters in one mole, the molar volume. That means dividing by R, which includes molar volume. Notice in the setup so far, liters cancel.

Now, if you remember from Equation 13.24 that molar volume is RT/P, you can complete the setup by putting temperature in the denominator and pressure in the numerator. If you have not memorized the equation, you can look at the uncancelled pressure and temperature units in the setup so far and place the values where they will cancel. Complete the setup, but do not calculate the answer.

$$3.85\ \cancel{L}\ H_2 \times \frac{mol \cdot \cancel{K}}{0.0821\ \cancel{L} \cdot \cancel{atm}} \times \frac{15.0\ \cancel{atm}}{(273 + 85)\cancel{K}} \times \underline{\qquad\qquad} \times \underline{\qquad\qquad} =$$

The pressure unit in R is in the denominator, so the pressure value must be in the numerator. The temperature unit in R is in the numerator, so the temperature value must be in the denominator. With all units cancelled, the setup to this point is moles of hydrogen.

Step 1 of the stoichiometry pattern is complete. The next two steps are routine: change moles of hydrogen to moles of ammonia, using the equation coefficients, and then change moles of ammonia to grams by means of its molar mass. Calculate the answer.

$$3.85\ \cancel{L}\ \cancel{H_2} \times \frac{\cancel{mol} \cdot \cancel{K}}{0.0821\ \cancel{L} \cdot atm} \times \frac{15.0\ \cancel{atm}}{(273 + 85)\cancel{K}} \times \frac{2\ \cancel{mol\ NH_3}}{3\ \cancel{mol\ H_2}} \times \frac{17.0\ g\ NH_3}{1\ \cancel{mol\ NH_3}} = 22.3\ g\ NH_3$$

EXAMPLE 13.17 How many liters of CO_2, measured at 740 torr and 130°C, will be produced by the complete burning of 16.2 g of butane, C_4H_{10}? The equation is $2\ C_4H_{10}(g) + 13\ O_2(g) \rightarrow 8\ CO_2(g) + 10\ H_2O(g)$.

Two steps of the stoichiometry pattern take you from grams of given species to moles of wanted species, CO_2. The last step takes you from moles of CO_2 to liters of CO_2 at 740 torr and 130°C. Will you multiply by molar volume or divide in the last step? Will you multiply by R or divide by R? Set up the problem completely and calculate the answer.

$$16.2\ \cancel{g\ C_4H_{10}} \times \frac{1\ \cancel{mol\ C_4H_{10}}}{58.0\ \cancel{g\ C_4H_{10}}} \times \frac{8\ \cancel{mol}\ CO_2}{2\ \cancel{mol\ C_4H_{10}}} \times \frac{62.4\ L \cdot \cancel{torr}}{\cancel{mol} \cdot \cancel{K}} \times \frac{403\cancel{K}}{740\ \cancel{torr}} = 38.0\ L\ CO_2$$

In this case you must multiply by molar volume—multiply by R—to change moles to volume.

13.13 AVOGADRO'S LAW: VOLUME–VOLUME STOICHIOMETRY

PG 13Q State Avogadro's Law regarding gas volumes and number of molecules.

Solving the ideal gas equation, $PV = nRT$, for n gives

$$n = \frac{PV}{RT} \tag{13.25}$$

This equation is true for *all gases*. If you have equal volumes of two gas samples—they may be the same gas or different gases—at the same temperature and pressure, all four factors on the right side of Equation 13.25 are the same for both samples. Therefore, the two samples must contain the same number of moles and the same number of molecules. In other words, **equal volumes of all gases at the same temperature and pressure contain the same number of molecules.** This statement is **Avogadro's Law;** it is illustrated in Figure 13.10.

Avogadro did not derive his hypothesis from the ideal gas equation. Instead, he suggested it as an explanation for the experimental observation that when gases react with each other, if the reacting volumes are measured at the same temperature and pressure, those volumes are in a ratio of small whole numbers. This is known as the **law of combining volumes.** It may be demonstrated in the laboratory by these and other reactions:

$$H_2(g) + Cl_2(g) \rightarrow 2\,HCl(g) \tag{13.26}$$

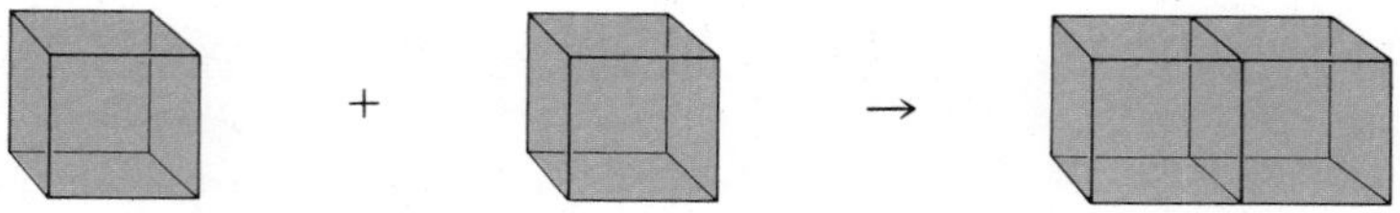

$$1\text{ L } H_2 + 1\text{ L } Cl_2 \rightarrow 2\text{ L HCl}$$

$$2\,H_2(g) + O_2(g) \rightarrow 2\,H_2O(g) \tag{13.27}$$

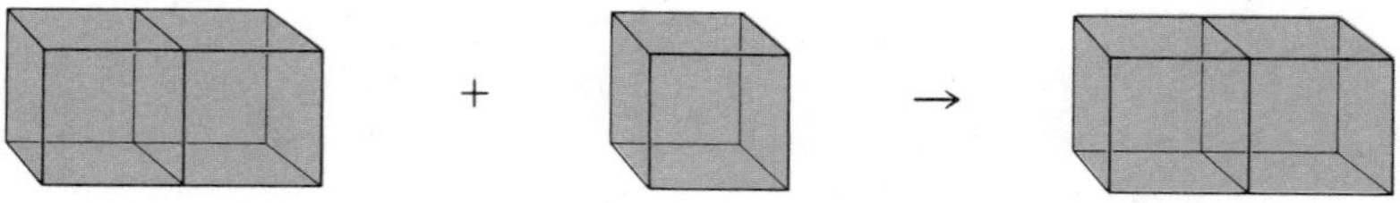

$$2\text{ L } H_2 + 1\text{ L } O_2 \rightarrow 2\text{ L } H_2O$$

$$N_2(g) + 3\,H_2(g) \rightarrow 2\,NH_3(g) \tag{13.28}$$

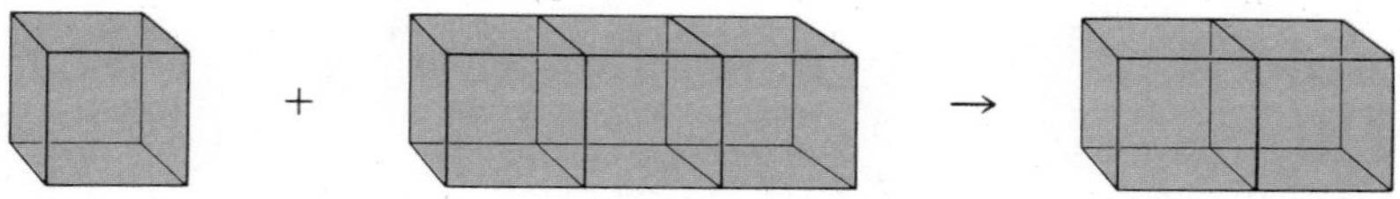

$$1\text{ L } N_2 + 3\text{ L } H_2 \rightarrow 2\text{ L } NH_3$$

The interesting fact derived from these observations is that the ratio of volumes in the above reactions is identical to the ratio of moles, represented by the coefficients in the equation. Numbers demonstrate that this is necessary

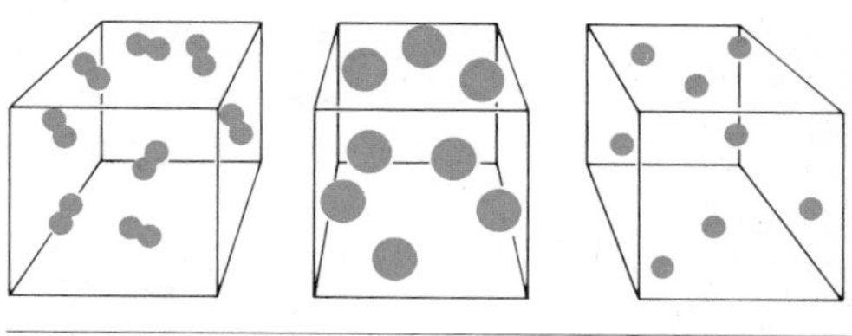

Figure 13.10
Avogadro's Law. Equal volumes of gases, measured at the same temperature and pressure, contain the same number of molecules.

if Avogadro's Law is correct. In Equation 13.28, for example, the coefficients show that the ratio of reacting moles of hydrogen to moles of nitrogen is 3/1. Suppose the reacting gas volumes are measured at STP. What will be the ratio of volumes of hydrogen and nitrogen? At STP there are 22.4 liters per mole. Three moles of hydrogen occupy 3 × 22.4 liters; one mole of nitrogen fills 1 × 22.4 liters. The ratio of volumes is

$$\frac{3 \times 22.4 \text{ L } H_2}{1 \times 22.4 \text{ L } N_2} = \frac{3 \text{ L } H_2}{1 \text{ L } N_2}$$

This is the same as the coefficients in the equation. Therefore, if 3 mol $H_2 \simeq$ 1 mol N_2 for Equation 13.28, then 3 L $H_2 \simeq$ 1 L N_2. This gives a direct volume-given-gas → volume-wanted-gas shortcut in solving stoichiometry problems involving two gases—*provided the gas volumes are measured at the same temperature and pressure.*

EXAMPLE 13.18 1.30 liters of ethene, C_2H_4, are burned completely. What volume of oxygen is required if both gas volumes are measured at STP? The equation is

$$C_2H_4(g) + 3\ O_2(g) \rightarrow 2\ CO_2(g) + 2\ H_2O(\ell)$$

The equivalence 1 liter $C_2H_4 \simeq$ 3 liters O_2 tells us that the volume of oxygen is three times the volume of ethene. Set up and solve.

$$1.30 \cancel{\text{ L } C_2H_4} \times \frac{3 \text{ L } O_2}{1 \cancel{\text{ L } C_2H_4}} = 3.90 \text{ L } O_2$$

This result is confirmed if the problem is solved by the full three steps of the stoichiometric pattern:

$$1.30 \cancel{\text{ L } C_2H_4} \times \frac{1 \cancel{\text{ mol } C_2H_4}}{\cancel{22.4} \cancel{\text{ L } C_2H_4}} \times \frac{3 \cancel{\text{ mol } O_2}}{1 \cancel{\text{ mol } C_2H_4}} \times \frac{\cancel{22.4} \text{ L } O_2}{1 \cancel{\text{ mol } O_2}} = 3.90 \text{ L } O_2$$

Be sure to recognize the restriction on this simplified solution process: both gas volumes *must be measured at the same temperature and pressure*, but not necessarily STP, as in the above example.

If the given and wanted gas volumes are at different temperatures and pressures, use pressure and temperature correction ratios to convert the volume of the *given* species from its temperature and pressure to the volume it *would occupy* at the temperature and pressure specified for the *wanted* species. From that point the solution is as in Example 13.18.

EXAMPLE 13.19 1.75 liters of oxygen, measured at 24°C and 755 torr, are consumed in burning sulfur. At one point in the exhaust hood the sulfur dioxide produced is at 165°C and a pressure of 785 torr. Find the volume of the sulfur dioxide at those conditions. The equation is

$$S(s) + O_2(g) \rightarrow SO_2(g)$$

First use pressure and temperature ratios to change the given volume of oxygen (1.75 L) from the starting conditions (24°C and 755 torr) to the volume it would occupy at the hood conditions (165°C and 785 torr). Example 13.6 is similar, so you may find it helpful to look back. Set up that far, but do not solve.

$$1.75\ \text{L}\ O_2 \times \frac{755\ \cancel{\text{torr}}}{785\ \cancel{\text{torr}}} \times \frac{438\ \cancel{\text{K}}}{297\ \cancel{\text{K}}} \times \underline{\hspace{3cm}}$$

The pressure change is from 755 torr to 785 torr. An increase in pressure reduces volume, so the pressure correction is less than 1. Temperature increases from 297 K to 438 K. A temperature rise increases volume, so the temperature correction is more than 1.

From here the problem is of the form, "How many liters of SO_2 are equivalent to the volume of O_2 in the above setup, both gases measured at the same temperature and pressure?" Extend the setup and complete the problem.

$$1.75\ \cancel{\text{L}\ O_2} \times \frac{755\ \cancel{\text{torr}}}{785\ \cancel{\text{torr}}} \times \frac{438\ \cancel{\text{K}}}{297\ \cancel{\text{K}}} \times \frac{1\ \text{L}\ SO_2}{1\ \cancel{\text{L}\ O_2}} = 2.48\ \text{L}\ SO_2$$

13.14 DALTON'S LAW OF PARTIAL PRESSURES

PG 13R Given the partial pressure of each component in a mixture of gases, find the total pressure.

13S Given the total pressure of a gaseous mixture and the partial pressures of all components except one, or information from which those partial pressures can be obtained, find the partial pressure of the remaining component.

Figure 13.11 shows the apparatus for an experiment. Suppose that n_a moles of gas A occupy volume V at absolute temperature T. The pressure exerted by A, p_a, may be found by solving the ideal gas equation for pressure: $p_a = (n_aRT)/V$. Now suppose that a container of equal volume holds n_b moles of gas B at the same temperature. The pressure of the second gas is $p_b = (n_bRT)/V$.

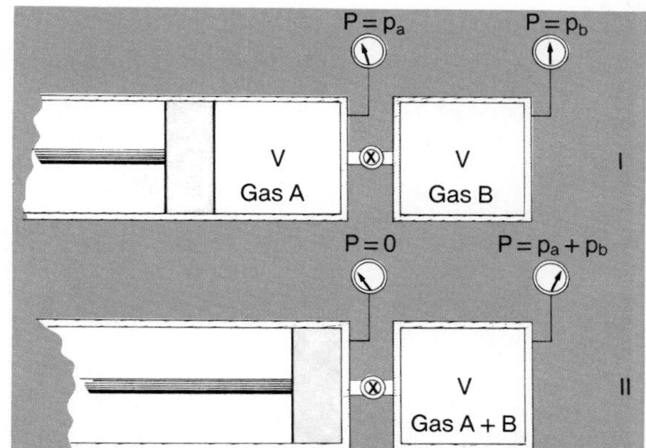

Figure 13.11
Experimental apparatus to demonstrate Dalton's Law of Partial Pressures.

Let both gases be forced into one of the containers, all at constant temperature. The mixture of gases consists of $n_a + n_b$ moles. They occupy volume V at temperature T. What is the pressure of the mixture?

The ideal gas equation is valid for a mixture of gases as well as for a pure gas. Therefore, if n is the total number of moles, $n_a + n_b$, the pressure of the mixture, P, is

$$P = \frac{nRT}{V} = \frac{(n_a + n_b)RT}{V} = \frac{n_aRT}{V} + \frac{n_bRT}{V} = p_a + p_b$$

The ideal gas equation thus predicts that the pressure of the mixed gases is equal to the sum of the pressures of the individual gases. Experiments confirm the prediction.

Observations such as these are summed up in **Dalton's Law of Partial Pressures: The total pressure exerted by a mixture of gases is the sum of the partial pressures of the components. The partial pressure of a component is the pressure that component would exert if it alone occupied the total volume at the same temperature.** Mathematically, this is

$$P = p_1 + p_2 + p_3 + \cdots \tag{13.29}$$

where P is the total pressure and p_1, p_2, and p_3 . . . are the partial pressures of components 1, 2, 3. . . .

EXAMPLE 13.20 In a gas mixture the partial pressure of methane is 150 torr, of ethane, 180 torr, and of propane, 450 torr. Find the total pressure exerted by the mixture.

This is a straightforward application of Equation 13.29.

$$P = 150 \text{ torr} + 180 \text{ torr} + 450 \text{ torr} = 780 \text{ torr}$$

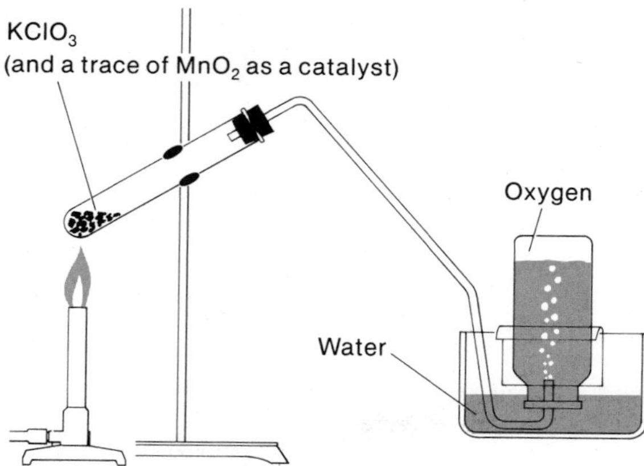

Figure 13.12
Laboratory preparation of oxygen.

A laboratory application of Dalton's Law of Partial Pressures has to do with mixtures of gases with water vapor. Gases may be prepared and collected by an apparatus such as that shown in Figure 13.12. The oxygen formed in the test tube is bubbled through water, at which time it becomes "saturated" with water vapor. The "gas" collected is therefore actually a mixture of the oxygen being generated and water vapor. The pressure exerted by the mixture is the sum of the partial pressure of the oxygen and the partial pressure of the water vapor. The latter is the equilibrium vapor pressure of water (Section 14.4), which depends only upon temperature. Values for this quantity at various temperatures are given in Table 13.1.

EXAMPLE 13.21 If the total gas pressure in an oxygen generator such as that shown in Figure 13.14 is 755 torr, and the temperature of the system is 22°C, find the partial pressure of the oxygen.

First, to see clearly where you are headed in this problem, write the partial pressure equation (Equation 13.29) as it applies specifically to this problem.

----- -----

$P = p_{O_2} + p_{H_2O}$

The water vapor pressure, p_{H_2O}, may be found from Table 13.1. It is . . .

----- -----

19.8 torr

Solve the partial pressure equation for p_{O_2}, substitute the known values, and calculate the answer.

----- -----

$p_{O_2} = P - p_{H_2O} = 755 \text{ torr} - 19.8 \text{ torr} = 735 \text{ torr}$

Table 13.1 Water Vapor Pressure

Temperature (°C)	Vapor Pressure (torr)	Temperature (°C)	Vapor Pressure (torr)
15	12.8	27	26.7
20	17.5	28	28.3
21	18.6	29	30.0
22	19.8	30	31.8
23	21.1	40	55.3
24	22.4	60	149.4
25	23.8	80	355.1
26	25.2	100	760.0

Hydrogen may also be collected over water, as in Figure 13.12. A popular laboratory experiment combines partial pressure and ideal gas law calculations.

EXAMPLE 13.22 85.5 mL of a mixture of hydrogen and water vapor are collected at 29°C. If the total pressure of the mixture is 748 torr, how many moles of hydrogen are present?

Your strategy is to solve the ideal gas equation for moles of hydrogen, n_{H_2} (see Equation 13.25):

$$n_{H_2} = \frac{p_{H_2} V}{RT}$$

The partial pressure of hydrogen is the only unknown. But you do know the total pressure of hydrogen plus water vapor, and the vapor pressure of water at 29°C is listed in Table 11.1. As in Example 13.23, calculate the partial pressure of hydrogen in the mixture.

$$p_{H_2} = P - p_{H_2O} = 748 \text{ torr} - 30 \text{ torr} = 718 \text{ torr}$$

Now you have everything you need to calculate n from Equation 13.25. Complete the problem. (Watch your volume figure.)

$$n = \frac{PV}{RT} = 718 \text{ torr} \times \frac{0.0855 \text{ L}}{302 \text{ K}} \times \frac{\text{mol K}}{62.4 \text{ L torr}} = 3.26 \times 10^{-3} \text{ mol } H_2$$

CHAPTER 13 IN REVIEW

13.1 Properties of Gases

13.2 The Kinetic Theory of Gases and the Ideal Gas Model

13A Explain physical properties of gases, or physical phenomena related to gases, in terms of the ideal gas model.

13.3 Gas Measurements

13B List the measurable properties of a gas.

13C Given a gas pressure in atmospheres, torr, millimeters of mercury, centimeters of mercury, inches of mercury, pounds per square inch, pascals, or kilopascals, express that pressure in each of the other units.

13.4 Boyle's Law

13D Given the initial volume (or pressure) and initial and final pressures (or volumes) of a fixed quantity of gas at constant temperature, calculate the final volume (or pressure).

13.5 Absolute Temperature

13E Given a temperature in degrees Celsius (or kelvins), convert it to kelvins (or degrees Celsius).

13.6 Charles' Law

13F Given the initial volume, and initial and final temperatures of a fixed quantity of gas at constant pressure, calculate the final volume.

13.7 The Ideal Gas Equation

13G Write the equation that shows how the measurable properties of an ideal gas are related to each other (the ideal gas equation).

13.8 Pressure, Volume, and Temperature Changes With a Fixed Sample of Gas

13H For a fixed quantity of a confined gas, given the initial volume, pressure, and temperature and the final pressure and temperature, calculate the final volume.

13.9 Standard Temperature and Pressure

13I Given the volume of a gas at one temperature and pressure (or at STP), find the volume it would occupy at STP (or at a stated temperature and pressure).

13.10 Applications of the Ideal Gas Equation

13J Write the variation of the ideal gas equation that includes the mass of the gas sample and its molar mass.

13K Given values for all except one of the variables in either form of the ideal gas equation, calculate the value of that remaining variable.

13.11 Molar Volume at Standard Temperature and Pressure

13L Define molar volume. State the molar volume of any gas at STP.

13M Given the volume (or number of moles) of any gas at STP, find the number of moles (or volume).

13N Given two of the following for any gas at STP, find the third: grams, volume, molar mass.

13O Given gas density at STP (or molar mass), find molar mass (or gas density at STP).

13.12 Gas Stoichiometry

13P Given a chemical equation, or a reaction for which an equation can be written, and the number of grams of one species, or the volume of any gaseous species at specified temperature and pressure, find the number of grams of any other species, or volume of any other gaseous species at specified temperature and pressure.

13.13 Avogadro's Law: Volume–Volume Stoichiometry

13Q Explain Avogadro's Law regarding gas volume and number of molecules.

13.14 Dalton's Law of Partial Pressures

13R Given the partial pressure of each component in a mixture of gases, find the total pressure.

13S Given the total pressure of a gaseous mixture and the partial pressures of all components except one, or information from which those partial pressures can be obtained, find the partial pressure of the remaining component.

TERMS AND CONCEPTS

13.2 Kinetic theory of gases
Kinetic molecular theory
Ideal gas model

13.3 Pressure
Pascal (pressure unit)
Millimeter of mercury (pressure unit)

Torr (pressure unit)
Atmosphere (pressure unit)
Barometer
Barometric (atmospheric) pressure
Manometer
Gauge pressure
13.4 Boyle's Law
Pressure correction
13.5 Absolute temperature, absolute zero
Kelvin temperature scale
Kelvin
13.6 Charles' Law
Temperature correction
13.7 Universal gas constant, R
Ideal gas equation, ideal gas law
13.9 Standard temperature and pressure, STP
13.11 Molar volume
13.13 Avogadro's Law
Law of combining volumes
13.14 Dalton's Law of Partial Pressures

Most of these terms and many others are defined in the Glossary. Use your Glossary regularly.

QUESTIONS AND PROBLEMS

Section 13.2

(*1*) What is the kinetic theory of gases? How does it describe an "ideal gas"?

(2) What causes pressure in a gas?

(3) List some properties of air that make it suitable for use in automobile tires. Explain how each property is related to the ideal gas model.

(55) What is kinetic energy? What "properties" of gases are involved in their kinetic energies?

(56) State how and explain why the pressure exerted by a gas is different from the pressure exerted by a liquid or solid.

(57) What are the desirable properties of air in an air mattress used by a camper? Show which part of the ideal gas model is related to each property.

For the next five questions in each column, explain how each physical phenomenon described is related to one or more of the features of the ideal gas model.

(4) Gases with a distinctive odor can be detected some distance from their source.

(5) Steam bubbles rise to the top in boiling water.

(6) The density of liquid oxygen is about 1.4 g/cm^3. Vaporized at 0°C and 760 torr, this same 1.4 g occupy 980 cm^3, an expansion of nearly 1000 times.

(7)* Properties of gases become less "ideal"—the substance adopts behavior patterns not typical of gases—when subjected to very high pressures such that the individual molecules are close to each other.

(8) Any container, regardless of size, will be completely filled by one gram of hydrogen.

(58) Pressure is exerted on the top of a tank holding a gas, as well as on its sides and bottom.

(59) Balloons expand in all directions when blown up, not just at the bottom as when filled with water.

(60) Even though an automobile tire is "filled" with air, more air can always be added without increasing the volume of the tire significantly.

(61) Very small dust particles, seen in a beam of light passing through a darkened room, appear to be moving about erratically.

(62) Gas bubbles always rise through a liquid and become larger as they move upward (see Prologue).

Section 13.3

(*9*) Four properties of gases may be measured. Name them.

(*10*) Define and/or distinguish between the atmosphere, millimeter of mercury, and the torr as units of pressure.

(11) Complete the following table:

atm	PSI	inches Hg	cm Hg	mm Hg	torr	Pa	kPa
1.84							
	13.9						
		28.7					
			74.8				
				785			
					124		
						1.18×10^5	
							91.4

(12) A manometer is connected to a gas storage tank, as in Figure 13.4. The mercury level difference is 284 mm, with the open leg level higher than the closed leg. If atmospheric pressure is 752 torr, what is the pressure of the gas in the tank?

(63) What is the meaning of *pressure*?

(64) Distinguish between a barometer and a manometer. Explain how each measures the pressure of a gas.

(65) A jet eductor is used to create low pressures in equipment such as evaporators and condensers. When used on a condenser in a steam plant, the closed leg of an open end manometer is 692 mm above the open leg. What pressure is being maintained in the condenser if atmospheric pressure is 743 torr?

Section 13.4

(13) Squeezing a balloon is one way to burst it. Why?

(14) A gas has a volume of 5.83 L at 2.18 atm. What will its volume be if the pressure is changed to 5.03 atm?

(15) A cylindrical gas chamber has a piston at one end that can be used to compress or expand the gas. If the gas is initially at 664 torr when the volume is 3.19 L, what will be the volume if the pressure is reduced to 529 torr?

(66) If you squeeze the bulb of a dropping pipet (eye dropper) when the tip is below the surface of a liquid, bubbles appear. On releasing the pipet, liquid flows into it. Explain why in terms of Boyle's Law.

(67) The pressure on 482 mL of a gas is changed from 772 torr to 695 torr. What is the volume at the new pressure?

(68) Calculate the gas volume in the system described in Problem 15 if the gas begins at 7.26 L and 1.22 atm and the pressure is increased to 2.12 atm.

(16)* A 1.91-L gas chamber contains air at 959 torr. It is connected through a closed valve to another chamber whose volume is 2.45 L. The larger chamber is evacuated to negligible pressure. What pressure will be reached in both chambers if the valve between them is opened and the air occupies the total volume?

(69)* The volume of the air chamber of a bicycle pump is 0.26 L. The volume of a bicycle tire, including the hose between the pump and the tire, is 1.80 L. If both the tire and the air in the pump chamber begin at 754 torr, what will be the pressure in the tire after a single stroke of the pump?

Section 13.5

(17) What does the term *absolute zero* mean? What physical condition is presumed to exist at absolute zero?

(70) A student records a temperature of −18 K in an experiment. What is the nature of things at that temperature? What would you guess the student meant to record? What is the absolute temperature that corresponds to the temperature the student meant to record?

(18) Can you ice skate on a river if the temperature is −8°C? What is this temperature in kelvins?

(71) If the temperature in the room is 21°C, what is the equivalent absolute temperature?

(19) Sodium chloride (table salt) melts at 801°C. Express this temperature in kelvins.

(72) Hydrogen remains a gas at very low temperatures. It does not condense to a liquid until the temperature is −253°C, and it freezes shortly thereafter at −259°C. What are the equivalent absolute temperatures?

(20) The temperature outside is 246 K. Is this a good day for swimming? Explain.

(73) The melting point of nickel is 1726 K. What is this in Celsius degrees?

Section 13.6

(21) 1.20 L of a gas are heated from 15°C to 40°C at constant pressure. What will be the volume at the higher temperature?

(74) A variable-volume container holds 17.7 L of gas at 55°C. If pressure remains constant, what will the volume be if the temperature falls to 15°C?

(22) A large industrial gas storage tank delivers fuel to a boiler at constant pressure. The pressure is maintained by a piston that rises or falls to adjust gas volume. The volume of the tank is 14.2 m^3 at the beginning of a weekend when the temperature is 42°C. What will the volume be for the same quantity of gas on Monday morning when the temperature has dropped to 18°C?

(75) A spring-loaded closure maintains constant pressure on a gas system that holds a fixed quantity of gas, but a bellows allows the volume to adjust for temperature changes. From a starting point of 1.26 L at 19°C, to what volume will the system change if the temperature rises to 28°C?

(23) Air in a steel cylinder is heated from 19°C to 42°C. If the initial pressure was 4.26 atm, what is the final pressure?

(76) A hydrogen cylinder holds gas at 3.67 atm in a laboratory where the temperature is 25°C. To what will the pressure change when it is placed in a storeroom where the temperature drops to 13°C?

(24)* A gas in a steel cylinder shows a gauge pressure of 355 psi while sitting on a loading dock in winter when the temperature is −18°C. What pressure will the gauge show when the tank is brought inside and its contents warm up to 23°C?

(77)* A gas storage tank is designed to hold a fixed volume and quantity of gas at 1.74 atm and 35°C. To prevent excessive pressure due to overheating, the tank is fitted with a relief valve that opens at 2.00 atm. To what temperature must the gas rise in order to open the valve?

Sections 13.8, 13.9

(25) What is the meaning of STP?

(26) An industrial process yields 8.42 L of NO at 35°C and 725 torr. What would this volume be if adjusted to STP?

(27) If the STP volume of a gas is 6.29 L, what would it be at 1.86 atm and −35°C?

(28) The stack gases of a certain industrial process are discharged to the atmosphere at 135°C and 844 torr. To what volume will 1.00 L of stack gas change as it adjusts to atmospheric conditions of 14°C and 748 torr?

(29)* The pressure gauge reads 125 psi on a 0.140-m^3 compressed air tank when the gas is at 33°C. To what volume will the contents of the tank expand if they are released to an atmospheric pressure of 751 torr and the temperature is 13°C?

(78) Why have the arbitrary conditions of STP been established? Are they realistic?

(79) If one cubic foot—28.4 L—of air at common room conditions of 23°C and 749 torr were adjusted to STP, what would the volume become?

(80) An experiment is designed to yield 75.0 mL O_2 measured at STP. If the actual temperature is 28°C, and the actual pressure is 0.894 atm, what volume of oxygen will result?

(81) A collapsible balloon for carrying meteorological testing instruments aloft is partly filled with 282 L of helium, measured at 25°C and 756 torr. Assuming the volume of the balloon is free to expand or contract according to changes in pressure and temperature, what will be its volume at an altitude where temperature is −58°C and pressure is 0.641 atm?

(82)* If 1.62 m^3 of air at 18°C and 738 torr are compressed into the 0.140-m^3 tank described in Problem 29, and the temperature is raised to 28°C, what pressure will be read on the gauge?

Section 13.10

(30) A 9.81-L cylinder contains 23.5 mol of nitrogen at 23°C. What pressure is exerted by the gas? Answer in atmospheres.

(31) A pressure of 850 torr is exerted by 28.6 g of sulfur dioxide at a temperature of 40°C. Calculate the volume of the vessel holding the gas.

(32) How many moles of NO_2 are in a 5.24-L cylinder if the pressure is 1.62 atm at 17°C?

(33) At what Celsius temperature will argon have a density of 10.3 g/L and a pressure of 6.43 atm?

(34) Calculate the mass of ammonia in a 6.64-L cylinder if the pressure is 4.76 atm at a temperature of 25°C.

(35) The density of an unknown gas at 20°C and 749 torr is 1.31 g/L. Estimate the molar mass of the gas.

(83) Find the pressure in torr produced by 4.27 g of carbon dioxide in a 5.00-L vessel at 36°C.

(84) 20.3 atm is the pressure caused by 6.04 mol of nitric oxide, NO, at a temperature of 18°C. What is the volume of the gas in liters?

(85) A 721-mL hydrogen lecture bottle is left with the valve slightly open. Assuming no air has mixed with the hydrogen, how many moles of hydrogen are left in the bottle after the pressure has become equal to an atmospheric pressure of 752 torr at a temperature of 22°C?

(86) At what temperature will 0.750 mol of chlorine in a 15.7-L vessel exert a pressure of 756 torr?

(87) How many grams of carbon monoxide must be placed into a 40.0-L tank to develop a pressure of 965 torr at 23°C?

(88) 0.710 g of an unidentified gas is placed in a 390-ml cylinder. At 52°C the gas yields a pressure of 520 torr. Find the molar mass of the gas.

(36) Air, which is about 21% oxygen (MM = 32) and 79% nitrogen (MM = 28) by moles, behaves as if its "molar mass" is 29. Calculate the density of air at 25°C and 1 atm.

(37) Just above its boiling point at 445°C, sulfur appears to be a mixture of polyatomic molecules. Above 1000°C, however, there is but one structure. Determine the formula of molecular sulfur if its vapor density is 0.625 g/L at 1.10 atm and 1100°C.

(38) A pure, unknown liquid is found to be 85.7% carbon and 14.3% hydrogen by weight. If 29.4 g of this liquid are vaporized in a 3.60-L cylinder at 260°C, the pressure is 2.84 atm. Determine the molecular formula of the compound.

Section 13.11

(39) What is the meaning of "molar volume?"

(40) What volume will be occupied by 28.4 g of propane, C_3H_8, at STP?

(41) How many grams of nitrogen will be in a 1.50-L flask at STP?

(42) Find the STP density of neon.

(43) If the density of an unknown gas at STP is 1.63 g/L, what is the molar mass of the gas?

(44) When filled with an unknown gas at STP, a 2.10-L vessel weighs 2.63 g more than it does when evacuated. Find the molar mass of the gas.

Section 13.12

(45) Carbon dioxide is released when cream of tartar reacts with baking powder: $KHC_4H_6O_6(s) + NaHCO_3(s) \rightarrow NaKC_4H_4O_6(s) + H_2O(g) + CO_2(g)$. If 8.55 g of $NaHCO_3$ react, what volume of CO_2 will be released at 0.949 atm in an oven that is at 325°C?

(46) Considering natural gas in a laboratory burner to be pure methane, CH_4, calculate the number of grams of carbon dioxide that would result from the complete burning of 35.0 L of methane, measured at 749 torr and 22°C.

(89) What is the density of neon at 40°C and 1.23 atm?

(90) Phosphorus vapor apparently consists of polyatomic molecules, the number of atoms in the molecule depending on the temperature. Measured at 790 torr, the vapor density is 2.74 g/L at 300°C and 0.617 g/L at 1000°C. Determine the molecular formulas at the two temperatures.

(91) An organic compound has the following percentage composition: 55.8% carbon, 7.0% hydrogen, and 37.2% oxygen. 3.26 g of the compound occupy 1.47 L at 160°C and 0.914 atm. Find the molecular formula of the compound.

(92) Explain the restrictions placed on the statement that 22.4 L/mol is the molar volume of any gas.

(93) Calculate the volume of 3.07 g of argon at STP.

(94) Calculate the mass of 162 L of chlorine, measured at STP.

(95) Calculate the density of nitric oxide, NO, at STP.

(96) The STP density of an unidentified gas is 2.14 g/L. Calculate the molar mass of the gas.

(97)* A student evacuates a gas-weighing bottle and finds its mass to be 135.821 g. She then fills the bottle with an unknown gas, adjusts the temperature to 0°C and the pressure to 1.00 atm and weighs it again at 136.201 g. She then fills the bottle with water and finds its mass to be 385.42 g. Find the molar mass of the gas.

(98) One source of sulfur dioxide used in making sulfuric acid comes from sulfide ores by the reaction $4\ FeS_2(s) + 11\ O_2(g) \rightarrow 2\ Fe_2O_3(s) + 8\ SO_2(g)$. Calculate the grams of FeS_2 that must react to produce 423 L of SO_2, measured at 1.36 atm and 384°C.

(99) How many liters of hydrogen, measured at 0.940 atm and 32°C, will result from the electrolytic decomposition of 10.0 g of water?

(47) Dolomite, used in the manufacture of refractory brick for lining very high temperature furnaces, is processed through a rotary kiln in which carbon dioxide is driven off: $CaCO_3 \cdot MgCO_3 \rightarrow CaO \cdot MgO + 2\ CO_2$. For each kilogram of dolomite processed, how many liters of carbon dioxide escape to the atmosphere at 225°C and 825 torr?

(48)* Solder is an alloy of lead and tin. 3.54 g of solder are treated with nitric acid, causing the tin to react: $Sn(s) + 4\ HNO_3(aq) \rightarrow SnO_2(s) + 4\ NO_2(g) + 2\ H_2O(\ell)$. The NO_2 produced has a volume of 1.39 L at 751 torr and 23°C. Calculate (a) the grams of tin that reacted and (b) the percentage composition of the solder.

(100) The reaction chamber in a modified Haber process for making ammonia by direct combination of its elements is operated at 550°C and 250 atm. How many grams of ammonia will be produced by the reaction of 75.0 L of nitrogen if introduced at the temperature and pressure of the chamber?

(101)* When properly detonated, ammonium nitrate explodes violently, releasing hot gases: $NH_4NO_3(s) \rightarrow N_2O(g) + 2\ H_2O$. Calculate the total volume of gas released at 975°C and 1.22 atm by the explosion of 46.4 g NH_4NO_3.

Section 13.13

(49) What is Avogadro's Law? Why do the volumes of gaseous reactants and products in a reaction have the same ratio as the coefficients in the reaction equation, provided the volumes are measured at the same temperature and pressure?

(50) Nitrogen dioxide is used in the chamber process for manufacturing sulfuric acid. It is made by direct reaction of oxygen with nitric oxide: $2\ NO(g) + O_2(g) \rightarrow 2\ NO_2(g)$. How many liters of nitrogen dioxide will be produced by the reaction of 155 L of oxygen, both gases being measured at atmospheric conditions, 24°C and 752 torr?

(51) Carbon monoxide is the gaseous reactant in a blast furnace that reduces iron ore to iron. It is produced by the reaction of coke with oxygen from preheated air. How many liters of atmospheric oxygen at an effective pressure of 160 torr and 23°C are required to produce 525 L of carbon monoxide at 440 torr and 1700°C? The equation is $2\ C + O_2 \rightarrow 2\ CO$.

(102) Suggest a reason why Avogadro's Law is acceptable for gases but not for liquids and solids.

(103) One of the methods for making sodium sulfate, used largely in the production of kraft paper for grocery bags, involves passing air and sulfur dioxide from a furnace over lumps of salt. The equation is $4\ NaCl + 2\ SO_2 + 2\ H_2O + O_2 \rightarrow 2\ Na_2SO_4 + 4\ HCl$. (Note the hydrochloric acid, an important by-product of the process.) If 2000 cubic feet of oxygen at 400°C and 835 torr react in a given period of time, how many cubic feet of sulfur dioxide react if measured at the same conditions?

(104) In the natural oxidation of hydrogen sulfide released by decaying organic matter, the following reaction occurs: $2\ H_2S + 3\ O_2 \rightarrow 2\ SO_2 + 2\ H_2O$. How many milliliters of oxygen at 4.52 atm and 18°C are required to react with 2.09 L of hydrogen sulfide measured at 31°C and 0.923 atm in a laboratory reproduction of the reaction?

Section 13.14

(52) State Dalton's Law of Partial Pressures, either in words or as an equation. Explain how this law "fits" the ideal gas model.

(105)* A gaseous mixture contains only nitrogen and hydrogen at a total pressure of 1.00 atm. If the partial pressure of H_2 is 0.50 atm, what is p_{N_2}? Which gas, if either, is present in greatest mass? Explain.

(53) What is the pressure of a gaseous mixture if the partial pressures of its components are: methane, 0.319 atm; ethane, 0.605 atm; propane, 0.456 atm?

(54) The properties of oxygen are studied in the laboratory by collecting samples of the gas over water. In one experiment the total pressure of the oxygen and water vapor was 762 torr. The temperature was 26°C, at which the vapor pressure of water is 25 torr. What was the partial pressure of the oxygen?

(106) Atmospheric pressure is the total pressure of the gaseous mixture called air. Atmospheric pressure is 749 torr on a day that the partial pressures of nitrogen, oxygen, and carbon dioxide are 584 torr, 144 torr, and 15 torr, respectively. What is the partial pressure of all miscellaneous gases in the air on that day?

(107)* The total volume of the gas collected in Problem 54 was 83.9 mL. The moisture was removed from the mixture, and the dry oxygen adjusted to STP. What volume did it occupy then, and what was its mass?

MISCELLANEOUS QUESTIONS

(*108*) Distinguish precisely and in scientific terms the differences between items in each of the following groups:

(a) Kinetic theory of gases, kinetic molecular theory
(b) Pascal, mm Hg, torr, atmosphere, psi
(c) Barometer, manometer
(d) Pressure, gauge pressure
(e) Boyle's Law, Charles' Law
(f) Celsius and Kelvin temperature scales
(g) 0°C, 0 K
(h) Ideal gas equation, ideal gas law
(i) Temperature and pressure, standard temperature and pressure
(j) Molar volume, molar mass, density (of a gas)
(k) Avogadro's Law, law of combining volumes
(l) Total pressure, partial pressure

(109) Determine whether each statement that follows is true or false:

(a) Gas molecules change speed when they collide with each other.
(b) A gas molecule may gain or lose kinetic energy in colliding with another molecule.
(c) The total kinetic energy of two molecules is the same before and after they collide with each other.
(d) Gas molecules are strongly attracted to each other.
(e) Gauge pressure is always greater than absolute pressure except in a vacuum.
(f) For a fixed amount of gas at constant temperature, if volume increases, pressure decreases.
(g) For a fixed amount of gas at constant pressure, if temperature increases, volume decreases.
(h) For a fixed amount of gas at constant volume, if temperature increases, pressure increases.
(i) At a given temperature, the number of degrees Celsius is larger than the number of kelvins.
(j) Both temperature and pressure correction ratios are larger than 1 when calculating the gas volume as conditions change from STP to 15°C and 0.894 atm.
(k) The molar volume of a gas at 750 torr and 25°C is more than 22.4 L/mol.
(l) To change liters of a gas to moles, multiply by RT/P.
(m) The mass of 5.00 L of ammonia is the same as the mass of 5.00 L of carbon monoxide if both volumes are measured at the same temperature and pressure.
(n) In a mixture of two gases, the gas with the higher molar mass exerts the higher partial pressure.

(110)* The compression ratio in an automobile engine is the ratio of gas pressure at the end of the compression stroke to the pressure at the beginning. Assume that the ideal gas law is obeyed and that compression occurs at constant temperature. The total volume of the cylinder in a compact automobile is 350 cm^3, and the displacement (the reduction in volume during the compression stroke) is 309 cm^3. What is the compression ratio in that engine?

(111)* The volume of a bicycle tire is 1.5 L; assume it to be the same when "empty" or "full." The tire is to be pumped up to 60 psi gauge, using a bicycle tire pump that forces 0.39 L of air at 1.0 atm into the tire with each stroke. Assume that temperature remains constant at 22°C (295 K) during the process. Work in two significant figures, and use the rounded-off result from each step in the step that follows. Also, use 15 psi = 1 atm.

(a) What is the gas pressure (atmospheres) in the tire when "empty?"
(b) What will be the gas pressure (atmospheres) in the tire when "full?"
(c) How many moles of "air" (gas mixture) are forced into the tire with each stroke of the pump?
(d) How many moles of air are in the tire when "empty?"
(e) How many moles of air are in the tire when "full?"
(f) How many strokes of the pump are required to pump up the tire?
(g) Why does each stroke of the pump become more difficult as you pump up a tire? (Assume you are tireless—and no pun is intended!)

(112)* Suppose you were to use the bicycle pump described in Problem 111 to pump up a 41-L automobile tire to 30 psi gauge, with all other conditions being the same. How many strokes would be required?

14 Liquids and Solids

LOOKING BACK	LOOKING AHEAD IN CHAPTER 14
2.2 According to the kinetic molecular theory, particles of a gas move randomly, filling and taking the shape of the container that holds them; particles of a liquid move freely among themselves at the bottom of the container that holds them; and particles of a solid vibrate in fixed positions in a fixed shape.	A general understanding of the kinetic theory of matter and the arrangement and movement of particles in gases, liquids, and solids is essential to an understanding of the particle activity described in this chapter.
2.7 Particles with the same charge attract; those with unlike charges repel.	Attractions and repulsions between molecules are electrostatic in character.
3.8 Energy is measured in joules (calories) and kilojoules (kilocalories).	Heats of vaporization and fusion are expressed in these units.
3.9 Specific heat is used in calculating heat flow as the temperature of a substance changes.	Specific heat calculations are combined with change-of-state calculations to get total heat flow for a process involving both changes.
10.5 Electronegativity determines the polarity of a bond.	The large electronegativity difference between hydrogen and nitrogen, oxygen, or fluorine underlies hydrogen bonding.
11.9 The polarity of a molecule depends on its shape and the polarity of bonds in the molecule.	Molecular polarity is responsible for intermolecular attractions, and in turn, the physical properties of substances.
13.5 Temperature is a measure of average translational kinetic energy.	Rate of evaporation depends on kinetic energy and, therefore, on temperature.
13.7 and 13.14 $pV = nRT$, the ideal gas equation, relates the partial pressure, quantity, volume, and temperature of a component of a gaseous mixture.	Solved for partial pressure, the ideal gas equation shows why the partial pressure of a vapor above its liquid phase depends on vapor concentration and temperature.

14.1 THE NATURE OF THE LIQUID STATE

PG 14A Explain the differences between the physical behavior of liquids and gases in terms of the relative distances between molecules and the effect of those distances on intermolecular forces.

The behavior of gases is neatly summarized in the ideal gas equation (Section 13.7), which may be applied to nearly all gases. No similar relationship is available for liquids. Two things about liquids are quite different from gases. First, liquid molecules are very close to each other, instead of being widely separated as in a gas. Second, because of this closeness, attractions between molecules are fairly strong and play a major role in determining the physical properties of a liquid. The properties of a gas can be explained only if we assume that intermolecular attractions are negligible.

The fact that gases can be compressed and liquids cannot is explained by these two differences. To compress a gas you need only to push the molecules closer to each other. In a liquid the molecules are already "touchingly close," so to compress a liquid you would have to crush or distort the actual molecules. This does not appear to occur. If you heat a gas at constant pressure or reduce the pressure on a gas at constant temperature, the gas expands readily. Neither an increase in temperature nor a reduction in pressure causes a major increase in the volume of a liquid. Evidently intermolecular attractions in a liquid hold the molecules together in an almost fixed total volume.

A gas may be condensed to a liquid by reducing its temperature, by compressing it, or by a combination of these actions.* Lowering the temperature slows the molecules in their random movement so that the intermolecular collisions are no longer elastic. Instead, the attractions between molecules cause them to stay together. This leads eventually to the formation of a liquid drop, and ultimately to nearly complete condensation. When a gas is compressed, the molecules are pushed closer to each other. At the smaller intermolecular distances the attractive forces are strong enough to cause the molecules to stick together, leading to condensation.

Quick Check 14.1
What is the main difference between gases and liquids that accounts for the large differences between their properties?

14.2 PHYSICAL PROPERTIES OF LIQUIDS

PG 14B For two liquids, given comparative values of physical properties that depend on intermolecular attractions, predict the relative strengths of those attractions; or, given a comparison of the strengths of intermolecular attractions, predict the relative values of physical properties that depend on them.

Many physical properties of liquids are directly related to the strength of intermolecular attractions. Among them are the following:

Vapor Pressure. Because of evaporation, the open space above any liquid contains some molecules in the gaseous, or vapor, state. The partial pressure

*There is a temperature, called critical temperature, above which a gas cannot be liquified by compression alone.

exerted by these gaseous molecules is called **vapor pressure.** If the gas space above the liquid is closed, the vapor pressure increases to a definite value, referred to as the **equilibrium vapor pressure** (see Section 14.4). Equilibrium vapor pressure is related to the strength of intermolecular forces. At a given temperature *liquids with relatively weak intermolecular attractions evaporate more readily, yielding higher vapor concentrations and, therefore, higher vapor pressures than liquids with strong intermolecular forces.*

Boiling Point. Liquids may be changed to gases by boiling. A liquid must be heated to make it boil. When it boils, the average kinetic energy of the liquid particles is high enough to overcome the forces of attraction that hold molecules in the liquid state, and it becomes a gas. *Liquids with strong intermolecular forces require higher temperatures for boiling than liquids with weak intermolecular attractions.*

Molar Heat of Vaporization. Even after the temperature of a liquid has been raised to its boiling point, additional energy is required to separate the liquid molecules from each other and keep them apart. The energy required to vaporize one mole of liquid is called the **molar heat of vaporization.** *A mole of a liquid with strong intermolecular attractions requires more energy to vaporize it than a mole of a liquid with weak intermolecular attractions.*

The trends in vapor pressure, boiling point, and molar heat of vaporization are correlated with strength of intermolecular attractions in Table 14.1.

Viscosity. One of the characteristics of liquids, according to the kinetic molecular theory, is that the liquid molecules are free to move about relative to each other within the body of the liquid. That freedom to move is not the same in all liquids. You are aware, for example, that water may be poured much more freely than syrup, and syrup more readily than honey. The unique pouring characteristic of each liquid is the result of its **viscosity.** Viscosity may be thought of as an internal resistance to flow. *Liquids with strong intermolecular forces are generally more viscous than liquids with weak intermolecular attractions.*

Surface Tension. When a liquid is broken into "small pieces" it forms drops. A liquid drop that is not falling has the shape of a sphere. A sphere has the smallest surface area possible for a drop of any given volume. This tendency toward a minimum surface is the result of **surface tension.** Within a liquid, each molecule is attracted in all directions by the molecules that surround it. At the

Table 14.1 Physical Properties of Liquids

Substance	Vapor Pressure at 20°C	Normal Boiling Point	Heat of Vaporization	Intermolecular Forces
Mercury	0.0012 torr	357°C	59 kJ/mol	Strongest
Water	17.5	100	41	↑
Benzene	75	80	31	
Ether	442	35	26	
Ethane	27 000	−89	15	Weakest

surface, however, the attraction is nearly all inward, pulling the surface molecules into a sort of tight skin over a standing liquid or around a drop. This is surface tension. The effect of surface tension in water may be seen when a needle floats if placed gently on a still surface, or when small bugs run across the surface of a quiet pond (Fig. 14.1). *Liquids with strong intermolecular attractions have higher surface tension than liquids with weak intermolecular forces.*

We now summarize these relationships between intermolecular attractions and physical properties. If you think in terms of the "stick togetherness" of molecules with strong attractive forces, you can usually reason to correct conclusions. With all other influences being equal, strong intermolecular attractions generally lead to:

1. Low vapor pressure. This is an "inverse" relationship because the vapor pressure is a result of the *gas* released by evaporation when the intermolecular attractions between liquid molecules are overcome. If the *stick togetherness* is high between liquid molecules, not much gas escapes, so the vapor pressure is low.
2. High boiling point. When *stick togetherness* is high, it takes a lot of agitation (high temperature) before the molecules can even *begin* to tear loose from each other within the liquid, where boiling occurs.
3. High heat of vaporization. Again, if *stick togetherness* is high, it takes still more energy to separate the molecules, even after they have reached the boiling point.
4. High viscosity. High *stick togetherness* means more "gooey-ness."
5. High surface tension. High *stick togetherness* at the surface means more resistance to anything that would break through or stretch that surface (high surface tension).

Figure 14.1
Surface tension. Unbalanced downward attractive forces at the surface of a liquid pull molecules into a difficult-to-penetrate skin capable of supporting small bugs or thin pieces of steel, such as a needle or razor blade. A bug literally runs *on* the water; it does not float *in* it. Molecules within the water are attracted in all directions, as shown.

Quick Check 14.2

1. Intermolecular attractions are stronger in A than they are in B. Which do you expect will have the higher surface tension, molar heat of vaporization, vapor pressure, boiling point, and viscosity?
2. X has a higher molar heat of vaporization than Y. Which do you expect will have a higher vapor pressure? Why?

14.3 TYPES OF INTERMOLECULAR FORCES

PG 14C Identify and describe or explain dipole forces, dispersion forces, and hydrogen bonds.

14D Given the structure of a molecule, or information from which it may be determined, identify the significant intermolecular forces present.

14E Given the molecular structures of two substances, or information from which they may be obtained, compare or predict relative values of physical properties that are related to them.

As a group, the intermolecular attractions that are responsible for so many physical properties of liquids are called **van der Waals forces.** These forces are electrostatic in character; the attractions are between negative and positive charges. But molecules, you may protest, are electrically neutral. How can there be electrostatic attractions between them? The explanation lies in molecular polarity, the *distribution* of electrical charge within the molecule. Molecules with a symmetrical distribution of charge are nonpolar; if the charge distribution is unbalanced, the molecule is polar (Section 11.9).

There are three kinds of intermolecular forces that can be traced to electrostatic attractions: dipole forces, dispersion forces, and hydrogen bonds.

1. Dipole Forces. A polar molecule is sometimes described as a **dipole.** This means that the molecule has two "poles," positive in one region and negative in another. Depending on their size and shape, dipoles can interact with each other in a number of ways (Fig. 14.2).

Iodine chloride is a substance whose molecules are dipoles. In the solid state they are held as in Figure 14.2A. The forces between molecules are overcome by molecular motion at 27°C, and the substance melts. As a liquid the molecules still attract each other, but they "flow" freely. These attractions are overcome at 97°C, the boiling point of iodine chloride.

The effect of dipole forces is seen when the melting and boiling points of iodine chloride are compared with those of bromine. The molecules are similar in size, shape, and molecular mass. The main difference between them is that ICl molecules are polar and Br_2 molecules are not. Bromine melts at −7°C and boils at 59°C, both temperatures lower than the corresponding temperatures for iodine chloride. This shows that intermolecular forces are weaker in nonpolar bromine than in polar iodine chloride.

Table 14.2 shows additional polar–nonpolar boiling point comparisons between compounds having otherwise similar molecules. In each case the com-

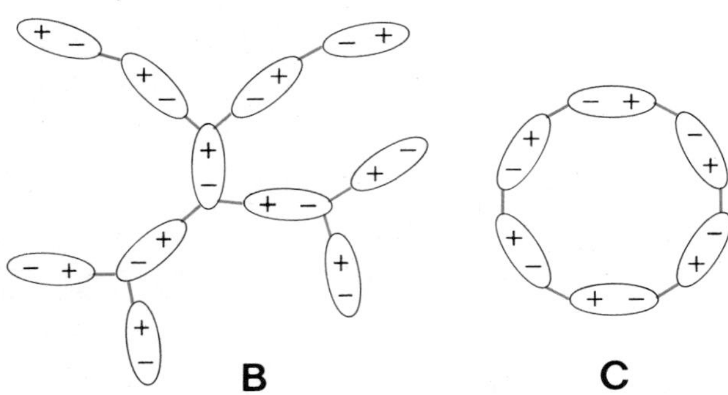

Figure 14.2
Dipole forces. A. The polar molecules are arranged in a regular geometric pattern, as they might be found in a molecular solid. B. Similar polar molecules are arranged in a random fashion, such as they might be found in a liquid. C. The molecules have grouped themselves into a larger unit, evidence of which is known to all three states of matter. In all cases the positive region of one polar molecule is attracted to the negative region of the second polar molecule.

pound with polar molecules has the higher boiling point, supporting the explanation of stronger intermolecular forces between dipoles.

2. Dispersion (London) Forces. Returning to bromine, Br_2, what type of intermolecular attractions hold it in the liquid state at temperatures up to 59°C? Why is it not, like chlorine, Cl_2, a gas at room temperature? And why is iodine, I_2, a solid at room temperature? The forces responsible for bromine being a liquid and iodine a solid are called **dispersion forces,** or **London forces.**

Dispersion forces are believed to be the result of "temporary dipoles" that are formed by the shifting electron clouds within molecules that are, on the average, nonpolar (Fig. 14.3). These dipoles attract or repel the electron clouds of nearby nonpolar molecules, thereby inducing them to become dipoles temporarily. As long as these dipoles exist—a very small fraction of a second in each individual case—there is an attraction between them.

The strength of dispersion forces depends on the ease with which electron distributions can be distorted, or "polarized." Large molecules, with many

Table 14.2 Boiling Points of Polar vs. Nonpolar Substances

Formulas	Polar or Nonpolar	Molecular Mass	Boiling Point (°C)	Formulas	Polar or Nonpolar	Molecular Mass	Boiling Point (°C)
N_2	Nonpolar	28	−196	GeH_4	Nonpolar	77	−90
CO	Polar	28	−192	AsH_3	Polar	78	−55
SiH_4	Nonpolar	32	−112	Br_2	Nonpolar	160	59
PH_3	Polar	34	−85	ICl	Polar	162	97

Molecules

I II

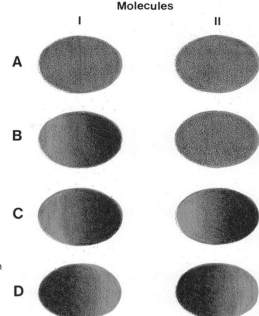

A

B

C

D

Figure 14.3
The origin of dispersion forces. A. Molecules I and II are temporarily close to each other. Uniform shading indicates overall uniform distribution of electron cloud in molecule. B. Electron cloud in Molecule I has shifted to right, forming a temporary dipole with concentration of negative charge to right, positive charge to left. C. Negative charge concentration at the right of Molecule I repels electron cloud in Molecule II to right side of molecule, causing it to become a temporary dipole similar to Molecule I. At this time there is a weak dipole–dipole attraction between the two "instantaneous dipoles." D. A small fraction of a second later the electron clouds may shift to opposite sides of the molecules. The instantaneous dipoles still attract each other weakly. Again a small fraction of a second later the molecules interact similarly with outer nearby molecules forming new weak intermolecular attractions.

electrons, or with electrons far removed from the nucleus, are more easily polarized than small, compact molecules in which the nuclei hold electrons in position more firmly. Larger molecules are generally heavier. As a consequence, intermolecular forces tend to increase with increasing molecular mass among otherwise similar substances. Figure 14.4 shows this for the halogens and for a series of organic compounds known as normal alkanes.

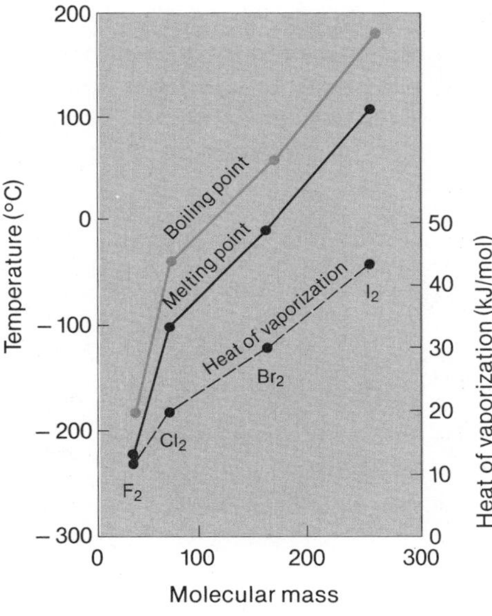

Boiling Points of Normal Alkanes

Formula	Boiling Point (°C)
CH_4	– 162
C_2H_6	– 88
C_3H_8	– 42
C_4H_{10}	0
C_5H_{12}	36
C_6H_{14}	69
C_7H_{16}	98
C_8H_{18}	126
C_9H_{20}	151
$C_{10}H_{22}$	174

Figure 14.4
Physical properties as a function of molecular size. The graph shows the boiling points, melting points, and heats of vaporization of the halogens. The table lists the boiling points of normal alkane hydrocarbons, which have the general formula C_nH_{2n+2}. When molecular size increases, as it generally does with increasing molecular mass, intermolecular attractions due to dispersion forces increase. This causes an increase in the magnitude of physical properties related to these attractions.

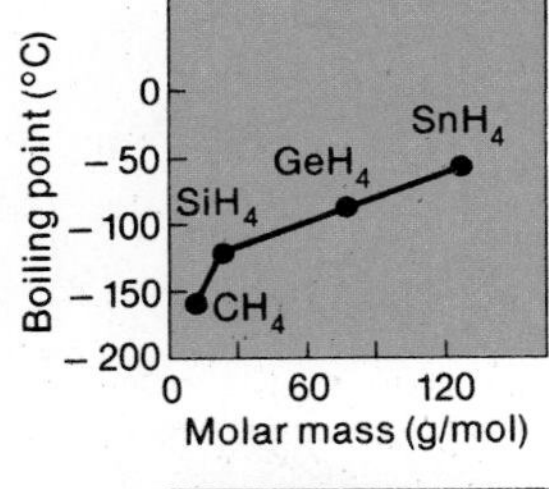

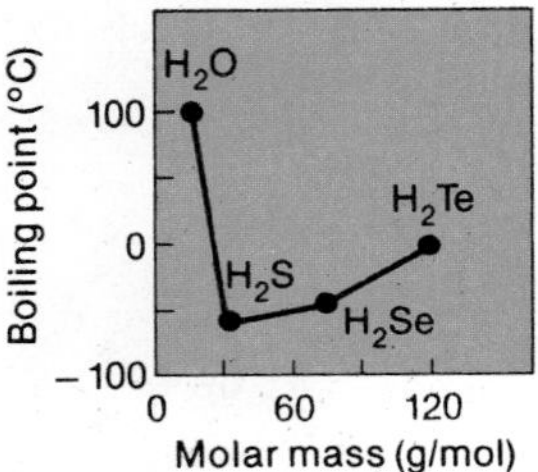

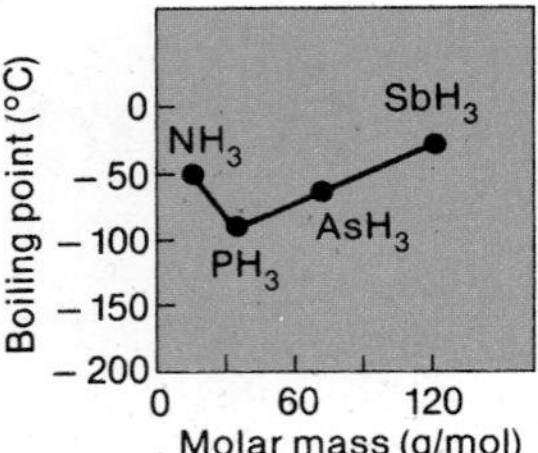

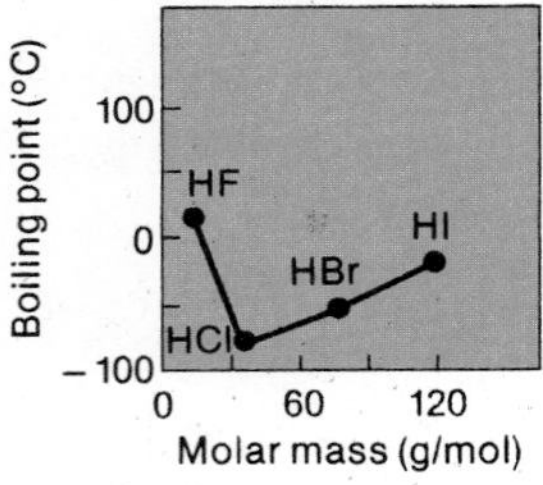

Figure 14.5
Boiling points of different hydrides.

Dispersion forces exist between *all* kinds of molecules, polar or nonpolar. Among small, low molecular mass molecules they are weak, and important only when there are no dipole forces. But among compounds with large molecules and higher molecular mass, these forces can become quite strong—strong enough to make liquids and solids out of nonpolar substances like bromine and iodine.

3. Hydrogen Bonds. Four hydrides of the Group 4A elements, in order of decreasing molecular mass, are SnH_4, GeH_4, SiH_4, and CH_4. These molecules are all tetrahedral in shape; they are nonpolar. The only intermolecular forces are dispersion forces. We would expect their boiling points to decrease in the order shown, matching the decreasing molecular sizes. They conform to this expectation, as shown in the first graph in Figure 14.5.

With some modification, similar reasoning would lead us to expect that ammonia, NH_3, should have the lowest boiling point of the hydrides of the Group 5A elements. Also, water should have the lowest boiling point of the Group 6A hydrides, and hydrogen fluoride, HF, should be the lowest boiling of the Group 7A hydrides. Obviously, from Figure 14.5, this is not the case. Why not? What is responsible for the unexpectedly strong intermolecular attractions in ammonia, water, and hydrogen fluoride?

The name **hydrogen bond** is given to the abnormally strong intermolecular forces in NH_3, H_2O, and HF. The attraction is between the hydrogen atom of one molecule and the nitrogen, oxygen, or fluorine atom of another. Figure 14.6 shows the hydrogen-bonding effect in water. The electronegativity of an oxygen atom (3.5) is considerably higher than that of a hydrogen atom (2.1). The electron pairs that form bonds between the oxygen atom and each hydrogen atom are drawn closely to the oxygen atom, giving that region of the molecule a negative charge. This leaves each hydrogen nucleus—nothing more than a proton—as a small, highly concentrated region of positive charge. The negatively charged oxygen region of one water molecule can get quite close to the small, positively charged region of a neighboring molecule. The result is a hydrogen bond, a dipole–dipole attraction that is much stronger than ordinary dipole forces. Hydrogen bonds between molecules are roughly one tenth as strong as covalent bonds between atoms within a molecule.

To recognize the possibility of hydrogen bonding in a substance, examine its structure. Figure 14.7 shows the structures present in most hydrogen bonding situations, and the caption explains why these structures produce hydrogen bonding.

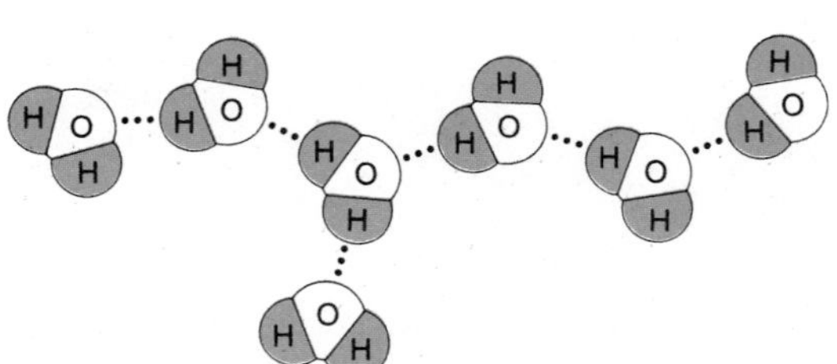

Figure 14.6
Hydrogen bonding in water. Intermolecular hydrogen bonds are present between the electronegative oxygen region of one molecule and the electropositive hydrogen region of a second molecule.

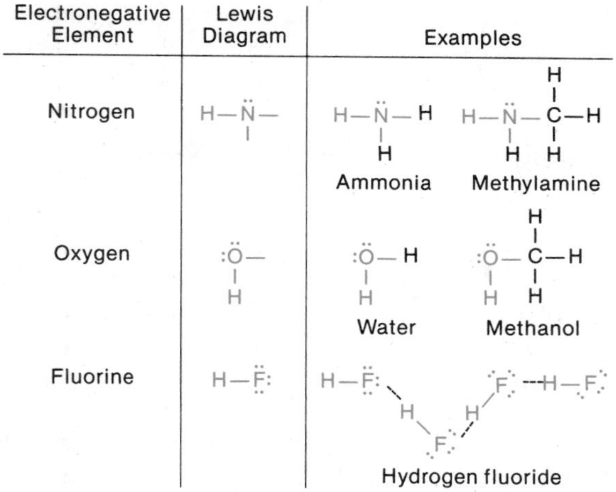

Electronegative Element	Lewis Diagram	Examples
Nitrogen	H—N̈— (with —H below)	H—N̈—H (with H below) Ammonia; H—N̈—CH₃ (H—N̈—C(H)(H)—H, with H below N) Methylamine
Oxygen	:Ö— (with H below)	:Ö—H (with H below) Water; :Ö—C(H)(H)—H (with H below O) Methanol
Fluorine	H—F̤̈:	H—F:···H—F:···H—F: Hydrogen fluoride

Figure 14.7
Recognizing hydrogen bonding. Hydrogen bonding most frequently occurs when hydrogen is covalently bonded to an atom of a highly electronegative element that has one or more unshared electron pairs. Fluorine, oxygen, and nitrogen are the most common. The hydrogen bond is established between one of these strongly electronegative atoms in one molecule and a hydrogen atom that is bonded to another highly electronegative atom in a nearby molecule. Hydrogen atoms bonded to carbon in methanol and methylamine do not engage in hydrogen bonding because a carbon atom does not attract the bonding electron pair away from hydrogen strongly enough. There is evidence that hydrogen flouride molecules form chains of variable length, as shown, and even closed rings as in Figure 14.2C.

The abnormal boiling point of water—abnormal in the sense that it departs from what would be expected when compared to other Group 6A hydrides—is but one example of a long list of "abnormal" physical properties of water. Water is so common a substance we take it much for granted. But, in relative trends of physical properties predicted from the periodic table, water is one of the most "uncommon" substances known. Nearly all of its unique properties are at least partly the result of hydrogen bonding.

Quick Check 14.3: Are the following true or false:

1. Dispersion forces are present only with nonpolar molecules.
2. All other things being equal, hydrogen bonds are stronger than dipole–dipole attractions.
3. Dipoles have a net electrical charge.
4. Intermolecular forces are magnetic in character.
5. H_2O displays hydrogen bonding, but H_2S does not.

14.4 LIQUID–VAPOR EQUILIBRIUM

PG 14F Describe or explain the equilibrium between a liquid and its own vapor, and the process by which the equilibrium is reached.

In Section 14.2, vapor pressure and boiling point were identified as physical properties that are related to intermolecular attractions. In this section and the next we will examine these properties in greater detail and, thereby, learn a bit more about the liquid state.

In Section 13.5, temperature was described as a measure of the average kinetic energy of the particles in a sample of matter. The word *average* suggests that at a given temperature the particles in the sample do not all have the same kinetic energy but rather a range of energies. Kinetic energy is expressed mathematically as $\frac{1}{2}mv^2$, where m is the mass of the particle and v is its velocity. If the sample is a pure substance, all particles have the same mass, except for minor variations due to isotopes. Their different kinetic energies therefore indicate that they have different velocities.

What happens at the molecular level when a liquid evaporates? Intermolecular forces tend to keep the substance in the liquid state. But a few of the faster moving molecules near the surface have enough kinetic energy to overcome these attractions and escape (evaporate) from the liquid. At a given temperature the fraction, or percentage, of the total sample that has this much energy is constant. As a consequence the rate of evaporation per unit of surface area is also constant at that temperature. If the temperature rises, a larger fraction of the sample has enough kinetic energy to evaporate, and the evaporation rate is greater.

If a liquid with weak intermolecular forces, such as benzene, is placed in an Erlenmeyer flask, which is then stoppered as in Figure 14.8, the benzene will begin to evaporate. If temperature is held constant, the *rate* of evaporation will remain constant throughout the experiment. The constant evaporation rate is represented by the fixed length of black arrows pointing upward from the liquid in each view of the flask, and also as the horizontal black line in a graph of evaporation rate versus time in Figure 14.8.

At first (Time 0) the movement of molecules is entirely in one direction, from the liquid to the vapor. However, as the concentration of molecules in the vapor builds up, an occasional molecule hits the surface and reenters the liquid. The change of state from a gas to a liquid is called **condensation.** The *rate* of condensation depends on the concentration of molecules in the vapor state. At Time 1 there will be a small number of molecules in the vapor state, so the condensation rate will be more than zero, but much less than the evaporation rate. This is shown by the arrow lengths in the Time 1 flask of Figure 14.8.

As long as the rate of evaporation is greater than the rate of condensation, the vapor concentration will rise over the next interval of time. Therefore, the rate of return from vapor to liquid rises with time (Time 2). Eventually the rates of vaporization and condensation become equal (Time 3). The number of molecules moving from vapor to liquid in unit time just balances the number moving in the opposite direction. We describe this situation, **when opposing rates of change are equal,** as a condition of **dynamic equilibrium** between liquid and vapor. Once equilibrium is reached, the concentration of molecules in the vapor has a certain fixed value which does not change. The rates therefore remain equal (Time 4).

Changes that occur in either direction, such as the change from a liquid to a vapor and the opposite change from a vapor to a liquid, are called **reversible**

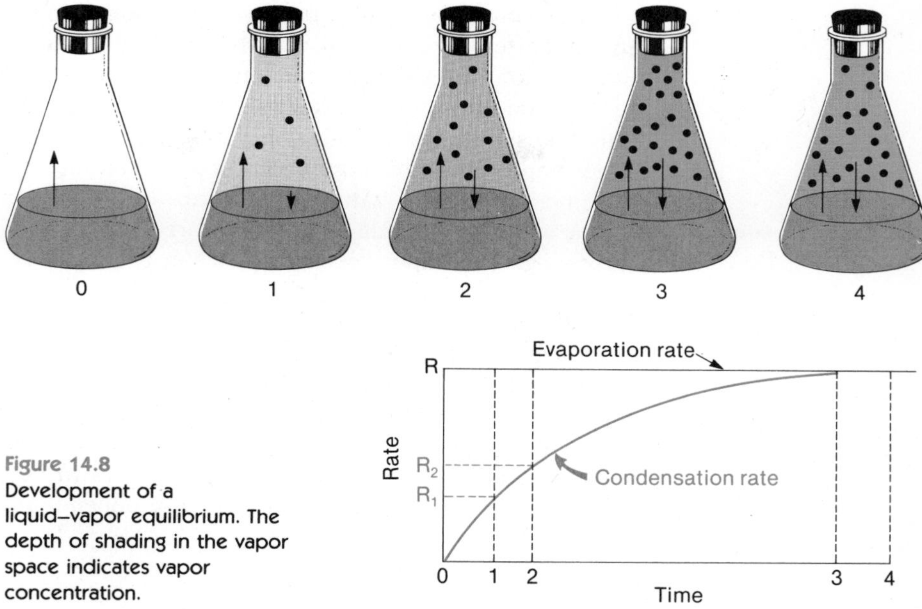

Figure 14.8
Development of a liquid–vapor equilibrium. The depth of shading in the vapor space indicates vapor concentration.

changes; if the change is chemical it is a **reversible reaction.** Chemists write equations describing reversible changes with a double arrow, one pointing in each direction. For example, the reversible change between liquid benzene, $C_6H_6(\ell)$, and benzene vapor, $C_6H_6(g)$, is represented by the equation

$$C_6H_6(\ell) \rightleftarrows C_6H_6(g) \tag{14.1}$$

If the ideal gas equation is solved for the partial pressure of a vapor,

$$p = \frac{nRT}{V} = RT \times \frac{n}{V} \tag{14.2}$$

we see that at constant temperature, the partial pressure is proportional to vapor concentration, n/V. When the vapor concentration becomes constant at equilibrium, the vapor pressure also becomes constant. **The partial pressure exerted by a vapor in equilibrium with its liquid phase at a given temperature is the equilibrium vapor pressure of the substance at that temperature.**

The Effect of Temperature

PG 14G Describe the relationship between vapor pressure and temperature for a liquid–vapor system in equilibrium; explain this relationship in terms of the kinetic molecular theory.

Equation 14.2 predicts that vapor pressure increases as temperature rises. This prediction is confirmed in the laboratory. This is understood in terms of how equilibrium is reached in a liquid–vapor system. It was noted earlier that only a fraction of the molecules present have enough kinetic energy to evaporate at any given temperature and that this fraction is larger at higher temperatures. Evaporation rate therefore increases with temperature. If the system reaches equilibrium, the condensation rate must also be greater at higher temperatures. This takes a higher equilibrium vapor concentration, which in turn exerts a larger vapor pressure. Figure 14.9 shows how the equilibrium vapor pressures of some liquids vary with temperature.

The extent to which vapor pressure increases with temperature may be surprising. The vapor pressure of water at 25°C, for example, is 24 torr, but at 60°C it is 149 torr, about 500% greater. The absolute temperature increase from 298 K to 333 K is only 35°, about 12%. Clearly the vapor pressure change is not a gas law pressure increase, in which pressure is proportional to absolute temperature. It is a fact, however, that the percentage of water molecules with enough kinetic energy to evaporate increases by about 600% over the same temperature range, matching much more closely the percentage increase in vapor pressure. It is the *number of particles with enough energy to evaporate* that is responsible for the vapor pressure dependence on temperature.

Quick Check 14.4: Are the following true or false:

1. A liquid–vapor equilibrium is reached when the amount of liquid is equal to the amount of vapor.
2. Rate of evaporation depends on temperature.
3. Equilibrium vapor pressure is higher at higher temperatures.

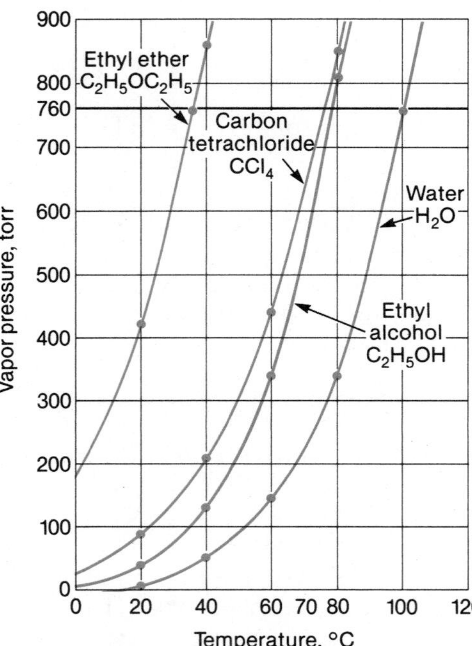

Figure 14.9
Vapor pressures of some common liquids at different temperatures.

14.5 THE BOILING PROCESS

PG 14H Describe the process of boiling and the relationships among boiling point, vapor pressure, and surrounding pressure.

When a liquid is heated in an open container, bubbles form, usually at the base of the container where heat is being applied. The first bubbles are often air, driven out of solution by an increase in temperature. Eventually, when a certain temperature is reached, vapor bubbles form throughout the liquid, rise to the surface, and break. When this happens we say the liquid is boiling.

In order for a stable bubble to form in a boiling liquid, the vapor pressure within the bubble must be high enough to push back the surrounding liquid and the atmosphere above the liquid. The minimum temperature at which this can occur is called the **boiling point: the boiling point is that temperature at which the vapor pressure of the liquid is equal to the pressure above its surface.** Actually the vapor pressure within a bubble must be a tiny bit greater than the surrounding pressure, which suggests that bubbles probably form in local "hot spots" within the boiling liquid. The boiling temperature at one atmosphere—the temperature at which the vapor pressure is equal to one atmosphere—is called the **normal boiling point.** Figure 14.9 shows that the normal boiling point of water is 100°C; of ethyl alcohol, 78°C; of carbon tetrachloride, 77°C, and of ethyl ether, 35°C.

According to the definition, the boiling point of a liquid depends on the pressure above it. If that pressure is reduced, the temperature at which the vapor pressure equals the lower surrounding pressure comes down also, and the liquid will boil at that lower temperature. This is why liquids at higher altitudes boil at reduced temperatures. In mile-high Denver, where atmospheric pressure is typically about 630 torr, water boils at 95°C. It is possible to boil water at room temperature by creating a vacuum in the space above it. When pressure is reduced to 20 torr, water boils at 22°C, about 72°F. A method for purifying a compound that might decompose or oxidize at its normal boiling point is to boil it at reduced temperature in a vacuum, and then condense the vapor.

It is also possible to *raise* the boiling point of a liquid by *increasing* the pressure above it. The pressure cooker used in the kitchen takes advantage of this effect. By allowing the pressure to build up within the cooker, it is possible to reach temperatures as high as 110°C without boiling off the water. At this temperature, foods cook in about half the time required at 100°C.

Quick Check 14.5: Are the following true or false:

1. Water can be made to boil at 60°F.
2. Bubbles can form anyplace in a boiling liquid.

14.6 THE NATURE OF THE SOLID STATE

PG 14I Distinguish between crystalline and amorphous solids.

Figure 14.10
Crystals of NaCl, $NH_4H_2PO_4$, and $CuSO_4 \cdot 5H_2O$.

A solid whose particles are arranged in a geometric pattern that repeats itself over and over in three dimensions is a **crystalline solid.** Each particle occupies a fixed position in the crystal. It can vibrate about that site but cannot move past its neighbors. The high degree of order often leads to large crystals which have a precise geometric shape. In ordinary table salt we can distinguish small cubic crystals of sodium chloride. Large, beautifully formed crystals of such minerals as quartz (SiO_2) and fluorite (CaF_2) are found in nature. Figure 14.10 has photographs of three crystalline solids.

In an **amorphous solid** such as glass, rubber, or plastic, there is no long-range order. Even though the arrangement around a particular site may resemble that in a crystal, the pattern does not repeat itself throughout the solid (Fig. 14.11). From a structural standpoint, we may regard an amorphous solid as intermediate between the crystalline and the liquid states. In many amorphous solids, the particles have some freedom to move with respect to one another. The elasticity of rubber and the tendency of glass to flow when subjected to stress over a long period of time suggest that the particles in these materials are not rigidly fixed in position.

Crystalline solids have characteristic physical properties that can serve to identify them. Sodium chloride, for example, melts sharply at 801°C. This is in striking contrast to glass, which first softens and then slowly liquifies over a wide range of temperatures.

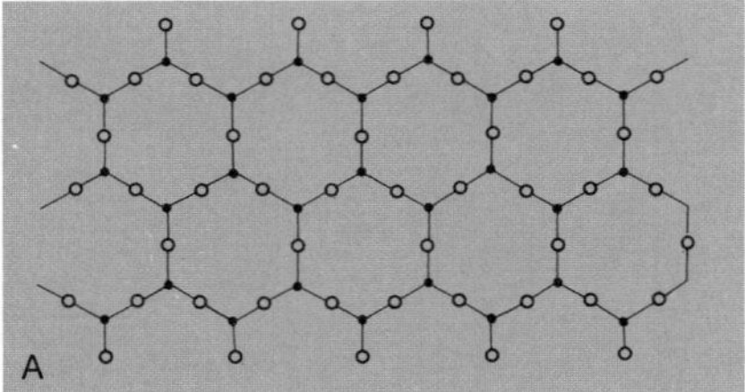

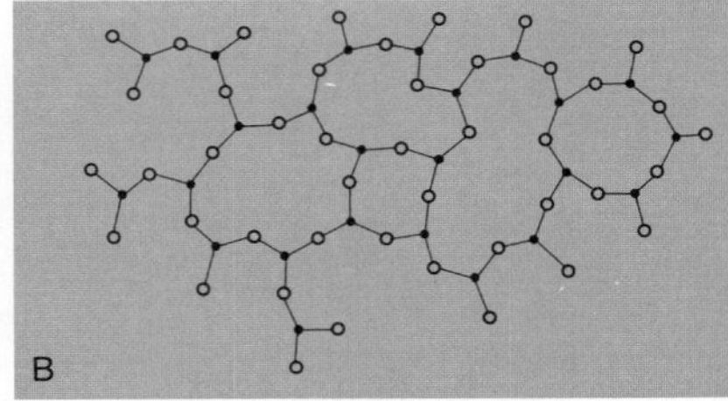

Figure 14.11
Crystalline (A) and amorphous (B) solids.

14.7 TYPES OF CRYSTALLINE SOLIDS

PG 14J Distinguish among the following types of crystalline solids: ionic, molecular, macromolecular, and metallic.

Solids can be divided into four classes on the basis of their particle structure and the type of forces that hold these particles together in the crystal lattice.

Ionic Crystals. Examples are NaF, $CaCO_3$, AgCl, and NH_4Br. Oppositely charged ions are held together by strong electrostatic forces (recall Figs. 10.3 and 10.4). As pointed out earlier, ionic crystals are typically high melting, frequently water-soluble, and have very low electrical conductivities. Their melts and water solutions, in which the ions can move around, conduct electricity readily.

Molecular Crystals. Examples are I_2 and ICl. Small, discrete molecules are held together by relatively weak intermolecular forces of the types discussed in Section 14.3. Molecular crystals are typically soft, low melting, and generally (but not always) insoluble in water. They usually dissolve in nonpolar or slightly polar organic solvents such as carbon tetrachloride or chloroform. Molecular substances, with rare exceptions, are nonconductors when pure, even in the liquid state.

Macromolecular Crystals. Examples are diamond, C, and quartz, SiO_2. Atoms are covalently bonded to each other to form one "huge molecule" making up the entire crystal. The prefix *macro-* means large.

There are no small, discrete molecules in macromolecular crystals. In diamonds each carbon atom is covalently bonded to four other carbon atoms to give a structure that repeats throughout the entire crystal (Fig. 14.12A). The structure of silicon dioxide resembles that of diamond in that the atoms are held together by a continuous series of covalent bonds. Each silicon atom is bonded to four oxygen atoms, each oxygen to two silicons, as shown in Figure 14.12B.

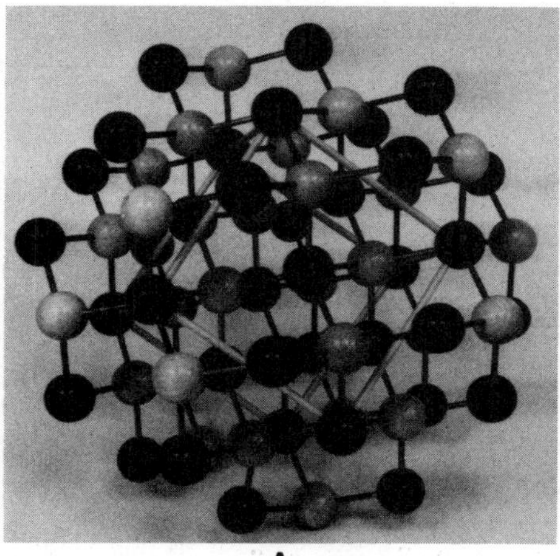

A

B

Figure 14.12
A. Model of diamond (carbon) crystal. B. Model of quartz (SiO_2) crystal. (Quartz crystal, courtesy of Klinger Scientific Apparatus Corporation, Jamaica, NY)

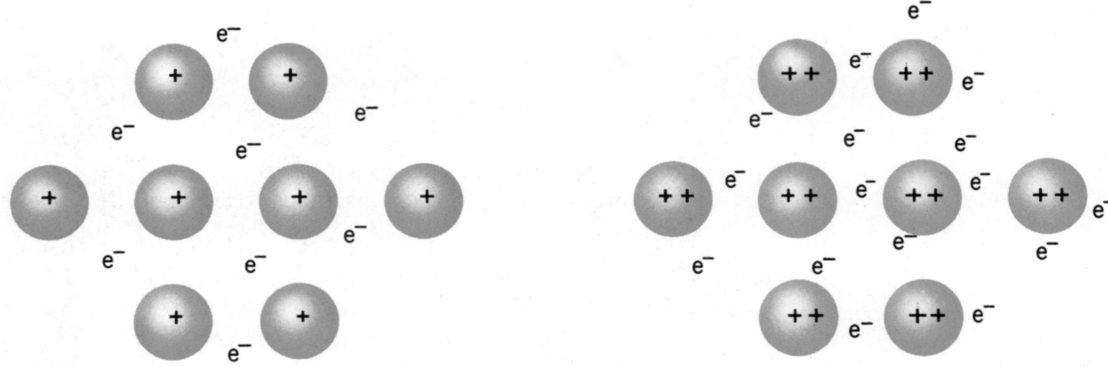

Figure 14.13
In a metal, the metal ions remain fixed in a definite crystal pattern, but the electrons are relatively free to move. This explains the high electrical conductivity of metals. The drawing on the left might represent metallic sodium, and that on the right, magnesium.

Macromolecular solids, like ionic crystals, are high melting. A very high temperature is needed to break the covalent bonds in crystals of quartz (1710°C) or diamond (3500°C). Macromolecular crystals are almost always insoluble in water or any common solvent. They are generally poor conductors of electricity in either the solid or liquid state.

Metallic Crystals. A simple model of bonding in a metal consists of a crystal of positive ions through which valence electrons move freely. This so-called "electron sea" model of a metallic crystal is illustrated in Figure 14.13 for silver. The positively charged ions form the backbone of the crystal; the electrons surrounding these ions are not tied down to any particular ion and hence are not restricted to a particular location. It is because of these freely moving electrons that metals are excellent conductors of electricity.

The general properties of the four kinds of crystalline solids are summarized in Table 14.3.

Table 14.3 Types of Crystals

Type	Examples	Properties
Ionic	KNO_3, NaCl, MgO	High melting; generally water-soluble; brittle; conduct only when melted or dissolved in water.
Molecular	$C_{10}H_8$, I_2	Low melting; usually more soluble in organic solvents than in water; nonconductors in pure state.
Macromolecular	SiO_2, C	Very high melting; insoluble in all common solvents; brittle; non- or semiconductors.
Metallic	Cu, Fe	Wide range of melting points; insoluble in all common solvents; malleable; ductile; good electrical conductors.

14.8 ENERGY AND CHANGE OF STATE

Heat flows into or out of a substance when it changes state. It takes heat to melt a solid, an endothermic change; and heat is lost by a liquid when it freezes, an exothermic change. In the change between a liquid and a gas, vaporization is endothermic and condensation is exothermic. In this section you will learn how to calculate the heat flow that accompanies a change of state.

Heat of Vaporization

PG 14K Given two of the following, calculate the third: (a) mass of a pure substance changing between the liquid and vapor (gaseous) states; (b) heat of vaporization; (c) heat flow.

In Section 14.2, molar heat of vaporization was defined as the energy required to vaporize one mole of a liquid. It is the energy needed to overcome the intermolecular attractions in the liquid, separating the particles into a gas. From the definition, molar heat of vaporization is measured in energy units per mole. This is not a convenient quantity to work with, however, because moles cannot be measured on a balance. Grams can. Therefore, the more common measurement of **heat of vaporization** (omitting *molar*) is **the energy required to vaporize one gram of a substance,** usually expressed in kilojoules per gram. Heats of vaporization for several substances are given in Table 14.4.

When a vapor condenses to a liquid at the boiling point, the reverse energy change occurs. The heat flow is then referred to as **heat of condensation.** Values are identical to heats of vaporization, except that they are negative, indicating that heat is flowing *from* the substance (an exothermic change), rather than into it.

The energy required to vaporize a given quantity of a liquid may be calculated from the equation

$$Q = m \times \Delta H_{vap} \qquad (14.3)$$

where m is mass in grams and ΔH_{vap} is heat of vaporization in kilojoules per gram. Applications are straightforward "substitute and solve" problems.

EXAMPLE 14.1 Calculate the heat of vaporization of a substance if 241 kJ (57.6 kcal) are required to vaporize a 78.2-g sample.

We recommend that you solve Equation 14.3 for ΔH_{vap} first, and then substitute numbers and units and calculate the answer. Equally correct, you may substitute numbers and units directly into the equation and solve for the missing item. Handling units is slightly more difficult by the second method. Take your choice and solve the problem completely.

----- -----

$$\Delta H_{vap} = \frac{Q}{m} = \frac{241 \text{ kJ}}{78.2 \text{ g}} = 3.08 \text{ kJ/g} \quad [737 \text{ cal/g}]$$

Table 14.4 Heats of Fusion and Heats of Vaporization

Substance	Melting Point (°C)	Boiling Point (°C)	Heat of Fusion		Heat of Vaporization	
			(J/g)	(cal/g)	(kJ/g)	(cal/g)
$H_2O(s)$	0		335	80		
$H_2O(\ell)$		100			2.26	540
Na	98	892	113	27	4.27	1020
NaCl	801	1413	519	124		
Cu(s)	1083		205	49		
Cu(ℓ)		2595			4.81	1150
Zn	419	907	100	24	1.76	420
Bi	271	1560	54	13		
Pb	327	1744	23	5.5		
Ni	1453	2732	310	74		
Au	1063	2807	64	15		
Ag	961	2212	105	25		
Fe	1535	2750	27	64		
Cd	321	765	54	13		

EXAMPLE 14.2 Calculate the energy required to vaporize 250 g of water at its boiling point.

$$Q = m \times \Delta H_{vap} = 250\ \not{g} \times \frac{2.26\text{ kJ}}{\not{g}} = 565\text{ kJ}\quad [135\text{ kcal}]$$

Heat of Fusion

PG 14L Given two of the following, calculate the third: (a) mass of a pure substance changing between the solid and liquid states; (b) heat of fusion; (c) heat flow.

To melt a solid, energy must be applied to break down the crystal structure. Conceptually similar to heat of vaporization, the **heat of fusion, ΔH_{fus}, of a substance is the energy required to melt one gram of that substance.** Heats of fusion are generally much smaller than heats of vaporization, so the usual units are joules per gram. Some typical heats of fusion are given in Table 14.4.

Just as condensation is the opposite of vaporization, freezing is the opposite of melting. The quantity of heat released in freezing a sample is identical to the heat required to melt that sample. Accordingly, **heat of solidification** is numerically equal to heat of fusion, but the sign is negative.

The heat flow equation for the change between solid and liquid is

$$Q = m \times \Delta H_{fus} \qquad (14.4)$$

Calculation methods are identical to those in vaporization problems.

EXAMPLE 14.3 Calculate the energy required to melt 135 g of sodium at its melting point. Express the answer in both joules and kilojoules.

The heat of fusion of sodium may be found in Table 14.4. That and equation 14.4 are all you need.

$$Q = m \times \Delta H_{fus} = 135\ \cancel{g} \times \frac{113\ J}{\cancel{g}} = 1.53 \times 10^4\ J = 15.3\ kJ \quad [3.6 \times 10^3\ cal = 3.6\ kcal]$$

14.9 CHANGE IN TEMPERATURE PLUS CHANGE OF STATE

PG 14M Sketch, interpret, and/or identify regions in a graph of temperature versus energy for a pure substance over a temperature range from below the melting point to above the boiling point.

14N Given (1) the mass of a pure substance, (2) ΔH_{vap} and/or ΔH_{fus} of the substance, and (3) the average specific heat of the substance in the solid, liquid, and/or vapor state, calculate the total heat flow in going from one state and temperature to another state and temperature.

In Section 3.9 you learned how to calculate heat flow when a solid, liquid, or gas changes temperature as it is heated or cooled without a change of state. The equation for that calculation is

$$Q = m \times c \times \Delta T \qquad (3.21)$$

where m is the mass of the sample in grams, c is specific heat in joules per gram degree Celsius, and ΔT is the temperature change in degrees Celsius.

Heat flows for temperature changes and for changes of state occur independently of each other, and sometimes one after the other. The total heat flow when a sample of matter is changed from one temperature and state to another temperature and state is the sum of the individual steps. Figure 14.14 traces the series of steps when 1.00 g of ice at −10°C is heated until it becomes steam at 120°C.

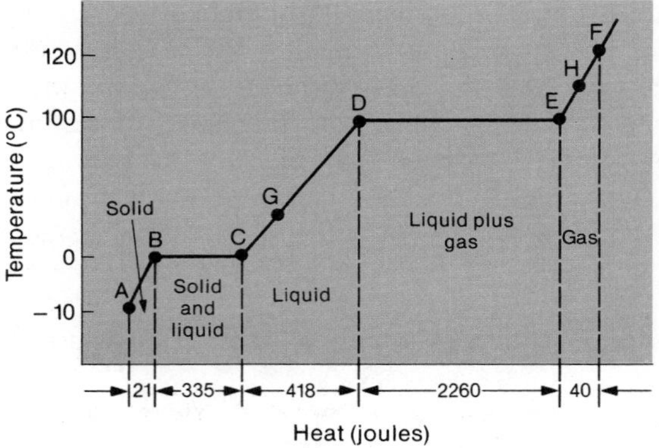

Figure 14.14
Temperature vs. energy absorbed as one gram of ice at −10°C is warmed, melted to the liquid state, heated to the boiling point, boiled, and then heated to 120°C. The curve is typical of most pure substances passing between the solid and vapor states. (Axes are not drawn to scale.)

Point A to Point B, Warming Ice from −10°C to 0°C. Warming ice from −10°C to 0°C is a change in temperature without a change of state. The energy absorbed may be calculated from Equation 3.21, $Q = m \times c \times \Delta T$. From Table 3.4 the specific heat of ice, $H_2O(s)$, is 2.1 J/g·°C. The heat absorbed from point A to point B, $Q_{A\ to\ B}$, is

$$Q_{A\ to\ B} = 1.00\ \not{g} \times \frac{2.1\ J}{\not{g}\cdot{}^\circ\not{C}} \times [0 - (-10)]^\circ\not{C} = 21\ J$$

Point B to Point C, Melting Ice at 0°C. This is a change of state at constant temperature. The energy absorbed may be calculated from Equation 14.4, $Q = m \times \Delta H_{fus}$. From Table 14.4, the heat of fusion of water is 335 J/g. The heat absorbed from Point B to Point C, $Q_{B\ to\ C}$, is

$$Q_{B\ to\ C} = 1.00\ \not{g} \times \frac{335\ J}{\not{g}} = 335\ J$$

Point C to Point D, Warming Water from 0°C to 100°C. Like A to B, this is a temperature change without a change of state. Using the specific heat of $H_2O(\ell)$,

$$Q_{C\ to\ D} = 1.00\ \not{g} \times \frac{4.18\ J}{\not{g}\cdot{}^\circ\not{C}} \times (100 - 0)^\circ\not{C} = 418\ J$$

Point D to Point E, Boiling Water at 100°C. Like B to C, this is a change of state at constant temperature. Using the heat of vaporization of water,

$$Q_{D\ to\ E} = 1.00\ \not{g} \times \frac{2.26\ kJ}{\not{g}} = 2.26\ kJ = 2260\ J$$

Point E to Point F, Heating Steam from 100°C to 120°C. Like A to B and D to E, this is a temperature change without a change of state. Using the specific heat of $H_2O(g)$,

$$Q_{E\ to\ F} = 1.00\ \not{g} \times \frac{2.0\ J}{\not{g}\cdot{}^\circ\not{C}} \times (120 - 100)^\circ\not{C} = 40\ J$$

To find the heat flow from any point on the graph to any other point you need only to calculate the heat flows of the individual steps in the process and add them. For example, the heat required to melt one gram of ice at 0°C (point B to C), heat it to boiling (C to D), and change it to steam at the boiling point (D to E) is

$$\begin{aligned} Q_{B\text{ to }E} &= Q_{B\text{ to }C} + Q_{C\text{ to }D} + Q_{D\text{ to }E} \\ &= 335\text{ J} + 418\text{ J} + 2260\text{ J} = 3013\text{ J} = 3.01\text{ kJ} \end{aligned}$$

(rounded off because of 2.26 kJ for $Q_{D\text{ to }E}$).

EXAMPLE 14.4 Calculate the heat flow when 45.0 g of steam, initially at 110°C, are changed to water at 25°C.

This problem may be related to Figure 14.14 if point G is at a temperature of 25°C and point H is at 110°C. The 45.0 g of steam are to be cooled from H to G. Three heat calculations are needed: from H to E, from E to D, and from D to G. Calculate the first heat flow, when the steam cools to 100°C.

----- -----

$$45.0\text{ g} \times \frac{2.0\text{ J}}{\text{g}\cdot{}^\circ\text{C}} \times (100 - 110)^\circ\text{C} = -900\text{ J} = -0.90\text{ kJ} \quad [-0.22\text{ kcal}]$$

The minus sign indicates that energy is being lost; the change is exothermic.

Now find the heat flow as the steam condenses at 100°C.

----- -----

$$45.0\text{ g} \times \frac{-2.26\text{ kJ}}{\text{g}} = -102\text{ kJ} \quad [-24.3\text{ kcal}]$$

The sign of heat of vaporization is changed to negative to get the heat of condensation.

Now the heat flow as the water cools to 25°C:

----- -----

$$45.0\text{ g} \times \frac{4.18\text{ J}}{\text{g}\cdot{}^\circ\text{C}} \times (25 - 100)^\circ\text{C} = -1.4 \times 10^4\text{ J} = -14\text{ kJ} \quad [-3.38\text{ kcal}]$$

You now have the energy for each of the three steps. What is the total energy for the whole process?

----- -----

$$Q = -0.90\text{ kJ} - 102\text{ kJ} - 14\text{ kJ} = -117\text{ kJ} \quad [-27.9\text{ kcal}]$$

CHAPTER 14 IN REVIEW

14.1 The Nature of the Liquid State

14A Explain the differences between the physical behavior of liquids and gases in terms of the relative distances between molecules and the effect of those distances on intermolecular forces.

14.2 Physical Properties of Liquids

14B For two liquids, given comparative values of physical properties that depend on intermolecular attractions, predict the relative strengths of those attractions; or, given a comparison of the strengths of intermolecular attractions, predict the relative values of physical properties that depend on them.

14.3 Types of Intermolecular Forces

14C Identify and describe or explain dipole forces, dispersion forces, and hydrogen bonds.

14D Given the structure of a molecule, or information from which it may be determined, identify the significant intermolecular forces present.

14E Given the molecular structures of two substances, or information from which they may be obtained, compare or predict relative values of physical properties that are related to them.

14.4 Liquid–Vapor Equilibrium

14F Describe or explain the equilibrium between a liquid and its own vapor, and the process by which the equilibrium is reached.

14G Describe the relationship between vapor pressure and temperature for a liquid–vapor system in equilibrium; explain this relationship in terms of the kinetic molecular theory.

14.5 The Boiling Process

14H Describe the process of boiling and the relationships among boiling point, vapor pressure, and surrounding pressure.

14.6 The Nature of the Solid State

14I Distinguish between crystalline and amorphous solids.

14.7 Types of Crystalline Solids

14J Distinguish among the following types of crystalline solids: ionic, molecular, macromolecular, metallic.

14.8 Energy and Change of State

14K Given two of the following, calculate the third: (a) mass of a pure substance changing between the liquid and vapor (gaseous) states; (b) heat of vaporization; (c) heat flow.

14L Given two of the following, calculate the third: (a) mass of a pure substance changing between the solid and liquid states; (b) heat of fusion; (c) heat flow.

14.9 Change in Temperature Plus Change of State

14M Sketch, interpret, and/or identify regions in a graph of temperature versus energy for a pure substance over a temperature range from below the melting point to above the boiling point.

14N Given (a) the mass of a pure substance, (b) ΔH_{vap} and/or ΔH_{fus} of the substance, and (c) the average specific heat of the substance in the solid, liquid, and/or vapor state, calculate the total heat flow in going from one state and temperature to another state and temperature.

TERMS AND CONCEPTS

14.2 Evaporate; evaporation
Vapor pressure
Equilibrium vapor pressure
Boiling point
Molar heat of vaporization
Viscosity
Surface tension

14.3 van der Waals forces
Dipole; dipole forces
Dispersion (London) forces
Hydrogen bond

14.4 Condense, condensation
Dynamic equilibrium
Reversible change, reaction

14.5 Boiling
Normal boiling point

14.6 Crystalline solid
Amorphous solid

14.7 Ionic crystal
Molecular crystal
Macromolecular crystal
Metallic crystal

14.8 Heat of vaporization
Heat of condensation
Heat of fusion
Heat of solidification

Most of these terms and many others are defined in the Glossary. Use your Glossary regularly.

QUESTIONS AND PROBLEMS

Section 14.1

(*1*) Explain why gases are less dense than liquids.

(2) Explain why water is less compressible than air.

(46) Explain why two gases will mix with each other more rapidly than two liquids.

(47) Explain why intermolecular attractions are stronger in the liquid state than in the gas state.

Section 14.2

(3) Identify and explain the relationship between intermolecular attractions and equilibrium vapor pressure.

(*4*) What is meant by molar heat of vaporization?

(5) Which liquid is more viscous, water or motor oil? In which liquid do you suppose the intermolecular attractions are stronger? Explain.

(6) A falling drop of any liquid tends to be spherical in shape. If the liquid is water there is a visible elongation of the drop into a "teardrop" form. A drop of mercury, however, remains more spherical. Compare the surface tension of water with that of mercury. What does this suggest about the strength of intermolecular attractions in water compared to the interatomic attractions in mercury?

(7) One of the functions of soap is to change the surface tension of water. Considering the purpose of laundry soap, do you think soap increases or decreases intermolecular attractions in water? Explain.

(48) How do intermolecular attractions influence the boiling point of a pure substance?

(49) Why does molar heat of vaporization depend on the strength of intermolecular attractions?

(50) A tall glass cylinder is filled to a depth of 1 m with water. Another tall glass cylinder is filled to a depth of 1 m with syrup. Identical ball bearings are dropped into each tube at the same instant. In which tube will the ball bearing reach the bottom first? Explain your prediction in terms of viscosity and intermolecular attractions.

(51) If water is spilled on a laboratory desk top, it usually spreads over the surface, wetting any papers or books that may be in its path. If mercury is spilled, it neither spreads nor makes paper it contacts wet, but rather forms little drops that are easily combined into pools by pushing them together. Suggest an explanation for these facts in terms of the apparent surface tension and intermolecular attraction in mercury and in water.

(52) The level at which a duck floats on water is determined more by the thin oil film that covers its feathers than by a lower density of its body compared to water. The water does not "mix" with the oil, and therefore does not penetrate the feathers. If, however, a few drops of "wetting agent" are placed in the water near the duck, the poor bird will sink. State the effect of wetting agent on surface tension and intermolecular attractions of water.

The normal boiling and melting points for three nitrogen oxides are given at the right. Refer to this table in answering the next two questions in each column.

	NO	*N_2O*	*NO_2*
Boiling point	*−152°C*	*−88.5°C*	*+21.2°C*
Melting point	*−164°C*	*−90.8°C*	*−11.2°C*

(8)* In which physical state, gas, liquid, or solid, would you find NO at −90.0°C?

(53)* Which of the three oxides would you expect to have the highest molar heat of vaporization? Explain how you reached your conclusion.

(9)* Which of the three oxides would you expect to have the highest viscosity at −90°C? Explain how you reached your conclusion.

(54)* Which of the three oxides would you expect to have a measurable vapor pressure at −90.0°C? Explain your answer.

Section 14.3

(*10*) Identify the three major types of intermolecular forces and explain why they exist.

(11) Identify the principal intermolecular forces in each of the following compounds: HBr; C_2H_2; NF_3; C_2H_5OH.

(12) Given that ionic compounds generally have higher melting points than molecular compounds of similar molar mass, compare dipole forces and forces between ions. How are they alike and how are they different?

(55) Under what circumstances are dispersion forces likely to produce stronger intermolecular attractions than dipole forces, and when are dispersion forces likely to be weaker?

(56) Identify the principal intermolecular forces in each of the following compounds: $NH(CH_3)_2$; CH_2F_2; C_3H_8.

(57) Compare dipole forces and hydrogen bonds. How are they different, and how are they similar?

For the next two questions in each column, predict, on the basis of molecular size, molecular polarity, and hydrogen bonding, which member of each of the following pairs has the higher boiling point and state the reason for your choice. Assume that molecular size is related to molar mass.

(13) CH_4 and CCl_4.

(14) H_2S and PH_3.

(15) Hydrogen is usually covalently bonded to one of what three elements in order for hydrogen bonding to be present? What unique feature do these elements share that sets them apart from other elements?

(16) Of the three types of intermolecular forces, which one(s) (a) operate in all molecular substances, and (b) operate between all polar molecules?

(17) In which of the following substances would you expect dipole forces to operate?

(a) H—C≡N (b) O=C=O

(c) P bonded to H, H, H

(d) C bonded to H, Cl, Cl, Cl

(e)

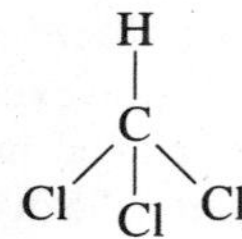

(58) CH_4 and NH_3.

(59) Ar and Ne.

(60) What physical feature of the hydrogen atom, when covalently bonded to an appropriate second element, is largely responsible for the strength of hydrogen bonding between molecules?

(61) Of the three types of intermolecular forces, which one(s) (a) account for the high melting point, boiling point, and other abnormal properties of water, and (b) increase with molecular size?

(62) In which of the following substances would you expect hydrogen bonds to form?

(a) H—C(H)(H)—C(H)(H)—O—H

(b) H—C(H)(H)—O—C(H)(H)—H

(c)

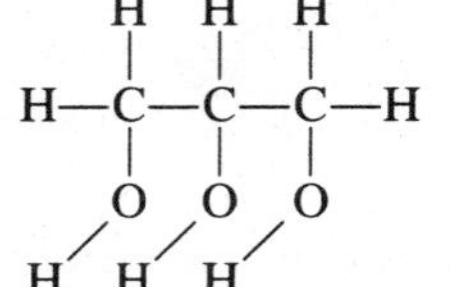

(d) H—N(H)—C(H)(H)—H

(e) H—C(H)(H)—C(H)(H)—H

(18) Predict which compound, C_3H_8 or C_6H_{14}, has the higher melting and boiling points. Explain your prediction.

(19) Predict which compound, SO_2 or CO_2, has the higher vapor pressure as a liquid at a given temperature. Explain your prediction.

(63) Predict which compound, CO_2 or CS_2, has the higher melting and boiling points. Explain your prediction.

(64) Predict which compound, CH_4 or CH_3F, has the higher vapor pressure as a liquid at a given temperature. Explain your prediction.

Section 14.4

(*20*) What essential condition exists when a system is in a state of *dynamic* equilibrium?

(65) Explain why the rate of evaporation from a liquid depends on temperature. Explain why the rate of condensation depends on concentration in the vapor state.

The next two questions in each column are based on the apparatus shown in Figure 14.15. Study the caption that describes how vapor pressure is measured, and then answer the questions.

(21)* A student uses the apparatus in Figure 14.15 to determine the equilibrium vapor pressure of a volatile liquid. He observes that the pressure increases rapidly at first, but more slowly as equilibrium is approached. Suggest a reason for this.

(22)* Suppose in making the vapor pressure measurement described in Figure 14.15 that all of the liquid introduced into the flask evaporates. Explain what this means in terms of evaporation and condensation rates. How does the vapor pressure in the flask compare with the equilibrium vapor pressure at the existing temperature?

(23) Using the ideal gas equation, show why the partial pressure of a gaseous component depends on its vapor concentration at a given temperature.

(24)* Three closed boxes have identical volumes. A beaker containing a small quantity of acetone, an easily vaporized liquid, is placed in one. Over

(66)* Why would the apparatus in Figure 14.15 be of little or no value in determining the equilibrium vapor pressure of water if used as described? Under what conditions might the apparatus give acceptable values for water vapor pressure?

(67)* After the system has come to equilibrium, as in Figure 14.15B an additional volume of liquid is introduced into the flask. Describe and explain what will happen to the pressure indicated by the manometer. Disregard any Boyle's Law effect; assume that the change in gas volume resulting from increased liquid volume is negligible.

(68) Using the ideal gas equation, show why equilibrium vapor pressure is dependent upon temperature.

(69)* Three closed boxes have different volumes: one is small, one medium-sized, and one large. Beakers containing equal quantities of acetone are

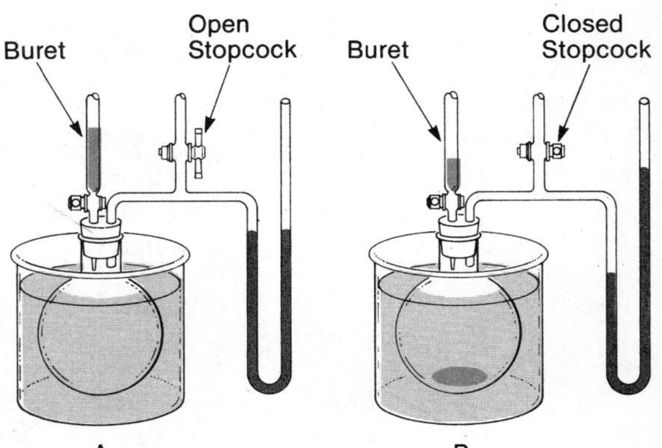

Figure 14.15
Measurement of vapor pressure. A. The buret contains the liquid whose vapor pressure is to be measured. The flask, tubes, and manometer above the mercury in the left leg are all at atmospheric pressure through the open stopcock. The mercury in the open right leg of the manometer is also at atmospheric pressure, so the mercury levels are the same in the two legs. B. To measure vapor pressure, stopcock is closed, trapping air in flask in space above the left mercury level. Liquid is introduced to flask from buret. Evaporation occurs until equilibrium is reached. Vapor causes increase in pressure, which is measured directly by the difference in mercury levels.

a period of time it evaporates completely. A medium quantity of acetone is placed in the second. Eventually it evaporates until only a small portion of liquid remains. A larger quantity is placed in the third, and about half of it eventually evaporates. (a) Which box or boxes develop the greatest acetone vapor pressure? (b) Which probably has the least? (c) Explain both answers.

(25)* The equilibrium vapor pressure of water at 50°C is 93 torr. A sealed flask under vacuum—assume the pressure inside is zero—contains a vial of liquid water. The vial is broken, and some of the water vaporizes. What is the maximum pressure that could be reached in this system?

placed in the boxes. Eventually all the acetone evaporates in one box, but equilibrium is reached in the other two. (a) In which box does complete evaporation occur? (b) Compare the eventual vapor pressures in the three boxes. (c) Explain both answers.

(70)* Suppose, in Question 25, the sealed flask contained air at 760 torr instead of a vacuum, and the same vial of liquid water. What then would be the maximum that could be reached when some of the liquid vaporized?

Section 14.5

(26) Define boiling point. Draw a vapor pressure–temperature curve and locate the boiling point on it.

(27) The vapor pressure of a certain compound at 20°C is 906 torr. Is the substance a gas or a liquid at 760 torr? Explain.

(28) Explain why high boiling liquids usually have low vapor pressures.

(29) The molar heat of vaporization of substance X is 34 kJ/mol; of substance Y, 27 kJ/mol. Which substance would be expected to have the higher normal boiling point? the higher vapor pressure at 25°C?

(71) An industrial process requires boiling a liquid whose boiling point is so high that maintenance costs on associated pumping equipment are prohibitive. Suggest a way this problem might be solved.

(72) Normally a gas may be condensed by cooling it. Suggest a second method, and explain why it will work.

(73) Explain why low-boiling liquids usually have low molar heats of vaporization.

(74) At 20°C the vapor pressure of substance M is 520 torr; of substance N, 634 torr. Which substance will have the lower boiling point? the lower molar heat of vaporization?

Section 14.6

(30) Compare amorphous and crystalline solids in terms of structure. How do crystalline and amorphous solids differ in physical properties? Explain the difference.

(75) Is ice a crystalline solid or an amorphous solid? On what properties do you base your conclusion?

Section 14.7

Problems (31) and (76): For each solid whose physical properties are tabulated below, state whether it is most likely to be ionic, molecular, macromolecular, or metallic:

	SOLID	MELTING POINT	WATER SOLUBILITY	CONDUCTIVITY (PURE)	TYPE OF SOLID
(31)	A	150°C	Insoluble	Nonconductor	________
	B	1450°C	Insoluble	Excellent	________
(76)	C	2000°C	Insoluble	Nonconductor	________
	D	1050°C	Soluble	Nonconductor	________

Section 14.8

When necessary for problems in this section, use Table 14.4 as a source of heats of fusion and vaporization.

(32) A calorimetry experiment is performed in which it is found that 29.3 kJ are given off when 6.04 g of a substance condenses. What is the heat of vaporization of that substance?

(33) How much energy is needed to vaporize 16 g of copper at its normal boiling point?

(34) What mass of hexane, a solvent used in rubber cement, can be boiled by 18.3 kJ if its heat of vaporization is 0.371 kJ/g.

(35) Dichlorodifluoromethane, CCl_2F_2, commonly known as Freon-12, is the refrigerant used in many freezers. Calculate the amount of energy absorbed as 744 g of CCl_2F_2 vaporize. Its molar heat of vaporization is 35 kJ/mol.

(36) How much energy is required to melt 35.4 g of gold?

(37) 7.08 kJ are required to melt 46.9 g of naphthalene, which is used in mothballs. What is the heat of fusion of naphthalene?

(38) Calculate the number of grams of silver that can be changed from a solid to a liquid by 11.3 kJ.

(77) A student is to find the heat of vaporization of isopropyl alcohol (rubbing alcohol). She vaporizes 63.6 g of the liquid at its boiling point and measures the energy required at 44.8 kJ. What heat of vaporization does she report?

(78) Calculate the energy released as 585 g of sodium vapor condense.

(79) 96.4 kJ were released by the condensation of a sample of ethyl alcohol. If $\Delta H_{vap} = 0.880$ kJ/g, what was the mass of the sample?

(80) Acetone, C_3H_6O, is a highly volatile solvent sometimes used as a cleansing agent prior to vaccination. It evaporates quickly from the skin, making it feel cold. How much heat is absorbed by 9.34 g of acetone as it evaporates if its molar heat of vaporization is 32.0 kJ/mol?

(81) Calculate the heat flow when 5.54 kg of lead freeze.

(82) 32.1 g of an unknown metal release 2.51 kJ of energy in freezing. What is the heat of fusion of that metal?

(83) A piece of zinc releases 2.04 kJ while freezing. What is the mass of the sample?

Section 14.9

Figure 14.16 is a graph of energy versus temperature for a sample of a pure substance. Assume that letters J through P on the horizontal and vertical axes represent numbers, and that expressions such as R − S or X + Y + Z represent arithmetic operations to be performed with those numbers. The next five questions in each column are related to Figure 14.16.

(39) What values are plotted, both horizontally and vertically?

(40) Identify in Figure 14.16 all points on the curve where the substance is entirely liquid.

(84) Identify by letter the boiling and freezing points in Figure 14.16.

(85) Identify all points on the curve in Figure 14.16 where the substance is entirely gas.

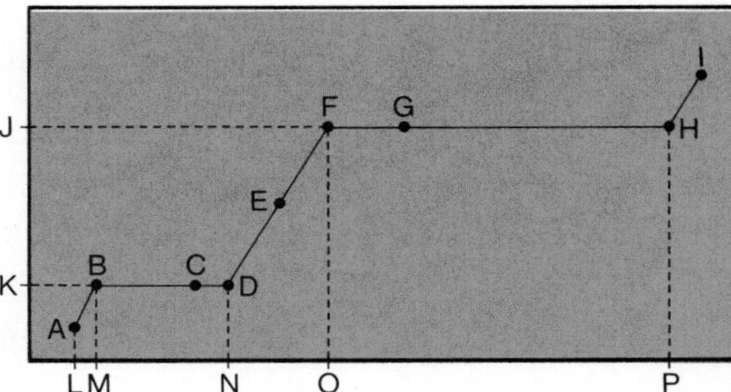

Figure 14.16

(41) Identify all points on the curve in Figure 14.16 where the substance is partly liquid and partly gas.

(42) Describe what happens physically as the energy represented by N − M is added to the sample.

(43) Using letters from the graph, write the expression for the energy required to raise the temperature of the liquid from the freezing point to the boiling point.

(44) A 127-g piece of ice is removed from a refrigerator at −11°C. It is placed in a bowl where it melts and eventually warms to room temperature, 21°C. Calculate the amount of heat the sample has absorbed from the atmosphere.

(45)* Find the heat flow when 25.1 kg of iron are drawn from a blast furnace at 1645°C and poured into a mold where it cools, freezes, and cools further to a shop temperature of 33°C. The average specific heats of iron are 0.452 J/g·°C over the liquid temperature range, and 0.444 J/g·°C over the solid range.

(86) Identify in Figure 14.16 all points on the curve where the substance is partly solid and partly liquid.

(87) Describe the physical changes that occur as energy N − P is removed from the sample.

(88) Using letters from the graph, show how you would calculate the energy required to boil the liquid at its boiling point.

(89)* A 54.1-g aluminum ice tray in a home refrigerator holds 312 g of water. How much energy must be removed from the tray and its contents to reduce their temperature from 17°C to 0°C, freeze the water, and further reduce the temperature of the tray and ice to −9°C? The specific heat of aluminum is 0.88 J/g·°C.

(90) A certain "white metal" alloy of lead, antimony, and bismuth melts at 264°C, and its heat of fusion is 29 J/g. Its average specific heat is 0.21 J/g·°C as a liquid and 0.27 J/g·°C as a solid. How much energy is required to heat the 818 kg of that alloy in a metal pot from a starting temperature of 26°C to its operating temperature, 339°C?

Miscellaneous Questions

(91) Distinguish precisely and in scientific terms the differences between items in each of the following groups:
(a) Intermolecular forces, chemical bonds
(b) Vapor pressure, equilibrium vapor pressure
(c) Molar heat of vaporization, heat of vaporization
(d) van der Waals forces, dipole forces, dispersion forces, London forces, hydrogen bonds
(e) Evaporation, vaporization, boiling
(f) Evaporation, condensation
(g) Fusion, solidification
(h) Boiling point, normal boiling point
(i) Amorphous solid, crystalline solid
(j) Ionic, molecular, macromolecular, and metallic crystals
(k) Heat of vaporization, heat of condensation
(l) Heat of fusion, heat of solidification

(92) Classify each of the following statements as true or false:
(a) Intermolecular attractions are stronger in liquids than in gases.
(b) Substances with weak intermolecular attractions generally have low vapor pressures.
(c) Liquids with high molar heats of vaporization usually are more viscous than liquids with low molar heats of vaporization.
(d) A substance with a relatively high surface tension usually has a very low boiling point.
(e) All other things being equal, hydrogen bonds are the weakest of the van der Waals forces.
(f) Dispersion forces become very strong between large molecules.
(g) Other things being equal, nonpolar molecules have stronger intermolecular attractions than polar molecules.
(h) The essential feature of a dynamic equilibrium is that the rates of opposing change are equal.
(i) Equilibrium vapor pressure depends on the concentration of a vapor above its own liquid.
(j) The heat of vaporization is equal to the heat of fusion, but with opposite sign.
(k) The boiling point of a liquid is a fixed property of the liquid.

(l) If you break (shatter) an amorphous solid, it will break in straight lines, but if you break a crystal, it will break in curved lines.
(m) Ionic crystals are seldom soluble in water.
(n) Macromolecular crystals are nearly always soluble in water.
(o) The numerical value of molar heat of vaporization is always larger than the numerical value of heat of vaporization.
(p) The units of heat of fusion are kJ/g·°C.
(q) The temperature of water drops while it is freezing.

(93)* It is a hot summer day and John wants a glass of lemonade. There is none in the refrigerator, so he makes it from freshly squeezed lemons. When finished, he finds he has 175 g of lemonade at 23°C. That is not a very refreshing temperature, so it must be cooled with ice. But John doesn't like ice in his lemonade! He therefore decides to use just enough ice to cool the lemonade to 5°C. Of course, the ice will melt and reach the same temperature. If the ice starts at −8°C, and if the specific heat of lemonade is the same as that of water, and if there is no heat transfer to or from the surroundings, how many grams of ice must John use? Answer in two significant figures.

(94)* The labels have come off the bottles of two white crystalline solids. You know one is sugar and the other is potassium sulfate. Suggest a test by which you could determine which is which.

(95) Identify the intermolecular attractions in CH_3OH and CH_3F. Which of the two substances do you expect will have the higher boiling point and which will have the higher equilibrium vapor pressure? Justify your choices.

(96)* The melting point of an amorphous solid is not always a definite value as it should be for a pure substance. Suggest a reason for this.

(97) Under what circumstances might you find that a substance having only dispersion forces is more viscous than a substance that exhibits hydrogen bonding?

(98) Why does dew form overnight?

15 Solutions

LOOKING BACK	LOOKING AHEAD IN CHAPTER 15
8.7 Equations were written for reactions that occur in water solutions.	Equations for reactions that occur in solution are required for solution stoichiometry problems.
9.2, 9.4, 9.5, and 13.12 A three-step pattern for solving stoichiometry problems gives the following unit path: quantity given substance → moles given substance → moles wanted substance → quantity wanted substance. Quantity was measured in grams or in volume of gas at specified temperature and pressure.	Volume of solution at specified molarity is added as another way to express quantity and used as a conversion relationship in the first and third steps of the stoichiometry pattern.
13.14 The partial pressure of one component in a gaseous mixture is the pressure that component would exert if it alone occupied the same volume at the same temperature.	The equilibrium between a gas and its solution in a liquid, and hence its solubility in that liquid, depends on the partial pressure of the gas over the liquid.
14.4 Evaporation and condensation are reversible changes, shown by an equation with a double arrow: $X(\ell) \rightleftharpoons X(g)$. When rates of change in the opposite directions are equal, equilibrium is reached.	Dissolving and crystallization are reversible changes that are shown by the equation $X(s) \rightleftharpoons X(aq)$. Equilibrium is reached when the opposing rates are equal, at which time the solution is saturated.

15.1 THE CHARACTERISTICS OF A SOLUTION

Solutions abound in nature. We are surrounded by the gaseous solution known as air. The oceans are a water solution of sodium chloride and other substances. Some of these are present in sufficient concentration to make it commercially profitable to extract them. Magnesium is a notable example. Even what we call ''fresh'' water is a solution, although the concentrations are so low we tend to think of the water we drink as ''pure.'' ''Hard'' water may be sufficiently pure for human consumption, but there are enough calcium and magnesium salts present to form solid deposits in hot water pipes and boilers. Even rain water is a solution, containing dissolved gases. Oxygen is not very soluble in water but what little there is in solution is mighty important to fish, who cannot survive without it.

A solution is a homogeneous mixture. This implies uniform distribution of solution components, so that a sample taken from any part of the solution will have the same composition. Two solutions made up of the same substances, however, may have different compositions. A solution of ammonia in water, for example, may contain 1% ammonia by weight, or 2%, 5%, 20.3% . . . up to the 29% solution called "concentrated ammonia." This leads to variable physical properties, which are determined by the composition of a mixture (see Section 2.4). By contrast, the composition of a compound is always the same (the Law of Constant Compositions, Section 2.3). The composition of ammonia is always 82.4% nitrogen and 17.6% hydrogen. As a consequence, the physical properties of ammonia, or any compound, do not vary.

A solution may exist in any of the three states, gas, liquid, or solid. Air is a gaseous solution, made up of nitrogen, oxygen, carbon dioxide, and other gases in small amounts. In addition to oxygen in water, dissolved carbon dioxide in carbonated beverages is a familiar liquid solution of a gas. Alcohol in water is an example of the solution of two liquids, and the oceans of the world are natural liquid solutions of solids. Solid state solutions are common in the form of metal alloys.

Particle size distinguishes solutions from other mixtures. Dispersed particles in solutions, which may be atoms, ions, or molecules, are very small—generally less than 5×10^{-7} cm in diameter. Particles of this size do not settle on standing, and they are too small to be seen.

Quick Check 15.1: Which among the following are properties of a solution:
(a) Definite percentage composition
(b) Variable physical properties
(c) Always made up of two pure substances
(d) Different parts can be detected visually

15.2 SOLUTION TERMINOLOGY

PG 15A Distinguish among terms in the following groups:

Solute and solvent
Concentrated and dilute
Solubility, saturated, unsaturated, and supersaturated
Miscible and immiscible

In discussing solutions, a language of closely related and sometimes overlapping terms is used. We will now identify and define these terms.

Solute and Solvent. When solids or gases are dissolved in liquids, the solid or gas is said to be the **solute** and the liquid the **solvent.** More generally, the solute is taken to be the substance present in a relatively small amount. The medium in which the solute is dissolved is called the solvent. The distinction is not precise, however. Water is capable of dissolving more than its own weight of some solids, but the water continues to be called the solvent. In alcohol–water solutions, either liquid may be the more abundant and, in a given context, either might be called the solute or solvent.

Concentrated and Dilute. A **concentrated** solution has a *relatively* large quantity of a specific solute per unit amount of solution, and a **dilute** solution has a *relatively* small quantity of the same solute per unit amount of solution. The terms compare concentrations of two solutions of the *same solute and solvent.* They carry no other quantitative meaning.

Solubility, Saturated, and Unsaturated. **Solubility** is a measure of how much solute will dissolve in a given amount of solvent at a given temperature. It is sometimes expressed by giving the number of grams of solute that will dissolve in 100 g of solvent. A solution that can exist in equilibrium with undissolved solute is a **saturated** solution. A solution whose concentration corresponds to the solubility limit is therefore saturated. If the concentration of a solute is less than the solubility limit it is **unsaturated.**

Supersaturated Solutions. Under carefully controlled conditions, a solution can be produced in which the concentration of solute is greater than the normal solubility limit. Such a solution is said to be **supersaturated.** A supersaturated solution of sodium acetate, for example, may be prepared by dissolving 80 g of the salt in 100 g of water at about 50°C. If the solution is then cooled to 20°C without stirring, shaking, or other disturbance, all 80 g of solute will remain in solution even though the solubility at 20°C is only 46.5 g/100 g of water. The supersaturated solution can be maintained indefinitely so long as there are no tiny particles upon which crystallization can start. If a small seed crystal of sodium acetate is added, crystallization takes place until equilibrium is attained by the formation of a saturated solution.

Miscible and Immiscible. Miscible and immiscible are terms customarily limited to solutions of liquids in liquids. If two liquids dissolve in each other in all proportions they are said to be **miscible** in each other. Alcohol and water, for example, are miscible liquids. Liquids that are insoluble in each other, as oil and water, are **immiscible.** Some liquid pairs will mix appreciably with each other, but in limited proportions; they are said to be *partially miscible*.

Quick Check 15.2

0.1 g of A is dissolved in 1000 mL of water, and 10 g of B are dissolved in 500 mL of water. Identify:

(a) The solute and solvent in each solution.
(b) The solution that is more apt to be saturated.
(c) Which solution is "concentrated," and which is "dilute."

15.3 THE FORMATION OF A SOLUTION

PG 15B Describe the formation of a saturated solution from the time excess solid solute is first placed into a liquid solvent.

15C Identify and explain the factors that determine the time required to dissolve a given amount of solute, or to reach equilibrium.

When a soluble ionic crystal is placed in water, the negatively charged ions at the surface are attracted by the positive region of the polar water molecules (Fig. 15.1). A "tug of war" for the negative ion begins; water molecules tend to pull them from the crystal, while neighboring positive ions tend to hold them in the crystal. In a similar way, positive ions at the surface are attacked by the negative portion of the water molecules and are torn from the crystal. Once released, the ions are surrounded by the polar water molecules. Such ions are said to be **hydrated.**

The dissolving process is reversible. As the dissolved solute particles move randomly through the solution, they come into contact with the undissolved solute and crystallize—return to the solid state. The rate at which crystallization occurs depends on the concentration of solute at the surface of the undissolved solid. If, before all the solute is dissolved, the concentration increases to the point that the crystallization rate is equal to the rate at which the solid is dissolving, an equilibrium is established. For NaCl, this equilibrium may be represented by the "reversible reaction" equation (Section 14.4),

$$NaCl(s) \rightleftharpoons Na^+(aq) + Cl^-(aq) \tag{15.1}$$

The concentration at equilibrium identifies a saturated solution, as noted in the previous section, and represents the solubility at the existing temperature (Fig. 15.2).

The time required to dissolve a given amount of solute—or to reach equilibrium concentration, if excess solute is present—depends upon several factors:

1. The dissolving process depends on surface area. Therefore, a finely divided solid, which offers more surface per unit of mass than a coarsely divided solid, will dissolve more rapidly.

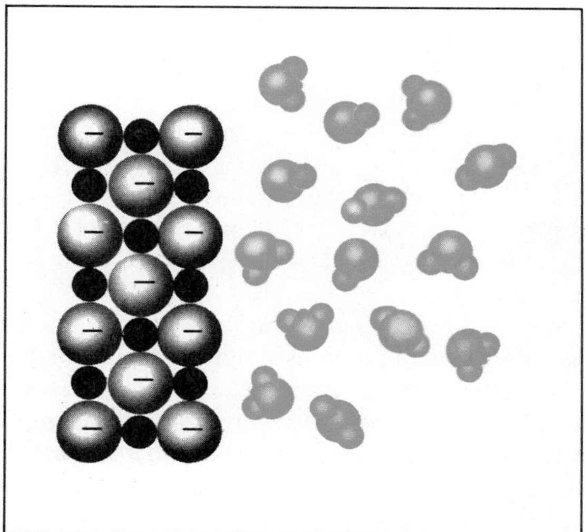

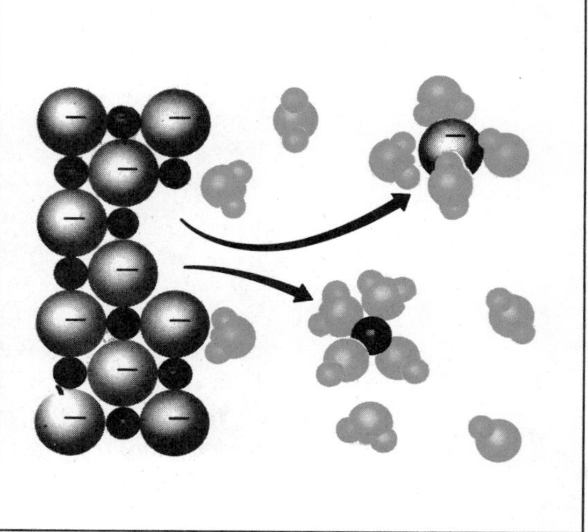

Figure 15.1
Dissolving of an ionic solute in water.

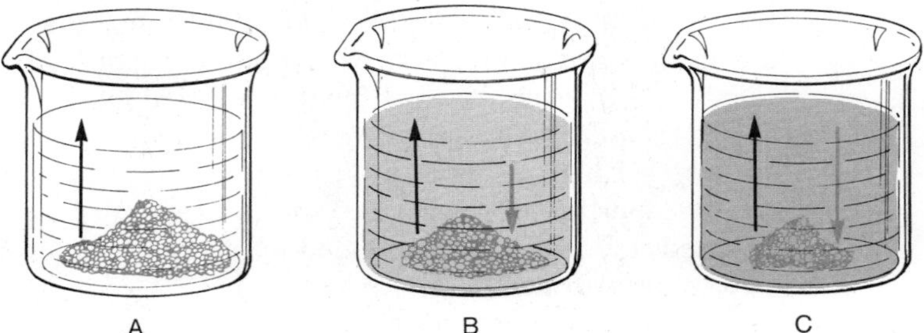

Figure 15.2
Development of equilibrium in producing a saturated solution. The solution dissolves at a constant rate, shown by the black arrows. In A, when dissolving has just begun, the solute concentration in the solvent is zero, so no crystallization can occur. In B the solute concentration has risen to yield a crystallization rate indicated by the colored arrow—still less than the dissolving rate. Eventually, in C, the concentration has increased to the point that the dissolving and crystallization rates are equal. Equilibrium has been reached, and the solution is saturated.

2. Stirring or agitating the solution prevents a localized build-up of concentration to near-saturation level at the solute surface, thereby minimizing the rate of crystallization. The *net* dissolving rate is therefore maximized.
3. At higher temperatures, particle movement is more rapid, thereby speeding up all physical processes.

15.4 FACTORS THAT DETERMINE SOLUBILITY

The extent to which a particular solute will dissolve in a given solvent is related to three factors: the strength of intermolecular forces between solvent molecules and solute molecules, the temperature and, in cases of gases dissolved in liquids, the partial pressure of the solute gas.

Intermolecular Forces

PG 15D Given the structural formulas of two molecular substances, or other information from which the strength of their intermolecular forces may be estimated, predict if they will dissolve appreciably in each other, and state the criteria on which your prediction is based.

You saw in Chapter 14 that physical properties of a molecular substance usually may be associated with the intermolecular forces resulting from the geometry of its molecules. Solubility is among these properties. Generally speaking, *if the forces between molecules of substance A are roughly the same as the intermolecular forces of substance B, substances A and B will probably dissolve in each other*. From the standpoint of these forces, the molecules appear to be able to replace each other. On the other hand, if the intermolecular forces between solute molecules are quite different from the forces between solvent molecules, it is unlikely that the substances will dissolve in each other.

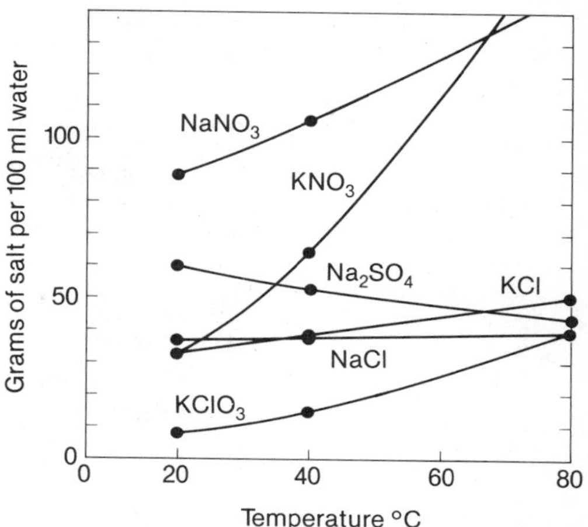

Figure 15.3
Temperature–solubility curves for various salts in water.

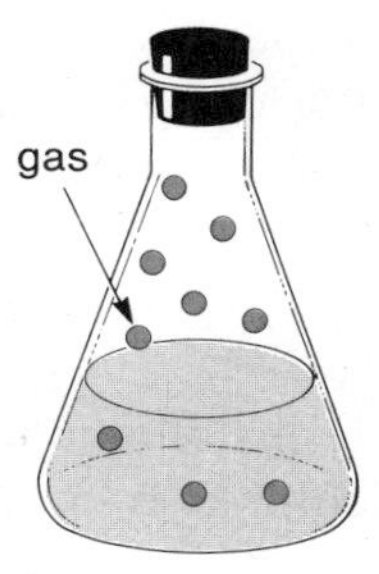

If we consider, for example, intermolecular attractions between such substances as hexane, C_6H_{14}, and decane, $C_{10}H_{22}$, we find that each substance has only dispersion forces. These forces are roughly the same for the two substances, which are soluble in each other. Neither, however, is soluble in water or methanol, CH_3OH, two liquids that exhibit strong hydrogen bonding. But water and methanol are soluble in each other, again supporting the correlation between solubility and similarity of intermolecular forces.

Temperature

Temperature exerts a major influence on most chemical equilibria, including solution equilibria. Consequently, solubility is temperature-dependent. Figure 15.3 indicates that solubility of most solids increases with rising temperature, but there are notable exceptions. The solubilities of gases in liquids, on the other hand, are generally lower at higher temperatures. The explanation of the interrelationship between temperature and solubility involves energy changes in the solution process, as well as other factors. We will look into this matter qualitatively in Chapter 19.

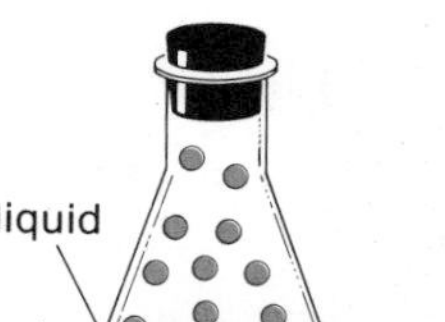

Effect of partial pressure of a gas on its solubility in a liquid. Solute gas concentration, and therefore its partial pressure, is lower in the first flask than in the second. Consequently, the solute concentration in solution is lower in the first flask.

Pressure

PG 15E Predict how the solubility of a gas in a liquid will be affected by a change in the partial pressure of that gas over the liquid.

Changes in partial pressure (Section 13.14) of a solute gas over a liquid solution have a pronounced effect on the solubility of the gas (Fig. 15.4). This is sometimes startlingly apparent on opening a bottle or can of a carbonated beverage. Such beverages are bottled or canned under carbon dioxide partial pressure slightly greater than one atmosphere, which increases the solubility of the gas. This is what is meant by ''carbonated.'' As the pressure is released on opening, solubility decreases, resulting in bubbles of carbon dioxide escaping from the solution.

In an "ideal" solution, the solubility of a gaseous solute in a liquid is directly proportional to the partial pressure of the gas over the surface of the liquid. A state of equilibrium is reached that is very similar to the vapor pressure equilibrium described by Figure 14.9 (Section 14.4) and the solid-in-liquid saturated solution shown in Figure 15.2. It is important to note that neither the partial pressure nor the total pressure caused by other gases has a significant effect on the solubility of the solute gas. This is what would be expected for an ideal gas, where all molecules are widely separated and completely independent.

Pressure has little or no effect on the solubility of solids or liquids in a liquid solvent. None of the events described in Figure 15.2 are influenced by gas pressure above the liquid surface.

Quick Check 15.4
If given the structural formulas of two substances, list the things you would look for to predict whether or not one would dissolve in the other. For each item listed, state the conditions under which solubility would be more probable.

15.5 SOLUTION CONCENTRATION: PERCENTAGE BY WEIGHT

The concentration of a solution is ordinarily expressed in terms of amount of solute present in a given quantity of solvent or of total solution. Amount of solute may be given in grams or moles; quantity of solvent or solution may be stated in mass or volume units. In this and the next three sections we will examine four ways to express solution concentration. We begin with percentage by weight.

PG 15F Given grams of solute and grams of solvent or solution, calculate percentage concentration.

15G Given grams of solution and percentage concentration, calculate grams of solute and grams of solvent.

If the concentration of a solute in a solution is given in percent, you may assume it to be percent by weight unless specifically stated otherwise. As always, percent is $\frac{\text{part quantity}}{\text{total quantity}} \times 100$. Accordingly, the percentage by weight of solute is defined by the following equations:

$$\% \text{ by weight of solute} = \frac{\text{grams solute}}{\text{grams solution}} \times 100 \qquad (15.2)$$

$$= \frac{\text{grams solute}}{\text{grams solute} + \text{grams solvent}} \times 100 \qquad (15.3)$$

EXAMPLE 15.1 125 g of solution, when evaporated to dryness, was found to contain 42.3 g of solute. What was the percentage by weight of the solute?

This involves only a direct substitution into one of the defining equations.

$$\frac{\text{g solute}}{\text{g solution}} \times 100 = \frac{42.3}{125} \times 100 = 33.8\%$$

There are 42.3 g of solute; the total quantity is 125 g of solution.

EXAMPLE 15.2 3.50 g of potassium nitrate are dissolved in 25.0 g of water. Calculate the percentage concentration of potassium nitrate.

Again, one of the defining equations may be used.

$$\frac{\text{g solute}}{\text{g solute} + \text{g solvent}} \times 100 = \frac{3.50}{3.50 + 25.0} \times 100 = 12.3\%$$

EXAMPLE 15.3 You are to prepare 250 g of 7.00% sodium carbonate solution. How many grams of sodium carbonate and how many milliliters of water do you use? (Recall that the density of water is 1.00 g/mL.)

You may approach this as a typical percentage problem, or from a dimensional analysis viewpoint, where percentage is grams of solute per 100 g of solution. On this basis, a 7.00% solution means 7.00 g of solute ≏ 100 g of solution (Appendix I, Part E).

$$250 \text{ g solution} \times \frac{7.00 \text{ g } Na_2CO_3}{100 \text{ g solution}} = 17.5 \text{ g } Na_2CO_3; \textit{ or}$$

7.00% of 250 = 0.0700 × 250 = 17.5 g Na_2CO_3

grams water = grams solution − grams solute = 250 − 17.5 = 232 g water

At 1.00 g/mL, 232 g of water = 232 mL.

15.6 SOLUTION CONCENTRATION: MOLALITY (OPTIONAL)

Many physical properties of solutions are related to the solution concentration expressed as **molality. Molality is the number of moles of solute dissolved in one kilogram of solvent.** Mathematically,

$$m = \frac{\text{moles of solute}}{\text{kilograms of solvent}} \tag{15.4}$$

where m is the symbol of molality. In essence, the molality of a solution is *the number of moles of solute that are equivalent to one kilogram of solvent*. In a 1.5 molal solution, for example, 1.5 moles of solute $\simeq$ 1 kilogram of solvent.

The quantity of solvent in molality problems is usually given in *grams*, or in volume, which may be converted to grams by multiplying by density. Molality, however, is based on *kilograms* of solvent. To convert grams to kilograms you divide by 1000—or more simply, move the decimal three places to the left. It is convenient to do this mentally in the initial setup of the problem.

EXAMPLE 15.4 Calculate the molality of a solution prepared by dissolving 15.0 g of sugar, $C_{12}H_{22}O_{11}$ (MM = 342 g/mol), in 350 mL of water.

SOLUTION: The data of the problem may be interpreted as stating the concentration of the solution in grams of solute per milliliter of solvent: 15.0 g solute per 350 mL H_2O. Because the density of water is 1.00 g/mL, the concentration is 15.0 g solute per 350 g H_2O, or

$$\frac{15.0 \text{ g } C_{12}H_{22}O_{11}}{0.350 \text{ kg } H_2O}$$

Molality is moles of solute per kilogram of solvent. To convert the above expression to molality it is necessary only to change 15.0 g of $C_{12}H_{22}O_{11}$ to moles:

$$\frac{15.0 \cancel{\text{g } C_{12}H_{22}O_{11}}}{0.350 \text{ kg } H_2O} \times \frac{1 \text{ mol } C_{12}H_{22}O_{11}}{342 \cancel{\text{g } C_{12}H_{22}O_{11}}} = 0.125 \text{ m}$$

EXAMPLE 15.5 How many grams of KCl (MM = 74.6 g/mol) must be dissolved in 250 g of water to make a 0.400 m solution?

SOLUTION: Molality provides a unit conversion between moles of solute and kilograms of solvent. Beginning with the given quantity, 250 g of water, or 0.250 kg, the unit path becomes kilograms water → moles solute → grams solute. The second conversion is by molar mass. The entire setup is

$$0.250 \cancel{\text{kg } H_2O} \times \frac{0.400 \cancel{\text{mol KCl}}}{1 \cancel{\text{kg } H_2O}} \times \frac{74.6 \text{ g KCl}}{1 \cancel{\text{mol KCl}}} = 7.46 \text{ g KCl}$$

15.7 SOLUTION CONCENTRATION: MOLARITY

PG 15H Given two of the following, calculate the third: volume of solution, molarity, and moles (or grams with known or calculable molar mass) of solute.

Percentage and molality concentrations are both based on mass of solute and mass of solvent, each considered separately. In using liquids, volume is much more easily measured than mass. Therefore, a solution's concentration is frequently expressed in terms of its **molarity,** which relates to a certain volume of solution. Abbreviated M, molarity is defined as **moles of solute per liter of solution.** In the form of an equation,

$$M = \frac{\text{moles solute}}{\text{liter solution}} = \frac{\text{mol}}{\text{L}} \tag{15.5}$$

In essence, the molarity of a solution is *the number of moles of solute that are equivalent to one liter of solution.* In an X molar solution, X moles of solute are equivalent to one liter of solution:

$$\text{X mol solute} \simeq 1 \text{ L solution}^* \tag{15.6}$$

The preparation of a solution of known molarity is a common procedure in a laboratory. The volume of solution to be prepared is often measured in milliliters. Because molarity is defined in liters, milliliters must be changed to match the larger units. This means dividing by 1000, which requires moving the decimal three places to the left. Thus, 750 mL is 0.750 L:

$$750 \cancel{\text{mL}} \times \frac{1 \text{ L}}{1000 \cancel{\text{mL}}} = 0.750 \text{ L}$$

This milliliters-to-liters conversion will be performed without comment in the examples that follow.

EXAMPLE 15.6 How many grams of silver nitrate, $AgNO_3$ (MM = 170 g/mol), must be dissolved to prepare 5.00×10^2 mL of 0.150 M $AgNO_3$?

The given quantity is 5.00×10^2 mL. If this is expressed in liters, the unit path becomes liters → moles $AgNO_3$ → grams $AgNO_3$. Complete the problem.

$$0.500 \cancel{\text{L}} \times \frac{0.150 \cancel{\text{mol } AgNO_3}}{1 \cancel{\text{L}}} \times \frac{170 \text{ g } AgNO_3}{1 \cancel{\text{mol } AgNO_3}} = 12.8 \text{ g } AgNO_3$$

*While molarity is easier to work with because liquid volumes are measured more readily than their masses, it has the disadvantage of being temperature dependent. The volume occupied by a fixed amount of solution changes with temperature, thereby changing the molarity of the solution. The variation is small, however, and we regard it as negligible for all problems in this text. Percentage concentration and molality are independent of temperature.

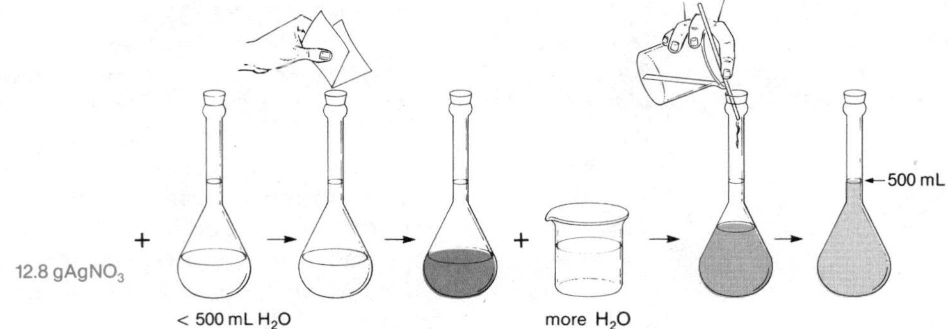

Figure 15.5 Preparation of 500 mL 0.150 M $AgNO_3$.

A clearer idea of the meaning of molarity may be gained by imagining the preparation of the solution in Example 15.6. The first step is to weigh out 12.8 g of silver nitrate (Fig. 15.5). This is transferred to a 500-mL volumetric flask containing *less than* 500 mL water. After dissolving the solute, water is added to the 500-mL mark on the neck of the flask. Notice the definition of molarity is based on the volume of *solution,* not the volume of *solvent.* This is why the solute is dissolved in less than 500 mL of water and then diluted to that volume. If it were dissolved *in* 500 mL of water, the final solution volume would be slightly more than 500 mL.

EXAMPLE 15.7 15.8 g of sodium hydroxide (MM = 40.0 g/mol) are dissolved in water and diluted to 1.00×10^2 mL. Calculate the molarity.

From the data presented, the concentration may be expressed in grams of solute per liter of solution. Write the concentration in these units.

$$\frac{15.8 \text{ g NaOH}}{100 \text{ mL solution}} = \frac{15.8 \text{ g NaOH}}{0.100 \text{ L solution}} \times \underline{\qquad\qquad}$$

The only requirement for changing the above expression to molarity, or moles per liter, is to convert grams of solute to moles. Complete the problem.

$$\frac{15.8 \cancel{\text{g NaOH}}}{0.100 \text{ L}} \times \frac{1 \text{ mol NaOH}}{40.0 \cancel{\text{g NaOH}}} = 3.95 \text{ mol NaOH/L} = 3.95 \text{ M}$$

One of the more important functions of molarity is the conversion between volume of solution and moles of solute. To change from volume to moles, multiply by molarity:

$$\cancel{\text{liters}} \times \frac{\text{moles}}{\cancel{\text{liter}}} = \text{liters} \times \text{molarity} = \text{moles}; \ \text{V} \times \text{M} = \cancel{\text{L}} \times \frac{\text{mol}}{\cancel{\text{L}}} = \text{mol} \quad (15.7)$$

EXAMPLE 15.8 How many moles of solute are there in 45.3 mL 0.550 M solution?

From the given quantity it is a one-step conversion to moles. Set up and solve.

$$0.0453\ \cancel{L} \times \frac{0.550\ \text{mol}}{1\ \cancel{L}} = 0.0249\ \text{mol}$$

The reverse problem is also significant. To change from moles to volume, divide by molarity:

$$\text{mol} \div \text{M} = \text{mol} \times \frac{1}{\text{M}} = \cancel{\text{mol}} \times \frac{\text{L}}{\cancel{\text{mol}}} = \text{L} \qquad (15.8)$$

EXAMPLE 15.9 Find the number of milliliters of 1.40 M solution that contain 0.287 mol of solute.

Converting the given quantity to volume is easy enough, but be sure you answer the question in the units required.

$$0.287\ \cancel{\text{mol}} \times \frac{1\ \text{L}}{1.40\ \cancel{\text{mol}}} = 0.205\ \text{L} = 205\ \text{mL}$$

Recall that to convert from units to milliunits, multiply by 1000, or move the decimal three places to the right.

15.8 SOLUTION CONCENTRATION: NORMALITY (OPTIONAL)

PG 15I Define an equivalent of an acid or base.

15J Given an equation for an acid–base reaction, state the number of equivalents of acid or base per mole and calculate the equivalent mass of the acid or base.

15K Given an equation for an acid–base reaction and two of the following, calculate the third: volume of solution, normality, and equivalents of acid or base (or grams with known or calculable molar mass).

Normality is a particularly convenient way to express concentration in analytical work. Designated N, it is **the number of equivalents of solute per liter of solution.** Mathematically,

$$\text{N} = \frac{\text{equivalents solute}}{\text{liter solution}} = \frac{\text{eq}}{\text{L}} \qquad (15.9)$$

In essence, the normality of a solution is the number of equivalents of solute that are the same as one liter of solution. In an X normal solution,

$$\text{X equivalents of solute} \simeq \text{1 liter of solution} \quad (15.10)$$

The concept of normality leads immediately to the meaning of *equivalent*. **One equivalent of an acid is that quantity that yields one mole of hydrogen ions in a chemical reaction; one equivalent of a base is that quantity that reacts with one mole of hydrogen ions.** Because hydrogen and hydroxide ions combine on a one-to-one basis, one mole of hydroxide ions is one equivalent. According to these definitions, both one mole of HCl and one mole of NaOH are one equivalent because they yield, respectively, one mole of hydrogen ions and one mole of hydroxide ions in a reaction. H_2SO_4, on the other hand, has two equivalents per mole because each mole of the compound can release two moles of hydrogen ions. Similarly, one mole of $Al(OH)_3$ may represent three equivalents because three moles of hydroxide ion may react.

Notice that the number of equivalents in a mole of an acid or base depends on a specific reaction, not simply on the number of moles of hydrogen or hydroxide ions present in one mole of solute. The number of moles of these ions that actually react may or may not be the same as those present. For example, we might expect that phosphoric acid, H_3PO_4, has three equivalents per mole—and indeed it does in the reaction

$$H_3PO_4(aq) + 3\ NaOH(aq) \rightarrow Na_3PO_4(aq) + 3\ HOH(\ell) \quad (15.11)$$

For reasons beyond the scope of this discussion, this reaction is difficult to perform quantitatively in the laboratory. It is possible, however, to control the reaction in a titration experiment (Section 15.10) so that 1.00 mol of H_3PO_4 reacts with 1.00 mol of NaOH in

$$H_3PO_4(aq) + NaOH(aq) \rightarrow NaH_2PO_4(aq) + HOH(\ell) \quad (15.12)$$

In Equation 15.10, one mole of phosphoric acid yields one mole of hydrogen ions to react with one mole of sodium hydroxide. This is one equivalent of sodium hydroxide. Phosphoric acid has one equivalent per mole in Equation 15.12.

Under other conditions the phosphoric acid–sodium hydroxide reaction yields a one-to-two mole ratio between the acid and the base.

$$H_3PO_4(aq) + 2\ NaOH(aq) \rightarrow Na_2HPO_4(aq) + 2\ HOH(\ell) \quad (15.13)$$

In this reaction one mole of phosphoric acid releases two moles of hydrogen ions, so there are two equivalents per mole.

EXAMPLE 15.10 For each of the following equations state the number of equivalents of acid and base per mole:

	eq acid/mol	**eq base/mol**
$2\ HBr + Ba(OH)_2 \rightarrow BaBr_2 + 2\ HOH$		
$H_3C_6H_5O_7 + 2\ KOH \rightarrow K_2HC_6H_5O_7 + 2\ HOH$		

Remember, you are interested only in the number of moles of $H+$ or OH^- that *react*, not the number present, in one mole of acid or base. The formula of citric acid, $H_3C_6H_5O_7$, is written as the inorganic chemist is most apt to write it, with three ionizable hydrogens first.

	eq acid/mol	eq base/mol
$2\ HBr + Ba(OH)_2 \rightarrow BaBr_2 + 2\ HOH$	1	2
$H_3C_6H_5O_7 + 2\ KOH \rightarrow K_2HC_6H_5O_7 + 2\ HOH$	2	1

In the first equation each mole of $Ba(OH)_2$ yields two OH^- ions, so there are two equivalents per mole. Each mole of HBr produces one H^+, so there is one equivalent per mole. In the second equation there could be one, two, or three equivalents per mole of $H_3C_6H_5O_7$, depending on how many ionizable hydrogens are released in the reaction. That number is two: $H_3C_6H_5O_7 \rightarrow 2\ H^+ + HC_6H_5O_7^{2-}$. The two H^+ ions released combine with the two OH^- ions from two moles of KOH to form two HOH molecules. There are therefore two equivalents of acid per mole. In potassium hydroxide there is only one OH^- in a formula unit, so there can be only one equivalent per mole.

Example 15.10 illustrates an important fact we will identify here but not use until Section 15.11. In both reactions, notice that *the total number of equivalents of each reacting species is the same*. In the first reaction, 2 mol HBr are 2 eq of HBr and 1 mol $Ba(OH)_2$ is 2 eq of $Ba(OH)_2$. In the second equation, 1 mol $H_3C_6H_5O_7$ is 2 eq of $H_3C_6H_5O_7$ and 2 mol KOH are 2 eq of KOH. The "same number of equivalents of all reactants" idea extends to the product species, too. Once you find the number of equivalents of one species in a reaction, you have the number of equivalents of *all* species. It is this fact that makes normality such a useful tool in quantitative work.

It is sometimes convenient to find the **equivalent mass** of a substance, **the number of grams per equivalent.** Equivalent mass, g/eq, is similar to molar mass, g/mol. Equivalent mass (EM) is readily calculated by dividing molar mass by equivalents per mole:

$$\frac{\text{g/mol}}{\text{eq/mol}} = \frac{\text{g}}{\cancel{\text{mol}}} \times \frac{\cancel{\text{mol}}}{\text{eq}} = \text{g/eq} \tag{15.14}$$

In ordinary acid-base reactions there are one, two, or three equivalents per mole. It follows that the equivalent mass of an acid or base is the same as, one half of, or one third of the molar mass. The molar mass of phosphoric acid is 98.0 g/mol. For the three reactions of phosphoric acid (Equations 15.11, 15.12, and 15.13) the equivalent masses are:

$$\text{Equation 15.12:}\quad \frac{98.0\ \text{g}\ H_3PO_4/\cancel{\text{mol}}}{1\ \text{eq}\ H_3PO_4/\cancel{\text{mol}}} = \frac{98.0\ \text{g}\ H_3PO_4}{1\ \text{eq}\ H_3PO_4} = 98.0\ \text{g}\ H_3PO_4/\text{eq}\ H_3PO_4$$

$$\text{Equation 15.13:}\quad \frac{98.0\ \text{g}\ H_3PO_4/\cancel{\text{mol}}}{2\ \text{eq}\ H_3PO_4/\cancel{\text{mol}}} = \frac{98.0\ \text{g}\ H_3PO_4}{2\ \text{eq}\ H_3PO_4} = 49.0\ \text{g}\ H_3PO_4/\text{eq}\ H_3PO_4$$

$$\text{Equation 15.11:}\quad \frac{98.0\ \text{g}\ H_3PO_4/\cancel{\text{mol}}}{3\ \text{eq}\ H_3PO_4/\cancel{\text{mol}}} = \frac{98.0\ \text{g}\ H_3PO_4}{3\ \text{eq}\ H_3PO_4} = 32.7\ \text{g}\ H_3PO_4/\text{eq}\ H_3PO_4$$

EXAMPLE 15.11 Calculate the equivalent masses of KOH (56.1 g/mol), $Ba(OH)_2$ (171 g/mol), and $H_3C_6H_5O_7$ (192 g/mol) for the reactions in Example 15.12.

KOH $\qquad$ $Ba(OH)_2$ $\qquad$ $H_3C_6H_5O_7$

KOH: $\frac{56.1 \text{ g KOH}}{1 \text{ eq KOH}} = 56.1 \text{ g KOH/eq}$; $Ba(OH)_2$: $\frac{171 \text{ g } Ba(OH)_2}{2 \text{ eq } Ba(OH)_2} = 85.5 \text{ g } Ba(OH)_2\text{/eq}$

$H_3C_6H_5O_7$: $\frac{192 \text{ g } H_3C_6H_5O_7}{2 \text{ eq } H_3C_6H_5O_7} = 96.0 \text{ g } H_3C_6H_5O_7\text{/eq}$

Just as molar mass makes it possible to convert in either direction between grams and moles, equivalent mass sets the path between grams and equivalents. Equation 7.4 for molar mass has its counterpart for equivalent mass

$$\text{EM grams} \triangleq 1 \text{ equivalent} \tag{15.15}$$

In practice it is often more convenient to use the fractional form for equivalent mass—the molar mass over the number of equivalents per mole, which are boxed in above. We will use both setups in the next example, but only the fractional setup thereafter. If your instructor emphasizes equivalent mass as a quantity, you should, of course, follow those instructions.

EXAMPLE 15.12 Calculate the number of equivalents in 68.5 g $Ba(OH)_2$.

The numbers you need are in Example 15.11. Complete the problem.

Using equivalent mass, $68.5 \cancel{\text{ g } Ba(OH)_2} \times \frac{1 \text{ eq } Ba(OH)_2}{85.5 \cancel{\text{ g } Ba(OH)_2}} = 0.801 \text{ eq } Ba(OH)_2$

Using the fractional setup, $68.5 \cancel{\text{ g } Ba(OH)_2} \times \frac{2 \text{ eq } Ba(OH)_2}{171 \cancel{\text{ g } Ba(OH)_2}} = 0.801 \text{ eq } Ba(OH)_2$

You are now ready to use the equivalent concept in normality problems.

EXAMPLE 15.13 Calculate the normality of a solution prepared by dissolving 2.50 g NaOH in 5.00×10^2 mL of solution.

SOLUTION: This is just like Example 15.7, except that moles in moles per liter, mol/L, have been replaced by equivalents in equivalents per liter, eq/L. After converting 500 mL to liters, the concentration may be expressed in grams of solute per liter, g/L. It must be changed to eq/L. In other words, grams must be changed to equivalents. This is done by equivalent weight, g/eq:

$$\frac{2.50\ \cancel{g\ NaOH}}{0.500\ L} \times \frac{1\ eq\ NaOH}{40.0\ \cancel{g\ NaOH}} = 0.125\ eq\ NaOH/L = 0.125\ N\ NaOH$$

In essence, grams per liter has been divided by equivalent mass, grams per equivalent.

EXAMPLE 15.14 2.50×10^2 mL of a sulfuric acid solution contains 10.5 g H_2SO_4. Calculate its normality for the reaction $H_2SO_4 + 2\ NaOH \rightarrow Na_2SO_4 + 2\ HOH$.

The procedure is the same. Express the concentration in grams per liter and convert to equivalents per liter. In doing so you must determine the number of equivalents in one mole of sulfuric acid. Complete the problem.

$$\frac{10.5\ \cancel{g\ H_2SO_4}}{0.250\ L} \times \frac{2\ eq\ H_2SO_4}{98.1\ \cancel{g\ H_2SO_4}} = 0.856\ eq\ H_2SO_4/L = 0.856\ N\ H_2SO_4$$

The setup is the same as dividing the concentration in grams per liter by the equivalent mass of the acid:

$$\frac{10.5\ \cancel{g\ H_2SO_4}}{0.250\ L} \times \frac{1\ eq\ H_2SO_4}{49.1\ \cancel{g\ H_2SO_4}} = 0.856\ eq\ H_2SO_4/L = 0.856\ N\ H_2SO_4$$

Just as molarity provides a way to convert in either direction between moles of solute and volume of solution, normality offers a unit path between equivalents of solute and volume of solution. Equation 15.8 is the link.

EXAMPLE 15.15 How many equivalents are in 18.6 mL 0.856 N H_2SO_4?

The unit path is L → eq H_2SO_4. 0.865 is the number of equivalents in 1 L. Set up and solve.

$$0.0186\ \cancel{L} \times \frac{0.856\ eq\ H_2SO_4}{1\ \cancel{L}} = 0.0159\ eq\ H_2SO_4$$

Example 15.14 presents an important relationship involving normality—the product of volume (L) times normality (eq/L) is equivalents of solute:

$$V \times N = \text{equivalents} \tag{15.16}$$

We will use this relationship in Section 15.11.

Naturally, it would be nice to know how to prepare a solution of specified normality.

EXAMPLE 15.16 How many grams of phosphoric acid must be used to prepare 1.00×10^2 mL 0.350 N H_3PO_4 to be used in the reaction $H_3PO_4 + NaOH \rightarrow NaH_2PO_4 + HOH$?

This is exactly like Example 15.6, except that moles have been replaced by equivalents. After volume is changed to liters, the unit path becomes L $\rightarrow$ eq H_3PO_4 $\rightarrow$ g H_3PO_4. Complete the problem.

$$0.100\ \cancel{L} \times \frac{0.350\ \cancel{eq\ H_3PO_4}}{1\ \cancel{L}} \times \frac{98.0\ g\ H_3PO_4}{1\ \cancel{eq\ H_3PO_4}} = 3.43\ g\ H_3PO_4$$

Only one of the three available hydrogen ions in phosphoric acid reacts, so there is only one equivalent per mole.

EXAMPLE 15.17 How many grams of oxalic acid, $H_2C_2O_4$, must you use to prepare 2.50 L of 0.440 N solution for the reaction $H_2C_2O_4 + 2\ KOH \rightarrow K_2C_2O_4 + 2\ HOH$?

This problem is similar to Example 15.16. Set up and solve.

$$2.50\ \cancel{L} \times \frac{0.440\ \cancel{eq\ H_2C_2O_4}}{1\ \cancel{L}} \times \frac{90.0\ g\ H_2C_2O_4}{2\ \cancel{eq\ H_2C_2O_4}} = 49.5\ g\ H_2C_2O_4$$

The equation shows that both hydrogens in $H_2C_2O_4$ react, so there are two equivalents per mole.

15.9 SOLUTION STOICHIOMETRY

PG 15L Given the quantity of any species participating in a chemical reaction for which the equation may be written, find the quantity of any other species, either quantity being measured in (a) grams, (b) volume of gas at specified temperature and pressure, or (c) volume of solution at specified molarity.

The pattern for solving stoichiometry problems was introduced in Section 9.2. It is repeated here for ready reference:

1. Convert the quantity of given species to moles.
2. Convert the moles of given species to moles of wanted species.
3. Convert the moles of wanted species to the quantity of units required.

Equation 9.2 (Section 9.2) established a unit path matching the stoichiometric pattern for quantities measured in grams. It was expanded to include volumes of gases at specified temperature and pressure in Equation 13.23 (Section 13.12). In Examples 15.8 and 15.9 you learned how to convert in either direction between volume of solution of known molarity and moles of solute. You now have three ways to perform Steps 1 and 3 of the stoichiometric pattern. They are summarized in this unit path equation:

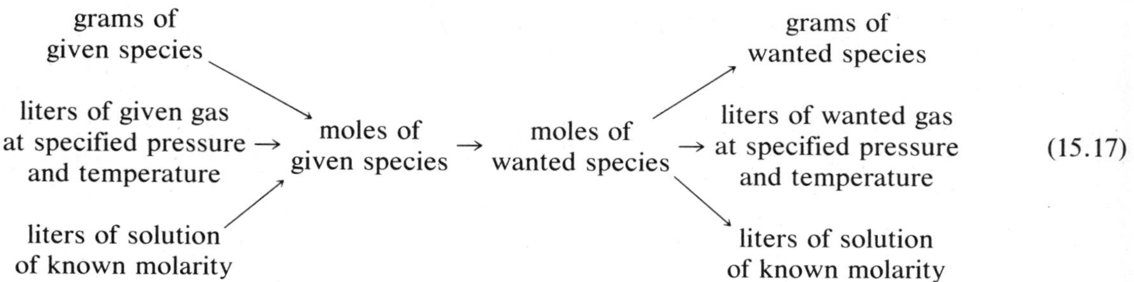

From Equations 15.7 and 15.8 and Examples 15.8 and 15.9, you can calculate any quantity on the right from any quantity on the left.

EXAMPLE 15.18 How many grams of lead(II) iodide will be precipitated from the addition of excess potassium iodide solution to 50.0 mL 1.22 M $Pb(NO_3)_2$? The equation is:

$$Pb(NO_3)_2(aq) + 2\ KI(aq) \rightarrow PbI_2(s) + 2\ KNO_3(aq)$$

Begin with the given quantity from the statement of the problem.

Write the unit path, ending with grams of PbI_2.

L $Pb(NO_3)_2(aq)$ → mol $Pb(NO_3)_2$ (Example 15.8) → mol PbI_2 → g PbI_2

Set up the first step to moles of given species, but do not solve.

$$0.0500\ \cancel{L} \times \frac{1.22\ \text{mol}\ Pb(NO_3)_2}{1\ \cancel{L}} \times \underline{\hspace{3cm}}$$

The remainder of the problem is like the first stoichiometry problems: convert moles of given to moles of wanted, and then to grams. Complete the setup and the solution.

$$0.0500\ \cancel{L} \times \frac{1.22\ \cancel{\text{mol}\ Pb(NO_3)_2}}{1\ \cancel{L}} \times \frac{1\ \cancel{\text{mol}\ PbI_2}}{1\ \cancel{\text{mol}\ Pb(NO_3)_2}} \times \frac{461\ \text{g}\ PbI_2}{1\ \cancel{\text{mol}\ PbI_2}} = 28.1\ \text{g}\ PbI_2$$

EXAMPLE 15.19 Calculate the number of milliliters of 0.842 M NaOH that will be required to precipitate as $Cu(OH)_2$ all the copper ion in 30.0 mL 0.635 M $CuSO_4$. The equation is

$$2\ NaOH(aq) + CuSO_4(aq) \rightarrow Cu(OH)_2(s) + Na_2SO_4(aq)$$

As before, write the unit path.

$$L\ CuSO_4(aq) \rightarrow mol\ CuSO_4 \rightarrow mol\ NaOH \rightarrow L\ NaOH(aq)$$

It is necessary, of course, to convert between milliliters and liters at both ends of the unit path.

The first two steps in the unit path this time are as in the last example. Set up that far, but do not solve.

$$0.0300\ \cancel{L} \times \frac{0.635\ \cancel{mol\ CuSO_4}}{1\ \cancel{L}} \times \frac{2\ mol\ NaOH}{1\ \cancel{mol\ CuSO_4}} \times \underline{\qquad\qquad}$$

At this point you have the number of moles of sodium hydroxide required. It is "packaged" at 0.842 mol/L. How many such "packages" are required to obtain the number of moles indicated by the above setup? Recall Example 15.9, in which you performed the identical operation. Complete the setup and solve the problem—remembering, of course, that the answer is required in milliliters.

$$0.0300\ \cancel{L} \times \frac{0.635\ \cancel{mol\ CuSO_4}}{1\ \cancel{L}} \times \frac{2\ \cancel{mol\ NaOH}}{1\ \cancel{mol\ CuSO_4}} \times \frac{1\ \cancel{L}}{0.842\ \cancel{mol\ NaOH}} = 0.0452\ L = 45.2\ mL\ NaOH$$

EXAMPLE 15.20 How many liters of dry hydrogen, measured at STP, will be released by the complete reaction of 45.0 mL 0.486 M H_2SO_4 with excess granular zinc? The equation is

$$Zn(s) + H_2SO_4(aq) \rightarrow ZnSO_4(aq) + H_2(g)$$

Recalling that the molar volume of any ideal gas at STP is 22.4 L/mol, set up and solve this problem completely.

$$0.0450\ \cancel{L} \times \frac{0.486\ \cancel{mol\ H_2SO_4}}{1\ \cancel{L}} \times \frac{1\ \cancel{mol\ H_2}}{1\ \cancel{mol\ H_2SO_4}} \times \frac{22.4\ L\ H_2}{1\ \cancel{mol\ H_2}} = 0.490\ L\ H_2$$

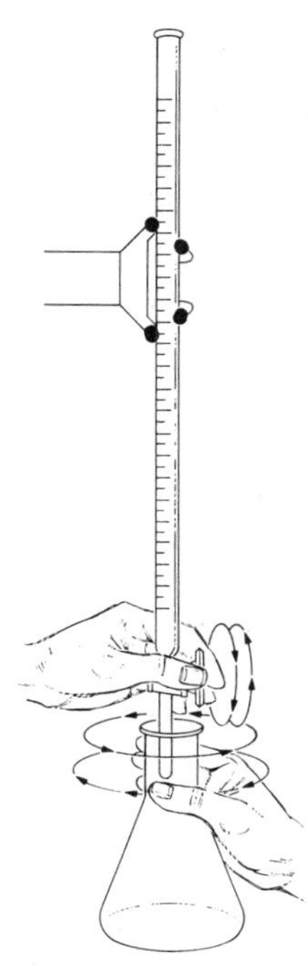

Figure 15.6
Titrating from a buret into a flask.

15.10 TITRATION USING MOLARITY

PG 15M Given the volume of a solution that reacts with a known mass of a primary standard and the equation for the reaction, calculate the molarity of the solution.

15N Given the volumes of two solutions that react with each other in a titration, the molarity of one solution, and the equation for the reaction, calculate the molarity of the second solution.

One of the more important laboratory operations in analytical chemistry is called **titration.** Titration is the very careful addition of one solution into another by means of a buret (Fig. 15.6). The buret accurately measures the volume of a solution required to react with a certain quantity of another dissolved substance. When precisely that volume has been reached, an **indicator** changes color, and the operator stops the flow from the buret. Phenolphthalein is a typical indicator for acid-base titrations. It is colorless in an acid solution and pink in a basic solution.

Titration can be used to **standardize** a solution, which means finding its precise concentration for use in later titrations. Sodium hydroxide cannot be weighed accurately because it absorbs moisture from the air and increases in weight during the weighing process. Therefore, it is not possible to prepare a sodium hydroxide solution whose molarity is known precisely. Instead the solution is standardized against a weighed quantity of something that can be weighed accurately. Such a substance is called a **primary standard.** Oxalic acid, $H_2C_2O_4$, is commonly used.* When used to standardize sodium hydroxide the equation is

$$H_2C_2O_4(s) + 2\ NaOH(aq) \rightarrow Na_2C_2O_4(aq) + 2\ HOH(\ell)$$

The following example illustrates the standardization process.

EXAMPLE 15.21 0.839 g $H_2C_2O_4$ is dissolved in water, and the solution is titrated with a solution of NaOH of unknown concentration. 28.3 mL NaOH(aq) are required to neutralize the acid. Calculate the molarity of the NaOH.

SOLUTION: Your goal in this problem is to find the moles of NaOH per liter of solution. The reaction involves 28.3 mL, or 0.0283 L. If you can find the number of moles in 0.0283 L, all you have to do is to divide that number of moles by 0.0283 L to get molarity.

How do you find the number of moles of NaOH in 0.0283 L? Whatever that number is, you know it is the number that reacts with 0.839 g $H_2C_2O_4$. We need a unit path from grams of $H_2C_2O_4$ to moles of NaOH. This path is the first two steps of the stoichiometric pattern: g $H_2C_2O_4 \rightarrow$ mol $H_2C_2O_4 \rightarrow$ mol NaOH. The setup is

$$0.839\ \cancel{g\ H_2C_2O_4} \times \frac{1\ \cancel{mol\ H_2C_2O_4}}{90.0\ \cancel{g\ H_2C_2O_4}} \times \frac{2\ mol\ NaOH}{1\ \cancel{mol\ H_2C_2O_4}} \times \underline{\qquad\qquad}$$

*Actually oxalic acid dihydrate, $H_2C_2O_4 \cdot 2\ H_2O$, is used, but to avoid confusion we will assume that the anhydrous compound is suitable.

Now that you have moles of NaOH, all that remains is to divide by volume, 0.0283 L. To divide by 0.0283 L you multiply by its inverse:

$$0.839 \text{ g } H_2C_2O_4 \times \frac{1 \text{ mol } H_2C_2O_4}{90.0 \text{ g } H_2C_2O_4} \times \frac{2 \text{ mol NaOH}}{1 \text{ mol } H_2C_2O_4} \times \frac{1}{0.0283 \text{ L}} = 0.659 \text{ M NaOH}$$

EXAMPLE 15.22 A potassium hydroxide solution is standardized by titrating against sulfamic acid, HSO_3NH_2 (97.1 g/mol). The equation is

$$HSO_3NH_2 + KOH \rightarrow KSO_3NH_2 + HOH$$

34.2 mL of solution are required to neutralize 0.395 g HSO_3NH_2. Find the molarity of the KOH.

----- -----

$$0.395 \text{ g } HSO_3NH_2 \times \frac{1 \text{ mol } HSO_3NH_2}{97.1 \text{ g } HSO_3NH_2} \times \frac{1 \text{ mol KOH}}{1 \text{ mol } HSO_3NH_2} \times \frac{1}{0.0342 \text{ L}} = 0.119 \text{ M KOH}$$

Once a solution is standardized it may be used to find the concentration of other solutions. This is a widely used procedure in industrial laboratories.

EXAMPLE 15.23 25.0 mL of an electroplating solution are analyzed for the sulfuric acid concentration. 46.8 mL 0.659 M NaOH are required to neutralize the sample. Calculate the molarity of H_2SO_4 in the bath. The equation is

$$2 \text{ NaOH} + H_2SO_4 \rightarrow Na_2SO_4 + 2 \text{ HOH}.$$

As in Example 15.22, you are seeking a molarity, moles per liter. If you can find the number of moles of acid in the 25.0-mL (0.0250 L) sample, you can divide moles by liters to find concentration. The problem appears to have two given quantities, both volumes of solution. Only one, 46.8 mL NaOH, is accompanied by a molarity that permits you to convert to moles. This is your starting point. Think in terms of the unit path from L NaOH to mol H_2SO_4, the first two steps of the stoichiometric pattern. Set up the problem that far, but do not calculate the answer.

----- -----

$$0.0468\ \cancel{L} \times \frac{0.659\ \cancel{mol\ NaOH}}{1\ \cancel{L}} \times \frac{1\ mol\ H_2SO_4}{2\ \cancel{mol\ NaOH}} \times \frac{\quad}{\quad} =$$

The unit path is L NaOH → mol NaOH → mol H_2SO_4.

The final step is to divide by the volume of solution that has the above number of moles of sulfuric acid, as in the last two examples. Complete the problem.

$$0.0468\ \cancel{L} \times \frac{0.659\ \cancel{mol\ NaOH}}{1\ \cancel{L}} \times \frac{1\ mol\ H_2SO_4}{2\ \cancel{mol\ NaOH}} \times \frac{1}{0.0250\ L} = 0.617\ M\ H_2SO_4$$

15.11 TITRATION USING NORMALITY (OPTIONAL)

PG 15O Given the volume of a solution that reacts with a known mass of a primary standard and the equation for the reaction, calculate the normality of the solution.

15P Given the volumes of two solutions that react with each other in titration, the normality of one solution, and the equation for the reaction, calculate the normality of the second solution.

It was noted earlier that normality is a convenient concentration unit in analytical work. This is particularly true in commercial laboratories where the same titration is performed again and again. The laboratory operations are the same, but the calculations are somewhat different. To illustrate this we will repeat Examples 15.21 and 15.23, using the same titration data with equivalents and normality. In solving these problems, recall the important fact that was pointed out in Section 15.8: *the number of equivalents of all species in a reaction is the same*. If you find the number of equivalents of one substance, you have found the number of equivalents of all species.

EXAMPLE 15.24 0.839 g $H_2C_2O_4$ is dissolved in water and the solution is titrated with 28.3 mL of a solution of NaOH of unknown concentration. Calculate the normality of the NaOH for the reaction.

$$H_2C_2O_4 + 2\ NaOH \rightarrow Na_2C_2O_4 + 2\ HOH$$

Your goal this time is to find the number of equivalents of NaOH per liter. Find first the number of equivalents of NaOH in 0.0283 liters. Then divide the equivalents by the volume and you have normality.

The quantity you know about is 0.839 g $H_2C_2O_4$. Look at the equation and determine the number of equivalents of acid per mole. Put these together and calculate the number of equivalents of $H_2C_2O_4$ in 0.839 g. Try to figure it out without looking back, but if you have trouble check Example 15.12.

$$0.839 \text{ g } \cancel{H_2C_2O_4} \times \frac{2 \text{ eq } H_2C_2O_4}{90.0 \text{ g } \cancel{H_2C_2O_4}} = 0.0186 \text{ eq } H_2C_2O_4$$

The equation shows that each mole of $H_2C_2O_4$ releases two moles of H^+, so oxalic acid has two equivalents per mole.

How many equivalents of sodium hydroxide react with 0.0186 eq $H_2C_2O_4$?

----- -----

0.0186 eq NaOH

Once you have found the equivalents of one species in a reaction you have found the equivalents of all species.

Dividing equivalents by volume (28.3 L) in liters gives normality. Complete the problem.

----- -----

$$\frac{0.0186 \text{ eq NaOH}}{0.0283 \text{ L}} = 0.657 \text{ eq NaOH/L} = 0.657 \text{ N NaOH}$$

The answer to Example 15.21 was 0.659 M NaOH. The molarity and normality should be the same when there is one equivalent per mole. The difference is caused in the rounding off that occurred in calculating the number of equivalents. The same problem in a single calculation setup is

$$0.839 \text{ g } \cancel{H_2C_2O_4} \times \frac{2 \text{ eq } \cancel{H_2C_2O_4}}{90.0 \text{ g } \cancel{H_2C_2O_4}} \times \frac{1 \text{ eq NaOH}}{1 \text{ eq } \cancel{H_2C_2O_4}} \times \frac{1}{0.0283 \text{ L}} = 0.659 \text{ N NaOH}$$

Compare this setup with the setup for Example 15.21. The numbers are the same, but the arrangement and units differ slightly.

Once you have the normality of a solution, you can use it to find the normality of other solutions. Equation 15.15 (Section 15.8) indicates that the number of equivalents of a species in a reaction is the product of the solution volume times normality. If the number of equivalents of all species in the reaction is the same, then

$$V_1N_1 = V_2N_2 \tag{15.18}$$

where subscripts 1 and 2 identify the reacting solutions. Solving for the second normality,

$$N_2 = \frac{V_1N_1}{V_2} \tag{15.19}$$

That's all it takes to calculate normality in this follow-up to Example 15.23.

EXAMPLE 15.25 25.0 mL of an electroplating solution are analyzed for its sulfuric acid concentration. 46.8 mL 0.659 N NaOH are required to neutralize the sample. Calculate the normality of H_2SO_4 in the bath.

Substitute directly into Equation 15.19 and solve. You may omit units. They can be included, but they do not make the problem easier and the setup becomes needlessly cluttered.

$$N_2 = \frac{46.8 \times 0.659}{25.0} = 1.23 \text{ N } H_2SO_4$$

Notice that volume in milliliters was used in the setup. There is a volume factor in both the numerator and denominator of Equation 15.19, so the conversion to liters cancels.

The normality of sulfuric acid in Example 15.25 is twice the molarity calculated in Example 15.23. This is as it should be, since

$$\frac{0.617 \cancel{\text{mol } H_2SO_4}}{1 \text{ L}} \times \frac{2 \text{ eq } H_2SO_4}{1 \cancel{\text{mol } H_2SO_4}} = 1.23 \text{ eq } H_2SO_4/\text{L} = 1.23 \text{ N } H_2SO_4$$

15.12 COLLIGATIVE PROPERTIES OF SOLUTIONS (OPTIONAL)

A pure solvent has certain distinct, definite physical properties, as does any pure substance. The introduction of a solute into the solvent affects these properties. The properties of the solution, a mixture, depend on the relative quantities of solvent and solute. It has been found experimentally that, in *dilute* solutions of certain solutes, the *change* in some of these properties is proportional to the molal concentration of the solute particles. The proportionality constant is independent of the solute; it is a property of the solvent. Solution properties that are determined solely by the *number* of solute particles dissolved in a fixed quantity of solvent are called **colligative properties.**

Freezing and boiling points of solutions are colligative properties. Perhaps the most common example is the antifreeze used in the cooling systems of automobiles. The solute dissolved in the radiator water reduces the freezing temperature to a level well below the normal freezing point of pure water, and also raises the boiling point above the normal boiling point.

The mathematical relationship for the change in freezing point between a solution and a pure solvent is

$$\Delta T_f = K_f m \qquad (15.20)$$

where ΔT_f is the **freezing point depression,** as it is called, K_f is a proportionality constant known as the *molal freezing point constant,* and m is the molality of the solution. Similarly,

$$\Delta T_b = K_b m \qquad (15.21)$$

in which ΔT_b is the **boiling point elevation** and K_b is the **molal boiling point constant.** For water $K_f = -1.86$, and $K_b = 0.52$. Boiling and freezing point problems may be solved algebraically by direct substitution into one or the other of the above equations.

EXAMPLE 15.26 Determine the freezing point of a solution of 12.0 grams of urea, $CO(NH_2)_2$, in 250 grams of water.

SOLUTION: To use Equation 15.20 it is necessary to express the solution concentration in molality:

$$\frac{12.0\ \cancel{\text{g } CO(NH_2)_2}}{0.250\ \text{kg } H_2O} \times \frac{1\ \text{mol } CO(NH_2)_2}{60.0\ \cancel{\text{g } CO(NH_2)_2}} = 0.800\ \text{m}$$

If $K_f = -1.86$ for water,

$$\Delta T_f = K_f m = -1.86 \times 0.800 = -1.49°C$$

This indicates the freezing point of the solution is 1.49°C below the normal freezing point of water, 0°C. The solution freezes at -1.49°C.

Notice that Equations 15.20 and 15.21 do not yield freezing or boiling points, but indicate *changes* in these properties. ΔT_f and the freezing points of the solution are the same in Example 15.26 only because the solvent happens to freeze at 0°C, a convenience that will not be true for any solvent other than water. Notice also that we have not used units in the usual way. They can be included, but they are awkward and tend to confuse the calculation rather than aid it.

Freezing point depression and/or boiling point elevation can be used to find the approximate molar mass of an unknown solute. The solution is prepared with measured masses of the solute and a solvent whose freezing or boiling point constant is known. From the mass data the solution concentration can be expressed in grams of solute per kilogram of solvent. Call it X grams of solute per kilogram of solvent, or

$$X\ \text{g solvent} \triangleq 1\ \text{kg solvent} \qquad (15.22)$$

The freezing point depression or boiling point elevation is determined in the laboratory, and the molality calculated from Equation 15.19 or 15.20. Let its value be Y. As noted in Section 15.6, molality represents ". . . the number of moles of solute that are equivalent to one kilogram of solvent":

$$Y\ \text{mol solute} \triangleq 1\ \text{kg solvent} \qquad (15.23)$$

Combining Equations 15.22 and 15.23 gives a single expression that includes the mass and moles of the same amount of solute:

$$X\ \text{g solute} \triangleq 1\ \text{kg solvent} \triangleq Y\ \text{mol solute} \qquad (15.24)$$

Dividing X and Y gives g solute/mol solute, which is the molar mass.

EXAMPLE 15.27 The molal boiling point constant of benzene is 2.5. A solution of 15.2 g of unknown solute in 91.1 g of benzene boils at a temperature 2.1°C higher than the boiling point of pure benzene. Estimate the molar mass of the solute.

SOLUTION: The concentration of the solution from mass data is

$$\frac{15.2 \text{ g solute}}{91.1 \text{ g solvent}} = \frac{15.2 \text{ g solute}}{0.0911 \text{ kg solvent}} = 167 \text{ g solute/kg solvent} = X$$

The molality of the solution comes from Equation 15.19, solved for m:

$$m = \frac{\Delta T_b}{K_b} = \frac{2.1}{2.5} = 0.84 \text{ mol solute/kg solvent} = Y$$

Thus 167 g solute (X) = 0.84 mol solute (Y). Dividing X by Y gives

$$\frac{167 \text{ g solute}}{0.84 \text{ mol solute}} = 2.0 \times 10^2 \text{ g/mol}$$

CHAPTER 15 IN REVIEW

15.1 The Characteristics of a Solution

15.2 Solution Terminology

15A Distinguish between terms in the following groups: solute and solvent; concentrated and dilute; solubility, saturated, unsaturated, and supersaturated; miscible and immiscible.

15.3 The Formation of a Solution

15B Describe the formation of a saturated solution from the time excess solid solute is first placed into a liquid solvent.

15C Identify and explain the factors that determine the time required to dissolve a given amount of solute, or to reach equilibrium.

15.4 Factors That Determine Solubility

15D Given the structural formulas of two molecular substances, or other information from which the strength of their intermolecular forces may be estimated, predict if they will dissolve appreciably in each other and state the criteria on which your prediction is based.

15E Predict how the solubility of a gas in a liquid will be affected by a change in the partial pressure of that gas over the liquid.

15.5 Solution Concentration: Percentage by Weight

15F Given grams of solute and grams of solvent or solution, calculate percentage concentration.

15G Given grams of solution and percentage concentration, calculate grams of solute and grams of solvent.

15.6 Solution Concentration: Molality (Optional)

15.7 Solution Concentration: Molarity

15H Given two of the following, calculate the third: volume of solution, molarity, and moles (or grams with known or calculable molar weight) of solute.

15.8 Solution Concentration: Normality (Optional)

15I Define an equivalent of an acid or a base.

15J Given an equation for an acid–base reaction, state the number of equivalents of acid or base per mole and calculate the equivalent mass of the acid or base.

15K Given an equation for an acid–base reaction and two of the following, calculate the third: volume of solution, normality, and equivalents of acid or base (or grams with known or calculable molar mass).

15.9 Solution Stoichiometry

15L Given the quantity of any species participating in a chemical reaction for which the equation may be written, find the quantity of any other species, either quantity being measured in (a) grams, (b) volume of gas at specified temperature and pressure, or (c) volume of solution at specified molarity.

15.10 Titration Using Molarity

15M Given the volume of a solution that reacts with a known mass of a primary standard and the equation for the reaction, calculate the molarity of the solution.

15N Given the volumes of two solutions that react with each other in a titration, the molarity of one solution, and the equation for the reaction, calculate the molarity of the second solution.

15.11 Titration Using Normality (Optional)

15O Given the volume of a solution that reacts with a known mass of a primary standard and the equation for the reaction, calculate the normality of the solution.

15P Given the volumes of two solutions that react with each other in a titration, the normality of one solution, and the equation for the reaction, calculate the normality of the second solution.

15.12 Colligative Properties of Solutions (Optional)

TERMS AND CONCEPTS

15.1 Solution
15.2 Solute
Solvent
Concentrated
Dilute
Solubility
Saturated, unsaturated
Supersaturated
Miscible, immiscible
15.3 Hydrated
15.5 Percentage (concentration)
15.6 Molality (concentration)
15.7 Molarity (concentration)
15.8 Normality (concentration)
Equivalent
Equivalent mass
15.10 Titration
Indicator
Standardize
Primary standard
15.12 Colligative properties
Freezing point depression
Molal freezing point constant
Boiling point elevation
Molal boiling point constant

Most of these terms and many others appear in the Glossary. Use the Glossary regularly.

QUESTIONS AND PROBLEMS

Section 15.1

(*1*) Explain why the physical properties of solutions do not have fixed values, as they have for pure substances.

(67) Mixtures of gases are always true solutions. True or false? Explain why.

(*2*) What kinds of solute particles are present in a solution of an ionic compound? Of a molecular compound?

(68) Into what kinds of particles is the solute divided in a solution? Can they usually be divided into smaller particles?

Section 15.2

(*3*) Explain why the distinction between solute and solvent is not clearly defined in many solutions.

(69) Identify the *solute* and the *solvent* in each of the following solutions: (a) salt water [NaCl(aq)]; (b) sterling silver (92.5% Ag, 7.5% Cu); (c) air (about 80% N_2, 20% O_2).

(4) Solution A contains 10 g of solute dissolved in 100 g of solvent, while solution B has only 5 g of a different solute per 100 g of solvent. Under what circumstances can solution A be classified as dilute and solution B as concentrated?

(70) Would it be proper to say that a saturated solution is a concentrated solution? or that a concentrated solution is a saturated solution? Point out the distinctions between these sometimes confused terms.

(5) Suggest simple laboratory tests by which you could determine if a solution is unsaturated, saturated, or supersaturated. Explain why your suggestions would distinguish between the different classifications.

(6) Suggest units in which solubility might be expressed other than grams per 100 g of solvent.

(7) Contrast the terms miscibility, miscible, and immiscible with their counterparts, solubility, soluble, and insoluble.

(71) What happens if you add a very small amount of solid salt (NaCl) to each of the beakers described below? Include a statement about the *amount* of solid eventually found in the beaker, compared with the amount you added: (a) a beaker containing *saturated* NaCl solution; (b) a beaker with *unsaturated* NaCl solution; (c) a beaker containing *supersaturated* NaCl solution.

(72) In stating solubility, an important variable must be specified. What is that variable, and how does solubility of a solid solute *usually* depend on it?

(73) Give an example of two immiscible substances other than oil and water.

Section 15.3

(8) Describe the forces that promote the dissolving of a solid solute in a liquid solvent.

(9) "A dynamic equilibrium exists when a saturated solution is in contact with excess solute." Explain the meaning of that statement. Is it possible to have a saturated solution *without* excess solute? Explain.

(10) Why is it customary to stir coffee or tea after putting sugar into it?

(74) Describe the forces that oppose the dissolving of a solute in a liquid solvent.

(75) Bakers use confectioner's sugar because it is more finely powdered than the crystals of table (granulated) sugar. Do you think confectioner's sugar would dissolve more or less quickly than table sugar? Why?

(76) Explain the effect of heating on the rate at which a solid dissolves in a liquid.

Section 15.4

(11) Would water or carbon tetrachloride be a better solvent for benzene, C_6H_6? Why? The structural formula of benzene may be represented as

```
          H
          |
          C
        /   \\
   H—C         C—H
     ||        |
   H—C         C—H
        \   //
          C
          |
          H
```

(12) Suppose you have a spot on some clothing, and water will not take it out. If you have available

(77) Suggest why water and liquid HF are good solvents for many ionic salts, but not for waxes and oils having structures such as

```
     H   H   H         H
     |   |   |         |
  H—C—C—C · · · · · C—H
     |   |   |         |
     H   H   H         H
```

(78) Glycerin and normal hexane (see structures below) are organic compounds of approximately

cyclopentane and methanol (see structures below), which would you choose as the more promising solvent to try? Why?

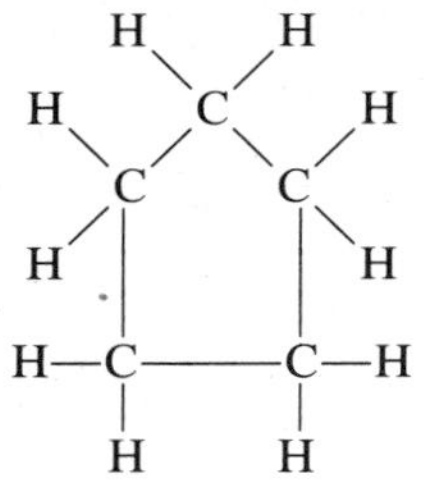

Cyclopentane

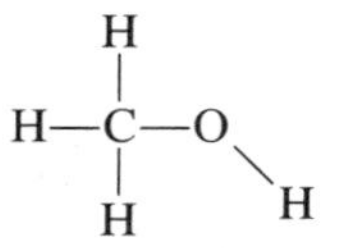

Methanol

the same molecular weight. Which of these is more apt to be miscible with carbon tetrabromide? Why?

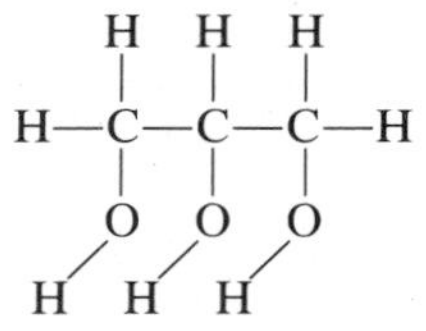

Glycerin

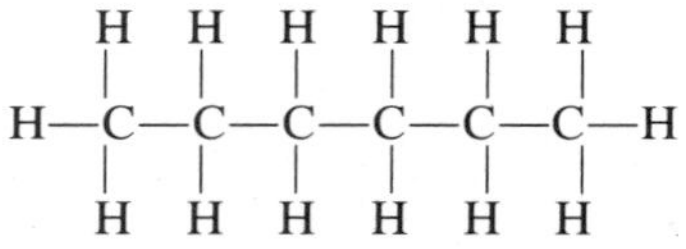

Normal hexane

(13) "The solubility of carbon dioxide in a carbonated beverage may be increased by raising the air pressure under which it is bottled." Criticize this statement.

(79) On opening a bottle of carbonated beverage, many bubbles are released, suggesting that the beverage is bottled under high pressure. Yet, for safety reasons, the pressure cannot be much more than one atmosphere. How, then, do you account for the substantial reduction of CO_2 solubility in an opened bottle?

Section 15.5

(14) 135 g of solution contain 18.5 g of dissolved salt. What is the percentage of salt in the solution?

(15) How many grams of solute are in 65.0 g of a 13.0% solution?

(80) Calculate the percentage concentration of a solution prepared by dissolving 5.29 g of calcium chloride in 81.0 g of water.

(81) How many grams of ammonium nitrate must be weighed out to make 445 g of a 58.0% solution? In how many milliliters of water should it be dissolved?

Section 15.6

(16) If 20.0 g of sugar, $C_{12}H_{22}O_{11}$, are dissolved in 1.00×10^2 mL of water, what is the molality of the solution?

(17) If you are to prepare a 4.00-molal solution of urea in water, how many grams of urea, $CO(NH_2)_2$, would you dissolve in 80.0 mL of water?

(18) Into what volume of water must 90.9 g of acetic acid, $HC_2H_3O_2$, be dissolved to produce 1.40 m $HC_2H_3O_2$?

(82) Calculate the molal concentration of a solution of 26.9 g of naphthalene, $C_{10}H_8$, in 125 g of benzene, C_6H_6.

(83) Diethylamine, $(CH_3CH_2)_2NH$, is highly soluble in ethanol, C_2H_5OH. Calculate the number of grams of diethylamine that would be dissolved in 400 g of ethanol to produce 4.50 m $(CH_3CH_2)_2NH$.

(84)* Methyl ethyl ketone, C_4H_8O, is a solvent popularly known as MEK that is used to cement plastics. How many grams of MEK must be dissolved in 1.00×10^2 mL of benzene, specific gravity 0.879, to yield a 0.720 molal solution?

Section 15.7

(19) 6.00×10^2 mL of a solution contains 23.5 g of sodium sulfate, a chemical used in dyeing operations. What is the molarity of this solution?

(20) The chemical name for the "hypo" used in photographic developing is sodium thiosulfate. It is sold as a pentahydrate, $Na_2S_2O_3 \cdot 5\ H_2O$. What is the molarity of a solution prepared by dissolving 1.2×10^2 g of this compound in water and diluting the solution to 1250 mL?

(21) Sodium carbonate is one of the most widely used sodium compounds. How many grams are required by an analytical chemist to prepare 4.00×10^2 mL 0.800 M Na_2CO_3?

(22) How many grams of acetic acid, the odor and taste producer in vinegar, must be dissolved in water and diluted to 7.50×10^2 mL to yield 0.600 M $HC_2H_3O_2$?

(23) A laboratory reaction requires 0.0150 moles of HCl. How many milliliters of 0.850 M HCl would you use?

(24) Calculate the volume of concentrated ammonia solution, which is 15 molar, that contains 75.0 g of NH_3.

(25) How many moles of solute are in 65.0 mL 2.20 M NaOH?

(26) 29.3 mL 0.482 M H_2SO_4 are used to titrate a base of unknown concentration. How many moles of sulfuric acid react?

(27) Concentrated HCl is 12 molar. Find the molarity of the solution prepared by diluting 1.00×10^2 mL to 2.00 L.

(28) How many milliliters of concentrated ammonia, 15 M NH_3, would you dilute to 5.00×10^2 mL to produce 6.0 M NH_3?

(29)* The density of 18.0% HCl is 1.09 g/mL. Calculate its molarity.

(85) Potassium iodide is the additive in "iodized" table salt. Calculate the molarity of a solution prepared by dissolving 3.59 g of potassium iodide in water and diluting to 50.0 mL.

(86) 15.0 g of anhydrous nickel chloride are dissolved in water and diluted to 75.0 mL. 25.0 g of nickel chloride hexahydrate are also dissolved in water and diluted to 75.0 mL. Identify the solution with the higher molar concentration and calculate its molarity.

(87) Large quantities of silver nitrate are used in making photographic chemicals. Find the mass that must be used in preparing 2.50×10^2 mL 0.125 M $AgNO_3$.

(88) Potassium hydroxide is used in making liquid soap, as well as many other things. How many grams would you use to prepare 2.50 L 2.25 M KOH?

(89) What volume of concentrated sulfuric acid, which is 18 molar, is required to obtain 2.8 mol of the acid?

(90) 0.150 M NaCl is to be the source of 8.33 g of dissolved solute. What volume of solution is needed?

(91) Calculate the moles of silver nitrate in 18.6 mL 0.204 M $AgNO_3$.

(92) Despite its intense purple color, potassium permanganate is used in bleaching operations. How many moles are in 25.0 mL 0.0904 M $KMnO_4$?

(93) What is the molarity of the acetic acid solution if 45.0 mL 17 M $HC_2H_3O_2$ are diluted to 1.2 L?

(94) How many milliliters of concentrated nitric acid, 16 M HNO_3, will you use to prepare 7.50×10^2 mL 0.50 M HNO_3?

(95)* The density of 2.50 M KNO_3 is 1.15 g/mL. What is its percentage concentration?

Section 15.8

(30) Explain why the number of equivalents in a mole of acid or base is not always the same.

(96) What is equivalent mass? Why can you state positively the equivalent mass of LiOH, but not H_2SO_4?

(31) How many equivalents are in one mole of each of the following: HF; $H_2C_2O_4$ in $H_2C_2O_4 \rightarrow 2\ H^+ + C_2O_4^{2-}$?

(32) What is the number of equivalents in one mole of $Zn(OH)_2$; in one mole of RbOH?

(33) What is the equivalent mass of HF and $H_2C_2O_4$ in Problem 31?

(34) Find the equivalent mass of $Zn(OH)_2$ and RbOH (Problem 32; Rb is rubidium, Z = 37).

(35) Calculate the normality of a solution prepared by dissolving 17.2 g of $HC_2H_3O_2$ in 3.00×10^2 mL of solution.

(36) 9.79 g of $NaHCO_3$ are dissolved in 5.00×10^2 mL of solution. What is the normality in the reaction $HCO_3^- + H^+ \rightarrow H_2O + CO_2$?

(37) How many grams of KOH must be used to prepare 6.00×10^2 mL 2.00 N KOH?

(38)* Calculate the mass of oxalic acid dihydrate, $H_2C_2O_4 \cdot 2\ H_2O$, required for 2.50×10^2 mL 0.500 N $H_2C_2O_4$ for the reaction $H_2C_2O_4 + 2\ OH^- \rightarrow C_2O_4^{2-} + 2\ HOH$.

(39) What is the normality of (a) 0.423 M HCl; (b) 0.423 M H_2SO_4 (assume complete ionization of H_2SO_4)?

(40) How many milliliters of 12 M HCl must be diluted to 2.0 L to produce 0.50 N HCl?

(41) 25.0 mL 15 M HNO_3 are diluted to 4.00×10^2 mL. What is the normality of the diluted solution?

(42) How many equivalents are in 2.25 L 0.871 N H_2SO_4?

(43) Calculate the volume of 0.371 N HCl that contains 0.0385 eq.

(97) State the number of equivalents in one mole of HNO_2; in one mole of H_2SeO_4. (Se is selenium, Z = 34.)

(98) How many equivalents are in one mole of $Cu(OH)_2$; in one mole of $Fe(OH)_3$?

(99) Calculate the equivalent mass of HNO_2 and H_2SeO_4 in Problem 97.

(100) What are the equivalent masses of $Cu(OH)_2$ and $Fe(OH)_3$ (Problem 98)?

(101) What is the normality of the solution made when 8.09 g of KOH are dissolved in water and diluted to 2.50×10^2 mL?

(102) 3.48 g of $H_2C_2O_4$ are dissolved in water, diluted to 1.00×10^2 mL, and used in a reaction in which it ionizes as follows: $H_2C_2O_4 \rightarrow H^+ + HC_2O_4^-$. What is the normality of the solution?

(103) $NaHSO_4$ is used as an acid in the reaction $HSO_4^- \rightarrow H^+ + SO_4^{2-}$. What mass of $NaHSO_4$ must be dissolved in 5.00×10^2 mL of solution to produce 0.200 N $NaHSO_4$?

(104)* Sodium carbonate 10-hydrate, $Na_2CO_3 \cdot 10\ H_2O$, is used as a base in the reaction $CO_3^{2-} + 2\ H^+ \rightarrow CO_2 + H_2O$. Calculate the mass of hydrate needed to prepare 1.00×10^2 mL 0.400 N Na_2CO_3.

(105) What is the molarity of (a) 0.767 N NaOH; (b) 0.401 N H_3PO_4 in $H_3PO_4 + 2\ NaOH \rightarrow Na_2HPO_4 + 2\ HOH$?

(106) Calculate the volume of 18 M H_2SO_4 required to prepare 3.0 L 3.0 N H_2SO_4 for reactions in which the sulfuric acid is completely ionized.

(107) Calculate the normality of a solution prepared by diluting 10.0 mL 15 M H_3PO_4 to 2.50×10^2 mL. The solution will be used in the same reaction as that in Problem 105.

(108) 45.8 mL 0.834 N NaOH has how many equivalents of solute?

(109) What volume of 0.492 N $KMnO_4$ contains 0.0758 eq?

Section 15.9

(44) How many grams of AgCl can be precipitated by adding excess NaCl to 50.0 mL 0.855 M $AgNO_3$?

(110) Calculate the grams of magnesium hydroxide that will precipitate from 25.0 mL 0.611 M $MgCl_2$ by the addition of excess NaOH solution.

(45) What mass of barium fluoride can be precipitated from 40.0 mL 0.436 M NaF by adding excess barium nitrate solution?

(46)* 25.0 mL 0.350 M NaOH are added to 45.0 mL 0.125 M $CuSO_4$. How many grams of copper(II) hydroxide will precipitate?

(47) Calculate the volume of chlorine, measured at STP, that can be recovered from 50.0 mL 1.20 M HCl by the reaction $MnO_2 + 4\ HCl \rightarrow MnCl_2 + 2\ H_2O + Cl_2$, assuming complete conversion of reactants to products.

(48) How many milliliters of 0.715 M HCl are needed to neutralize 1.24 g of sodium carbonate.

(111)* The iron(III) ion content of a solution may be found by precipitating it as $Fe(OH)_3$, and then decomposing the hydroxide to Fe_2O_3 by heat. How many grams of iron(III) oxide can be collected from 35.0 mL 0.345 M $Fe(NO_3)_3$?

(112)* 25.0 mL of 0.235 M $NiCl_2$ are combined with 30.0 mL 0.260 M KOH. How many grams of nickel hydroxide will precipitate?

(113) How many milliliters of 1.50 M NaOH must react with aluminum to yield 2.00 L of hydrogen, measured at 22°C and 710 torr, by the reaction $2\ Al + 6\ NaOH \rightarrow 2\ Na_3AlO_3 + 3\ H_2$, assuming complete conversion of reactants to products?

(114) What volume of 0.842 M NaOH would react with 2.14 g of sulfamic acid, NH_2SO_3H, a solid acid with one replaceable hydrogen?

Section 15.10

(49) Potassium hydrogen iodate is used as a primary standard in finding the concentration of a solution of potassium hydroxide by the reaction $KH(IO_3)_2 + KOH \rightarrow 2\ KIO_3 + HOH$. What is the molarity of the base if 32.14 mL are required to titrate 1.359 g of the primary standard?

(50)* Oxalic acid dihydrate, $H_2C_2O_4 \cdot 2\ H_2O$, is the primary standard used for finding the molarity of a sodium hydroxide solution. 3.290 g are dissolved in water and diluted to 500.0 mL. A 25.00 mL sample of that solution is titrated with the NaOH solution; 30.10 mL are required. Find (a) the molarity of the acid and (b) the molarity of the base. (Both replaceable hydrogens in the oxalic acid react.)

(51)* What minimum number of grams of oxalic acid dihyrate would you specify for a student experiment involving a titration of no fewer than 15.0 mL 0.100 M NaOH? The reaction is the same as in Example 50.

(52) An analytical procedure for finding the chloride ion concentration in a solution involves the precipitation of silver chloride: $Ag^+ + Cl^- \rightarrow AgCl$. What is the molarity of the chloride ion if 16.80 mL 0.629 M $AgNO_3$ (the source of Ag^+) are needed to precipitate all of the chloride in a 25.00 mL sample of the unknown?

(115) $2\ HC_7H_5O_2 + Na_2CO_3 \rightarrow 2\ NaC_7H_5O_2 + H_2O + CO_2$ is the equation for a reaction by which a solution of sodium carbonate may be standardized. 5.038 g of $HC_7H_5O_2$ uses 41.28 mL of the solution in the titration. Find the molarity of the sodium carbonate.

(116) A student is to titrate solid maleic acid, $H_2C_4H_2O_4$ (two replaceable hydrogens), with 0.500 M NaOH. What is the maximum number of grams of maleic acid that can be used if the titration is not to exceed 25.0 mL?

(117)* 16.06 g of $NaHCO_3$ are dissolved in water and diluted to 500.0 mL in a volumetric flask. 37.80 mL of that solution are required to titrate a 20.00 mL sample of sulfuric acid solution. What is the molarity of the acid? The reaction equation is $H_2SO_4 + 2\ NaHCO_3 \rightarrow Na_2SO_4 + 2\ H_2O + 2\ CO_2$.

(118)* Calculate the hydroxide ion concentration in a 20.00 mL sample of an unknown if 14.30 mL 0.248 M H_2SO_4 are used in a neutralization titration.

(53)* A 694 mg sample of impure Na_2CO_3 was titrated with 41.24 mL 0.244 M HCl. Calculate the percent Na_2CO_3 in the sample.

(119)* A student received a 599-mg sample of a mixture of Na_2HPO_4 and NaH_2PO_4. She is to find the percentage of each compound in the sample. After dissolving the mixture, she titrated it with 15.70 mL 0.201 M NaOH. If the only reaction is $NaH_2PO_4 + NaOH \rightarrow Na_2HPO_4 + HOH$, find the required percentages.

Section 15.11

(54) What is the normality of the base in Problem 49?

(120) What is the normality of the sodium carbonate solution in Problem 115?

(55) Calculate the normalities of the acid and the base in Problem 50. Set up the problem completely from the data, not from the answers to Problem 50.

(121) What is the normality of the acid in Problem 117?

(56) What is the normality of an acid if 12.8 mL are required to titrate 15.0 mL 0.882 N NaOH?

(122) Calculate the normality of a solution of sodium carbonate if a 25.0 mL sample requires 22.8 mL 0.405 N H_2SO_4 in a titration.

(57) 28.4 mL 0.424 N $AgNO_3$ are required to titrate the chloride ion in a 25.0 mL sample of nickel chloride solution. Find the normality of the nickel chloride.

(123) 28.9 mL 0.402 N NaOH are required to titrate 50.0 mL of a solution of tartaric acid ($H_2C_4H_4O_6$) of unknown concentration. Find the normality of the acid.

(58) Using a particular acid–base indicator, 20.0 mL of a phosphoric acid solution require 32.6 mL 0.208 N NaOH in a titration reaction. Find the normality of the acid.

(124) Repeating the titration described in Problem 58, but with a different indicator, it is found that only 16.3 mL 0.208 N NaOH are required for 20.0 mL of the phosphoric acid solution. Calculate the normality of the acid and account for the difference in the answers in Problem 58 and this problem.

(59) 15.6 mL 0.562 N NaOH are required to titrate a solution prepared by dissolving 0.631 g of an unknown acid. What is the equivalent mass of the acid?

(125) 1.21 g of an organic compound that functions as a base in reaction with sulfuric acid are dissolved in water and titrated with 0.294 N H_2SO_4. What is the equivalent mass of the base if 30.7 mL of acid are required in the titration?

Section 15.12

(60) The specific gravity of a solution of KCl is greater than 1.00. The specific gravity of a solution of NH_3 is less than 1.00. Is specific gravity a colligative property? Why, or why not?

(126) Is the partial pressure exerted by one component of a gaseous mixture at a given temperature and volume a colligative property? Justify your answer, pointing out in the process what classifies a property as "colligative."

(61) Calculate the boiling and freezing points of a solution of 50.0 g of glucose, $C_6H_{12}O_6$, in 1.00×10^2 g of water.

(127) 20.0 g of analine, $C_6H_5NH_2$, are dissolved in 1.20×10^2 g of water. At what temperatures will the solution freeze and boil?

(62) 4.34 g of paradichlorobenzene, $C_6H_4Cl_2$, are dissolved in 65.0 g of naphthalene, $C_{10}H_8$. Calculate the freezing point of the solution if pure naphthalene freezes at 80.2°C and K_f is -6.9.

(128) Calculate the freezing point of a solution of 1.99 g of naphthalene, $C_{10}H_8$, in 32.0 g of benzene, C_6H_6. Pure benzene freezes at 5.50°C, and its $K_f = -5.10$.

(63) Calculate the molal concentration of an aqueous solution that boils at 100.84°C.

(64) If a solution prepared by dissolving 26.0 g of an unknown solute in 3.80×10^2 g of water freezes at −1.18°C, what is the approximate molar mass of the solute?

(65) A solution of 12.0 g of an unknown solute dissolved in 80.0 g $C_{10}H_8$ freezes at 71.3°C. The normal freezing point of $C_{10}H_8$ is 80.2°C, and $K_f = -6.9$. Estimate the molar mass of the solute.

(66) The boiling point of a solution of 1.40 g of urea, NH_2CONH_2, in 16.3 g of a certain solvent is 3.92°C higher than the boiling point of the pure solvent. What is the molal boiling point constant of the solvent?

(129) What is the molality of a solution of an unknown solute in acetic acid if it freezes at 13.4°C? The normal freezing point of acetic acid is 16.6°C, and $K_f = -3.90$.

(130) A solution of 15.1 g of an unknown solute in 6.00×10^2 g of water boils at 100.28°C. Find the molar mass of the solute.

(131) When 11.0 g of an unknown solute are dissolved in 90.0 g of phenol, the freezing point depression is 9.6°C. Calculate the molar mass of the solute if $K_f = -3.56$ for phenol.

(132) The normal freezing point of an unknown solvent is 28.7°C. A solution of 11.4 g of ethanol, C_2H_5OH, in 2.00×10^2 g of the solvent freezes at 21.5°C. What is the molal freezing point constant of the solvent?

Miscellaneous Questions

(*133*) Distinguish precisely and in scientific terms the differences between items in each of the following groups:
(a) Solute, solvent
(b) Concentrated, dilute
(c) Saturated, unsaturated, supersaturated
(d) Soluble, miscible
(e) (Optional) molality, molarity, normality
(f) (Optional) molar mass, equivalent mass
(g) (Optional) freezing point depression, boiling point elevation
(h) (Optional) molal freezing point constant, molal boiling point constant

(134) Determine whether each statement that follows is true or false:
(a) The solute particles in a colorless solution are colorless; the solute particles in a colored solution have the same color as the solution.
(b) The concentration is the same throughout a beaker of solution.
(c) A saturated solution of solute A is always more concentrated than an unsaturated solution of solute B.
(d) A solution can never have a concentration greater than its solubility at a given temperature.
(e) A finely divided solute dissolves faster because more surface area is exposed to the solvent.
(f) Stirring a solution increases the rate of crystallization.
(g) Crystallization ceases when equilibrium is reached.
(h) All solubilities increase at higher temperatures.
(i) Increasing air pressure over water increases the solubility of nitrogen in the water.
(j) An ionic solute is more apt to dissolve in a nonpolar solvent than in a polar solvent.
(k) (Optional) The molarity of a solution changes slightly with temperature, but the molality does not.
(l) (Optional) If an acid and a base react on a two-to-one mole ratio, there are twice as many equivalents of acid as there are base in the reaction.
(m) The concentration of a primary standard is found by titration.
(n) Colligative properties of a solution are independent of the kinds of solute particles, but they are dependent on particle concentration.

(135)* Sketch a graph on which rate of dissolving and rate of crystallization are plotted vertically and time is plotted horizontally. Let time be 0 when solute X is placed in water. Carry the graph to the point that equilibrium is reached and allow for its extension beyond that time.

(a) At what time is the rate of crystallization at its maximum?
(b) At what time is the absolute rate of dissolving at its maximum?
(c) At what time is the net rate of dissolving at its maximum?
(d) How would the shape of the graph change if
 (1) larger crystals of X were dissolved?
 (2) the mixture was stirred up to the time equilibrium was reached?
 (3) the dissolving was performed at a lower temperature?
(e) At time P, after equilibrium is reached, more X is poured into the mixture. Show on the graph what happens.
(f) At time Q, after equilibrium is reached, a little more water is poured into the mixture. Show on the graph what happens.
(g) At time R, after equilibrium is reached, a lot more water is poured into the mixture. Show on the graph what ultimately happens.

(136) When you heat water on a stove, small bubbles appear long before the water begins to boil. What are they? Explain why they appear.

(137) Antifreeze is put into the water in an automobile to prevent it from freezing in winter. What does the antifreeze do to the boiling point of the water, if anything?

(138)* 60.0 mL 0.322 M KI are combined with 20.0 mL 0.530 M $Pb(NO_3)_2$. (a) How many grams of PbI_2 will precipitate? (b) What is the final molarity of the K^+ ion? (c) What is the final molarity of the Pb^{2+} or I^- ion, whichever one is in excess?

(139)* A solution has been defined as a homogeneous mixture. Pure air is a solution. Does it follow that the atmosphere is a solution? Explain.

(140) Does percentage concentration of a solution depend on temperature?

(141)* If you know either the percentage concentration of a solution or its molarity, what additional information must you have before you can convert to the other concentration?

16 Reactions That Occur in Water Solutions: Net Ionic Equations

LOOKING BACK

8.7 "Conventional" equations were written for "single replacement" redox, precipitation, and neutralization reactions that occur in water solution. A neutralization reaction between an acid and a base yields a salt and water.

12.3 The ionization of a strong acid by water is shown by an ionization equation in which the acid molecule separates into a hydrogen ion and an anion.

LOOKING AHEAD IN CHAPTER 16

Net ionic equations are written for single replacement redox, precipitation, and molecule formation reactions, including neutralization reactions. The terms *acid, base,* and *salt* are used throughout the chapter.

The solution inventory of a strong acid is a hydrogen ion and the anion that comes from its complete ionization.

In Chapter 8 you wrote chemical equations for some reactions that occur in water solutions. At that time you were promised a description of these reactions in the form of net ionic equations. That promise will be fulfilled in this chapter.

Net ionic equations give us an understanding of what really happens when solutions react. To write a net ionic equation we must identify precisely what is in a solution, the exact nature of the tiniest solute particle. Our study therefore begins by examining the electrical properties of solutions and asking what makes them what they are.

16.1 ELECTROLYTES AND SOLUTION CONDUCTIVITY

PG 16A Distinguish among strong electrolytes, weak electrolytes, and nonelectrolytes.

16B Describe or explain electrical conductivity through a solution.

Suppose two metal strips, called **electrodes,** are placed in a liquid and wired to a battery and an electric light bulb (Fig. 16.1). To light the bulb a current must pass through the liquid from one metal strip to the other. If the liquid is pure water, no appreciable current passes; the bulb does not light up. If the electrodes are in a solution of sodium chloride, the bulb glows brightly—almost as much as if the metal strips were "shorted" with a screwdriver. A salt solution is an

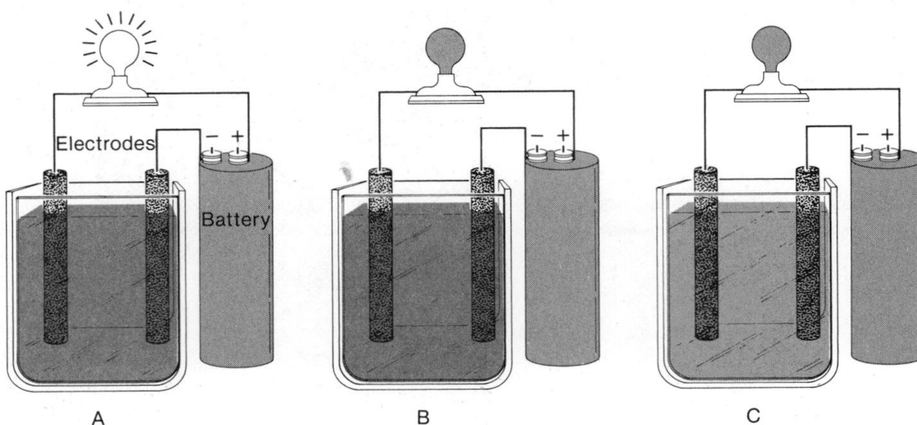

Figure 16.1 Electrolytes and nonelectrolytes. A. Solution of table salt, a strong electrolyte, is a good conductor, as indicated by the glowing light bulb. B. Solution of table sugar, a nonelectrolyte, is a nonconductor, as indicated by the unlit bulb. C. Pure water, a nonelectrolyte, is a nonconductor.

excellent **conductor** of electricity. A solution of sugar produces no such effect; like pure water, a sugar solution is a **nonconductor.**

Passage of electricity through a liquid is called **electrolysis.** A solute that yields a solution that is a good conductor is called a **strong electrolyte.** Sugar, whose solution is a nonconductor, is a **nonelectrolyte.** Some solutes, such as acetic acid, are **weak electrolytes.** Their solutions conduct electricity, but poorly, permitting only a dim glow of the lamp in Figure 16.1. The term *electrolyte* is also used to refer to the *solution* through which the current passes. The acid solution in an automobile storage battery is an electrolyte in this sense. In commercial electroplating, *electrolyte* is almost always used in this way.

What is called an electric **current** is a movement of electric charge. In a metal wire it is a movement of electrons that makes up a current. In a solution the charge is carried by ions rather than electrons. When an ionic compound dissolves, its positive ions (cations) and negative ions (anions) are free to move. When a pair of electrodes is connected to a source of direct current, such as a storage battery, one electrode, called an **anode,** has a positive charge, and the other, the **cathode,** has a negative charge. The positively charged *an*ode attracts the negatively charged *an*ions in the solution, while the positively charged *cat*ions move toward the negatively charged *ca*thode (Fig. 16.2).* This movement of ions is the electric current that flows through the solution. *Solution conductivity is positive evidence of the presence of ions.*

When sugar, a nonelectrolyte, dissolves, the crystal breaks into individual sugar molecules. These molecules are electrically neutral and do not move toward either electrode. Even if there was movement there would be no current, because the molecules have no charge. If the solute is a weak electrolyte, such as acetic acid, only a small fraction of the molecules separate into ions. This allows a weak current flow, which accounts for the dim glow of the bulb in Figure 16.1. This is the essence of *weak* in this sense: a large majority of solute particles in a weak electrolyte are unionized (un-ionized, not union-ized) molecules.

*Formally, the anode is the electrode where *oxidation* occurs, and *reduction* occurs at the cathode. Oxidation and reduction are described in Chapter 18.

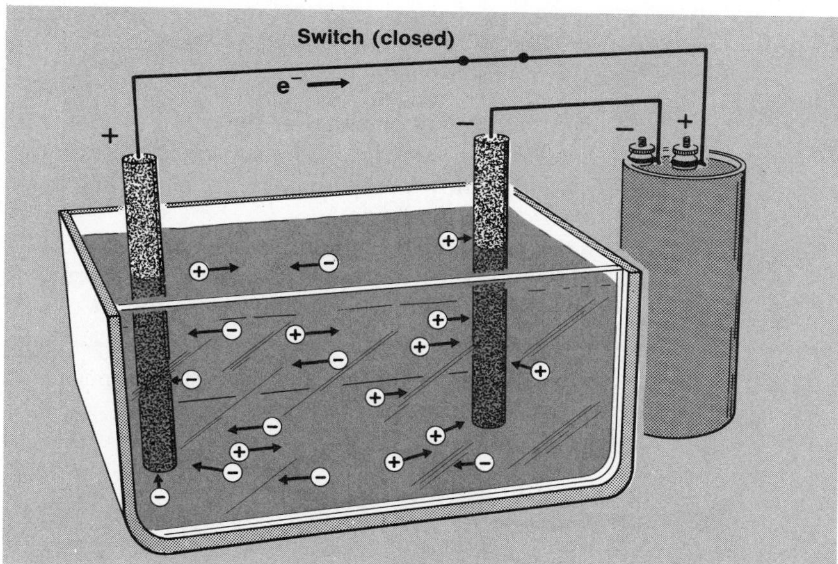

Figure 16.2
Conductivity in an ionic solution. The conductivity of a solution is positive evidence that mobile ions are present.

The equations that describe solution reactions include the ions and/or molecules that are the actual solute particles in the solution. These particles must be identified by their chemical formulas. A list of the particles in any solution is its **solution inventory.** You will learn how to write solution inventories for different kinds of solutes in the next two sections.

Quick Check 16.1
Three compounds, A, B, and C, are dissolved in water, and the solutions are tested for electrical conductivity. Solution A is a poor conductor, B is a good conductor, and C does not conduct. Classify A, B, and C as electrolytes (strong, weak, non-) and state the significance of the conductivities of their solutions.

16.2 SOLUTION INVENTORIES OF IONIC COMPOUNDS

PG 16C Given the formula of an ionic compound, write the solution inventory when it is dissolved in water.

If an ionic compound dissolves, its solution inventory always consists of ions. These ions are identified simply by separating the compound into its ions. When sodium chloride dissolves, the solution inventory is sodium ions and chloride ions:

$$NaCl(s) \xrightarrow{H_2O} Na^+(aq) + Cl^-(aq) \qquad (16.1)$$

If the solute is barium chloride, the inventory is barium ions and chloride ions:

$$BaCl_2(s) \xrightarrow{H_2O} Ba^{2+}(aq) + 2\ Cl^-(aq) \quad (16.2)$$

Notice that no matter where a chloride ion comes from, its formula is *always* Cl^-, never Cl_2^- or Cl_2^{2-}. The subscript after an ion in a formula, as the 2 in $BaCl_2$, tell us how many ions are present in the formula unit. The subscript is not part of the ion formula.

In this chapter you should include state symbols for all species in equations. It helps in writing net ionic equations. Also remember to include the charge every time you write the formula of an ion.

EXAMPLE 16.1 Write solution inventories for the following ionic compounds by writing their dissolving equations.

$$NaOH(s) \xrightarrow{H_2O}$$

$$K_2SO_4(s) \xrightarrow{H_2O}$$

$$(NH_4)_2CO_3(s) \xrightarrow{H_2O}$$

$$NaOH(s) \xrightarrow{H_2O} Na^+(aq) + OH^-(aq)$$

$$K_2SO_4(s) \xrightarrow{H_2O} 2\ K^+(aq) + SO_4^{2-}(aq)$$

$$(NH_4)_2CO_3(s) \xrightarrow{H_2O} 2\ NH_4^+(aq) + CO_3^{2-}(aq)$$

Polyatomic and monatomic ions are handled in exactly the same way in writing solution inventories. In $(NH_4)_2CO_3$, notice that the subscript outside the parentheses tells us how many ammonium ions are present, while the subscript 4 inside the parentheses is part of the polyatomic ion formula.

EXAMPLE 16.2 Write the solution inventories of the following ionic solutes *without* writing the dissolving equations: $MgSO_4$; $Ca(NO_3)_2$; $AlBr_3$; $Fe_2(SO_4)_3$.

This question is the same as Example 16.1. It asks simply for the products of the reaction without writing an equation. This is how you will write these inventories later in the chapter. Be sure to show the number of each kind of ion released by a formula unit—the coefficient if the equation were written. Also remember state symbols.

$MgSO_4$:

$Ca(NO_3)_2$:

$AlBr_3$:

$Fe_2(SO_4)_3$:

$MgSO_4$: $Mg^{2+}(aq) + SO_4^{2-}(aq)$

$Ca(NO_3)_2$: $Ca^{2+}(aq) + 2\ NO_3^-(aq)$

$AlBr_3$: $Al^{3+}(aq) + 3\ Br^-(aq)$

$Fe_2(SO_4)_3$: $2\ Fe^{3+}(aq) + 3\ SO_4^{2-}(aq)$

Quick Check 16.2
Which ionic compounds have ions in their solution inventories, and which do not?

16.3 STRONG ACIDS AND WEAK ACIDS

PG 16D Explain why the solution of an acid may be a good conductor or a poor conductor of electricity.

16E Given the formula of a soluble acid, write the solution inventory when it is dissolved in water.

To write the solution inventory of an acid, you must be able to do two things: first, recognize the compound as an acid, and second, classify it as a strong acid or weak acid. An acid, as we have used the term so far, is a hydrogen-bearing compound that releases hydrogen ions in water solution. Its formula has the form HX, H_2X, or H_3X, where X represents anything that becomes an anion when the acid ionizes. In HCl, X is Cl^-; in H_2SO_4, X is SO_4^{2-}. Notice that the anion may or may not contain oxygen. Even hydrogen may be present; acetic acid, $HC_2H_3O_2$, for example, yields $C_2H_3O_2^-$, the acetate ion, and H_3PO_4 yields $H_2PO_4^-$. Do not be concerned if the anion is unfamiliar. It will behave in exactly the same way as the ion from any other acid, and should be treated accordingly.

Hydrochloric acid is the water solution of hydrogen chloride, a gaseous molecular compound. Hydrochloric acid is a **strong acid.** This means it is almost completely ionized in water and is an excellent conductor. The ionization equation (Section 12.3) is

$$HCl(g) \xrightarrow{H_2O} HCl(aq) \rightarrow H^+(aq) + Cl^-(aq) \qquad (16.3)$$

Hydrofluoric acid is the water solution of hydrogen fluoride, another gaseous molecular compound. Hydrofluoric acid is a **weak acid.** It is only slightly ionized in water and is a poor conductor. The ionization process may be described by an equation similar to that for hydrochloric acid, but with an important difference:

$$HF(g) \xrightarrow{H_2O} HF(aq) \rightleftharpoons H^+(aq) + F^-(aq) \qquad (16.4)$$

The double arrow shows that the ionization process is reversible (Section 14.4). Its greater length from right to left says it is much more likely for hydrogen and fluoride ions to form dissolved hydrogen fluoride molecules than for the molecules to form ions. In other words, the dissolved particles in hydrofluoric acid are mostly unionized HF molecules, and a relatively small number of ions. This is why the solution is a poor conductor.

In describing acids, the words *strong* and *weak* are used exactly the same way they are used in describing electrolytes. A strong acid is almost completely ionized in water, and a weak acid is ionized only slightly. *The solution inventory of a strong acid is the ions it forms; the solution inventory of a weak acid is the unionized molecule.*

Now, how do you determine if an acid is strong or weak? The answer is, by knowing the acids that are strong. We will consider only six:

HCl	hydrochloric acid	HNO_3	nitric acid
HBr	hydrobromic acid	H_2SO_4	sulfuric acid
HI	hydroiodic acid	$HClO_4$	perchloric acid

These acids are easily memorized. HCl, HNO_3, and H_2SO_4 are the three best known acids. HBr and HI are, like HCl, made up of hydrogen and a halogen. $HClO_4$ is not widely discussed in an introductory course; it simply must be remembered as a strong acid. In classifying an acid, you first check to see if it is one of the strong acids. If not, it must be weak.*

EXAMPLE 16.3 Write the solution inventories of the following acids: HNO_2; H_2SO_4; HI; $HC_3H_5O_2$.

You could write these inventories with equations, as in Example 16.1. It is better just to identify the ions—or molecules—as in Example 16.2. If a molecule produces more than one ion of a kind, show it. Use state symbols.

HNO_2: HI:

H_2SO_4: $HC_3H_5O_2$:

HNO_2: $HNO_2(aq)$ HI: $H^+(aq) + I^-(aq)$

H_2SO_4: $2\ H^+(aq) + SO_4^{2-}(aq)$ $HC_3H_5O_2$: $HC_3H_5O_2(aq)$

HNO_2 and $HC_3H_5O_2$ are not among the six strong acids, so their solution inventories are unionized molecules. HI and H_2SO_4 are strong acids; they break up into ions. A common mistake with H_2SO_4 is to write $H_2^+(aq)$ or something like that for the hydrogen ion, $H^+(aq)$. Like the chloride ion in Equations 16.1 and 16.2, the hydrogen ion has the same formula no matter where it comes from.

Quick Check 16.3

What is the difference between a strong acid and a weak acid? How do you identify a weak acid? What two classes of solutes contain ions in their solution inventories?

*The dividing line between strong and weak acids is arbitrary, and at least three acids are marginal in their classifications. Sulfuric acid is definitely strong in its first ionization step, but questionable in the second. In all reactions in this book the second ionization does occur, so we regard it as a strong acid that releases both hydrogen ions. Both oxalic acid, $H_2C_2O_4$, and phosphoric acid, H_3PO_4, are marginal in their first ionizations, but definitely weak in the second and, in the case of phosphoric acid, third. We regard them as weak, but we will avoid questions in which they might have to be classified.

16.4 NET IONIC EQUATIONS: WHAT THEY ARE AND HOW TO WRITE THEM

If a solution of lead nitrate is added to a solution of sodium chloride, lead chloride precipitates. In Chapter 8 you learned how to write the equation for a precipitation reaction. In this case it is

$$Pb(NO_3)_2(aq) + 2\ NaCl(aq) \rightarrow PbCl_2(s) + 2\ NaNO_3(aq) \qquad (16.5)$$

In this chapter we will identify this kind of equation as a *conventional* equation.

A conventional equation serves many useful purposes, including its essential role in solving stoichiometry problems. For a reaction that occurs in water solutions, however, it is not always satisfactory. Usually it does not identify the reactants and/or products correctly, and it generally does not tell exactly what chemical changes takes place. A conventional equation has no value when it comes to solving chemical equilibrium problems, such as those in Chapter 19 of this book.

To illustrate the shortcomings of a conventional equation, the lead nitrate solution in the above reaction contains no substance whose formula is $Pb(NO_3)_2$. Actually present in that solution are the solution inventory species, lead ion, $Pb^{2+}(aq)$, and nitrate ion, $NO_3^-(aq)$. Similarly, the sodium chloride solution really contains $Na^+(aq)$ and $Cl^-(aq)$. The reaction does not produce anything whose formula is $NaNO_3$. The ions, $Na^+(aq)$ and $NO_3^-(aq)$, are present in the solution inventory of the combined solutions. The only substance in Equation 16.5 that is *really there* is solid lead chloride, $PbCl_2(s)$. If you perform the reaction you can see the precipitate.

To write an equation that describes the above reaction more accurately, the formulas of the dissolved substances are replaced with their solution inventories. This produces the **ionic equation:**

$$Pb^{2+}(aq) + 2\ NO_3^-(aq) + 2\ Na^+(aq) + 2\ Cl^-(aq) \rightarrow PbCl_2(s) + 2\ Na^+(aq) + 2\ NO_3^-(aq) \qquad (16.6)$$

Notice that, in order to keep the ionic equation balanced, the coefficients from the conventional equation are repeated. One formula unit of $Pb(NO_3)_2$ gives one Pb^{2+} ion and two NO_3^- ions, two formula units of NaCl give two Na^+ ions and two Cl^- ions, and two formula units of $NaNO_3$ give two Na^+ ions and two NO_3^- ions.

An ionic equation tells more than just what happens in a chemical change. It includes **spectator ions,** or simply **spectators.** A spectator is an ion that is present at the scene of a reaction, but experiences no chemical change. It appears on both sides of the ionic equation. $Na^+(aq)$ and $NO_3^-(aq)$ are spectators in Equation 16.6. To change an ionic equation into a **net ionic equation** you simply remove the spectators:

$$Pb^{2+}(aq) + 2\ Cl^-(aq) \rightarrow PbCl_2(s) \qquad (16.7)$$

A net ionic equation tells exactly what chemical change took place, and nothing else.

In Equations 16.5 to 16.7 you have the three steps to be followed in writing a net ionic equation. They are:

1. Write the conventional equation, including designations of state—(g), (ℓ), (s), or (aq). Balance the equation.
2. Write the ionic equation by replacing each *dissolved* substance, designated (aq) in the conventional equation, with its solution inventory species. Never change solids (s), liquids (ℓ), or gases (g) in this step. Be sure the equation is balanced in both atoms and charge. (Charge balance is discussed in the next section.)
3. Write the net ionic equation by removing the spectators from the ionic equation. Reduce coefficients to lowest terms, if necessary. Be sure the equation is balanced in both atoms and charge.

Quick Check 16.4
After writing a conventional equation, including state designations (s), (ℓ), (g), or (aq), which species do you consider breaking into ions in writing the ionic equation? Of those, which *do* you break into ions, and which do you *not* break into ions?

16.5 REDOX REACTIONS THAT ARE DESCRIBED BY "SINGLE REPLACEMENT" EQUATIONS

PG 16F Given two substances that may engage in a redox reaction and an activity series by which the reaction may be predicted, write the conventional, ionic, and net ionic equations for the reaction that will occur, if any.

A piece of zinc is dropped into sulfuric acid. Hydrogen gas bubbles out vigorously. When the reaction ends, the vessel contains a solution of zinc sulfate. The question is, what happened?

To "answer" the question by the conventional equation (Step 1):

$$Zn(s) + H_2SO_4(aq) \rightarrow H_2(g) + ZnSO_4(aq)$$

But this doesn't really answer the question. There were no H_2SO_4 molecules in the sulfuric acid. Instead, each H_2SO_4 molecule released two $H^+(aq)$ ions and one $SO_4^{2-}(aq)$ ion. At the end of the reaction there were no white $ZnSO_4$ crystals present. There was only a clear colorless solution of $Zn^{2+}(aq)$ and $SO_4^{2-}(aq)$ ions. So the ionic equation (Step 2), in which the dissolved substances in the conventional equation are replaced by their solution inventories, provides a better explanation of what happened:

$$Zn(s) + 2\ H^+(aq) + SO_4^{2-} \rightarrow H_2(g) + Zn^{2+} + SO_4^{2-} \qquad (16.8)$$

The ionic equation remains balanced in both atoms and charge. The net charge is zero on both sides.

At this point the answer to what happened is cluttered with something that was present, but was no part of the chemical change. The sulfate ion, SO_4^{2-}, appears on both sides of the ionic equation. It is a spectator. It may be taken away (Step 3):

$$Zn(s) + 2\ H^+(aq) \rightarrow H_2(g) + Zn^{2+}(aq) \qquad (16.9)$$

This is the net ionic equation. This is what happened—and no more.

Notice that all equations are balanced.* You have been balancing atoms since Chapter 8, but charge is something new. Neither protons nor electrons are created or destroyed in a chemical change, so the total charge among the reactants must be equal to the total charge among the products. In Equation 16.8, two plus charges in $2\ H^+(aq)$ added to two negative charges in $SO_4^{2-}(aq)$ give a net zero charge on the left. Two plus charges in $Zn^{2+}(aq)$ and two negative charges in $SO_4^{2-}(aq)$ on the right also total zero. The equation is balanced in charge. Equation 16.9 is also balanced, this time with a net charge of +2 on each side. Notice that the net charge does not have to be zero for the equation to be balanced.

EXAMPLE 16.4 A reaction occurs when a piece of zinc is dipped into $Cu(NO_3)_2(aq)$. Write the conventional, ionic, and net ionic equations. See Color Plate 3 opposite page 85.

Begin by writing the conventional equation. Do you know what the products are? What might they be? You know that the equation for this kind of redox reaction is a single replacement type (Section 8.7). Something must replace something else in a compound. Don't forget the state designations.

----- -----

$$Zn(s) + Cu(NO_3)_2(aq) \rightarrow Cu(s) + Zn(NO_3)_2(aq)$$

In the single replacement equation, zinc appears to replace copper in $Cu(NO_3)_2$.

Now write the ionic equation, replacing formulas of dissolved substances with their solution inventories.

----- -----

$$Zn(s) + Cu^{2+}(aq) + 2\ NO_3^-(aq) \rightarrow Cu(s) + Zn^{2+}(aq) + 2\ NO_3^-(aq)$$

The ionic equation will always be balanced if the solution inventories are written with coefficients to show how many ions come from all formula units. You should always check, though. The total charge is zero on both sides.

Examine the ionic equation. Are there any spectators? If so, remove them and write the net ionic equation.

----- -----

$$Zn(s) + Cu^{2+}(aq) \rightarrow Cu(s) + Zn^{2+}(aq)$$

*If the purpose is simply to write the net ionic equation, it is not necessary to balance the conventional and ionic equations. Most instructors, however, recommend balancing all three equations, both in atoms and charge.

Table 16.1 Activity Series

Li
K
Ba
Ca
Na
Mg
Al
Zn
Ni
Pb
H_2
Cu
Ag

If you were to place a strip of copper into a solution of zinc nitrate and go through the identical thought process, the equations would be exactly the reverse of those in Example 16.4. Does the reaction occur in both directions? If not, in which way does it occur? And how can you tell?

Technically both reactions occur, or can be made to occur. Practically, only the reaction in Example 16.4 takes place. The best way to find out which of two reversible reactions occurs is to try them and see. These experiments have been done, and the results are summarized in the **activity series** in Table 16.1. In solution, under normal conditions, any element in the table will replace the ions of any element beneath it. Zinc is above copper in the table, so zinc will replace $Cu^{2+}(aq)$ ions in solution. Copper, being below zinc, will not replace $Zn^{2+}(aq)$ in solution. You may use Table 16.1 to predict whether or not a redox reaction will take place. If asked to write the equation for a reaction that does not occur, write NR for "no reaction" on the product side: $Cu(s) + Zn^{2+}(aq) \rightarrow NR$.

EXAMPLE 16.5 Write the conventional, ionic, and net ionic equations for the reaction that will occur, if any, between calcium and hydrochloric acid.

First, will a reaction occur? Check the activity series; then answer.

----- -----

Yes. Calcium is above hydrogen in the series, so calcium will replace hydrogen ions from solution.

Write the conventional equation. Remember the state designation.

----- -----

$Ca(s) + 2\ HCl(aq) \rightarrow H_2(g) + CaCl_2(aq)$

Now write the ionic equation.

----- -----

$Ca(s) + 2\ H^+(aq) + 2\ Cl^-(aq) \rightarrow H_2(g) + Ca^{2+} + 2\ Cl^-(aq)$

Each HCl(aq) yields one $H^+(aq)$ and one $Cl^-(aq)$. The conventional equation has two HCl(aq), so there will be 2 $H^+(aq)$ and 2 $Cl^-(aq)$ in the ionic equation.

Now eliminate the spectators and write the net ionic equation.

----- -----

$Ca(s) + 2\ H^+(aq) \rightarrow H_2(g) + Ca^{2+}(aq)$

Chloride ion, $Cl^-(aq)$, is the only spectator.

EXAMPLE 16.6 Write the three equations for the reaction between nickel and a solution of $MgCl_2$, if any.

See what you can do this time without hints.

$$Ni(s) + MgCl_2(aq) \rightarrow NR$$

Nickel is beneath magnesium in the activity series, so no redox reaction occurs.

EXAMPLE 16.7 Copper is placed into a solution of silver nitrate. Write the three equations.

$$Cu(s) + 2\ AgNO_3(aq) \rightarrow 2\ Ag(s) + Cu(NO_3)_2(aq)$$

$$Cu(s) + 2\ Ag^+(aq) + 2\ NO_3^-(aq) \rightarrow 2\ Ag(s) + Cu^{2+}(aq) + 2\ NO_3^-(aq)$$

$$Cu(s) + 2\ Ag^+(aq) \rightarrow 2\ Ag(s) + Cu^{2+}(aq)$$

EXAMPLE 16.8 If potassium is placed into water, hydrogen gas bubbles out. Write all three equations.

This reaction is more clearly seen if you write the formula of water as HOH. Treat it as a weak acid, with only the first hydrogen ionizable. Go for all three equations, but watch those state symbols.

$$2\ K(s) + 2\ HOH(\ell) \rightarrow H_2(g) + 2\ KOH(aq)$$

$$2\ K(s) + 2\ HOH(\ell) \rightarrow H_2(g) + 2\ K^+(aq) + 2\ OH^-(aq)$$

Water is a liquid molecular compound, $HOH(\ell)$. It does not break into ions. Never separate into ions a gas (g), liquid (ℓ), or solid (s). This particular ionic equation has no spectators, so it is also the net ionic equation.

16.6 ION COMBINATIONS THAT FORM PRECIPITATES

PG 16G Predict whether or not a precipitate will form when known solutions are combined; if a precipitate forms, write the net ionic equation. (Reference to a solubility table may or may not be allowed.)

16H Given the product of a precipitation reaction, write the net ionic equation.

An **ion-combination reaction** occurs when the cation from one reactant combines with the anion from another to form a particular kind of product compound. The conventional equation is a double displacement type in which the ions appear to "change partners" (Section 8.7): $MY + NX \rightarrow MX + NY$. In this section the product is an insoluble ionic compound that settles to the bottom of the mixed solutions. A solid formed this way is called a **precipitate;** the reaction is a **precipitation reaction.**

EXAMPLE 16.9 When hydrochloric acid and silver nitrate solutions are mixed, a white precipitate of silver chloride is produced. Develop the net ionic equation for the reaction.

The same three steps you used on redox reactions are applied to precipitation reactions. Write all three equations.

$$HCl(aq) + AgNO_3(aq) \rightarrow HNO_3(aq) + AgCl(s)$$

$$H^+(aq) + Cl^-(aq) + Ag^+(aq) + NO_3^-(aq) \rightarrow H^+(aq) + NO_3^-(aq) + AgCl(s)$$

$$Ag^+(aq) + Cl^-(aq) \rightarrow AgCl(s)$$

$H^+(aq)$ and $NO_3^-(aq)$ are spectators in the ionic equation.

Let's try another that has two interesting features at the end.

EXAMPLE 16.10 When solutions of silver chlorate, $AgClO_3(aq)$, and aluminum chloride, $AlCl_3(aq)$, are combined, silver chloride precipitates. Write the three equations.

Proceed as usual. When you get to the net ionic equation, which will have something you haven't seen so far, see if you can figure out what to do about it. If necessary, read Step 3 of the procedure in Section 16.4.

$$3\ AgClO_3(aq) + AlCl_3(aq) \rightarrow 3\ AgCl(s) + Al(ClO_3)_3(aq)$$

$$3\ Ag^+(aq) + 3\ ClO_3^-(aq) + Al^{3+}(aq) + 3\ Cl^-(aq) \rightarrow 3\ AgCl(s) + Al^{3+}(aq) + 3\ ClO_3^-(aq)$$

$$3\ Ag^+(aq) + 3\ Cl^-(aq) \rightarrow 3\ AgCl(s);$$

$$Ag^+(aq) + Cl^-(aq) \rightarrow AgCl(s)$$

The net ionic equation this time has 3 for the coefficient of all species. The third step of the procedure says to reduce coefficients to lowest terms. The equation may be divided by 3, as shown. That leads to the second interesting feature.

Examples 16.9 and 16.10 produced the same net ionic equation even though the reacting solutions were completely different. Actually only the spectators were different, but they are not part of the chemical change. Getting them out of the way shows that both reactions are exactly the same. It will be this way whenever a solution containing $Ag^+(aq)$ ion is added to a solution containing $Cl^-(aq)$ ion. Therefore, if asked to write the net ionic equation for the reaction between such solutions, you can go directly to $Ag^+(aq) + Cl^-(aq) \rightarrow AgCl(s)$. (It is possible that a second reaction may occur between the other pair of ions, yielding a second net ionic equation.)

This simple and direct procedure may be used to write the net ionic equation for the precipitation of any insoluble ionic compound. The compound is the product, and the reactants are the ions in the compound. It is like writing a solution inventory equation (Equations 16.1 and 16.2, and Example 16.1) in reverse. Try it on the following:

EXAMPLE 16.11 Write the net ionic equations for the precipitation of the following from aqueous solutions:

CuS:

$Mg(OH)_2$:

Li_3PO_4:

$$Cu^{2+}(aq) + S^{2-}(aq) \rightarrow CuS(s)$$

$$Mg^{2+}(aq) + 2\ OH^-(aq) \rightarrow Mg(OH)_2(s)$$

$$3\ Li^+(aq) + PO_4^{3-}(aq) \rightarrow Li_3PO_4(s)$$

If we knew in advance those combinations of ions that yield insoluble compounds, we could predict precipitation reactions. These compounds have been identified in the laboratory. Table 16.2 shows the result of such experiments for a large number of ionic compounds. These solubilities have also been summarized in a set of "solubility rules" that your instructor may ask you to memorize. These rules are in Table 16.3.

Table 16.2 Solubilities of Ionic Compounds*

IONS	Acetate	Bromide	Carbonate	Chlorate	Chloride	Fluoride	Hydrogen Carbonate	Hydroxide	Iodide	Nitrate	Nitrite	Phosphate	Sulfate	Sulfide	Sulfite
Aluminum	I	S		S	S	I		I	—	S		I	S	—	
Ammonium	S	S	S	S	S	S	S	—	S	S	S	S	S	S	S
Barium	—	S	I	S	S	I		S	S	S	S	I	I	—	I
Calcium	S	S	I	S	S	I		I	S	S	S	I	I	—	I
Cobalt(II)	S	S	I	S	S	—		I	S	S		I	S	I	I
Copper(II)	S	S			S	S		I		S		I	S	I	
Iron(II)	S	S	I		S	I		I	S	S		I	S	I	I
Iron(III)	—	S			S	I		I	S	S		I	S	—	
Lead(II)	S	I	I	S	I	I		I	I	S	S	I	I	I	I
Lithium	S	S	S	S	S	S	S	S	S	S	S	I	S	S	
Magnesium	S	S	I	S	S	I		I	S	S	S	I	S	—	S
Nickel		S	I	S	S	S		I	S	S		I	S	I	I
Potassium	S	S	S	S	S	S	S	S	S	S	S	S	S	S	S
Silver	I	I	I	S	I	S		—	I	S	I	I	I	I	I
Sodium	S	S	S	S	S	S	S	S	S	S	S	S	S	S	S
Zinc	S	S	I	S	S	S		I	S	S		I	S	I	I

*Compounds having solubilities of 0.1 M or more at 20°C are listed as soluble (S); if the solubility is less than 0.1 M, the compound is listed as insoluble (I). A dash (—) identifies an unstable species in aqueous solution, and a blank space indicates lack of data. In writing equations for reactions that occur in water solution, insoluble substances, shown by I in this table, have the state symbol for a solid, (s). Dissolved substances, S in the table, are designated by (aq) in an equation.

In the next four examples use either table to predict precipitation reactions. We recommend that you use each table at least once.

EXAMPLE 16.12 Solutions of lead(II) nitrate and sodium fluoride are combined. Write the net ionic equation for any precipitation reaction that may occur.

Start with the conventional equation.

----- -----

$Pb(NO_3)_2(aq) + 2\ NaF(aq) \rightarrow PbF_2(\quad) + 2\ NaNO_3(\quad)$

The spaces between parentheses after the products have been left blank. You must determine if the compound is soluble or insoluble. If soluble, the state symbol should be (aq); if insoluble, (s). You will have to use Table 16.2 to find out about the solubility of lead(II) fluoride; fluoride solubilities are not easily included in solubility rules. The solubility of sodium nitrate is readily found from either table. Fill in the state symbols.

----- -----

$Pb(NO_3)_2(aq) + 2\ NaF(aq) \rightarrow PbF_2(s) + 2\ NaNO_3(aq)$

In Table 16.2 the intersection of the lead(II) ion line and the fluoride column shows that lead(II) fluoride is insoluble, so the designation (s) follows PbF_2. The intersection of the sodium ion line and nitrate ion column shows that sodium nitrate is soluble in water. This is confirmed by the solubility rules, which indicate that all sodium compounds and all nitrates are soluble. The designation (aq) therefore follows $NaNO_3$.

Complete the example by writing the ionic and net ionic equations.

----- -----

$$Pb^{2+}(aq) + 2\ NO_3^-(aq) + 2\ Na^+(aq) + 2\ F^-(aq) \rightarrow PbF_2(s) + 2\ Na^+(aq) + 2\ NO_3^-(aq)$$

$$Pb^{2+}(aq) + 2\ F^-(aq) \rightarrow PbF_2(s)$$

Table 16.3 Solubility Rules for Ionic Compounds*

The following compounds are **Soluble**	**Except**	*The following compounds are* **Insoluble**	**Except**
Ammonium salts		Carbonates	NH_4^+ and alkali metals
Alkali metal salts (Column 1A)	Some Li^+	Phosphates	Na^+, K^+, and NH_4^+
		Hydroxides	Ba^{2+} and alkali metals
Nitrates		Sulfides	NH_4^+ and salts of alkali metals
Halides (Column 7A)	Ag^+, Hg_2^{2+}, Pb^{2+}		
Acetates	Ag^+, Al^{3+}		
Chlorates			
Sulfates	Ba^{2+}, Sr^{2+}, Ca^{2+}, Pb^{2+}, Ag^+, Hg^{2+}, Hg_2^{2+}		

*For purposes of these rules, a compound is considered to be soluble if it dissolves to a concentration of 0.1 M or more at 20°C.

EXAMPLE 16.13 Write the net ionic equation for any reaction that will occur when solutions of nickel chloride and ammonium carbonate are combined.

This example does not ask for all three equations, but only the net ionic equation. You might wish to try for the net ionic equation without the other two. Can you, simply by looking at the names of the reactants, identify the names of the products as the ions rearrange themselves? If so, determine from Table 16.2 or 16.3 which, if either, is insoluble. For any insoluble compound you can write the net ionic equation directly, as in Example 16.11. If you are not ready for this step, write all three equations, as before. The three steps are shown in the answer.

----- -----

$$NiCl_2(aq) + (NH_4)_2CO_3(aq) \rightarrow NiCO_3(s) + 2\ NH_4Cl(aq)$$

$$Ni^{2+}(aq) + 2\ Cl^-(aq) + 2\ NH_4^+(aq) + CO_3^{2-}(aq) \rightarrow NiCO_3(s) + 2\ NH_4^+(aq) + 2\ Cl^-(aq)$$

$$Ni^{2+}(aq) + CO_3^{2-}(aq) \rightarrow NiCO_3(s)$$

EXAMPLE 16.14 Write the net ionic equation for any reaction that occurs between solutions of aluminum sulfate and calcium acetate, $Al_2(SO_4)_3(aq)$ and $Ca(C_2H_3O_2)_2(aq)$.

Be careful here!

----- -----

$$Al_2(SO_4)_3(aq) + 3\ Ca(C_2H_3O_2)_2(aq) \rightarrow 3\ CaSO_4(s) + 2\ Al(C_2H_3O_2)_3(s)$$

$$2\ Al^{3+}(aq) + 3\ SO_4^{2-}(aq) + 3\ Ca^{2+}(aq) + 6\ C_2H_3O_2^-(aq) \rightarrow 3\ CaSO_4(s) + 2\ Al(C_2H_3O_2)_3(s)$$

This time *both* new combinations of ions precipitate. There are no spectators, so the ionic equation is the net ionic equation. Actually there are two separate reactions taking place at the same time:

$$Al^{3+}(aq) + 3\ C_2H_3O_2^-(aq) \rightarrow Al(C_2H_3O_2)_3(s)$$

$$Ca^{2+}(aq) + SO_4^{2-}(aq) \rightarrow CaSO_4(s)$$

EXAMPLE 16.15 Write the net ionic equation for any reaction that occurs when solutions of ammonium sulfate and potassium nitrate are combined.

Again, be careful.

$$(NH_4)_2SO_4(aq) + KNO_3(aq) \rightarrow NR$$

This time both new combinations of ions are soluble, so there is no precipitation reaction. If you complete the conventional equation, as you would have done in Chapter 8, and from it write the ionic equation, you will find that *all* ions are spectators.

PG 16I Given reactants that yield a molecular product, write the net ionic equation.

The reaction of an acid often leads to an ion combination that yields a molecular product instead of an insoluble ionic compound. Except for the difference in the product, the equations are written in exactly the same way. Just as you had to recognize an insoluble product and not break it up in the ionic equations, you must now recognize a molecular product and not break it into ions. Water or a weak acid are the two kinds of molecular products you will find.

The neutralization reaction between an acid and a hydroxide to produce water and a salt (Section 8.7) is the most common molecular product reaction.

EXAMPLE 16.16 Write the conventional, ionic, and net ionic equations for the reaction between hydrochloric acid and a solution of sodium hydroxide.

Proceed just as you did for precipitation reactions. Watch your state designations.

$$HCl(aq) + NaOH(aq) \rightarrow HOH(\ell) + NaCl(aq)$$

$$H^+(aq) + Cl^-(aq) + Na^+(aq) + OH^-(aq) \rightarrow HOH(\ell) + Na^+(aq) + Cl^-(aq)$$

$$H^+(aq) + OH^-(aq) \rightarrow HOH(\ell)$$

Water is the molecular product. It is unionized and is in the liquid state.

The acid in a neutralization may be a weak acid. You must then recall that the solution inventory of a weak acid is the acid molecule; it is not broken into ions.

EXAMPLE 16.17 Write the three equations leading to the net ionic equation for the reaction between acetic acid, $HC_2H_3O_2(aq)$, and a solution of sodium hydroxide.

$$HC_2H_3O_2(aq) + NaOH(aq) \rightarrow HOH(\ell) + NaC_2H_3O_2(aq)$$

$$HC_2H_3O_2(aq) + Na^+(aq) + OH^-(aq) \rightarrow HOH(\ell) + Na^+(aq) + C_2H_3O_2^-(aq)$$

$$HC_2H_3O_2(aq) + OH^-(aq) \rightarrow HOH(\ell) + C_2H_3O_2^-(aq)$$

The weak acid molecule appears in its molecular form in the net ionic equation.

When the reactants are a strong acid and the salt of a weak acid, that weak acid is formed as the molecular product. You must recognize it as a weak acid and leave it in molecular form in the solution inventory.

EXAMPLE 16.18 Develop the net ionic equation for the reaction between hydrochloric acid and a solution of sodium acetate, $NaC_2H_3O_2(aq)$.

$$HCl(aq) + NaC_2H_3O_2(aq) \rightarrow HC_2H_3O_2(aq) + NaCl(aq)$$

$$H^+(aq) + Cl^-(aq) + Na^+(aq) + C_2H_3O_2^-(aq) \rightarrow HC_2H_3O_2(aq) + Na^+(aq) + Cl^-(aq)$$

$$H^+(aq) + C_2H_3O_2^-(aq) \rightarrow HC_2H_3O_2(aq)$$

Compare the reactions in Examples 16.16 and 16.18. The only difference between them is that Example 16.16 has the hydroxide ion as a reactant and Example 16.18 has the acetate ion as a reactant. In the first case the molecular product is water, formed when the hydrogen ion bonds to the hydroxide ion. In the second case the molecular product is acetic acid, a weak acid, formed when the hydrogen ion bonds to the acetate ion. Actually water is a weak acid too, weaker than acetic acid or any other acid that dissolves in water. "Weak," you recall, means that the acid does not ionize to a large extent. It is because water ionizes less than acetic acid that weak acids can neutralize hydroxides, as in Example 16.17.

Just as you can write the net ionic equation for a precipitation reaction without the conventional and ionic equations, so you can write the net ionic equation for a molecule formation reaction. Again you must recognize the product from the formulas of the reactants. The acid will contribute a hydrogen ion to the molecular product. It will form a molecule with the anion from the other reactant. If the molecule is water or a weak acid, you have the reactants and products of the net ionic equation.

EXAMPLE 16.19 Without writing the conventional and ionic equations, write the net ionic equations for each of the following pairs of reactants:

H_2SO_4 and LiOH:

KNO_2 and HBr:

$H^+(aq) + OH^-(aq) \rightarrow HOH(\ell)$ $\quad$ $H^+(aq) + NO_2^-(aq) \rightarrow HNO_2(aq)$

$Li^+(aq)$ and $SO_4^{2-}(aq)$ are spectators in the first equation for neutralization reaction. HNO_2 is recognized as a weak acid in the second reaction because it is not one of the six strong acids. $K^+(aq)$ and $Br^-(aq)$ are spectators.

There are two points by which you identify a molecular product reaction: (1) one reactant is an acid, usually strong; (2) one product is water or a weak acid. One of the most common mistakes in writing net ionic equations is the failure to recognize a weak acid as a molecular product. If one reactant is a strong acid, you can be sure there will be a molecular product. If it isn't water, look for a weak acid.

16.8 ION COMBINATIONS THAT FORM UNSTABLE PRODUCTS

PG 16J Given reactants that form H_2CO_3, H_2SO_3, or "NH_4OH" by ion combination, write the net ionic equation for the reaction.

Three ion combinations yield molecular products that are not the products you would expect. Two of the expected products are carbonic and sulfurous acids. If hydrogen ions from one reactant reach carbonate ions from another, H_2CO_3, carbonic acid, should form:

$$2\ H^+(aq) + CO_3^{2-}(aq) \rightarrow H_2CO_3(aq)$$

But carbonic acid is unstable and decomposes to carbon dioxide gas and water. The correct net ionic equation is therefore

$$2\ H^+(aq) + CO_3^{2-}(aq) \rightarrow CO_2(g) + H_2O(\ell)$$

Sulfurous acid, H_2SO_3, decomposes in the same way to sulfur dioxide and water, but the sulfur dioxide remains in solution:

$$2\ H^+(aq) + SO_3^{2-}(aq) \rightarrow SO_2(aq) + H_2O(\ell)$$

The third ion combination that yields unexpected molecular products occurs when ammonium and hydroxide ions meet:

$$NH_4^+(aq) + OH^-(aq) \rightarrow \text{``}NH_4OH\text{''}$$

In spite of printed labels and laboratory bottles with NH_4OH etched on them, and wide use of the name "ammonium hydroxide," no substance having the formula NH_4OH exists at ordinary temperatures. The actual product is a solution of ammonia molecules, $NH_3(aq)$. The proper net ionic equation is therefore

$$NH_4^+(aq) + OH^-(aq) \rightarrow NH_3(aq) + H_2O(\ell)$$

The reaction is reversible, and actually reaches an equilibrium in which the solution inventory is overwhelmingly dissolved ammonia molecules.

There is no system by which these three "different" molecular product reactions can be recognized. You simply must be alert to them and catch them when they appear. Once again, the predicted but unstable formulas are H_2CO_3, H_2SO_3, and NH_4OH.

EXAMPLE 16.20 Write the conventional, ionic, and net ionic equations for the reaction between solutions of sodium carbonate and hydrochloric acid.

$$2\ HCl(aq) + Na_2CO_3(aq) \rightarrow 2\ NaCl(aq) + CO_2(g) + H_2O(\ell)$$

$$2\ H^+(aq) + 2\ Cl^-(aq) + 2\ Na^+(aq) + CO_3^{2-}(aq) \rightarrow 2\ Na^+(aq) + 2\ Cl^-(aq) + CO_2(g) + H_2O(\ell)$$

$$2\ H^+(aq) + CO_3^{2-}(aq) \rightarrow CO_2(g) + H_2O(\ell)$$

16.9 ION-COMBINATION REACTIONS WITH UNDISSOLVED SOLUTES

In every ion-combination reaction considered so far, it has been assumed that both reactants are in solution. This is not always the case. Sometimes the description of the reaction will indicate that a reactant is a solid, liquid, or gas, even though it may be soluble in water. In such a case write the correct state symbol after the formula in the conventional equation and carry the formula through all three equations unchanged.

A common example of this kind of reaction occurs with a compound that the handbooks say is insoluble in water but soluble in acids. The net ionic equation shows why.

EXAMPLE 16.21 Write the net ionic equation to describe the reaction when solid aluminum hydroxide dissolves in hydrochloric acid, nitric acid, and sulfuric acid.

This is a neutralization reaction between a strong acid and a solid hydroxide. Write the conventional equation for hydrochloric acid only.

----- -----

$$3\ HCl(aq) + Al(OH)_3(s) \rightarrow 3\ HOH(\ell) + AlCl_3(aq)$$

Now write the ionic and net ionic equations. Remember that you replace only dissolved substances, designated (aq), with their solution inventories.

----- -----

$$3\ H^+(aq) + 3\ Cl^-(aq) + Al(OH)_3(s) \rightarrow 3\ HOH(\ell) + Al^{3+}(aq) + 3\ Cl^-(aq)$$
$$3\ H^+(aq) + Al(OH)_3(s) \rightarrow 3\ HOH(\ell) + Al^{3+}(aq)$$

Now think a bit before writing the net ionic equation for the reaction between solid aluminum hydroxide and nitric acid. The question asks only for the net ionic equation. Can you write it directly, without the conventional and ionic equations? If not, write all three equations.

----- -----

$$3\ H^+(aq) + Al(OH)_3(s) \rightarrow 3\ HOH(\ell) + Al^{3+}(aq)$$

This is the same as the equation for hydrochloric acid. Sulfuric acid will also produce the same equation. The chloride, nitrate, and sulfate ions are spectators in the three reactions.

16.10 SUMMARY

In this chapter you have learned how to write net ionic equations for three kinds of reactions. The process is simpler if you recognize the kind of reaction and know what to expect. This summary should help:

If the reactants are an *element* and a *compound,* write the single replacement equation for what must be a redox reaction. Proceed with solution inventories, the ionic equation, and the net ionic equation as you learned in Section 16.5.

If the reactants are *two ionic compounds,* write the double displacement equation for what must be an ion-combination–precipitation reaction. Identify the insoluble product and proceed as in Section 16.6.

If the reactants are an *acid* and an *ionic compound,* write the double displacement equation for what must be an ion-combination–molecule-formation reaction. The hydrogen-bearing product is the unionized molecule; it will be water or a weak acid. Proceed as in Section 16.7.

There are four pitfalls to avoid. They are:

1. Be sure your reactant states are correct. If the question says a reactant is a solid, liquid, or gas, it must appear as such in the net ionic equation, even if it is soluble in water.
2. Be sure there is a reaction. Check the activity series to be sure a possible redox reaction will occur. Check ionic products to be sure one will precipitate. Check molecular products to be sure one is water or a *weak* acid.
3. Watch for double reactions. A double replacement equation can have two insoluble products, or an insoluble product and an unionized molecule.
4. Watch for the unstable ion-combination products H_2CO_3, H_2SO_3, or "NH_4OH." If you have one, proceed as in Section 16.8.

CHAPTER 16 IN REVIEW

16.1 Electrolytes and Solution Conductivity
- **16A** Distinguish among strong electrolytes, weak electrolytes, and nonelectrolytes.
- **16B** Describe or explain electrical conductivity through a solution.

16.2 Solution Inventories of Ionic Compounds
- **16C** Given the formula of an ionic compound, write the solution inventory when it is dissolved in water.

16.3 Strong Acids and Weak Acids
- **16D** Explain why the solution of an acid may be a good conductor or a poor conductor of electricity.
- **16E** Given the formula of a soluble acid, write the solution inventory when it is dissolved in water.

16.4 Net Ionic Equations: What They Are and How to Write Them

16.5 Redox Reactions That Are Described by "Single Replacement" Equations
- **16F** Given two substances that may engage in a redox reaction and an activity series by which the reaction may be predicted, write the conventional, ionic, and net ionic equations for the reaction that will occur, if any.

16.6 Ion Combinations That Form Precipitates
- **16G** Predict whether or not a precipitate will form when known solutions are combined; if a precipitate forms, write the net ionic equation. (Reference to a solubility table may or may not be allowed.)
- **16H** Given the product of a precipitation reaction, write the net ionic equation.

16.7 Ion Combinations That Form Molecules
- **16I** Given reactants that yield a molecular product, write the net ionic equation.

16.8 Ion Combinations That Form Unstable Products
- **16J** Given reactants that form H_2CO_3, H_2SO_3, or "NH_4OH" by ion combination, write the net ionic equation for the reaction.

16.9 Ion-Combination Reactions with Undissolved Solutes

16.10 Summary

TERMS AND CONCEPTS

16.1 Electrode
Conductor, nonconductor
Electrolysis
Electrolyte: strong, weak, nonelectrolyte
Anode; cathode
Solution inventory

16.3 Strong acid; weak acid

16.4 Conventional equation
Ionic equation
Spectator ion; spectator
Net ionic equation

16.5 Activity series

16.6 Ion-combination reaction
Precipitate; precipitation reaction

Most of these terms and many others appear in the Glossary. Use the Glossary regularly.

QUESTIONS AND PROBLEMS

Section 16.1

(*1*) If solid solute A is a strong electrolyte and solute B is a nonelectrolyte, state how these solutes differ.

(2) How do you explain the passage of "electricity" through a solution? The ability of a solution to conduct is considered evidence of what?

(35) How does a weak electrolyte differ from a strong electrolyte?

(36) All soluble ionic compounds are electrolytes. A molecular solute may or may not be an electrolyte. Show how both of these statements are true.

For the next four questions in each column, write the solution inventory for the water solution of each substance given.

Section 16.2

(3) LiF; $Mg(NO_3)_2$; $FeCl_3$.

(4) NH_4NO_3; $Ca(OH)_2$; Li_2SO_3.

(37) Na_2S; $(NH_4)_2SO_4$; $MnCl_2$.

(38) $NiSO_4$; K_3PO_4; KNO_2.

Section 16.3

(5) $HCHO_2$; HCl; $HC_4H_4O_6$.

(6) HI; H_2SO_4; $H_3C_2H_5O_7$.

(39) HNO_3; $HC_2H_3O_2$; HBr.

(40) $HClO_4$; $H_2C_4H_4O_4$; HF.

Section 16.5

For each pair of reactants in the next three questions in each column, write the net ionic equation for any redox reaction that may be predicted by Table 16.1 (Section 16.5). If no redox reaction occurs, write NR.

(7) $Ba(s)$, $Zn(NO_3)_2(aq)$.

(8) $Pb(s)$, $Ca(NO_3)_2(aq)$.

(9) $Mg(s)$, $Al_2(SO_4)_3(aq)$.

(41) $Cu(s)$, $Li_2SO_4(aq)$.

(42) $Ba(s)$, $HCl(aq)$.

(43) $Ni(s)$, $AgNO_3(aq)$.

Section 16.6

For each pair of reactants given in the next four questions in each column, write the net ionic equation for any precipitation reaction that may be predicted by Tables 16.2 and 16.3 (Section 16.6). If no precipitation occurs, write NR.

(10) $BaCl_2(aq)$, $Na_2CO_3(aq)$.

(11) $CoSO_4(aq)$, $NaOH(aq)$.

(12) $FeCl_2(aq)$, $(NH_4)_2S(aq)$.

(13) $NiCl_2(aq)$, $CuSO_4(aq)$.

(14) Write the net ionic equations for the precipitation of each of the following insoluble ionic compounds: MgF_2; $Zn_3(PO_4)_2$.

(44) $Pb(NO_3)_2(aq)$, $KI(aq)$.

(45) $KClO_3(aq)$, $Mg(NO_2)_2(aq)$.

(46) $AgNO_3(aq)$, $LiBr(aq)$.

(47) $ZnCl_2(aq)$, $Na_2SO_3(aq)$.

(48) Write the net ionic equations for the precipitation of each of the following insoluble ionic compounds: $PbCO_3$; $Ca(OH)_2$.

Section 16.7

For each pair of reactants given in the next three questions in each column, write the net ionic equation for the molecule formation reaction that will occur.

(15) $NaC_6H_5O_7(aq)$, $HNO_3(aq)$.

(16) $Ca(OH)_2(aq)$, $HBr(aq)$.

(17) $KC_4H_4O_6(aq)$, $HCl(aq)$.

(49) $NaNO_2(aq)$, $HI(aq)$.

(50) $KC_3H_5O_3(aq)$, $HClO_4(aq)$.

(51) $RbOH(aq)$, $H_2SO_4(aq)$.

Section 16.8

For each pair of reactants given in the next two questions in each column, write the net ionic equation for the reaction that will occur.

(18) $K_2CO_3(aq)$, $HNO_3(aq)$.

(19) $LiOH(aq)$, $(NH_4)_2SO_4(aq)$.

(52) $(NH_4)_2SO_3(aq)$, $HBr(aq)$.

(53) $MgSO_3(aq)$, $H_3PO_4(aq)$.

The remaining questions include all types of reactions covered in this chapter. Use the suggestions in Section 16.10 to identify the reaction type, and then develop the net ionic equation. Use the activity series and solubility tables to predict whether or not redox or precipitation reactions will take place. If you find a "no reaction" combination, mark it NR.

(20) Silver nitrate solution is added to a solution of ammonium carbonate.

(21) Aluminum nitrate solution is added to potassium phosphate solution.

(22) Metallic calcium is dropped into a solution of silver nitrate.

(23) When combined in an oxygen-free atmosphere, sodium sulfite solution and hydrochloric acid react.

(24) Sulfuric acid neutralizes potassium hydroxide.

(25) A solution of potassium iodide is able to dissolve solid silver chloride and form a new precipitate.

(26) Lithium metal is immersed in a solution of copper(II) sulfate.

(27) Sodium benzoate solution, $NaC_7H_5O_2(aq)$, is treated with hydrochloric acid.

(28) Ammonium chloride solution is added to a solution of potassium hydroxide.

(29) A solution of sodium fluoride is poured into a solution of calcium nitrate.

(30) A piece of magnesium is dropped into a solution of zinc nitrate.

(31) Dilute nitric acid is poured over solid barium hydroxide.

(54) Barium chloride and sodium sulfate solutions are combined in an oxygen-free atmosphere.

(55) Copper(II) sulfate and sodium hydroxide solutions are combined.

(56) Carbon dioxide bubbles appear as hydrochloric acid is poured onto solid magnesium carbonate.

(57) Nitric acid is able to dissolve solid lead(II) hydroxide.

(58) Oxalic acid, $H_2C_2O_4(s)$, is neutralized by sodium hydroxide solution.

(59) What happens if barium is placed into hydrochloric acid?

(60) Hydrochloric acid is poured into a solution of sodium hydrogen sulfite.

(61) Solutions of magnesium sulfate and ammonium bromide are combined.

(62) Magnesium ribbon is placed in hydrochloric acid.

(63) Solid nickel hydroxide is readily dissolved by hydrobromic acid.

(64) Sodium fluoride solution is poured into nitric acid.

(65) Silver wire is dropped into hydrochloric acid.

(32) Sulfuric acid is able to dissolve copper(II) hydroxide.

(33) Aluminum reacts with a solution of sodium hydroxide to produce hydrogen gas and a solution of sodium aluminate, Na_3AlO_3.

(34) Consider H_3PO_4 to be a weak acid and write the net ionic equations for Equations 15.12 and 15.13, p. 342.

(66) When metallic lithium is added to water, hydrogen is released.

(67) Aluminum shavings are dropped into a solution of copper(II) nitrate.

(68) (a) Hydrochloric acid reacts with a solution of sodium hydrogen carbonate. (b) Hydrochloric acid reacts with a solution of sodium carbonate. (c) A limited amount of hydrochloric acid reacts with a solution of sodium carbonate, yielding $NaHCO_3(aq)$ as one product.

Miscellaneous Questions

(69) Distinguish precisely and in scientific terms the differences among items in each of the following groups:
- (a) Strong electrolyte, weak electrolyte
- (b) Electrolyte, nonelectrolyte
- (c) Anode, cathode
- (d) Strong acid, weak acid
- (e) Conventional, ionic, and net ionic equations
- (f) Ion combination, precipitation, and molecule formation reactions
- (g) Molecule formation and neutralization reactions
- (h) Acid, base, salt

(70) Classify each of the following statements as true or false:
- (a) Electrons carry electricity through a solution.
- (b) The cathode of an electrolytic cell and cations have opposite charges.
- (c) Ions must be present if a solution conducts electricity.
- (d) Ions make up the solution inventory of a solution of an ionic compound.
- (e) There are no ions present in the solution of a weak acid.
- (f) Only six important acids are weak.
- (g) Hydrofluoric acid, which is used to etch glass, is a strong acid.
- (h) Spectators are included in a net ionic equation.
- (i) A net ionic equation for a reaction between an element and an ion is the equation for a redox reaction.
- (j) A compound that is insoluble forms a precipitate when its ions are combined.
- (k) Precipitation and molecule formation reactions are both ion-combination reactions having double displacement conventional equations.
- (l) Neutralization is a special case of a molecule-formation reaction.
- (m) One product of a molecule-formation reaction is a strong acid.
- (n) Ammonium hydroxide is a possible product of a molecule-formation reaction.

17 Acid–Base (Proton-Transfer) Reactions

LOOKING BACK

6.7, 12.3 A hydrogen-bearing compound that yields hydronium (hydrogen) ions in water solution is one kind of acid. A base is something that reacts with these ions. Hydroxide ion is the most common base.

12.3, 12.4 When polyprotic acids ionize, they do so in stepwise fashion, losing one proton in each step.

14.4 A chemical or physical change that can proceed in either direction is a reversible change. When the rates of change in the two directions are equal, a state of equilibrium is reached.

16.1, 16.3 The solution inventory of a strong acid is hydrogen ions and the anions that come from the ionization of the acid; the solution inventory of a weak acid is the unionized acid molecule.

16.4 Net ionic equations exclude all spectators and show only those solution inventory species that are a part of a chemical change.

Appendix I, Parts B and G. The mathematics of logarithms and exponentials is reviewed.

LOOKING AHEAD IN CHAPTER 17

Two broader acid–base concepts, both of which include the hydrogen and hydroxide ions, are introduced.

In the stepwise ionization of a weak acid, each acid in the sequence is weaker than the acid that preceded it.

The ionization of weak acids is reversible and eventually reaches a state of equilibrium.

Both strong and weak acids are considered in this chapter, with greater emphasis going to the weak acids.

Most of the study of acid–base reactions centers around net ionic equations.

The pH concept is based on the mathematics of logarithms and exponentials.

17.1 ACID–BASE THEORIES

Originally, the word **acid** was used to describe something that had a sour, biting taste. The tastes of vinegar (acetic acid) and lemon juice (citric acid) are typical examples. Substances with such a taste have other common properties too. Among them are the ability to impart certain colors to organic substances, such as litmus, which is red in acid solutions; the ability to react with carbonate ions and release carbon dioxide; the ability to react with and neutralize a base; and the ability to react with certain metals and release hydrogen.

Traditionally, a **base** is also characterized by a taste sensation: bitterness. A base is the opposite of an acid in its ability to impart a different color to some organic compounds (blue to litmus) and to react with and neutralize an acid. Bases feel slippery, or "soapy." Bases form precipitates when added to solutions of nearly all salts.

The Arrhenius Theory of Acids and Bases

PG 17A Distinguish between an acid and a base in terms of their general properties and the ions associated with those properties.

In 1884 a brilliant young chemist, Svante Arrhenius, observed that all substances classified as acids contain hydrogen ions, H^+. An acid was thus identified as **a solution that has a high concentration of hydrogen ions.** The hydroxide ion is present in all solutions then known as bases, so a base became **a solution that has a high concentration of hydroxide ions.** From the Arrhenius standpoint, the properties of an acid are the properties of the hydrogen ion, and the properties of a base are the properties of the hydroxide ion.

Particularly noteworthy among the properties of acids and bases is their ability to neutralize each other. In Example 16.16, the net ionic equation for the reaction between a strong acid (HCl) and a strong base (NaOH) is a molecule-forming ion combination that yields water:

$$H^+(aq) + OH^-(aq) \rightarrow HOH(\ell)$$

If the acid is weak, as in Example 16.17, the hydroxide ion essentially pulls a hydrogen ion right out of the unionized acid molecule:

$$HC_2H_3O_2(aq) + OH^-(aq) \rightarrow HOH(\ell) + C_2H_3O_2{}^-(aq)$$

The carbon dioxide from the reaction of an acid with a carbonate is the end result of the combination of hydrogen and carbonate ions. The expected product is carbonic acid, H_2CO_3. However, in Section 16.8 you learned that this is one of the "unstable products" of an ion combination. It decomposes into carbon dioxide and water (Example 16.20):

$$2\ H^+(aq) + CO_3{}^{2-}(aq) \rightarrow CO_2(g) + H_2O(\ell)$$

Hydrogen ions of acids and hydrogen gas occupy a unique position in the activity series (Section 16.5). All other members of this series are metals and metal ions that engage in single-replacement-type redox reactions. The reaction of the hydrogen ion appears in the net ionic equation for the reaction of calcium with hydrochloric acid (Example 16.5):

$$Ca(s) + 2\ H^+(aq) \rightarrow H_2(g) + Ca^{2+}(aq)$$

Table 16.2 shows that only the hydroxides of barium and the alkali metals are soluble. This is why these solutions are the only strong Arrhenius bases. It also explains the basic property that precipitates form when a strong base is combined with the salt of most metals. The net ionic equation for the precipitation of magnesium hydroxide is typical (Example 16.11):

$$Mg^{2+}(aq) + 2\ OH^-(aq) \rightarrow Mg(OH)_2(s)$$

Quick Check 17.1A
According to the Arrhenius theory of acids and bases, how do you recognize an acid and a base?

The Brönsted–Lowry Theory of Acids and Bases

PG 17B State and illustrate the Brönstead–Lowry concept of acids and bases.

When the simple hydrogen–hydroxide ion concept of acids and bases is examined from the standpoint of bond breaking and bond forming, we find two things. First, all such reactions can be interpreted as a transfer of a proton (hydrogen ion) from the acid to a hydroxide ion. Second, substances other than the hydroxide ion can receive protons. These observations were announced independently by Johannes N. Brönsted and Thomas M. Lowry in 1923. Their proposal is known by both their names, the Brönsted–Lowry theory of acids and bases. By this concept, **an acid–base reaction is a proton transfer reaction in which a proton is transferred from the acid to the base. The acid is a proton donor, and the base is a proton acceptor.** According to this theory, anything that can receive a proton is a base. The hydroxide ion is only the most common example.

The Lewis diagram showing the formation of the hydronium ion when a hydrogen-bearing molecule reacts with water (Section 12.3) illustrates a proton-transfer reaction. Using hydrogen chloride as an example,

$$H_2O(\ell) + HCl(g) \rightarrow H_2O^+(aq) + Cl^-(aq) \qquad (17.1)$$

$$\mathrm{H{-}\ddot{\underset{|}{O}}\!: \; + \; H{-}\ddot{\underset{\cdot\cdot}{Cl}}\!: \;\rightarrow\; \left[H{-}\ddot{\underset{|}{O}}{-}H\right]^+ + \left[:\ddot{\underset{\cdot\cdot}{Cl}}:\right]^-}$$

H (below O) ... H (below O)

Base proton receiver — Acid proton donor

The proton from the hydrogen chloride molecule is clearly transferred to one of the unshared electron pairs of the water molecule. The HCl molecule is the acid, the proton donor. The acceptor of the proton is the water molecule; water is a base in Equation 17.1.

Another proton-transfer reaction occurs when ammonia reacts with water:

$$NH_3(aq) + HOH(\ell) \rightleftharpoons NH_4^+(aq) + OH^-(aq) \qquad (17.2)$$

$$\mathrm{H{-}\overset{H}{\underset{H}{N}}\!: \; + \; H{-}\ddot{\underset{H}{O}}\!: \;\rightleftharpoons\; \left[H{-}\overset{H}{\underset{H}{N}}{-}H\right]^+ + \left[:\ddot{\underset{H}{O}}:\right]^-}$$

Base proton receiver — Acid proton donor

This time water is an acid. It donates a proton to ammonia, the base. A substance that can behave as an acid in one case and a base in another, as water does in Equations 17.1 and 17.2, is said to be **amphoteric.**

The double arrow in Equation 17.2 indicates that the reaction is reversible (Section 14.4). It suggests correctly that the reaction reaches a chemical equilibrium. When a reversible reaction is read from left to right, the **forward reaction,** or the reaction in the **forward direction,** is described; from right to left, the change is the **reverse reaction,** or in the **reverse direction.** In 1 M $NH_3(aq)$, less than 1% of the ammonia is changed to ammonium ions. The solution inventory is mainly ammonia molecules. The reverse reaction is therefore said to be **favored.** The favored direction of an equilibrium resulting from the ionization of an acid or a base points to the more abundant species.

HCl is an acid, a proton donor, in Equation 17.1. NH_3 is a base, a proton acceptor, in Equation 17.2. Do you suppose that if we were to put HCl and NH_3 together, the HCl would give its proton to NH_3? The answer, yes, is readily evident if bottles of concentrated ammonia and hydrochloric acid are opened next to each other. The solid product is often found on the outside of reagent bottles in the laboratory. Conventional and Lewis diagram equations describe the reaction:

$$NH_3(g) + HCl(g) \rightarrow NH_4Cl(s) \qquad (17.3)$$

$$H{-}\overset{\displaystyle H}{\underset{\displaystyle H}{\overset{|}{\underset{|}{N}}}}{:} + H{-}\ddot{\underset{..}{Cl}}{:} \rightarrow \left[H{-}\overset{\displaystyle H}{\underset{\displaystyle H}{\overset{|}{\underset{|}{N}}}}{-}H\right]^+ + \left[{:}\ddot{\underset{..}{Cl}}{:}\right]^-$$

Base proton receiver (NH₃) — Acid proton donor (HCl)

$NH_4Cl(s)$ is an assembly of NH_4^+ and Cl^- ions. Equation 17.3 is an acid–base reaction in the absence of both water and the hydroxide ion.

Reviewing Equations 17.1 to 17.3, we find that all fit into the general equation for a Brönsted–Lowry proton-transfer reaction:

$$B + HA \rightarrow HB^+ + A^- \qquad (17.4)$$

B	HA
Base proton receiver	Acid proton donor

In this equation, note that the charges are not "absolute" charges. They indicate, rather, that the acid species, in losing a proton, leaves a species having a charge one less than the acid, and that the base, in gaining a proton, increases by 1 in charge.

Quick Check 17.1B

1. What is the difference between a Brönsted–Lowry acid and an Arrhenius acid? between a Brönsted–Lowry base and an Arrhenius base?
2. Are all Brönsted–Lowry bases also Arrhenius bases? Are all Arrhenius bases also Brönsted–Lowry bases? Explain.

The Lewis Theory of Acids and Bases (Optional)

> **PG 17C** Describe the Lewis theory of acids and bases and the structural features associated with it; identify potential Lewis acids and Lewis bases.

Look at the Lewis diagrams of the bases in Equations 17.1 to 17.3. In each case, the structural feature of the base that permits it to receive the proton is an unshared pair of electrons. A hydrogen ion has no electron to contribute to the bond. Both bonding electrons must come from the base. The Lewis theory of acids and bases expands the concept of an acid to include anything that can accept an electron pair in forming a covalent bond. By this theory, **a Lewis base is an electron-pair donor,** and a **Lewis acid is an electron-pair acceptor.**

A Lewis acid is not limited to a hydrogen ion. A common example is boron trifluoride, which behaves as a Lewis acid by accepting an unshared pair of electrons from ammonia:

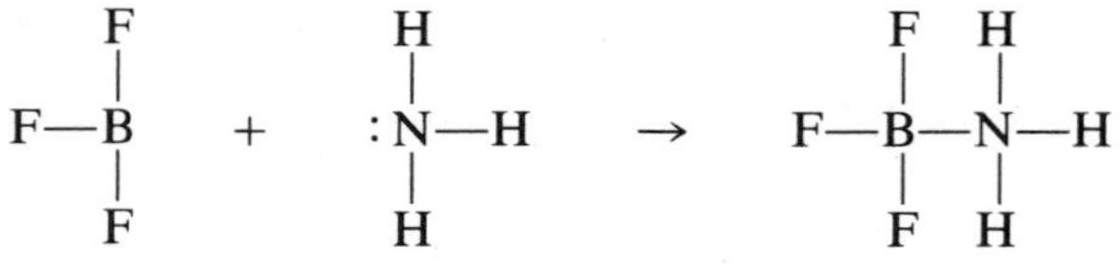

$$\begin{array}{ccccccc} & F & & & H & & & F\ \ H \\ & | & & & | & & & |\ \ \ \ | \\ F- & B & + & :N-H & & \rightarrow & & F-B-N-H \\ & | & & & | & & & |\ \ \ \ | \\ & F & & & H & & & F\ \ H \\ & \text{Acid} & & & \text{Base} & & & \end{array}$$

Other examples of Lewis acid–base reactions appear in the formation of some complex ions and in organic reactions. You will study these in more advanced courses.

> **Quick Check 17.1C**
> By the Brönsted–Lowry theory, water can be either an acid or a base. Can water be a Lewis acid? a Lewis base? Explain.

Summary

The identifying features of acids and bases according to the three acid–base theories are summarized below:

THEORY	ACID	BASE
Arrhenius	Hydrogen ion	Hydroxide ion
Brönsted–Lowry	Proton donor	Proton acceptor
Lewis	Electron-pair acceptor	Electron-pair donor

Our principal interest in acid–base chemistry is in aqueous solutions, where the Brönsted–Lowry theory prevails. The balance of the chapter is limited to the proton-transfer concept of acids and bases.

17.2 CONJUGATE ACID–BASE PAIRS

PG 17D Define or identify conjugate acid–base pairs.

It was noted that Equation 17.2 reaches equilibrium. The fact is that most acid–base reactions reach a state of equilibrium. Accordingly, Equation 17.5 is a rewrite of Equation 17.4, except that a double arrow is used to show the reversible character of the reaction. Look carefully at the reverse reaction.

$$B + HA \rightleftharpoons \underset{\substack{\text{Acid}\\\text{proton}\\\text{donor}}}{HB^+} + \underset{\substack{\text{Base}\\\text{proton}\\\text{receiver}}}{A^-} \qquad (17.5)$$

Is not HB^+ donating a proton to A^- in the reverse reaction? In other words, HB^+ is an acid in the reverse reaction, and A^- is a base. From this we see that the products of any proton-transfer acid–base reaction are *another* acid and base for the reverse reaction.

Combinations such as acid HA and base A^- that result from an acid losing a proton or a base gaining one are called **conjugate acid–base pairs.** In the forward reaction of Equation 17.2, $HOH(\ell)$ releases a hydrogen ion, leaving the hydroxide ion:

$$HOH(\ell) \rightarrow H^+(aq) + OH^-(aq)$$

Water is an acid, a proton donor. What is left after the proton is gone, $OH^-(aq)$, is the **conjugate base** of water. In the reverse direction $OH^-(aq)$ is a base because it gains a proton. The **conjugate acid** of $OH^-(aq)$ is $HOH(\ell)$, the species formed when the base gains a proton. In the other part of the reaction,

$$NH_3(aq) + H^+(aq) \rightleftharpoons NH_4^+(aq)$$

the base, $NH_3(aq)$, in the forward direction gains a proton to form its conjugate acid, $NH_4^+(aq)$; and $NH_3(aq)$ is the conjugate base of acid $NH_4^+(aq)$ in the reverse direction. For the whole reaction, two conjugate acid–base pairs can be identified:

conjugate acid–base pair (HOH(ℓ) — OH⁻(aq))

$$NH_3(aq) + HOH(\ell) \rightleftharpoons NH_4^+(aq) + OH^-(aq) \qquad (17.2)$$

conjugate acid–base pair (NH₃(aq) — NH₄⁺(aq))

You can write the formula of the conjugate base of any acid simply by removing a proton. If HCO_3^- acts as an acid, its conjugate base is CO_3^{2-}. You can also write the formula of the conjugate acid of any base by adding a proton. If HCO_3^- is a base, its conjugate acid is H_2CO_3.

EXAMPLE 17.1 (a) Write the formula of the conjugate base of H_3PO_4.
(b) Write the formula of the conjugate acid of $C_7H_5O_2^-$.

In (b) don't let $C_7H_5O_2^-$ confuse you just because it is unfamiliar. Just do what must be done to find the formula of a conjugate acid.

------ ------

(a) $H_3PO_4 \rightarrow H^+ + H_2PO_4^-$, the conjugate base of H_3PO_4

(b) $C_7H_5O_2^- + H^+ \rightarrow HC_7H_5O_2$, the conjugate acid of $C_7H_5O_2^-$

Remove a proton to get a conjugate base, and add one to get a conjugate acid.

EXAMPLE 17.2 Nitrous acid engages in a proton-transfer reaction with sulfite ion:

$$HNO_2(aq) + SO_3^{2-}(aq) \rightarrow NO_2^-(aq) + HSO_3^-(aq)$$

Answer the questions about this reaction in the steps that follow:

For the forward reaction, identify the acid and the base.

------ ------

HNO_2 is the acid; it donates a proton to SO_3^{2-}, the base, or proton receiver.

Identify the acid and base for the reverse reaction.

------ ------

HSO_3^- is the acid; it donates a proton to NO_2^-, the base, or proton receiver.

Identify the conjugate of HNO_2. Is it a conjugate acid or a conjugate base?

------ ------

NO_2^- is the conjugate *base* of HNO_2. It is the base that remains after the proton has been donated by the acid.

Identify the other conjugate acid–base pair, and classify each species as the acid or the base.

------ ------

SO_3^{2-} and HSO_3^- is the other conjugate acid–base pair. SO_3^{2-} is the base, and HSO_3^- is the conjugate acid—the species produced when the base accepts a proton.

EXAMPLE 17.3 Identify the conjugate acid–base pairs in

$$HCHO_2 + PO_4^{3-} \rightleftharpoons HPO_4^{2-} + CHO_2^-$$

$HCHO_2$ and CHO_2^- are one conjugate acid–base pair; PO_4^{3-} and HPO_4^{2-} are the second pair.

17.3 RELATIVE STRENGTHS OF ACIDS AND BASES

PG 17E Distinguish between a strong acid and a weak acid; between a strong base and a weak base.

17F Given a table of relative strengths of acids and bases, identify the stronger and weaker of two acids or two bases.

In Section 16.3, the distinction was made between the relatively few *strong* acids and the many *weak* acids. Strong acids are those that ionize almost completely, whereas weak acids ionize but slightly. Hydrochloric acid is a strong acid; 0.10 M HCl is almost 100% ionized. Acetic acid is a weak acid; only 1.3% of the molecules ionize in 0.10 M $HC_2H_3O_2$. In a Brönsted–Lowry sense, acid strength is a measure of the tendency of an acid to lose protons. *A strong acid loses protons readily—ionizes more completely; a weak acid clings to its protons—does not ionize significantly.*

Table 17.1 is a list of acids arranged in order of decreasing strength. The strongest acids are at the top and the weakest are at the bottom. Each ionization is shown as an equilibrium equation of the general form $HA(aq) \rightleftharpoons H^+(aq) + A^-(aq)$. Notice the ionization of **polyprotic acids,** those able to release more than one proton. Sulfuric acid loses its first proton almost completely, according to the sixth equation in Table 17.1. This places H_2SO_4 among the strong acids. The HSO_4^- ion, four lines lower in the table, releases the second proton less readily. The second ionization step occurs to about 29% in 0.10 M HSO_4^-. **Diprotic** (two-proton) and **triprotic** (three-proton) acids always ionize in a stepwise fashion like this, and each succeeding acid is weaker than the one before.

The species on the right side of each equation in Table 17.1 is the conjugate base of the acid on the left. Being a base, it is able to accept a proton from a potential donor. *A strong base is one that has a strong attraction for protons; a weak base has a weak attraction for protons.* It follows that the conjugate base of a strong acid is a weak base, and that a weak acid has a strong conjugate base. HCl, for example, a strong acid, loses its proton readily to yield Cl^-. The chloride ion has no tendency to gain a proton to form HCl by the reverse reaction, so Cl^- is a weak base. Conversely, the carbonate ion, CO_3^{2-}, has a strong attraction for protons to form weak acid HCO_3^-. Thus CO_3^{2-} is a strong

Table 17.1 Relative Strengths of Acids and Bases

Acid Name	Acid Formula		Base Formula
Perchloric	$HClO_4$	⇌	$H^+ + ClO_4^-$
Hydroiodic	HI	⇌	$H^+ + I^-$
Hydrobromic	HBr	⇌	$H^+ + Br^-$
Hydrochloric	HCl	⇌	$H^+ + Cl^-$
Nitric	HNO_3	⇌	$H^+ + NO_3^-$
Sulfuric	H_2SO_4	⇌	$H^+ + HSO_4^-$
Hydronium ion	H_3O^+	⇌	$H^+ + H_2O$
Oxalic	$H_2C_2O_4$	⇌	$H^+ + HC_2O_4^-$
Sulfurous	H_2SO_3	⇌	$H^+ + HSO_3^-$
Hydrogen sulfate ion	HSO_4^-	⇌	$H^+ + SO_4^{2-}$
Phosphoric	H_3PO_4	⇌	$H^+ + H_2PO_4^-$
Hydrofluoric	HF	⇌	$H^+ + F^-$
Nitrous	HNO_2	⇌	$H^+ + NO_2^-$
Formic (methanoic)	$HCHO_2$	⇌	$H^+ + CHO_2^-$
Benzoic	$HC_7H_5O_2$	⇌	$H^+ + C_7H_5O_2^-$
Hydrogen oxalate ion	$HC_2O_4^-$	⇌	$H^+ + C_2O_4^{2-}$
Acetic (ethanoic)	$HC_2H_3O_2$	⇌	$H^+ + C_2H_3O_2^-$
Propionic (propanoic)	$HC_3H_5O_2$	⇌	$H^+ + C_3H_5O_2^-$
Carbonic	H_2CO_3	⇌	$H^+ + HCO_3^-$
Hydrosulfuric	H_2S	⇌	$H^+ + HS^-$
Dihydrogen phosphate ion	$H_2PO_4^-$	⇌	$H^+ + HPO_4^{2-}$
Hydrogen sulfite ion	HSO_3^-	⇌	$H^+ + SO_3^{2-}$
Hypochlorous	$HClO$	⇌	$H^+ + ClO^-$
Boric	H_3BO_3	⇌	$H^+ + H_2BO_3^-$
Ammonium ion	NH_4^+	⇌	$H^+ + NH_3$
Hydrocyanic	HCN	⇌	$H^+ + CN^-$
Hydrogen carbonate ion	HCO_3^-	⇌	$H^+ + CO_3^{2-}$
Monohydrogen phosphate ion	HPO_4^{2-}	⇌	$H^+ + PO_4^{3-}$
Hydrogen sulfide ion	HS^-	⇌	$H^+ + S^{2-}$
Water	HOH	⇌	$H^+ + OH^-$
Hydroxide ion	OH^-	⇌	$H^+ + O^{2-}$

(Left side, acids: ↑ Increasing — STRENGTH — Decreasing ↓; right side, bases: ↑ Decreasing — STRENGTH — Increasing ↓)

base. By this reasoning we conclude that the bases in Table 17.1 are listed in order of *increasing* strength; the weaker bases are at the top, and the stronger bases are at the bottom.

EXAMPLE 17.4 Using Table 17.1, list the following acids in order of decreasing strength (strongest first): $HC_2O_4^-$; NH_4^+; H_3PO_4.

Find the three acids among those listed in the table, and list them from strongest (first) to weakest (last).

H_3PO_4; $HC_2O_4^-$; NH_4^+

EXAMPLE 17.5 Using Table 17.1, list the following bases in order of decreasing strength (strongest first): $HC_2O_4^-$; SO_3^{2-}; F^-.

SO_3^{2-}; F^-; $HC_2O_4^-$

The ion $HC_2O_4^-$ appears in both Example 17.4 and 17.5, first as an acid and second as a base. It is the intermediate ion in the two-step ionization of oxalic acid, $H_2C_2O_4$. The $HC_2O_4^-$ ion is amphoteric, as are most intermediate ions in the stepwise ionization of polyprotic acids.

17.4 PREDICTING ACID–BASE REACTIONS

PG 17G Given a table of the relative strengths of acids and bases and the identity of an acid and a base from the table, (a) write the equation for the single-proton transfer reaction between them, and (b) predict which direction, forward or reverse, will be favored at equilibrium.

A chemist likes to know if an acid-base reaction will occur if certain reactants are brought together. Obviously, there must be a potential proton donor and acceptor—there can be no proton-transfer reaction without both. From there the decision is based on the relative strengths of the conjugate acid–base pairs. The stronger acid and base are the most reactive. They *do* what they must do to behave as an acid and a base. The weaker acid and base are more stable—less reactive. It follows that *the stronger acid will always transfer a proton to the stronger base, yielding the weaker acid and base as favored species at equilibrium.* Figure 17.1 summarizes the proton transfer from the stronger acid to the stronger base from the standpoint of positions in Table 17.1.

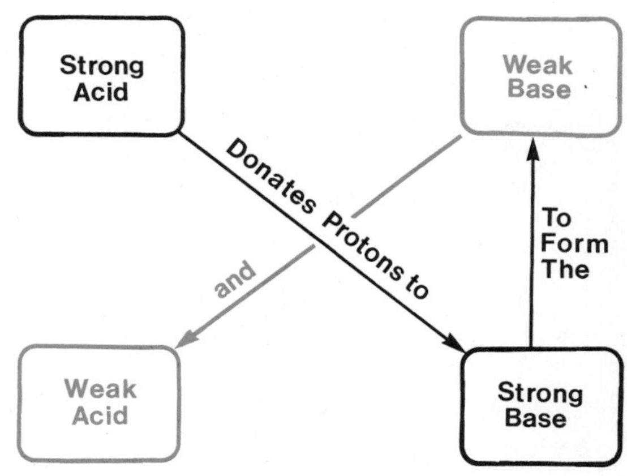

Figure 17.1
Correlation between prediction of an acid–base reaction and position in Table 17.1. Of the two conjugate acid–base pairs in the proton-transfer equation, the stronger acid (*upper left*) will donate protons to the stronger base (*lower right*), yielding the weaker base (*upper right*) and weaker acid (*lower left*). That direction, forward or reverse, will be favored that has the weaker acid and weaker base as products.

If a hydrogen sulfate ion, HSO_4^-, a relatively strong acid trying to lose a proton, finds a hydroxide ion, OH^-, a strong base seeking a proton, both tendencies can be satisfied by the transfer of the proton from HSO_4^- to OH^-:

$$HSO_4^-(aq) + OH^-(aq) \rightleftharpoons SO_4^{2-}(aq) + HOH(\ell) \qquad (17.6)$$

Now identify the conjugate acid–base pairs. In the forward direction the acid is HSO_4^-. Its conjugate base for the reverse reaction is SO_4^{2-}. Similarly OH^- is the base in the forward reaction, and HOH is the conjugate acid for the reverse reaction. In Equation 17.7 the acid–base roles for the different directions are shown with the letters A for acid and B for base, and colors identify the conjugate pairs.

$$\underset{A}{HSO_4^-(aq)} + \underset{B}{OH^-(aq)} \rightleftharpoons \underset{B}{SO_4^{2-}(aq)} + \underset{A}{HOH(\ell)} \qquad (17.7)$$

Now compare the two acids in strength. HSO_4^- is near the top of the list, a much stronger acid than water. We therefore label HSO_4^- with SA for strong acid, and water with WA for weak acid. Similarly, compare the bases: OH^- is a stronger base (SB) than SO_4^{2-} (WB).

$$\underset{SA}{HSO_4^-(aq)} + \underset{SB}{OH^-(aq)} \rightleftharpoons \underset{WB}{SO_4^{2-}(aq)} + \underset{WA}{HOH(\ell)} \qquad (17.8)$$

As you see, the weaker combination is on the right in this equation. This indicates that the reaction is favored in the forward direction: the proton transfers spontaneously from the strong proton donor to the strong proton receiver, yielding as products in greater abundance the weaker conjugate base and conjugate acid.

The following procedure is recommended in the prediction of acid–base reactions:

1. For a given pair of reactants, write the equation for the transfer of *one* proton from one species to the other. (Do not transfer two protons.)
2. Identify the acid and base on each side of the equation.
3. Determine which side of the equation has *both* the weaker acid and the weaker base (they must both be on the same side). That side identifies the products in the favored direction.

EXAMPLE 17.6 Write the net ionic equation for the reaction between hydrofluoric acid, HF, and the sulfite ion, SO_3^{2-}, and predict which side will be favored at equilibrium.

The first step is to write the equation for the single-proton transfer reaction between HF and SO_3^{2-}. Complete this step.

----- -----

$HF(aq) + SO_3^{2-}(aq) \rightleftharpoons F^-(aq) + HSO_3^-(aq)$

Next we identify the acid and base on each side of the equation. Do so with letters A and B, as in the preceding discussion.

----- -----

$$\underset{A}{HF(aq)} + \underset{B}{SO_3^{2-}(aq)} \rightleftharpoons \underset{B}{F^-(aq)} + \underset{A}{HSO_3^-(aq)}$$

In each case the acid is the species with a proton to donate. It is transferred from acid HF to base SO_3^{2-} in the forward reaction, and from acid HSO_3^- to base F^- in the reverse reaction.

Finally, determine which reaction, forward or reverse, is favored at equilibrium. It is the side with the weaker acid and base. Refer to Table 17.1.

----- -----

The forward reaction is favored at equilibrium.

HSO_3^- is a weaker acid than HF, and F^- is a weaker base than SO_3^{2-}. These species are the products in the favored direction.

EXAMPLE 17.7 Write the net ionic equation for the acid–base reaction between HCO_3^- and ClO^- and predict which side will be favored at equilibrium.

Complete the example as before.

----- -----

$HCO_3^-(aq) + ClO^-(aq) \rightleftharpoons CO_3^{2-}(aq) + HClO(aq)$

The reverse reaction will be favored. This conclusion is based on HCO_3^- being a weaker acid than HClO, and ClO^- being a weaker base than CO_3^{2-}.

Up to this point most attention has been given to the direction in which an equilibrium is favored. This does not mean that we can ignore the unfavored direction. Consider, for example, the reaction described by Equation 17.2: $NH_3(aq) + HOH(\ell) \rightleftharpoons NH_4^+(aq) + OH^-(aq)$. Although this reaction proceeds only slightly in the forward direction, many of the properties of household ammonia, in particular its cleaning power, depend on the presence of OH^- ions.

17.5 THE WATER EQUILIBRIUM: pH

PG 17H Write the equation for the ionization of water, the equation for K_w, and state the value of K_w at 25°C.

17I Distinguish among an acidic solution, a basic solution, and a neutral solution in terms of pH and $[H^+]$.

17J Given one of the following, calculate any or all of the others: pH, $[H^+]$, $[OH^-]$, pOH.

17K Given the pH, $[H^+]$, $[OH^-]$, or pOH of two or more solutions, rank them in order of acidity, basicity, pH, $[H^+]$, $[OH^-]$, or pOH.

In this section and the next you will be multiplying and dividing exponentials, taking the square root of an exponential, and working with logarithms. You may wish to review these procedures in Appendix I, Parts B and G.

One of the most critical equilibria in all of chemistry is represented by the next-to-last line in Table 17.1, the ionization of water. Careful control of tiny traces of hydrogen and hydroxide ions marks the difference between success and failure in an untold number of industrial chemical processes; and in biochemical systems, these concentrations are vital to survival itself.

Though pure water is generally regarded as a nonconductor, a sufficiently sensitive detector shows that even water contains a tiny concentration of ions. These ions come from the ionization of the water molecule:

$$HOH(\ell) \rightleftharpoons H^+(aq) + OH^-(aq) \qquad (17.9)$$

Each chemical equilibrium has an **equilibrium constant** that is calculated from the concentrations of one or more species in the equation.* The equilibrium constant for water, K_w, at 25°C is

$$K_w = [H^+][OH^-] = 1.0 \times 10^{-14} \qquad (17.10)$$

Enclosing a chemical symbol in square brackets is one way to represent the moles-per-liter concentration of that species. Thus, $[H^+]$ and $[OH^-]$ are the concentrations of the hydrogen and hydroxide ions. At first we will consider $K_w = 10^{-14}$ and work only with concentrations that can be expressed with whole-number exponents. In the next section, after you have become familiar with the mathematical procedures, concentrations will be written in the usual exponential notation form, including coefficients.

The stoichiometry of Equation 17 indicates that the theoretical concentrations of hydrogen and hydroxide ions in pure water must be equal.† If $x = [H^+] = [OH^-]$, then substituting into Equation 17.10 gives

$$x^2 = 10^{-14}$$

$$x = \sqrt{10^{-14}} = 10^{-7} \text{ moles/liter}$$

*Equilibrium constants are covered in more detail in Chapter 19.

†Natural water is not pure. Dissolved minerals from the ground and gases from the atmosphere may cause variations as large as two orders of magnitude from expected hydrogen and hydroxide ion concentrations.

Water or water solutions in which $[H^+] = [OH^-] = 10^{-7}$ are said to be neutral solutions, neither acidic nor basic. A solution in which $[H^+]$ is greater than $[OH^-]$ is acidic; a solution in which $[OH^-]$ is greater than $[H^+]$ is basic.

Equation 17.10 indicates an inverse relationship between $[H^+]$ and $[OH^-]$; if one concentration goes up, the other must go down. In fact, if we know either the hydrogen or hydroxide ion concentration, we can, by Equation 17.10, find the other.

EXAMPLE 17.8 Find the hydroxide ion concentration in a solution in which $[H^+] = 10^{-5}$.

This problem requires little explanation. Solve Equation 17.10 for $[OH^-]$, substitute, and calculate.

----- -----

From Equation 17.10, $[OH^-] = \frac{10^{-14}}{[H^+]} = \frac{10^{-14}}{10^{-5}} = 10^{-14-(-5)} = 10^{-9}$

Chemists use base 10 logarithms in the form of ''p'' values to express the very small values of $[H^+]$ and $[OH^-]$. By this system, if Q is a number, then

$$pQ = -\log Q \qquad (17.11)$$

Applied to $[H^+]$ and $[OH^-]$, it follows that

$$pH = -\log [H^+] \qquad pOH = -\log [OH^-] \qquad (17.12)$$

$$[H^+] = \text{antilog } -pH = 10^{-pH} \qquad [OH^-] = \text{antilog } -pOH = 10^{-pOH} \qquad (17.13)$$

The following procedures result:

1. If hydrogen ion concentration in moles per liter is expressed as 10 raised to some negative exponent, pH is the exponent with its sign changed to positive. Thus, if $[H^+] = 10^{-x}$, then $pH = x$.
2. Given the pOH of a solution, the hydroxide ion concentration is 10 raised to a power that is the negative of the pOH. Thus, if $pOH = y$, then $[OH^-] = 10^{-y}$. Also, $[OH^-] = \text{antilog} - pOH$.*

EXAMPLE 17.9 (a) The hydrogen ion concentration of a solution is 10^{-9}. What is its pH? (b) The pOH of a certain electroplating bath is 4. What is the hydroxide ion concentration?

----- -----

(a) $[H^+] = 10^{-pH} = 10^{-9}$. pH = 9, the negative of the exponent of 10.

(b) $[OH^-] = 10^{-pOH} = 10^{-4}$. The exponent of 10 is the negative of the pOH.

*Mathematically, pH and pOH are negative numbers when the corresponding molarities are 1 or more. The ''p'' value is not ordinarily used for concentrations greater than 1 molar, so we need not be concerned with negative values of pH or pOH.

The nature of the equilibrium constant and the logarithmic relationship between pH and pOH yield a simple equation that ties the two together:

$$pH + pOH = 14.0 \qquad (17.14)$$

Between Equations 17.10, 17.12, 17.13, and 17.14, if you know any one of the group consisting of pH, pOH, $[OH^-]$, or $[H^+]$, you can calculate the others. Figure 17.2 is a "pH loop" that summarizes these calculations.

EXAMPLE 17.10 Assuming complete ionization of 0.01 M NaOH, find its pH, pOH, $[OH^-]$, and $[H^+]$.

SOLUTION: (a) Starting with $[OH^-]$, if 0.01 mol of NaOH is dissolved in 1 L of solution, the concentration of the hydroxide ion is 0.01 molar: $[OH^-] = 0.01 = 10^{-2}$.

(b) If $[OH^-] = 10^{-2}$, $pOH = -\log 10^{-2} = 2$.

(c) $pH = 14 - pOH = 14 - 2 = 12$.

(d) If $pH = 12$, $[H^+] = 10^{-pH} = 10^{-12}$.

Two things are worth noting about Example 17.10. First, if we extend the problem by one more step, we complete the full pH loop. We began with $[OH^-]$ and went counterclockwise through pOH, pH, and $[H^+]$. The $[H^+]$ of 10^{-12} can be converted to $[OH^-]$ by Equation 17.10:

$$[OH^-] = \frac{K_w}{[H^+]} = \frac{10^{-14}}{10^{-12}} = 10^{-2}$$

This is the same as the starting $[OH^-]$. Completing the loop may therefore be used to check the correctness of the other steps in the process.

The second observation from Example 17.10 is that the loop may be circled in either direction. Starting with $[OH^-] = 10^{-2}$ and moving clockwise,

$$[H^+] = \frac{K_w}{[OH^-]} = \frac{10^{-14}}{10^{-2}} = 10^{-12}$$

It follows that $pH = 12$ and $pOH = 14 - 12 = 2$, the same results reached by circling the loop in the opposite direction.

You should now be able to make a complete trip around the pH loop.

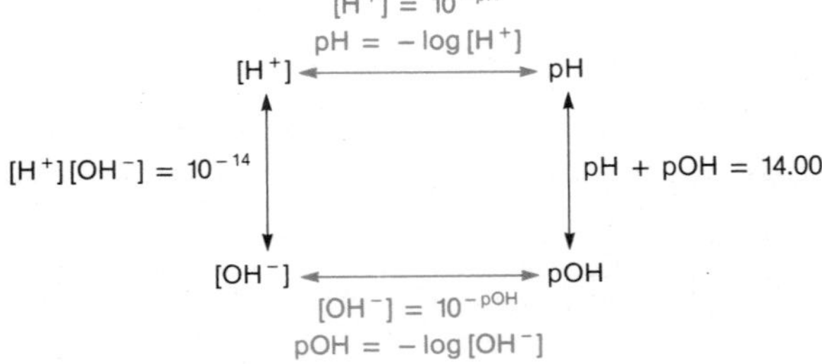

Figure 17.2
The "pH loop." Given the value for any corner of the pH loop, all other values may be calculated by progressing around the loop in either direction. Conversion equations are shown for each step.

EXAMPLE 17.11 The pH of a solution is 3. Calculate the pOH, $[H^+]$, and $[OH^-]$ in any order. Confirm your result by calculating the starting pH—by completing the loop.

You may go either way around the loop, but complete it whichever way you choose, making sure you return to the starting point.

COUNTERCLOCKWISE

From pH = 3, $[H^+] = 10^{-3}$

$$[OH^-] = \frac{10^{-14}}{10^{-3}} = 10^{-11}$$

From $[OH^-] = 10^{-11}$, pOH = 11

pH = 14 − 11 = 3

CLOCKWISE

pOH = 14 − 3 = 11

From pOH = 11, $[OH^-] = 10^{-11}$

$$[H^+] = \frac{10^{-14}}{10^{-11}} = 10^{-3}$$

From $[H^+] = 10^{-3}$, pH = 3

Most of the solutions we work with in the laboratory and all of those involved in biochemical systems have pH values between 1 and 14. This corresponds to H^+ concentrations between 10^{-1} and 10^{-14}, as shown in Table 17.2.

Let's pause for a moment to develop a "feeling" for pH—what it means. pH is a measure of acidity, at least in the sense that hydrogen ion concentration expresses acidity. It is an inverse sort of measurement; the higher the pH, the lower the acidity, and vice versa. Table 17.3 brings out this relationship.

On examining Table 17.3, we see that each pH unit represents a factor of 10. Thus a solution of pH 2 is 10 times as acidic as a solution with pH = 3, and 100 times as acidic as the solution of pH 4. In general, the relative acidity in terms of $[H^+]$ is 10^x, where x is the difference between the two pH measurements. From this we conclude that a 0.1 M solution of a strong acid, with pH = 1, is one million times as acidic as a neutral solution, with pH = 7. (One million is based on the pH difference, 7 − 1 = 6. As an exponential, 10^6 = 1 000 000.)

If you understand the idea behind pH, you should be able to make some comparisons.

EXAMPLE 17.12 Arrange the following solutions in order of decreasing acidity (i.e., highest $[H^+]$ first, lowest last): Solution A, pH = 8; Solution B, pOH = 4; Solution C, $[H^+] = 10^{-6}$; Solution D, $[OH^-] = 10^{-5}$.

To make comparisons, all values should be converted to the same basis, pH, pOH, $[H^+]$, or $[OH^-]$. Because the question asks for a list based on acidity, find the $[H^+]$ for each solution.

$[H^+] = 10^{-8}$ for A; 10^{-10} for B; 10^{-6} for C; 10^{-9} for D.

Now arrange these $[H^+]$ values in decreasing order. Remember that the exponents are negative.

C, A, D, B $10^{-6} > 10^{-8} > 10^{-9} > 10^{-10}$

Various methods are used to measure pH in the laboratory (Fig. 17.3). Acid–base indicators have been mentioned already. Each indicator is effective over a specific pH range. Paper strips, impregnated with an indicator dye that functions over a range selected for the pH being measured, are widely used for rough pH readings. More accurate measurements are made with pH meters.

17.6 NONINTEGER pH–[H⁺] AND pOH–[OH⁻] CONVERSIONS [OPTIONAL]

It is sad but true that real-world solutions do not come neatly packaged in concentrations that can be expressed as whole-number powers of ten. $[H^+]$ is more apt to have a value such as 2.7×10^{-4}, or the pH of a solution is more

Table 17.2 pH Values of Common Liquids

Liquid	pH
Human gastric juices	1.0–3.0
Lemon juice	2.2–2.4
Vinegar	2.4–3.4
Carbonated drinks	2.0–4.0
Orange juice	3.0–4.0
Black coffee	3.7–4.1
Tomato juice	4.0–4.4
Cow's milk	6.3–6.6
Human blood	7.3–7.5
Sea water	7.8–8.3
Saturated $Mg(OH)_2$	10.5
Household ammonia (1–5%)	10.5–11.5
0.1 M Na_2CO_3	11.7
1 M NaOH	14.0

Table 17.3 pH and Hydrogen Ion Concentration

$[H^+]$	$[H^+]$	pH	Acidity or Basicity*
1.0	10^0	0	Strongly acid $pH < 4$
0.1	10^{-1}	1	
0.01	10^{-2}	2	
0.001	10^{-3}	3	
0.000 1	10^{-4}	4	
0.000 1	10^{-4}	4	Weakly acid $4 \le pH < 6$
0.000 01	10^{-5}	5	
0.000 001	10^{-6}	6	
0.000 001	10^{-6}	6	Neutral (or near neutral) $6 \le pH < 8$
0.000 000 1	10^{-7}	7	
0.000 000 01	10^{-8}	8	
0.000 000 01	10^{-8}	8	Weakly basic $8 \le pH < 11$
0.000 000 001	10^{-9}	9	
0.000 000 000 1	10^{-10}	10	
0.000 000 000 1	10^{-10}	10	Strongly basic $11 \le pH$
0.000 000 000 01	10^{-11}	11	
0.000 000 000 001	10^{-12}	12	
0.000 000 000 000 1	10^{-13}	13	
0.000 000 000 000 01	10^{-14}	14	

* Ranges of acidity and basicity are arbitrary.

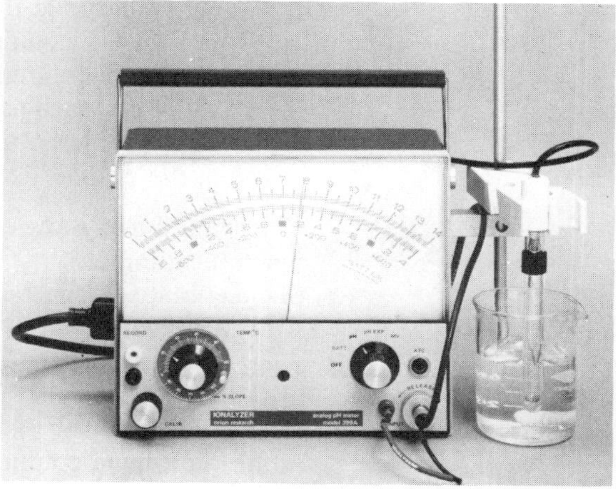

Figure 17.3

likely to be 6.24. The chemist must be able to convert from each of these to the other.

A pH number is a logarithm. To find a pH on a calculator, you must therefore find a logarithm. This procedure is discussed in Appendix I, Part B; you may wish to review it at this time.

Table 17.4 shows the logarithms of 3.45 multiplied by five different powers of ten: 0, 1, 2, 8, and 12. One column shows the value of the logarithm to seven decimals. Another shows how that value is presented in the display of a calculator. A third column has the logarithms rounded off to the correct number of significant figures. Notice three things:

1. The mantissa of the logarithm—the number to the right of the decimal in the VALUE column—is always the same, 0.5378191. This is the logarithm of 3.45, the coefficient for each entry in the EXPONENTIAL NOTATION column in Table 17.4.

Table 17.4 Logarithms and Exponential Notation

Number		Logarithm		
DECIMAL FORM	EXPONENTIAL NOTATION	VALUE	CALCULATOR DISPLAY	ROUNDED OFF
3.45	3.45×10^0	0.5378191	0.5378191	0.538
34.5	3.45×10^1	1.5378191	1.5378191	1.538
345	3.45×10^2	2.5378191	2.5378191	2.538
345 000 000	3.45×10^8	8.5378191	8.5378 00	8.538
3 450 000 000 000	3.45×10^{12}	12.5378191	1.2538 01 (12.5378 00)	12.538

2. The characteristic of the logarithm—the number to the left of the decimal in the VALUE column—is the same as the exponent in the EXPONENTIAL NOTATION column.
3. All numbers in the first two columns are three-significant-figure numbers. This appears in both the decimal form of the number and in the coefficient when the number is written in exponential notation.

Item 2 above shows that the characteristic of a logarithm and the exponent when the number is expressed in scientific notation are related to the magnitude of the number—the location of the decimal point. They have nothing to do with significant figures. Significant figures in the number appear only in the *coefficient* in exponential notation. These show up in the *mantissa* of the logarithm. To be correct in significant figures, the coefficient and the mantissa must have the same number of digits. The correctly rounded off logarithms are in the right-hand column of Table 17.4. All numbers in that column are written in three significant figures. Only the digits after the decimal point are significant.

In working with pH, you will be finding logarithms of numbers smaller than 1. These logarithms are negative. The sign is changed to positive when the logarithm is written as a ''p'' value. Try one. Find log 3.45×10^{-6}. Enter the number into your calculator, and then press the ''log'' key. The display should read $-4.4622\ 00$. If 3.45×10^{-6} represented $[H^+]$, the pH would be the opposite of -5.4622, or 5.462 rounded off to three significant figures and with the sign changed.*

In the example that follows, the calculator sequence is given in detail. Although calculator keys may be marked differently, the procedure is essentially the same for AOS logic and RPN logic.

EXAMPLE 17.13 Calculate the pH of a solution if $[H^+] = 2.7 \times 10^{-4}$.

SOLUTION: $\text{pH} = -\log(2.7 \times 10^{-4}) = 3.57$

PRESS	DISPLAY
2.7	2.7
EE	2.7 00
4	2.7 04
+/−	2.7 - 04
log	− 3.5686 00
+/−	3.5686 00

The answer should be rounded off to two significant figures, 3.57.

*The mantissa appears to be different here than it was in Table 17.4, but really it is not when you trace its origin:

$$\log(3.45 \times 10^{-6}) = \log 3.45 + \log 10^{-6} = 0.538 + (-6) = -5.462$$

EXAMPLE 17.14 Find the pOH of a solution if its hydroxide ion concentration is 7.9×10^{-5} moles per liter.

$$pOH = -\log(7.9 \times 10^{-5}) = 4.10 \text{ (two significant figures)}$$

There are two ways to change pH to $[H^+]$. Both are based on the fact that pH is the negative of a logarithm, which is an exponent of 10. The first is simply to raise 10 to the negative pH power. The second is to enter the negative of the pH into the calculator, and then find its antilogarithm, or inverse log, as it is sometimes called. This is done by pressing the INV key, for those calculators so equipped. The mathematics is summarized in Equation 17.15, and the procedure is detailed in Example 17.15:

$$10^{-pH} = [H^+] = \text{antilog } -pH \tag{17.15}$$

EXAMPLE 17.15 The pOH of a solution is 6.24. Find $[OH^-]$.

SOLUTION: $[OH^-] = \text{antilog } -6.24 = 10^{-6.24} = 5.8 \times 10^{-7}$

10^{-pOH} SEQUENCE

PRESS	DISPLAY
10	10
y^x	10
6.24	6.24
+/−	−6.24
EE	−6.24 00
=	5.7544 - 07

ANTILOG SEQUENCE

PRESS	DISPLAY
6.24	6.24
+/−	−6.24
EE	−6.24 00
INV	−6.24 00
log	5.7544 - 07

EXAMPLE 17.16 Find the hydrogen ion concentration of a solution if its pH is 11.62.

$$[H^+] = \text{antilog } -11.62 = 10^{-11.62} = 2.4 \times 10^{-12}$$

EXAMPLE 17.17 $[OH^-] = 5.2 \times 10^{-9}$ for a certain solution. Calculate in order pOH, pH, and $[H^+]$, and then complete the pH loop by recalculating $[OH^-]$ from $[H^+]$.

$$pOH = -\log (5.2 \times 10^{-9}) = 8.28$$

$$pH = 14.00 - 8.28 = 5.72$$

$$[H^+] = \text{antilog} - 5.72 = 10^{-5.72} = 1.9 \times 10^{-6}$$

$$[OH^-] = \frac{1.0 \times 10^{-14}}{1.9 \times 10^{-6}} = 5.3 \times 10^{-9}$$

The variation between 5.2×10^{-9} and 5.3×10^{-9} comes from rounding off in expressing intermediate answers. If the calculator sequence is completed without rounding off, the loop returns to 5.2×10^{-9} for $[OH^-]$.

CHAPTER 17 IN REVIEW

17.1 Acid–Base Theories

17A Distinguish between an acid and a base in terms of their general properties and the ions associated with those properties.

17B State and illustrate the Brönsted–Lowry concept of acids and bases.

17C Describe the Lewis theory of acids and bases and the structural features associated with it; identify potential Lewis acids and Lewis bases.

17.2 Conjugate Acid–Base Pairs

17D Define or identify conjugate acid–base pairs.

17.3 Relative Strengths of Acids and Bases

17E Distinguish between a strong acid and a weak acid; between a strong base and a weak base.

17F Given a table of relative strengths of acids and bases, identify the stronger and weaker of two acids or two bases.

17.4 Predicting Acid–Base Reactions

17G Given a table of the relative strengths of acids and bases and the identity of an acid and a base from the table, (a) write the equation for the single-proton transfer reaction between them, and (b) predict which direction, forward or reverse, will be favored at equilibrium.

17.5 The Water Equilibrium: pH

17H Write the equation for the ionization of water, the equation for K_w, and state the value of K_w at 25°C.

17I Distinguish among an acidic solution, a basic solution, and a neutral solution in terms of pH and $[H^+]$.

17J Given one of the following, calculate any or all of the others: pH; $[H^+]$; $[OH^-]$; pOH.

17K Given the pH, $[H^+]$, $[OH^-]$, or pOH of two or more solutions, rank them in order of acidity, basicity, pH, $[H^+]$, $[OH^-]$, or pOH.

17.6 Noninteger pH-$[H^+]$ and pOH-$[OH^-]$ Conversions (Optional)

TERMS AND CONCEPTS

17.1 Acid
Base
Arrhenius theory
Brönsted–Lowry theory
Proton-transfer reaction
Proton donor or acceptor
Amphoteric
Forward or reverse reaction or direction
Favored (equilibrium)
Lewis theory

Electron pair donor or acceptor
17.2 Conjugate acid–base pair
Conjugate acid; conjugate base
17.3 Strong or weak acid, base
Polyprotic, diprotic, triprotic

17.5 Equilibrium constant
Water constant, K_w
$[H^+]$, $[OH^-]$
Acidic, neutral, or basic solution
pH, pOH

Most of these terms and many others appear in the Glossary. Use the Glossary regularly.

QUESTIONS AND PROBLEMS

Section 17.1

(*1*) Identify the ions traditionally present in solutions called acids and bases. List three compounds that are commonly regarded as acids and three that are regarded as bases which contain these ions in their water solutions.

(2) Distinguish between an Arrhenius acid and a Brönsted–Lowry acid. Are the two concepts in agreement? Justify your answer.

(*3*) What structural feature must be present in a compound for it to qualify as a Lewis acid? Lewis base?

(4) Is the Arrhenius theory of acids and bases consistent with the Lewis acid–base theory? Explain.

(5)* Aluminum chloride, $AlCl_3$, behaves more as a molecular compound than an ionic one. This is illustrated in its ability to form a fourth covalent bond with a chloride ion:

$$AlCl_3 + Cl^- \rightarrow AlCl_4^-.$$

From the Lewis diagram of the aluminum chloride molecule and the electron configuration of the chloride ion, show that this is an acid–base reaction in the Lewis sense, and identify the Lewis acid and the Lewis base.

(30) Identify at least two of the classical properties of acids and two of bases. For one acid property and one base property, show how it is related to the ion associated with an acid or a base.

(31) Distinguish between an Arrhenius base and a Brönsted–Lowry base. Are the two concepts in agreement? Justify your answer.

(32) Explain and/or illustrate by an example what is meant by identifying a Lewis acid as an electron-pair acceptor, and a Lewis base as an electron-pair donor.

(33) Is the Brönsted–Lowry acid–base theory consistent with the Lewis acid–base theory? Explain.

(34)* Diethyl ether reacts with boron trifluoride by forming a covalent bond between the molecules. Describe the reaction from the standpoint of the Lewis acid–base theory, based on the following "structural" equation:

$$BF_3 + :\ddot{O}(C_2H_5)_2 \rightarrow F_3B\text{—}\ddot{O}(C_2H_5)_2$$

Section 17.2

(6) What are the conjugate acids of OH^- and HCO_3^-? Write the formulas of the conjugate bases of H_3O^+ and HCO_3^-.

(7) For the reaction

$$HNO_2(aq) + CN^-(aq) \rightleftharpoons NO_2^-(aq) + HCN(aq)$$

identify the acid and base on each side of the equation—i.e., the acid and base for the forward reaction and the acid and base for the reverse reaction.

(35) Give the formula of the conjugate base of HF; of $H_2PO_4^-$. Give the formula of the conjugate acid of NO_2^-; of $H_2PO_4^-$.

(36) For the reaction

$$HSO_4^-(aq) + C_2O_4^{2-}(aq) \rightleftharpoons SO_4^{2-}(aq) + HC_2O_4^-(aq)$$

identify the acid and the base on each side of the equation—i.e., the acid and base for the forward reaction and the acid and base for the reverse reaction.

(8) Identify the conjugate acid–base pairs in Question 7.

(9) Identify both conjugate acid–base pairs in the reaction

$$HSO_4^-(aq) + C_2O_4^{2-}(aq) \rightleftharpoons SO_4^{2-}(aq) + HC_2O_4^-(aq)$$

(10) Identify the conjugate acid–base pairs in

$$H_2PO_4^-(aq) + HCO_3^-(aq) \rightleftharpoons HPO_4^{2-}(aq) + H_2CO_3(aq)$$

(37) Identify the conjugate acid–base pairs in Question 36.

(38) For the reaction

$$HNO_2(aq) + CN^-(aq) \rightleftharpoons NO_2^-(aq) + HCN(aq)$$

identify both conjugate acid–base pairs.

(39) Identify the conjugate acid–base pairs in

$$NH_4^+(aq) + HPO_4^{2-}(aq) \rightleftharpoons NH_3(aq) + H_2PO_4^-(aq)$$

Section 17.3

Refer to Table 17.1 when answering questions in this section.

(*11*) What is the difference between a strong acid and a weak acid, according to the Brönsted–Lowry concept? Identify two examples of strong acids and two examples of weak acids.

(12) List the following bases in order of their decreasing strength (strongest base first): SO_4^{2-}; Br^-; $H_2PO_4^-$; CO_3^{2-}.

(13) List the following acids in order of their decreasing strength (strongest acid first): $H_2C_2O_4$; HSO_3^-; H_2O; HI; NH_4^+.

(40) What is the difference between a strong base and a weak base, according to the Brönsted–Lowry concept? Identify two examples of strong bases and two examples of weak bases.

(41) List the following acids in order of their increasing strength (weakest acid first): $HC_2O_4^-$, H_2SO_3; HOH; HClO.

(42) List the following bases in order of their decreasing strength (strongest base first): CN^-; H_2O; HSO_3^-; ClO^-; Cl^-.

Section 17.4

For each acid and base given in this section, complete a proton-transfer equation for the transfer of one proton. Using Table 17.1 predict the direction in which the resulting equilibrium will be favored.

(14) $HC_7H_5O_2(aq) + SO_4^{2-}(aq) \rightleftharpoons$

(15) $H_2C_2O_4(aq) + NH_3(aq) \rightleftharpoons$

(16) $H_3PO_4(aq) + CN^-(aq) \rightleftharpoons$

(17) $H_2BO_3^-(aq) + NH_4^+(aq) \rightleftharpoons$

(18) $HPO_4^{2-}(aq) + HC_2H_3O_2(aq) \rightleftharpoons$

(43) $HC_3H_5O_2(aq) + PO_4^{3-}(aq) \rightleftharpoons$

(44) $HSO_4^-(aq) + CO_3^{2-}(aq) \rightleftharpoons$

(45) $H_2CO_3(aq) + NO_3^-(aq) \rightleftharpoons$

(46) $NO_2^-(aq) + H_3O^+(aq) \rightleftharpoons$

(47) $HSO_4^-(aq) + HC_2O_4^-(aq) \rightleftharpoons$

Section 17.5

(19) How is it that we can classify water as a nonconductor of electricity and yet talk about the ionization of water? If it ionizes, why does it not conduct?

(*20*) Identify the ranges of the pH scale that we classify as strongly acidic, weakly acidic, strongly basic, weakly basic, and neutral, or close to neutral.

(48) Of what significance is the very small value of 10^{-14} for K_w, the ionization equilibrium constant for water?

(49) What is meant by saying that one solution is acidic, another is neutral, and another is basic?

(*21*) Select any integer from 1 to 14 and explain what is meant by saying that this number is the pH of a certain solution.

(50) If the pH of a solution is 9.4, is the solution acidic or basic? How do you reach your conclusion? List in order the pH values of a solution that is neutral, one that is basic, and one that is acidic.

In the next four questions the pH, pOH, [OH⁻], or [H⁺] of a solution is given. Find each of the other values. Also classify each solution as strongly acidic, weakly acidic, neutral (or close to neutral), weakly basic, or strongly basic, as these terms are used in Table 17.3.

(22) pOH = 6

(23) $[H^+] = 0.1$

(24) $[OH^-] = 10^{-2}$

(25) pH = 4

(51) pH = 3

(52) $[OH^-] = 10^{-5}$

(53) pOH = 1

(54) $[H^+] = 10^{-6}$

Section 17.7

In the questions in this section the pH, pOH, [OH⁻], or [H⁺] of a solution is given. Find each of the other values.

(26) pH = 6.62

(27) $[OH^-] = 1.1 \times 10^{-11}$

(28) pOH = 5.54

(29) $[H^+] = 7.2 \times 10^{-2}$

(55) $[OH^-] = 5.6 \times 10^{-10}$

(56) pH = 2.28

(57) $[H^+] = 9.1 \times 10^{-1}$

(58) pOH = 6.82

Miscellaneous Questions

(*59*) Distinguish precisely and in scientific terms the differences between items in each of the following groups:
- (a) Acid, base—by Arrhenius theory
- (b) Acid, base—by Brönsted–Lowry theory
- (c) Acid, base—by Lewis theory
- (d) Forward reaction, reverse reaction
- (e) Acid, conjugate base; base, conjugate acid
- (f) Strong acid, weak acid
- (g) Strong base, weak base
- (h) $[H^+]$, $[OH^-]$
- (i) pH, pOH

(60) Classify each of the following statements as true or false:
- (a) All Brönsted–Lowry acids are Arrhenius acids.
- (b) All Arrhenius bases are Brönsted–Lowry bases, but not all Brönsted–Lowry bases are Arrhenius bases.
- (c) HCO_3^- is capable of being amphoteric.
- (d) HS^- is the conjugate base of S^{2-}.
- (e) If the species on the right side of an ionization equilibrium are present in greater abundance than those on the left, the equilibrium is favored in the forward direction.
- (f) NH_4^+ cannot act as a Lewis base.
- (g) Weak bases have a weak attraction for protons.
- (h) The stronger acid and the stronger base are always on the same side of a proton-transfer reaction equation.
- (i) A proton-transfer reaction is always favored in the direction that yields the stronger acid.
- (j) A solution with pH = 9 is more acidic than one with pH = 4.
- (k) A solution with pH = 3 is twice as acidic as one with pH = 6.
- (l) A pOH of 4.65 expresses the hydroxide ion concentration of a solution in three significant figures.

(61) Theoretically, can there be a Brönsted–Lowry acid–base reaction between OH^- and NH_3? If not, why not? If yes, write the equation.

(*62*) Explain what amphoteric means. Give an example of an amphoteric substance, other than water, that does not contain carbon.

(63) Very small concentrations of ions other than hydrogen and hydroxide are sometimes expressed with "p" numbers. Calculate $[Cl^-]$ in a solution for which pCl = 7.49×10^{-8}.

(64)* Suggest a reason why, in a polyprotic acid, the acid strength decreases with each step in the ionization.

(65) Theoretically, can there be a Brönsted–Lowry acid–base reaction between SO_4^{2-} and F^-? If not, why not? If yes, write the equation.

(66) Sodium carbonate is among the most important industrial bases. How can it be a base when it does not contain a hydroxide ion? Write the equation for a reaction that demonstrates its character as a base.

(67) Which of the three theories of acids and bases described in this chapter is the broadest, including the largest number of compounds that can be considered as acids and bases? Justify your choice.

18 Oxidation–Reduction (Electron Transfer) Reactions

LOOKING BACK	LOOKING AHEAD IN CHAPTER 18
8.7, 16.5 Conventional and net ionic equations were written for redox reactions of the "single replacement" type.	"Single replacement" redox reactions are electron-transfer reactions that are used as a basis for the study of redox chemistry.
12.1 Oxidation state (number) was used in chemical nomenclature.	Oxidation state (number) is used to identify an oxidation–reduction reaction and as a means for keeping track of the transfer of electrons.
16.1 An electrolytic system was used to introduce solution inventory, with emphasis on the electrolyte.	An electrolytic system is used to examine the oxidation–reduction reactions that occur at the electrodes.
17.1 Acid–base reactions are proton transfer reactions.	Many redox reactions are, and most redox reactions can be regarded as, electron-transfer reactions.
17.3, 17.4 Acids and bases can be listed according to their strength, and acid–base reactions can be predicted from that list.	Oxidizing and reducing agents can be listed according to their strength, and redox reactions can be predicted from that list.

18.1 ELECTRON-TRANSFER REACTIONS

PG 18A Distinguish between oxidation and reduction in terms of electrons gained or lost.

18B Given an oxidation half-reaction equation and a reduction half-reaction equation, add them to obtain a balanced redox reaction equation.

Figure 18.1 illustrates a **voltaic cell** of the type that was used to operate telegraph relays and doorbells in the 19th century. A strip of zinc is immersed in a solution of zinc ions, and a piece of copper is placed in a solution of copper ions. The solutions are connected by a "salt bridge," an electrolyte whose ions are not involved in the net chemical change. The two electrodes are connected by a wire. A voltmeter in the external circuit detects a flow of electrons from the zinc electrode to the copper electrode and also measures the "force" that moves the electrons through the circuit.

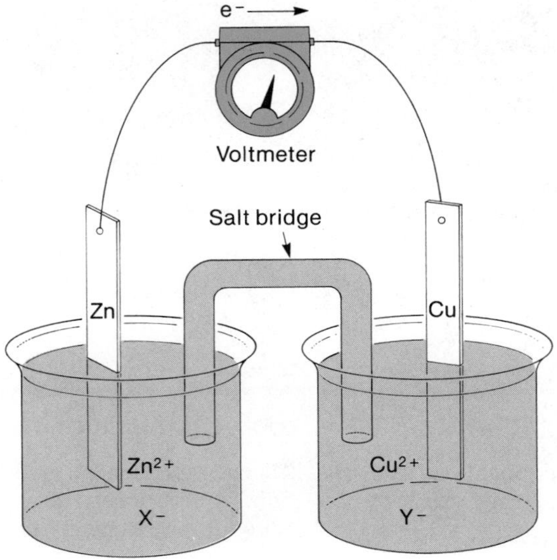

Figure 18.1
Voltaic cell.

Where do the electrons entering the voltmeter come from, and where do they go on leaving? Four measurable observations give the answer to that question. After the cell has operated for a period of time, (a) the mass of the zinc electrode decreases; (b) the Zn^{2+} concentration increases; (c) the mass of the copper electrode increases; (d) the Cu^{2+} concentration decreases. The first two observations indicate that neutral zinc atoms lose two electrons to become zinc ions. Another way of putting it is that zinc atoms are being divided into zinc ions and two electrons:

$$Zn(s) \rightarrow Zn^{2+}(aq) + 2\ e^- \qquad (18.1)$$

The electrons flow through the wire and the voltmeter to the copper electrode, where they join a copper ion to become a copper atom:

$$Cu^{2+}(aq) + 2\ e^- \rightarrow Cu(s) \qquad (18.2)$$

The chemical change that occurs at the zinc electrode is oxidation. **Oxidation is defined as the loss of electrons.** The reaction is described as a **half-reaction** because it cannot occur by itself. There must be a second half-reaction. The electrons lost by the substance **oxidized** must have some place to go. In this case they go to the copper ion, which is **reduced. Reduction is a gain of electrons.**

Equations 18.1 and 18.2 are **half-reaction equations.** If the half-reaction equations are combined—added algebraically—the result is the net ionic equation for the oxidation–reduction (redox) reaction:

$$Zn(s) \rightarrow Zn^{2+}(aq) + \cancel{2e^-} \qquad (18.1)$$

$$Cu^{2+}(aq) + \cancel{2e^-} \rightarrow Cu(s) \qquad (18.2)$$

$$Zn(s) + Cu^{2+}(aq) \rightarrow Cu(s) + Zn^{2+}(aq) \qquad (18.3)$$

This chemical change is an **electron-transfer reaction.** Electrons have been transferred from zinc atoms to copper(II) ions. Notice that, while no electrons

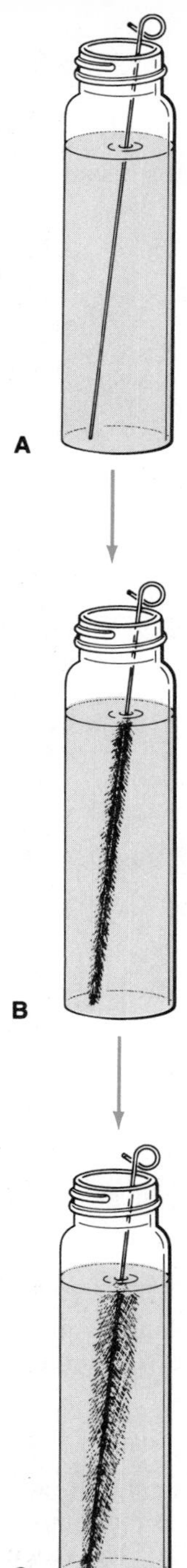

Figure 18.2
"Chemical pine tree." Copper wire is placed in a silver nitrate solution (A). Silver begins to appear (B), finally taking the shape of silver needles (C).

appear in the final equation, the electron-transfer character of the reaction is quite clear in the half-reactions. Notice also that the number of electrons lost by one species is exactly equal to the number of electrons gained by the other.

If there is no need for the electrical energy that can be derived from this cell, the same reaction can be performed by simply dipping a strip of zinc into a solution of copper(II) ions. A coating of copper atoms quickly forms on the surface of the zinc. If the copper atoms are washed off of the zinc and the zinc weighed, its mass will be less than it was at the beginning. The concentration of copper ions in the solution goes down, and zinc ions appear. The half-reaction and net ionic equations are exactly as they are for the voltaic cell.

This same reaction was used in Example 16.4 as, "A reaction occurs when a piece of zinc is dipped into $Cu(NO_3)_2(aq)$. See Color Plate 3. Write the conventional, ionic, and net ionic equations." The conventional equation is a "single replacement" equation:

$$Zn(s) + Cu(NO_3)_2(aq) \rightarrow Cu(s) + Zn(NO_3)_2(aq) \qquad (18.4)$$

Equation 18.3 is the net ionic equation produced in Example 16.4.

All of the single replacement redox reactions encountered in Chapters 8 and 16 can be analyzed in terms of half-reactions. For example:

1. The evolution of hydrogen gas on adding zinc to sulfuric acid (Sec. 16.5):

Reduction: $2\ H^+(aq) + \cancel{2e^-} \rightarrow H_2(g)$

Oxidation: $Zn(s) \rightarrow Zn^{2+}(aq) + \cancel{2e^-}$

Redox: $2\ H^+(aq) + Zn(s) \rightarrow H_2(g) + Zn^{2+}(aq)$ (18.5)

2. The preparation of bromine by bubbling chlorine gas through a solution of NaBr (Example 8.12):

Reduction: $Cl_2(g) + 2\ e^- \rightarrow 2\ Cl^-$

Oxidation: $2\ Br^-(aq) \rightarrow Br_2(\ell) + 2\ e^-$

Redox: $Cl_2(g) + 2\ Br^-(aq) \rightarrow 2\ Cl^-(aq) + Br_2(\ell)$ (18.6)

3. The formation of a "chemical pine tree" with needles of silver (Example 8.11 and Fig. 18.2) by placing a copper wire into a silver nitrate solution:

Reduction: $2\ Ag^+(aq) + \cancel{2e^-} \rightarrow 2\ Ag(s)$

Oxidation: $Cu(s) \rightarrow Cu^{2+}(aq) + \cancel{2e^-}$

Redox: $2\ Ag^+(aq) + Cu(s) \rightarrow 2\ Ag(s) + Cu^{2+}(aq)$ (18.7)

The development of Equation 18.7 needs special comment. The usual reduction equation for silver ion is $Ag^+(aq) + e^- \rightarrow Ag(s)$. Because two moles of electrons are lost in the oxidation reaction, *two moles of electrons must be gained in the reduction reaction*. As has already been noted, the number of electrons lost by one species must equal the number gained by the other species. It is therefore necessary to multiply the usual Ag^+ reduction equation by 2 to bring about this equality in electrons gained and lost. They then cancel when the half-reaction equations are added.

EXAMPLE 18.1 Combine the following half-reactions to produce a balanced redox reaction equation. Indicate which half-reaction is an oxidation reaction, and which is a reduction.

$$Co^{2+}(aq) + 2\ e^- \rightarrow Co(s)$$

$$Sn(s) \rightarrow Sn^{2+}(aq) + 2\ e^-$$

Reduction:	$Co^{2+}(aq) + \cancel{2e^-} \rightarrow Co(s)$
Oxidation:	$Sn(s) \rightarrow Sn^{2+}(aq) + \cancel{2e^-}$
Redox:	$Co^{2+}(aq) + Sn(s) \rightarrow Co(s) + Sn^{2+}(aq)$

EXAMPLE 18.2 Combine the following half-reactions to produce a balanced redox equation. Identify the oxidation half-reaction and reduction half-reaction.

$$Fe^{2+}(aq) \rightarrow Fe^{3+}(aq) + e^-$$

$$Al^{3+}(aq) + 3\ e^- \rightarrow Al(s)$$

Oxidation:	$3\ Fe^{2+}(aq) \rightarrow 3\ Fe^{3+}(aq) + \cancel{3e^-}$
Reduction:	$Al^{3+}(aq) + \cancel{3e^-} \rightarrow Al(s)$
Redox:	$Al^{3+}(aq) + 3\ Fe^{2+}(aq) \rightarrow Al(s) + 3\ Fe^{3+}(aq)$

In this example it is necessary to multiply the oxidation half-reaction equation by 3 in order to balance the electrons gained and lost.

Another reaction involving iron and aluminum introduces an additional technique.

EXAMPLE 18.3 Arrange and modify the following half-reactions as necessary, so they add up to produce a balanced redox equation. Identify the oxidation half-reaction and the reduction half-reaction.

$Fe^{2+}(aq) + 2\ e^- \rightarrow Fe(s)$; $Al(s) \rightarrow Al^{3+}(aq) + 3\ e^-$

This will extend you a bit when it comes to balancing electrons. Two electrons are transferred for each atom of iron and three per atom of aluminum. In what ratio must the atoms be used to equate the electrons gained and lost? Multiply and add the half-reaction equations accordingly.

Reduction:	$3\ Fe^{2+}(aq) + 6e^- \rightarrow 3\ Fe(s)$
Oxidation:	$2\ Al(s) \rightarrow 2\ Al^{3+}(aq) + 6e^-$
Redox:	$3\ Fe^{2+}(aq) + 2\ Al(s) \rightarrow 3\ Fe(s) + 2\ Al^{3+}(aq)$

In this example electrons are transferred two at a time in the iron half-reaction and three at a time in the aluminum half-reaction. The simplest way to equate these is to take the iron half-reaction three times and the aluminum half-reaction twice. This gives six electrons for both half-reactions—just as two Al^{3+} and three O^{2-} balance the positive and negative charges in the ions making up the formula of Al_2O_3.

18.2 OXIDATION NUMBERS AND REDOX REACTIONS

PG 18C Given the formula of a chemical species, determine the oxidation number of each element it contains.

18D Distinguish between oxidation and reduction in terms of oxidation number change.

The redox reactions that we have discussed up to this point have been rather simple ones involving only two reactants. With Equations 18.3 and 18.5–18.7 we can see at a glance which species has gained and which has lost electrons. Some oxidation–reduction reactions are not so readily analyzed. Consider, for example, a reaction that is sometimes used in the general chemistry laboratory to prepare chlorine gas from hydrochloric acid:

$$MnO_2(s) + 4\ H^+(aq) + 2\ Cl^-(aq) \rightarrow Mn^{2+}(aq) + Cl_2(g) + 2\ H_2O(\ell); \quad (18.8)$$

or the reaction, taking place in a lead storage battery, that produces the electrical spark to start an automobile:

$$Pb(s) + PbO_2(s) + 4\ H^+(aq) + 2\ SO_4{}^{2-}(aq) \rightarrow 2\ PbSO_4(s) + 2\ H_2O(\ell) \quad (18.9)$$

Looking at these equations, it is by no means obvious which species are gaining and which are losing electrons.

"Electron bookkeeping" in redox reactions like Equations 18.8 and 18.9 is accomplished by using **oxidation numbers,** which were introduced in Section 12.1. By following a set of rules, oxidation numbers may be assigned to each element in a molecule or ion. The rules are repeated here for your convenience:

1. The oxidation number of any elemental substance is 0 (zero).
2. The oxidation number of an element as a monatomic ion is the same as the charge on the ion.
3. The oxidation number of combined oxygen is -2, except in peroxides (-1), superoxides $(-\frac{1}{2})$, and OF_2 $(+2)$.
4. The oxidation number of combined hydrogen is $+1$, except in hydrides (-1).
5. In any molecular or ionic species, the sum of the oxidation numbers of all atoms in a formula unit is equal to the charge on the unit.

EXAMPLE 18.4 What are the oxidation numbers of the elements in MnO_2?

Oxidation Rules 1, 2, and 4 do not apply. Rule 3 gives one of the two oxidation numbers that are required. What is it?

Oxygen, -2

Manganese can exist in several different oxidation states. You can decide which one by applying Rule 5. The thought process is the same as in naming a compound, as in distinguishing between FeO and Fe_2O_3. In fact, an acceptable name for MnO_2 is manganese( ) oxide, where the oxidation number goes into the parentheses. What is that oxidation number?

$+4$—manganese(IV) oxide

Note that Rule 5 requires the sum of the oxidation numbers of *atoms* in the formula unit. There are two oxygen atoms, each at -2. The total contribution of oxygen is 2×-2, or -4. The sum of -4 plus the oxidation number of manganese is equal to 0, the total charge on the species. Manganese must therefore be $+4$.

There is a mechanical way to reach the same conclusion that you might find helpful in more complicated examples. Applied to MnO_2, it is:

Write the formula with space between the symbols of elements or ions. Place the oxidation number of each element or ion beneath its symbol. Use n for the unknown oxidation number.	Mn O$_2$ n -2
Multiply each oxidation number by the number of atoms of that element in the formula unit.	Mn O$_2$ n $2(-2)$
Add the oxidation numbers, set them equal to the charge on the species, and solve for the unknown oxidation number. In this case, n = +4.	Mn O$_2$ $n + 2(-2) = 0$ $n = +4$

EXAMPLE 18.5 Find the oxidation number of:

(a) S in SO_4^{2-} (b) Cr in $HCr_2O_7^-$ (c) Fe in $Fe_2(SO_4)_3$

(a) S O$_4$
$n + 4(-2) = -2$
$n = +6$

(b) H Cr$_2$ O$_7$
$1 + 2(n) + 7(-2) = -1$
$n = +6$

(c) Fe$_2$ (SO$_4$)$_3$
$2(n) + 3(-2) = 0$
$n = +3$

Let's glance back to some of the equations in Section 18.1 to see what happens to oxidation numbers during a redox reaction. Looking first at the oxidation half-reaction from Equation 18.5, $Zn \rightarrow Zn^{2+} + 2e^-$, the oxidation number of Zn is 0, by Rule 1; the oxidation number of Zn^{2+} is +2, by Rule 2. The oxidation state of zinc *increased* from 0 to +2 as it was *oxidized*. Similarly, from

1. Equation 18.6, bromine was oxidized from −1 to 0: $2\ Br^- \rightarrow Br_2$. The oxidation number increased.
2. Example 18.2, Fe^{2+} was oxidized from +2 to +3: $Fe^{2+} \rightarrow Fe^{3+}$. The oxidation number increased.

If we examine an endless number of *oxidation* half-reactions we find that in each case there is an *increase* in the oxidation number of the element oxidized.

At this point you probably suspect that reduction is accompanied by a reduction in oxidation number. So it is. For the same reactions:

Equation 18.5: $H^+ \rightarrow H_2$, a reduction from +1 to 0

Equation 18.6: $Cl_2 \rightarrow 2\ Cl^-$, a reduction from 0 to −1

Example 18.2: $Al^{3+} \rightarrow Al$, a reduction from +3 to 0

We can now state a second and broader definition of oxidation and reduction. **Oxidation is an increase in oxidation number; reduction is a reduction in oxidation number.** These definitions are more useful in identifying the elements oxidized and reduced in a redox reaction when electron transfer is not apparent. All you must do is find the elements that change oxidation number and determine the direction of the change.

There are some techniques that enable you to spot quickly an element that changes oxidation number, or to dismiss quickly some elements that do not change. These are:

1. An element that is in its elemental state must change. As an element on one side of the equation, its oxidation number is 0; as anything other than an element on the other side, it is *not* 0.
2. In other than elemental form, hydrogen is +1 and oxygen is −2. Unless they are elements on one side, they do not change. In more advanced courses you will have to be alert to the hydride, peroxide, and superoxide exceptions noted in the oxidation number rules.
3. A Group 1A or 2A element has only one oxidation state other than 0. If it does not appear as an element, it does not change. This observation is helpful when you must find the element oxidized or reduced in a conventional equation.

We will now use these ideas to find the elements oxidized and reduced in Equation 18.8,

$$MnO_2(s) + 4\ H^+(aq) + 2\ Cl^-(aq) \rightarrow Mn^{2+}(aq) + Cl_2(g) + 2\ H_2O(\ell)$$

Chlorine is an element on the right, so it must be something else on the left. It is—the chloride ion, Cl^-. The oxidation number change is −1 to 0, an *increase*. Chlorine is *oxidized*.

Neither hydrogen nor oxygen appears as elements, so we conclude that they do not change oxidation state. That leaves manganese. Its oxidation state is +4 in MnO_2 on the left, and +2 as Mn^{2+} on the right. This is a *decrease,* from +4 to +2; manganese is *reduced.*

EXAMPLE 18.6 Determine the element oxidized and the element reduced in a lead storage battery, Equation 18.9:

$$Pb(s) + PbO_2(s) + 4\ H^+(aq) + 2\ SO_4^{2-}(aq) \rightarrow 2\ PbSO_4(s) + 2\ H_2O(\ell)$$

Assign oxidation numbers to as many elements as necessary until you come up with the pair that changed. Then identify the oxidation and reduction changes. (Careful. This one is a bit tricky.)

----- -----

Lead is both oxidized (0 in Pb to +2 in $PbSO_4$) and reduced (+4 in PbO_2 to +2 in $PbSO_4$).

The oxidation of lead can be spotted quickly, as it is an element on the left. You might have thought sulfur to be the element reduced, but its oxidation state is +6 in the sulfate ion whether the ion is by itself on the left, or part of a solid ionic compound on the right.

While oxidation number is a very useful device to keep track of what the electrons are up to in a redox reaction, we should emphasize that it is a manmade concept that has no real physical basis. Unlike the charge of a monatomic ion, the oxidation number of an atom in a molecule or polyatomic ion cannot be measured in the laboratory. It is all very well to talk about "+4 manganese" in MnO_2 or "+6 sulfur" in the SO_4^{2-} ion, but take care not to fall into the trap of thinking that the elements in these species actually carry positive charges equal to their oxidation numbers.

Quick Check 18.2

Oxidation and reduction can be defined as a gain or loss of electrons, or an increase or decrease in oxidation number. Give these definitions by inserting *gain, loss, increase,* or *decrease* in the following table:

	OXIDATION	REDUCTION
Change in Electrons		
Change in Oxidation Number		

18.3 OXIDIZING AGENTS (OXIDIZERS); REDUCING AGENTS (REDUCERS)

PG 18E Distinguish between an oxidizing agent and a reducing agent.

The two essential reactants in a redox reaction are given special names to indicate the role they play. The species that accepts electrons is referred to as an **oxidizing agent,** or **oxidizer;** the species that donates the electrons so reduction can occur is called a **reducing agent,** or **reducer.** For example, in Equation 18.5,

$$2\ H^+(aq) + Zn(s) \rightarrow H_2(g) + Zn^{2+}(aq)$$

H^+ has accepted electrons from Zn—it has *oxidized* Zn to Zn^{2+}—and is therefore the oxidizing agent. Conversely, Zn has donated electrons to H^+—it has reduced H^+ to H_2—and is therefore the reducing agent. In Equation 18.8,

$$MnO_2(s) + 4\ H^+(aq) + 2\ Cl^-(aq) \rightarrow Mn^{2+}(aq) + Cl_2(g) + 2\ H_2O(\ell)$$

Cl^- is the reducer, reducing manganese from +4 to +2. The oxidizer is MnO_2—the whole compound, not just the Mn; it oxidizes chlorine from −1 to 0.

The following example summarizes the redox concepts:

EXAMPLE 18.7 Consider the redox equation

$$5\ NO_3^-(aq) + 3\ As(s) + 2\ H_2O(\ell) \rightarrow 5\ NO(g) + 3\ AsO_4^{3-}(aq) + 4\ H^+(aq)$$

(a) Determine the oxidation number in each species:

N: _____ in NO_3^-, and _____ in NO

O: _____ in NO_3^-, _____ in H_2O, _____ in NO, and _____ in AsO_4^{3-}

As: _____ in As, and _____ in AsO_4^{3-}

H: _____ in H_2O, and _____ in H^+

(b) Identify (1) the element oxidized _____ (2) the element reduced _____
(3) the oxidizing agent _____ (4) the reducing agent _____

(a) N: +5 in NO_3^-, and +2 in NO
O: −2 in all species
As: 0 in As, and +5 in AsO_4^{3-}
H: +1 in both H_2O and H^+.

(b) (1) As is oxidized, increasing in oxidation number from 0 to +5.
(2) N is reduced, decreasing in oxidation number from +5 to +2.
(3) NO_3^- is the oxidizing agent, removing electrons from As.
(4) As is the reducing agent, furnishing electrons to NO_3^-.

18.4 REDOX REACTIONS AND ACID–BASE REACTIONS COMPARED

PG 18F Identify the essential difference between a redox reaction and an acid–base reaction.

18G Distinguish between a strong oxidizing agent and a weak oxidizing agent; between a strong reducing agent and a weak reducing agent.

At this point it may be useful to pause briefly and point out how redox reactions resemble acid–base reactions.

1. Acid–base reactions involve a transfer of protons; redox reactions, a transfer of electrons.
2. In both cases, the reactants are given special names to indicate their role in the transfer process. An acid is a proton donor; a base is a proton acceptor. A reducing agent is an electron donor; an oxidizing agent is an electron acceptor.
3. Just as certain species (e.g., HCO_3^-, H_2O) can either donate or accept protons and thereby behave as an acid in one reaction and a base in another, certain species can either accept or donate electrons, acting as an oxidizing agent in one reaction and a reducing agent in another. An example is the Fe^{2+} ion, which can oxidize Zn atoms to Zn^{2+} in the reaction

$$Fe^{2+}(aq) + Zn(s) \rightarrow Fe(s) + Zn^{2+}(aq)$$

Fe^{2+} can also reduce Cl_2 molecules to Cl^- ions in another reaction:

$$Cl_2(g) + 2\ Fe^{2+}(aq) \rightarrow 2\ Cl^-(aq) + 2\ Fe^{3+}(aq)$$

4. Just as acids and bases may be classified as "strong" or "weak" depending on how readily they donate or accept protons, the strengths of oxidizing and reducing agents may be compared according to their tendencies to attract or release electrons. A "strong" oxidizing agent is one that has a strong attraction for electrons; a "strong" reducing agent gives up electrons readily.
5. Just as most acid–base reactions in solution reach a state of equilibrium, so most aqueous redox reactions reach equilibrium. Just as the favored side of an acid–base equilibrium can be predicted from acid–base strength, so the favored side of a redox equilibrium can be predicted from oxidizer–reducer strength.

Let's look more closely into the matter of the strength of oxidizing and reducing agents.

18.5 STRENGTHS OF OXIDIZING AGENTS AND REDUCING AGENTS

PG 18H Given a table of relative strengths of oxidizing and reducing agents, identify the stronger and weaker of two oxidizing or two reducing agents.

With a proper selection of electrodes, a pH meter (Section 17.5) can be used to measure electrical potential, or voltage, in a redox reaction. The instrument

Table 18.1 Relative Strengths of Oxidizing and Reducing Agents

	Oxidizing Agent		Reducing Agent	
↓Decreasing——STRENGTH——Increasing↑	$F_2(g) + 2e^-$	→	$2\ F^-$	↓Increasing——STRENGTH——Decreasing↑
	$Cl_2(g) + 2e^-$	→	$2\ Cl^-$	
	$\frac{1}{2}\ O_2(g) + 2\ H^+ + 2e^-$	→	H_2O	
	$Br_2(\ell) + 2e^-$	→	$2\ Br^-$	
	$NO_3^- + 4\ H^+ + 3e^-$	→	$NO(g) + 2\ H_2O$	
	$Ag^+ + e^-$	→	$Ag(s)$	
	$Fe^{3+} + e^-$	→	Fe^{2+}	
	$I_2(s) + 2e^-$	→	$2\ I^-$	
	$Cu^{2+} + 2e^-$	→	$Cu(s)$	
	$2\ H^+ + 2e^-$	→	$H_2(g)$	
	$Ni^{2+} + 2e^-$	→	$Ni(s)$	
	$Co^{2+} + 2e^-$	→	$Co(s)$	
	$Cd^{2+} + 2e^-$	→	$Cd(s)$	
	$Fe^{2+} + 2e^-$	→	$Fe(s)$	
	$Zn^{2+} + 2e^-$	→	$Zn(s)$	
	$Al^{3+} + 3e^-$	→	$Al(s)$	
	$Na^+ + e^-$	→	$Na(s)$	
	$Ca^{2+} + 2e^-$	→	$Ca(s)$	
	$Li^+ + e^-$	→	$Li(s)$	

is, in fact, the voltmeter that appears in Figure 18.1. Voltages from such cells enable us to determine the relative strengths of oxidizing and reducing agents.

Table 18.1 is a list of oxidizing agents in order of decreasing strength on the left side of the equation, and of reducing agents in order of increasing strength on the right. (Compare to Table 17.1, Section 17.3, which lists acids in order of decreasing strength on the left and bases in order of increasing strength on the right.) The strongest oxidizing agent in the table is the F_2 molecule, located at the upper left. Chlorine, Cl_2, listed just below F_2, is used as a disinfectant in water supplies because of its ability to oxidize harmful organic matter.

Perhaps you recognize Table 18.1 as the source of the activity series (Table 16.1, Section 16.5) that you used to predict simple redox reactions in writing net ionic equations. Such predictions are based on the relative strengths of the reducing agents. The activity series corresponds with the right side of Table 18.1, from the bottom to the top.

18.6 PREDICTING REDOX REACTIONS

PG 18I Given a table of relative strengths of oxidizing and reducing agents and the identity of an oxidizer and a reducer from the table, (a) write the net ionic equation for the redox reaction between them, and (b) predict whether the forward or reverse reaction will be favored at equilibrium.

As Table 17.1 enables us to write acid–base reaction equations and predict the direction that will be favored at equilibrium, Table 18.1 enables us to do the same things for redox reactions. The redox table has a limitation, however. Acid–base reactions are all *single*-proton transfer reactions and equations are automatically balanced if taken directly from the table. Redox half-reactions, on the other hand, frequently involve unequal numbers of electrons. They must be balanced as in Equation 18.7 and Example 18.2. The next five examples illustrate the process.

EXAMPLE 18.8 Write the net ionic equation for the redox reaction between the cobalt(II) ion, Co^{2+}, and metallic silver, Ag.

SOLUTION: First, as in a Brönsted–Lowry acid–base reaction there must be a proton giver and a proton taker, so in a redox reaction there must be an electron giver (reducer) and an electron taker (oxidizer). Consulting Table 18.1, we find Co^{2+} among the oxidizers and Ag among the reducers. The reduction half-reaction is

Reduction: $Co^{2+}(aq) + 2e^- \rightarrow Co(s)$

To obtain the oxidation half-reaction for silver it is necessary to *reverse* the reduction half-reaction found in the table:

Oxidation: $Ag(s) \rightarrow Ag^+(aq) + e^-$

Multiplying the oxidation equation by 2 to equalize electrons gained and lost, and adding to the reduction equation yields

Reduction: $Co^{2+}(aq) + \cancel{2e^-} \rightarrow Co(s)$

2 × Oxidation: $2\,Ag(s) \rightarrow 2\,Ag^+(aq) + \cancel{2e^-}$

Redox: $Co^{2+}(aq) + 2\,Ag(s) \rightarrow Co(s) + 2\,Ag^+(aq)$

The principle underlying the prediction of the favored direction of a redox reaction is the same as for an acid–base reaction. The stronger oxidizing agent—strong in its attraction for electrons—will take the electrons from a strong reducing agent—strong in its tendency to donate electrons—to produce the weaker oxidizer and reducer. This is shown in Figure 18.3, which is strikingly similar to Figure 17.1, Section 17.5. In the reaction $2\,H^+(aq) + Zn(s) \rightleftharpoons H_2(g) + Zn^{2+}(aq)$ (Equation 18.2), the positions in the table establish H^+ and Zn as the stronger oxidizer and reducer. The reaction is favored in the forward direction, yielding the weaker oxidizer and reducer, Zn^{2+} and H_2. (Note: We will use double arrows when necessary to indicate the equilibrium character of redox reactions.)

EXAMPLE 18.9 In which direction, forward (__) or reverse (__) will the redox reaction in Example 18.8 be favored?

Reverse

Ag^+ is a stronger oxidizer than Co^{2+}, and is therefore able to take electrons from cobalt atoms. Also, cobalt atoms are a stronger reducer than silver atoms, and therefore readily release electrons to Ag^+. The weaker reducer and oxidizer, Ag and Co^{2+}, are favored.

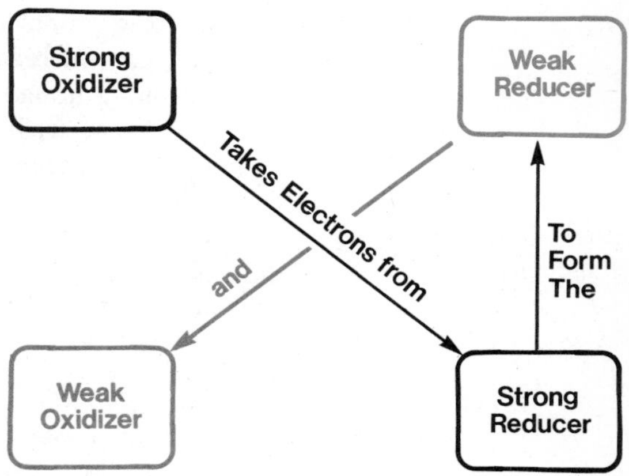

Figure 18.3
Correlation between prediction of a redox reaction and position in Table 18.1. The stronger oxidizing agent (*upper left*) takes electrons from the stronger reducing agent (*lower right*), yielding the weaker reducing agent (*upper right*) and weaker oxidizing agent (*lower left*). That direction, foward or reverse, will be favored that has the weaker reducer and oxidizer as products. The similarity between acid–base and redox reaction predictions may be seen by comparing this illustration with Figure 17.1.

EXAMPLE 18.10 Write the redox reaction equation between metallic copper and a strong acid, H^+, and indicate the direction that is favored.

Reduction: $2\ H^+(aq) + \cancel{2}e^- \rightleftharpoons H_2(g)$

Oxidation: $Cu(s) \rightleftharpoons Cu^{2+}(aq) + \cancel{2e^-}$

Redox: $2\ H^+(aq) + Cu(s) \rightleftharpoons H_2(g) + Cu^{2+}(aq)$

The reverse reaction is favored.

EXAMPLE 18.11 Write the net ionic equation for the redox reaction between Al(s) and $Ni^{2+}(aq)$, and predict the favored direction, forward or reverse.

Reduction half-reaction: $Ni^{2+}(aq) + 2e^- \rightleftharpoons Ni(s)$

Oxidation half-reaction: $Al(s) \rightleftharpoons Al^{3+}(aq) + 3e^-$

3 × Reduction: $3\ Ni^{2+}(aq) + \cancel{6e^-} \rightleftharpoons 3\ Ni(s)$

2 × Oxidation: $2\ Al(s) \rightleftharpoons 2\ Al^{3+}(aq) + \cancel{6e^-}$

$$3\ Ni^{2+}(aq) + 2\ Al(s) \rightleftharpoons 3\ Ni(s) + 2\ Al^{3+}(aq)$$

The forward reaction is favored.

One of the properties of acids listed in Section 17.1 is their ability to release hydrogen gas on reaction with certain metals. Judging from Example 18.10, copper is not among those metals. The metals that do release hydrogen are the reducers below hydrogen in Table 18.1. But there is more to the reactions between metals and acids than meets the eye.

EXAMPLE 18.12 Write the equation for the reaction between copper and nitric acid, and predict which direction is favored.

Copper is in Table 18.1, but you will search in vain for HNO_3. The solution inventory species of nitric acid (NO_3^- and H^+) are present, though. We will comment on the imbalance between hydrogen ions and nitrate ions shortly. This reaction summarizes our equation writing methods to this point. Take it all the way.

2 × Reduction: $2\ NO_3^-(aq) + 8\ H^+(aq) + \cancel{6e^-} \rightleftharpoons 2\ NO(g) + 4\ H_2O(\ell)$

3 × Oxidation: $3\ Cu(s) \rightleftharpoons 3\ Cu^{2+}(aq) + \cancel{6e^-}$

Redox: $2\ NO_3^-(aq) + 8\ H^+(aq) + 3\ Cu(s) \rightleftharpoons 2\ NO(g) + 4\ H_2O(\ell) + 3\ Cu^{2+}(aq)$

The forward reaction is favored.

Don't worry about those missing nitrate ions, the six unaccounted for from the eight moles of HNO_3 that furnished the 8 H^+. They're in there as spectators, just enough to balance the 3 Cu^{2+}.

18.7 WRITING REDOX EQUATIONS

PG 18J given the identify of an oxidizer and reducer, and the identify of their reduced and oxidized products in an acidic solution, write the net ionic equation for the reaction.

Thus far we have considered only redox reactions for which the oxidation and reduction half-reactions are known. We are not always this fortunate. Sometimes we know only the reactants and products. Considering nitric acid, for example, suppose we know only that the product of the reduction of nitric acid is NO(g). How do we get from this information to the reduction half-reaction given in Table 18.1?

The steps for writing a half-reaction equation in an acidic solution are listed below. Each step is illustrated for the NO_3^- to NO change in Example 18.12.

1. *After identifying the element oxidized or reduced, write a partial half-reaction equation with the element in its original form (element, monatomic*

ion, or part of a polyatomic ion or compound) on the left, and in its final form on the right.

$$NO_3^-(aq) \rightarrow NO(g)$$

2. *Balance the element oxidized or reduced.*
   Nitrogen is already balanced.
3. *Balance elements other than hydrogen or oxygen, if any.*
   There are none.
4. *Balance oxygen by adding water molecules where necessary.*
   There are three oxygens on the left and one on the right. Two water molecules are needed on the right.

$$NO_3^-(aq) \rightarrow NO(g) + 2\ H_2O(\ell)$$

5. *Balance hydrogen by adding H^+ where necessary.*
   There are four hydrogens on the right and none on the left. Four hydrogen ions are needed on the left.

$$4\ H^+(aq) + NO_3^-(aq) \rightarrow NO(g) + 2\ H_2O(\ell)$$

6. *Balance charge by adding electrons to the more positive side.*
   Total charge on the left is $+4 + (-1) = +3$; on the right, zero. Three electrons are needed on the left.

$$3\ e^- + 4\ H^+(aq) + NO_3^-(aq) \rightarrow NO(g) + 2\ H_2O(\ell)$$

7. *Recheck the equation to be sure it is balanced in both atoms and charge.*

Notice that the above instructions are for redox half-reactions in *acidic* solutions. The procedure is somewhat different with basic solutions, but we will omit that procedure in this introductory text.

When you have both half-reaction equations, proceed as in the earlier examples.

EXAMPLE 18.13 Write the net ionic equation for the redox reaction between iodide and sulfate ions in an acidic solution. The products are iodine and sulfur: $I^-(aq) + SO_4^{2-}(aq) \rightarrow I_2(s) + S(s)$.

First, identify the element reduced and the element oxidized.

Sulfur is reduced (ox. no. change $+6$ to 0) and iodine is oxidized (-1 to 0).

Balance atoms first, and then charges, in the oxidation half-reaction:

$$I^-(aq) \rightarrow I_2(s)$$

$2\ I^-(aq) \rightarrow I_2(s) + 2e^-$. (This one happens to be in Table 18.1.)

Now for the reduction half-reaction:

$$SO_4^{2-}(aq) \rightarrow S(s)$$

Sulfur is already in balance. The only other element is oxygen. According to Step 4, oxygen is balanced by adding the necessary water molecules. Complete that step.

----- -----

$$SO_4^{2-}(aq) \rightarrow S(s) + 4\ H_2O(\ell)$$

Four oxygen atoms in a sulfate ion require four water molecules.

Next comes the hydrogen balancing, using H^+ ions.

----- -----

$$8\ H^+(aq) + SO_4^{2-}(aq) \rightarrow S(s) + 4\ H_2O(\ell)$$

Finally, add to the positive side the electrons that will bring the charges into balance.

----- -----

$$6\ e^- + 8\ H^+(aq) + SO_4^{2-}(aq) \rightarrow S(s) + 4\ H_2O(\ell)$$

On the left there are 8 + charges from hydrogen ion and 2 − charges from sulfate ion, a net of 6 +. On the right the net charge is zero. Charge is balanced by adding 6 electrons to the left (positive) side.

Now that you have the two half-reaction equations, finish writing the net ionic equation as you did before.

$$2\ I^-(aq) \rightarrow I_2(s) + 2e^-$$
$$8H^+ + SO_4^{2-}(aq) + 6e^- \rightarrow S(s) + 4\ H_2O(\ell)$$

----- -----

Reduction: $8\ H+(aq) + SO_4^{2-}(aq) + \cancel{6e^-} \rightarrow S(s) + 4\ H_2O(\ell)$

3 × Oxidation: $6\ I^-(aq) \rightarrow 3\ I_2(s) + \cancel{6e^-}$

Redox: $8\ H^+(aq) + SO_4^{2+}(aq) + 6\ I^-(aq) \rightarrow 4\ H_2O(\ell) + 3\ I_2(s)$

Let's check to make sure the equation is, indeed, balanced:

—the atoms balance (six I, one S, four O, and eight H atoms on each side);
—the charge balances $(+8\ -2\ -6 = 0 + 0 + 0)$.

EXAMPLE 18.14 The permanganate ion, MnO_4^-, is a strong oxidizing agent that oxidizes chloride ion to chlorine in an acidic solution. Manganese ends up as a monatomic manganese(II) ion. Write the net ionic equation for the redox reaction.

This is a challenging example, but watch how it falls into place following the procedure that has been outlined. To be sure the question has been interpreted correctly, begin by writing an unbalanced skeleton equation. Put the identified reactants on the left and the identified products on the right.

$$MnO_4^-(aq) + Cl^-(aq) \rightarrow Cl_2(g) + Mn^{2+}(aq)$$

The oxidation half-reaction is easiest. Write it next.

$$2\ Cl^-(aq) \rightarrow Cl_2(g) + 2e^-$$

With a switch from iodine to chlorine, this is the same oxidation half-reaction as in the last example.

Now write the formulas of the starting and ending species for the reduction reaction on opposite sides of the arrow.

$$MnO_4^-(aq) \rightarrow Mn^{2+}(aq)$$

Can you take the reduction to a complete half-reaction equation? First do oxygen, then hydrogen, and finally charge.

$$8\ H^+(aq) + MnO_4^-(aq) + 5e^- \rightarrow Mn^{2+}(aq) + 4\ H_2O(\ell)$$

By steps:

- Oxygen: $MnO_4^-(aq) \rightarrow Mn^{2+}(aq) + 4\ H_2O(\ell)$ (four waters for four oxygen atoms).
- Hydrogen: $8\ H^+(aq) + MnO_4^-(aq) \rightarrow Mn^{2+}(aq) + 4\ H_2O(\ell)$ (eight H^+ for four waters).
- Charge: $+8 - 1$ on left is $+7$; $+2 + 0$ on right is $+2$. Charge is balanced by adding 5 negatives on the left, or five electrons, as in the final answer.

Now rewrite and combine the half-reaction equations for the net ionic equation.

2 × Reduction:	$16\ H^+(aq) + 2\ MnO_4^-(aq) + 10e^- \rightarrow 2\ Mn^{2+}(aq) + 8\ H_2O(\ell)$
5 × Oxidation:	$10\ Cl^-(aq) \rightarrow 5\ Cl_2(g) + 10e^-$
Redox:	$16\ H^+(aq) + 2\ MnO_4^-(aq) + 10\ Cl^-(aq) \rightarrow 2\ Mn^{2+}(aq) + 8\ H_2O(\ell) + 5\ Cl_2(g)$

Checking:
—atoms balance (sixteen H, two Mn, eight O, and ten Cl);
—charges balance ($+16 - 2 - 10 = +4$).
Can you imagine a trial and error approach to an equation such as this?

CHAPTER 18 IN REVIEW

18.1 Electron-Transfer Reactions

18A Distinguish between oxidation and reduction in terms of electrons gained or lost.

18B Given an oxidation half-reaction equation and a reduction half-reaction equation, add them to obtain a balanced redox reaction equation.

18.2 Oxidation Numbers and Redox Reactions

18C Given the formula of a chemical species, determine the oxidation number of each element it contains.

18D Distinguish between oxidation and reduction in terms of oxidation number change.

18.3 Oxidizing Agents (Oxidizers); Reducing Agents (Reducers)

18E Distinguish between an oxidizing agent and a reducing agent.

18.4 Redox Reactions and Acid–Base Reactions Compared

18F Identify the essential difference between a redox reaction and an acid–base reaction.

18G Distinguish between a strong oxidizing agent and a weak oxidizing agent; between a strong reducing agent and a weak reducing agent.

18.5 Strengths of Oxidizing Agents and Reducing Agents

18H Given a table of relative strengths of oxidizing and reducing agents, identify the stronger and weaker of two oxidizing or two reducing agents.

18.6 Predicting Redox Reactions

18I Given a table of relative strengths of oxidizing and reducing agents and the identity of an oxidizer and a reducer from the table, (a) write the net ionic equation for the redox reaction between them, and (b) predict whether the forward or reverse reaction will be favored at equilibrium.

18.7 Writing Redox Equations

18J Given the identity of an oxidizer and reducer, and the identity of their reduced and oxidized products in an acidic solution, write the net ionic equation for the reaction.

TERMS AND CONCEPTS

18.1 Voltaic cell
Oxidation; oxidize
Half-reaction; half-reaction equation
Reduction; reduce
Electron-transfer reaction

18.2 Oxidation number

18.3 Oxidizing agent; oxidizer
Reducing agent; reducer

Most of these terms and many others appear in the Glossary. Use the Glossary regularly.

QUESTIONS AND PROBLEMS

Section 18.1

(*1*) Define oxidation; define reduction. It is sometimes said that oxidation and reduction are simultaneous processes—that you cannot have one without the other. From the standpoint of your definitions, explain why.

(25) Using any example of a redox reaction, explain why such reactions are described as electron transfer reactions.

(2) Classify each of the following half-reaction equations as oxidation or reduction half-reactions:
(a) $2\ Cl^- \rightarrow Cl_2 + 2e^-$
(b) $Na \rightarrow Na^+ + e^-$
(c) $Sn^{2+} \rightarrow Sn^{4+} + 2e^-$
(d) $O_2 + 4\ H^+ + 4e^- \rightarrow 2\ H_2O$

(26) Classify each of the following half-reaction equations as oxidation or reduction half-reactions:
(a) $Zn \rightarrow Zn^{2+} + 2e^-$
(b) $2\ H^+ + 2e^- \rightarrow H_2$
(c) $Fe^{2+} \rightarrow Fe^{3+} + e^-$
(d) $NO + 2\ H_2O \rightarrow NO_3^- + 4\ H^+ + 3e^-$

For the next two questions in each column, classify the equation given as an oxidation or a reduction half-reaction equation.

(3) Tarnishing of silver:

$$2\ Ag + S^{2-} \rightarrow Ag_2S + 2e^-$$

(27) Dissolving ozone, O_3, in water:

$$O_3 + H_2O + 2e^- \rightarrow O_2 + 2\ OH^-$$

(4) One side of an automobile battery:

$$PbO_2 + SO_4^{2-} + 4\,H^+ + 2e^- \rightarrow PbSO_4 + 2\,H_2O$$

(5) Combine the following half-reaction equations to produce a balanced redox equation:

$Ni^{2+} + 2e^- \rightarrow Ni$; $Mg \rightarrow Mg^{2+} + 2e^-$.

(6) The half-reaction at one electrode of a lead storage battery, used in automobiles and boats, is given in Question 18.4. The reaction at the other electrode is

$$Pb + SO_4^{2-} \rightarrow PbSO_4 + 2e^-$$

Write the equation for the overall battery reaction.

(28) Dissolving gold (Z = 79):

$$Au + 4\,Cl^- \rightarrow AuCl_4^- + 3e^-$$

(29) Combine the following half-reaction equations to produce a balanced redox equation:

$Cr \rightarrow Cr^{3+} + 3e^-$; $Cl_2 + 2e^- \rightarrow 2\,Cl^-$.

(30) The half-reactions that take place at the electrodes of an alkaline cell, widely used in flashlights, calculators, etc. are

$$Ni_2O_3 + 3\,H_2O + 2e^- \rightarrow 2\,Ni(OH)_2 + 2\,OH^-$$
$$Cd + 2\,OH^- \rightarrow Cd(OH)_2 + 2e^-$$

Which equation is for the oxidation half-reaction? Write the overall equation for the cell.

Section 18.2

In the next two questions in each column, give the oxidation state of the element whose symbol is underlined in each formula.

(7) $\underline{Mg}^{2+}$; $\underline{Cl}^-$; $\underline{Cl}O^-$; $K\underline{Cl}O_3$

(8) $\underline{N}_2O_5$; $\underline{N}H_4^+$; $\underline{Mn}O_4^-$; $Na_2H\underline{P}O_3$

(31) $\underline{Al}^{3-}$; $\underline{S}^{2-}$; $\underline{S}O_3^{2-}$; $Na_2\underline{S}O_4$

(32) $\underline{N}_2O_3$; $\underline{N}O_3^-$; $\underline{Cr}O_4^{2-}$; $NaH_2\underline{P}O_4$

In the next three questions in each column, (1) identify the element experiencing oxidation or reduction; (2) state "oxidized" or "reduced"; and (3) show the change in oxidation number. Example: $2\,Cl^- \rightarrow Cl_2 + 2e^-$. Chlorine oxidized from −1 to 0.

(9) (a) $Cu^{2+} + 2e^- \rightarrow Cu$
(b) $Co^{3+} + e^- \rightarrow Co^{2+}$

(10) (a) $H_2O + SO_3^{2-} \rightarrow SO_4^{2-} + 2\,H^+ + 2e^-$
(b) $PH_3 \rightarrow P + 3\,H^+ + 3e^-$

(11) (a) $2\,HF \rightarrow F_2 + 2\,H^+ + 2e^-$
(b) $MnO_4^{2-} + 2\,H_2O + 2e^- \rightarrow MnO_2 + 4\,OH^-$

(33) (a) $Br_2 + 2e^- \rightarrow 2\,Br^-$
(b) $Pb^{2+} + 2\,H_2O \rightarrow PbO_2 + 4\,H^+ + 2e^-$

(34) (a) $8\,H^+ + IO_4^- + 8e^- \rightarrow I^- + 4\,H_2O$
(b) $4\,H^+ + O_2 + 4e^- \rightarrow 2\,H_2O$

(35) (a) $NO_2 + H_2O \rightarrow NO_3^- + 2\,H^+ + e^-$
(b) $2\,Cr^{3+} + 7\,H_2O \rightarrow Cr_2O_7^{2-} + 14\,H^+ + 6e^-$

Section 18.3

(12) In the reaction between copper(II) oxide and hydrogen, identify the oxidizing agent and the reducing agent:

$$CuO + H_2 \rightarrow Cu + H_2O$$

(13) Name the oxidizing and reducing agents in the reaction:

$$BrO_3^- + 3\,HNO_2 \rightarrow Br^- + 3\,NO_3^- + 3\,H^+$$

(36) Identify the oxidizing and reducing agents in

$$Cl_2 + 2\,Br^- \rightarrow 2\,Cl^- + Br_2$$

(37) What is the oxidizing agent in the equation for the storage battery, $Pb + PbO_2 + 4\,H^+ + 2\,SO_4^{2-} \rightarrow 2\,PbSO_4 + 2\,H_2O$? What does it oxidize? Also name the reducing agent and the species it reduces.

Section 18.4

(14) Show how redox and acid-base reactions parallel each other—how they are similar, but also what makes them different.

(38) Explain how a strong acid is similar to a strong reducer. Also, explain how a strong base compares to a strong oxidizer.

Section 18.5

(15) Which is the stronger oxidizing agent, Ag^+ or H^+? What is the basis of your selection? What is the meaning of the statement that one substance is a stronger oxidizer than another?

(16) Arrange the following reducing agents in order of *decreasing* strength—i.e, strongest reducer first: H_2; Al; Cl^-; Fe^{2+}.

(39) Identify the stronger reducer between Zn and Fe^{2+}. On what basis do you make your decision? What is the significance of one reducer being stronger than another?

(40) Arrange the following oxidizers in order of *increasing* strength—i.e., weakest oxidizing agent first: Na^+; Br_2; Fe^{2+}; Cu^{2+}.

Section 18.6

In this section, write the redox equation for the two redox reactants given, using Table 18.1 as a source of the required half-reactions. Then predict the direction in which the reaction will be favored at equilibrium.

(17) $Ni + Zn^{2+} \rightleftharpoons$

(18) $Fe^{3+} + Co \rightleftharpoons$

(19) $O_2 + H^+ + Ca \rightleftharpoons$

(41) $Br_2 + I^- \rightleftharpoons$

(42) $H^+ + Br^- \rightleftharpoons$

(43) $NO + H_2O + Fe^{2+} \rightleftharpoons$

Section 18.7

In this section, each "equation" identifies an oxidizer and a reducer, as well as the oxidized and reduced products of the redox reaction. Write the separate oxidation and reduction half-reaction equations, assuming that the reaction takes place in an acidic solution, and add them to produce a balanced redox equation.

(20) $Ag + SO_4^{2-} \rightarrow Ag^+ + SO_2$

(21) $NO_3^- + Zn \rightarrow NH_4^+ + Zn^{2+}$

(22) $Cr_2O_7^{2-} + Fe^{2+} \rightarrow Cr^{3+} + Fe^{3+}$

(23) $I^- + MnO_4^- \rightarrow I_2 + MnO_2$

(24) $BrO_3^- + Br^- \rightarrow Br_2$

(44) $S_2O_3^{2-} + Cl_2 \rightarrow SO_4^{2-} + Cl^-$

(45) $Sn + NO_3^- \rightarrow H_2SnO_3 + NO_2$

(46) $C_2O_4^{2-} + MnO_4^- \rightarrow CO_2 + Mn^{2+}$

(47) $Cr_2O_7^{2-} + NH_4^+ \rightarrow Cr_2O_3 + N_2$

(48) $As_2O_3 + NO_3^- \rightarrow AsO_4^{3-} + NO$

Miscellaneous Questions

(49) Distinguish precisely and in scientific terms the difference between items in each of the following groups:

(a) Oxidation, reduction (in terms of electrons)
(b) Half-reaction equation, net ionic equation
(c) Oxidation, reduction (in terms of oxidation number)
(d) Oxidizing agent (oxidizer), reducing agent (reducer)
(e) Electron-transfer reaction, proton-transfer reaction
(f) Strong oxidizing agent, weak oxidizing agent
(g) Strong reducing agent, weak reducing agent
(h) Atom balance, charge balance (in equations)

(50) Classify each of the following statements as true or false:

(a) Oxidation and reduction occur at the electrodes in a voltaic cell.
(b) The sum of the oxidation numbers in a molecular compound is zero, but in an ionic compound that sum may or may not be zero.
(c) The oxidation number of oxygen is the same in all of the following: O^{2-}; $HClO_3^-$; $S_2O_3^{2-}$; NO_2.
(d) The oxidation number of an alkali metal is always -1.
(e) A substance that gains electrons and increases oxidation number is oxidized.
(f) A strong reducing agent has a strong attraction for electrons.
(g) The favored side of a redox equilibrium equation is the side with the weaker oxidizer and reducer.

(51) One of the properties of acids listed in Section 17.1 is "the ability to react with certain metals and release hydrogen." An acid is an acid, and a metal is a metal. Why is the property of acids limited to certain metals? Identify two metals that would release hydrogen from an acid and two that would not.

(52)* When writing a half-reaction equation that takes place in acidic solution, why is it permissible to use hydrogen ions and water molecules without regard to their source?

(53)* There is a fundamental difference between the electrolytic cell in Figure 16.2 and the voltaic cell in Figure 18.1. Examine both illustrations and see if you can identify that difference.

(54) *Anode* and *cathode* were defined in Section 16.2, with reference to Figure 16.2. Which electrode is the anode and which is the cathode in Figure 18.2? On what do you base your choice? Which half-reaction, oxidation or reduction, occurs at the cathode in Figure 16.2? in Figure 18.1?

(55)* In Figure 16.2, electrons appear to be flowing through the external (metal) part of the circuit from the positively charged electrode (anode) of the electrolytic cell to the negatively charged electrode (cathode). This is strange behavior for negatively charged electrons! Can you explain it?

(56) It is sometimes said that, in a redox reaction, the oxidizing agent is reduced and the reducing agent is oxidized. Is this statement (a) always correct, (b) never correct, or (c) sometimes correct. If you select (b) or (c), give an example in which the statement is incorrect.

19 Chemical Equilibrium

LOOKING BACK	LOOKING AHEAD IN CHAPTER 19
9.1 Coefficients in a chemical equation are used to convert from moles of one substance to moles of any other substance.	Equilibrium calculations (optional) include converting from moles of one species in a reaction to moles of another.
14.4 Equilibrium vapor pressure is greater at higher temperatures because a larger fraction of the liquid sample has enough kinetic energy to evaporate.	The rate of a chemical reaction is greater at higher temperatures because a larger fraction of the reactants has enough kinetic energy to engage in an effective particle collision.
14.4, 15.3 Two physical equilibria, the liquid–vapor equilibrium of a volatile liquid and the solute–solution equilibrium of a soluble salt, show that a reversible change reaches equilibrium when the rates of change in the forward and reverse directions become equal.	The dynamic character of an equilibrium is reviewed, analyzed, and summarized in a list of conditions that are satisfied by all equilibrium systems.
15.7 Molarity is the concentration of a solution in moles per liter.	Molarity is used for solution concentrations in equilibrium calculations (optional).
17.6 (Optional) Given either $[H^+]$ or pH, the other may be calculated.	Equilibrium calculations (optional) include conversions between pH and $[H^+]$.
Appendix I, Parts E and F Percentage and calculations with exponents are reviewed.	Equilibrium calculations (optional) include percentage and calculations with exponents.

19.1 THE CHARACTER OF AN EQUILIBRIUM

PG 19A Identify the essential characteristics of an equilibrium system.

A careful review of a liquid-vapor equilibrium (Section 14.4) and a solute-solution equilibrium (Section 15.3) reveals four conditions that are found in every equilibrium.

1. *The change is reversible and can be represented by an equation with a double arrow.* In a reversible change, the substances on the left side of the equation produce the substances on the right, and those on the right are changed back into the substances on the left.

2. *The equilibrium system is "closed"—closed in the sense that no substance can enter or leave the immediate vicinity of the equilibrium.* All substances on either side of an equilibrium equation must remain to form the substances on the other side.
3. *The equilibrium is dynamic.* The reversible changes occur continuously, even though there is no appearance of change. By contrast, items in a static equilibrium are stationary, without motion, as an object hanging on a spring.
4. *The things that are* equal *in an equilibrium are the forward rate of change (from left to right in the equation) and the reverse rate of change (from right to left).* Specifically, amounts of substances present in an equilibrium are *not* equal.

Quick Check 19.1
Can a solution in a beaker that is open to the atmosphere contain a chemical equilibrium? Explain.

19.2 THE COLLISION THEORY OF CHEMICAL REACTIONS

PG 19B Explain why, according to the collision theory of chemical reactions, some molecular collisions result in a chemical reaction and others do not.

If two molecules are to react chemically, it is reasonable to expect that they must come into contact with each other. What we see as a chemical reaction is the overall effect of a huge number of individual collisions between reacting particles. This view of chemical change is the **collision theory of chemical reactions.**

Figure 19.1 examines three kinds of molecular collisions for the imaginary reaction $A_2 + B_2 \rightarrow 2\ AB$. If the collision is to produce a reaction, the bond between A atoms in A_2 must be broken; similarly, the bond in the B_2 molecules must be broken. It takes energy to break these bonds. This energy comes from the kinetic energy of the molecules just before they collide. In other words, it must be a violent, bond-breaking collision. This is most apt to occur if the molecules are moving at high speed. Figure 19.1A pictures a reaction-producing collision.

Not all collisions result in a reaction; in fact, most of them do not. If the colliding molecules do not have enough kinetic energy to break the bonds, the original molecules simply bounce off each other with the same identity they had before the collision. Sometimes they may have enough kinetic energy, but only "sideswipe" each other and glance off unchanged (Fig. 19.1B). Other sufficiently energetic collisions may have an orientation that pushes atoms in the original molecules closer together, rather than pulling them apart (Fig. 19.1C).

To summarize: In order for an individual collision to result in a reaction, the particles must have (1) enough kinetic energy and (2) the proper orientation. The rate of a particular reaction depends on the frequency of effective collisions.

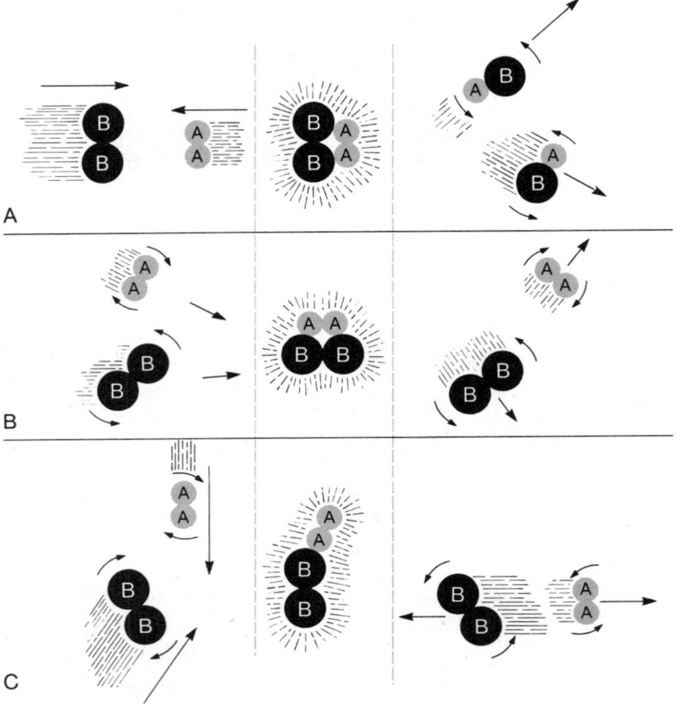

Figure 19.1
Molecular collisions and chemical reactions. (A.) Reaction-producing collision between molecules A_2 and B_2, which have sufficient kinetic energy and proper orientation. (B.) Glancing collision has proper orientation but insufficient kinetic energy. There is no reaction. (C.) Collision has sufficient kinetic energy, but poor orientation. There is no reaction.

Quick Check 19.2
Describe three kinds of molecular collisions that will not result in a chemical change.

19.3 ENERGY CHANGES DURING A MOLECULAR COLLISION

PG 19C Sketch and/or interpret an energy-reaction graph. Identify (a) the activated complex region, (b) activation energy, and (c) ΔE for the reaction.

When a rubber ball is dropped to the floor, it bounces back up. Just before it hits the floor, it has a certain amount of kinetic energy. It also has kinetic energy as it leaves the floor on the rebound. But during the time the ball is in contact with the floor, it slows down, even stops, and then builds velocity in an upward direction. During the period of reduced velocity, the kinetic energy, $\frac{1}{2}mv^2$, is reduced, and even reaches zero at the turnaround instant. While the collision is in progress, the initial kinetic energy is changed to potential energy in the partially flattened ball. The potential energy is changed back into kinetic energy as the ball bounces upward.

It is believed that there is a similar conversion of kinetic energy to potential energy during a collision between molecules. This can be shown by a graph of energy versus a reaction coordinate that traces the energy of the system before,

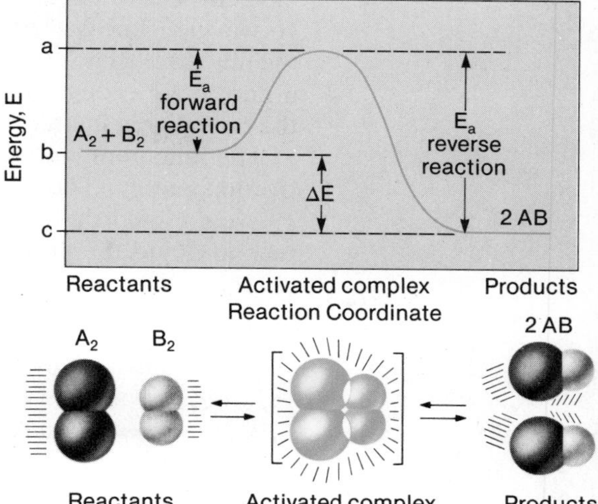

Figure 19.2
Energy-reaction graph for the reaction $A_2 + B_2/2\ AB$. E_a represents the activation energy for the forward and reverse reactions, as indicated. (Energy values a, b, and c are reference points in an end-of-chapter question.)

during, and after the collision (Fig. 19.2). The product energy minus the reactant energy is the ΔE for the reaction.

When two molecules are colliding, they form an **activated complex** that has a high potential energy in the hump of the curve. The increase in potential energy comes from the loss of kinetic energy during the collision. The activated complex is unstable. It quickly separates into two parts. If the collision is "effective" in producing a reaction, the parts will be product molecules; if the collision is ineffective, they will be the original reactant molecules.

The hump in Figure 19.2 is a potential energy barrier that must be surpassed before a collision can be effective. It is like rolling a ball over a hill. If the ball has enough kinetic energy to get to the top, it will roll down the other side (Fig. 19.3A). This corresponds to an effective collision in which the colliding

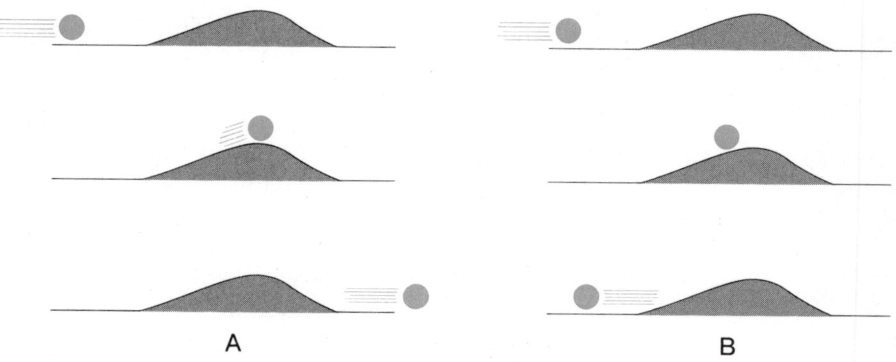

Figure 19.3
Potential energy barrier. As the ball rolls toward the hill (potential energy barrier), its speed (kinetic energy) determines whether or not it will pass over the hill. Kinetic energy is changed to potential energy as the ball climbs the hill. In A, it has more than enough kinetic energy to reach the top and roll down the other side. In B, all of the original kinetic energy is changed to potential energy before the ball reaches the top, so it rolls back down the same side.

molecules have enough kinetic energy to get over the potential energy barrier. However, if the ball does not have enough kinetic energy to reach the top of the hill, it rolls back to where it came from (Fig. 19.3B). If the colliding particles do not have enough kinetic energy to meet the potential energy requirement, the collision is ineffective.

The minimum energy needed to produce an effective collision is called **activation energy.** The difference between the energy at the peak of the Figure 19.2 curve and the reactant energies is the activation energy for the forward reaction, and the difference between the energy at the peak and the product energies is the activation energy for the reverse reaction.

Quick Check 19.3
In what sense is activation energy a *barrier?*

19.4 CONDITIONS THAT AFFECT THE RATE OF A CHEMICAL REACTION

The Effect of Temperature on Reaction Rate

PG 19D State and explain the relationship between reaction rate and temperature.

Chemical reactions are faster at higher temperatures. This is seen in two simple facts from the kitchen. The changes that occur in cooking are chemical reactions. One way to speed up cooking reactions that are performed in boiling water is to use a pressure cooker. Under pressure, the boiling point of water is higher, so the time required for a given cooking operation is reduced. The opposite effect is seen in open cooking at high altitudes, where reduced atmospheric pressure allows water to boil at lower temperatures. Here cooking is slower, the result of reduced reaction rates.

To understand the effect of temperature on reaction rates, recall that temperature is a measure of the *average* kinetic energy of the molecules in a sample (Section 13.5). This means that some molecules in the sample move slowly, some very fast, and the majority fall in between. We have already seen that collisions between molecules with low kinetic energies are ineffective collisions, and only high-energy collisions yield reactions. This means that only a fraction of the sample, that fraction with high kinetic energy, can engage in reaction-producing collisions.

If the temperature of the reactants is increased, the molecules will have a higher average kinetic energy. It follows that a larger fraction of the sample will have enough kinetic energy to convert to potential energy and cross the potential energy barrier. Quite often a temperature increase of about 10°C will double the rate of a reaction, suggesting that the number of particles with sufficient kinetic energy to react has been doubled. Is this reasoning familiar? You have encountered it before; it is the same reasoning that explains the sharp increase in vapor pressure at higher temperatures (Section 14.4).

It can be argued that a higher temperature increases the frequency of all collisions, and that this alone would increase reaction rate. This is correct, but this explanation cannot by itself account for the increase in reaction rate of about 100% when the average absolute temperature is raised by about 3%. We must conclude that *the main reason a chemical change occurs more rapidly at higher temperatures is that a larger portion of the total number of particles has sufficient kinetic energy for effective collisions.*

The Effect of a Catalyst on Reaction Rate

PG 19E Using an energy-reaction graph, explain how a catalyst affects reaction rate.

In driving from one city to another you usually have a choice of two or more routes. One of these would be the quickest, and in all probability your choice. If a superhighway were to be built between the cities you would have another route that would be faster than any of the earlier choices. If your purpose was to make the trip in the minimum time, you would no doubt turn to the superhighway for future trips.

A **catalyst** for a chemical reaction is similar to the superhighway in furnishing an alternative "route" for the reactants to change to products. It does this by changing the reaction path. Specifically, a lower-energy activated complex is found, with a lower activation energy than the uncatalyzed reaction. Figure 19.4 compares the reaction paths for a catalyzed and an uncatalyzed reaction. E_a is the activation energy for the reaction without a catalyst, and E_a' represents the activation energy in the presence of a catalyst. Many molecules that don't have enough kinetic energy to overcome the E_a potential energy barrier are able to cross the lower E_a' hump. With a larger fraction of molecules engaging in reaction-producing collisions, the reaction is faster.

Catalysts exist in several different forms, and the precise function of many catalysts is not clearly understood. Some catalysts are mixed intimately in the reacting chemicals, while others seem to do no more than provide a surface upon which the reaction may occur. In either case, the catalyst is not permanently affected by the reaction. Many times the catalyst is only a bystander,

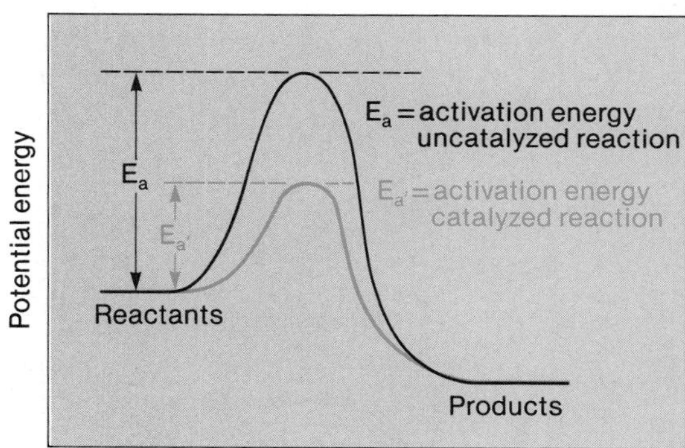

Figure 19.4
Lowering of activation energy by adding a catalyst.

although an important one from the standpoint of the reaction. In other cases, the catalyst may actually participate in the reaction and undergo chemical change, but eventually it will be regenerated in the same amount as at the start.

The catalytic reaction that appears most often in beginning chemistry laboratories is the making of oxygen by decomposing potassium chlorate. Manganese dioxide is mixed in as a catalyst. A well-known industrial process is the catalytic cracking of crude oil, in which large hydrocarbon molecules are broken down into simpler and more useful products in the presence of a catalyst. Biological reactions are controlled by catalysts called *enzymes*.

Some substances interfere with a normal reaction path from reactants to products, forcing the reaction to a higher activation energy route that is slower. Such substances are called **negative catalysts,** or **inhibitors.** Inhibitors are used to control the rates of certain industrial reactions. Sometimes negative catalysts can have disastrous results, as when mercury poisoning prevents the normal biological function of enzymes.

The Effect of Concentration on Reaction Rate

PG 19F Explain the relationship between reactant concentration and reaction rate.

If a reaction rate depends on frequency of effective collisions, the influence of concentration is readily predictable. The more particles there are in a given space, the more frequently collisions will occur, and the more rapidly the reaction will take place.

The effect of concentration on reaction rate is easily seen in the rate at which objects burn in air compared to the rate of burning in an atmosphere of pure oxygen. If a burning splint is thrust into pure oxygen the burning is brighter, more vigorous, and much faster. In fact, the typical laboratory test for oxygen is to ignite a splint, blow it out, and then, while there is still a faint glow, place it in oxygen. It immediately bursts back into full flame and burns vigorously. Charcoal, phosphorus, and other substances behave similarly.

* * * * *

Three factors that influence the rate of a chemical reaction have been identified. In relating these factors to equilibrium considerations, we will consider only concentration and temperature. These variables affect forward and reverse reaction rates differently. A catalyst, on the other hand, has the same effect on both forward and reverse rates and, therefore, does not alter a chemical equilibrium. A catalyst does cause a system to reach equilibrium more quickly.

Quick Check 19.4

1. What happens to a reaction rate as temperature drops? Give two explanations for the change. State which one is more important and explain why.
2. How does a catalyst affect reaction rates?
3. Compare the reaction rates when a certain reactant is at high concentration and at low concentration. Explain the difference.

19.5 THE DEVELOPMENT OF A CHEMICAL EQUILIBRIUM

PG 19G Trace the changes in concentrations of reacting species that lead to a chemical equilibrium.

The role of concentration in chemical equilibrium may be illustrated by tracing the development of an equilibrium. We will do this in two ways, graphically and by means of the reversible-reaction equation. For our purpose, the hypothetical reaction $A_2 + B_2 \rightarrow 2\ AB$ is assumed to take place by the simple collision of A_2 and B_2 molecules, which separate as two AB molecules (Fig. 19.1). The reverse reaction is exactly the reverse process: two AB molecules collide and separate as one A_2 molecule and one B_2 molecule.

Figure 19.5 is a graph of forward and reverse reaction rates vs. time. Initially, at Time 0, pure A_2 and B_2 are introduced to the reaction chamber. At the initial concentrations of A_2 and B_2 the forward reaction begins at a certain rate, F_0. Initially there are no AB molecules present, so the reverse reaction cannot occur. At Time 0 the reverse reaction rate, R_0, is zero. These points are plotted on the graph.

As soon as the reaction begins, A_2 and B_2 are consumed, thereby reducing their concentrations in the reaction vessel. As these reactant concentrations decrease, the forward reaction rate declines. Consequently, at Time 1, the forward reaction rate drops to F_1. During the same interval, some AB molecules are produced by the forward reaction, and the concentration of AB becomes greater than zero. Therefore, the reverse reaction begins with the reverse rate rising to R_1 at Time 1.

At Time 1 the forward rate is greater than the reverse rate. Therefore A_2 and B_2 are consumed by the forward reaction more rapidly than they are produced by the reverse reaction. The net change in the concentrations of A_2 and B_2 is therefore downward, causing a further reduction in the forward rate at Time 2. Conversely, the forward reaction produces AB more rapidly than the reverse reaction uses it. The net change in the concentration of AB is therefore an increase. This, in turn, raises the reverse reaction rate at Time 2.

Similar changes occur over successive intervals until the forward and reverse rates eventually become equal. At this point a dynamic equilibrium is established. From this analysis we may state the following generalization: *For any reversible reaction in a closed system, whenever the opposing reactions are*

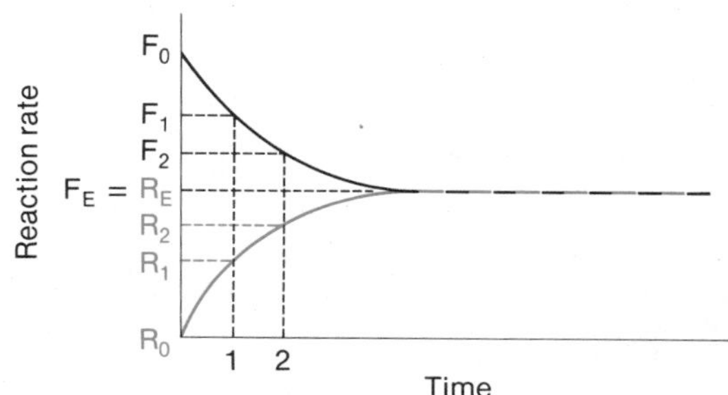

Figure 19.5
Changes in reaction rates during the development of a chemical equilibrium.

occurring at different rates, the faster reaction will gradually become slower, and the slower reaction will become faster. Finally they become equal, and equilibrium is established.

Quick Check 19.5
Concentration, temperature, and the presence of a catalyst have been identified as three factors that determine the rate of a reaction. Which is the only important one in explaining how a system reaches equilibrium at constant temperature? Why?

19.6 Le CHATELIER'S PRINCIPLE

In the last section you saw how concentrations and reaction rates jockey with each other until the rates become equal and equilibrium is reached. In this section we will start with a system already at equilibrium and see what happens to it when the equilibrium is upset. To ''upset'' an equilibrium you must somehow make the forward and reverse reactions unequal, at least temporarily. One way to do this is to add something to or remove something from the system. A gaseous equilibrium can often be upset simply by changing the volume of the container.

How an equilibrium responds to a disturbance can be predicted from the concentration and temperature effects already considered. The predictions may be summarized in **Le Chatelier's Principle,** which says that **if an equilibrium system is subjected to a change, processes occur that tend to partially counteract the initial change, thereby bringing the system to a new position of equilibrium.**

We will now apply this system and see why it works.

The Concentration Effect

PG 19H Given the equation for a chemical equilibrium, predict the direction in which the equilibrium will shift because of a change in the concentration of one species.

The reaction of hydrogen and iodine to produce hydrogen iodide comes to equilibrium with hydrogen iodide as the favored species. The equal forward and reverse reaction rates are shown by the equal length arrows in the equation

$$H_2(g) + I_2(g) \rightleftharpoons 2\ HI(g)$$

The sizes of the formulas represent the relative concentrations of the different species in the reaction. If more HI is forced into the system, [HI] is increased. (Recall from Section 17.6 that a formula enclosed in square brackets refers to the concentration of the substance in moles per liter.) This raises the rate of the reverse reaction, shown by the longer arrow in the equation, in which HI is a reactant:

$$H_2(g) + I_2(g) \rightleftharpoons 2\ HI_{(g)}$$

The rates no longer being equal, the equilibrium is destroyed.

At this point the unequal reaction rates cause the system to "shift" in the direction of the faster rate—to the left, or in the reverse direction. As a result H_2 and I_2 are made by the reverse reaction faster than they are used by the forward reaction. Their concentrations increase, so the forward rate increases. Simultaneously HI is used faster than it is produced, reducing both [HI] and the reverse reaction rate. Eventually the rates become equal at an intermediate value (note arrow lengths) and a new equilibrium is reached:

$$H_2(g) + I_2(g) \rightleftharpoons 2\ HI_{(g)}$$

The sizes of the formulas indicate that all three concentrations are larger than they were originally, although [HI] has come back from the maximum it reached just after it was added.

The above example shows *why* an equilibrium shifts when it is disturbed in terms of concentrations and reaction rates. Le Chatelier's Principle makes it possible to predict the direction of a shift, forward or reverse, without such a detailed analysis. Just remember that the shift is always in the direction that tries to return the substance disturbed to its original condition.

EXAMPLE 19.1 The system $N_2(g) + 3\ H_2(g) \rightleftharpoons 2\ NH_3(g)$ is at equilibrium. Use Le Chatelier's Principle to predict the direction in which the equilibrium will shift if ammonia is withdrawn from the reaction chamber.

The equilibrium disturbance is clearly stated: ammonia is withdrawn. In which direction, forward (__) or reverse (__), must the reaction shift to *counteract* the removal of ammonia—that is, to *produce* more ammonia to replace some of what has been taken away?

The shift will be in the *forward* direction.

Ammonia is the product of the forward reaction and therefore will be partially restored to its original concentration by a shift in the forward direction.

EXAMPLE 19.2 Predict the direction, forward or reverse, of a Le Chatelier shift in the equilibrium $CH_4(g) + 2\ H_2S(g) \rightleftharpoons 4\ H_2(g) + CS_2(g)$ caused by each of the following:

(a) increase $[H_2S]$: (b) reduce $[CS_2]$: (c) increase $[H_2]$:

(a) Forward, counteracting partially the increase in $[H_2S]$ by consuming some of it.

(b) Forward, restoring some of the CS_2 removed.

(c) Reverse, consuming some of the added H_2.

The Volume Effect

PG 191 Given the equation for a chemical equilibrium involving one or more gases, predict the direction in which the equilibrium will shift because of a change in the volume of the system.

A change in the volume of an equilibrium that includes one or more gases changes the concentrations of those gases. Usually—there is one exception, as you will see shortly—there is a Le Chatelier shift that partially offsets the initial change. Both the change and the adjustment involve the pressure exerted by the entire system.

From the ideal gas equation, PV = nRT, the pressure exerted by a gas at constant temperature is proportional to the concentration of the gas, n/V, measured in moles per liter:

$$P = RT \times \frac{n}{V} = \text{constant} \times \frac{n}{V}; \qquad P \propto \frac{n}{V} \tag{19.1}$$

If volume is reduced, the denominator in n/V becomes smaller and the fraction becomes larger. Pressure increases. Le Chatelier's Principle calls for a shift that will partially counteract the change—to reduce pressure. What change can reduce the pressure of the system at the new volume? The numerator in Equation 19.1 must be reduced; there must be fewer gaseous molecules in the system. In general, *if a gaseous equilibrium is compressed, the increased pressure will be partially relieved by a shift in the direction of fewer gaseous molecules; if the system is expanded, the reduced pressure will be partially restored by a shift in the direction of more gaseous molecules.*

Realizing that the coefficients of gases in an equation are in the same proportion as the number of gaseous molecules (Section 13.13), let's try some examples.

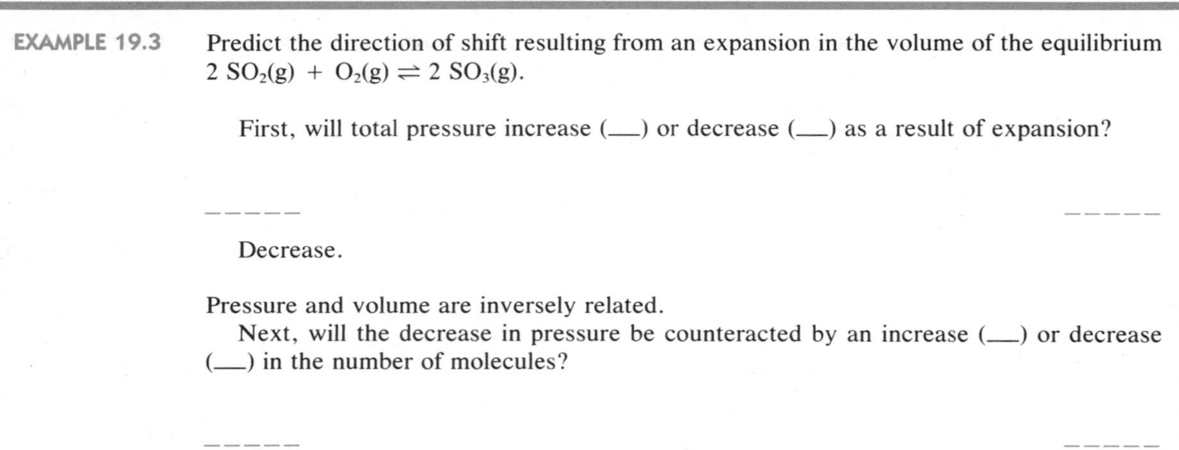

EXAMPLE 19.3 Predict the direction of shift resulting from an expansion in the volume of the equilibrium $2\ SO_2(g) + O_2(g) \rightleftharpoons 2\ SO_3(g)$.

First, will total pressure increase (__) or decrease (__) as a result of expansion?

Decrease.

Pressure and volume are inversely related.

Next, will the decrease in pressure be counteracted by an increase (__) or decrease (__) in the number of molecules?

Increase.

Pressure is directly related to number of molecules.

Finally, examining the equation, notice that a forward shift finds *three* reactant molecules, two SO_2 and one O_2, forming *two* SO_3 product molecules, whereas a reverse shift has *two* reactant molecules yielding *three* product molecules. In which direction will the reaction shift to increase the total number of molecules?

----- -----

Reverse.

Three product molecules replace two reactant molecules in a reverse shift. The increase in number of molecules partially restores the total pressure lost in expansion.

EXAMPLE 19.4 The volume occupied by the equilibrium

$$SiF_4(g) + 2\ H_2O(g) \rightleftharpoons SiO_2(s) + 4\ HF(g)$$

is reduced. Predict the direction of shift in the position of equilibrium.

Will the shift be in the direction of more (__) gaseous molecules or fewer (__)?

----- -----

Fewer.

If volume is reduced, pressure increases. Increased pressure will be counteracted by fewer molecules.

Now predict the direction of shift and justify your prediction by stating the numerical loss in molecules predicted by the equation.

----- -----

The shift will be in the reverse direction. The reverse reaction has *four gaseous* molecules being reduced to three. Note the number is not five reduced to three, but four. Only the gaseous molecules are involved in pressure adjustments.

EXAMPLE 19.5 Returning to the familiar $H_2(g) + I_2(g) \rightleftharpoons 2\ HI(g)$, predict the direction of the shift that will occur because of a volume increase.

Take it all the way, but be careful.

----- -----

There will be no shift. In this case there are two gaseous molecules on *each* side of the equation. Therefore, there is no gain or loss in number of molecules by a shift in either direction. When the number of gaseous molecules on the two sides of the equilibrium equation is the same, increasing or reducing volume has no effect on the equilibrium.

The Temperature Effect

PG 19J Given a thermochemical equation for a chemical equilibrium, predict the direction in which the equilibrium will shift because of a change in temperature.

A change in temperature of an equilibrium will change both forward and reverse reaction rates, but the rate changes are not equal. The equilibrium is therefore destroyed temporarily. The events that follow are again predictable by Le Chatelier's Principle.

EXAMPLE 19.6

$$PCl_5(g) \rightleftharpoons PCl_3(g) + Cl_2(g) + 92.5 \text{ kJ}$$

is increased, predict the direction of the Le Chatelier shift.

In order to raise the temperature of something, it must be heated. We therefore interpret an increase in temperature as the "addition of heat," and a lowering of temperature as the "removal of heat." In applying Le Chatelier's Principle to a thermochemical equation, heat may be regarded in much the same manner as any chemical species in the equation. Accordingly, if heat is *added* to the equilibrium system shown, in which direction must it shift to *use up*, or *consume*, some of the heat that was added? (If chlorine was *added*, in which direction would the equilibrium shift to use up some of the chlorine? Forward (__); reverse (__).

The equilibrium must shift in the *reverse* direction to use up some of the added heat. An endothermic reaction consumes heat. As the equation is written, the reverse reaction is endothermic.

EXAMPLE 19.7 The thermal decomposition of limestone reaches the following equilibrium: $CaCO_3(s) + 176 \text{ kJ} \rightleftharpoons CaO(s) + CO_2(g)$. Predict the direction this equilibrium will shift if the temperature is reduced: Forward (__); reverse (__).

The shift will be in the *reverse* direction. Reduction in temperature is interpreted as the removal of heat. The reaction will respond to replace some of the heat removed—as an exothermic reaction. Heat is produced as the reaction proceeds in the reverse direction.

19.7 THE LAW OF CHEMICAL EQUILIBRIUM: THE EQUILIBRIUM CONSTANT

PG 19K Given any chemical equilibrium equation, write the equilibrium constant expression.

Gas Phase Equilibria

If 1.000 mole of H_2(g) and 1.000 mole of I_2(g) are introduced into a 1.000-liter reaction vessel under pressure, they will react to produce the following equilibrium:

$$H_2(g) + I_2(g) \rightleftharpoons 2\,HI(g) \tag{19.2}$$

At a temperature of 440°C, analysis at equilibrium will show hydrogen and iodine concentrations of 0.218 mol/L, and a hydrogen iodide concentration of 1.564 mol/L. Development of these concentrations is shown in black in Figure 19.6A.

Working in the opposite direction, if gaseous hydrogen iodide is introduced to a reaction chamber at 440°C at an initial concentration of 2.000 mol/L, it will decompose by the reverse reaction. The system will eventually come to equilibrium with $[H_2] = [I_2] = 0.218$, and $[HI] = 1.564$, exactly the same equilibrium concentrations as in the first example. This illustrates the experimental fact that the position of an equilibrium is the same, regardless of the direction from which it is approached (Fig. 19.6A, white lines).

As a third example, if the initial hydrogen iodide concentration is half as much, 1.000 mol/L, the equilibrium concentrations will be half what they are above: $[H_2] = [I_2] = 0.109$ and $[HI] = 0.782$.

If the experiment is repeated at the same temperature, starting this time with hydrogen at 1.000 mol/L and iodine at 0.500 mol/L, the equilibrium concentrations will be $[H_2] = 0.532$, $[I_2] = 0.032$, and $[HI] = 0.936$, as shown in Figure 19.6B.

The data of these four equilibria are summarized in Table 19.1. Analysis of these data leads to the observation that, at equilibrium, the ratio $\frac{[HI]^2}{[H_2][I_2]}$, has a value of 51.5 for all four equilibria. For that matter, this ratio of equilibrium concentrations is always the same regardless of the initial concentrations of hydrogen, iodine, and hydrogen iodide, as long as the temperature is held at 440°C.

Data from countless other equilibrium systems show a similar regularity. Expressed in words, it is known as the **Law of Chemical Equilibrium: For any**

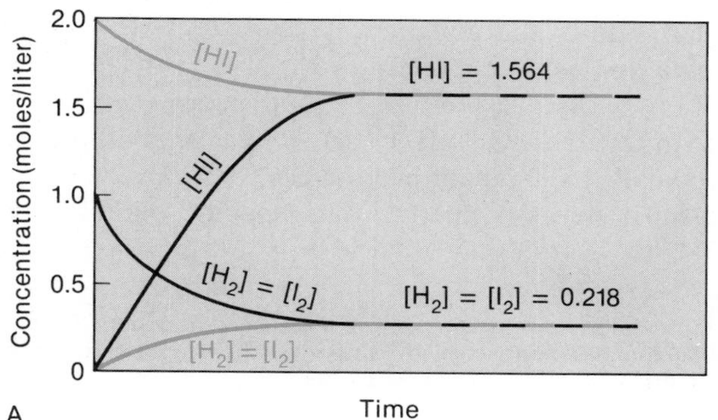

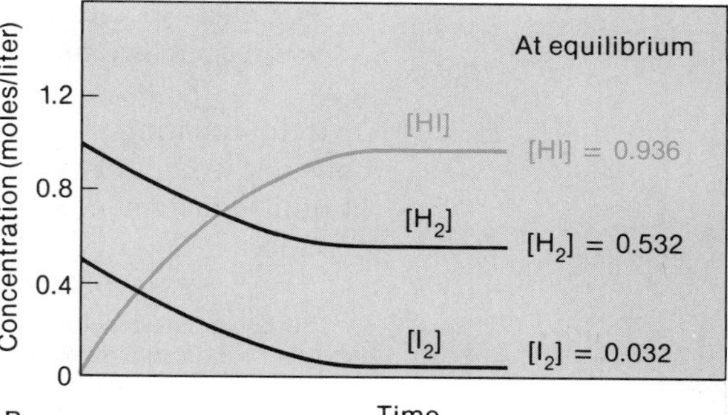

Figure 19.6
Changes in concentrations during the development of a chemical equilibrium.

Table 19.1

Equi-librium	Initial			Equilibrium			
	$[H_2]$	$[I_2]$	$[HI]$	$[H_2]$	$[I_2]$	$[HI]$	$\frac{[HI]^2}{[H_2][I_2]}$
1	1.000	1.000	0	0.218	0.218	1.564	51.5
2	0	0	2.000	0.218	0.218	1.564	51.5
3	0	0	1.000	0.109	0.109	0.782	51.5
4	1.000	0.500	0	0.532	0.032	0.936	51.5

equilibrium at a given temperature, the ratio of the product of the concentrations of the species on the right side of the equilibrium equation, each raised to a power equal to its coefficient in the equation, to the corresponding product of the concentrations of the species on the left side of the equation is a constant. The constant is called the **equilibrium constant, K.** Thus for Equation 19.2 at 440°C,

$$K = \frac{[HI]^2}{[H_2][I_2]} = 51.5$$

The statement of the equilibrium law sets the procedure by which any equilibrium constant expression may be written. For the general equilibrium $aA + bB \rightleftharpoons cC + dD$, where A, B, C, and D are chemical formulas, and a, b, c, and d are their coefficients in the equilibrium equation

1. Write in the *numerator* the concentration of each species on the *right* side of the equation. $K = \frac{[C][D]}{}$
2. For each species on the right side of the equation, use its coefficient in the equation as an exponent. $K = \frac{[C]^c[D]^d}{}$
3. Write in the *denominator* the concentration of each species on the *left* side of the equation. $K = \frac{[C]^c[D]^d}{[A][B]}$
4. For each species on the left side of the equation, use its coefficient in the equation as an exponent. $K = \frac{[C]^c[D]^d}{[A]^a[B]^b}$* (19.3)

If the equation were written in reverse, $cC + dD \rightleftharpoons aA + bB$, the equilibrium constant would be the reciprocal of the constant in Equation 19.3. Every equilibrium constant expression must be associated with a specific equilibrium equation.

*Although this development of the equilibrium constant expression has been from hypothetical laboratory data, the same expression can also be reached by a rigorous theoretical derivation. Theory and experiment support each other completely in this area.

EXAMPLE 19.8 Write equilibrium constant expressions for each of the following equilibria:

(a) $2\ HI(g) \rightleftharpoons H_2(g) + I_2(g)$

(b) $HI(g) \rightleftharpoons \frac{1}{2}\ H_2(g) + \frac{1}{2}\ I_2(g)$

(c) $2\ Cl_2(g) + 2\ H_2O(g) \rightleftharpoons 4\ HCl(g) + O_2(g)$

(a) $K = \frac{[H_2][I_2]}{[HI]^2}$. Notice that this is the reciprocal of the equilibrium constant developed in the text, when the equation was written $H_2(g) + I_2(g) \rightleftharpoons 2\ HI(g)$. Its numerical value at 440°C is 1/51.5, or 0.0194.

(b) $K = \frac{[H_2]^{1/2}[I_2]^{1/2}}{[HI]}$. The value of this equilibrium constant is *not* the same as above, 0.0194. It is, in fact, the square root of 0.0194, or 0.139. This emphasizes the importance of associating any equilibrium constant expression with a specific chemical equation.

(c) $K = \frac{[HCl]^4[O_2]}{[Cl_2]^2[H_2O]^2}$. The procedure for writing the equilibrium constant expression is the same no matter how complex the equation is.

Solids in Equilibrium Constant Expressions

In Chapter 16 silver chloride was identified as an "insoluble" ionic compound. It is more accurately described as a very, very slightly soluble compound. Its solution becomes saturated at an extremely low concentration, roughly 0.000 01 mol/L. The equilibrium equation is $AgCl(s) \rightleftharpoons Ag^+(aq) + Cl^-(aq)$. The equilibrium constant expression, written according to the above procedure, is

$$K = \frac{[Ag^+][Cl^-]}{[AgCl]}$$

The "concentration" of the solid, [AgCl], in the above expression has no effect on the equilibrium and, for all practical purposes, no meaning. The fact is that, as long as there is undissolved solute in contact with the saturated solution, the *amount* of solute does not influence the ion concentrations. It is the indicated product of the ionic concentrations that is constant at equilibrium:

$$K_{sp} = [Ag^+][Cl^-] \qquad (19.4)$$

The subscript is K_{sp} identifies a **solubility product constant,** the special name given to equilibrium constants for low-solubility solids. Accordingly, *when writing the equilibrium constant expression for an equilibrium that includes a solid, the concentration of the solid is omitted.* The resulting expression includes only those substances whose concentrations are variable. These are the concentrations that affect forward and reverse reaction rates.

EXAMPLE 19.9 Write the equilibrium constant expressions for each of the following equilibria:

(a) $CaCO_3(s) \rightleftharpoons CaO(s) + CO_2(g)$

(b) $Li_2CO_3(s) \rightleftharpoons 2\ Li^+(aq) + CO_3^{2-}(aq)$

(c) $4\ H_2O(g) + 3\ Fe(s) \rightleftharpoons 4\ H_2(g) + Fe_3O_4(s)$

(a) $K = [CO_2]$, (b) $K = [Li^+]^2[CO_3^{2-}]$; (c) $K = \dfrac{[H_2]^4}{[H_2O]^4}$

In all cases the equilibrium constant expression is written in the usual way, except that the solids are omitted.

Liquid Water in Equilibrium Constant Expressions

Pure water ionizes to a very small extent, yielding a small concentration of hydrogen and hydroxide ions, just like a low-solubility solid dissolves only slightly to yield a low concentration of its ions. The equilibrium equation for water is

$$HOH(\ell) \rightleftharpoons H^+(aq) + OH^-(aq)$$

The equilibrium constant expression follows:

$$K = \frac{[H^+][OH^-]}{[HOH]}$$

To two significant figures, [HOH] is constant both in water and in all dilute aqueous solutions. Therefore, multiplying both sides of the equation by [HOH] yields the product of two constants on the left, which is another constant called the **water constant, K_w:**

$$K[HOH] = K_w = [H^+][OH^-] \qquad (19.5)$$

(In Section 17.5 you learned that $K_w = 1.0 \times 10^{-14}$ at 25°C.)

The liquid water effect is the same in all dilute aqueous solution equilibria. Therefore, *when we write the equilibrium constant expression for a dilute* *aqueous solution equilibrium equation that includes liquid water, the concentration of water, $[H_2O]$, is omitted.* Again, the resulting expression includes only those species whose concentrations are variable.

EXAMPLE 19.10 Write the equilibrium constant expression for each equilibrium equation:

(a) $HF(aq) + H_2O(\ell) \rightleftharpoons H_3O^+(aq) + F^-(aq)$

(b) $NH_3(aq) + H_2O(\ell) \rightleftharpoons NH_4^+(aq) + OH^-(aq)$

(a) $K = \frac{[H_3O^+][F^-]}{[HF]}$ (b) $K = \frac{[NH_4^+][OH^-]}{[NH_3]}$

19.8 THE SIGNIFICANCE OF THE VALUE OF K

PG 19L Given an equilibrium equation and the value of the equilibrium constant, identify which direction is favored.

By definition, an equilibrium constant is a ratio—a fraction. The numerical value of an equilibrium constant may be very large, very small, or any place in between. While there is no defined intermediate range, equilibria with constants between 0.01 and 100 (10^{-2} to 10^2) will have appreciable quantities of all species present at equilibrium.

To see what is meant by "very large" or "very small" K values, consider an equilibrium similar to the hydrogen iodide system studied earlier. Substituting chlorine for iodine, the equilibrium equation is $H_2(g) + Cl_2(g) \rightleftharpoons 2\,HCl(g)$. At 25°C

$$K = \frac{[HCl]^2}{[H_2][Cl_2]} = 2.4 \times 10^{33}$$

This is a very large number—ten billion times larger than the number in a mole! The only way an equilibrium constant ratio can become so huge is for the concentration of one or more reacting species to be very close to zero. If the denominator of a ratio is nearly zero, the value of the ratio will be very large. A near-zero denominator and large K means the equilibrium is favored in the forward direction.

By contrast, if the equilibrium constant is very small, it means the concentration of one or more of the species on the right side of the equation is nearly zero. This puts a near-zero number in the numerator of K, and the equilibrium is strongly favored in the reverse direction.

To summarize: If an equilibrium constant is very large, the forward reaction is favored; if the constant is very small, the reverse reaction is favored. If the constant is neither large nor small, appreciable quantities of all species are present at equilibrium.

19.9 EQUILIBRIUM CALCULATIONS [OPTIONAL]

Equilibrium calculations cover a wide range of problem types. A thorough understanding of these is essential to the understanding of many chemical phenomena in the laboratory, in industry, and in living organisms. We will sample only a few of these in this section.*

Two things should be written before attempting to solve any equilibrium problem. First, is the equilibrium equation; second is the equilibrium constant expression.

Solubility Equilibria

The ion concentration of a low solubility solid can be used to calculate a solubility product constant. The following example illustrates the method:

EXAMPLE 19.11 The chloride ion concentration of saturated silver chloride is 1.3×10^{-5} molar. Calculate K_{sp} for AgCl.

SOLUTION: First the equilibrium equation and K_{sp} expression are written:

$$AgCl(s) \rightleftharpoons Ag^+(aq) + Cl^-(aq) \qquad K_{sp} = [Ag^+][Cl^-]$$

Stoichiometric reasoning plays a large part in solving equilibrium problems. Given information about the concentration of one species in a problem, the concentrations of other substances can be determined. In this case, silver chloride has dissolved to the extent of releasing 1.3×10^{-5} moles of Cl^- per liter. According to the reaction equation, the moles of Ag^+ produced is the same as the moles of Cl^-. The concentrations at equilibrium must therefore both be 1.3×10^{-5}. Thus

$$K_{sp} = [Ag^+][Cl^-] = (1.3 \times 10^{-5})^2 = 1.7 \times 10^{-10}$$

EXAMPLE 19.12 The solubility of magnesium fluoride is 73 mg/L. What is the indicated K_{sp}?

SOLUTION: Again, the two equations come first:

$$MgF_2(s) \rightleftharpoons Mg^{2+}(aq) + 2\,F^-(aq) \qquad K_{sp} = [Mg^{2+}][F^-]^2$$

*For a more complete consideration of equilibrium problems see Chapters 13 to 17 of Peters, *Problem Solving for Chemistry,* Second Edition, Saunders College Publishing, 1976.

The K_{sp} equation requires concentrations in moles per liter. These can be calculated from the solubility. For the magnesium ion,

$$\frac{0.073 \text{ g } MgF_2}{L} \times \frac{1 \text{ mol } Mg^{2+}}{62.3 \text{ g } MgF_2} = 1.17 \times 10^{-3} \text{ mol } Mg^{2+}/L = [Mg^{2+}]$$

(We have carried an extra significant figure at this point to approach more closely the digits that will normally be retained in a calculator.) The dissolving equation shows that twice as many fluoride ions as magnesium ions are released. Therefore

$$[F^-] = 2 \times [Mg^{2+}] = 2 \times 1.17 \times 10^{-3} = 2.34 \times 10^{-3}$$

The two concentrations are now substituted into the K_{sp} expression:

$$K_{sp} = [Mg^{2+}][F^-]^2 = (1.17 \times 10^{-3})(2.34 \times 10^{-3})^2 = 6.4 \times 10^{-9}$$

Solubility product constants have already been determined for most common salts, and their values may be found in handbooks. They are used in all sorts of problems, one of which is the reverse of the last two examples.

EXAMPLE 19.13 Calculate the solubility of zinc carbonate, $ZnCO_3$, in (a) mol/L and (b) g/100 mL if $K_{sp} = 1.4 \times 10^{-11}$

SOLUTION:

$$ZnCO_3(s) \rightleftharpoons Zn^{2+}(aq) + CO_3^{2-}(aq) \qquad K_{sp} = 1.4 \times 10^{-11}$$

If the $ZnCO_3$ is the only source of Zn^{2+} and CO_3^{2-} ions, stoichiometry dictates that they must be present in equal numbers and equal concentration. Furthermore, that concentration is the number of moles of solute that dissolve in 1 L of solution, or the solubility in moles per liter. Let the algebraic symbol s be solubility: s = solubility = $[Zn^{2+}]$ = $[CO_3^{2-}]$. Use the value of K_{sp} to find s:

$$K_{sp} = [Zn^{2+}][CO_3^{2-}] = s^2 = 1.4 \times 10^{-11} \qquad s = 3.7 \times 10^{-6}$$

The solubility can now be used to find the number of grams per 100 milliliters, or the number of grams in 0.100 L:

$$0.100 \text{ L} \times \frac{3.7 \times 10^{-6} \text{ mol}}{L} \times \frac{125 \text{ g } ZnCO_3}{1 \text{ mol } ZnCO_3} = 4.6 \times 10^{-5} \text{ g } ZnCO_3 \text{ per 100 mL}$$

Suppose a soluble carbonate, such as Na_2CO_3, were to be dissolved in the saturated solution of zinc carbonate in Example 19.13. What would happen to the solubility of $ZnCO_3$? $[CO_3^{2-}]$ would increase. It would no longer be the same as $[Zn^{2+}]$ because the two ions would now be coming from different sources. According to Le Chatelier's Principle, the equilibrium should shift in the direction that would reduce $[CO_3^{2-}]$, which is in the reverse direction. In other words, less $ZnCO_3$ would dissolve; the solubility would be reduced.

EXAMPLE 19.14 Calculate the solubility of $ZnCO_3$ in 0.010 M Na_2CO_3. Answer in moles per liter.

SOLUTION: The product of the concentration of the two ions is the same, 1.4×10^{-11}, whether the concentrations are the same or different. Assuming the Na_2CO_3 is completely ionized, $[CO_3^{2-}]$ is 0.010. Substituting into the K_{sp} expression,

$$K_{sp} = [Zn^{2+}][CO_3^{2-}] = 0.010[Zn^{2+}] = 1.4 \times 10^{-11}$$

$$[Zn^{2+}] = 1.4 \times 10^{-9}$$

$[Zn^{2+}]$ is equal to the solubility of zinc carbonate in 0.010 M Na_2CO_3. As predicted by Le Chatelier's Principle, it is less than the solubility in water that was calculated in Example 19.13.

Example 19.14 illustrates the **common ion effect,** in which an ion already present in the equilibrium is added from a second, independent source. The common ion produces a Le Chatelier shift away from that ion in the equilibrium, reducing the concentration of any other species on the same side of the equilibrium equation. In the $AgCl \rightleftharpoons Ag^+ + Cl^-$ equilibrium, addition of Cl^- from HCl or a soluble chloride, such as NaCl, would force the equilibrium left, reducing $[Ag^+]$ and the solubility of AgCl. Similarly, adding Ag^+ from a soluble silver compound, such as $AgNO_3$, would also push the equilibrium in the reverse direction, reducing $[Cl^-]$ and the solubility of AgCl.

Ionization Equilibria

When a strong acid ionizes, it is like a salt that dissolves. The solute is changed completely to ions, as was the Na_2CO_3 in Example 19.14. Thus, in 0.010 M HCl, $[Cl] = [H^+] = 0.010$. Most of the time the interest is in $[H^+]$ and the pH it represents. In 0.010 M HCl, pH = 2.0. In general, for a strong monoprotic acid whose molarity is M,

$$[H^+] = M \tag{19.6}$$

Only a small percentage of a weak acid ionizes before a state of equilibrium is reached. If HA is a typical acid, its ionization and equilibrium equations are

$$HA(aq) \rightleftharpoons H^+(aq) + A^-(aq) \tag{19.7}$$

$$K_a = \frac{[H^+][A^-]}{[HA]} \tag{19.8}$$

The subscript "a" is used for the **acid constant.** If (a) the ionization of the weak acid, HA, is so slight that the reduction in [HA] is negligible, and if (b) HA is the only appreciable source of either H^+* or A^-, then $[H^+] = [A^-]$, just as

*These two assumptions are generally valid for acids whose molarities are 0.1 or more, and whose K_a values fall between 10^{-6} and 10^{-4}. We will limit problems in this book to those for which the assumptions are valid. In more advanced courses you will learn how to check their validity and what to do if they are not acceptable.

the silver and chloride ion concentrations were equal in Example 19.11. This makes it possible to calculate the percentage ionization and K_a from the pH of a weak acid of known molarity:

EXAMPLE 19.15 A 0.13 M solution of an unknown acid has a pH of 3.12. Calculate the percent ionization and K_a.

SOLUTION: From Equation 17.15, $[H^+] = 10^{-3.12} = 7.6 \times 10^{-4}$. This is also $[A^-]$. This means that only 0.00076 of the 0.13 mol of acid in 1 L of solution are ionized. Expressed as percent,

$$\frac{7.6 \times 10^{-4}}{0.13} \times 100 = 0.58\% \text{ ionized}$$

Substituting values of $[H^+]$ and $[A^-]$ into Equation 19.8 gives K_a:

$$K_a = \frac{[H^+][A^-]}{[HA]} = \frac{(7.6 \times 10^{-4})^2}{0.13} = 4.4 \times 10^{-6}.$$

As the K_{sp} values of low-solubility solids are listed in handbooks, so are the K_a values of most weak acids. They can be used to determine $[H^+]$ and the pH of solutions of those acids. If the ionization of the acid is the only source of H^+ and A^-, multiplying both sides of Equation 19.8 by [HA] and substituting $[H^+]$ for its equal, $[A^-]$ yields

$$[H^+][A^-] = [H^+]^2 = K_a[HA] \qquad (19.9)$$

$$[H^+] = \sqrt{K_a[HA]} \qquad (19.10)$$

EXAMPLE 19.16 What is the pH of a 0.20 molar solution of the acid in Example 19.15, for which $K_a = 4.4 \times 10^{-6}$?

SOLUTION: $[H^+]$ may be found by direct substitution into Equation 19.10. Its negative logarithm is the pH of the solution.

$$[H^+] = \sqrt{K_a[HA]} = \sqrt{(4.4 \times 10^{-6})(0.20)} = 9.4 \times 10^{-4} \qquad pH = 3.03$$

Just as solubility equilibria can be forced in the reverse direction by the addition of a common ion, so can a weak acid equilibrium by adding a soluble salt of the acid. If A^- is added to the equilibrium in Equation 19.7, $[A^-]$ no longer is equal to $[H^+]$. To find the pH of such a solution, Equation 19.8 may be solved for $[H^+]$:

$$[H^+] = K_a \times \frac{[HA]}{[A^-]} \qquad (19.11)$$

It is assumed that neither [HA] nor $[A^-]$ are changed significantly by the ionization of HA. This assumption is almost always valid because of the Le Chatelier effect of the common ion.

EXAMPLE 19.17 Find the pH of a 0.20 M solution of the acid in Examples 19.15 and 19.16 ($K_a = 4.4 \times 10^{-6}$) if the solution is also 0.15 M in A^-.

SOLUTION: Direct substitution into Equation 19.11 gives $[H^+]$. pH follows.

$$[H^+] = K_a \times \frac{[HA]}{[A^-]} = 4.4 \times 10^{-6} \times \frac{0.20}{0.15} = 5.9 \times 10^{-6} \qquad pH = 5.23$$

The solution in Example 19.17 is a **buffer solution,** or, more simply, a **buffer.** A buffer is a solution that resists changes in pH because it contains relatively high concentrations of both a weak acid and a weak base. The acid is able to consume any OH^- that may be added, and the base can absorb H^+, both without significant change in either [HA] or $[A^-]$. For example, if 0.001 mol of HCl was dissolved in one liter of water, the pH would be 3. If the same amount of HCl was added to a liter of the buffer in Example 19.11, it would react with 0.001 mol of the A^- present. The new concentration of A^- would be $0.15 - 0.001 = 0.15$, unchanged according to the rules of significant figures. An additional 0.001 mol of HA would be formed in the reaction, but that added to 0.20 is still 0.20. In other words, the $[HA]/[A^-]$ ratio would be unchanged, so the $[H^+]$ and pH would also be unchanged.

This all suggests that a buffer can be tailor-made for any pH simply by adjusting the $[HA]/[A^-]$ ratio to the proper value. Solving Equation 19.8 for that ratio gives

$$\frac{[HA]}{[A^-]} = \frac{[H^+]}{K_a} \qquad (19.12)$$

EXAMPLE 19.18 What $[HA]/[A^-]$ ratio is necessary to produce a buffer with a pH of 5.00, using HA and A^- from Example 19.15? $K_a = 4.4 \times 10^{-6}$.

SOLUTION: If pH is 5.00, $[H^+] = 10^{-5.00}$, or simply 10^{-5}. Substituting into Equation 19.12,

$$\frac{[HA]}{[A^-]} = \frac{[H^+]}{K_a} = \frac{10^{-5}}{4.4 \times 10^{-6}} = 2.3$$

Homogeneous Equilibria

All the equilibria considered thus far in this section have been in aqueous solutions. A **homogeneous equilibrium** is one that involves compounds in the gas or liquid phase. The calculation principles and stoichimetric reasoning are the same.

EXAMPLE 19.19 0.052 mol of NO and 0.054 mol of O_2 are placed in a 1.00 L vessel at a certain temperature. They react until equilibrium is reached according to the equation $2\ NO(g) + O_2(g) \rightleftharpoons 2\ NO_2(g)$. At equilibrium, $[NO_2] = 0.028$. Calculate K.

SOLUTION:

$$K = \frac{[NO_2]^2}{[NO]^2[O_2]}$$

In order to find K, all concentrations must be found at equilibrium. The volume of the vessel is 1.00 L, so the moles present are numerically equal to the concentration in moles per liter. The thought processes are:

$[NO_2]$ Initially the vessel contained no NO_2. By the time equilibrium was reached, 0.028 mol of NO_2 was present in the 1.00 L, produced according to the reaction equation.

$$[NO_2] = 0.028.$$

$[NO]$ Each mole of NO_2 formed uses 1 mol of NO. Initially there was 0.052 mol of NO, and 0.028 mol was used. Therefore:

$$[NO] = 0.052 - 0.028 = 0.024.$$

$[O_2]$ Each mole of NO_2 formed uses 0.5 mol of O_2. Initially there was 0.054 mol of O_2, and 0.014 mol was used. Therefore:

$$[O_2] = 0.054 - 0.014 = 0.040.$$

Sometimes it is helpful to develop the equilibrium concentrations by tabulating the initial, reacting, and equilibrium moles per liter under the equation. The starting information is given in italics, and the results of the above reasoning in roman type.

	$2\ NO(g)$	+	$O_2(g)$	$\rightleftharpoons$	$2\ NO_2(g)$
INITIAL	*0.052*		*0.054*		*0.000*
REACTING	−0.028		−0.014		+0.028
EQUILIBRIUM	0.024				*0.028*

K may now be calculated by substituting the equilibrium concentrations into the equilibrium constant expression:

$$K = \frac{[NO_2]^2}{[NO]^2[O_2]} = \frac{0.028^2}{(0.024)^2(0.040)} = 34$$

CHAPTER 19 IN REVIEW

19.1 The Character of an Equilibrium

19.2 The Collision Theory of Chemical Reactions

19B Explain why, according to the collision theory of chemical reactions, some molecular collisions result in a chemical reaction and others do not.

19.3 Energy Changes during a Molecular Collision

19C Sketch and/or interpret an energy-reaction graph. Identify (a) the activated complex region; (b) activation energy, and (c) ΔE for the reaction.

19.4 Conditions That Affect the Rate of a Chemical Reaction

19D State and explain the relationship between reaction rate and temperature.

19E Using an energy-reaction graph, explain how a catalyst affects reaction rate.

19F Explain the relationship between reactant concentration and reaction rate.

19.5 The Development of a Chemical Equilibrium

19G Trace the changes in concentrations of reacting species that lead to a chemical equilibrium.

19.6 Le Chatelier's Principle

19H Given the equation for a chemical equilibrium, predict the direction in which the equilibrium will shift because of a change in the concentration of one species.

19I Given the equation for a chemical equilibrium involving one or more gases, predict the direction in which the equilibrium will shift because of a change in the volume of the system.

19J Given a thermochemical equation for a chemical equilibrium, predict the direction in which the equilibrium will shift because of a change in temperature.

19.7 The Law of Chemical Equilibrium: The Equilibrium Constant

19K Given any chemical equilibrium equation, write the equilibrium constant expression.

19.8 The Significance of the Value of K

19L Given an equilibrium equation and the value of the equilibrium constant, identify which direction is favored.

19.9 Equilibrium Constant Calculations (Optional)

TERMS AND CONCEPTS

19.1 "Closed" system
Dynamic, static
Forward, reverse direction
Favored direction
19.2 Collision theory of chemical reactions
19.3 Activated complex
Potential energy barrier
Activation energy
19.4 Catalyst
Negative catalyst
Inhibitor
19.6 Le Chatelier's Principle
"Shift" of an equilibrium
19.7 Law of Chemical Equilibrium
Equilibrium constant, K
Solubility product constant, K_{sp}
Water constant, K_w
19.9 Common ion effect
Acid constant, K_a
Buffer solution; buffer
Homogeneous equilibrium

Most of these terms and many others appear in the Glossary. Use the Glossary regularly.

QUESTIONS AND PROBLEMS

Section 19.1

(*1*) What things are equal in an equilibrium? Give an example.

(40) What does it mean when an equilibrium is described as **dynamic?** Compare a dynamic equilibrium with one that is static.

(*2*) What is meant by saying that an equilibrium is confined to a "closed system?"

(41) Undissolved table salt is in contact with a saturated salt solution in (a) a sealed container and (b) an open beaker. Which system, if either, can reach equilibrium? Explain your answer.

(3) With regard to the equilibrium

$$F^-(aq) + H_2O(\ell) \rightleftarrows HF(aq) + OH^-(aq)$$

what is meant by saying that the equilibrium is favored in the reverse direction?

(42) The equilibrium equation for the ionization of formic acid is

$$HCHO_2(aq) \rightleftarrows H^+(aq) + CHO_2^-(aq)$$

and the equilibrium is favored in the reverse direction. If you were to write a net ionic equation in which formic acid is a reactant, would you show formic acid in its ionized form or molecular form? Why?

Section 19.2

(4) According to the collision theory of chemical reactions, what two conditions must be satisfied if a molecular collision is to result in a reaction?

(43) Explain why a molecular collision can be sufficiently energetic to cause a reaction, yet no reaction occurs as a result of that collision.

Section 19.3

Assume heat to be the only form of reaction energy in the following questions. This makes ΔE equal to the ΔH discussed in Section 9.6.

(5) In the reaction for which Figure 19.2 is the energy-reaction graph, is ΔE for the reaction positive or negative? Is the reaction exothermic or endothermic? Use the letters a, b, and c on the vertical axis of Figure 19.2 to state algebraically the ΔE and activation energy of the reaction.

(6) Assume the reaction described by Figure 19.2 is reversible. Compare the magnitude of the activation energies for the forward and reverse reactions described by Figure 19.2. Which is greater, or are they equal?

(7) Explain the significance of activation energy. For two reactions that are identical in all respects except activation energy, identify the reaction that would have the higher rate and tell why.

(44) Sketch an energy-reaction graph for which the answers to the first two questions in Question 5 would be the opposite of what they were in Question 5. Include a, b, and c points on the vertical axis and use them for the algebraic expressions of ΔE and the activation energy.

(45) Assuming the reaction described by Figure 19.2 to be reversible, compare the signs of the activation energies for the forward and reverse reactions. Which is positive, which is negative, or are they the same, and if so, are they positive or negative?

(46) What is an "activated complex"? Why is it that we cannot list the physical properties of the species represented as an activated complex?

Section 19.4

(8) "At a given temperature, only a small fraction of the molecules in a sample has sufficient kinetic energy to engage in chemical reaction." What is the meaning of that statement?

(9) What is a catalyst? Explain how a catalyst affects reaction rates.

(10) For the hypothetical reaction $A + B \rightarrow C$, what will happen to the rate of reaction if the concentration of A is increased? What will happen if the concentration of B is decreased? Explain why in both cases.

(47) State the effect of a temperature increase and a temperature decrease on the rate of a chemical reaction. Explain each effect.

(48) Suppose that two substances are brought together under conditions that cause them to react and reach equilibrium. Suppose that in another vessel the same substances and a catalyst are brought together, and again equilibrium is reached. How are the processes alike, and how are they different?

(49) For the hypothetical reaction $A + B \rightarrow C$, what will happen to the reaction rate if the concentration of A is increased *and* the concentration of B is decreased? Explain.

Section 19.5

If nitrogen and hydrogen are brought together at the proper temperature and pressure, they will react until they reach equilibrium: $N_2(g) + 3\ H_2(g) \rightleftharpoons 2\ NH_3(g)$. *Answer the following two questions in each column with regard to the establishment of that equilibrium:*

(11) When will the forward reaction rate be at a maximum: at the start of the reaction, after equilibrium has been reached, or at some point in between?

(12) What happens to the concentration of each of the three species between the start of the reaction and the time equilibrium is reached?

(50) When will the reverse reaction rate be at a maximum: at the start of the reaction, after equilibrium has been reached, or at some point in between?

(51) On a single set of coordinate axes, sketch graphs of the forward reaction rate vs. time and the reverse reaction rate vs. time from the moment the reactants are mixed to a point beyond the establishment of equilibrium.

Section 19.6

(13) Predict the direction in which the equilibrium $SO_2(g) + NO_2(g) \rightleftharpoons SO_3(g) + NO(g)$ will shift if the concentration of NO is increased. Explain or justify your prediction.

(14) The equilibrium system

$$COCl_2(g) \rightleftharpoons CO(g) + Cl_2(g)$$

has some of the chlorine removed. Predict the direction in which the equilibrium will shift, and explain your prediction.

(15) In which direction would the equilibrium $HC_2H_3O_2(aq) \rightleftharpoons H^+(aq) + C_2H_3O_2^-(aq)$ shift if the concentration of the $C_2H_3O_2^-$ ion were increased? Explain your prediction.

(16) Which direction will be favored if you reduce the volume of the equilibrium

$$N_2O_3(g) \rightleftharpoons N_2O(g) + O_2(g)?$$

Explain.

(17) What shift will occur, according to Le Chatelier's Principle, to the equilibrium

$$C(s) + H_2O(g) \rightleftharpoons CO(g) + H_2(g)$$

if volume is increased? Explain.

(18) The equilibrium $CO(g) + H_2O(g) \rightleftharpoons CO_2(g) + H_2(g) + 41.4\ kJ$ is heated. Predict the direction in which the equilibrium will be favored as a result, and justify your prediction in terms of Le Chatelier's Principle.

(52) If the system $2\ SO_2(g) + O_2(g) \rightleftharpoons 2\ SO_3(g)$ is at equilibrium and the concentration of O_2 is reduced, predict the direction in which the equilibrium will shift. Justify or explain your prediction.

(53) If additional oxygen is pumped into the equilibrium system

$$4\ NH_3(g) + 5\ O_2(g) \rightleftharpoons 4\ NO(g) + 6\ H_2O(g),$$

in which direction will the reaction shift? Justify your answer.

(54) Predict the direction of shift for the equilibrium

$$Cu(NH_3)_4^{2+}(aq) \rightleftharpoons Cu^{2+}(aq) + 4\ NH_3(aq)$$

if the concentration of ammonia were reduced. Explain your prediction.

(55) A container holding the equilibrium

$$4\ H_2(g) + CS_2(g) \rightleftharpoons CH_4(g) + 2\ H_2S(g)$$

is enlarged. Predict the direction of the Le Chatelier shift. Explain.

(56) In what direction will

$$CO(g) + H_2O(g) \rightleftharpoons CO_2(g) + H_2(g)$$

shift as a result of a reduction in volume? Explain.

(57) Which direction of the equilibrium

$$2\ NO_2(g) \rightleftharpoons N_2O_4(g) + 59.0\ kJ$$

will be favored if the system is cooled? Explain.

(19) If you wished to increase the relative amount of HI in the equilibrium

$$H_2(g) + I_2(g) + 25.9\ kJ \rightleftharpoons 2\ HI(g)$$

would you heat or cool the system? Explain your decision.

(20) Consider the equilibrium:

$$4\ NH_3(g) + 5\ O_2(g) \rightleftharpoons 4\ NO(g) + 6\ H_2O(g) + 905\ kJ$$

Determine the direction of the Le Chatelier shift, forward or reverse, for each of the following actions: (a) add ammonia; (b) raise temperature; (c) reduce volume; (d) remove $H_2O(g)$.

(58) If your purpose were to increase the yield of SO_3 in the equilibrium

$$SO_2(g) + NO_2(g) \rightleftharpoons SO_3(g) + NO(g) + 41.8\ kJ$$

would you use the highest or lowest operating temperature possible? Explain.

(59)* The solubility of calcium hydroxide is low; it reaches about 2.4×10^{-2} M at saturation. In acid solutions, with many H^+ ions present, calcium hydroxide is quite soluble. Explain this fact in terms of Le Chatelier's Principle. (Hint: Recall what you know of reactions in which molecular products are formed.)

Section 19.7

For each equilibrium equation shown, write the equilibrium constant expression.

(21) $2\ SO_2(g) + O_2(g) \rightleftharpoons 2\ SO_3(g)$

(22) $4\ H_2(g) + CS_2(g) \rightleftharpoons CH_4(g) + 2\ H_2S(g)$

(23) $Cd(OH)_2(s) \rightleftharpoons Cd^{2+}(aq) + 2\ OH^-(aq)$

(24) $HNO_2(aq) \rightleftharpoons H^+(aq) + NO_2^-(aq)$

(25) $Ag(CN)_2^-(aq) \rightleftharpoons Ag^+(aq) + 2\ CN^-(aq)$

(26) "The equilibrium constant expression for a given reaction depends on how the equilibrium equation is written." Explain the meaning of that statement. You may, if you wish, use the equilibrium equation $N_2(g) + 3\ H_2(g) \rightleftharpoons 2\ NH_3(g)$ to illustrate your explanation.

(60) $CO(g) + H_2O(g) \rightleftharpoons CO_2(g) + H_2(g)$

(61) $C(s) + H_2O(g) \rightleftharpoons CO(g) + H_2(g)$

(62) $Zn_3(PO_4)_2(s) \rightleftharpoons 3\ Zn^{2+}(aq) + 2\ PO_4^{3-}(aq)$

(63) $HNO_2(aq) + H_2O(\ell) \rightleftharpoons H_3O^+(aq) + NO_2^-(aq)$

(64) $Cu(NH_3)_4^{2+}(aq) \rightleftharpoons Cu^{2+}(aq) + 4\ NH_3(aq)$

(65) The equilibrium between nitrogen monoxide, oxygen, and nitrogen dioxide may be expressed in the equation

$$2\ NO(g) + O_2(g) \rightleftharpoons 2\ NO_2(g)$$

Write the equilibrium constant expression for this equation. Then express the same equilibrium in at least two other ways, and write the equilibrium constant expression for each. Are the constants numerically equal? Cite some evidence to support your yes or no answer.

Section 19.8

(27) The equilibrium constant is 2.3×10^{-8} for the equilibrium

$$HCO_3^-(aq) + HOH(\ell) \rightleftharpoons H_2CO_3(aq) + OH^-(aq)$$

In which direction is the reaction favored at equilibrium? State the basis for your answer.

(66) If sodium cyanide solution is added to silver nitrate solution the following equilibrium will be reached:

$$Ag^+(aq) + 2\ CN^-(aq) \rightleftharpoons Ag(CN)_2^-(aq)$$

For this equilibrium $K = 5.6 \times 10^{18}$. In which direction is the equilibrium favored? Justify your answer.

(28) Acetic acid, $HC_2H_3O_2$, is a soluble weak acid. When placed in water it partially ionizes and reaches equilibrium. Write the equilibrium equation for the ionization. Will the equilibrium constant be large or small? Justify your answer.

(67) A certain equilibrium has a very small equilibrium constant. In which direction, forward or reverse, is the equilibrium favored? Explain.

In Chapter 16 you learned how to write solution inventories and net ionic equations, based on the solubility of ionic compounds, the strength of acids, and the stability of certain ion combinations. Use these ideas to predict the favored direction of each equilibrium below. In each case, state whether you expect the equilibrium constant to be large or small.

(29)* (a) $HCl(aq) \rightleftharpoons H^+(aq) + Cl^-(aq)$
(b) $BaSO_4(s) \rightleftharpoons Ba^{2+}(aq) + SO_4{}^{2-}(aq)$

(68)* (a) $H_2SO_3(aq) \rightleftharpoons H_2O(\ell) + SO_2(aq)$
(b) $H^+(aq) + C_2H_3O_2{}^-(aq) \rightleftharpoons HC_2H_3O_2(aq)$

Section 19.9

(30) The solubility of cadmium sulfide, CdS, is 8.8×10^{-14} mol/L. Calculate K_{sp}.

(31) If 1.0×10^{-3} g of CuBr dissolved in 100 mL of water yields a saturated solution, calculate the K_{sp} of CuBr.

(32) $K_{sp} = 8.7 \times 10^{-9}$ for $CaCO_3$. Calculate its solubility in (a) moles per liter and (b) grams per 100 mL.

(33)* $K_{sp} = 1.7 \times 10^{-6}$ for BaF_2. Calculate its solubility in moles per liter.

(34) $K_{sp} = 8.1 \times 10^{-9}$ for barium carbonate. What is the solubility (mol/L) of $BaCO_3$ in 0.10 M $BaCl_2$?

(35) pH = 1.93 for 1.0 M $HC_3H_5O_3$ (lactic acid). Calculate its K_a and percent ionization.

(36) Calculate the pH of 0.1 M HNO_2. ($K_a = 4.6 \times 10^{-4}$)

(37) What is the pH of a solution of 0.25 M $NaNO_2$ in 0.75 M HNO_2? ($K_a = 4.6 \times 10^{-4}$ for HNO_2)

(38) What concentration ratio of benzoic acid to benzoate ion ($[HC_7H_5O_2]/[C_7H_5O_2{}^-]$ will produce a buffer having a pH of 4.80 if $K_a = 6.5 \times 10^{-5}$ for benzoic acid?

(39) 0.069 mol PCl_3 and 0.058 mol Cl_2 are placed into a vessel whose volume is 1.0 L. They react to establish the equilibrium $PCl_3(g) + Cl_2(g) \rightleftharpoons PCl_5$. $[Cl_2] = 0.031$ when equilibrium is reached. Calculate K at the temperature of the system.

(69) $Co(OH)_2$ dissolves in water to the extent of 3.7×10^{-6} mol/L. Find its K_{sp}.

(70) 250 mL of water will dissolve only 8.7 mg of silver carbonate. What is the K_{sp} of Ag_2CO_3?

(71) Find the moles per liter and grams per 100 mL solubility of silver iodate, $AgIO_3$, if its $K_{sp} = 2.0 \times 10^{-8}$.

(72)* Find the solubility (mol/L) of $Mn(OH)_2$ if its $K_{sp} = 1.0 \times 10^{-13}$

(73)* How many grams of calcium oxalate will dissolve in 5.0×10^2 mL 0.22 M $Na_2C_2O_4$ if $K_{sp} = 2.4 \times 10^{-9}$ for CaC_2O_4?

(74) The pH of 0.22 M $HC_4H_5O_3$ (acetoacetic acid) is 2.12. Find its K_a and percent ionization.

(75) Find the pH of 0.30 M $HC_2H_3O_2$. ($K_a = 1.8 \times 10^{-5}$)

(76)* 28.0 g of sodium acetate, $NaC_2H_3O_2$, are dissolved in 5.00×10^2 mL of 0.12 M $HC_2H_3O_2$ ($K_a = 1.8 \times 10^{-5}$). Calculate the pH of the solution.

(77) Find the ratio $[HC_2H_3O_2]/[C_2H_3O_2{}^-]$ that will yield a buffer in which pH = 4.15. ($K_a = 1.8 \times 10^{-5}$)

(78)* 0.351 mol of CO and 1.340 mol of Cl_2 are introduced into a reaction chamber having a volume of 3.00 L. When equilibrium is reached according to the equation $CO(g) + Cl_2(g) \rightleftharpoons COCl_2(g)$, there are 1.050 mol of Cl_2 in the chamber. Calculate K.

Miscellaneous Questions

(*79*) Distinguish precisely and in scientific terms the differences between items in each of the following groups:
- (a) Reaction and reversible reaction
- (b) Open system, closed system
- (c) Dynamic equilibrium, and static equilibrium
- (d) Activated complex and activation energy
- (e) Catalyzed reaction, uncatalyzed reaction
- (f) Catalyst, inhibitor
- (g) Buffered solution, unbuffered solution

(80) Classify each of the following statements as true or false:
- (a) Some equilibria depend on a steady supply of a reactant in order to maintain the equilibrium.
- (b) Both forward and reverse reactions continue after equilibrium is reached.
- (c) Every time reaction molecules collide there is a reaction.
- (d) Potential energy during a collision is greater than potential energy before or after the collision.
- (e) The properties of an activated complex are between those of the reactants and the products.
- (f) Activation energy is positive for both the forward and reverse reactions.
- (g) Kinetic energy is changed to potential energy during a collision.
- (h) An increase in temperature speeds the forward reaction, but slows the reverse reaction.
- (i) A catalyst changes the steps by which a reaction is completed.
- (j) An increase in concentration of a substance on the right side of an equation speeds the reverse reaction rate.
- (k) An increase in the concentration of a substance in an equilibrium increases the reaction rate in which the substance is a product.
- (l) Reducing the volume of a gaseous equilibrium shifts the equilibrium in the direction of fewer gaseous molecules.
- (m) Raising temperature results in a shift in the forward direction of an endothermic equilibrium.
- (n) The value of an equilibrium constant depends on temperature.
- (o) Liquid water is never included in an equilibrium constant expression.
- (p) A large K indicates an equilibrium is favored in the reverse direction.

(81) At Time 1, two molecules are about to collide. At Time 2, they are in the process of colliding, and their form is that of the activated complex. Compare the sum of their kinetic energies at Time 1 with the kinetic energy of the activated complex at Time 2. Explain your conclusions.

(82) List three things you might do to increase the rate of the reverse reaction for which Figure 19.2 is the energy-reaction graph.

(83)* The Haber process for making ammonia by direct combination of the elements is described by the equation $N_2(g) + 3\ H_2(g) \rightarrow 2\ NH_3(g) +$ 92 kJ. If the purpose of the manufacturer is to make the greatest amount of ammonia in the least time, is he most apt to conduct the reaction at (a) high pressure or low pressure; (b) high temperature or low temperature? Explain your choice in each case.

(84) Under proper conditions, the reaction in Question 83 will reach equilibrium. Is the manufacturer apt to conduct the reaction under those conditions, i.e, at equilibrium? Why or why not?

(85)* The reaction in Question 83 has a yield of about 98% at 200°C and 1000 atm. Commercially the reaction is performed at about 500°C and 350 atm, where the yield is only about 30%. Suggest reasons why operation at the lower yield is economically more favorable.

(86)* $K_{sp} = 1.6 \times 10^{-52}$ for HgS. Calculate the number of pounds of HgS that will dissolve in a cubic mile of water.

(87)* The solubility of calcium hydroxide is low enough to be listed as "insoluble" in Table 16.2, but it is much more soluble than most of the other salts that are similarly classified. Its K_{sp} is 5.5×10^{-6}.
- (a) Write the equation for the equilibrium to which the K_{sp} is related.
- (b) If you had such an equilibrium, name at least two chemicals or general classes of chemicals that might be added to (1) reduce the solubility of $Ca(OH)_2$; (2) increase its solubility. Justify your choices.
- (c) Without adding a calcium or hydroxide ion, name a chemical or class of chemicals that

would, if added, (1) increase $[OH^-]$; (2) reduce $[OH^-]$. Justify your choices.

(88)* $4\ NH_3(g) + 7\ O_2(g) \rightleftharpoons 6\ H_2O(g) + 4\ NO_2(g) +$ energy. The table below lists several "disturbances" that may or may not produce a Le Chatelier shift in the foregoing equilibrium. If the disturbance is an immediate change in the concentration of any species in the equilibrium, place in the concentration column of that substance an "I" if the change is an increase, and a "D" if it is a decrease. If a shift will result, place F in the SHIFT column if the shift is in the forward direction, and R if it is in the reverse direction. Then determine what will happen to the concentrations of the other species because of the shift, and insert "I" or "D" for increase or decrease. If there is no Le Chatelier shift, write "None" in the SHIFT column, and leave the other columns blank.

DISTURBANCE	SHIFT	$[NH_3]$	$[O_2]$	$[H_2O]$	$[NO_2]$
Add NO_2					
Reduce temperature					
Add N_2					
Remove NH_3					
Add a catalyst					

20 Nuclear Chemistry

LOOKING BACK

4.3 The proton, neutron, and electron are the three major subatomic particles. Their masses are measured in atomic mass units, amu.

4.4 Protons and neutrons that account for most of the mass of an atom make up its nucleus, while electrons thinly populate the almost empty space around the nucleus.

4.5 Nuclei of different atoms of the same element all have the same number of protons, but they may have different numbers of neutrons, giving them different masses. These are isotopes, and they are distinguished by their nuclear symbols:

$$_{\text{atomic number}}^{\text{mass number}}\text{Sy} = {}_{Z}^{A}\text{Sy} = {}^{A}\text{Sy}$$

LOOKING AHEAD IN CHAPTER 20

Individually and in groups, protons, neutrons, and electrons participate in nuclear reactions.

Nuclear chemistry is concerned with the nucleus of the atom and almost disregards the electrons outside the nucleus.

Nuclear reactions are concerned with the isotopes of an element. Nuclear symbols are used to write equations for nuclear reactions. The atomic number (Z) and mass number (A) are used in balancing nuclear equations.

20.1 THE DAWN OF NUCLEAR CHEMISTRY

Serendipity. This pleasant sounding word refers to finding valuable things you are not looking for—an accidental discovery, in other words. Serendipity has played an important part in many scientific discoveries, but rarely has an unplanned observation led to a discovery as far-reaching as that of Henri Becquerel. What he stumbled across now affects the lives of you and me and potentially of every creature on this planet. Becquerel discovered nuclear chemistry.

Becquerel became interested in X-rays soon after they were discovered, also accidentally, by Wilhelm Roentgen in 1895. Becquerel was also interested in fluorescence and wondered if the two were related. To produce fluorescence you shine one color of light on an object, and it glows with another color. X-rays are like light. They are part of the electromagnetic spectrum (Section 5.1), but they are outside the visible range. They are high-energy radiations that are able to go through human tissue, black paper, and many other solids that visible light cannot penetrate.

Becquerel's plan was to put some fluorescent material, a uranium salt, on top of unexposed photographic film wrapped in black paper and place them in the sunlight. The sunlight could not expose the film by itself, but if the film were exposed it would show that fluorescent rays from the uranium salt had passed through the paper like X-rays.

The day Becquerel picked for his experiment was cloudy. After waiting in vain for the sun to come out, he put the uranium salt and wrapped film into a drawer, to be used the next sunny day. The weather continued to be dull for several days. Becquerel finally decided to develop the film to see if any fluroescent light had gotten through, even on the cloudy day. To his amazement, the film was strongly exposed. The only possible conclusion was that rays were being emitted from the salt all the time in the darkness of the drawer, and that they had nothing whatever to do with sunlight. It was later shown that, although their penetrating ability is like that of X-rays, the emissions were not X-rays but a form of energy that had never been observed before.

This is how Becquerel discovered **radioactivity,** a natural, spontaneous process that has been going on in the nuclei of atoms since the beginning of time.

Once the door to nuclear chemistry was opened, progress was rapid. While working with a uranium-bearing mineral called *pitchblende,* Marie and Pierre Curie discovered that a second element, thorium, gave off penetrating radiations. They also found two new radioactive elements: polonium, which is about 400 times more radioactive than uranium, and radium, which is again many times more radioactive than polonium.

20.2 NATURAL RADIOACTIVITY

PG 20A Define radioactivity. Name, identify from a description, or describe the three types of radioactive emission.

Radioactivity is the spontaneous emission of rays resulting from the decay, or breaking up, of an atomic nucleus. "Natural" radioactivity begins with a radioactive isotope that is found in nature. Three kinds of rays can be identified. If a beam consisting of all three rays is directed into an electric field, as in Figure 20.1A, the individual rays are separated. One, called an alpha ray, or α-ray,* is attracted toward the negatively charged plate, indicating that it has a positive charge. The α-ray has little penetrating power; it can be stopped by the outer layer of skin or a few sheets of paper (Fig. 20.1B). Alpha rays are now known to be nuclei of helium atoms, having the nuclear symbol ${}^{4}_{2}He$. They are commonly called alpha particles, or α particles.

The second kind of ray also turns out to be a beam of particles, but they are negatively charged and are therefore attracted to the positively charged plate (Fig. 20.1A). Called beta rays, or β-rays, they have been identified as electrons. The nuclear symbol for a β-particle is ${}^{0}_{-1}e$, indicating zero atomic mass and a -1 charge. β-particles have considerably more penetrating power than α-particles, but they can be stopped by a sheet of aluminum about 4-mm thick (Fig. 20.1B).

*α, β, and γ are the Greek letters alpha, beta, and gamma.

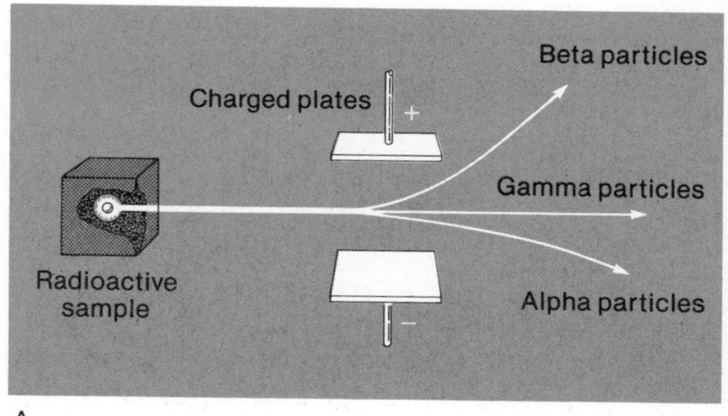

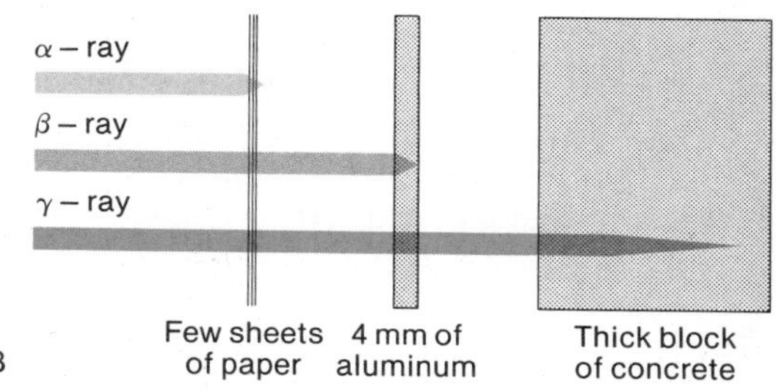

Figure 20.1
Alpha (α), beta (β) and gamma (γ) radioactive emissions. A. Alpha and beta particles are deflected by an electric field, but gamma rays are not. B. The penetrating power of radioactive emissions increases in the order of alpha, beta, and gamma.

The third kind of radiation is the gamma ray, or γ-ray. Gamma rays are not particles, but very high energy electromagnetic rays, similar to X-rays. Because of their high energy, gamma rays have high penetrating power. They can be stopped only by thick layers of lead or heavy concrete walls, as shown in Fig. 20.1B. Not having an electric charge, gamma rays are not deflected by an electric field (Fig. 20.1A).

Quick Check 20.2

1. Write the nuclear symbol and electrical charge of an alpha particle and a beta particle.
2. List alpha, beta, and gamma rays in order of decreasing penetrating power.

20.3 INTERACTION OF RADIOACTIVE EMISSIONS WITH MATTER

PG 20B Describe how emissions from radioactive substances affect gases and living organisms.

When α, β, or γ radiation collides with an atom or molecule, some of its energy is given to the target particle. The collision changes the electron arrangement in the species hit. An electron may be excited to a higher energy level, leading to electromagnetic radiation as the electron drops back to its ground state level. The electron may be knocked all the way out of an atom or molecule, ionizing it. Air, or any gas, can be ionized by a radioactive substance. If the radiation strikes chemically bonded atoms, it often breaks those bonds, thereby causing a chemical reaction.

The ionization of molecules in air by radioactive emission has a present-day application in one kind of home smoke detector. These "ionization" detectors generally use a tiny chip of americium-241, a radioactive isotope of an element not found in nature. The ionization of air causes a small current to flow through the air inside the detector. When smoke enters, it breaks the circuit and sets off the alarm.

It is the ability of radiation to start chemical change that is responsible for its effect on body tissue, leading to illness and even death in extreme cases. These effects have been studied among the survivors of the atomic bombs dropped at Hiroshima and Nagasaki at the close of World War II and among workers in nuclear plants and laboratories who have been overly exposed to radiation. People working in such plants wear a small device called a dosimeter that monitors the amount of radiation that falls on them. By this and other safety measures, the danger to workers associated with industrial radiation has been minimized.

20.4 DETECTION OF EMISSIONS FROM RADIOACTIVE SUBSTANCES

PG 20C Define or describe a Geiger counter. Explain how it operates.

There are several ways to detect radioactivity. Perhaps the most obvious, but not necessarily the most convenient, is through its effect on photographic film, the very property that led to its discovery. Another is based on its ability to produce visible emissions in fluorescent materials, as in watch dials that can be seen in the dark. The path of a radioactive particle can be observed in a **cloud chamber,** an enclosed air space that is supersaturated with some vapor, often water. The particle ionizes the air as it travels through the chamber. The vapor condenses on the ions, leaving a visible track that can be photographed.

The **Geiger–Müller counter** (often called a **Geiger counter**) is probably the best known instrument for detecting and measuring radiation. It consists of a tube filled with a gas, as shown in Figure 20.2. The gas is ionized by radiation

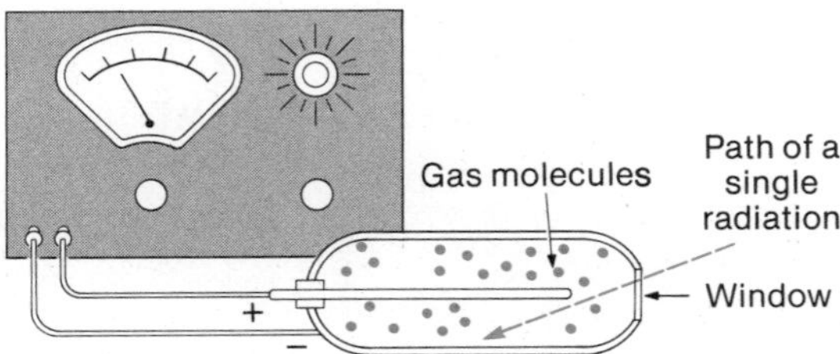

Figure 20.2
Geiger–Müller counter. The tube contains argon at low pressure. A high electrical potential is established between a positively charged wire in the center and the negatively charged case. Radiations enter the window and ionize the argon. The ions move to the electrodes, creating a measurable electric discharge and an audible "click."

entering the tube through a window, permitting an electrical discharge between two electrodes. The current is related to the quantity of radiation received, and can be measured on a meter. Many Geiger counters indicate radiation by a clicking sound.

20.5 NATURAL RADIOACTIVE DECAY SERIES—NUCLEAR EQUATIONS

PG 20D Describe a natural radioactive decay series.

21E Given the identity of a radioactive isotope and the particle it emits, write a nuclear equation for the emission.

When a nucleus emits an α- or β-particle, there is a change in the make-up of the nucleus. It has a new identity directly related to the particle emitted. The emission of one or more gamma rays does not, by itself, change the composition of the nucleus. For the balance of the chapter, therefore, we will consider only alpha and beta particles in nuclear reactions.

When a radioactive nucleus emits an alpha or beta particle, there is a **transmutation** of an element, or a change from one element to another. Recall that the elemental identity of an atom depends on the number of protons in the nucleus. If an atom changes from one element to another, there must be a change in the number of protons. A change of this kind is represented by a nuclear equation showing the nuclear symbols of the reactant and product isotopes. The emission observed by Becquerel was an alpha particle emission, also called an **alpha decay reaction.** The nuclear product remaining after a ${}^{238}_{92}U$ nucleus emits an α-particle is a thorium nucleus, ${}^{234}_{90}Th$. In other words, a ${}^{238}_{92}U$ nucleus has *disintegrated,* or *decayed,* into a ${}^{4}_{2}He$ nucleus and a ${}^{234}_{90}Th$ nucleus. This may be shown in a nuclear equation:

$$ {}^{238}_{92}U \rightarrow {}^{4}_{2}He + {}^{234}_{90}Th \qquad (20.1)$$

Notice that the above equation is "balanced" in both neutrons and protons. The total number of neutrons and protons is 238, the mass number of the uranium isotope. The total mass number of the two products is 234 + 4, again 238. In terms of protons, the 92 in a uranium nucleus are accounted for by 90 in the thorium nucleus plus 2 in the helium nucleus. A nuclear equation is balanced if the sums of the mass numbers on the two sides of the equation are equal, and if the sums of the atomic numbers are equal.

The ${}^{234}_{90}Th$ nucleus resulting from the disintegration of uranium-238 is also radioactive. It is a **beta decay reaction;** it emits a beta particle, ${}^{0}_{-1}e$, and produces an isotope of protactinium, ${}^{234}_{91}Pa$:

$$ {}^{234}_{90}Th \rightarrow {}^{234}_{91}Pa + {}^{0}_{-1}e \qquad (20.2)$$

In a β-particle emission, the mass numbers of the reactant and product isotopes are the same, while the atomic number increases by 1. Though the actual process is more complex, it appears as if a neutron divides into a proton and an electron, and the electron is ejected.

The two radioactive disintegration steps described by Equations 20.1 and 20.2 are only the first two of fourteen steps that begin with ${}^{238}_{92}U$. There are

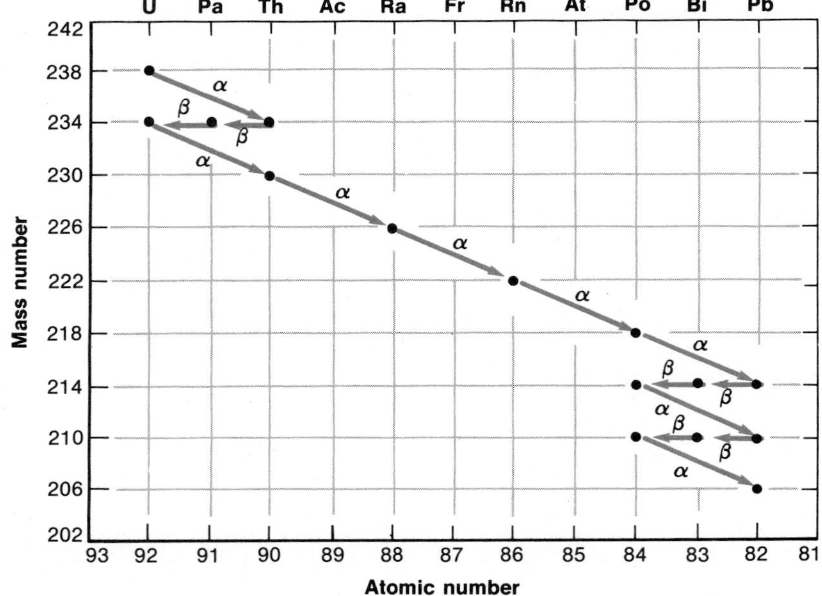

Figure 20.3
Radioactive decay series. This series begins with $^{238}_{92}U$, and after eight alpha emissions and six beta emissions, produces $^{206}_{82}Pb$ as a stable end product.

eight α-particle emissions and six β-particle emissions, leading ultimately to a stable isotope of lead, $^{206}_{82}Pb$. This entire **natural radioactive decay series** is described in Figure 20.3. There are two other natural disintegration series. One begins with $^{232}_{90}Th$ and ends with $^{208}_{82}Pb$, and the other passes from $^{235}_{92}U$ to $^{207}_{82}Pb$.

EXAMPLE 20.1 Write the nuclear equation for the changes that occur in the uranium-238 disintegration series when $^{226}_{88}Ra$ ejects an α-particle. Ra is the symbol for radium, one of the elements discovered by Pierre and Marie Curie in their study of radioactivity.

In writing a nuclear equation, one product will be the particle ejected. The mass number of the other product will be such that, when added to the mass number of the ejected particle, the total will be the mass number of the original isotope. What is the mass number of the second product of the emission of an alpha particle from a $^{226}_{88}Ra$ nucleus?

222.

The reactant isotope has a mass number of 226. It emits a particle having a mass number of 4. This leaves $226 - 4 = 222$ as the mass number of the remaining particle.

Now find the atomic number of the second product particle. The atomic number of the starting isotope is 88. It emitted a particle having two protons. How many protons are left in the nucleus of the other product?

86.

If two protons are emitted from a nucleus having 88 protons, 86 remain.

You now know the mass number and the atomic number of the second product of an alpha particle emission from $^{226}_{88}Ra$. Using a periodic table you can find the elemental symbol of this product, and assemble all three symbols into the required nuclear equation.

$$^{226}_{88}Ra \longrightarrow ^{4}_{2}He + ^{222}_{86}Rn$$

The second product is a radioactive isotope of the noble gas radon, whose atomic number is 86. This isotope continues the natural emission series by emitting another α-particle.

EXAMPLE 20.2 Write the nuclear equation for the emission of a β-particle from $^{210}_{83}Bi$.

The method is the same. Remember the beta particle, $^{0}_{-1}e$, has zero mass number, and an effective atomic number of -1. Both mass number and atomic number must be conserved in the equation.

$$^{210}_{83}Bi \longrightarrow ^{0}_{-1}e + ^{210}_{84}Po$$

In the emission of a β-particle, which has effectively no mass, the mass number of the radioactive isotope and the product iosotope are the same. The product isotope has an atomic number greater by one than the radioactive isotope, an increase of one proton. Po is the symbol that corresponds to atomic number 84. The element is polonium, the other element discovered by the Curies in their investigation of radioactivity. The name of the element was selected to honor Mme. Curie's native Poland.

20.6 HALF-LIFE

PG 20F Describe or illustrate what is meant by the half-life of a radioactive isotope.

20G Given the starting quantity of a radioactive iostope, Figure 20.4, and two of the following, calculate the third: half-life; elapsed time; quantity of isotope remaining.

The rate at which a single step in nuclear disintegration occurs is measured by its **half-life, the time required for the disintegration of one half of the radioactive atoms in a sample.** Each radioactive isotope has its own unique half-life, commonly written $\mathbf{t_{1/2}}$. The half-lives of some isotopes are very long; the half-life of $^{238}_{92}U$ is 4.5 billion years. Other half-lives are very short. The half-life of $^{234}_{90}Th$ (Equation 20.2) is a comfortable 24.1 days, but the half-life of the particle that appears after $^{234}_{90}Th$ emits a β-particle, $^{234}_{91}Pa$, is only 1.18 minutes. The half-life of one species in the same radioactive disintegration series, $^{214}_{84}Po$, is only

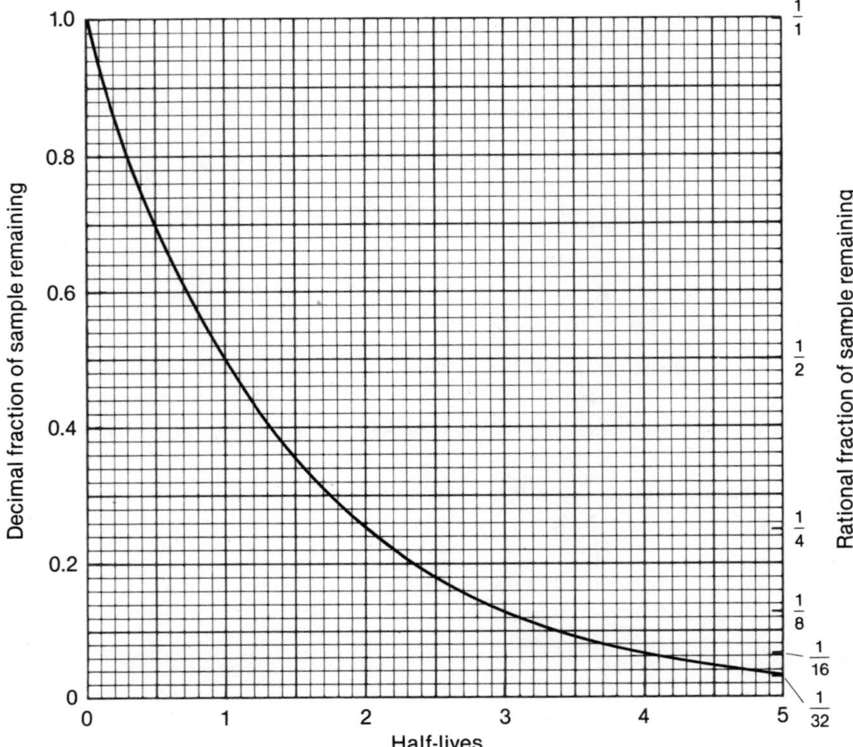

Figure 20.4
Half-life decay Curve for a Radioactive substance.

0.000 16 second. Needless to say, people who work with radioactive substances with such short half-lives are rather pressed for time.

Figure 20.4 is a graph of the fraction of an original sample that remains (vertical axis) after a number of half-lives (horizontal axis). The vertical axis at the left has a conventional scale, giving values in decimal fractions. The scale values on the axis on the right are the fractions that remain after each half-life period: $\frac{1}{2}$ after the first half-life, $\frac{1}{2}$ of $\frac{1}{2}$, or $\frac{1}{4}$ after the second, $\frac{1}{2}$ of $\frac{1}{4}$, or $\frac{1}{8}$ after the third, and so forth. Thus, the fraction of a sample that is still present after n half-lives is $(\frac{1}{2})^n$. If S is the starting quantity and R is the amount that remains after n half-lives, then

$$R = S \times (\tfrac{1}{2})^n \tag{20.3}$$

EXAMPLE 20.3 The half-life of $^{210}_{83}Bi$ is 5.0 days. If you begin with 16 grams of $^{210}_{83}Bi$ how many grams will you have 25 days (5 half-lives) later?

This problem is solved by a straightforward substitution into Equation 20.3. Recall that, to raise a number to a power on a calculator, enter that number (0.5 for ½), press y^x, then the power (5), and finally =.

————— —————

$$R = 16 \times (0.5)^5 = 0.50 \text{ grams}$$

EXAMPLE 20.4 The half-life of $^{45}_{19}K$ is 20 minutes. If you have a sample containing 2.1×10^3 micrograms (μg) of this isotope at noon, how many micrograms will remain at 3 o'clock in the afternoon?

First of all, through how many half-lives will the decay process pass between noon and 3 o'clock, at 20 minutes for each half-life?

----- -----

$$3 \text{ hours} \times \frac{60 \text{ minutes}}{1 \text{ hour}} \times \frac{1 \text{ half-life}}{20 \text{ minutes}} = 9 \text{ half-lives}$$

From here the solution comes by direct substitution into Equation 21.3.

----- -----

$$R = 2.1 \times 10^3\ \mu g \times (1/2)^9 = 4.1\ \mu g$$

To find the half-life of a radioactive isotope you must determine starting and remaining quantities over a measured period of time. One way to interpret these data is to express the numbers as the fraction of the sample remaining. R/S, which is the vertical axis in Figure 20.4. Locate the fraction on the curve and read on the horizontal axis the number of half-lives in the measured time period. Divide time by half-lives to get the half-life of the substance.

EXAMPLE 20.5 $^{125}_{52}Sb$ has a longer half-life than many man-made radioisotopes. The mass of a sample is found to be 8.623 grams. The sample is set aside for 146 days—a convenient period equal to 0.40 year—and then weighed again. The mass is 7.782 grams. Find $t_{1/2}$.

Begin by calculating the decimal fraction of the sample that remains after 0.40 year.

----- -----

7.782/8.623 = 0.9025

Use Figure 20.4 to estimate the number of half-lives when 90.25% of the original sample remains after 0.40 year.

----- -----

Either 0.14 or 0.15 would be a reasonable estimate. We will use 0.15.

The half-life in *years* per *half-life* is the quotient of those two numbers.

----- -----

0.40 year/0.15 half-live = 2.7 years in one half-life.

This half-life rate of decay of radioactive substances is the basis of **radiocarbon dating,** by which the age of fossils is estimated. Carbon is the principal chemical element in all living organisms, both plant and animal. Most carbon atoms are carbon-12; but a small portion of the carbon in atmospheric carbon dioxide is carbon-14, a radioactive isotope with a half-life of 5.72×10^3 years. When a plant or animal is alive, it takes in this isotope from the atmosphere, while the same isotope in the organism is disappearing by nuclear disintegration. A "steady state" situation exists while the system lives, maintaining a constant ratio of carbon-14 to carbon-12. When the organism dies, the disintegration of carbon-14 continues, but its intake stops. This leads to a gradual reduction in the ratio of ${}^{14}_{6}C$ to ${}^{12}_{6}C$. By measuring the ratio and the amount of ${}^{14}_{6}C$ now present in a sample it is possible to calculate the ${}^{14}_{6}C$ present when the organism died. Figure 20.4 is then used to calculate the age of the sample.

EXAMPLE 20.6 An archaeologist analyzes an organic fossil and finds that for every 14 units of ${}^{14}_{6}C$ present in the sample at death, 9.3 remain today. Calculate the age of the sample.

The quantities 14 and 9.3 refer to radioactive disintegrations detected by a Geiger counter or some other instrument. They express initial and final amounts, just as if they were expressed in grams. Use these quantities to find the number of half-lives as you did in Example 20.5.

$$\frac{9.3}{14} = 0.66 \text{ of sample remains} = 0.59 \text{ half-life}$$

If each half-life is 5.72×10^3 years, find the age of the fossil.

$$0.59 \text{ half-life} \times \frac{5.72 \times 10^3 \text{ years}}{1 \text{ half-life}} = 3.4 \times 10^3 \text{ years}$$

Carbon dating has produced evidence of modern man's presence on earth as long ago as 14 000 to 15 000 years. Similar dating techniques are also applied to mineral deposits. Analyses of geological deposits have yielded rocks with an estimated age of 3.0 to 4.5 billion years, the latter figure being the scientist's estimate of the age of the earth. The oldest moon rocks analyzed to date indicate an age of about 3.5 billion years.

20.7 NUCLEAR REACTIONS AND ORDINARY CHEMICAL REACTIONS COMPARED

PG 20H List or identify four ways in which nuclear reactions differ from ordinary chemical reactions.

Now that you have seen the nature of a nuclear change and the type of equation by which it is described, we will pause to compare nuclear reactions with the others you have studied. There are four areas of comparison:

1. In ordinary chemical reactions, the chemical properties of an element depend only on the electrons outside the nucleus, and the properties are essentially the same for all isotopes of the element. The nuclear properties of the various isotopes of an element are quite different, however. In the radioactive decay series beginning with uranium-238, $^{234}_{90}Th$ emits a β-particle, whereas a bit farther down the line $^{230}_{90}Th$ ejects an α-particle. Both $^{214}_{82}Pb$ and $^{210}_{82}Pb$ are β-particle emitters toward the end of the series, while the final product, $^{206}_{82}Pb$, has a stable nucleus, emitting neither alpha nor beta particles, nor gamma rays.
2. Radioactivity is independent of the state of chemical combination of the radioactive isotope. The reaction of $^{210}_{83}Bi$ occurs for atoms of that particular isotope whether they are in pure chemical bismuth, combined in bismuth chloride, $BiCl_3$, bismuth sulfate, $Bi_2(SO_4)_3$, or any other bismuth compound, or if they happen to be present in the low melting bismuth alloy used for fire protection in sprinkler systems in large buildings.
3. Nuclear reactions usually result in the formation of different elements because of changes in the number of protons in the nucleus of an atom. In ordinary chemical reactions the atoms keep their identity while changing from one compound as a reactant to another as a product.
4. Both nuclear and ordinary chemical changes involve energy, but the amount of energy for a given amount of reactant in a nuclear change is enormous—greater by several orders of magnitude, or multiples of ten—compared to the energies of ordinary chemical reactions. Further comment appears in Sections 20.11 and 20.13.

20.8 NUCLEAR BOMBARDMENT AND INDUCED RADIOACTIVITY

PG 20I Define or identify nuclear bombardment, reactions.

20J Distinguish natural radioactivty from induced radioactivity produced by bombardment reactions.

20K Define or identify transuranium elements.

In natural radioactive decay, we find an example of the alchemist's get-rich-quick dream of converting one element to another. But the natural process for uranium did not yield the gold coveted by the alchemist; rather, it produced the element lead, with which the dreamer wanted to begin his transmutation. The question was still present after the discovery of radioactivty: can man cause the transmutation of one element into another?

In 1919, Rutherford produced a "Yes" answer to that question. He found that he could "bombard" the nucleus of a nitrogen atom with a beam of alpha particles from a radioactive source, knocking a proton out of the nucleus and producing an atom of oxygen-17:

$$^{14}_{7}N + ^{4}_{2}He \rightarrow ^{17}_{8}O + ^{1}_{1}H$$

The oxygen isotope produced is stable; the experiment did not yield any man-made radioactive isotopes. Similar experiments were conducted with other elements, using high-speed alpha particles as atomic "bullets." It was found that most of the elements up to potassium can be changed to other elements by nuclear bombardment. None of the isotopes produced were radioactive.

One experiment during this period was first thought to yield a nuclear particle that emitted some sort of high-energy radiation, perhaps a gamma ray. In 1932 James Chadwick correctly interpreted the experiment and, in doing so, he became the first person to identify the neutron. The reaction comes from bombarding a beryllium atom with a high-energy α-particle:

$$^{9}_{4}\text{Be} + ^{4}_{2}\text{He} \rightarrow ^{12}_{6}\text{C} + ^{1}_{0}\text{n}$$

$^{1}_{0}\text{n}$ is the nuclear symbol for the neutron, with a mass number of 1 and zero charge.

Two years later, in 1934, Irene Curie, the daughter of Pierre and Marie Curie, and her husband, Frederic Joliot, used high-energy α-particles to produce the first man-made radioactive isotope. Their target was boron-5; the product was a radioactive nitrogen nucleus

$$^{10}_{5}\text{B} + ^{4}_{2}\text{He} \rightarrow ^{13}_{7}\text{N} + ^{1}_{0}\text{n}$$

Because this radioactive isotope is not found in nature, its decay is an example of **induced** or **artificial radioactivity.** When $^{13}_{7}\text{N}$ decays, it emits a particle having the mass of an electron and a charge equal to that of an electron, except that it is positive. This "positive electron" is called a **positron,** and it is represented by the symbol $^{0}_{1}\text{e}$. The decay equation is

$$^{13}_{7}\text{N} \rightarrow ^{13}_{6}\text{C} + ^{0}_{1}\text{e}$$

Today hundreds of **radioisotopes** have been produced in laboratories all over the world, and they find broad use in medicine, industry, and research. Many of these isotopes have been made in different kinds of *particle accelerators,* which use electric fields to increase the kinetic energy of the charged particles that bombard nuclei. Among the earliest and best known accelerators is the cyclotron (Fig. 20.5), designed by E. O. Lawrence at the University of California, Berkeley. Other more powerful accelerators are approximately circular in shape (over a mile in diameter), or linear (two miles long).

One of the more exciting areas of research with bombardment reactions has been the production of elements that do not exist in nature. Except for trace quantities, no natural elements having atomic numbers greater than 92 have ever been discovered. In 1940, however, it was found that $^{238}_{92}\text{U}$ is capable of capturing a neutron:

$$^{238}_{92}\text{U} + ^{1}_{0}\text{n} \rightarrow ^{239}_{92}\text{U} \qquad (20.4)$$

The newly formed isotope is unstable, progressing through two successive β-particle emissions, yielding isotopes of the elements having atomic numbers 93 and 94:

$$^{239}_{92}\text{U} \rightarrow ^{0}_{-1}\text{e} + ^{239}_{93}\text{Np (neptunium)} \qquad (20.5)$$

$$^{239}_{93}\text{Np} \rightarrow ^{0}_{-1}\text{e} + ^{239}_{94}\text{Pu (plutonium)} \qquad (20.6)$$

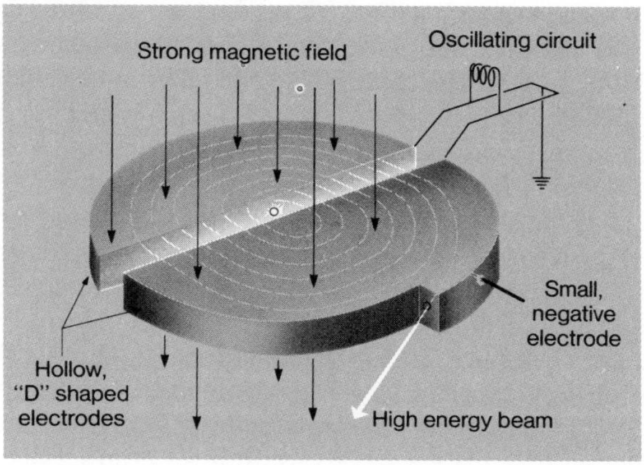

Figure 20.5
Cyclotron. The cyclotron consists of two oppositely charged, evacuated "dees" placed between the poles of a powerful electromagnet. Positive ions, originating at the center, enter the upper dee, which is originally at a negative charge. They pass through this dee in a curved path. At the instant they reenter the central corridor, the polarity of the dees is reversed, and the particles enter the lower dee at an increased velocity. The procedure is repeated over and over; the particles move at higher and higher velocities in paths of greater and greater radius. Eventually they are deflected from the outer dee to strike the target.

Neptunium, plutonium, and all the other man-made elements having atomic numbers greater than 92 are called the **transuranium elements.** All the transuranium isotopes are radioactive, and a few have been isolated only in isotopes with very short half-lives and in extremely small quantities. Some of the bombardments yielding transuranium products use relatively heavy isotopes as bullets. For example, einsteinium-247 is produced by bombarding uranium 238 with ordinary nitrogen nuclei:

$$^{238}_{92}\text{U} + ^{14}_{7}\text{N} \rightarrow ^{247}_{99}\text{Es} + 5\,^{1}_{0}\text{n}$$

Quick Check 20.8

1. What is produced in a nuclear bombardment reaction?
2. What property must a particle have if it is to be used in a particle accelerator?

20.9 USES OF RADIOISOTOPES

Shortly after radioactivity was discovered, it was thought that the radiations had certain curative powers. Radium compounds were made, and radium solutions were bottled and sold for drinking and bathing, before the harmful effects of radiation exposure were well understood. Today's medical practitioners are much wiser, and they have devised sophisticated ways to examine their patients, diagnose their illnesses, and treat their disorders, all using man-made radioisotopes.

One means of radioisotope examination involves injecting a radioactive sodium isotope into the bloodstream, and then tracing its progress through the body with a suitable detector, perhaps a computer-controlled body scan technique. If some portion of the body shows a low radiation count, it is an indication of a circulatory problem in that area. The normal absorption of iodine by the thyroid glands is checked by adding a radioactive iodine isotope to drinking water, followed by monitoring the radiation of that isotope. The same isotope is used to treat cancer of the thyroid glands. Radioactive isotopes of cobalt, phosphorus, radium, yttrium, strontium, and cesium are used to destroy cancerous tissue in different forms of radiation therapy.

Industrial applications of radioisotopes include studies of piston wear and corrosion resistance. Petroleum companies use radioisotopes to monitor the progress of certain oils through pipelines. The thickness of thin sheets of metal, plastic, and paper is subject to continuous production control by using a Geiger counter to measure the amount of radiation that passes through the sheet; the thinner the sheet, the more radiation that will be detected by the counter. Quality control laboratories can detect small traces of radioactive elements in a metal part.

Scientific research is another major application of radioisotopes. Chemists use "tagged" atoms as *radioactive tracers* to study the mechanism, or series of individual steps, in complicated reactions. For example, by using water containing radioactive oxygen it has been determined that the oxygen in the glycose, $C_6H_{12}O_6$, formed in photosynthesis,

$$6\ CO_2(g) + 6\ H_2O(\ell) \rightarrow C_6H_{12}O_6(s) + 6\ O_2(g)$$

comes entirely from the carbon dioxide, and all oxygen from water is released as oxygen gas. Archaeologists use neutron bombardment to produce radioactive isotopes in an artifact, which makes it possible to analyze the item without destroying it. Biologists employ radioactive tracers in the water absorbed by the roots of plants to study the rate at which the water is distributed throughout the plant system. These are but a few of the many ingenious applications that have been devised for this useful tool of science.

20.10 NUCLEAR FISSION

PG 20L Define or identify a nuclear fission reaction.

20M Define or identify a chain reaction.

In 1938, during the period when Nazi Germany was moving steadily toward war, dramatic and far-reaching events were taking place in her laboratories. A team made up of Otto Hahn, Fritz Strassman, and Lise Meitner was working with neutron bombardment of uranium. In the products of the reaction they were finding, surprisingly, atoms of barium and krypton, and other elements far removed in both atomic mass and atomic number from the uranium atoms and neutrons used to produce them. The only explanation was, at that time, unbelievable: the nucleus must be splitting into two nuclei of smaller mass. This kind of reaction is called **nuclear fission.**

Only 0.7% of all naturally occurring uranium is $^{235}_{92}U$, one isotope that is capable of the **chain reaction** required to keep a fission reaction going. In the fission of uranium-235 there are many products; it is not possible to write a single equation to show what happens. A representative equation is

$$^{235}_{92}U + ^{1}_{0}n \rightarrow ^{94}_{38}Sr + ^{139}_{54}Xe + 3\ ^{1}_{0}n$$

Notice that it takes a neutron to initiate the reaction. Notice also that the reaction produces *three* neutrons. If one or two of these collide with other fissionable uranium nuclei, there is the possibility of another fission or two. And the neutrons from those reactions can trigger others, repeatedly, as long as the supply of nuclei lasts. This is what is meant by a *chain reaction* (Fig. 20.6), in which a nuclear product of the reaction becomes a nuclear reactant in the next step, thereby continuing the process.

The number of neutrons produced in the fission of $^{235}_{92}U$ varies with each reaction. Some reactions yield two neutrons per uranium atom, others, like the above, yield three, and still others produce four or more. The average is about 2.5. If the quantity of uranium, or any other fissionable isotope, is large enough that most of the neutrons produced are captured within the sample, rather than escaping to the surroundings, the chain reaction will continue. The minimum quantity required for this purpose is called the **critical mass.**

Because uranium-235 makes up less than 1% of all naturally occurring uranium, it is not a very satisfactory source of nuclear fuel. An alternate is the

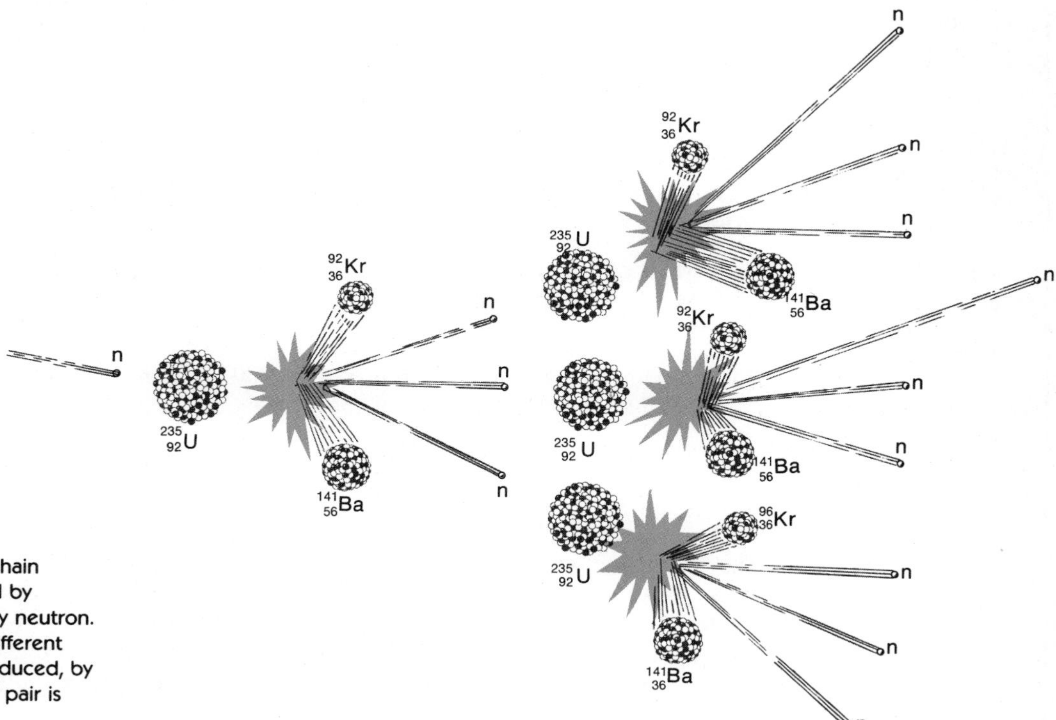

Figure 20.6
Illustration of a chain reaction initiated by capture of a stray neutron. (Many pairs of different isotopes are produced, by only one kind of pair is shown.)

plutonium isotope, $^{239}_{94}Pu$, produced from $^{238}_{92}U$, the more abundant uranium isotope (Equations 20.4 to 20.6). $^{234}_{94}Pu$ has a long half-life (24 360 years) and is fissionable. It has been used in the production of atomic bombs and is also used in some nuclear power plants to generate electrical energy. It is made in a **breeder reactor,** the name given to a device whose purpose is to produce fissionable fuel from nonfissionable isotopes.

Quick Check 20.10
How do the products of a fission reaction compare with the reactants?

20.11 NUCLEAR ENERGY

PG 20N Explain the source of energy in a nuclear reaction.

The term *atomic energy* is widely used; a better term is *nuclear energy,* for it is the nucleus that is the source of the enormous energy associated with a nuclear reaction. In Chapter 2 it was pointed out that the conservation of mass and energy laws must be combined to account for nuclear energy, in which matter is converted to energy. The conversion is expressed by Einstein's equation, $\Delta E = \Delta mc^2$, in which ΔE is the change in energy, Δm is the change in mass, and c is the speed of light. The amount of energy for a given quantity of matter is huge. For example, it can be shown that the energy released by changing a given mass of matter to energy is about 2.5 *billion* times greater than the energy released by burning the same mass of coal (see Fig. 20.7).

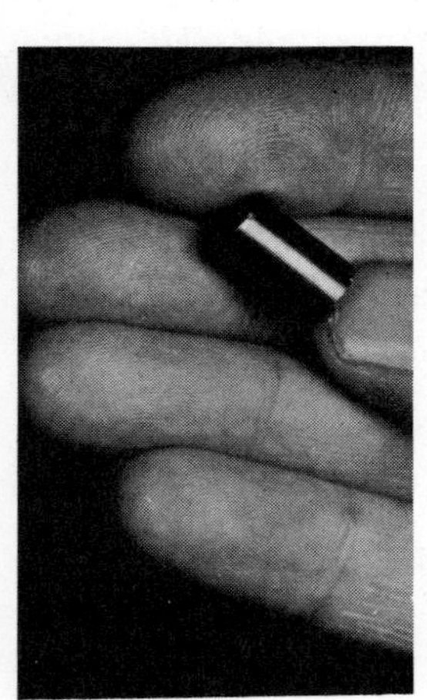

Figure 20.7
This uranium pellet, which provides nuclear fuel for a reactor, produces the energy equivalent of one ton of coal. (Courtesy of Duke Power Company.)

An atom of carbon-12 can be used to illustrate the mass-energy relationship. By the definition of an atomic mass unit, a carbon-12 atom has a mass of exactly 12 atomic mass units. The atom consists of a nucleus and six electrons. If the mass of the electrons (0.000 549 amu each) is subtracted from the mass of the atom, the difference is the mass of the nucleus:

Mass of atom:	12.000 000 amu
Mass of electrons (6 × 0.000 549):	−0.003 294 amu
Mass of nucleus:	11.996 706 amu

The nucleus is made up of six protons (1.007 27 amu each) and six neutrons (1.008 67 amu each). The sum of the masses of these component parts is

Mass of protons (6 × 1.007 28):	6.043 68 amu
Mass of neutrons (6 × 1.008 67):	+6.052 02 amu
Total	12.095 70 amu

The difference between the total mass of the *parts* of the nucleus and the actual mass of the *whole* nucleus is called the **mass defect.** For carbon-12 it is 12.095 70 − 11.996 706 = 0.098 99 amu. This 0.098 99 amu of mass is lost—converted to energy—when six protons and six neutrons combine to form a nucleus of $^{12}_{6}C$. The energy represented by this difference is called the **binding energy;** it is the energy that holds the nucleus together against the tremendous repulsion forces between the tightly packed positively charged protons.

In nuclear changes some of this binding energy is lost as one nucleus is destroyed, and another quantity of energy is absorbed as new nuclei form. The differences between these binding energies for all atoms involved represent the energy of a nuclear reaction.

20.12 ELECTRICAL ENERGY FROM NUCLEAR FISSION

Aside from hydroelectric plants located on major rivers, most of the electrical energy consumed in the world comes from generators driven by steam. Traditionally the steam comes from boilers fueled by oil, gas, or coal. The fast dwindling supplies of these fossil fuels, and the uncertainties surrounding the availability and cost of petroleum from the countries where it is so abundant, have turned attention to nuclear fission as an alternative energy source.

A diagram of a nuclear power plant is shown in Figure 20.8. The turbine, generator, and condenser are similar to those found in any fuel-burning power plant. The nuclear fission reactor has three main components: the fuel elements, control rods, and moderator. The fuel elements are simply long trays that hold fissionable material in the reactor. As the fission reaction proceeds, fast-moving neutrons are released. These neutrons are slowed down by a moderator, which is water in the reactor illustrated. When the slower neutrons collide with more

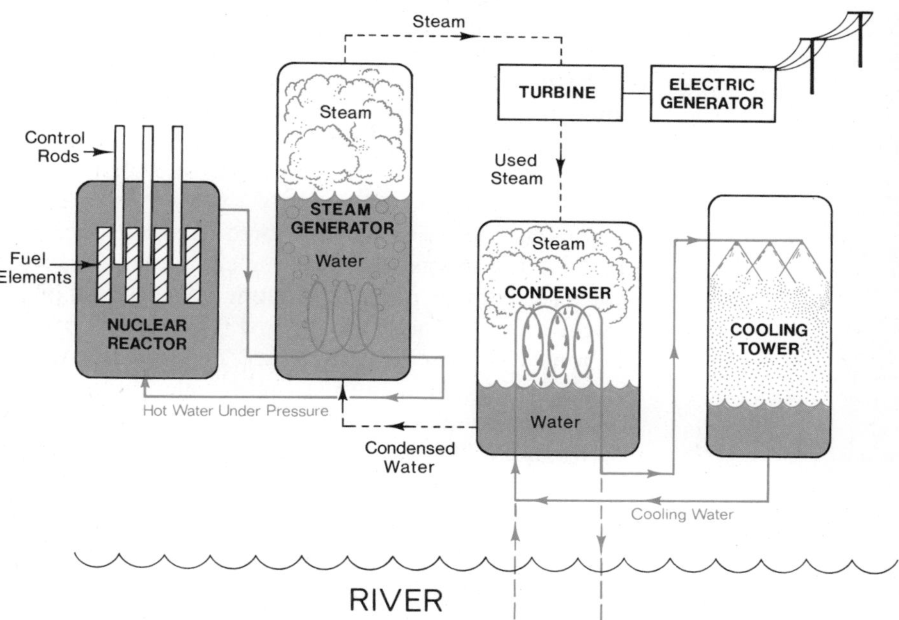

Figure 20.8
Schematic diagram of a nuclear power plant. Nuclear fission occurs in the reactor. Fission energy is used to convert water to steam in the steam generator. High pressure steam is delivered to the turbine, where it drives the generator that produces electrical energy, which is distributed to the users. Spent steam from the turbine is converted to liquid water in the condenser and then recycled back to the steam generator. Cooling water for the condenser comes from a cooling tower, to which it is recycled. Makeup cooling water, and sometimes the cooling water itself, is drawn from a river, lake, or ocean.

fissionable material the reaction is continued. The reaction rate is governed by cadmium control rods, which absorb excess neutrons. At times of peak power demand, the control rods are largely withdrawn from the reactor, permitting as many neutrons as necessary to find fissionable nuclei. When demand drops, the control rods are pushed in, absorbing neutrons and limiting the reaction.

The energy released in a nuclear reactor appears as heat. This heat is transferred to the water, which is in the liquid state under a pressure of 50 atm or more. The water from the reactor enters a heat exchanger at about 300°C. After giving up some of its heat in the heat exchanger, it is returned to the reactor, as shown.

Low pressure water that runs through a coil in the heat exchanger absorbs the heat coming from the high-temperature reactor water. The low pressure water forms steam, which is delivered to the turbines that drive the generators. Spent steam from the turbines is sent through a condenser where residual heat is removed and the steam condenses to water. The water is then cycled through the reactor heat exchanger again.

The building, and even the continued use, of nuclear power plants faces stiff opposition from both individuals and public groups. The threat of an accident that might release large amounts of radiation over a densely populated area is the major concern. To the date of this writing, the safety record of nuclear power plants is unblemished, but there have been a number of "close calls." The best known of these is the Three Mile Island accident in 1979. These incidents point out the need for absolutely foolproof safeguards in the design, construction, inspection, and operation of nuclear power facilities.

Even if a nuclear accident never occurs, there is still the problem of how and where to dispose of the dangerous radioactive wastes from nuclear reactors. One method is to collect them in large containers that may be buried in the earth. People who live and work near such disposal sites are seldom enthusiastic about this solution.

Finally there is fear that some irresponsible government may use nuclear fuel for the manufacture of nuclear weapons, spreading the threat of atomic warfare. Almost more frightening is the possibility that some terrorist group might steal the materials needed to build a bomb. While these threats cannot be removed from the earth today, perhaps they would be minimized if the large scale production of nuclear fuel for electric power were eliminated.

On the other side of all these concerns is, of course, the question, "If we do not build and operate nuclear power plants, how else will the energy needs of the coming decades be met?" Perhaps the next section offers one answer, if it can only be reached in time.

20.13 NUCLEAR FUSION

PG 200 Define or identify a nuclear fusion reaction.

There is nothing new about nuclear energy. Man didn't invent it. In fact, without knowing it, man has been enjoying its benefits since the beginning of recorded time, and before. In its common form, though, we do not call it nuclear energy. We call it solar energy. It is the energy that comes from the sun.

The energy the earth derives from the sun comes from another type of nuclear reaction called **nuclear fusion, in which two small nuclei combine to form a larger nucleus.** The smaller nuclei are "fused" together, you might say. The typical fusion reaction believed to be responsible for the heat energy radiated by the sun is represented by the equation

$$^{2}_{1}H + ^{3}_{1}H \rightarrow ^{4}_{2}He + ^{1}_{0}n$$

Fusion processes are, in general, more energetic than fission reactions. The fusion of one gram of hydrogen in the above reaction yields about four times as much energy as the fission of an equal mass of uranium-235. So far man has been able to produce only one kind of fusion reaction, and that has been the explosion of a hydrogen bomb.

Much research effort is being made to develop nuclear fusion as a source of useful energy. It has several advantages over a fission reactor. It presents more energy per given quantity of fuel. The isotopes required for fusion are far more abundant that those needed for fission. Perhaps the biggest advantage is that fusion yields no radioactive waste, removing both the need for extensive disposal systems and the danger of accidental release of radiation to the atmosphere.

The main obstacle to be overcome before energy can be obtained from fusion is the extremely high temperature required to start and sustain the reaction. This is no problem on the sun, where temperatures are more than one million degrees. On earth, no substance known can hold the reactants at the required temperature. Experiments are now in progress on "magnetic containment," in which the fuel is suspended in a magnetic field. The only way now known to reach the necessary temperature in a hydrogen bomb is to explode a small fission bomb first. New research is investigating "energy pellets" that react when "hit" by a laser or ion beam.

Even if the technological obstacles to energy from fusion are overcome, time remains a serious problem. Only the most optimistic predictions forsee an operating plant in this century, and it will probably be well into the next before a significant portion of our energy needs can be supplied by fusion.

Quick Check 20.13
How do the products of a fusion reaction compare with the reactants?

CHAPTER 20 IN REVIEW

20.1 The Dawn of Nuclear Chemistry

20.2 Natural Radioactivity

20A Define radioactivity. Name, identify from a description, or describe the three types of radioactive emission.

20.3 Interaction of Radioactive Emissions With Matter

20B Describe how emissions from radioactive substances affect gases and living organisms.

20.4 Detection of Emissions from Radioactive Substances

20C Define or describe a Geiger counter. Explain how it operates.

20.5 Natural Radioactive Decay Series—Nuclear Equations

20D Describe a natural radioactive decay series.

20E Given the identity of a radioactive isotope

and the particle it emits, write a nuclear equation for the emission.

20.6 Half-Life

20F Describe or illustrate what is meant by the half-life of a radioactive isotope.

20G Given the starting quantity of a radioactive isotope, Figure 20.4, and two of the following, calculate the third: half-life; elapsed time; quantity of isotope remaining.

20.7 Nuclear Reactions and Ordinary Chemical Reactions Compared

20H List or identify four ways in which nuclear reactions differ from ordinary chemical reactions.

20.8 Nuclear Bombardment and Induced Radioactivity

20I Define or identify nuclear bombardment reactions.

20J Distinguish natural radioactivity from induced radioactivity produced by bombardment reactions.

20K Define or identify transuranium elements.

20.9 Uses of Radioisotopes

20.10 Nuclear Fission

20L Define or identify a nuclear fission reaction.

20M Define or identify a chain reaction.

20.11 Nuclear Energy

20N Explain the source of energy in a nuclear reaction.

20.12 Electrical Energy from Nuclear Fission

20.13 Nuclear Fusion

20O Define or identify a nuclear fusion reaction.

TERMS AND CONCEPTS

20.1 X-ray
Radioactivity
20.2 Alpha particle, α-particle; alpha ray, α-ray
Beta particle, β-particle; beta ray, β-ray
Gamma ray, γ-ray
20.4 Cloud chamber
Geiger—Müller counter
20.5 Transmutation
Alpha decay reaction
Beta decay reaction
Natural radioactive decay series
20.6 Half-life
Radiocarbon dating
20.8 Nuclear bombardment
Induced (artificial) radioactivity
Positron
Radioisotope
Particle accelerator
Transuranium element
20.9 Radioactive tracer
20.11 Nuclear fission
Chain reaction
Critical mass
Breeder reactor
20.12 Nuclear energy
Mass defect
Binding energy
20.13 Nuclear fusion

Most of these terms and many others appear in the Glossary. Use the Glossary regularly.

QUESTIONS AND PROBLEMS

Section 20.2

(*1*) What is meant by radioactivity?

(*2*) Identify the three types of emission normally associated with radioactive decay. What is each type of emission made of; that is, what is its "structure?" Write the nuclear symbol, if any, for each kind of emission.

(26) What is the meaning of the terms *disintegration* and *decay* in relation to radioactivity?

(27) Compare the three principal forms of radioactive emission in terms of mass, electrical charge, and penetrating power.

Section 20.3

(*3*) When an emission from a radioactive substance passes through air or any other gas, what effect does it have?

(28) Explain how rays from radioactive substances can cause injury or damage to internal tissue in a living organism.

Section 20.4

(*4*) What properties of radioactive decay are used in detecting and measuring it?

(29) What is a Geiger counter? How does it work?

Section 20.5

(*5*) What is meant by the expression *transmutation of an element?*

(*6*) Describe the change in mass number and atomic number that accompanies an alpha particle emission from a radioactive nucleus.

(7) Write nuclear equations for beta emissions from $^{212}_{82}Pb$ and from $^{231}_{90}Th$.

(8) Write nuclear equations for the ejection of an alpha particle from $^{228}_{90}Th$ and from $^{222}_{86}Rn$.

(30) What is a *natural radioactive decay series?* How many such series have been found?

(31) If a radioactive nucleus emits a beta particle, what change will occur in the atomic number and mass number of the nucleus that remains?

(32) Write nuclear equations for beta emissions from $^{228}_{89}Ac$ and from $^{214}_{83}Bi$.

(33) Write nuclear equations for alpha decay of $^{216}_{84}Po$; of $^{234}_{92}U$.

Section 20.6

(9) What is meant by the half-life of a radioactive substance? What fraction of an original sample remains after the passage of six half-lives?

(10) One of the man-made isotopes used in radiotherapy is $^{60}_{27}Co$, which has a half-life of 5.2 years. (a) What percentage of the stored amount of this material is lost annually due to radioactive decay? (b) How many grams of a 125-g sample of this material will remain after 18 years?

(11) One of the more hazardous radioactive isotopes in the fallout of an atomic bomb is strontium-90, $^{90}_{38}Sr$. A 227-g sample taken in 1947 was carefully stored for 20.0 years. In 1967 its mass was down to 138 g. Using Figure 20.4, find the half-life of $^{90}_{38}Sr$.

(12)* An isotope of radon has a half-life of 3.8 days. The emission of a sample was measured at 7.1×10^4 disintegrations per second (dps). (The number of disintegrations is proportional to the mass of a radioactive isotope present in a sample, so dps values may be treated in the same way as mass values.) At a later time the decay is measured at 1.9×10^4 dps. How much time elapsed between the measurements?

(34) Suggest some of the difficulties that might surround the determination of physical properties of the radioactive isotope of an element that has a short half-life. What fraction of the sample would remain after the passage of four half-lives?

(35) 3.1 minutes is the half life of $^{208}_{81}Tl$. A 48.4-g sample is studied in the laboratory. (a) How many grams of the sample will remain after 8.7 minutes? (b) In how many minutes will the mass that remains be only 4.38 g?

(36) Uranium-235, the uranium isotope used in making the first atomic bomb, is the starting point of one of the natural radioactivity series. The next isotope in the series is $^{231}_{90}Th$. At the beginning of a test period a sample contained 9.53 g of the isotope. At the end of the period, 63.2 hr later, the mass of the sample was 1.71 g. What is the half-life of the isotope in hours?

(37) A bottle of radioactive material that was purchased by a radiotherapy laboratory was lost by being placed on the wrong shelf in a storeroom. In a general cleaning of the storeroom some time later, the sample was found. Analysis showed that the sample contained 21.8 g of the radioactive isotope. The label on the bottle indicated that the mass was 50.0 g at the time of delivery. How long was the bottle lost if the half-life of the radioisotope is 198 days?

Section 20.7

(*13*) Explain why the chemical properties of an element are the same for all isotopes in an ordinary chemical change, but depend on the particular isotope for a nuclear change.

(14)* If the uranium in pure UCl_4 and UBr_4 has all isotopes in their normal percentage distribution in nature, which will exhibit the greatest amount of radioactivity, 100 g of UCl_4 or 100 g of UBr_4? How about 0.10 mol of UCl_4 or 0.10 mol of UBr_4? Explain both answers.

(15) A fundamental idea in Dalton's atomic theory is that atoms of an element can be neither created nor destroyed. How, then, can you account for the fact that the number of lead atoms in the world is constantly increasing?

(38) Two bottles of the same lead compound are carelessly left exposed. Explain the circumstances under which one of these bottles might be hazardous, but not the other.

(39) An ore sample containing a certain quantity of radioactive uranium disintegrates at 7000 counts per minute, a way of expressing rate of radioactive decay when measured with a Geiger counter. If all the uranium in the sample is extracted and isolated as a pure element, would you expect the rate of decay to remain at 7000 counts per minute, would it be less than 7000, or would it be more? (Disregard any loss of the radioactive isotope because of disintegration during the extraction process.) Explain your answer.

(40) A radiochemical laboratory prepares a sample of pure KCl containing a measurable amount of $^{43}_{19}K$, a radioactive isotope that emits a beta particle and has a half-life of 22.4 hr. The compound is securely stored overnight in an inert atmosphere. The next day the compound will no longer be pure. Why? With what element would you expect it to be contaminated?

Section 20.8

(16) What is the meaning of the expression *nuclear bombardment?*

(*17*) Particles used for nuclear bombardment reactions frequently do not have sufficient kinetic energy when obtained from their natural sources. Identify some of the "particle accelerators" that have been developed to increase their kinetic energy.

(*18*) What are transuranium elements? Where are they located on the periodic table?

(41) How does induced radioactivity differ from natural radioactivity?

(42)* Describe the principle by which a particle accelerator increases the kinetic energy of a particle used in nuclear bombardment. What major nuclear particle cannot be accelerated? Why?

(43)* Why are transuranium elements not found in nature, except for trace quantities? Did they ever exist in nature? Explain why or why not.

Complete each of the following nuclear bombardment equations by supplying the nuclear symbol for the missing species:

(19) $^{98}_{42}Mo + ^{2}_{1}H \rightarrow ? + ^{1}_{0}n$

(20) $^{238}_{92}U + ^{4}_{2}He \rightarrow 3\ ^{1}_{0}n + ?$

(21) $? + ^{2}_{1}H \rightarrow ^{60}_{27}Co + ^{1}_{1}H$

(44) $^{44}_{20}Ca + ^{1}_{1}H \rightarrow ? + ^{1}_{0}n$

(45) $^{252}_{98}Cf + ^{10}_{5}B \rightarrow 5\ ^{1}_{0}n + ?$

(46) $^{106}_{46}Pd + ^{4}_{2}He \rightarrow ^{109}_{47}Ag + ?$

Section 20.10

(*22*) Explain what is meant by a *fission* reaction.

(**47**) What is a *chain reaction?* What essential feature must be present in a reaction before it can become a chain reation?

Section 20.11

(*23*) What is the source of the enormous energy released in a nuclear reaction?

(**48**) For all practical purposes, the Law of Conservation of Mass is obeyed in ordinary chemical reactions, but not in nuclear reactions. Would the mass of the products of an atomic bomb explosion be more than or less than the mass of the reactants? Explain your answer.

Section 20.12

(*24*) List some of the principal advantages that are associated with nuclear power plants as a source of electrical energy.

(**49**)* In September, 1984, when this question was written, there were serious concerns about using nuclear reactors as a source of electrical energy. List those concerns. In the period since 1984, have there been changes that have removed or reduced the earlier objections to nuclear energy? If so, identify them. Has anything happened over the same period to show that the worries of 1981 were justified, and that continuing to build nuclear power plants is an unwise procedure? If so, identify the events. How do you feel about nuclear power sources today?

Section 20.13

(*25*) What is a *fusion* reaction? How does it differ from a fission reaction?

(**50**) Why is nuclear fusion more promising as a source of electrical energy than nuclear fission? What major obstacle prevents us from building nuclear fusion power plants?

Miscellaneous Questions

(*51*) Distinguish precisely and in scientific terms the differences between items in each of the following groups:
- (a) Alpha, beta, and gamma radiation
- (b) X-rays and γ-rays
- (c) α-particle, β-particle
- (d) Natural and induced radioactivity
- (e) Chemical reaction, nuclear reaction
- (f) Isotope, radioisotope
- (g) Element, transuranium element
- (h) Nuclear fission, nuclear fusion
- (i) Atom bomb, hydrogen bomb

(**52**) Classify each of the following statements as true or false:
- (a) A radioactive atom decays in the same way whether or not the atom is chemically bonded in a compound.
- (b) The chemical properties of a radioactive atom of an element are different from the chemical properties of a nonradioactive atom of the same element.
- (c) α-rays have more penetrating power than β-rays.
- (d) Alpha and beta rays are particles, but a gamma ray is an "energy ray."
- (e) Radioactivity is a nuclear change that has no effect on the electrons in nearby atoms.
- (f) The number of protons in a nucleus changes when it emits a beta particle.

(g) The mass number of a nucleus changes in an alpha emission but not in a beta emission.
(h) The higher the atomic number of an isotope, the longer its half-life.
(i) The first transmutations were achieved by alchemists.
(j) Radioisotopes are made by bombarding a nonradioactive isotope with atomic nuclei or subatomic particles.
(k) The atomic numbers of products of a fission reaction are smaller than the atomic number of the original nucleus.
(l) The mass of an atom is equal to the sum of the masses of electrons, protons and neutrons that make up the atom.
(m) Nuclear power plants are a safe source of electrical energy.
(n) The main obstacle to developing nuclear fusion as a source of electrical energy is a shortage of nuclei to serve as "fuel."

(53) What is the function of particle accelerators? What kinds of particles do they accelerate? What kinds of particles cannot be accelerated?

(54)* The examples and most of the problems having to do with half-life compared initial and final quantities of a radioactive "substance" in terms of mass. This simplification is usually unrealistic. Can you suggest a reason why? If mass is not a suitable measure of an amount of radioactive matter, what is?

(55) A major form of fuel for nuclear reactors used to produce electrical energy is a fissionable isotope of plutonium. Plutonium is a transuranium element. Why is this element used instead of a fissionable isotope that occurs in nature?

(56) Why is half-life used for measuring rate of decay rather than the time required for the complete decay of a radioactive isotope?

(57) A ton of high grade coal has an energy output of about 2.5×10^7 kJ. The energy released in the fission of one mole of $^{235}_{92}U$ is about 2.0×10^{10} kJ. How many tons of coal could be replaced by one pound of uranium-235, assuming the material and the technology were available?

21 Introduction to Organic Chemistry

LOOKING BACK	LOOKING AHEAD IN CHAPTER 21
10.4 Lewis diagrams are drawn to show how atoms are covalently bonded to each other.	Lewis diagrams are used to represent the structures of organic molecules.
11.5 to 11.9 Linear, planar triangular, tetrahedral, trigonal pyramidal, and bent molecular geometry develop around a central atom, causing a molecule to be polar or nonpolar.	The structure of organic molecules around carbon, nitrogen, and oxygen atoms is linear, planar triangular, tetrahedral, trigonal pyramidal, or bent.
14.2 The physical properties of compounds depend on intermolecular forces, which, in turn, depend on molecular structure.	Physical properties of organic substances are related to their molecular geometries.
14.3 The size and polarity of molecules, and the presence or absence of hydrogen bonding, determine the strength of intermolecular forces.	Intermolecular attractions between organic molecules depend on the size and polarity of those molecules and the presence or absence of hydrogen bonding.

Chapter 21 is a brief survey of organic chemistry. It is unlike the first twenty chapters, in which we could expect to find mastery of specific chemical concepts. While we might seek similar achievement in selected areas of this chapter, most topics are presented so briefly it is unrealistic to set for them the kind of performance goals used earlier. In their place the following chapter-wide performance goals are offered for your guidance:

1. Distinguish between organic and inorganic chemistry.
2. Define a hydrocarbon.
3. Distinguish between saturated and unsaturated hydrocarbons.
4. Write, recognize, or otherwise identify (1) the structural unit, or functional group, (2) the general formula, and (3) molecular or structural formulas and/or names of specific examples of the following classes of organic compounds: alkanes, alkenes, alkynes, aromatic hydrocarbons, alcohols, ethers, aldehydes, ketones, carboxylic acids, and esters.
5. Define and give examples of isomerism.

21.1 THE NATURE OF ORGANIC CHEMISTRY

In the early development of chemistry, the logical starting point was a study of substances that occur in nature. As in the organization of any body of knowledge, substances were grouped by certain common characteristics. One system assigns substances to groups labeled animal, vegetable, or mineral. So far in this text, attention has been directed almost entirely to minerals and the compounds that may be derived from them. These substances make up that area of chemistry commonly known as *inorganic chemistry*.

Animal and vegetable substances are, or at one time were, composed of living organisms, a distinction that sets them apart from inorganic substances. **Organic chemistry** was originally defined as the chemistry of living organisms, including those compounds directly derived from living organisms by natural processes of decay. It was learned, however, that compounds called ''organic'' can be produced from inorganic chemicals. Ultimately it was recognized that all organic compounds contain the element **carbon.** We now consider organic chemistry to be **the chemistry of carbon compounds.** Carbonates, cyanides, oxides of carbon, and a few other compounds are exceptions that are still classified as inorganic.

The only really unique feature of organic chemistry is the huge and rapidly growing number of organic compounds—many times more than the total number of known compounds that do not contain carbon. All of the chemical principles we have studied for inorganic chemicals, such as bonding, reaction rates, equilibrium, and others, apply equally to organic compounds. In particular, a clear picture of bonding and the structure of molecules is the cornerstone of all that we understand about organic chemistry, as the remaining pages will show.

21.2 THE MOLECULAR STRUCTURE OF ORGANIC COMPOUNDS

Table 21.1 summarizes the covalent bonding properties of carbon, hydrogen, oxygen, nitrogen, and the halogens, the elements most frequently found in organic compounds. All the bond geometries are indicated—the linear, planar, or three-dimensional shapes and, where constant, the actual bond angles. Of particular significance is the number of covalent bonds that atoms of the different elements can form. This is determined by the electron configuration of the atom. The bonding relationships of these elements are basic to your understanding of the structure of organic compounds.

When carbon forms four single bonds, they are arranged tetrahedrally around the carbon atom; the molecular geometry is tetrahedral (see Fig. 11.2). Recall from Chapter 11 that it is not possible to represent this three dimensional shape accurately in a two dimensional sketch. Thus the four bonds radiating from each carbon atom in

```
    x   z   y
    |   |   |
x—C—C—C—y
    |   |   |
    x   z   y
```

Table 21.1 Bonding in Organic Compounds

Element	Number of Bonds*	Bond Geometry: SINGLE BOND	Bond Geometry: DOUBLE BOND	Bond Geometry: DOUBLE BONDS	Bond Geometry: TRIPLE BOND
Carbon	4	—C— (four single bonds) Tetrahedral: 109.5° angles	>C= Planar: 120° angles	=C= Linear: 180° angle	—C≡ Linear: 180° angle
Hydrogen	1	H—			
Halogens	1	:X— (with lone pairs)			
Oxygen	2	O (two bonds, lone pairs) Bent structure	O= (lone pairs)		
Nitrogen	3	N (three bonds, lone pair) Pyramidal structure	N (single and double bond, lone pair) Bent structure		:N≡

*Number of bonds to which an atom of the element shown can contribute *one* electron.

all form tetrahedral bond angles (109.5°). Furthermore, all three x positions are geometrically equal, three y positions are equal, and two z positions are equal. It follows that

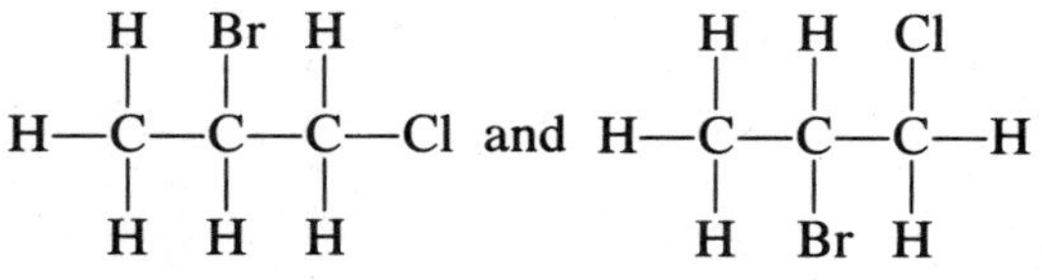

$$\begin{array}{ccccccc} & H & & Br & & H & \\ & | & & | & & | & \\ H— & C & — & C & — & C & —Cl \\ & | & & | & & | & \\ & H & & H & & H & \end{array} \quad \text{and} \quad \begin{array}{ccccccc} & H & & H & & Cl & \\ & | & & | & & | & \\ H— & C & — & C & — & C & —H \\ & | & & | & & | & \\ & H & & Br & & H & \end{array}$$

are the same compound.

Hydrocarbons

The simplest type of organic compound is the hydrocarbon. As the name suggests, **hydrocarbons** consist of two elements, hydrogen and carbon. A hydrocarbon may be classified into one of several categories based upon its structure: (1) alkanes, (2) alkenes, (3) alkynes, and (4) aromatic hydrocarbons. The first three of these are sometimes grouped together as the *aliphatic* hydrocarbons,

Table 21.2 Aliphatic Hydrocarbons

Number of Carbon Atoms	Alkane	Alkene	Alkyne
1	H H—C—H H CH_4 Methane		
2	H H H—C—C—H H H C_2H_6 Ethane	H H H—C=C—H C_2H_4 Ethylene ethene	H—C≡C—H C_2H_2 Acetylene ethyne
3	H H H H—C—C—C—H H H H C_3H_8 Propane	H H H H—C=C—C—H H C_3H_6 Propylene propene	H H—C≡C—C—H H C_3H_4 Propyne
Structural Unit or Functional Group	—C— (with four single bonds)	>C=C<	—C≡C—

in which the carbon atoms are arranged in chains. Some aliphatic hydrocarbons are shown in Table 21.2; the aromatic hydrocarbons will be considered separately.

21.3 SATURATED HYDROCARBONS: THE ALKANES

The **alkanes** are known as **saturated hydrocarbons.** This means that each carbon atom is bonded by four single covalent bonds to four other atoms, the maximum number possible according to the octet rule (Section 10.4). Structural models of the first three alkane molecules, methane, ethane, and propane, are shown in Figure 21.1. Notice the tetrahedral orientation of atoms bonded to carbon in all three molecules. Because of this tetrahedral arrangement any continuous chain of three or more carbon atoms through a saturated hydrocarbon has a crooked path. Furthermore there is free rotation around single carbon–carbon

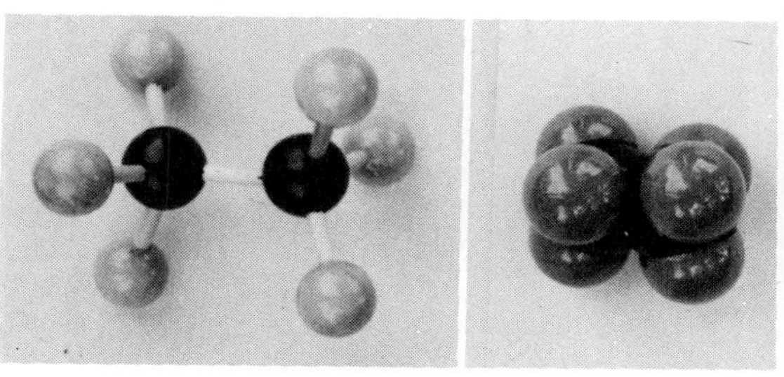

Figure 21.1
Ball-and-stick and space-filling models of the first three members of the alkane hydrocarbon series. Methane (above left), ethane (above), and propane (left).

bonds. The shape of the molecule varies. At one moment the molecule can curl up so the end carbons are close to each other, and later the molecule is stretched out with the end carbons far apart. The straight lines we draw on paper are not true representations of carbon chains.

The General Formula for the Alkanes—A Homologous Series

If you examine the molecular and structural formulas of methane, ethane, and propane, you will find a pattern. Each additional carbon atom is accompanied by two more hydrogen atoms. The alkane with four carbon atoms is butane, C_4H_{10}, five carbon atoms yield pentane, C_5H_{12}, and so forth, with each additional step extending the chain by a —CH_2— structural unit.

A series of compounds in which each member differs from the members before and after it by the same structural unit is called a **homologous series.** The alkane series may be represented by the general formula C_nH_{2n+2}, where n is the number of carbon atoms in the molecule. With this general formula you can produce the molecular formula for any member of the series. For octane, the alkane with 8 carbon atoms, n = 8. The number of hydrogen atoms is 2(8) + 2 = 18. The formula of octane is therefore C_8H_{18}.

The formulas and names of the first ten alkanes are shown in Table 21.3. Also shown are the melting and boiling points of the so-called **normal** alkanes, those in which the carbon atoms form a continuous chain. You will recall from Chapter 14 that intermolecular forces between nonpolar molecules increase with increasing molecular size, and that stronger intermolecular attractions yield higher boiling points. Accordingly, alkanes having fewer than five carbons have the weakest intermolecular attractions, low boiling points, and are gases at normal temperatures. All are used as fuels for stoves. Methane is the main

Table 21.3 The Alkane Series

Molecular Formula	Name	Number of Carbon Atoms	Prefix	Melting Point (°C)	Boiling Point (°C)
CH_4	Methane	1	Meth-	−183	−162
C_2H_6	Ethane	2	Eth-	−172	−89
C_3H_8	Propane	3	Prop-	−187	−42
C_4H_{10}	Butane	4	But-	−138	0
C_5H_{12}	Pentane	5	Pent-	−130	36
C_6H_{14}	Hexane	6	Hex-	−95	69
C_7H_{16}	Heptane	7	Hept-	−91	98
C_8H_{18}	Octane	8	Oct-	−57	126
C_9H_{20}	Nonane	9	Non-	−54	151
$C_{10}H_{22}$	Decane	10	Dec-	−30	174

constituent of "natural gas." It is also found in large quantities in the forbidding atmosphere of planets like Jupiter and Saturn.

Intermolecular forces are stronger between the larger alkane molecules from C_5H_{12} to $C_{17}H_{36}$. These higher boiling compounds are liquids at room temperature. Several of the lower molecular weight liquid alkanes, notably octane, are present in gasoline. Diesel fuel and lubricating oils are made up largely of higher molecular weight liquid alkanes. Alkanes with molecular weights greater than 250 are normally solids at room temperature.

Because of isomerism (see below), a molecular formula does not identify a compound adequately. Structural formulas, on the other hand, are complex, although frequently the only satisfactory representation of a compound. A compromise sometimes used is the **condensed formula,** or **line formula,** in which the structure is indicated by repeating the groups it contains. Line formulas for a few sample alkanes are shown below:

Ethane, C_2H_6	CH_3CH_3
Propane, C_3H_8	$CH_3CH_2CH_3$
Butane, C_4H_{10}	$CH_3CH_2CH_2CH_3$
Octane, C_8H_{18}	$CH_3CH_2CH_2CH_2CH_2CH_2CH_2CH_3$

When line formulas become long, as in the case of octane, they are sometimes shortened by grouping the CH_2 units: $CH_3(CH_2)_6CH_3$.

The alkane hydrocarbons also serve to introduce organic nomenclature. Table 21.3 illustrates the system for the first ten alkanes. Each alkane is named by combining a prefix and a suffix. The prefix indicates the number of carbons in the chain. The first ten prefixes used in this nomenclature system appear in the fourth column of Table 21.3. The suffix identifying an *alkane* is *-ane*. Thus the name of methane comes from combining the prefix *meth-,* indicating one carbon atom, with the suffix *-ane* indicating an alkane. We will see shortly that these prefixes are used in naming other organic compounds and groups as well.

Isomerism in the Alkane Series

Not all alkanes have their carbons bonded in a continuous chain; some have branches. The smallest alkane in which this is possible is butane, which has two possible structures:*

```
    H  H  H  H            H      H      H
    |  |  |  |            |      |      |
 H—C—C—C—C—H        H—C——C——C—H
    |  |  |  |            |      |      |
    H  H  H  H            H      |      H
                                 |
                              H—C—H
                                 |
                                 H
```

n-Butane ("normal butane")   Isobutane

You will observe that in the compound at the left, called normal butane, or *n*-butane, the four carbons are in a single chain. In isobutane, the structure at the right, there are only three carbons in the chain, with the fourth carbon branching off the middle carbon of the three. Both compounds have the same molecular formula, C_4H_{10}. These compounds are **isomers: two compounds having the same molecular formula but different structures are called isomers.** It should be realized that isomers are distinctly different compounds, having different physical and chemical properties.

The number of isomers that are possible increases rapidly with the number of carbon atoms in the compound. There are three different pentanes; their structures, showing only the carbon skeletons to make the diagrams less "cluttered," are

```
                                                  |
                                                —C—
  |  |  |  |  |        |  |  |  |              |  |  |
—C—C—C—C—C—      —C—C—C—C—          —C—C—C—
  |  |  |  |  |        |  |  |  |              |  |  |
                            —C—                 —C—
                              |                   |
```

n-Pentane   Isopentane   Neopentane

There are five isomeric hexanes, nine heptanes, and 75 possible decanes. It is possible to draw over 300 000 isomeric structures for $C_{20}H_{42}$, and more than 100 million for $C_{30}H_{62}$. Obviously not all of them have been prepared and identified! This does give us some idea, though, why there are so many organic compounds.

(You might like to try your hand at writing isomers of the alkanes. See if you can sketch the five isomers of hexane. In doing so, be sure they are all

*The length of lines representing bonds has no significance. Different lengths are used only to separate on paper those parts of the molecule that are not bonded to each other.

different. You would be wise to start with the longest chain possible, then shorten it by one, and again shorten it by one, drawing all possible structures each time until all combinations are exhausted. The correct diagrams are shown on page 501.)

Distinguishing between isomeric structures requires some extension of nomenclature rules. As noted earlier, a continuous straight chain alkane is called a normal alkane—hence the name "normal butane," written *n*-butane. In a normal alkane each carbon atom is bonded to no more than two other carbon atoms. The branched chain isomer of butane is called isobutane. In it, one carbon atom is bonded to three other carbon atoms. The *normal* and *iso-* terminology is carried forth to the isomers of pentane, and is expanded to *neo*pentane to provide for the third isomer in which one carbon atom is bonded to four other carbon atoms.

Beyond pentane the number of isomers becomes so large it is necessary to develop a systematic nomenclature. In fact, many chemists prefer to drop the *iso-* and *neo-* prefixes and use a single system for all hydrocarbons. Several different systems exist, but the one most widely adopted is the IUPAC* system. But before we can consider this system we must define an *alkyl group*.

Alkyl Groups

In inorganic chemistry we found it convenient to assign names to certain groups of atoms, such as sulfate, nitrate, and ammonium ions, that function as units in forming chemical compounds. Similarly, in organic chemistry, it is convenient to identify **alkyl groups** that may be derived from an alkane. If, on paper, we remove a hydrogen atom from methane, CH_4, we get $—CH_3$. This $—CH_3$ group, appearing in the structural formula of a compound, is called a **methyl** group, the term being made up of the prefix *meth-* for one carbon (Table 21.3) and the suffix *-yl* applied to all alkyl groups. If we compare two compounds,

$$\begin{array}{ccccccccccc} & & H & & H & & H & & H & & H & \\ & & | & & | & & | & & | & & | & \\ H & — & C & — & C & — & C & — & C & — & C & — & H \\ & & | & & | & & | & & | & & | & \\ & & H & & H & & H & & H & & H & \end{array} \quad \text{and} \quad \begin{array}{ccccccccccccc} & & H & & H & & H & & H & & H & & \\ & & | & & | & & | & & | & & | & & \\ H & — & C & — & C & — & C & — & C & — & C & — & H \\ & & | & & | & & | & & | & & | & & \\ & & H & & H & & | & & H & & H & & \\ & & & & & H & — C — & H & & & & & \\ & & & & & & | & & & & & & \\ & & & & & & H & & & & & & \end{array}$$

we see that the colored H in the first compound has been replaced by a $—CH_3$ group, or methyl group, in the second. If the replacement group has two carbon atoms,

*International Union of Pure and Applied Chemistry (p. 243).

```
      H   H      H      H   H
      |   |      |      |   |
  H—C—C——C——C—C—H
      |   |      |      |   |
      H   H      |      H   H
                 |
             H—C—H
                 |
             H—C—H
                 |
                 H
```

it is an ethyl group, —C_2H_5, one hydrogen short of ethane, C_2H_6. All the alkyl groups are similarly named.

Frequently we wish to show a bonding situation in which *any* alkyl group may appear. The letter R is used for this purpose. Thus R—OH could be CH_3OH, C_2H_5OH, C_3H_7OH, or any other alkyl group attached to an —OH group.

Some chemists consider alkyl groups as **functional groups** also. A functional group is **an atom or group of atoms that both establishes the identity of a class of compounds and determines its chemical properties.** The structural units in the bottom row of Table 21.2 may be considered as the functional groups of the aliphatic hydrocarbons. Other functional groups will appear later in the chapter.

Naming the Alkanes by the IUPAC System

We are now ready to describe the IUPAC system of naming isomers of the alkanes, as well as other compounds we will encounter shortly. The system follows a set of rules:

1. **Identify as the parent alkane the longest continuous chain.** For example, in the compound having the structure

```
    |   |   |   |   |
  —C—C—C—C—C—
    |   |   |   |   |
              —C—
                |
              —C—
                |
```

the longest chain is six carbons long, not five as you might first expect. This is readily apparent if we number the carbon atoms in the original representation of the structure and an equivalent layout:

```
    |6  |5  |4  |3  |                          |
  —C—C—C—C—C—                          —C—
    |   |   |   |2  |            |6  |5  |4  |3  |2  |1
              —C—               —C—C—C—C—C—C—
                |1                 |   |   |   |   |   |
              —C—
                |
```

2. **Identify by number the carbon atom to which the alkyl group (or other species) is bonded to the chain.** In the example compound this is the *third* carbon, as shown. Notice that counting always begins at that end of the chain that places the branch on the *lowest* number carbon atom possible.
3. **Identify the branched group (or other species).** In this example the branch is a methyl group, —CH_3, shown in color.

```
                          H
  |   |   |   |           |
 —C—C—C—C————C—H
  |   |   |   |           |
            —C—           H
              |
            —C—
              |
```

These three items of information are combined to produce the name of the compound, *3-methylhexane*. The 3 comes from the third carbon (Step 2); methyl comes from the branch group (Step 3); and hexane is the parent alkane (Step 1).

Sometimes the same branch appears more than once in a single compound. This situation is governed by the following rule:

4. **If the same alkyl group, or other species, appears more than once, indicate the number of appearances by di-, tri-, tetra-, etc., and show the location of each branch by number.** For example,

```
         Cl  Cl
  |   |   |   |   |
 —C—C—C—C—C—
  |   |   |   |   |
```

would be 2,3-dichloropentane. In the other direction, to write the structural formula for 1,1,5-tribromohexane, we would establish a six-carbon skeleton and attach bromines as required, two to the first carbon and one to the fifth:

```
 Br              Br
  |   |   |   |   |   |
 —C—C—C—C—C—C—
  |   |   |   |   |   |
 Br
```

5. **If two or more different alkyl groups, or other species, are attached to the parent chain, they are named in order of increasing group size or in alphabetical order.** By this rule the compound

```
  |   |   |   |   |
 —C—C—C—C—C—
  |   |   |   |   |
     Cl  Br
```

would be named 3-bromo-2-chloropentane. The formula for 2,2-dibromo-4-chloroheptane would be

```
     Br     Cl
  |  |   |  |   |   |   |
—C—C—C—C—C—C—C—
  |  |   |  |   |   |   |
     Br
```

The five isomers of hexane further illustrate the application of these rules:

```
                                             |
                                           —C—
  |   |   |   |   |   |            |5  |4  |3  |2  |1
—C—C—C—C—C—C—        —C—C—C—C—C—
  |   |   |   |   |   |            |   |   |   |   |

        n-Hexane                     2-Methylpentane
```

```
        |                      |                         |
      —C—                    —C—                       —C—
  |1  |2  |3  |4  |5     |1  |2  |3  |4          |4  |3  |2  |1
—C—C—C—C—C—      —C—C—C—C—        —C—C—C—C—
  |   |   |   |   |      |   |   |   |           |   |   |   |
                               —C—                      —C—
                                 |                        |
   3-Methylpentane
                        2,3-Dimethylbutane       2,2-Dimethylbutane
```

Quick Check 21.3

1. Identify the alkanes among the following: C_7H_{16} C_5H_{10} $C_{11}H_{22}$ C_9H_{20}
2. Write the formula of the alkyl group derived from pentane.
3. Write a structural diagram of 3,3-difluoro-4-iododecane.

21.4 UNSATURATED HYDROCARBONS: THE ALKENES AND THE ALKYNES

Structure and Nomenclature

A saturated hydrocarbon has been identified as one in which each carbon atom is bonded to four other atoms. **Hydrocarbons in which two or more carbon atoms are (1) connected by a double or triple bond and (2) bonded to fewer than four other atoms are said to be unsaturated.**

If one hydrogen atom, complete with its electron, is removed from each of two adjacent carbon atoms in an alkane (A below), each carbon is left with a single unpaired electron (B). These electrons may then form a second bond between the two carbon atoms (C):

```
   H  H                      H  H                 H  H
   |  |                      |  |                 |  |
H—C—C—H   —2 H·→    H—C—C—H    →    H—C=C—H
   |  |                      ·  ·
   H  H

     A                         B                    C
```

Each carbon atom is now bonded to three other atoms. **An aliphatic hydrocarbon in which two carbon atoms are bonded to three other atoms and double-bonded to each other is called an alkene.** Figure 21.2 shows two models of the simplest alkene.

Removal of another hydrogen atom from each of the double-bonded carbon atoms in an alkene yields a triple bond:

$$\mathrm{H{-}\overset{\displaystyle H}{\overset{|}{C}}{=}\overset{\displaystyle H}{\overset{|}{C}}{-}H \xrightarrow{-2\,H\cdot} H{-}\dot{C}{=}\dot{C}{-}H \rightarrow H{-}C{\equiv}C{-}H}$$

Each carbon atom is now bonded to two other atoms. **An aliphatic hydrocarbon in which two carbon atoms are bonded to two other atoms and triple-bonded to each other is called an alkyne.** Models of acetylene, the most common alkyne, are shown in Figure 21.2.

Both the alkenes and the alkynes make up a new homologous series. Just as with the alkanes, each series may be extended by adding $—CH_2—$ units. Longer chains may have more than one multiple bond, but we will not consider such compounds in this text. The general formula for an alkene is C_nH_{2n}, and for an alkyne, C_nH_{2n-2}.

Table 21.4 gives the names and formulas of some of the simpler unsaturated hydrocarbons. The IUPAC nomenclature system for the alkenes matches that of the alkanes. The suffix designating the alkene hydrocarbon series is *-ene*, just as *-ane* identifies an alkane. The same prefixes are used to show the total number of carbon atoms in the molecule. For example, pentene is C_5H_{10}, hexene is C_6H_{12}, and octene is C_8H_{16}. The common names for the alkenes are produced similarly, except that the suffix is *-ylene*. These names are firmly entrenched in reference to the lower alkenes: C_2H_4 is almost always called ethylene, C_3H_6 is propylene and C_4H_8 is butylene.

Acetylene, C_2H_2, the first member of the alkyne series, is always called by its common name. The next alkynes are sometimes named as derivatives of acetylene, as methyl acetylene, C_3H_4, and ethyl acetylene, C_4H_6. The IUPAC system is more often employed for all alkynes except acetylene. The same prefixes are used, and the alkyne suffix is *-yne*. Thus, for the alkynes with two, three, and four carbon atoms, the formal names are ethyne, propyne and butyne, respectively.

Figure 21.2
Ball-and-stick and space-filling models of the first members of the alkene and alkyne hydrocarbon series, ethylene (left) and acetylene (right).

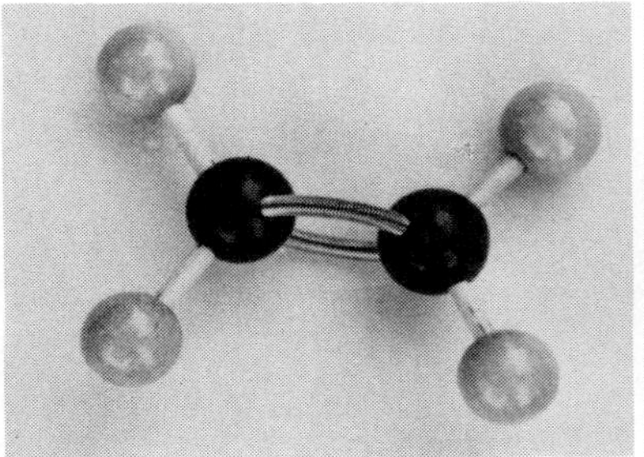

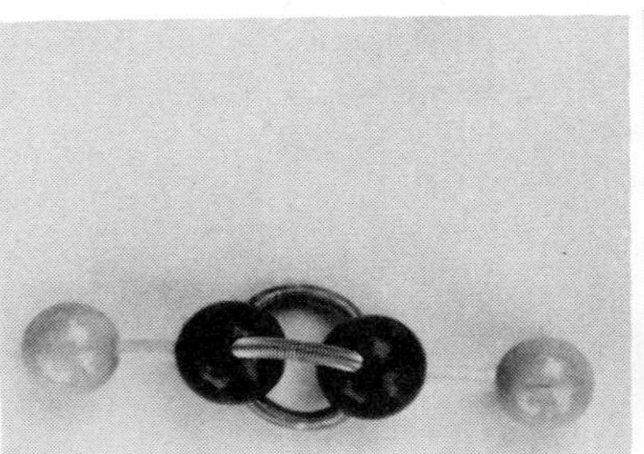
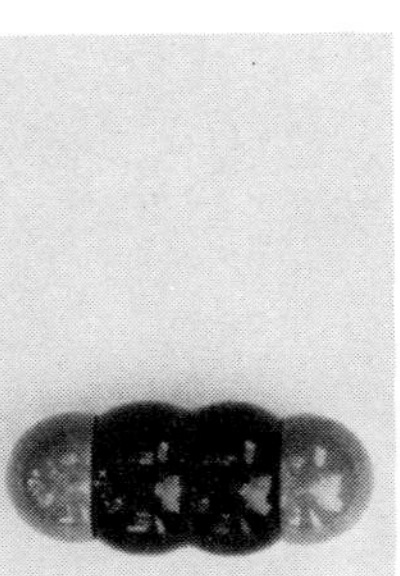

Table 21.4 Unsaturated Hydrocarbons

Hydrocarbon Series	n	Formulas		Names	
		MOLECULAR	STRUCTURAL	IUPAC	COMMON
Alkene, C_nH_{2n}	2	C_2H_4	H H C=C H H	Ethene	Ethylene
	3	C_3H_6	H H C=C—C—H H H H	Propene	Propylene
	4	C_4H_8	H H H C=C—C—C—H H H H H	Butene	Butylene
Alkynes, C_nH_{2n-2}	2	C_2H_2	H—C≡C—H	Ethyne	Acetylene
	3	C_3H_4	H H—C≡C—C—H H	Propyne	—

Isomerism among the Unsaturated Hydrocarbons

All possibilities for isomerism among the alkanes are duplicated in the alkenes and alkynes. Moreover, the unsaturated hydrocarbons introduce a second form of isomerism, and the alkenes even a third. The unique isomerism they both have concerns the location of the multiple bond, which can be anywhere in the chain. Double and triple bonds are handled similarly. In the simplest example, butene may have either of these two structures:

```
  |1  |2  |3  |4          |1  2   3  |4
—C=C—C—C—           —C—C=C—C—
          |   |           |   |   |   |
```

1-Butene　　　　2-Butene

The number appearing before the name is the lowest number possible to identify the carbon atom to which the double bond is attached. The compound

```
  |5  |4  |3  |2  |1
—C—C—C=C—C—
  |   |           |
```

is 2-pentene, because the double bond is attached to the *second* carbon atom counting from the right.

That part of a molecule that is on either side of a single bond may rotate freely around that bond as an axis. This is not so with a double bond. This leads to two possible arrangements around the double bond. The first alkene in which these options appear is butene:

$$\begin{array}{ccc} H_3C & & CH_3 \\ & C{=}C & \\ H & & H \end{array} \qquad \begin{array}{ccc} H_3C & & H \\ & C{=}C & \\ H & & CH_3 \end{array}$$

cis-2-Butene *trans*-2-Butene

The two methyl groups attached to the double-bonded carbons can be on the *same* side of the double bond, as in *cis*-2-butene, or on *opposite* sides, as in *trans*-2-butene. *Cis*- and *trans*- are prefixes meaning, respectively, *on this side* and *across*. The latter is perhaps most easily remembered by association with such words as transcontinental, suggesting across a continent.

Quick Check 21.4

1. Identify the alkenes among the following: C_4H_6 C_2H_6 C_7H_{12} C_8H_{16}.
2. Write a structural formula for *trans*-difluoroethylene.

21.5 SOURCES AND PREPARATION OF ALIPHATIC HYDROCARBONS

Petroleum Products. Alkanes and alkenes are natural products which have resulted from the decay of organic compounds from plants and animals that lived millions of years ago. They are found today as petroleum, mixtures of hydrocarbons containing up to 30 to 40 carbon atoms in the molecule. Different components of petroleum may be isolated by fractional distillation, a process that separates "fractions" that boil at different temperatures. The lower alkanes and alkenes, up to four or five carbons per molecule, may be obtained in pure form by this method. The boiling points of larger compounds are too close for their complete separation, so chemical methods must be employed to obtain pure samples.

Preparation of Alkenes. Two ways alkenes are produced are the *dehydration* of alcohols and the *dehydrohalogenation* of an alkyl halide. These two impressive terms describe very similar processes which are quite simple, at least in principle if not in practice. Dehydration is removal of water; dehydrohalogenation is removal of a hydrogen and a halogen. As an example, propanol is an alcohol (Section 21.9). Its formula is C_3H_7OH. Under certain conditions a molecule of water may be separated from a molecule of the alcohol, producing propene:

$$\begin{array}{c} \text{H}\ \ \text{H}\ \ \text{H} \\ \text{H—C—C—C—H} \\ \text{H}\ \ \boxed{\text{H}\ \ \text{OH}} \end{array} \xrightarrow{H_2SO_4} \begin{array}{c} \text{H}\ \ \text{H}\ \ \text{H} \\ \text{H—C—C=C—H} \\ \text{H} \end{array} + \text{HOH}$$

Propanol Propene Water

An alkyl halide is an alkane in which a halogen atom has been substituted for a hydrogen atom; or, viewed in another way, an alkyl halide is an alkyl group bonded to a halogen. The molecule is attacked with a base in the presence of an alcohol.

$$\begin{array}{c} \text{H}\ \ \text{H}\ \ \text{H} \\ \text{H—C—C—C—H} \\ \text{H}\ \ \boxed{\text{H}\ \ \text{X}} \end{array} + \text{KOH} \xrightarrow{\text{alcohol}} \begin{array}{c} \text{H}\ \ \text{H}\ \ \text{H} \\ \text{H—C—C=C—H} \\ \text{H} \end{array} + \text{KX} + \text{HOH}$$

Propyl halide Base Propene Salt Water

Preparation of Alkanes. There are several industrial and laboratory methods by which alkanes may be prepared. One of the more important is the catalytic **hydrogenation** of an alkene. Hydrogenation is the reaction of a substance with hydrogen. The general reaction of the hydrogenation of an alkene is

$$C_nH_{2n} + H_2 \xrightarrow{\text{catalyst}} C_nH_{2n+2}$$

Preparation of Acetylene. One alkyne is of major importance—acetylene. It is produced commercially in a two-step process in which calcium oxide reacts with coke (carbon) at high temperatures to produce calcium carbide and carbon monoxide:

$$CaO(s) + 3\ C(s) \rightarrow CaC_2(s) + CO(g)$$

Calcium carbide then reacts with water to produce acetylene:

$$CaC_2(s) + 2\ H_2O(\ell) \rightarrow C_2H_2(g) + Ca(OH)_2(s)$$

21.6 CHEMICAL PROPERTIES OF ALIPHATIC HYDROCARBONS

The combustibility—ability to burn in air—of the hydrocarbons is probably one of the most important of all chemical reactions to modern man. As components of liquid and gaseous fuels, hydrocarbons are among the most heavily processed and distributed chemical products in the world. When burned in an excess of air, the end products are water and carbon dioxide (Section 8.6).

One major distinction separates the chemical properties of saturated hydrocarbons from the unsaturated hydrocarbons. By opening a multiple bond in an alkene or alkyne, the compound is capable of reacting by **addition,** simply by adding atoms of some element to the molecule. By contrast, an alkane molecule is literally saturated; there is no room for an atom to join the molecule without first removing a hydrogen atom. A reaction in which a hydrogen atom in an alkane is replaced by an atom of another element is called a **substitution** reaction.

Both alkanes and alkenes undergo *halogenation* reactions—reaction with a halogen. These reactions serve to show the difference between addition and substitution reactions:

Addition Reaction.

$$\begin{array}{ccccccc} & \text{H} & & \text{H} & & \text{H} & \\ & | & & | & & | & \\ \text{H}- & \text{C} & - & \text{C} & = & \text{C} & -\text{H} \\ & | & & & & & \\ & \text{H} & & & & & \end{array} + \text{Cl}_2 \xrightarrow{\text{CCl}_4} \begin{array}{ccccccc} & \text{H} & & \text{H} & & \text{H} & \\ & | & & | & & | & \\ \text{H}- & \text{C} & - & \text{C} & - & \text{C} & -\text{H} \\ & | & & | & & | & \\ & \text{H} & & \text{Cl} & & \text{Cl} & \end{array}$$

Propene   Chlorine   1,2-Dichloropropane

Substitution Reaction.

$$\begin{array}{ccccccc} & \text{H} & & \text{H} & & \text{H} & \\ & | & & | & & | & \\ \text{H}- & \text{C} & - & \text{C} & - & \text{C} & -\text{H} \\ & | & & | & & | & \\ & \text{H} & & \text{H} & & \text{H} & \end{array} + \text{Cl}_2 \xrightarrow{\text{heat or light}} \begin{array}{ccccccc} & \text{H} & & \text{H} & & \text{H} & \\ & | & & | & & | & \\ \text{H}- & \text{C} & - & \text{C} & - & \text{C} & -\text{H} \\ & | & & | & & | & \\ & \text{H} & & \text{H} & & \text{Cl} & \end{array} + \text{HCl}$$

Propane   Chlorine   1-Chloropropane   Hydrogen chloride

The substituted chlorine atom may appear on either an end carbon atom or the middle carbon; the actual product is usually a mixture of 1-chloropropane and 2-chloropropane.

Normally, addition reactions are more readily accomplished than substitution reactions. This is hinted in the reaction conditions specified above. The addition of a halogen to an alkene will occur easily at room temperature, whereas the substitution of a halogen for a hydrogen in an alkane requires either high temperature or ultraviolet light. This shows that unsaturated hydrocarbons are more reactive than saturated hydrocarbons.

Hydrogenation is also an addition reaction. We have already indicated that the hydrogenation of an alkene may be used to produce an alkane. Hydrogenation of an alkyne is a stepwise process, which may often be controlled to give the intermediate alkene as a product:

$$\text{R}-\text{C}\equiv\text{C}-\text{R}' \xrightarrow[\text{catalyst}]{\text{H}_2} \begin{array}{ccccc} & \text{H} & & \text{H} & \\ & | & & | & \\ \text{R}- & \text{C} & = & \text{C} & -\text{R}' \end{array} \xrightarrow[\text{catalyst}]{\text{H}_2} \begin{array}{ccccc} & \text{H} & & \text{H} & \\ & | & & | & \\ \text{R}- & \text{C} & - & \text{C} & -\text{R}' \\ & | & & | & \\ & \text{H} & & \text{H} & \end{array}$$

Alkyne   Alkene   Alkane

A particularly interesting addition reaction is the addition of ethylene to itself. It is an example of **polymerization.** Polymerization is the process whereby small molecular units called **monomers** join together to form giant molecules called **polymers.** Ethylene polymerizes as follows:

$$\begin{array}{ccc} \text{H} & & \text{H} \\ | & & | \\ \text{C} & = & \text{C} \\ | & & | \\ \text{H} & & \text{H} \end{array} + \begin{array}{ccc} \text{H} & & \text{H} \\ | & & | \\ \text{C} & = & \text{C} \\ | & & | \\ \text{H} & & \text{H} \end{array} + \begin{array}{ccc} \text{H} & & \text{H} \\ | & & | \\ \text{C} & = & \text{C} \\ | & & | \\ \text{H} & & \text{H} \end{array} \rightarrow \begin{array}{ccccccccccccc} & \text{H} & & \text{H} & & \text{H} & & \text{H} & & \text{H} & & \text{H} & \\ & | & & | & & | & & | & & | & & | & \\ - & \text{C} & - & \text{C} & - & \text{C} & - & \text{C} & - & \text{C} & - & \text{C} & - \\ & | & & | & & | & & | & & | & & | & \\ & \text{H} & & \text{H} & & \text{H} & & \text{H} & & \text{H} & & \text{H} & \end{array}$$

Ethylene monomers   Segment of polyethylene

In this reaction the double bond of each ethylene molecule opens, and each carbon atom joins to a carbon atom of another molecule, producing the chain shown. The chains formed yield molecules of molecular weights in the area of 20 000. Most plastics are polymers: the example above is the familiar polyethylene used for squeeze bottles, toys, packaging, etc.

Pyrolysis is a decomposition by heat. The pyrolysis of petroleum, consisting primarily of high-molecular-weight alkanes, is called *cracking,* and is usually done in the presence of a catalyst. Pyrolysis is conducted at temperatures in the range of 400 to 600°C. The usual products are alkanes of fewer carbon atoms, alkenes, and hydrogen.

Cracking is one of the major operations in extracting gasoline from raw petroleum. It increases both the yield and quality of gasoline. Yield is increased because some of the longer chain hydrocarbons are reduced to acceptable length for use as gasoline (five to ten carbon atoms per molecule). Quality is improved because the alkenes resulting from the reaction have good antiknock properties.*

Quick Check 21.6 Complete the following equations:
(a) $C_4H_8 + Br_2 \rightarrow$ (b) $C_6H_{14} + Br_2 \rightarrow$

21.7 THE AROMATIC HYDROCARBONS

Historically the term **aromatic** was associated with a series of compounds found in such pleasant-smelling substances as oil of cloves, vanilla, wintergreen, cinnamon, and others. Ultimately, it was found that the key structure in these compounds is the benzene ring.

Benzene has been studied thoroughly in an attempt to determine its structure. Its molecular formula is C_6H_6. It is also known to be a ring compound. How this structure is to be represented in print is a problem, and a universally agreed upon answer has yet to be found. Two common forms are:

```
         H
         C
       /   \\
    HC       CH
    ||       |
    HC       CH
       \   //
         C
         H
         I                II
```

*"Knocking" in an internal combustion engine is a sharp detonation of the fuel–air mixture, rather than a smooth explosion. Knocking is heard when an automobile accelerates too quickly, or when climbing a hill. Knocking is rated on an arbitrary scale in which n-heptane is given an **octane number** of zero, and 2,2,4-trimethyl pentane ("isooctane") is rated at 100 in octane number.

Structure I is perhaps the most complete representation in that it shows all the carbon and hydrogen atoms, as well as alternating single and double bonds that would satisfy the requirements of the octet rule. This structure is not in agreement with experimental fact, however; among other things, all carbon–carbon bonds are known to be alike, rather than some being single, some double. Structure II is a compromise that suggests equality among these bonds. It is also understood that each "corner" of the hexagon represents a carbon atom that forms the equivalent of four covalent bonds, three within the benzene ring, and one without. If the fourth bonded atom is not shown, it is understood to be hydrogen.

An alkyl group, halogen, or other species may replace a hydrogen in the benzene ring.

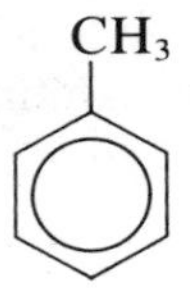

CH_3 Br

Toluene 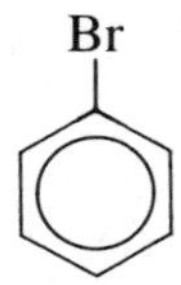 Bromobenzene

If two bromines substitute for hydrogens on the same ring we must consider three possible isomers:

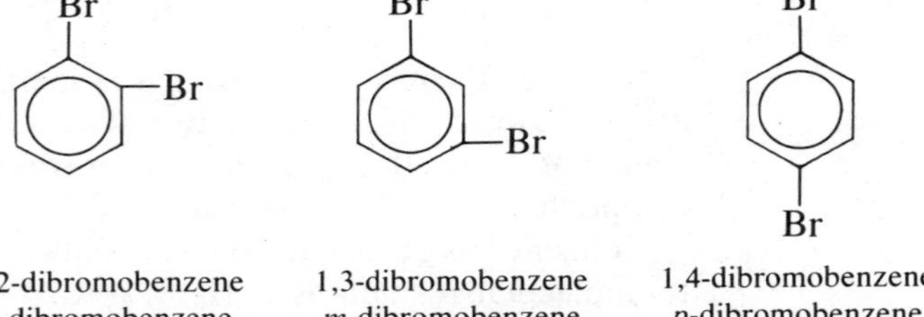

1,2-dibromobenzene
o-dibromobenzene

1,3-dibromobenzene
m-dibromobenzene

1,4-dibromobenzene
p-dibromobenzene

Two names are given for each isomer. The number system, which counts locations around the ring beginning at the substituted positon that yields the lowest numbers, is more formal and serves any number of substituents. The other names are pronounced *ortho*-dibromobenzene, *meta*-dibromobenzene, and *para*-dibromobenzene. *Ortho-*, *meta-*, and *para-* are prefixes commonly used when two hydrogens have been replaced from the benzene ring. Relative to position X, the other positions are shown:

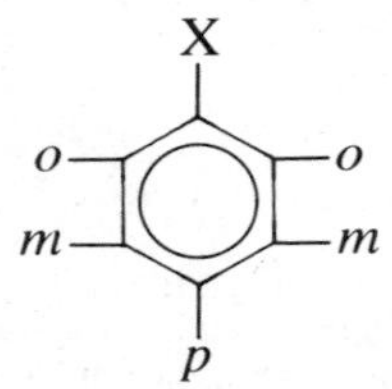

X
o *o*
m *m*
p

The physical properties of benzene and its derivatives are quite similar to those of other hydrocarbons. The compounds are nonpolar, insoluble in polar solvents such as water, but generally soluble in nonpolar solvents. In fact, benzene is widely used as the solvent for many nonpolar organic compounds. Like other hydrocarbons of comparable molecular weight, benzene is a liquid at room temperature. Members of the homologous series increase in boiling point in the usual manner as the number of carbon atoms increases.

The principal industrial source for benzene has been as a byproduct of the preparation of coke from coal. More recently, commercial methods have been developed by which certain petroleum products are converted to aromatic hydrocarbons. For example, toluene may be prepared from *n*-heptane:

$$\underset{n\text{-Heptane}}{CH_3(CH_2)_5CH_3} \xrightarrow[600°\text{ C} + \text{pressure}]{Cr_2O_3 + Al_2O_3} \underset{\text{Toluene}}{C_6H_5CH_3} + 4\ H_2$$

Perhaps the most significant—and surprising—chemical property of benzene is that, despite its high degree of unsaturation, it does not normally engage in addition reactions. The most important reaction of benzene itself is the substitution reaction in which one hydrogen is displaced from the benzene ring. Several substances may be used for substitution, including the halogens:

$$\underset{\text{Benzene}}{C_6H_6} + Cl_2 \xrightarrow{FeCl_3} \underset{\text{Chlorobenzene}}{C_6H_5Cl} + HCl$$

Substitutions with nitric and sulfuric acids yield, respectively, nitrobenzene and benzenesulfonic acid. Second substitutions on the same ring are possible though more difficult to bring about. Substitution reactions may also be performed on benzene derivatives, such as toluene, yielding isomers of nitrotoluene, for example. A triple nitro substitution produces 2,4,6-trinitrotoluene, better known simply as TNT.

Quick Check 21.7
Write the structural formulas of 1,3,5-trifluorobenzene and *p*-dichlorobenzene.

21.8 SUMMARY OF THE HYDROCARBONS

Four types of hydrocarbons we have considered are summarized in Table 21.5:

Table 21.5 Hydrocarbons

Type	Name	Formula	Saturation	Structure
Aliphatic Open Chain	Alkane	C_nH_{2n+2}	Saturated	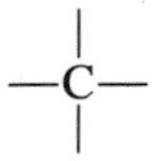 $-\overset{\vert}{\underset{\vert}{C}}-$
	Alkene	C_nH_{2n}	Unsaturated	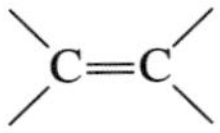 $>C=C<$
	Alkyne	C_nH_{2n-2}	Unsaturated	$-C\equiv C-$
Aromatic	—	—	Unsaturated	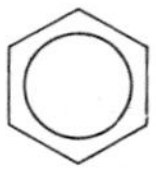 (benzene ring)

Organic Compounds with Oxygen

Thus far we have considered only the hydrocarbons and their derivatives. The third element most commonly found in organic chemicals is oxygen. Capable of forming two bonds (Table 21.1), oxygen serves as a connecting link between two other elements or, double-bonded, usually to carbon, it is a terminal point in a functional group. We will now examine functional groups that contain oxygen: the alcohols, ethers, aldehydes and ketones, acids, and esters.

21.9 THE ALCOHOLS

The Structure of Alcohols

Alcohol is the name given to a large class of compounds containing the **hydroxyl group,** —OH.* This functional group is not to be confused with the hydroxide ion of inorganic chemistry, which exists as an entity in ionic compounds and solutions. In alcohols the hydroxyl group is covalently bonded to an alkyl or other hydrocarbon group. Thus the general formula for an alcohol is R—OH, where R represents the alkyl group.

As shown in Table 21.1, and also in the two models of ethyl alcohol, C_2H_5OH in Figure 21.3, the bond angle around an oxygen atom is close to the tetrahedral angle—about 105°. Thus the alcohol molecule is a water molecule in which one hydrogen has been replaced by an alkyl group:

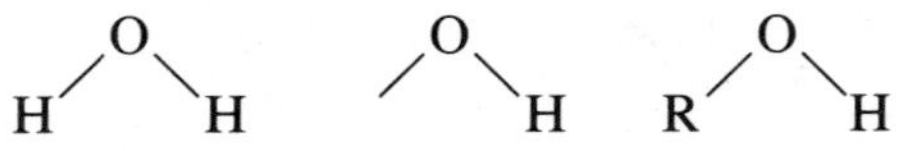

H—O—H (bent)	—O—H (bent)	R—O—H (bent)
Water	Functional group	Alcohol

*Some chemists refer to the hydroxyl group as the *hydroxy group*.

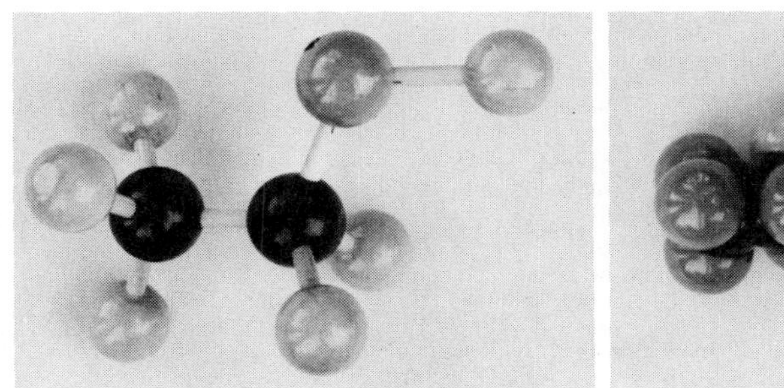

Figure 21.3
Ball-and-stick and space-filling models of ethanol (ethyl alcohol), C_2H_5OH.

This structural similarity correctly suggests similar intermolecular forces and therefore similar physical properties. The lower alcohols (one to three carbon atoms) are liquids with boiling points ranging from 65°C to 97°C, comparable with water but well above the boiling points of alkanes of about the same molecular mass. This is largely because of hydrogen bonding, very much in evidence in the lower alcohols (Fig. 21.4). Hydrogen bonding also accounts for the complete miscibility (solubility) between lower alcohols and water. As usual, boiling points rise with increasing molecular weight. Solubility drops off sharply as the alkyl chain lengthens and the molecule assumes more the character of the parent alkane.

When the hydroxyl group is attached to the end carbon in a chain, the compound is a *primary* alcohol. If the hydroxyl group is bonded to a carbon that is bonded to two other carbons, it is a *secondary* alcohol. Isopropyl alcohol (see below) is a secondary alcohol. A *tertiary* alcohol has the hydroxyl group attached to a carbon that is bonded to three other carbon atoms.

Names of the Alcohols

Alcohols are best known by their common names, which originate in the name of the alkyl group to which the hydroxyl group is bonded. This system names the alkyl group, followed by "alcohol." Thus CH_3OH is *methyl alcohol* and C_2H_5OH is *ethyl alcohol*. Under IUPAC nomenclature rules for alcohols, the *e* at the end of the corresponding alkane is replaced with the suffix *-ol* and the result is the name of the alcohol. Thus methyl alcohol becomes *methanol,* and ethyl alcohol is formally *ethanol*.

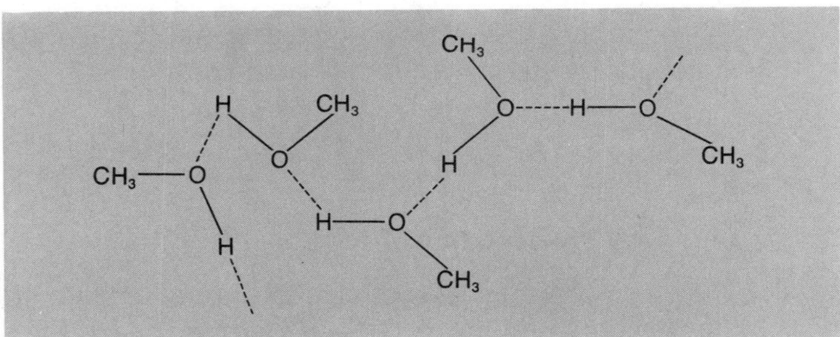

Figure 21.4
Hydrogen bonding in methanol.

Propyl alcohol has two isomers:

$$\begin{array}{ccccccc} & & \text{H} & & \text{H} & & \text{H} & \\ & & | & & | & & | & \\ \text{H} & — & \text{C} & — & \text{C} & — & \text{C} & — \text{H} \\ & & | & & | & & | & \\ & & \text{H} & & \text{H} & & \text{OH} & \end{array} \qquad \begin{array}{ccccccc} & & \text{H} & & \text{H} & & \text{H} & \\ & & | & & | & & | & \\ \text{H} & — & \text{C} & — & \text{C} & — & \text{C} & — \text{H} \\ & & | & & | & & | & \\ & & \text{H} & & \text{OH} & & \text{H} & \end{array}$$

n-Propyl alcohol
1-propanol

Isopropyl alcohol
2-propanol

These isomers are distinguished by stating the number of the carbon atom to which the hydroxyl group is bonded. Accordingly, *n*-propyl alcohol becomes 1-*propanol* and isopropyl alcohol is designated 2-*propanol*.

Sources and Preparation of Alcohols

Hydration of Alkenes. The major industrial source of several of our most important alcohols is the hydration of alkenes obtained from the cracking of petroleum. Beginning with ethylene, for example, the reaction may be summarized

$$\begin{array}{ccccc} & & \text{H} & & \text{H} & \\ & & | & & | & \\ \text{H} & — & \text{C} & = & \text{C} & — \text{H} \end{array} + \text{HOH} \rightarrow \begin{array}{ccccc} & & \text{H} & & \text{H} & \\ & & | & & | & \\ \text{H} & — & \text{C} & — & \text{C} & — \text{H} \\ & & | & & | & \\ & & \text{H} & & \text{OH} & \end{array}$$

Ethylene   Ethyl alcohol

Fermentation of Carbohydrates. Making ethyl alcohol by the fermentation of sugars in the presence of yeast is probably the oldest synthetic chemical process known:

$$\underset{\text{Glucose (sugar)}}{C_6H_{12}O_6} \xrightarrow{\text{yeast}} 2\ CO_2 + \underset{\text{Ethyl alcohol}}{2\ C_2H_5OH}$$

A solution that is 95% ethyl alcohol (190 proof) may be obtained from the final mixture by fractional distillation. The mixture also yields two other products that are of commercial importance today: *n*-butyl alcohol and acetone (Section 21.11).

Synthesis of Methyl Alcohol. Methyl alcohol is sometimes called wood alcohol because it was once made by the destructive distillation (heating in absence of air) of wood. It is now produced by the catalytic hydrogenation of carbon monoxide at high pressure and temperature:

$$CO + 2\ H_2 \xrightarrow[\text{250 atm + 300°C}]{ZnO + Cr_2O_3} CH_3OH$$

Chemical Properties of Alcohols

The chemical properties of alcohols are essentially the chemical properties of the functional group, —OH. In some reactions the C—OH bond is broken, separating the entire hydroxyl group. This is true in the dehydration of alcohols to form alkenes (Section 21.5). In other reactions of alcohols the O—H bond within the hydroxyl group is broken. We will postpone discussion of these reactions until the structures of the various products—aldehydes, ketones, carboxylic acids, and esters—have been examined.

Some Common Alcohols

Methyl alcohol is an important industrial chemical with production measured in the billions of pounds annually. It is a raw material for the production of many chemicals, particularly formaldehyde, which is widely used in the plastics industry. It is also used in antifreezes, commercial solvents, and as a denaturant, or additive to ethyl alcohol to make it unfit for human consumption. Taken internally, methyl alcohol is a deadly poison, frequently causing blindness in less-than-lethal doses.

In addition to its uses in beverages, ethyl alcohol is used in organic solvents and in the preparation of various organic compounds such as chloroform and ether. Its production is also measured in the billions of pounds annually.

Other widely used alcohols include isopropyl alcohol, which is sold as rubbing alcohol, and *n*-butyl alcohol, used in lacquers in the automobile industry. Alcohols containing more than one hydroxyl group are also common. Permanent antifreeze in automobiles is ethylene glycol, which has two hydroxyl groups in the molecule. Glycerine, or glycerol, a trihydroxyl alcohol, has many uses in the manufacture of drugs, cosmetics, explosives, and other chemicals.

Quick Check 21.9
Write the name and structural formula of the functional group that identifies an alcohol.

21.10 THE ETHERS

In the previous section we pointed out that structurally an alcohol might be considered as a water molecule in which one hydrogen has been replaced by an alkyl group. An **ether** may be similarly considered, except that *both* hydrogens are replaced by an alkyl group. The functional group that identifies an ether is simply the oxygen atom bonded to two alkyl groups:

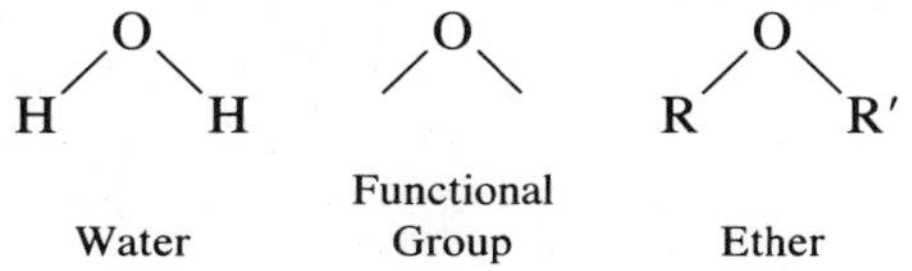

$H{-}O{-}H$	$-O-$	$R{-}O{-}R'$
Water	Functional Group	Ether

The R′ indicates that the functional groups may or may not be identical. For example, methyl ethyl ether and ethyl ether have the structures

$$\underset{\text{Methyl ethyl ether}}{CH_3{-}O{-}C_2H_5} \qquad \underset{\text{Ethyl ether}}{C_2H_5{-}O{-}C_2H_5}$$

Figure 21.5 shows models of methyl ether.

Ether molecules are less polar than alcohol molecules, and there is no opportunity for hydrogen bonding between them. Intermolecular attractions are therefore lower, as are the dependent boiling points. Up to three carbons, ethers are gases at room conditions, and the familiar ethyl ether is a volatile liquid that boils at 35°C. The solubility of an ether in water is about the same as the solubility of its isomeric alcohol, primarily because of hydrogen bonding between the ether molecule and water molecule:

$$\mathrm{R{-}O(R')\cdots H{-}O{-}H\cdots O(R)(R')}$$

All ethers are called "ether," and identified specifically by naming first the two alkyl groups that are bonded to the functional group. If the groups are identical, the prefix *di-* may be used, as in diethyl ether.

Under properly controlled conditions, ethers can be prepared by dehydrating alcohols. At 140°C, and with constant alcohol addition to replace the ether as it distills from the mixture, ethyl ether is formed from two molecules of ethanol:

$$\underset{\text{Ethyl alcohol}}{CH_3{-}CH_2{-}O{-}H} + \underset{\text{Ethyl alcohol}}{H{-}O{-}CH_2{-}CH_3} \rightarrow \underset{\text{Ethyl ether}}{CH_3{-}CH_2{-}O{-}CH_2{-}CH_3} + HOH$$

Aside from combustion, ethers are relatively unreactive compounds, being quite resistant to attack by active metals, strong bases, and oxidizing agents. They are, however, highly flammable and must be handled cautiously in the laboratory.

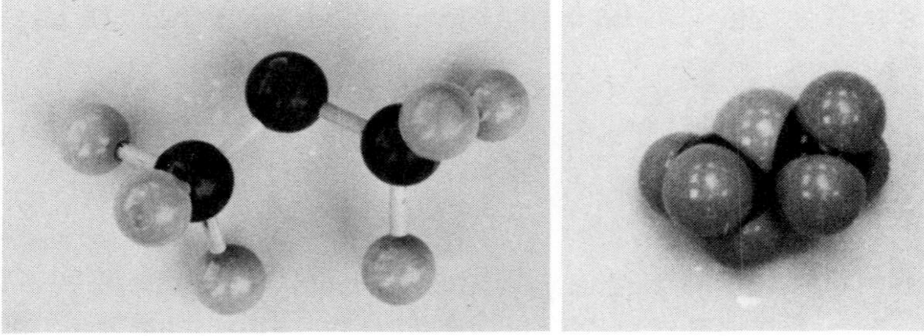

Figure 21.5
Ball-and-stick and space-filling models of methyl ether, CH_3OCH_3.

The isolated word "ether" generally prompts one to think of the anesthetic that is so identified. This compound is ethyl ether; its line formula is $C_2H_5—O—C_2H_5$. Recently its isomer, methyl propyl ether, $CH_3—O—C_3H_7$, has been gaining popularity as a substitute in this use; it has fewer objectionable after-effects than ethyl ether. Ethyl ether is also used as a solvent for fats from foods and animal tissue in the laboratory.

Quick Check 21.10
Write the structural formula of the functional group that identifies an ether.

21.11 THE ALDEHYDES AND KETONES

Aldehydes and **ketones** are characterized by the **carbonyl functional group,**

$$\overset{\displaystyle O}{\underset{\diagup\quad\diagdown}{\overset{\|}{C}}}$$

If at least one hydrogen atom is bonded to the carbonyl carbon, the compound is an aldehyde, RCHO; if two alkyl groups are attached, the compound is a ketone, R—CO—R′.

$$\underset{\text{Aldehyde}}{R-\overset{\displaystyle O}{\overset{\|}{C}}-H} \qquad \underset{\text{Ketone}}{R-\overset{\displaystyle O}{\overset{\|}{C}}-R'}$$

The simplest carbonyl compound is formaldehyde, HCHO, which has two hydrogen atoms bonded to the carbonyl carbon. If a methyl group replaces one of the hydrogens of formaldehyde, the result is acetaldehyde, CH_3CHO (Fig. 21.6). Replacement of both formaldehyde hydrogens with methyl groups yields acetone.

$$\underset{\text{Formaldehyde}}{H-\overset{\displaystyle O}{\overset{\|}{C}}-H} \qquad \underset{\text{Acetaldehyde}}{CH_3-\overset{\displaystyle O}{\overset{\|}{C}}-H} \qquad \underset{\text{Acetone}}{CH_3-\overset{\displaystyle O}{\overset{\|}{C}}-CH_3}$$

The carbonyl group is polar, thereby making ketone and aldehyde molecules polar, although not as polar as alcohols. Only formaldehyde is definitely a gas at room temperature (boiling point −21°C); acetaldehyde boils at 20°C. Aldehydes and ketones of up to about five carbons enjoy some solubility in water, no doubt because of the polarity of both the solvent and solute molecules and hydrogen bonding. The liquid "formaldehyde" we encounter in the laboratory is actually a water solution, sold under the trade name "Formalin."

Figure 21.6
Ball-stick and space-filling models of acetaldehyde (above) and acetone (below).

The lower aldehydes are best known by their common names. The IUPAC nomenclature system for aldehydes employs the name of the parent hydrocarbon, substituting the suffix *-al* for the final *e* to identify the compound as an aldehyde. Thus the IUPAC name for formaldehyde is methanal, for acetaldehyde, ethanal, and so forth.

Ketones are named by one of two systems. The first duplicates the method of naming ethers: identify each alkyl group attached to the carbonyl group, followed by the class name, ketone. Accordingly, methyl ethyl ketone has the structure

$$CH_3-\overset{\overset{\displaystyle O}{\|}}{C}-C_2H_5$$

Under the IUPAC system the number of carbons in the longest chain carrying the carbonyl carbon establishes the hydrocarbon base, which is followed by *-one* to identify the ketone as the class of compound. Methyl ethyl ketone, having four carbons, would be called *butanone*. Two isomers of pentanone would be 2-*pentanone* and 3-*pentanone,* the number being used to designate the carbonyl carbon:

$$CH_3-\overset{\overset{\displaystyle O}{\|}}{C}-CH_2CH_2CH_3 \qquad CH_3CH_2-\overset{\overset{\displaystyle O}{\|}}{C}-CH_2CH_3$$

2-Pentanone　　　　3-Pentanone

Aldehydes and ketones may be prepared by oxidation of alcohols. If the product is to be a ketone, the alcohol must be a *secondary* alcohol, in which the hydroxyl group is bonded to an interior carbon:

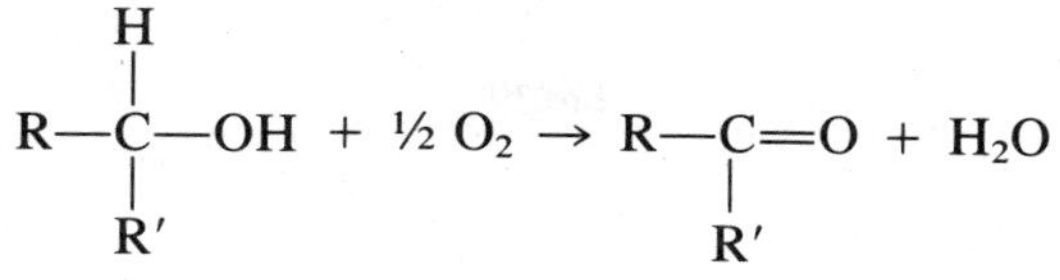

$$\underset{\text{Secondary alcohol}}{R-\underset{R'}{\overset{H}{\overset{|}{\underset{|}{C}}}}-OH} + \tfrac{1}{2}\,O_2 \rightarrow \underset{\text{Ketone}}{R-\underset{R'}{\underset{|}{C}}=O} + H_2O$$

Care must be taken not to overoxidize aldehyde preparations, since aldehydes are easily oxidized to carboxylic acids (see next section).

Aldehydes and ketones may also be produced by the hydration of alkynes. If the triple bond is on an end carbon, an aldehyde is produced; if between internal carbons, the result is a ketone. A typical reaction is the commercial preparation of acetaldehyde:

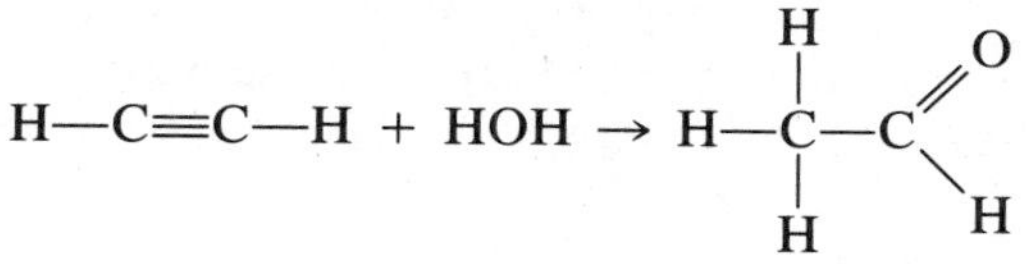

$$H-C\equiv C-H + HOH \rightarrow \underset{\text{Acetaldehyde}}{H-\underset{H}{\overset{H}{\overset{|}{\underset{|}{C}}}}-C\!\!\begin{smallmatrix}\nearrow O\\ \searrow H\end{smallmatrix}}$$

The double bond of the carbonyl group can engage in addition reactions, just like the double bond in the alkenes. One such reaction is the catalytic hydrogenation of ketones to secondary alcohols, in which the hydroxyl group is bonded to a carbon atom *within* the chain:

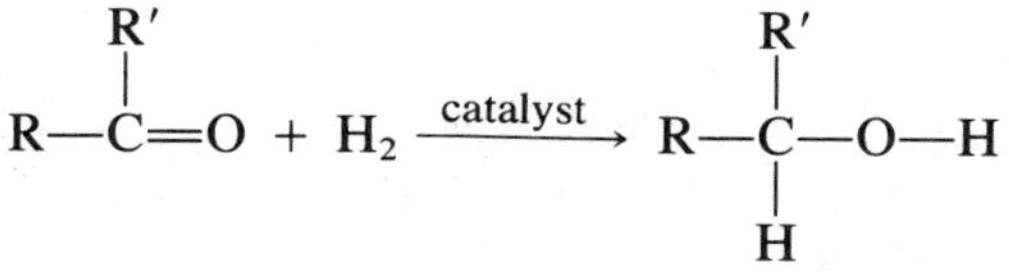

$$\underset{\text{Ketone}}{R-\overset{R'}{\overset{|}{C}}=O} + H_2 \xrightarrow{\text{catalyst}} \underset{\text{Secondary alcohol}}{R-\underset{H}{\overset{R'}{\overset{|}{\underset{|}{C}}}}-O-H}$$

Oxidation reactions occur quite readily with aldehydes, but are resisted by ketones. When an aldehyde is oxidized, the product is a carboxylic acid:

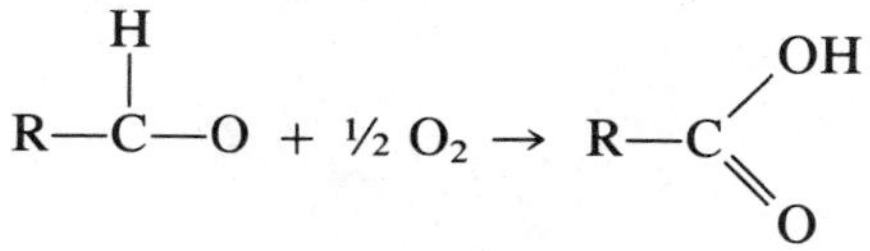

$$\underset{\text{Aldehyde}}{R-\overset{H}{\overset{|}{C}}-O} + \tfrac{1}{2}\,O_2 \rightarrow \underset{\text{Carboxylic acid}}{R-C\!\!\begin{smallmatrix}\nearrow OH\\ \searrow\!\!\searrow O\end{smallmatrix}}$$

Formaldehyde is probably the best known carbonyl compound. It is widely used as a preservative. Large quantities are consumed in the manufacture of resins and in the preparation of numerous organic compounds. Acetaldehyde finds use in the manufacture of acetic acid, ethyl acetate, and other organic products.

Acetone is the most important ketone. It is a solvent for many organic chemicals, including cellulose derivatives, varnish, lacquer, plastics, and resins. Methyl ethyl ketone finds application in the petroleum industry, and it is also familiar as an ingredient of fingernail polish remover.

Quick Check 21.11

1. Write the name and structural formula of the functional group that identifies an aldehyde or a ketone.
2. Write structural formulas that show the difference between an aldehyde and a ketone.

21.12 CARBOXYLIC ACIDS AND ESTERS

In the last section we saw that oxidation of an aldehyde produces a **carboxylic acid,** the general formula of which is frequently represented as RCOOH. The functional group, —COOH, shown at the left below, is a combination of a carbonyl group and a hydroxyl group, appropriately called the **carboxyl group.** You can probably pick out the carboxyl group in the formic and acetic acid models in Figure 21.7. In an **ester,** the carboxyl carbon may be bonded to a hydrogen atom or an alkyl group, and the carboxyl hydrogen is replaced by another alkyl group, as shown:

```
       O              O
      //             //
   —C         R—C
      \              \
       O—H            O—R'
 Carboxyl group      Ester
```

The geometry of the carboxyl group results in strong dipole attractions and hydrogen bonding between molecules. As a consequence, boiling points tend

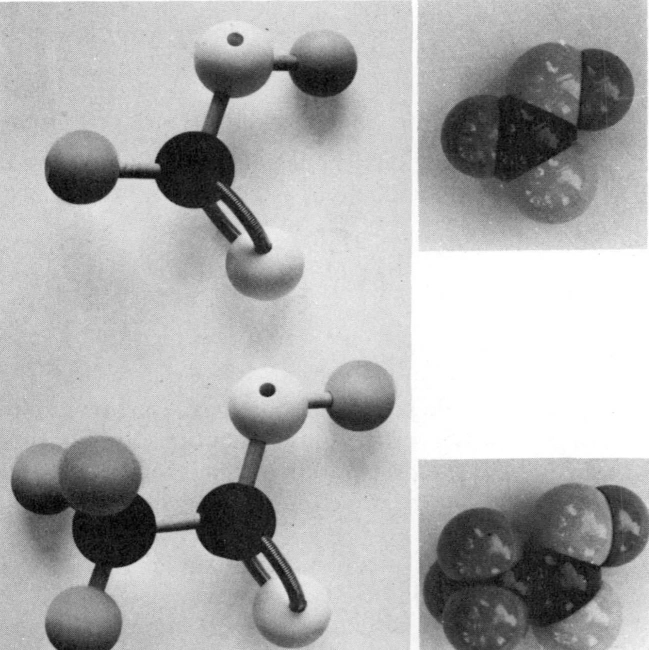

Figure 21.7
Ball-and-stick and space-filling models of formic acid (above) and acetic acid (below).

to be high compared to compounds of similar molecular weight. Formic acid, for example, boils at 100.5°C. Lower acids are completely miscible in water, but solubility drops off as the aliphatic chain lengthens and the molecule behaves more like a hydrocarbon.

Common names continue to be used for most acids. Formic acid, HCOOH, with only a hydrogen attached to the carboxyl group, is the simplest of the carboxylic acids. Next in the series is acetic acid, in which the methyl group is bonded to the carboxyl group: CH_3COOH. The names of many acids come from their sources or some physical property associated with them. Butyric acid, C_3H_7COOH, for example, is responsible for the odor of rancid butter, for which the Latin word is *butyrum*. The IUPAC system for naming carboxylic acids drops the *e* from the alkane of the same number of carbon atoms, and replaces it with *-oic*. Thus HCOOH is *methanoic acid;* CH_3COOH is *ethanoic acid;* C_2H_5COOH is *propanoic acid,* and so forth.

Formic acid is prepared commercially from sodium formate, produced by the reaction of carbon monoxide and sodium hydroxide. In a typical molecular product reaction (Section 16.7), sodium formate reacts with hydrochloric acid to yield formic acid and sodium chloride:

$$\mathrm{HCOONa + HCl \rightarrow HCOOH + NaCl}$$

Acetic acid, by far the most important of the carboxylic acids, is produced by the stepwise oxidation of ethanol, first to acetaldehyde and then to acetic acid:

$$\begin{array}{ccccc} \begin{array}{c} \mathrm{H}\ \ \mathrm{H} \\ |\ \ \ | \\ \mathrm{H{-}C{-}C{-}OH} \\ |\ \ \ | \\ \mathrm{H}\ \ \mathrm{H} \end{array} & \xrightarrow{\mathrm{KMnO_4}} & \begin{array}{c} \mathrm{H}\ \ \mathrm{H} \\ |\ \ \ | \\ \mathrm{H{-}C{-}C{=}O} \\ | \\ \mathrm{H} \end{array} & \xrightarrow{\mathrm{KMnO_4}} & \begin{array}{c} \mathrm{H}\quad\ \ \mathrm{OH} \\ |\quad\ / \\ \mathrm{H{-}C{-}C} \\ |\quad\ \backslash\backslash \\ \mathrm{H}\quad\ \ \mathrm{O} \end{array} \\ \text{Ethyl alcohol} & & \text{Acetaldehyde} & & \text{Acetic acid} \end{array}$$

Carboxylic acids are weak acids that release a proton from the carboxyl group on ionization.* Acetic acid, for example, ionizes in water as follows:

$$\mathrm{CH_3COOH(aq) \rightleftharpoons CH_3COO^-(aq) + H^+(aq)}$$

The ionization takes place but slightly; only about 1% of the acetic acid molecules ionize. The solution consists primarily of molecular CH_3COOH. This notwithstanding, acetic acid participates in typical acid reactions such as neutralization.

$$\mathrm{CH_3COOH(aq) + OH^-(aq) \rightarrow HOH(\ell) + CH_3COO^-(aq)}$$

and the release of hydrogen on reaction with a metal,

$$\mathrm{2\ CH_3COOH(aq) + Ca(s) \rightarrow 2\ CH_3COO^-(aq) + Ca^{2+}(aq) + H_2(g)}$$

Metal acetate salts may be obtained by evaporating the resulting solutions to dryness.

*In more advanced consideration of organic reactions, the term *acid* is also used in reference to Lewis acids (Section 17.1). This is why the adjective *carboxylic* is used to identify an organic acid containing the carboxyl group.

The reaction between an acid and an alcohol is called **esterification.** The products of the reaction are an ester and water. A typical esterification reaction is

$$\underset{\substack{\text{Acetic acid}\\ \text{Acid}}}{H-\overset{\overset{H}{|}}{\underset{\underset{H}{|}}{C}}-\overset{\overset{O}{\|}}{C}-O-H} + \underset{\substack{\text{Methanol}\\ \text{Alcohol}}}{H-O-\overset{\overset{H}{|}}{\underset{\underset{H}{|}}{C}}-H} \rightleftharpoons \underset{\substack{\text{Methyl acetate}\\ \text{Ester}}}{H-\overset{\overset{H}{|}}{\underset{\underset{H}{|}}{C}}-\overset{\overset{O}{\|}}{C}-O-\overset{\overset{H}{|}}{\underset{\underset{H}{|}}{C}}-H} + HOH$$

Notice how the water molecule is formed: the *acid contributes the entire hydroxyl group,* while the *alcohol furnishes only the hydrogen.*

The names of esters are derived from the parent alcohol and acid. The first term is the alkyl group associated with the alcohol; the second term is the name of the anion derived from the acid. In the example above, methyl alcohol (methanol) yields *methyl* as the first term, and acetic acid yields *acetate* as the second term.

Carboxylic acids engage in typical proton transfer acid–base type reactions with ammonia to produce salts. The ammonium salt so produced may then be heated, which causes it to lose a water molecule. The resulting product is called an *amide*. Compared to the original acid, an amide substitutes an $-NH_2$ group for the $-OH$ group of the acid (Section 21.14).

$$\underset{\text{Acid}}{R-\overset{\overset{O}{\|}}{C}-O-H} + NH_3 \rightarrow \underset{\text{Salt}}{\left[R-\overset{\overset{O}{\|}}{C}-O\right]^- NH_4^+} \xrightarrow{\Delta} \underset{\text{Amide}}{R-\overset{\overset{O}{\|}}{C}-NH_2} + H_2O$$

Formic acid and acetic acid are the two most important carboxylic acids. Formic acid is the source of irritation in the bite of ants and other insects, or the scratch of nettles. A liquid with a sharp, irritating odor, it is used in manufacturing esters, salts, plastics, and other chemicals. Acetic acid is present to about 4% to 5% in vinegar, and is responsible for its odor and taste. Acetic acid is among the least expensive organic acids, and is therefore a raw material in many commercial processes that require a carboxylic acid. Sodium acetate is one of several important salts of carboxylic acids. It is used to control the acidity of chemical processes and in the preparation of soaps and pharmaceutical agents.

Ethyl acetate and butyl acetate are two of the relatively few esters produced in large quantity. Both are used as solvents, particularly in the manufacture of lacquers. Other esters are involved in the plastics industry, and some find application in the medicinal fields. Esters are responsible for the odor of most fruit and flowers, leading to their use in the food and perfume industries.

Quick Check 21.12

1. Write the name and structural formula of the functional group that identifies a carboxylic acid.
2. Describe in words the reactants and products of an esterification reaction.

Organic Compounds with Nitrogen

Nitrogen is one of the important elements found in living organisms, so we might expect to find it in many organic compounds. We will mention briefly two classes of nitrogen-bearing organic compounds, the amines and the amides.

21.13 AMINES

Amines are organic derivatives of ammonia. In much the same way that an alcohol can be viewed structurally as a water molecule in which a hydrogen atom has been replaced by an alkyl group, and ether molecules are water molecules with both hydrogen atoms replaced by alkyl groups, an amine is an ammonia molecule with one, two, or three hydrogen atoms replaced by alkyl groups. The number of hydrogens so replaced distinguishes between a primary, secondary, and tertiary amine. Amines are named by identifying the alkyl groups that are bonded to the nitrogen atom, using appropriate prefixes if two or three identical groups are present, followed by the suffix *-amine*. Illustrative examples follow:

$$H—\ddot{N}(H)—H \qquad CH_3—\ddot{N}(H)—H \qquad CH_3—\ddot{N}(H)—C_2H_5 \qquad CH_3—\ddot{N}(CH_3)—C_2H_5$$

Ammonia	Methylamine	Ethylmethylamine	Dimethylethylamine
	Primary amine	Secondary amine	Tertiary amine

As ammonia is polar and capable of forming hydrogen bonds, so are the primary and secondary amines, but to a lesser extent. Tertiary amines are essentially nonpolar. These structural features contribute in the usual way to the physical properties of the amines. All three methylamines and ethylamines are gases, with boiling points in the range of $-6°C$ to $11°C$. Other amines are liquids with boiling points that increase with molecular weight and more complex structure. Amines can form hydrogen bonds with water molecules; therefore lower amines—particularly primary and secondary amines—are very soluble in water, and less soluble in nonpolar solvents.

Because of the unshared electron pair in the nitrogen atom, amines behave as Brönsted–Lowry or Lewis bases (Section 17.1). A typical reaction is

$$H—N(CH_3)_2: + HCl \rightarrow [H—N(CH_3)_2—H]^+ Cl^-$$

Dimethylamine — Dimethylammonium chloride

By reaction first with nitrous acid and then hydrogen, dimethylamine is made into dimethyl hydrazine, $(CH_3)_2NNH_2$, which is used as a rocket propellant. Dimethylamine and trimethylamine are both used in making anion exchange resins. Dyes, drugs, herbicides, fungicides, soaps, insecticides, and photographic developers are among the chemical products made from amines. Aniline, or phenylamine, an aromatic amine, is among the more important materials used in dye making.

Quick Check 21.13
Write the structural formula of the functional group that indentifies an amine.

21.14 AMIDES

In Section 21.12 an **amide** was shown to be a derivative of a carboxylic acid in which the hydroxyl part of the carboxyl group is replaced by an NH_2 group. For example,

$$\underset{\text{Acetic acid}}{CH_3{-}C{\overset{\displaystyle O}{\underset{\displaystyle OH}{\lessgtr}}}} \quad \text{becomes} \quad \underset{\text{Acetamide}}{CH_3{-}C{\overset{\displaystyle O}{\underset{\displaystyle NH_2}{\lessgtr}}}}$$

by substitution of the $—NH_2$ in place of the —OH, as shown. An amide is named by replacing the *-ic acid* name of the acid with *amide*.

Amides are polar compounds that are capable of strong hydrogen bonding between the electronegative oxygen of the carboxyl group of one molecule and the electropositive amide hydrogen of the next. As a consequence the amides as a group have higher melting and boiling points than otherwise similar compounds. Only formamide, $HCONH_2$, is a liquid at room temperature; all higher amides are solids. Polarity and hydrogen bonding predict accurately the solubility of the lower amides in water.

The amide structure appears in an important biochemical system, protein, as a connecting link between amino acids. The linkage is commonly called a *peptide linkage*. This linkage has the form

$$R{-}\overset{\displaystyle O}{\overset{\|}{C}}{-}\underset{\displaystyle H}{\underset{|}{N}}{-}C$$

Peptide linkage

An *amino acid* is an acid in which an amine group is substituted for a hydrogen atom in the molecule. The amino acids involved in the protein structure have the general formula

$$R{-}\overset{\displaystyle H}{\overset{|}{\underset{\displaystyle NH_2}{\underset{|}{C}}}}{-}C{\overset{\displaystyle O}{\underset{\displaystyle OH}{\lessgtr}}}$$

in which the amine and carboxyl groups are bonded to the same carbon atom. The peptide linkage is formed when the carboxyl group of one amino acid and the amine group of another combine by removing a water molecule:

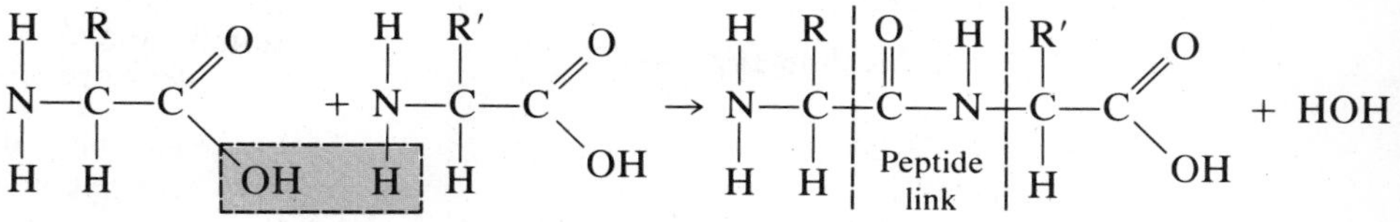

```
H  R      O      H  R'     O       H  R ¦ O     H ¦ R'     O
|  |     //      |  |     //       |  | ¦ ||    | ¦ |     //
N—C—C         + N—C—C          →  N—C—¦—C—N—¦—C—C            + HOH
|  |     \       |  |     \        |  | ¦       ¦ |     \
H  H     [OH   H] H       OH       H  H ¦Peptide¦ H      OH
                                        ¦ link  ¦
```

Proteins are chains of such links between as many as 18 different amino acids, producing huge molecules with molar weights ranging from about 34 500 to 50 000 000.

Quick Check 21.14
Write the structural formula of the functional group that identifies an amide.

21.15 SUMMARY OF THE ORGANIC COMPOUNDS OF CARBON, HYDROGEN, OXYGEN, AND NITROGEN

The eight types of organic compounds of carbon, hydrogen, oxygen and nitrogen we have considered are summarized in Table 21.6.

TERMS AND CONCEPTS

21.1 Organic chemistry
21.3 Hydrocarbon
Aliphatic hydrocarbon
Saturated hydrocarbon
Alkane
Homologous series
Normal alkane
Condensed (line) formula
Isomer; isomerism
Alkyl group
Functional group
21.4 Unsaturated hydrocarbon
Alkene
Alkyne
Cis-, *trans-* isomers
21.5 Fractional distillation
Dehydration
Dehydrohalogenation
Hydrogenation
21.6 Addition reaction
Substitution reaction
Polymerization
Monomer, polymer
Pyrolysis
Cracking
21.7 Aromatic hydrocarbon
Benzene ring
Ortho-, meta, para-
21.9 Alcohol
Hydroxyl (hydroxy) group
Primary, secondary, tertiary alcohol
21.10 Ether
21.11 Aldehyde
Ketone
Carbonyl group
21.12 Carboxylic acid
Carboxyl group
Ester
Esterification
21.13 Amine
Primary, secondary, tertiary amine
21.14 Amide

Most of these terms and many others appear in the Glossary. Use the Glossary regularly.

Table 21.6 Classes of Organic Compounds

Compound Class	General Formula	Functional Group	Names*
Alcohol	R—OH	—OH	Alkyl group + *alcohol;* methyl alcohol Alkane prefix + *-ol:* methanol
Ether	R—O—R′	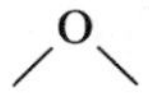 C—O—C	Name both alkyl groups + *ether:* ethyl methyl ether Alkyl group + *-oxy-* + alkane: methoxyethane
Aldehyde	R—CHO	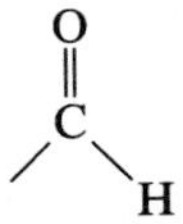 C(=O)H	Common prefix + *-aldehyde:* formaldehyde Alkane prefix + *-al:* methanal
Ketone	R—CO—R′	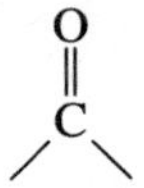 C(=O)	Name both alkyl groups + *ketone:* methyl ethyl ketone; methyl *n*-propyl ketone (Number carbonyl carbon) + alkane prefix + *-one:* butanone; 2-pentanone
Acid	R—COOH	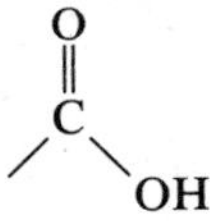 C(=O)OH	Common name + acid: formic acid Alkane prefix + *-oic* + *acid:* methanoic acid
Ester	R—CO—OR′	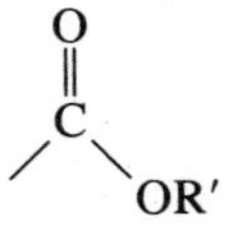 C(=O)OR′	Alcohol alkyl group + acid anion: methyl acetate Alcohol alkyl group + acid alkane prefix + *-oate;* methyl ethanoate
Amine	RNH_2 R_2NH R_3N	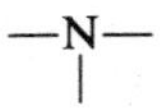 —N—	Name alkyl group(s) + *-amine:* methylamine *Amino-* + alkane: aminomethane
Amide	R—$CONH_2$	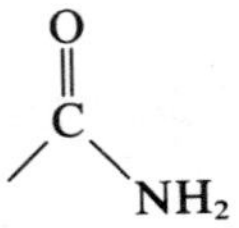 C(=O)NH_2	Common acid prefix + *-amide:* formamide Alkane prefix + *-amide;* methanamide

*Common name followed by IUPAC name.

QUESTIONS AND PROBLEMS

Section 21.1

(*1*) Compare the original and modern definitions of organic chemistry. Why was the definition changed? Which definition includes the other.

(43) Name some of the classes of carbon-bearing compounds that are not included in organic chemistry.

Section 21.3

(2) Define hydrocarbon.

(3) Define an alkane, and explain how it is an example of a homologous series.

(4) What is a condensed formula, or line formula? What advantage does it have over a molecular formula?

(5) Write both the structural formula and condensed (line) formula for the normal alkane having nine carbon atoms in its molecules.

(6) Define isomerism. From the following list of molecular and condensed formulas, select two that are isomers and explain why they may be so classified:
(a) $CH_3CH(CH_3)CH_2CH_2CH(CH_3)CH_2CH_3$
(b) $CH_3CH_2C(CH_3)_2CH_2CH_3$
(c) $CH_3(CH_2)_8CH_3$
(d) C_6H_{14}
(e) C_9H_{20}

(7) What is meant by an alkyl group? How are alkyl groups named? Give three examples of alkyl groups, both name and formula.

(8)

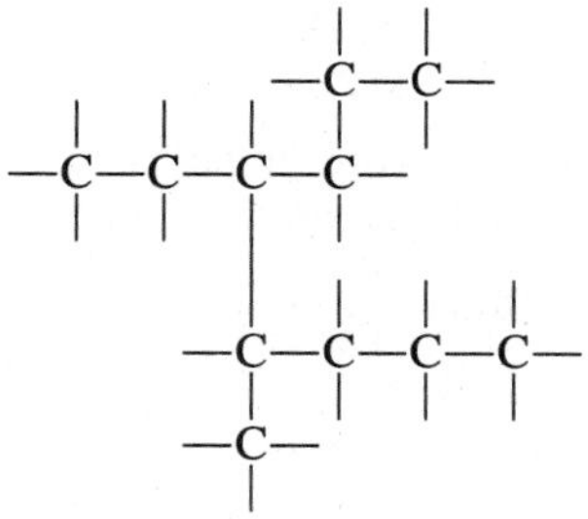

```
              |   |
             —C—C—
  |   |   |   |   |
 —C—C—C—C—
  |   |   |   |
          |
          |   |   |   |
         —C—C—C—C—
          |   |   |   |
         —C—
          |
```

is the skeleton structure for an alkane. Identify the alkane on which its IUPAC name would be based (e.g., propane, butane, etc.). Justify your choice.

(44) Explain the terms *saturated* and *unsaturated* as they are used in organic chemistry.

(45) Write the molecular formula for the alkane having 19 carbon atoms in its molecules. Explain how you determined this formula.

(46) Write a structural formula for the following compound: $CH_3(CH_2)_4CH_3$.

(47) Write the name of

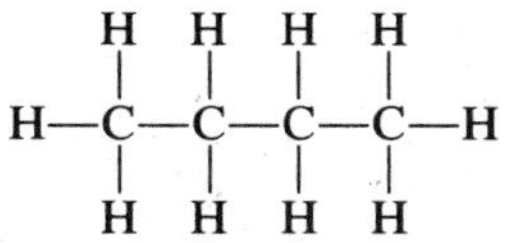

```
   H  H  H  H
   |  |  |  |
H—C—C—C—C—H
   |  |  |  |
   H  H  H  H
```

(48) Draw structural diagrams for all the isomers of heptane.

(49) What is the significance of the letter R in a formula such as R—Cl? Write two structural formulas that might be described by R—Cl.

(50) Redraw the skeleton structure of the alkane shown in Question 8 so its identification will be more evident.

Write the names of those compounds in Problems 9 to 16 and 51 to 58 whose structural formulas are given, and the structural formulas for the compounds whose names are given.

(9) 3-methylheptane

(10) 2,4-dimethyloctane

(51) 3-ethylhexane

(52) 2,2,4-trimethyl-3,6-diethyloctane

(11)
```
   |   |   |   |   |
 —C—C—C—C—C—
   |   |   |   |   |
       —C—
         |
       —C—
         |
```

(12)
```
   |   |   |   |   |   |   |
 —C—C—C—C—C—C—C—
   |   |   |   |   |   |   |
              Cl
```

(13)
```
  Br          Cl
   |   |   |   |   |   |
 —C—C—C—C—C—C—
   |   |   |   |   |   |
                  Cl
```

(14) 1,1-dibromoethane

(15) 1,1,4,5-tetrabromopentane

(16) 1,2,2-tribromo-1-chlorobutane

(53)
```
                 |
               —C—
                 |
               —C—
   |   |   |   |   |
 —C—C—C—C—C—
   |   |   |   |   |
               —C—
                 |
```

(54)
```
          Cl Cl
   |   |   |   |   |   |   |
 —C—C—C—C—C—C—C—
   |   |   |   |   |   |   |
              Cl
```

(55)
```
  Br  Cl
   |   |   |   |
 —C—C—C—C—Cl
   |   |   |   |
  Br      Cl
```

(56) 1,2,3-trichloropropane

(57) 2,3-dibromo-4-chlorohexane

(58) 1,1,1-tribromo-3-chloropentane

Section 21.4

(17) What structural feature identifies an alkene? Write the general formula for an alkene.

(18) Write structural formulas and names for the first three alkynes.

(19) Explain in words the difference among 1-hexene, 2-hexene, and 3-hexene.

(20) What are *cis-, trans-* isomers?

(21) What is the name of

```
CH3         C2H5
    \      /
     C=C
    /      \
   H        H
```

(59) What structural feature identifies an alkyne? Write the general formula for an alkyne.

(60) Write structural formulas and names for the first three alkenes.

(61) Write the structural formula for 3-hexyne.

(62) Write structural formulas for the *cis-* and *trans-* isomers of 3-hexene, and state which is which.

(63) What is the name of

```
   |   |   |   |        |   |
 —C—C—C—C≡C—C—C—
   |   |   |   |        |   |
```

Section 21.5

(22) Write the formula and name of the alkene that can be prepared by the dehydration of the following alcohol:

```
    H  H  H  H  H  H
    |  |  |  |  |  |
 H—C—C—C—C—C—C—OH
    |  |  |  |  |  |
    H  H  H  H  H  H
```

(64) Which of the two alcohol structures shown could be used to produce 2-butene? Explain.

(a)
```
      OH  H   H   H
       |  |   |   |
  H—C—C—C—C—H
       |  |   |   |
       H  H   H   H
```

(23) Write the structural formula and name of the alkene that can be prepared by the dehydrohalogenation of 1-chloropentane.

Section 21.6

(24) What structural feature allows a hydrocarbon to engage in a substitution reaction, but not an addition reaction? What must be present if there is to be an addition reaction?

(25) Write an equation for the hydrogenation of propene.

(*26*) What is polymerization?

(27) Chloroethylene is also known as vinyl chloride. When it polymerizes, it becomes the well-known polyvinyl chloride of which shower curtains, phonograph records, raincoats, and plastic pipe are made. Show how three vinyl chloride monomers would polymerize to form "PVC."

Section 21.7

(28) How do aromatic hydrocarbons differ from aliphatic hydrocarbons?

(29) Name the compounds represented by these three formulas:

(a) benzene ring with Cl at top and Cl at adjacent lower right position

(b) benzene ring with Cl at top and Cl at bottom

(c) benzene ring with Cl at top and Cl at lower left position

Section 21.9

(*30*) What is the functional group of an alcohol? Write both its name and formula. Write the general formula of an alcohol.

(31) Write the structural formula for 1-pentanol.

(b)

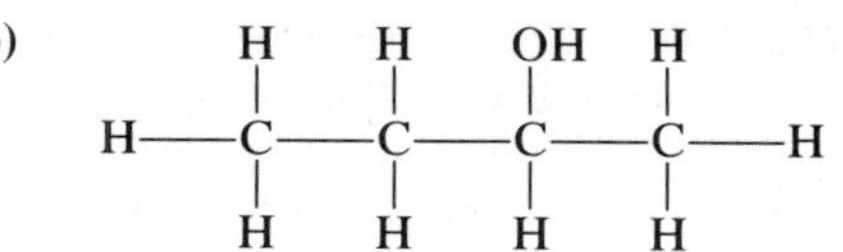

(65) Write the structural formula and name of two isomeric alkenes that could be formed by the dehydrohalogenation of 2-bromobutane.

(66) Explain why the chlorination of ethane yields chloroethane, whereas the chlorination of ethylene produces 1,2-dichloroethane.

(67) Write an equation for the hydrogenation of propyne.

(68) Distinguish between monomers and polymers.

(69) The popular kitchen material Saran Wrap is made from the copolymerization (polymerization of two monomers) of chloroethylene and vinylidene chloride, $CH_2{=}CCl_2$. Show how this copolymerization occurs, and write the structure through at least eight carbon atoms.

(70) Write the symbol used to represent the benzene ring and describe what is signified and/or understood about what it represents.

(71) Draw a structural formula for 1,3,5-trichlorobenzene.

(72) Account for the high boiling points and water solubility of alcohols compared to the same properties of alkanes of comparable molecular weight.

(73) Give the common and IUPAC name of

```
    H  H  H  H
    |  |  |  |
H—C—C—C—C—OH
    |  |  |  |
    H  H  H  H
```

(32) There must be at least three carbon atoms in a secondary alcohol. Explain why this is so. Write the structural formulas of two different secondary alcohols that have five carbon atoms.

(74) Write the structural formula of the tertiary alcohol that contains the smallest possible number of carbon atoms.

Section 21.10

(33) Show how water, alcohols, and ethers are structurally related.

(34) Write the structural formula for methyl propyl ether.

(75) Write a structural equation that shows how methyl ether might be prepared from an alcohol.

(76) "Each ether with two or more carbon atoms has an alcohol with which it is isomeric." Show that this statement is true.

Section 21.11

(35) Use structural formulas to show how an aldehyde is formed by oxidizing an alcohol.

(36) How are aldehydes and ketones alike, and how do they differ?

(77) Show why a ketone rather than an aldehyde is produced when a secondary alcohol is oxidized.

(78) Write structural formulas for butanal and butanone.

Section 21.12

(37) Write the general formula for a carboxylic acid, as well as the formula of the functional group.

(38) Account for the high boiling points of carboxylic acids compared to the boiling points of other compounds of comparable molecular weight.

(39) Demonstrate by equation the acid character of the carboxylic group.

(79) Write the structural formula for hexanoic acid. What is the name of C_4H_9COOH?

(80) Account for the high solubility of the lower carboxylic acids.

(81) Write the equation for the reaction between formic acid, HCOOH, and propanol, and name the ester formed.

Section 21.13

(40) Explain the relationship between ammonia and the amines.

(41) What is the name of

$$\begin{array}{c} CH_3{-}N{-}C_3H_7 \\ | \\ H \end{array}$$

(82) "Amines are bases in many chemical reactions." Explain why this is so.

(83) Write the structural formula of dibutylamine.

Section 21.14

(42) Compare the functional groups of carboxylic acids and amides.

(84) Explain by structural formula how a peptide link is formed.

Miscellaneous Questions

(85) Distinguish precisely and in scientific terms the differences among items in each of the following groups:

(a) Organic chemistry, inorganic chemistry
(b) Saturated and unsaturated hydrocarbons
(c) Alkanes, alkenes, alkynes
(d) Normal alkane, branched alkane
(e) *Cis-*, *trans-*, isomers
(f) Structural formula, condensed (line) formula, molecular formula

(g) Addition reaction, substitution reaction
(h) Alkane, alkyl group
(i) Monomer, polymer
(j) Aliphatic hydrocarbon, aromatic hydrocarbon
(k) *Ortho-, meta-, para-*
(l) Alcohol, aldehyde, carboxylic acid
(m) Hydroxyl group, carbonyl group, carboxyl group
(n) Primary, secondary, and tertiary alcohols
(o) Alcohol, ether
(p) Aldehyde, ketone
(q) Carboxylic acid, ester
(r) Carboxylic acid, amide
(s) Amine, amide
(t) Primary, secondary, tertiary amine

(86) Classify each of the following statements as true or false:
(a) To be classified as organic, a compound must be or have been a part of a living organism.
(b) Carbon atoms normally form four bonds in organic compounds.
(c) Only an unsaturated hydrocarbon can engage in an addition reaction.
(d) Members of a homologous series differ by a distinct structural unit.
(e) Alkanes, alkenes, and alkynes are unsaturated hydrocarbons.
(f) Alkyl groups are a class of organic compounds.
(g) Isomers have the same molecular formulas but different structural formulas.
(h) *Cis-, trans-* isomerism appears among alkenes but not alkynes.
(i) Aliphatic hydrocarbons are unsaturated.
(j) Aromatic hydrocarbons have a ring structure.
(k) An alcohol has one alkyl group bonded to an oxygen atom, and an ether has two.
(l) Carbonyl groups are found in alcohols and aldehydes.
(m) An ester is an aromatic hydrocarbon, made by the reaction of an alcohol with a carboxylic acid.
(n) An amine has one, two, or three alkyl groups substituted for hydrogens in an ammonia molecule.
(o) An amide has the structure of a carboxylic acid, except that $—NH_2$ replaces —OH in the carboxyl group.
(p) A peptide link arises when a water molecule forms from a hydrogen from the $—NH_2$ group of one amino acid molecule and an —OH from another amino acid molecule.

(87) Criticize the statement: "Carbon atoms in a normal alkane lie in a straight line."

(88) C_8H_{16} and C_5H_{12} are formulas of compounds in two homologous series. Give the formulas of the next members in each series. Name the class of hydrocarbon represented by each compound.

(89) What are the shapes of the following: (a) *cis*-dichloroethylene; (b) acetylene; (c) *n*-butane?

(90) Is a molecule of $HC{\equiv}CCH_3$ planar? Justify your answer.

(91) Which one or more among the following are most likely to be soluble in water: alkanes, alkenes, alkynes, alcohols? Why?

(92) In which acid is the H—O bond stronger, H_2SO_4 or CH_3COOH? How do you know?

(93) Use Lewis diagrams to show how methyl propyl ether might be formed from the dehydration of two alcohols.

(94) Write the name and formula of the alcohol formed by the combination of hydrogen with methyl ethyl ketone. Is it a primary or secondary alcohol?

(95)* Write the structural formula for alanylglycine, the dipeptide that forms by the combination of alanine, $CH_3CH(NH_2)COOH$, and glycine, $CH_2(NH_2)COOH$.

Appendix I
Chemical Calculations

A beginning student in chemistry is assumed to have developed calculation skills in his or her mathematics classes. Often chemistry is the first occasion for these skills to be put to the test of practical application. Experience shows that many students who may have learned these skills at one time, but have not used them regularly, can profit from a review of basic concepts. Others can benefit from a handy reference to the calculation techniques used in chemistry. This section of the Appendix is written to meet these needs.

PART A NUMBERS

When we count objects, the result is given in a **number.** One, two three, etc., are **counting numbers,** usually written as **numerals 1, 2, 3.** . . . Counting numbers are also called **whole numbers, integers,** and **integral numbers.** Zero is an integer. Numbers may be **positive** or **negative.** Numbers that lie between integers are called **fractions.** Fractions are sometimes written as a **ratio** of integers, such as ½, ¾, and 5/3. The number above the line is the **numerator,** and the number below the line is the **denominator.** In a **proper fraction** the numerator is smaller than the denominator; if the numerator is larger than the denominator, it is an **improper fraction.** Any fraction in which the numerator is the same as or equal to the denominator has a value of 1.

In chemistry, fractions are usually written as **decimal fractions.** Whole numbers appear to the left of the decimal point, and fractional parts on the right. The positions to the left and right of the decimal are **place values** or **digit values.** Reading left from the decimal point, the place values show, in order, the number of ones or units, tens, hundreds, etc.; from the decimal to the right are the tenths, hundredths, thousandths, etc. Each place value is ten times larger than the place value to its right and is said to be larger by one **order of magnitude** or a **multiple of 10.** For numbers less than 1, orders of magnitude are called **submultiples of 10.**

It has long been customary in the United States to write commas after every three orders of magnitude in a large number. One billion, for example, is written 1,000,000,000. Commas are not used to separate groups of three orders of magnitude in numbers less than 1; one millionth is 0.000001. Other nations use different "punctuation" in writing numbers. There is, at present, an international effort to standardize number writing by placing a space between each three orders of magnitude on both sides of the decimal point. By this procedure the number 32,845,702.83785 is written 32 845 702.837 85, and the number 0.04237986 is written 0.042 379 86. The space is often omitted for numbers that begin or end four spaces to the left or right of the decimal. Thus 4 968 is written 4968, and 0.589 3 is written 0.5893. This type of spacing in large and small numbers is used throughout this book.

Another standard practice in scientific writing is to place a zero before the decimal point of a decimal fraction. Thus .5893 is written 0.5893.

PART B THE HAND CALCULATOR

Today, every serious chemistry student uses a calculator to solve chemistry problems. A suitable calculator can (a) add, subtract, multiply, and divide; (b) perform these operations in exponential notation; (c) work with logarithms; and (d) raise any base to any power. Calculators that can do these things usually have other capabilities, too, such as squares and square roots, trigonometric functions, short cuts for pi and percentage, enclosures, statistical features, and different levels of storage and recall.

Most calculators operate with one of two *logic* systems, each with its own order of operations. One is called the Algebraic Operating System (AOS), and the other is Reverse Polish Notation (RPN). In the examples that follow, we will give *general* keyboard sequences for both systems as they are performed on calculators that are popularly used by students. Different brands may vary in details, particularly when some keys are used for more than one function. Some calculators offer an option on the number of digits to be displayed after the decimal point. With or without such an option, the number varies on different calculators. Accordingly, answers in this book may differ slightly from yours. Please consult the instruction book that accompanied your calculator for specific directions on these or other variations that may appear.

(You may wonder why you should not simply use your instruction book rather than the suggestions that follow. For complete mastery of your calculator, you should do just that. If your present purpose is to learn how to use the calculator for chemistry, these instructions will be much easier. They also include practical suggestions that do not appear in a formal instruction book.)

One precautionary note before we begin: *A calculator should* never *be used as a substitute for thinking*. If a problem is simple and can be solved mentally, do it "in your head." You will make fewer mistakes. If you use your calculator, *think* your way through each problem and estimate the answer mentally. Suggestions on approximating answers are given in Part H of this Appendix. If the calculator answer appears reasonable, round it off properly and write it down. Then run through the calculation again to be sure you haven't made a keyboard error. Your calculator is an obedient and faithful servant that will do exactly what you tell it to do, but it is not responsible for the mistakes you make in your instructions.

It was suggested above that you round off your answer properly before recording it. This is because calculator answers to many problems are limited in length only by the display. For example, 273 ÷ 45.6 = 5.986 842 1 on one calculator. Some calculators can show more numbers, and therefore do. Usually only the first three or four digits have meaning, and the others should be discarded. Procedures for deciding how many digits to write are given in Section 3.6, on significant figures.

Reciprocals, Square Roots, Squares, and Logarithms

Finding the reciprocal, square root, or square of a number are called one-number functions because only one number must be keyed into the calculator. In this case we are interested in finding $1/x$, $\sqrt{x}$, x^2, and log x. The procedures are the same for both operating systems:

1. Key in x.
2. Press function key ($1/x$, $\sqrt{x}$, x^2, or log x)

Example: Find 1/12.34, $\sqrt{12.34}$, 12.34^2, and log 12.34.

SOLUTION:

PROBLEM	PRESS	DISPLAY
1/12.34	12.34	*12.34*
	1/x	*0.0810373*
$\sqrt{12.34}$	12.34	*12.34*
	$\sqrt{\ }$	*3.5128336*
12.34^2	12.34	*12.34*
	x^2	*152.2756*
log 12.34	12.34	*12.34*
	log	*1.0913152*

Antilogarithms, 10^x, and y^x

If x is the base 10 logarithm of a number, N, then N = antilog x = 10^x. Some calculators have a

10^x key that makes finding an antilogarithm a one-number function. Other calculators use an inverse function key, sometimes marked INV, to reverse the logarithm function. Again, finding an antilogarithm is a one-number function, but two function keys are used.

Example: Find the antilogarithm of 3.19.

SOLUTION:

PRESS	DISPLAY
3.19	3.19
10^x	1548.8166

PRESS	DISPLAY
3.19	3.19
INV	3.19
log	1548.8166

The y^x key can be used to raise any base, y, to any power, x. The procedure differs on the two operating systems, as the following example shows:

Example: Calculate $8.25^{0.413}$

AOS LOGIC

PRESS	DISPLAY
8.25	8.25
y^x	8.25
.413	0.413
=	2.3905371

RPN LOGIC

PRESS	DISPLAY
8.25	8.25
ENTER	8.25
.413	0.413
y^x	2.3905371

Notice that, even though we always write a decimal fraction less than 1 with a zero before the decimal point, it is not necessary to enter such zeros into a calculator. The zeros are included in the display.

The y^x key can also be used to find an antilogarithm. In that case y = 10, and x is the given logarithm, the exponent to which 10 is to be raised.

Addition, Subtraction, Multiplication, and Division

For ordinary arithmetic operations, the procedures for AOS and RPN logics are as follows:

AOS LOGIC

The procedure for a common arithmetic operation is identical to the arithmetic equation for the same calculation. If you wish to add X to Y, the equation is X + Y = . The calculator procedure is:

1. Key in X.
2. Press the function key (+ for addition).
3. Key in Y.
4. Press =.

The display will show the calculated result.

Example: Solve 12.34 + 0.0567 = ?

SOLUTION:

PRESS	DISPLAY
12.34	12.34
+	12.34
.0567	0.0567
=	12.3967

RPN LOGIC

The procedure for a common arithmetic operation is to key in *both* numbers and then tell the calculator what to do with them. If you wish to add X to Y, you enter X, key in Y, and instruct the calculator to add. The procedure is:

1. Key in X.
2. Press ENTER.
3. Key in Y.
4. Press the function key (+ for addition).

The display will show the calculated result.

Example: Solve 12.34 + 0.0567 = ?

SOLUTION:

PRESS	DISPLAY
12.34	12.34
ENTER	12.34
.0567	0.0567
+	12.3967

The numbers in the PRESS column are, in order, Steps 1, 2, 3, and 4 of the procedure. In Step 2, the function key is − for subtraction, × for multiplication, and ÷ by division.

The numbers in the PRESS column are, in order, Steps 1, 2, 3, and 4 of the procedure. In Step 4, the function key is − for subtraction, × for multiplication, and ÷ for division.

You may wish to confirm the following results on your calculator: 12.34 − 0.0567 = 12.2833; 12.34 × 0.0567 = 0.699 678; 12.34 ÷ 0.0567 = 217.636 68.

Chain Calculations

A "chain calculation" is a series of two or more operations performed on three or more numbers. To the calculator, the sequence is a series of two-number operations in which the first number is always the result of all calculations completed to that point. For example, in X + Y − Z, the calculator first finds X + Y = A. The quantity A is already in and displayed by the calculator. All that needs to be done is to subtract Z from it. In the following example we deliberately begin with a negative number to illustrate the way such a number is introduced to the calculator. You simply press in the number, followed by the key that changes the sign, usually +/− or CHS. The example: −2.45 + 18.7 + 0.309 − 24.6 = ?

AOS LOGIC

PRESS	DISPLAY
2.45	2.45
+/−	−2.45
+	−2.45
18.7	18.7
+	16.25
.309	0.309
−	16.559
24.6	24.6
=	−8.041

RPN LOGIC

PRESS	DISPLAY
2.45	2.45
CHS	−2.45
ENTER	−2.45
18.7	18.7
+	16.25
.309	0.309
+	16.559
24.6	24.6
−	−8.04

Notice that it is not necessary to press the = key after each step *in a chain calculation involving only addition and/or subtraction*.

Combinations of multiplication and division are handled the same way. To solve 9.87 × 0.0654 ÷ 3.21:

AOS LOGIC

PRESS	DISPLAY
9.87	9.87
×	9.87
0.654	0.0654
+	0.645498
3.21	3.21
=	.20108972

RPN LOGIC

PRESS	DISPLAY
9.87	9.87
ENTER	9.87
.0654	0.0654
×	0.0645498
3.21	3.21
÷	0.20108972

Notice that it is not necessary to press the = key after each step *in a chain calculation involving only multiplication and/or division*.

Combination multiplication/division problems similar to the one above usually appear in the form of fractions in which all multipliers are in the numerator and all divisors are in the denominator. Thus 9.87 × 0.0654 ÷ 3.21 is the same as $\frac{9.87 \times 0.0654}{3.21}$. In chemistry there are often several numerator factors and several denominator factors. A simple calculation that is easily completed "in your head" brings out some important facts about using calculators for chain calculations. Mentally, right now, calculate $\frac{9 \times 4}{2 \times 6} = ?$

There are several ways to get the answer. The most probable one, if you do it mentally, is to multiply 9 × 4 = 36 in the numerator, and then multiply 2 × 6 = 12 in the denominator. This changes the problem to 36/12. Dividing 36 by 12 gives 3 for the answer. This perfectly correct approach is often followed by the beginning calculator user when faced with numbers that cannot be multiplied and divided mentally. It is not the best method, however. It is longer and there is greater probability of error than is necessary.

In solving a problem such as (9 × 4)/(2 × 6) you can begin with any number and perform the required operations with other numbers in any order. Logically you begin with one of the numerator factors. That gives you 12 different calculation sequences that yield the correct answer. They are

9 × 4 ÷ 2 ÷ 6	4 × 9 ÷ 2 ÷ 6
9 × 4 ÷ 6 ÷ 2	4 × 9 ÷ 6 ÷ 2
9 ÷ 2 ÷ 6 × 4	4 ÷ 2 ÷ 6 × 9
9 ÷ 2 × 4 ÷ 6	4 ÷ 2 × 9 ÷ 6
9 ÷ 6 × 4 ÷ 2	4 ÷ 6 × 9 ÷ 2
9 ÷ 6 ÷ 2 × 4	4 ÷ 6 ÷ 2 × 9

Practice a few of these sequences on your calculator to see how freely you may choose.

There is a common error in chain calculations that you should avoid. This is to interpret the above problem as 9 × 4 divided by 2 × 6, which is correct, but then punch it into the calculator as 9 × 4 ÷ 2 × 6, which is not correct. The calculator interprets these instructions as 9 × 4 = 36; 36 ÷ 2 = 18; 18 × 6 = 108. The last step should be 18 ÷ 6 = 3, as in the first setup in the list above. *In chain calculations you must always* divide *by each factor in the denominator*.

There are over a hundred different sequences by which $\frac{7.83 \times 86.4 \times 291}{445 \times 807 \times 0.302}$ can be calculated. Practice some of them and see if you can duplicate the answer, 1.815 214 7.

You have seen that in multiplication and division you can take the factors in any order. This is possible in addition and subtraction too, provided that you keep each positive and negative sign with the number that follows it and treat the problem as an algebraic addition of signed numbers. When you mix addition/subtraction with multiplication/division, however, you must obey the rules that govern the order in which arithmetic operations are performed. Briefly, these rules are:

1. Simplify all expressions enclosed in parentheses.
2. Complete all multiplications and divisions.
3. Complete all additions and subtractions.

If your calculator is able to store and recall numbers, it can solve problems with a very complex order of operations. In this book you will find no such problems, but only those that require the simplest application of the first rule. Our comments will be limited to that application, and we will not use the storage capacity of your calculator, as the instruction book would probably recommend.

A typical calculation is

$$6.02 \times (22.1 - 48.6) \times 0.134.$$

Recalling that factors in a multiplication problem may be taken in any order, you rearrange the numbers so the enclosed factor appears first:

$$(22.1 - 48.6) \times 6.02 \times 0.134.$$

You may then perform the calculation in the order in which the numbers appear:

AOS LOGIC		RPN LOGIC	
PRESS	DISPLAY	PRESS	DISPLAY
22.1	22.1	22.1	22.1
−	22.1	ENTER	22.1
48.6	48.6	48.6	48.6
=	−26.5	−	−26.5
×	−26.5	6.02	6.02
6.02	6.02	×	−159.53
×	−159.53	.134	0.134
.134	0.134	×	−21.37702
=	−21.37702		

Notice that, in a chain calculation involving *both* addition/subtraction *and* multiplication/division, *it is necessary to press the = key after each addition or subtraction sequence in the AOS logic system, before you proceed to a multiplication/division.*

Sometimes a factor in parentheses appears in the denominator of a fraction, where it is not easily taken as the first factor to be entered into the calculator. Again such problems in this book are relatively simple. They may be solved by working the problem upside down, and at the end using the 1/x key to turn it right side up. The process is demonstrated in calculating

$$\frac{13.3}{2.59(88.4 - 27.2)}$$

The procedure is to calculate $\frac{2.59(88.4 - 27.2)}{13.3}$ and find the reciprocal of the result.

AOS LOGIC

PRESS	DISPLAY
88.4	88.4
−	88.4
27.2	27.2
=	61.2
×	61.2
2.59	2.59
÷	158.508
13.3	13.3
=	11.9 17895
1/x	.08390744

RPN LOGIC

PRESS	DISPLAY
88.4	88.4
ENTER	88.4
27.2	27.2
−	61.2
2.59	2.59
×	158.508
13.3	13.3
÷	11.9179
1/x	0.0839

Exponential Notation

Modern calculators use exponential notation (Appendix I, Part F) for very large or very small numbers. Ordinarily, if numbers are entered as decimal numbers, the answer appears as a decimal number. If the answer is too large or too small to be displayed, it "overflows" or "underflows" into exponential notation automatically. If you want the answer in exponential notation, even if the calculator can display it as a decimal number, you instruct the machine accordingly. The symbol on the key for this instruction varies with different calculators. EE and EEX are common.

A number shown in exponential notation has a space in front of the last two digits, which are at the right side of the display. These last two digits are the exponent. Thus 4.68×10^{14} is displayed as 4.68 14; 2.39×10^{6} is 2.39 06. If the exponent is negative, a minus sign is present; 4.68×10^{-14} is 4.68 − 14.

To key 4.68×10^{14} and 4.68×10^{-14} into a calculator, proceed as follows:

AOS LOGIC

PRESS	DISPLAY
4.68	4.68
EE	4.68 00
14	4.68 14

RPN LOGIC

PRESS	DISPLAY
4.68	4.68
EEX	4.68 00
14	4.68 14

The calculator display now shows 4.68×10^{14}. Use the next step only if you wish to change to a negative exponent, 4.68×10^{-14}.

PRESS	DISPLAY
+/−	4.68 −14
Function key: +, −, ×, or ÷	4.68 −14

PRESS	DISPLAY
CHS	4.68 −14
ENTER	4.68 −14

The calculator is now ready for the next number to be keyed in, as in earlier examples.

In the section on exponential notation (Appendix I, Part F) you will find specific examples showing how to use your calculator to solve problems with these numbers. In Chapter 17 you will combine exponential notation with logarithms.

PART C ARITHMETIC AND ALGEBRA

We present here a brief review of arithmetic and algebra to the point that it is used or assumed in this text. Formal mathematical statement and development are avoided. The only purpose of this section is to refresh your memory in those areas where it may be needed.

1. ADDITION. a + b. Example: 2 + 3 = 5. The result of an addition is a **sum.**

2. SUBTRACTION. a − b. Example: 5 − 3 = 2. Subtraction may be thought of as the addition of a negative number. In that sense, a − b = a + (−b). Example:

$$5 - 3 = 5 + (-3) = 2.$$

The result of a subtraction is a **difference.**

3. MULTIPLICATION. a × b = ab = a · b = a(b) = (a)(b) = b × a = ba = b · a = b(a). The foregoing all mean that **factor** a is to be multiplied by factor b. Reversing the sequence of the factors, a × b = b × a, indicates that factors may be taken in any order when two or more are multiplied together. Examples:

$$2 \times 3 = 2 \cdot 3 = 2(3) = (2)(3) = 3 \times 2 = 3 \cdot 2 = 3(2) = 6.$$

The result of a multiplication is a **product.**

Grouping of factors: (a)(b)(c) = (ab)(c) = (a)(bc). Factors may be grouped in any way in multiplication. Example:

$(2)(3)(4) = (2 \times 3)(4) = (2)(3 \times 4) = 24.$

Multiplication by 1: $n \times 1 = n$. If any number is multiplied by 1, the product is the original number. Examples:

$6 \times 1 = 6; \quad 3.25 \times 1 = 3.25.$

Multiplication of fractions:

$$\frac{a}{b} \times \frac{c}{d} \times \frac{e}{f} = \frac{ace}{bdf}$$

If two or more fractions are to be multiplied, the product is equal to the product of the numerators divided by the product of the denominators. Example:

$$4 \times \frac{9}{2} \times \frac{1}{6} = \frac{4}{1} \times \frac{9}{2} \times \frac{1}{6}$$
$$= \frac{4 \times 9 \times 1}{1 \times 2 \times 6} = \frac{36}{12} = 3$$

4. DIVISION. $a \div b = a/b = \frac{a}{b}$. The foregoing all mean that a is to be divided by b. Example: $12 \div 4 = 12/4 = \frac{12}{4} = 3$. The result of a division is a **quotient.**

Special case: If any number is divided by the same or an equal number, the quotient is equal to 1. Examples: $\frac{4}{4} = 1; \frac{8-3}{4+1} = \frac{5}{5} = 1; \frac{n}{n} = 1.$

Division by 1. $\frac{n}{1} = n$. If any number is divided by 1, the quotient is the original number. Examples: $\frac{6}{1} = 6; \frac{3.25}{1} = 3.25$. From this it follows that any number may be expressed as a fraction having 1 as the denominator. Examples:

$$4 = \frac{4}{1}; 9.12 = \frac{9.12}{1}; m = \frac{m}{1}.$$

5. RECIPROCALS. If n is any number, the reciprocal of n is $\frac{1}{n}$; if $\frac{a}{b}$ is any fraction, the reciprocal of $\frac{a}{b}$ is $\frac{b}{a}$. The first part of the foregoing sentence is actually a special case of the second part: If n is any number, it is equal to $\frac{n}{1}$. Its reciprocal is therefore $\frac{1}{n}$.

A reciprocal is sometimes referred to as the **inverse** (more specifically, the multiplicative inverse) of a number. This is because the product of any number multiplied by its reciprocal equals 1. Examples:

$$2 \times \frac{1}{2} = \frac{2}{2} = 1 \qquad n \times \frac{1}{n} = \frac{n}{n} = 1$$

$$\frac{4}{3} \times \frac{3}{4} = \frac{12}{12} = 1 \qquad \frac{m}{n} \times \frac{n}{m} = \frac{mn}{mn} = 1$$

Division may be regarded as multiplication by a reciprocal:

$$a \div b = \frac{a}{b} = a \times \frac{1}{b}.$$

Example: $6 \div 2 = \frac{6}{2} = 6 \times \frac{1}{2} = 3.$

$$a \div b/c = \frac{a}{b/c} = a \times \frac{c}{b}.$$

Example: $6 \div \frac{2}{3} = \frac{6}{2/3} = 6 \times \frac{3}{2} = 9.$

6. SUBSTITUTION. If $d = b + c$, then $a(b + c) = ad$. Any number or expression may be substituted for its equal in any other expression. Example: $7 = 3 + 4$. Therefore $2(3 + 4) = 2 \times 7$.

7. "CANCELLATION." $\frac{ab}{ca} = \frac{\cancel{a}b}{c\cancel{a}} = \frac{b}{c}$. The process commonly called **cancellation** is actually a combination of grouping of factors (see 3), substitution (see 6) of 1 for a number divided by itself (see 4), and multiplication by 1 (see 3). Note the steps in the following examples:

$$\frac{xy}{yz} = \frac{yx}{yz} = \left(\frac{y}{y}\right)\left(\frac{x}{z}\right) = 1 \cdot \frac{x}{z} = \frac{x}{z}$$

$$\frac{24}{18} = \frac{6 \times 4}{6 \times 3} = \frac{6}{6} \times \frac{4}{3} = 1 \times \frac{4}{3} = \frac{4}{3}$$

Note that only *factors,* or *multipliers,* can be canceled. There is no cancellation in $\frac{a + b}{a + c}$.

8. EXPONENTIALS. An exponential has the form B^p, where B is the **base** and p is the **power** or **exponent.** An exponential indicates the number of times the base is used as a factor in multiplication. For example, 10^3 means 10 is to be used as a factor 3 times:

$$10^3 = 10 \times 10 \times 10 = 1000.$$

A negative exponent tells the number of times a base is used as a divisor. For example, 10^{-3} means 10 is used as a divisor 3 times:

$$10^{-3} = \frac{1}{10} \times \frac{1}{10} \times \frac{1}{10} = \frac{1}{10^3}$$

This also shows that an exponential may be moved in either direction between the numerator and denominator by changing the sign of the exponent:

$$\frac{1}{2^3} = 2^{-3} \qquad 3^4 = \frac{1}{3^{-4}}$$

Multiplication of exponentials having the same base: $a^m \times a^n = a^{m+n}$. To multiply exponentials, add the exponents. Example:

$$10^3 \times 10^4 = 10^7.$$

Division of exponentials having the same base: $a^m \div a^n = \frac{a^m}{a^n} = a^{m-n}$. To divide exponentials, subtract the denominator exponent from the numerator exponent. Example:

$$10^7 \div 10^4 = \frac{10^7}{10^4} = (10^7)(10^{-4}) = 10^{7-4} = 10^3.$$

Zero power: $a^0 = 1$. Any base raised to the zero power equals 1. The fraction $\frac{a^m}{a^m} = 1$ because the numerator is the same as the denominator. By division of exponentials, $\frac{a^m}{a^m} = a^{m-m} = a^0$.

Raising a product to a power: $(ab)^n = a^n \times b^n$. When the product of two or more factors is raised to some power, each factor is raised to that power. Example:

$$(2 \times 5y)^3 = 2^3 \times 5^3 \times y^3 = 8 \times 125 \times y^3 = 1000y^3$$

Raising a fraction to a power: $\left(\frac{a}{b}\right)^n = \frac{a^n}{b^n}$. When a fraction is raised to some power, the numerator and denominator are both raised to that power. Example:

$$\left(\frac{2x}{5}\right)^3 = \frac{2^3x^3}{5^3} = \frac{8x^3}{125} = 0.064x^3$$

Square root of exponentials: $\sqrt{a^{2n}} = a^n$. To find the square root of an exponential, divide the exponent by 2. Example: $\sqrt{10^6} = 10^3$. If the exponent is odd, see below.

Square root of a product: $\sqrt{ab} = \sqrt{a} \times \sqrt{b}$. The square root of the product of two numbers equals the product of the square roots of the numbers. Example:

$$\sqrt{9 \times 10^{-6}} = \sqrt{9} \times \sqrt{10^{-6}} = 3 \times 10^{-3}$$

Using this principle, by adjusting a decimal point you may take the square root of an exponential having an odd exponent. Example:

$$\sqrt{10^5} = \sqrt{10 \times 10^4} = \sqrt{10} \times \sqrt{10^4} = 3.16 \times 10^2$$

The same technique may be used in taking the square root of a number expressed in exponential notation. Example:

$$\sqrt{1.8 \times 10^{-5}} = \sqrt{18 \times 10^{-6}} = \sqrt{18} \times \sqrt{10^{-6}} = 4.2 \times 10^{-3}$$

9. SOLVING AN EQUATION FOR AN UNKNOWN QUANTITY: Most problems in this book can be solved by dimensional analysis methods. There are times, however, when algebra must be used, particularly in relation to the gas laws. Solving an equation for an unknown involves rearranging the equation so that the unknown is the only item on one side and only known quantities are on the other. "Rearranging" an equation may be done in several ways, but the important thing is that *whatever is done to one side*

of the equation must also be done to the other. The resulting relationship remains an equality, a true equation. Among the operations that may be performed on both sides of an equation are addition, subtraction, multiplication, division, and raising to a power, which includes taking square root.

In the following examples, a, b, and c represent known quantities, and x is the unknown. The object in each case is to solve the equation for x. The steps of the algebraic solution are shown, as well as the operation performed on both sides of the equation. Each example is accompanied by a practice problem that is solved by the same method. You should be able to solve the problem, even if you have not yet reached that point in the book where such a problem is likely to appear. Answers to these practice problems may be found at the end of Appendix I.

(1) $x + a = b$

$x + a - a = b - a$ Subtract a.

$x = b - a$ Simplify.

PRACTICE: *1) If* $P = p_{O_2} + p_{H_2O}$, *find* p_{O_2}, *when* $P = 748$ *torr and* $p_{H_2O} = 24$ *torr.*

(2) $ax = b$

$\frac{\not{a}x}{\not{a}} = \frac{b}{a}$ Divide by a which is called the **coefficient** of x.

$x = \frac{b}{a}$ Simplify.

PRACTICE: *2) At a certain temperature,* $PV = k$. *If* $P = 1.23$ *atm and* $k = 1.62$ *L atm, find V.*

(3) $\frac{x}{a} = b$

$\frac{\not{a}x}{\not{a}} = ba$ Multiply by a.

$x = ba$ Simplify.

PRACTICE: *3) In a fixed volume,* $\frac{P}{T} = k$. *Find P if* $k = \frac{2.4 \text{ torr}}{K}$ *and* $T = 300$ *K.*

(4) $\frac{a}{x} = b$

$\frac{\not{x}a}{\not{x}} = bx$ Multiply by x.

$a = bx$ Simplify.

$x = \frac{a}{b}$ Divide by b.

PRACTICE: *4) At what value of T will* $P = 760$ *torr when* $k = \frac{2.4 \text{ torr}}{K}$ *and* $\frac{P}{T} = k$?

(5) $\frac{a}{x} = \frac{b}{c}$

$\frac{c\not{x}}{b} \cdot \frac{a}{\not{x}} = \frac{\not{b}}{\not{c}} \cdot \frac{\not{c}x}{\not{b}}$ Multiply by $\frac{cx}{b}$.

$\frac{ac}{b} = x$ Simplify.

PRACTICE: *5) For gases at constant volume,* $\frac{P_1}{T_1} = \frac{P_2}{T_2}$. *If* $P_1 = 0.80$ *atm at* $T_1 = 320$ *K, at what value of* T_2 *will* $P_2 = 1.00$ *atm?*

(6) $\frac{a}{b + x} = c$

$(b + x)\frac{a}{(b + x)} = c(b + x)$ Multiply by $(b + x)$.

$a = c(b + x)$ Simplify.

$\frac{a}{c} = \frac{\not{c}(b + x)}{\not{c}}$ Divide by c.

$\frac{a}{c} = b + x$ Simplify.

$\frac{a}{c} - b = x$ Subtract b.

PRACTICE: *6) In how many grams of water must you dissolve 20.0 g of salt to make a 25% solution? The formula is*

$$\frac{g \text{ salt}}{g \text{ salt} + g \text{ water}} \times 100 = \%; \text{ or}$$

$$\frac{g \text{ salt}}{g \text{ salt} + g \text{ water}} = \frac{\%}{100}$$

PART D PROPORTIONALITIES

It is quite common, both in the laboratory and out, to find two quantities that are related to each other in a constant way. The weight of milk you carry out of a grocery store depends on the number of half-gallon cartons you buy. Four cartons weigh twice as much as two, and three weigh half as much as six. Mathematically we say that the weight is **directly proportional** to the number of cartons. Using W for weight, C for number of cartons, and the symbol $\propto$ for "is proportional to," the proportionality may be written

$$W \propto C \qquad \text{(AP.1)}$$

A proportionality may be changed to an equation by introducing a **proportionality constant,** k:

$$W = kC \qquad \text{(AP.2)}$$

Quite often a proportionality constant has units and physical significance. It does here. Solving Equation AP.2 for k gives

$$k = \frac{W}{c} \qquad \text{(AP.3)}$$

Equation AP.3 shows the "constant way" in which weight is related to the number of cartons: the ratio between them is constant. If three cartons weigh 14.4 lb, five cartons weigh 24.0 lb, and eight cartons weigh 38.4 lb, the ratios of pounds to cartons is

$$\frac{38.4 \text{ lb}}{8 \text{ cartons}} = \frac{24.0 \text{ lb}}{5 \text{ cartons}} = \frac{14.4 \text{ lb}}{3 \text{ cartons}},$$

which all reduce to $\frac{4.8 \text{ lb}}{1 \text{ carton}}$. Thus the value of k, its units, and its physical significance are 4.8 lb/carton.

Because the ratio of weight to number of cartons is constant, the ratios can be set equal to each other and the equation can be used to find the weight of any number of cartons. If you want to know the weight of seven cartons, you can set any known ratio equal to $\frac{x}{7}$, where x is the required weight, and solve. Using $\frac{24.0 \text{ lb}}{5 \text{ cartons}}$,

$$\frac{24.0 \text{ lb}}{5 \text{ cartons}} = \frac{x \text{ lb}}{7 \text{ cartons}}$$

$$x \text{ lb} = \frac{24.0 \text{ lb} \times 7 \text{ cartons}}{5 \text{ cartons}} = 33.6 \text{ lb}$$

This result is easily checked by reason. If each carton weighs 4.8 lb, then 7 cartons must weigh $7 \times 4.8 = 33.6$ lb.

Problems set up and solved this way are called "ratio and proportion" problems. This approach can be used in chemistry, and once was quite popular. Now dimensional analysis (Section 3.2) is much more common. Dimensional analysis can be used to change one quantity to another whenever the two quantities are proportional to each other. Proportional relationships commonly appear in experimental data.

It can be shown that whenever one variable is proportional to two others, it is also proportional to their product. The number of acres, A, in a rectangular field is proportional to both the length, L, and width, W, of the field:

$A \propto L$ and $A \propto W$; therefore, $A \propto L \times W$

Introducing a proportionality constant, k′,

$$A = k'LW$$

Solving for k′,

$$k' = \frac{A}{LW}$$

which has the physical meaning of acres per square foot. The value of k′ is $\frac{1}{43\,560}$, which is the inverse of the number of square feet in an acre.

In a direct proportionality, if one variable rises, the other rises too. An **inverse proportionality** is exactly the opposite; as one variable increases, the other decreases. A common example is the time it takes to drive from one point to another at different speeds. At 40 miles per hour (mph) it will take half as long as at 20 mph. As the speed is doubled, the time is cut in half. At 60 mph—three times as fast as 20 mph—the time is one-third. If r is speed and t is time, the inverse proportionality is written

$$r \propto \frac{1}{t}$$

This time we will use d for the proportionality constant:

$$r = d\left(\frac{1}{t}\right)$$

Solving for d gives

$$d = rt \qquad \text{(AP.4)}$$

which you may recognize as the familiar distance = rate × time equation. The physical significance of d in Equation AP.4 is distance. Notice that in an inverse proportionality it is the product of the variables that is constant.

Not all proportionality constants have both units and significance. The universal gas constant, R (Section 13.7), is made up of a combination of units that have no apparent meaning. A very familiar proportionality constant comes from the proportionality between the circumference and diameter of a circle: $c \propto d$. The constant is π, as in the equation $c = \pi d$. Solving for π gives a ratio of lengths. The units cancel, leaving a dimensionless, or pure, number.

Summary. Most of what appears here is background information to help you understand proportionality relationships as they appear in the text. The important things to recognize are:

1. A mathematical statement of proportionality may be changed to an equation by introducing a proportionality constant.
2. A proportionality constant may have units and physical meaning, which can be determined by solving the equation for the constant.
3. If one variable is proportional to two other variables, it is proportional to the product of those variables.
4. If A is *directly* proportional to B, then the ratio A/B is constant. If A increases, B increases; if A decreases, B decreases.
5. If A is *inversely* proportional to B, then the product A × B is constant. If A increases, B decreases; if A decreases, B increases.

PART E PER AND PERCENT

If you drive 100 miles in 4 hours, what is your average speed? Twenty-five miles per hour, you say. Fine. But how did you get it? Was it by *dividing* 100 by 4? In other words, 100 miles *per* 4 hours is the same as 25 miles *per* 1 hour, or $100 \div 4 = 25$, or $\frac{100}{4} = \frac{25}{1}$. The results of many division problems are expressed as some quantity *per unit*—per *one*—or something else. Thus the term *per* means *division*. Finally, *per* represents an equivalence. As long as speed is 25 miles per hour, 25 miles $\simeq$ 1 hour.

A familiar use of *per* is in **percent.** *Cent-* is an expression that refers to 100: there are 100 *centi*meters in a meter, and 100 *cents* in a dollar. Percent is commonly used to state the amount of one part of a mixture in 100 total parts of the mixture—parts per 100 parts total. If 78 of 100 cars in a parking lot are made in the United States, 78% are American made.

Suppose a laboratory analyzes 761 g of sea water and finds that it contains 21.3 g of salt. What percentage is salt? According to the data, there are 21.3 g of salt per 761 g of sea water, or $\frac{21.3\text{ g salt}}{761\text{ g sea water}}$. If X is the grams of salt in 100 g of sea water—X grams of salt per 100 grams of sea water, or $\frac{X\text{ g salt}}{100\text{ g sea water}}$ then, by definition, X is the percentage of sea water that is salt. The ratio of salt to sea water is the same regardless of the size of the sample, so the two ratios can be set equal to each other:

$$\frac{X\text{ g salt}}{100\text{ g sea water}} = \frac{21.3\text{ g salt}}{761\text{ g sea water}} \qquad \text{(AP.5)}$$

Solving this "ratio and proportion" problem for X gives

$$X = \frac{21.3\text{ g salt}}{761\text{ g sea water}} \times 100\text{ g sea water} \qquad \text{(AP.6)}$$

$$= 2.80\%\text{ salt}$$

The form of this equation shows how the percentage of one part, A, of any mixture may be found:

$$\%\text{ of A} = \frac{\text{parts of A}}{100\text{ parts total}} \times 100 \qquad \text{(AP.7)}$$

PRACTICE: *7) During a certain period, 89 Chinook salmon, 14 steelhead trout, and 21 fish of other kinds passed through a counting station in a fish ladder bypass around a hydro-electric dam. Calculate the percentage of Chinook salmon.**

In addition to methods you have already learned, percentage problems can be solved by dimensional analysis. Percentage establishes an equivalence between a given number of one part in a mixture and 100 total parts in the mixture. If sea water is 2.80% salt, then

$$2.80 \text{ grams salt} \simeq 100 \text{ grams water} \quad \text{(AP.8)}$$

This relationship can be used to calculate the number of grams of salt in 5000 g of sea water:

$$5000 \cancel{\text{g water}} \times \frac{2.80 \text{ g salt}}{100 \cancel{\text{g water}}} = 140 \text{ g salt}$$

The usual way to solve this problem is to calculate 2.80% of 5000. The percentage is changed to a decimal fraction by dividing by 100 (moving the decimal point two places left) and then multiplying by 5000: $0.028 \times 5000 = 140$.

PRACTICE: *8) 84% of the registered voters in a certain precinct cast ballots in a recent election. If there are 1482 registered voters in the district, how many ballots were cast?*

The dimensional analysis approach to percentage problems is even more useful in the reverse direction. For example, a state-supported university establishes a quota such that 27% of the freshmen admitted in a certain year will be from another state. If 2891 incoming freshmen were from out of state, what was the total number of freshmen enrolled last year? Dimensional analysis interprets 27% as 27 out-of-state students (OSS) $\simeq$ 100 students. Applied to the given quantity,

$$2891 \cancel{\text{OSS}} \times \frac{100 \text{ students}}{27 \cancel{\text{OSS}}} = 10707 \text{ students}$$

The algebraic approach to this problem is

$$27\% \text{ of total students} = \text{OSS}$$

$$0.27 \times \text{total students} = 2891$$

$$\text{total students} = \frac{2891}{0.27} = 10\,707$$

PRACTICE: *9) The total number of pages in a monthly magazine is planned on the number of advertising pages sold. If 58 pages of advertising sold for a forthcoming issue are to represent 46% of the pages in the magazine, what will be the total number of pages?*

PART F EXPONENTIAL NOTATION

Expressing Numbers in Exponential Notation

Problems in chemistry sometimes involve extremely large or very small numbers. For example, the mass of a single atom of hydrogen is 0.000 000 000 000 000 000 000 001 66 gram. In one liter of hydrogen, measured at one atmosphere pressure and 0°C, there are 53 770 000 000 000 000 000 000 hydrogen atoms.

Very large and very small numbers can be written in the form $C \times 10^n$, where C is called the **coefficient** and 10^n is an exponential. The exponent n may be a positive or negative integer. This form is called **exponential notation** or **scientific notation.** In **standard** exponential notation the coefficient is always equal to or greater than 1, but less than 10.* Unless there is reason for doing otherwise, C is written in that range.

To change an ordinary number to exponential notation, you first count the number of places, n, that the decimal point must be moved so it will appear immediately after the first nonzero digit. This number is the size of the exponent. If the decimal point in the original number is moved to the left, the exponent is positive; if the decimal moves right, the exponent is negative. The rule, left-positive, right-negative, is easily learned, but just as easily reversed in one's memory. A more certain way is to think of the original number multiplied by 1 in the form of 10^0. Two examples

*Answers to practice problems are at the end of Appendix I.

*Some books use the term *exponential* notation if C is not in the range from 1 to less than 10, and *scientific* notation if C is in the standard range.

show the thought process:

$$1234 = 1234 \times 10^0 = 1.234 \times 10^3$$

$$0.000\,567\,8 = 0.000\,567\,8 \times 10^0 = 5.678 \times 10^{-4}$$

In the first case, the coefficient became smaller (1234 became 1.234), while the exponent became larger, or *more positive* (0 became 3). In the second case, the coefficient was made larger (0.000 567 8 became 5.678), while the exponent was made smaller, or *more negative* (0 became −4). One factor always becomes larger, and the other becomes smaller.

PRACTICE: *10) Write each of the following numbers in exponential notation:**

10a) 3 762 199 *10c) 0.000 098* *10e) 0.004 60*
10b) 198.75 *10d) 10.080*

In converting a number expressed in exponential notation to an ordinary decimal number, the **absolute value** of the exponent (its *numerical* value, regardless of sign) tells you how many places the decimal point must be moved. If the exponent is positive, the number is larger than 1, so the decimal is moved to the right; if the exponent is negative, the number is smaller than 1, so the decimal goes to the left. For example, in 7.89×10^5 the exponent is positive. This indicates a large number, so the decimal is moved five places to the right, the same number of places as the value of the exponent: 789 000. In 5.37×10^{-4} the negative exponent identifies a small number, so the decimal point is moved four places to the left: 0.000 537.

PRACTICE: *11) Express each of the following numbers in ordinary decimal form:*

11a) 3.49×10^{-11} *11c)* 1.0×10^1
11b) 5.16×10^4 *11d)* 3.75×10^0

Sometimes calculations yield a result in exponential notation that is not in standard form, i.e., the coefficient is not in the range from 1 to 9.999. . . To change it to standard form, rewrite the coefficient itself in exponential notation, and then combine the exponentials. For example, if a calculation yields an answer of 345×10^{-8}, the coefficient can be written as 3.45×10^2. Then

$$345 \times 10^{-8} = (3.45 \times 10^2) \times 10^{-8} = 3.45 \times (10^2 \times 10^{-8}) = 3.45 \times 10^{-6}$$

PRACTICE: *12) Express each of the following numbers in the standard form of exponential notation:*

12a) 0.00427×10^7 *12b)* $55\,800 \times 10^{-9}$

Multiplication with Exponential Notation

In multiplying numbers expressed in exponential notation, the factors are rearranged, placing all the coefficients in one group and all the exponentials in a second group. The two groups are multiplied separately. The decimal result and the exponential result are then combined as the exponential notation expression of the product. For example:

$$(3.96 \times 10^4)(5.19 \times 10^{-7}) = 3.96 \times 10^4 \times 5.19 \times 10^{-7} = 3.96 \times 5.19 \times 10^4 \times 10^{-7} = (3.96 \times 5.19)(10^4 \times 10^{-7}) = 20.6 \times 10^{4-7} = 20.6 \times 10^{-3} = 2.06 \times 10^{-2}$$

If your calculator works in exponential notation, you will no doubt use it to solve problems like the one above. The factors are entered and the operations are performed as described in Part B of this Appendix. The answer is displayed in standard form.

AOS LOGIC		RPN LOGIC	
PRESS	DISPLAY	PRESS	DISPLAY
3.96	3.96	3.96	3.96
EE	3.96 00	EEX	3.96 00
4	3.96 04	4	3.96 04
×	3.96 04	ENTER	3.9600 04
5.19	5.19	5.19	5.19
EE	5.19 00	EEX	5.19 00
7	5.19 07	7	5.19 07
+/−	5.19 −07	CHS	5.19 −07
=	2.0552 −02	×	2.0552 −02

*Answers to practice problems are at the end of Appendix I.

PRACTICE: *13) Perform each of the following calculations and express the result in exponential notation:*

a) $(3.26 \times 10^4)(1.54 \times 10^6) =$
b) $(8.39 \times 10^{-7})(4.53 \times 10^9) =$
c) $(6.73 \times 10^{-3})(9.11 \times 10^{-3}) =$
d) $(2.93 \times 10^5)(4.85 \times 10^6)(5.58 \times 10^{-3}) =$
e) $(35\ 780)(761.9) =$
f) $(0.000\ 91)(235.6) =$
g) $(0.0890)(0.000\ 000\ 726) =$

Division with Exponential Notation

Division of numbers expressed in exponential notation is done the same way as multiplication. The coefficients and exponential factors are separated into two groups; the value of each group is calculated separately; and finally the decimal result is combined with the exponential result as the exponential notation form of the quotient. To illustrate,

$$3.96 \times 10^4 \div 5.19 \times 10^{-7} = \frac{3.96 \times 10^4}{5.19 \times 10^{-7}}$$
$$= \left(\frac{3.96}{5.19}\right)\left(\frac{10^4}{10^{-7}}\right)$$
$$= 0.763 \times 10^{11}$$
$$= 7.63 \times 10^{10}$$

PRACTICE: *14) Perform each of the following calculations and express the result in exponential notation.*

a) $\frac{8.94 \times 10^6}{4.35 \times 10^4} =$

b) $\frac{5.08 \times 10^{-3}}{7.23 \times 10^{-5}} =$

c) $\frac{(3.05 \times 10^{-6})(2.19 \times 10^{-3})}{5.48 \times 10^{-5}} =$

d) $\frac{(867)(7.2 \times 10^{-7})}{0.000\ 000\ 000\ 927} =$

e) $\frac{(7180)(9\ 420\ 000)}{(258\ 000)(0.000\ 618)} =$

Addition and Subtraction in Exponential Notation

As in arithmetic, addition and subtraction of exponentials require that digit values (hundreds, units, tenths, etc.) be aligned vertically. This may be achieved by adjusting coefficients and exponents so the exponents of all numbers are the same, and then proceeding as in ordinary arithmetic. (This adjustment is not necessary on a calculator. The numbers are keyed in just as they appear in the problem.) Shown below is the addition of $6.44 \times 10^{-7} + 1.3900 \times 10^{-5}$ in three ways: with exponents of 10^{-5}, with exponents of 10^{-7}, and as decimal numerals.

1.3900×10^{-5}	139.00×10^{-7}	0.000 013 900
0.0644×10^{-5}	6.44×10^{-7}	0.000 000 644
1.4544×10^{-5}	145.44×10^{-7}	0.000 014 544

PRACTICE: *15) Add or subtract the following numbers:*

a) $5.2 \times 10^5 + 3.98 \times 10^4 =$
b) $3.971 \times 10^2 + 1.98 \times 10^{-1} =$
c) $5.63 \times 10^4 + 2.93 \times 10^3 + 5.42 \times 10^5 =$
d) $1.05 \times 10^{-4} - 9.7 \times 10^{-5}$
e) $\frac{9.02 \times 10^7}{265.9} - \frac{7.89 \times 10^3}{2.37 \times 10^{-2}} =$

PART G LOGARITHMS

Sections 17.5 and 17.6 are the only places in this text that use logarithms, and most of what you need to know about logarithms is explained at that point. Comments here are limited to basic information needed to support the text explanations.

The common logarithm of a number is the power, or exponent, to which 10 must be raised to be equal to the number. Expressed mathematically,

$$\text{If } N = 10^x, \text{ then } \log N = \log 10^x = x \quad \text{(AP.9)}$$

The number 100 may be written as the base, 10, raised to the second power: $100 = 10^2$. According to Equation AP.9, 2, is the logarithm of 10^2, or 100. Similarly, if $1000 = 10^3$, $\log 1000 = \log 10^3 = 3$. And, if $0.0001 = 10^{-4}$, $\log 0.0001 = \log 10^{-4} = -4$.

Just as the powers to which 10 can be raised may be either positive or negative, so may logarithms be positive or negative. The changeover occurs at the value 1, which is 10^0. The logarithm of 1 is therefore 0. It follows that the logarithms of numbers greater than 1 are positive, and logarithms of numbers less than 1 are negative.

The powers to which 10 may be raised are not limited to integers. For example, 10 can be raised to the 2.45 power: $10^{2.45}$. The logarithm of $10^{2.45}$ is 2.45. Such a logarithm is made up of two parts. The digit or digits to the left of the decimal are the **characteristic.** The characteristic reflects the size of the number; it is related to the exponent of 10 when the number is expressed in exponential notation. The characteristic is 2 in 2.45. The digits to the right of the decimal make up the *mantissa,* which is the logarithm of the coefficient of the number when written in exponential notation. In 2.45, the mantissa is 0.45. These relationships are explained in greater detail at the point they are used in Section 17.6.

The number that corresponds to a given logarithm is its antilogarithm. In Equation AP.9, the antilogarithm of x is 10^x, or N. The antilogarithm of 2 is 10^2, or 100. The antilogarithm of 2.45 is $10^{2.45}$. The value of the antilogarithm of 2.45 can be found on a calculator, as described in Part B of Appendix I: antilog 2.45 = $10^{2.45} = 2.8 \times 10^2$. In this exponential form of the antilogarithm of 2.45, the characteristic, 2, is the exponent of 10, and the mantissa, 0.45, is the logarithm of the coefficient, 2.8. In terms of significant figures, the mantissa matches the coefficient. This is illustrated in Section 17.6.

Because logarithms are exponents, they are governed by the rules of exponents given in Secton 8 of Part C of this Appendix. For example, the product of two exponentials to the same base is the base raised to a power equal to the sum of the exponents: $a^m \times a^n = a^{m+n}$. The exponents are added. Similarly, exponents to the base 10 (logarithms) are added to get the logarithm of the product of two numbers: $10^m \times 10^n = 10^{m+n}$. Thus

$$\log ab = \log a + \log b \qquad \text{(AP.10)}$$

In a similar fashion, the logarithm of a quotient is the logarithm of the dividend minus the logarithm of the divisor (or the logarithm of the numerator minus the logarithm of the denominator if the expression is written as a fraction):

$$\log a/b = \log a - \log b \qquad \text{(AP.11)}$$

Equation AP.10 is the basis for converting between pH and hydrogen ion concentration in Section 17.6—or between any "p" number and its corresponding value in exponential notation. This is the only application of logarithms in this text. In more advanced chemistry courses you will encounter applications of Equation AP.11 and others that are beyond the scope of this discussion.

Ten is not the only base for logarithms. Many natural phenomena, both chemical and otherwise, involve logarithms to the **base e,** which is 2.718. . . Logarithms to base e are known as **natural logarithms.** Their value is 2.303 times greater than a base 10 logarithm. Physical chemistry relationships that appear in base e are often converted to base 10 by the 2.303 factor, although modern calculators make it just as easy to work in base e as in base 10. The "p" concept, however, uses base 10 by definition.

PART H ESTIMATING CALCULATION RESULTS

A large percentage of student calculation errors would never appear on homework or test papers if the student would estimate the answer before accepting the number displayed on a calculator. *Challenge every answer.* Be sure it is reasonable before you write it down.

There is no single "right" way to estimate an answer. As your mathematical skills grow, you will develop techniques that are best for you. You will also find that one method works best on one kind of problem, and another method on another problem. The ideas that follow should help you get started in this important practice.

In general, estimating a calculated result involves rounding off the given numbers and calculating the answer mentally. For example, if you multiply 325 by 8.36 on your calculator, you will get 2717. To see if this answer is reasonable, you might round off 325 to 300, and 8.36 to 8. The problem then becomes 300 × 8, which you can calculate mentally to 2400. This is reasonably close to your calculator answer. Even so, you should run your calculator again to be sure you haven't made a small "typing" error.

If your calculator answer for the above problem had been 1254.5, or 38.975 598, or 27 170, your estimated 2400 would signal that an error was made. These three numbers represent common calculation errors. The first comes from a mistake that may arise any time a number is transferred from one position (your paper) to another (your calculator). It is called transposition, and appears when two numerals are changed in position. In this case, 325 × 3.86 (instead of 8.36) = 1254.5.

The answer 38.875 598 comes from pressing the wrong function key: 325 ÷ 8.36 = 38.875 598. This answer is so unreasonable—325 × 8.36 = about 38!—that no mental arithmetic should be necessary to tell you it is wrong. But, like molten idols, calculators speak to some students with a mystic authority they would never dare to challenge. On one occasion neither student nor teacher could figure out why or how, on a test, the student used a calculator to divide 428 by 0.01, and then wrote down 7290 for the answer, when all he had to do was move the decimal two places!

The answer 27 170 is a decimal error, as might arise from transposing the decimal point and a number, or putting an incorrect number of zeros in numbers like 0.001 23 or 123 000. Decimal errors are also apt to appear through an incorrect use of exponential notation, either with or without a calculator.

Exponential notation is a valuable aid in estimating results. For example, in calculating 41 300 × 0.0524, you can regard both numbers as falling between 1 and 10 for a quick calculation of the coefficient: 4 × 5 = 20. Then, thinking of the exponents, changing 41 300 to 4 moves the decimal four places left, so the exponential is 10^4. Changing 0.0524 to 5 has the decimal moving two places right, so the exponential is 10^{-2}. Adding exponents gives 4 + (−2) = 2. The estimated answer is 20 × 10^2, or 2000. On the calculator it comes out to 2164.12.

Another "trick" that can be used is to move decimals in such a way that the moves cancel and at the same time the problem is simplified. In 41 300 × 0.0524, the decimal in the first factor can be moved two places left (divide by 100), and in the second factor two places right (multiply by 100). Dividing and multiplying by 100 is the same as multiplying by $\frac{100}{100}$, which is equal to 1. The problem simplifies to 413 × 5.24, which is easily estimated as 400 × 5 = 2000. The same technique can be used to simplify fractions, too. $\frac{371\ 000}{6240}$ can be simplified to $\frac{371}{6.24}$ by moving the decimal three places left in both numerator and denominator, which is the same as multiplying by 1 in the form $\frac{0.001}{0.001}$. An estimated $\frac{360}{6}$ gives 60 as the approximate answer. The calculator answer is 59.455 128. Similarly, $\frac{0.000\ 406}{0.000\ 839}$ becomes about $\frac{4}{8}$, or 0.5; by calculator, the answer is 0.483 904 2.

In Part F there are some practice problems (13 and 14) for exponential notation. Try estimating the answers to some of those problems, and then compare them with the calculated answers that follow immediately. Problems 13d, 14c, and 14e are good for this purpose, and use others if they will be helpful.

ANSWERS TO PRACTICE PROBLEMS IN APPENDIX I

1) $p_{O_2} = P - p_{H_2O} = 748 - 24 = 724$ torr

2) $V = \frac{k}{P} = \frac{1.62\ \text{L}\ \cancel{\text{atm}}}{1.23\ \cancel{\text{atm}}} = 1.32\ \text{L}$

3) $P = kT = \frac{2.4\ \text{torr}}{\cancel{\text{K}}} \times 300\ \cancel{\text{K}} = 720$ torr

4) $T = \frac{P}{k} = 760\ \cancel{\text{torr}} \times \frac{\text{K}}{2.4\ \cancel{\text{torr}}} = 317$ K

5) $T_2 = \frac{T_1P_2}{P_1} = \frac{(320\ \text{K})(1.00\ \cancel{\text{atm}})}{0.80\ \cancel{\text{atm}}} = 400$ K

6) Because of the complexity of this problem, it is easier to substitute the given values into the original equation and then solve for the unknown. The steps in the solution correspond to those on page APP-9.

$$\frac{\text{g salt}}{\text{g salt + g water}} = \frac{\%}{100}$$

$$\frac{20.0}{20.0 + \text{g water}} = \frac{25}{100}$$

$$20.0 = 0.25(20.0 + \text{g water})$$

$$= 5.0 + 0.25(\text{g water})$$

$$20.0 - 5.0 = 0.25(\text{g water})$$

$$\text{g water} = \frac{15.0}{0.25} = 60 \text{ g water}$$

7) total fish = 89 Chinook + 14 steelhead + 21 others = 124 fish

$\frac{\text{89 Chinook}}{\text{124 total}} \times 100 = 71.8\%$ (arbitrarily rounded off to three significant figures)

8) $1482 \text{ ~~registered~~} \times \frac{\text{84 voters}}{\text{100 ~~registered~~}}$ = 1245 voters; or $1482 \times 0.84 = 1245$

9) $58 \text{ ~~p sold~~} \times \frac{\text{100 pages total}}{\text{46 ~~pages sold~~}} = 126$ pages total; or $0.46 \times \text{total} = 58$
total = 126

10) a) $3.762\ 199 \times 10^6$ d) 1.0080×10^1
b) 1.9875×10^2 e) 4.60×10^{-3}
c) 9.8×10^{-5}

11) a) 0.000 000 000 034 9 c) 10
b) 51 600 d) 3.75

12) a) 4.27×10^4 b) 5.58×10^{-5}

13) *Answers are adjusted for significant figures.*
a) 5.02×10^{10}
b) $38.0 \times 10^2 = 3.80 \times 10^3$
c) $61.3 \times 10^{-6} = 6.13 \times 10^{-5}$
d) $79.3 \times 10^8 = 7.93 \times 10^9$
e) $27.26 \times 10^6 = 2.726 \times 10^7$
f) $21 \times 10^{-2} = 2.1 \times 10^{-1}$
g) $64.6 \times 10^{-9} = 6.46 \times 10^{-8}$

14) *Answers are adjusted for significant figures.*
a) 2.06×10^2
b) $0.703 \times 10^2 = 7.03 \times 10^1$
c) 1.22×10^{-4}
d) 6.7×10^5
e) 4.24×10^8

15) *Answers are adjusted for significant figures.*

a)
$$\begin{array}{r} 5.2 \quad \times 10^5 \\ \underline{0.398 \times 10^5} \\ 5.6 \quad \times 10^5 \end{array}$$

b)
$$\begin{array}{r} 3.971 \quad \times 10^2 \\ \underline{0.00198 \times 10^2} \\ 3.973 \quad \times 10^2 \end{array}$$

c)
$$\begin{array}{r} 0.563 \quad \times 10^5 \\ 0.0293 \times 10^5 \\ \underline{5.42 \quad \times 10^5} \\ 6.01 \quad \times 10^5 \end{array}$$

d)
$$\begin{array}{r} 10.5 \times 10^{-5} \\ \underline{-9.7 \times 10^{-5}} \\ 0.8 \times 10^{-5} \end{array} = 8 \times 10^{-6}$$

e)
$$\frac{9.02 \times 10^7}{265.9} = 3.39 \times 10^5$$

$$\frac{7.89 \times 10^3}{2.37 \times 10^{-2}} = 3.33 \times 10^5$$

$$\begin{array}{r} 3.39 \times 10^5 \\ \underline{-3.33 \times 10^5} \\ 0.06 \times 10^5 \end{array} = 6 \times 10^3$$

16) a) 1.43 g/cm^3
b) 643 cm^2
c) 0.007 46 or 7.46×10^{-3} kg
d) 2.10×10^4 cm
e) 4.59×10^4 cm^3
f) 0.0394 or 3.94×10^{-2} m
g) 3.61×10^{-7} cm
h) 3.50 sec

Appendix II
The SI System of Units

BASE UNITS

The International System of Units or *Système International* (SI), which represents an extension of the metric system, was adopted by the 11th General Conference of Weights and Measures in 1960. It is constructed from seven base units, each of which represents a particular physical quantity (Table AP-1).

Of the seven units listed in Table AP-1, the first five are particularly useful in general chemistry. They are defined as follows:

1. The *meter* was redefined in 1960 to be equal to 1 650 763.73 wavelengths of a certain line in the orange-red region of the emission spectrum of krypton-86.
2. The *kilogram* represents the mass of a platinum-iridium block kept at the International Bureau of Weights and Measures at Sevres, France.
3. The *second* was redefined in 1967 as the duration of 9 192 631 770 periods of a certain line in the microwave spectrum of cesium-133.
4. The *kelvin* is 1/273.16 of the temperature interval between the absolute zero and the triple point of water (0.01°C = 273.16 K).
5. The *mole* is the amount of substance which contains as many entities as there are atoms in exactly 0.012 kg of carbon-12.

Table AP-1 SI Base Units

Physical Quantity	Name of Unit	Symbol
1. Length	meter	m
2. Mass	kilogram	kg
3. Time	second	s
4. Temperature	kelvin	K
5. Amount of substance	mole	mol
6. Electric current	ampere	A
7. Luminous intensity	candela	cd

PREFIXES USED WITH SI UNITS

Decimal fractions and multiples of SI units are designated by using the prefixes listed in Table 3.1. Those which are most commonly used in general chemistry are underlined.

DERIVED UNITS

In the International System of Units, all physical quantities are expressed in the base units listed in Table AP-2 or in combination of those units. The combinations are called **derived units.** For example, the density of a substance is found by dividing the mass of a sample in kilograms by its volume in cubic meters. The resulting units are kilograms per cubic meter, or kg/m^3. Some of the derived units used in chemistry are given in Table AP-2.

If you have not studied physics, the SI units of force, pressure, and energy are probably new to you. Force is related to acceleration, which has to do with changing the velocity of an object. One **newton** is the force that, when applied for one second, will change the straight-line speed of a 1-kilogram object by 1 meter per second.

A **pascal** is defined as a pressure of one newton

Table AP-2 SI Derived Units

Physical Quantity	Name of Unit	Symbol	Definition
Area	square meter	m^2	
Volume	cubic meter	m^3	
Density	kilograms per cubic meter	kg/m^3	
Force	newton	N	$kg \cdot m/s^2$
Pressure	pascal	Pa	N/m^2
Energy	joule	J	$N \cdot m$

acting on an area of one square meter. A pascal is a small unit, so pressures are commonly expressed in kilopascals (kPa), 1000 times larger than the pascal.

A **joule** (pronounced jo͞ol, as in pool) is defined as the work done when a force of one newton acts through a distance of one meter. *Work* and *energy* have the same units. When speaking of large amounts of energy, it is often expressed in kilojoules, 1000 times larger than the joule.

SOME CHOICES

It is difficult to predict the extent to which SI units will replace traditional metric units in the coming years. This makes it difficult to select and use the units that will be most helpful to the readers of this textbook. Add to that the author's joy that, after nearly 200 years, the United States has finally begun to adopt the metric system, including some units that the SI system would eliminate, and his deep desire to encourage rather than complicate the use of metrics in his native land. With particular apologies to Canadian readers, who are more familiar with SI units than Americans are, we list the areas in which this book does not follow SI recommendations:

1. The SI unit of length is the *metre,* spelled in a way that corresponds to its French pronunciation. In America, and in this book, it is written *meter,* which matches the English pronunciation. "What's in a name?"
2. The SI volume unit, *cubic meter,* is huge for most everyday uses. The *cubic decimeter* is 1/1000 as large, and much more practical. When referring to liquids it is customary to replace this six-syllable name with the two-syllable *liter*—or *litre,* for the French spelling. (Which would you rather buy at the grocery store, two cubic decimeters of milk, or two liters?) In the laboratory the common units are again 1/1000 as large, the *cubic centimeter* for solids and the *milliliter* for liquids.
3. The *millimeter of mercury* has an advantage over the *pascal* or *kilopascal* as a pressure unit because the common laboratory instrument for "measuring" pressure literally measures millimeters of mercury. We lose some of the advantage of this "natural" pressure unit by us-

Mass	Length	Capacity
454 grams = 1 pound 28.3 grams = 1 ounce 1 kilogram = 2.20 pounds	2.54 centimeters = 1 inch* 30.5 centimeters = 1 foot 1 meter = 39.4 inches 1 meter = 3.28 feet 1 meter = 1.09 yards 1.61 kilometers = 1 mile	1 liter = 1.06 quarts 3.785 liters = 1 gallon 4.546 liters = 1 imperial gallon

Pressure	Energy
760.0 torr 760.0 mm mercury 76.00 cm mercury 101.3 kilopascals } = 1 atm = { $14.69\ lb/in^2$ 29.92 in mercury	4.184 joules = 1 calorie = 0.003 97 British thermal unit

*Exactly, by definition.

ing its other name, *torr*. Reducing eight syllables to one is worth the sacrifice. Again, "What's in a name?" For large pressures we continue to use the traditional *atmosphere,* which is 760 torr.

CONVERSIONS BETWEEN ENGLISH AND METRIC UNITS

Table AP-3 lists the conversion relationships between metric units and some of the more common English units.

Appendix III
Common Names of Chemicals

COMMON NAME	CHEMICAL NAME	FORMULA
Alumina	Aluminum oxide	Al_2O_3
Baking soda	Sodium hydrogen carbonate	$NaHCO_3$
Bluestone	Copper(II) sulfate pentahydrate	$CuSO_4 \cdot 5\ H_2O$
Borax	Sodium tetraborate decahydrate	$Na_2B_4O_7 \cdot 10\ H_2O$
Brimstone	Sulfur	S
Carbon tetrachloride	Tetrachloromethane	CCl_4
Chloroform	Trichloromethane	$CHCl_3$
Cream of tartar	Potassium hydrogen tartrate	$KHC_4H_4O_6$
Diamond	Carbon	C
Dolomite	Calcium magnesium carbonate	$CaCO_3 \cdot MgCO_3$
Epsom salts	Magnesium sulfate heptahydrate	$MgSO_4 \cdot 7\ H_2O$
Freon (refrigerant)	Dichlorodifluoromethane	CCl_2F_2
Galena	Lead(II) sulfide	PbS
Grain alcohol	Ethyl alcohol; ethanol	C_2H_5OH
Graphite	Carbon	C
Gypsum	Calcium sulfate dihydrate	$CaSO_4 \cdot 2H_2O$
Hypo	Sodium thiosulfate	$Na_2S_2O_3$
Laughing gas	Dinitrogen oxide	N_2O
Lime	Calcium oxide	CaO
Limestone	Calcium carbonate	$CaCO_3$
Lye	Sodium hydroxide	NaOH
Marble	Calcium carbonate	$CaCO_3$
MEK	Methyl ethyl ketone	$CH_3COC_2H_5$
Milk of magnesia	Magnesium hydroxide	$Mg(OH)_2$
Muriatic acid	Hydrochloric acid	HCl
Oil of vitriol	Sulfuric acid (conc.)	H_2SO_4
Plaster of Paris	Calcium sulfate hemihydrate	$CaSO_4 \cdot \frac{1}{2}\ H_2O$
Potash	Potassium carbonate	K_2CO_3
Pyrites (fool's gold)	Iron disulfide	FeS_2
Quartz	Silicon dioxide	SiO_2
Quicksilver	Mercury	Hg
Rubbing alcohol	Isopropyl alcohol	$(CH_3)_2CHOH$
Sal ammoniac	Ammonium chloride	NH_4Cl
Salt	Sodium chloride	NaCl
Saltpeter	Sodium nitrate	$NaNO_3$
Slaked lime	Calcium hydroxide	$Ca(OH)_2$
Sugar	Sucrose	$C_{12}H_{22}O_{11}$
Washing soda	Sodium carbonate decahydrate	$Na_2CO_3 \cdot 10\ H_2O$
Wood alcohol	Methyl alcohol; methanol	CH_3OH

Quick Check Answers

2.1 1 and 4: true; 2 and 3: false.

2.2 1: true; 2: false.

2.4 1: false; 2: true. Both questions are based on the fact that a compound is a pure substance, not a mixture.

2.5 a, b, and c are heterogeneous; d is homogeneous.

2.8 1: Potential energy; 2: Increase. The objects repel each other. Work must be done to push them closer together. This is an increase in potential energy.

3.2 a, d, and e can be given quantities; b, c, and f are conversion relationships.

3.3 1: 10 000, or 10^4; 2a: 1; 2b: 1000.

3.4 Centimeters: 100 cm = 1 m.

4.3 1 and 3: false; 2: true.

4.4 1: false; 2 and 3: true.

4.5 All true.

5.1 1: true; 2: false.

5.2 1, 2, and 4: false; 3: true.

5.5 1, 2, and 3: true; 4 and 5: false.

6.2 1: 2; 2: electrostatic; 3: by sharing electrons; 4: cation, positive; anion, negative.

6.5 By the periodic table group it is in.

6.6 Hydrogen ion, H^+, positively charged; hydride ion, H^-, negatively charged.

7.2 They have the same number—Avogadro's number—of atoms.

7.3 18.0 grams per mole. Molar mass is numerically equal to molecular mass.

8.1 Both false. In 1, oxygen is unbalanced. In 2, H_2O_2 is not the same chemical as H_2O.

8.6 Reactant: oxygen. Products: water and carbon dioxide.

8.7 1: An element and a compound. 2: Two compounds. 3: Precipitation and neutralization.

10.1 S^{2-} and Ba^{2+}.

10.2 Both are false.

10.4 1, 2, and 4: true; 3: false.

11.5 1: Linear; 2: linear and bent; 3: trigonal planar and trigonal pyramid; 4: tetrahedral.

11.6 1: linear, 180°; 2: trigonal planar, 120°; 3: tetrahedral, 109.5°.

14.1 Gas molecules are widely separated compared to liquid molecules.

14.2 1: A has the higher surface tension, molar heat of vaporization, boiling point and viscosity, and B has the higher vapor pressure; 2: If X has higher molar heat of vaporization, it has stronger intermolecular forces, and therefore lower vapor pressure.

14.3 1, 3, and 4: false; 2 and 5: true.

14.4 1: false; 2 and 3: true.

14.5 Both true.

15.1 Only b. There may be more than two pure substances in a solution.

15.2 (a) The solutes are A and B; water is the solvent. (b) Degree of saturation cannot be estimated without knowing the solubility of the compound. (c) The question has no meaning because *concentrated* and *dilute* compare solutions of the same solute, not different solutes.

15.4 Main criterion: do substances have similar intermolecular attractions? If yes, they are probably soluble. Look for similarities in polarity, hydrogen bonding, and size, each of which, if present, contributes to similar intermolecular forces.

16.1 A is a weak electrolyte because its solution is a poor conductor. B is a strong electrolyte because its solution is a good conductor. C is a nonelectrolyte because its solution does not conduct.

16.2 Soluble ionic compounds have ions in their solution inventories; insoluble compounds do not—they have no solution inventories.

16.3 Strong acids ionize nearly completely, weak acids only slightly. H_2SO_4, HNO_3, HCl, HBr, HI, and $HClO_4$ are strong acids. All others are weak.

16.4 Only species designated (aq) can be broken into ions. Of those, ionic compounds and strong acids are separated; weak acids are not.

17.1A An acid produces a H^+ ion, and a base yields an OH^- ion.

17.1B Brönsted–Lowry (B–L) and Arrhenius (Ar) acids both yield protons; they are the same. Ar bases all have hydroxide ions to receive protons; B–L bases are anything that can receive protons. All Ar bases are B–L bases, but not all B–L bases are Ar bases.

17.1C Water can be a Lewis base because it has unshared electron pairs. It cannot be an acid because it has no vacant valence orbitals to receive an electron pair from a Lewis base.

18.2 Oxidation is a loss of electrons and an increase in oxidation number. Reduction is a gain of electrons and a reduction in oxidation number.

19.1 An open beaker can contain a solution-type equilibrium if there is no significant evaporation of solvent.

19.2 "Glancing" collisions, collisions that have insufficient kinetic energy, and those that have improper orientation do not produce a chemical change.

19.3 A barrier is something that prevents or limits some event. Activation energy limits reaction to that fraction of the intermolecular collisions that have sufficient kinetic energy and proper orientation.

19.4 1: As temperature drops, the fraction of collisions with enough kinetic energy to meet the activation energy requirement drops significantly, which reduces reaction rate. Collision frequency also drops, which reduces reaction rate slightly. 2: A catalyst increases reaction rate by reducing the activation energy. 3: At higher concentrations there are more frequent collisions, so reaction rate increases.

19.5 Concentration is the only important factor. "Constant temperature" eliminates temperature from consideration. A catalyst speeds forward and reverse reaction rates equally so system reaches equilibrium more quickly, but it is the same equilibrium as that reached without a catalyst.

20.2 1: Alpha, ${}^{4}_{2}He$; beta, ${}^{0}_{-1}e$. 2: Gamma, beta, alpha.

20.8 1: Nuclear bombardment reactions produce isotopes that do not exist in nature. 2: To be accelerated, a particle must have an electric charge.

20.10 Isotopes produced by a fission reaction have smaller atomic numbers than the starting isotope.

20.13 An isotope produced by a fusion reaction has a larger atomic number than the starting isotopes.

21.3 1: C_7H_{16} and C_9H_{20} are alkanes. 2: $—C_5H_{11}$. 3:

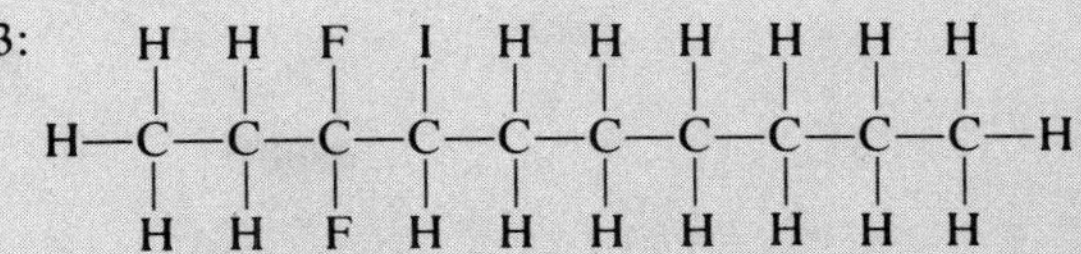

21.4 1: C_4H_6 and C_8H_{16} are alkenes. 2: F(C)C=C(C)F

21.6 (a) $C_4H_8 + Br_2 \rightarrow C_4H_8Br_2$ (b) $C_6H_{14} + Br_2 \rightarrow C_6H_{13}Br + HBr$

21.7 1,3,5-trifluorobenzene: *p*-dichlorobenzene:

21.9 Hydroxyl group, —OH.

21.10 —O—

21.11 1: Carbonyl group: >C=O 2: Aldehyde: R(H)C=O Ketone: R(R′)C=O

21.12 1: Carboxyl group: —C(=O)—O—H 2: Acid + alcohol → ester + water

21.13 $R_1—N(R_3)—R_2$ One or two of the Rs may be H.

21.14 $—C(=O)NH_2$

Answers to Questions and Problems

CHAPTER 2

(1) Physical: a, c, d; chemical: b, e. **(2)** Chemical: a, d, e; physical: b, c

(3) Toasting is a chemical change in which the surface of bread is partially burned or charred.

(4) Pick out balls or bearings based on size and appearance; use magnetic property of steel to pick up ball bearings with a magnet; use low density of balls and high density of steel to float the balls in water.

(5) Liquid volume usually slighly greater than solid; gas volume very much larger than liquid.

(6) As a solid, a snowman has a definite shape; as a liquid, the water takes the "shape" of its container, the ground.

(7) The particles in a solid occupy fixed positions relative to each other and cannot be poured, but different pieces of solids can move relative to each other.

(8) Elements: b, d, e; compounds: a, c.

(10) Only a compound with one taste can be decomposed into a solid having a different taste and a gas.

(11) Mixture of two elements.

(12) Pure substance. A mixture would have changed boiling temperature during distillation because of a change in composition of the mixture.

(13) New product. If both products are the same pure substance, they must be identical.

(14) Homogeneous: b, d; heterogeneous: a, c, e.

(15) Natural milk from the cow is a mixture of several chemicals that separate into layers. Milk is homogenized to give it consistency in properties, notably taste.

(16) Weights will be the same. The weight of the products of a chemical change is the same as the mass of the reactants.

(18) Exothermic: c, d, e; endothermic: a, b. **(19)** Kinetic energy increases.

(20) If the objects have opposite charge, there is an attraction between them. An increase in separation will be an increase in potential energy. If they have the same charge (repulsion), the greater distance will be lower in potential energy.

(21) Potential energy of water above dam to kinetic energy falling into turbine to kinetic energy of turbine and generator to electrical energy to heat and light energy given off by light bulb.

(44) True: e, f, i, j, l; false: a, b, c, d, g, h, k. **(45)** Nothing.

(49) Mercury, water, ice, carbon.

(50) Yes. Nitrogen and oxygen in air are a mixture of two elements . . .

(51) . . . that form at least six different compounds.

(52) Rain water is more pure. Ocean water is a solution of salt and other substances. The water is distilled by evaporation and condensed into rain.

(53) (a) The powder is neither an element nor a compound, both of which have a fixed composition. (b) The contents of the box are heterogeneous because samples from different locations have a different composition. (c) The contents must be a mixture of varying composition.

(54) The now-available sources of useable energy are limited. If we change them into forms that cannot be used, we are threatened with an energy shortage in the future.

CHAPTER 3

NOTE: Cancellation marks have been omitted in the answer section in order that problem setups may be seen more clearly.

(1) $3 \text{ miles} \times \dfrac{5280 \text{ ft}}{1 \text{ mile}} = 15\,840 \text{ ft}$

(2) $93\,000\,000 \text{ miles} \times \dfrac{1 \text{ sec}}{186\,000 \text{ miles}} \times \dfrac{1 \text{ min}}{60 \text{ sec}} = 8.33 \text{ min}$

(3) $63 \text{ hr} \times \dfrac{3 \text{ ft}}{24 \text{ hr}} = 7.9 \text{ ft}$

(4) $\$39.95 \text{ C} \times \dfrac{\$1.00 \text{ A}}{\$1.29 \text{ C}} = \$30.97 \text{ A}$

(5) $2 \text{ sem} \times \dfrac{900 \text{ students}}{\text{sem}} \times \dfrac{55 \text{ beakers}}{825 \text{ students}} = 120 \text{ beakers}$

(6) $125 \text{ con} \times \dfrac{5.0 \text{ lb chile}}{1 \text{ con}} \times \dfrac{0.054 \text{ oz p}}{1 \text{ lb chile}} \times \dfrac{1 \text{ lb}}{16 \text{ oz}} = 2.1 \text{ lb p}$

(7) 25 900 m; 0.00427 m; 94.6 mm; 9.37×10^{-7} m

(8) $399.9 \text{ m} \times \dfrac{1.09 \text{ yd}}{1 \text{ m}} \times \dfrac{3 \text{ ft}}{1 \text{ yd}} = 1.31 \times 10^3 \text{ ft (b)}$

$= 436 \text{ yd (a)}$

(9) $2350 \text{ miles} \times \dfrac{1.61 \text{ km}}{1 \text{ mile}} = 3.78 \times 10^3 \text{ km}$

(10) $29\,002 \text{ ft} \times \dfrac{1 \text{ m}}{3.28 \text{ ft}} \times \dfrac{1 \text{ km}}{1000 \text{ m}} = 8.84 \text{ km}$

(11) $46.2 \text{ m}^2 \times \dfrac{100^2 \text{ cm}^2}{1^2 \text{ m}^2} = 4.62 \times 10^5 \text{ cm}^2$

$751 \text{ mm}^3 \times \dfrac{1^3 \text{ m}^3}{1000^3 \text{ mm}^3} = 7.51 \times 10^{-7} \text{ m}^3$

$231 \text{ mL} \times \dfrac{1 \text{ L}}{1000 \text{ mL}} = 0.231 \text{ L}$

(12) $1.5 \text{ in} \times 3.5 \text{ in.} \times \dfrac{2.54^2 \text{ cm}^2}{1^2 \text{ in}^2} = 33.9 \text{ cm}^2$

(13) $20 \text{ ft} \times 36 \text{ in} \times 4 \text{ in} \times \dfrac{1 \text{ m}}{3.28 \text{ ft}} \times \dfrac{1^2 \text{ m}^2}{39.4 \text{ in}^2} = 0.566 \text{ m}^3$

(14) $50.0 \text{ L} \times \dfrac{1 \text{ gal}}{3.785 \text{ L}} = 13.2 \text{ gal}$

(15) $3.0 \text{ L} \times \dfrac{1000 \text{ cm}^3}{\text{L}} \times \dfrac{1^3 \text{ in}^3}{2.54^3 \text{ cm}^3} = 183 \text{ in}^3$

(16) $113 \text{ ft}^3 \times \dfrac{1^3 \text{ m}^3}{3.28^3 \text{ ft}^3} \times \dfrac{100^3 \text{ cm}^3}{1^3 \text{ m}^3} \times \dfrac{1 \text{ L}}{1000 \text{ cm}^3} = 3.20 \times 10^3 \text{ L}$

(17) $\dfrac{231 \text{ miles}}{1 \text{ hr}} \times \dfrac{1.61 \text{ km}}{1 \text{ mile}} \times \dfrac{1 \text{ hr}}{60 \text{ min}} = 6.20 \text{ km/min}$

(18) $\dfrac{\$1.23}{\text{gal}} \times \dfrac{1 \text{ gal}}{3.785 \text{ L}} = \$0.325/\text{L} > \$0.317/\text{L}$

(19) The mass of the rock is the same wherever it is. The weight of the rock is the same whether it be in the lake or on the beach—or anywhere in the gravitational field on the earth's surface. In the lake the upward buoyancy force cancels some of the downward weight force, making the rock seem "lighter."

(20) 595 g; 1490 cg; 871 000 mg; 9.13×10^{-14} g

(21) $26.4 \text{ g} \times \dfrac{1 \text{ oz}}{28.3 \text{ g}} = 0.933 \text{ oz}$

(22) $6 \text{ lb} \times \dfrac{454 \text{ g}}{1 \text{ lb}} + 7 \text{ oz} \times \dfrac{28.3 \text{ g}}{1 \text{ oz}} = 2922 \text{ g} = 2.92 \text{ kg}$

(23) $115 \text{ mg} \times \frac{1 \text{ g}}{1000 \text{ mg}} \times \frac{1 \text{ lb}}{454 \text{ g}} = 2.53 \times 10^{-4} \text{ lb}$

$= 0.115 \text{ g}$

(24) $922 \text{ lb} \times \frac{1 \text{ kg}}{2.20 \text{ lb}} = 419 \text{ kg}$

(25)

CELSIUS	FAHRENHEIT	KELVIN
69	156	342
−34	−29	239
−162	−260	111
2	36	275
85	185	358
−141	222	132

(26) 37°C 27) 26°C 28) 136°F

(29) $[243 - (-261)]°\text{F} \times \frac{1°\text{C}}{1.8°\text{F}} = 280°\text{C}$

(31) $\frac{1.51 \times 10^3 \text{ g}}{865 \text{ cm}^3} = 1.75 \text{ g/cm}^3$

(32) $\frac{4.92 \times 10^3 \text{ g}}{4.60 \text{ cm} \times 10.3 \text{ cm} \times 13.2 \text{ cm}} = 7.87 \text{ g/cm}^3$

(33) $35.3 \text{ mL} \times \frac{0.790 \text{ g}}{1 \text{ mL}} = 27.9 \text{ g}$

(34) $227 \text{ g} \times \frac{1 \text{ mL}}{0.92 \text{ g}} = 2.5 \times 10^3 \text{ mL}$

(35) $11.0 \text{ gal} \times \frac{3.785 \text{ L}}{1 \text{ gal}} \times \frac{1000 \text{ cm}^3}{1 \text{ L}} \times \frac{1^3 \text{ ft}^3}{30.5^3 \text{ cm}^3} \times \frac{42.0 \text{ lb}}{\text{ft}^3} \times \frac{1 \text{ kg}}{2.20 \text{ lb}} = 28.0 \text{ kg}$

(36) 5; 2; 3; uncertain—3 or 4; 3; 5; 3; 5

(37) 52.2 mL; 18.0 g; 78.5 mg; 2.36×10^7 μm; 4.20×10^{-3} kg

(38) 2.86 g + 3.9 g + 0.896 g + 246 g = 254 g

(39) 101.209 g − 94.33 g = 6.88 g

(40) $118 \text{ ft} \times \frac{1 \text{ m}}{3.28 \text{ ft}} = 36.0 \text{ ft}$

(41) (a) $60.5 \text{ cal} \times \frac{4.184 \text{ J}}{1 \text{ cal}} = 253 \text{ J}$

(b) $8.32 \text{ kJ} \times \frac{1 \text{ kcal}}{4.184 \text{ kJ}} = 1.99 \text{ kcal} = 1.99 \times 10^3 \text{ cal}$

(c) $0.753 \text{ kJ} \times \frac{1 \text{ kcal}}{4.184 \text{ kJ}} = 0.180 \text{ kcal}$

(42) $912 \text{ cal} \times \frac{4.184 \text{ J}}{1 \text{ cal}} = 3.82 \times 10^3 \text{ J} = 3.82 \text{ kJ}$

(43) $56.7 \text{ g S} \times \frac{3.94 \times 10^3 \text{ cal}}{1 \text{ g S}} \times \frac{4.184 \text{ J}}{1 \text{ cal}} = 9.35 \times 10^5 \text{ J} = 935 \text{ kJ}$

(44) Heat flow is proportional to ΔT. If all other factors are the same, less heat will be required—less time on the stove—for the smaller ΔT. Starting with hot water gives the smaller ΔT.

(45) $204 \text{ g} \times \frac{0.16 \text{ J}}{\text{g} \cdot °\text{C}} \times (64.9 - 22.8)°\text{C} = 1.37 \times 10^3 \text{ J} = 1.37 \text{ kJ}$

(46) $2.55 \times 10^3 \text{ g} \times \frac{0.444 \text{ J}}{\text{g} \cdot °\text{C}} \times (25 - 350)°\text{C} = -3.68 \times 10^5 \text{ J} = -368 \text{ kJ}$

(47) $\frac{3.14 \times 10^3 \text{ J}}{545 \text{ g}} \times \frac{\text{g} \cdot °\text{C}}{0.46 \text{ J}} = 12.5 = \Delta T = T_f - 25.0 \qquad T_f = 38°\text{C}$

(48) $72.0 \text{ g} \times \frac{4.184 \text{ J}}{\text{g} \cdot °\text{C}} \times (25.5 - 19.2)°\text{C} = 141 \text{ g} \times c \times -(25.5 - 89.0)°\text{C}$

$c = 0.21 \text{ J/g} \cdot °\text{C}$

(98) True: a, b, d, g, j, n, p, q, u. False: c, e, f, h, i, k, l, m, o, r, s, t.

(99) 0.72 mile^3.

(100) The height of a 5′8″ person is 173 cm, the centimeter being the closest metric unit to the American inch.

(101) The mass of a 140-pound person is 63.6 kg. The *kilogram* is probably the most suitable mass unit for humans because it is the standard unit of mass. The *hectogram* would avoid a decimal fraction—636 hg—but the unit is seldom used.

(102) 231.0 in^3.

(103) 250 mL is a "standard" metric size in volumetric glassware, but converting the fractions in which American recipes are written to metric units is awkward: ⅔ of 250, ¾ of 250; ⅜ of 250—a nightmare for most American cooks! Because 2, 3, 4, 8, and even 16 all divide evenly into 240, fractional conversions are much easier. Furthermore, the product would probably come closer to tasting how it should taste because 240 mL is closer to the real cup, which is 237 mL.

(104) 22 cm × 28 cm.

(105) 0.812 pound.

(106) 1.29 oz.

(107) 1.1. $\times 10^3$ cm^3.

CHAPTER 4

(4) 402 g of mercury combine with 71 g of chlorine in one compound, and 201 g of mercury combine with 71 g of chlorine in the other. $\frac{402}{201} = \frac{2}{1}$, a ratio of small, whole numbers.

(8) The Rutherford experiment showed that the heavier atomic particles were in the nucleus. When found, the proton and neutron proved to be the heavier particles.

(9) The number of protons is the same as the number of electrons. Neutrons are equal to or more than number of protons or electrons.

(10) Isotopes of an element cannot have same mass number. Mass number is sum of protons plus neutrons. Isotopes have different numbers of neutrons, but same number of protons. Sums must be different.

(11)

NAME OF ELEMENT	NUCLEAR SYMBOL	ATOMIC NUMBER	MASS NUMBER	NUMBER OF PROTONS	NUMBER OF NEUTRONS	NUMBER OF ELECTRONS
Arsenic	${}^{70}_{33}As$	33	70	33	37	33
Fluorine	${}^{19}_{9}F$	9	19	9	10	9
Chromium	${}^{52}_{24}Cr$	24	52	24	28	24
Gold	${}^{197}_{79}Au$	79	197	79	118	79
Iron	${}^{57}_{26}Fe$	26	57	26	31	26
Selenium	${}^{74}_{34}Se$	34	74	34	40	34

(13) Boron occurs in nature as a mixture of atoms that have different masses.

(14) Atomic mass is a *weighted* average that takes into account the percentage distribution of isotopes of an element, not an arithmetic average of the atomic masses of the isotopes.

(15) $0.1978 \times 10.0129 + 0.8022 \times 11.00931 = 1.981 + 8.832 = 10.813$ amu

(16)
$$\begin{aligned} 0.5182 \times 106.9041 \text{ amu} &= 55.40 \text{ amu} \\ 0.4818 \times 108.9047 \text{ amu} &= \underline{52.47} \text{ amu} \\ &\quad 107.87 \text{ amu} \end{aligned}$$

(17)
$$\begin{aligned} 0.5725 \times 120.9038 \text{ amu} &= 69.22 \text{ amu} \\ 0.4275 \times 122.9041 \text{ amu} &= \underline{52.54} \text{ amu} \\ &\quad 121.76 \text{ amu} \end{aligned}$$

(18)
$$\begin{aligned} 0.00193 \times 135.907 \text{ amu} &= 0.262 \text{ amu} \\ 0.00250 \times 137.9057 \text{ amu} &= 0.345 \text{ amu} \\ 0.8848 \times 139.9053 \text{ amu} &= 123.8 \text{ amu} \\ 0.1107 \times 141.9090 \text{ amu} &= \underline{15.71} \text{ amu} \\ &\quad 140.1 \end{aligned}$$

(19) Let x = fraction of $^{7}_{4}Li$; 1 − x = fraction of $^{6}_{4}Li$
(1 − x)(6.01512) + 7.01600x = 6.941 x = 0.925 = 92.5% $^{7}_{4}Li$; 7.5% $^{6}_{4}Li$

(20) Be, Mg, Ca, Sr, Ba, Ra; 11 to 18.

(21) (a) and (d), same period; (b) and (c), same family.

(22) Z = 24: 52.00; Z = 50: 118.7; Z = 77: 192.2.

(23) Potassium, 39.10; sulfur, 32.06.

(24) Check alphabetical list of elements inside back cover.

(48) True: b*, d, f, j, n, o, p; false, a, c, e, g, h, i, k, l, m.
*Regarding b, Dalton apparently did not make any specific comment about the diameter of an atom, but he did propose that all atoms of an element are identical in every respect. That would include diameters.

(49) What was left had to have a positive charge to account for the neutrality of the complete atom.

(50) Variable e/m ratio suggests positive portion of atom consists of at least two particles, one with positive charge and one neutral, present in varying number ratio, and/or with different masses.

(51) Planetary model of atom is similar to solar system in that it obeys the same rules of classical physics, both conceptually and quantitatively. Major difference is that electron energies are quantized, whereas planetary orbits are not.

(52) Chemical properties of isotopes of an element are identical.

(53) 12.09899 amu. The difference in masses of the nuclear parts and the sum of the masses of protons and neutrons is what is responsible for nuclear energy in an energy-mass conversion. See more complete explanation in Section 20.11.

(54) Electrons, 0.0272%; protons, 49.95%; neutrons, 50.02%.

(55) (a) $\dfrac{12.01 \text{ amu}}{1 \text{ C atom}} \times \dfrac{1.66 \times 10^{-24} \text{ g}}{1 \text{ amu}} \times \dfrac{1 \text{ C atom}}{1.9 \times 10^{-24} \text{ cm}^3} = 1.0 \times 10^1 \text{ g/cm}^3$

(b) In packing carbon atoms into a crystal there are void spaces between the atoms. There are no voids in a single atom. (In fact, voids in diamond account for 66% of the total volume, and in graphite, 78%.)

(c) $1.9 \times 10^{-24} \text{ cm}^3/(1 \times 10^5)^3 = 2 \times 10^{-39} \text{ cm}^3$

(d) $\dfrac{12.01 \text{ amu}}{1 \text{ C nucleus}} \times \dfrac{1.66 \times 10^{-24} \text{ g}}{1 \text{ amu}} \times \dfrac{1 \text{ C nucleus}}{2 \times 10^{-39} \text{ cm}^3} = 1 \times 10^{16} \text{ g/cm}^3$

(e) $4 \times 10^{-5} \text{ cm}^3 \times \dfrac{1 \times 10^{16} \text{ g}}{1 \text{ cm}^3} \times \dfrac{1 \text{ lb}}{454 \text{ g}} \times \dfrac{1 \text{ ton}}{2000 \text{ lb}} = 4 \times 10^5 \text{ tons}$

CHAPTER 5

(2) Something that behaves like a wave has properties normally associated with waves, some of which appear in Figure 5.3.

(3) (b) and (c) are quantized.

(6) An atom is in its ground state when all electrons are at their lowest possible energies.

(12) There is never more than one p sublevel at any value of n.

(13) An orbital is a region in space where there is a high probability of finding an electron. This implies correctly that there is a low but real probability of finding the ball outside that region—outside the ball.

(15) The quantum model of the atom gives no indication of the path of an electron. Furthermore, the orbital is three dimensional, not two as suggested by a "figure 8."

(16) If $n \geq 3$, the *actual* (not maximum) number of d orbitals is 5.

(17) Energies of principal energy levels increase in the order 1, 2, 3, . . . Energies of sublevels increase in the order s, p, d, f.

(18) The models are consistent in identifying quantized energy levels. The quantum model substitutes orbitals (regions in space) for orbits (electron paths) in the Bohr model. The quantum model goes beyond the Bohr model in identifying sublevels.

(19) $3p^4$ means there are four electrons in the 3p sublevel.

(20) Fluorine, Period 2, Group 7A.

(22) 4s and 3d orbitals are close in energy, so minor influences in the overall electron energies are sufficient to produce irregularities. Energies are even closer at higher values of n, so irregularities are even more likely.

(23) N: $1s^22s^22p^3$; Ti: $1s^22s^22p^63s^23p^64s^23d^2$.

(24) Ca: $1s^22s^22p^63s^23p^64s^2$; Cu: $1s^22s^22p^63s^23p^64s^13d^{10}$.

(25) Ti: $[Ar]4s^23d^2$; Ca: $[Ar]4s^2$; Cu: $[Ar]4s^13d^{10}$.

(26) Ge: $[Ar]4s^23d^{10}4p^2$.

(27) Ba: $[Xe]6s^2$. Tc: $[Kr]5s^24d^5$.

(30) Group 4A.

(32) Atoms in the same family become larger as atomic number increases. The highest energy electrons are farther from the nucleus and easier to remove. This appears as lower ionization energies.

(34) ns^1.

(36) Iodine, halogens. Rubidium, alkali metals.

(37) Chemical properties of elements are often determined by the number of valence electrons. Both magnesium and calcium have two: ns^2.

(38) Representative elements are in the A groups of the periodic table. Those to the left of the stair-step line are metals, and those to the right are nonmetals.

(39) Atoms of antimony (Z = 51) and bismuth (Z = 83) are larger than atoms of arsenic, and atoms of phosphorus and nitrogen are smaller.

(41) Outermost electrons of germanium are one energy level higher than outermost electrons of silicon. Germanium atoms are therefore larger.

(42) Aluminum atoms have a lower nuclear charge—fewer protons in the nucleus—than chlorine atoms. It is therefore easier to remove a 3d electron from an aluminum atom than from a chlorine atom.

(44) (a) Q M; (b) D E.

(45) J L W.

(46) E D G.

(94) True: b, e, g, i, k, m, n, p; false: a, c, d, f, h, j, l, o.

(95) The quantum and Bohr model explanations of atomic spectra are essentially the same.

(96) All species have a single electron. Species with two or more electrons are far more complex.

(97) The other lines are outside the visible spectrum, in the infrared or ultraviolet regions.

(98) Sc^{3+} is isoelectronic with an argon atom.

(99) To form a monatomic ion, carbon would have to lose or gain four electrons. The fourth ionization energy of any atom is very high, as is the energy required to add an electron to a C^+, C^{2+} or C^{3+} particle, the intermediate steps in reaching a C^{4+} ion.

(100) The smaller atoms in Group 5A tend to complete their octets by gaining or sharing electrons, which are characteristics of nonmetals. Larger atoms in the group tend to lose their highest energy s electrons and form positively charged ions, a characteristic of a metal.

(101) Xenon has the lowest ionization energy of the noble gases, and apparently the greatest reactivity. This is characteristic of the more active metals that form ionic compounds.

(102) Iron loses two electrons from the 4s orbital to form Fe^{2+}, and a third from the 3d orbital to form Fe^{3+}.

(103) (a) Ionization energy increases across a period of the periodic table because of increasing nuclear charge. (b) The breaks in ionization energy trends across Periods 2 and 3 occur just after the s orbital is filled, and just after the p orbitals are half filled.

CHAPTER 6

(1) Two nitrogen atoms, one oxygen atom.

(2) One sulfur atom, two oxygen atoms, two chlorine atoms.

(3) SiH_4.

(4) C_3H_8O.

(8) A crystal is "a solid in which the ions and/or molecules are arranged in a definite geometric pattern" (Glossary). Most metallic elements exist as crystals.

(10) Iron, Fe; chlorine, Cl_2; chromium, Cr; beryllium, Be.

(11) A molecular substance—compound or otherwise—has electrically neutral molecules as its fundamental structural unit.

(14) A cation has a positive charge, and an anion has a negative charge.

(16)

ATOMIC NUMBER	ION NAME	ION SYMBOL	ANION OR CATION
11	Sodium	Na^+	Cation
53	Iodide	I^-	Anion
3	Lithium	Li^+	Cation
20	Calcium	Ca^{2+}	Cation
16	Sulfide	S^{2-}	Anion

(17) Selenium forms the selenide ion, Se^{2-}, an anion. Rubidium forms the rubidium ion, Rb^+, a cation.

(18) F^-, fluoride ion; K^+, potassium ion; Al^{3+}, aluminum ion; O^{2-}, oxide ion.

(21) Hydrogen behaves as a monatomic cation, like other members of Group 1A.

(25) $HClO_3 \rightarrow H^+ + ClO_3^-$. Chlorate ion.

(27) Lithium chloride; CaS.

(28) Ammonium nitrate; $BaCO_3$.

(29) Barium bromide; K_3PO_4.

(30) Magnesium phosphate; $(NH_4)_2SO_4$.

(32) Two water molecules are in calcium chloride 2-hydrate.

(33) $Ba(OH)_2 \cdot 8\ H_2O$, barium hydroxide 8-hydrate.

(66) True: b, f, h, i, j, k, m, o, p; false, a, c, d, e, g, l, n.

(67) (a), SrS; (b) In_2Se_3; (c) gallium chloride; (d) sodium telluride; (e) iodic acid, HIO_3.

The following table contains the names and formulas of all compounds in Formula Writing and Nomenclature Exercises Numbers 1 and 2.

Aluminum
- bromide, $AlBr_3$
- carbonate, $Al_2(CO_3)_2$
- chloride, $AlCl_3$
- fluoride, AlF_3
- hydroxide, $Al(OH)_3$
- iodide, AlI_3
- nitrate, $Al(NO_3)_3$
- oxide, Al_2O_3
- phosphate, $AlPO_4$
- sulfate, $Al_2(SO_4)_3$
- sulfide, Al_2S_3

Ammonium
- bromide, NH_4Br
- carbonate, $(NH_4)_2CO_3$
- chloride, NH_4Cl
- fluoride, NH_4F
- hydroxide, NH_4OH
- iodide, NH_4I
- nitrate, NH_4NO_3
- oxide, $(NH_4)_2O$
- phosphate, $(NH_4)_3PO_4$
- sulfate, $(NH_4)_2SO_4$
- sulfide, $(NH_4)_2S$

Barium
- bromide, $BaBr_2$
- carbonate, $BaCO_3$
- chloride, $BaCl_2$
- fluoride, BaF_2
- hydroxide, $Ba(OH)_2$
- iodide, BaI_2
- nitrate, $Ba(NO_3)_2$
- oxide, BaO
- phosphate, $Ba_3(PO_4)_2$
- sulfate, $BaSO_4$
- sulfide, BaS

Calcium
- bromide, $CaBr_2$
- carbonate, $CaCO_3$
- chloride, $CaCl_2$
- fluoride, CaF_2
- hydroxide, $Ca(OH)_2$
- iodide, CaI_2
- nitrate, $Ca(NO_3)_2$
- oxide, CaO
- phosphate, $Ca_3(PO_4)_2$
- sulfate, $CaSO_4$
- sulfide, CaS

Lithium
- bromide, LiBr
- carbonate, Li_2CO_3
- chloride, LiCl
- fluoride, LiF
- hydroxide, LiOH
- iodide, LiI
- nitrate, $LiNO_3$
- oxide, Li_2O
- phosphate, Li_3PO_4
- sulfate, Li_2SO_4
- sulfide, Li_2S

Magnesium
- bromide, $MgBr_2$
- carbonate, $MgCO_3$
- chloride, $MgCl_2$
- fluoride, MgF_2
- hydroxide, $Mg(OH)_2$
- iodide, MgI_2
- nitrate, $Mg(NO_3)_2$
- oxide, MgO
- phosphate, $Mg_3(PO_4)_2$
- sulfate, $MgSO_4$
- sulfide, MgS

Potassium
- bromide, KBr
- carbonate, K_2CO_3
- chloride, KCl
- fluoride, KF
- hydroxide, KOH
- iodide, KI
- nitrate, KNO_3
- oxide, K_2O
- phosphate, K_3PO_4
- sulfate, K_2SO_4
- sulfide, K_2S

Sodium
- bromide, NaBr
- carbonate, Na_2CO_3
- chloride, NaCl
- fluoride, NaF
- hydroxide, NaOH
- iodide, NaI
- nitrate, $NaNO_3$
- oxide, Na_2O
- phosphate, Na_3PO_4
- sulfate, Na_2SO_4
- sulfide, Na_2S

CHAPTER 7

(3) To the four significant figures used in this book, the atomic masses are the same. The change was made to reconcile the atomic mass scales used by chemists and physicists (see Question 34). The carbon-12 scale reduced chemists' atomic weights, as they were called then, by 0.004%: $[(16.0000 - 15.9994)/16.0000] \times 100 = 0.04$.

(6) 64.12 grams of sulfur is just twice the atomic mass of sulfur, 32.06 g/mol. The sample contains two moles of atoms. Two moles of carbon atoms have a mass twice the atomic mass of carbon, or 2×12.01 g = 24.02 g.

(8) "Oxygen" may be oxygen atoms, O, or diatomic oxygen molecules, O_2. The molar mass of oxygen atoms is 16.00 g/mol, whereas the molar mass of oxygen molecules is 32.00 g/mol.

(9) (a) KI: $39.1 + 127 = 166$ g/mol
(b) H_2: $2 \times 1.01 = 2.02$ g/mol
(c) $NaNO_3$: $23.0 + 14.0 + 3(16.0) = 85.0$ g/mol
(d) $Mg_3(PO_4)_2$: $3(24.3) + 2(31.0) + 8(16.0) = 263$ g/mol
(e) $Ba(OH)_2 \cdot 8\,H_2O$: $137 + 10(16.0) + 18(1.01) = 315$ g/mol
(f) C_3H_7OH: $3(12.0) + 8(1.01) + 16.0 = 60.1$ g/mol
(g) $CuSO_4$: $63.6 + 32.1 + 4(16.0) = 159.7$ g/mol
(h) $CrCl_3$: $52.0 + 3(35.5) = 159$ g/mol
(i) $Al_2(C_2O_4)_3 \cdot 4\,H_2O$: $2(27.0) + 6(12.0) + 16(16.0) + 8(1.01) = 3.90 \times 10^2$ g/mol
(j) $NaC_2H_3O_2$: $23.0 + 2(12.0) + 3(1.01) + 2(16.0) = 82.0$ g/mol

(10) $454 \text{ g LiF} \times \dfrac{6.94 \text{ g Li}}{25.9 \text{ g LiF}} = 122 \text{ g Li}$

(11) $16.3 \text{ g } K_2SO_4 \times \dfrac{78.2 \text{ g K}}{174.3 \text{ g } K_2SO_4} = 7.31 \text{ g K}$

(12) $158 \text{ g Zn} \times \dfrac{117.4 \text{ g } Zn(CN)_2}{65.4 \text{ g Zn}} = 284 \text{ g } Zn(CN)_2$

(13) $7.80 \times 10^4 \text{ g } PbMoO_4 \times \dfrac{95.9 \text{ g Mo}}{367 \text{ g } PbMoO_4} = 2.04 \times 10^4 \text{ g} = 20.4 \text{ kg Mo}$

(14) $2.29 \text{ g Cl} \times \dfrac{207.1 \text{ g } Ca(ClO_3)_2}{71.0 \text{ g Cl}} = 6.68 \text{ g } Ca(ClO_3)_2$

(15) (a) (23.0/85.0)100 = 27.1% Na; (14.0/85.0)100 = 16.5% N; (48.0/85.0)100 = 56.5% O
(b) (72.9/263)100 = 30.5% Mg; (62.0/263)100 = 25.9% P; (128/263)100 = 53.6% O
(c) (52.0/159)100 = 32.7% Cr; (107/159)100 = 67.3% Cl
(d) (63.6/159.7)100 = 39.8% Cu; (32.1/159.7)100 = 20.1% S; (64.0/159.7)100 = 40.1% O

(16) $4\,H_2O = 4 \times 18.0 \text{ g } H_2O\text{/mol} = 72.0 \text{ g } H_2O\text{/mol hydrate}$; $(72.0/390)100 = 18.5\%\ H_2O$

(17) $\dfrac{\$104.00}{\text{kg } NiSO_4 \cdot 6\,H_2O} \times \dfrac{263 \text{ kg } NiSO_4 \cdot 6\,H_2O}{58.7 \text{ kg Ni}} = \$466\text{/kg Ni}$

$\dfrac{\$61.00}{\text{kg } NiCl_2 \cdot 6\,H_2O} \times \dfrac{238 \text{ kg } NiCl_2 \cdot 6\,H_2O}{58.7 \text{ kg Ni}} = \$247\text{/kg Ni}$

(18) (a) $8.59 \text{ g } H_2 \times \dfrac{1 \text{ mol } H_2}{2.02 \text{ g } H_2} = 4.25 \text{ mol } H_2$

(b) $9.85 \text{ g KI} \times \dfrac{1 \text{ mol KI}}{166 \text{ g KI}} = 0.0593 \text{ mol KI}$

(c) $427 \text{ g } CuSO_4 \times \dfrac{1 \text{ mol } CuSO_4}{160 \text{ g } CuSO_4} = 2.67 \text{ mol } CuSO_4$

(d) $22.8 \text{ g } Al_2(C_2O_4)_3 \cdot 4\,H_2O \times \dfrac{1 \text{ mol } Al_2(C_2O_4)_3 \cdot 4\,H_2O}{3.90 \times 10^2 \text{ g } Al_2(C_2O_4)_3 \cdot 4\,H_2O} = 0.0585 \text{ mol } Al_2(C_2O_4)_3 \cdot 4\,H_2O$

(19) (a) $0.581 \text{ mol } C_3H_7OH \times \dfrac{60.1 \text{ g } C_3H_7OH}{1 \text{ mol } C_3H_7OH} = 34.9 \text{ g } C_3H_7OH$

(b) $4.28 \text{ mol } NaC_2H_3O_2 \times \dfrac{82.0 \text{ g } NaC_2H_3O_2}{1 \text{ mol } NaC_2H_3O_2} = 351 \text{ g } NaC_2H_3O_2$

(c) $0.0913 \text{ mol } Mg_3(PO_4)_2 \times \dfrac{263 \text{ g } Mg_3(PO_4)_2}{1 \text{ mol } Mg_3(PO_4)_2} = 24.0 \text{ g } Mg_3(PO_4)_2$

(d) $0.148 \text{ mol Ba(OH)}_2 \cdot 8\ H_2O \times \frac{315 \text{ g Ba(OH)}_2 \cdot 8\ H_2O}{1 \text{ mol Ba(OH)}_2 \cdot 8\ H_2O} = 46.6 \text{ g Ba(OH)}_2 \cdot 8\ H_2O$

(20) (a) $0.818 \text{ mol K} \times \frac{6.02 \times 10^{23} \text{ K atoms}}{1 \text{ mol K}} = 4.92 \times 10^{23} \text{ K atoms}$

(b) $70.3 \text{ g } C_2H_6 \times \frac{1 \text{ mol } C_2H_6}{30.0 \text{ g } C_2H_6} \times \frac{6.02 \times 10^{23}\ C_2H_6 \text{ molecules}}{1 \text{ mol } C_2H_6} = 1.41 \times 10^{24}\ C_2H_6 \text{ molecules}$

(c) $3.78 \text{ g } NH_4Cl \times \frac{1 \text{ mol } NH_4Cl}{53.5 \text{ g } NH_4Cl} \times \frac{6.02 \times 10^{23} \text{ units } NH_4Cl}{1 \text{ mol } NH_4Cl} = 4.25 \times 10^{22} \text{ units } NH_4Cl$

(d) $0.629 \text{ mol Al} \times \frac{6.02 \times 10^{23} \text{ Al atoms}}{1 \text{ mol Al}} = 3.79 \times 10^{23} \text{ Al atoms}$

(e) $6.57 \text{ g Ca(NO}_3)_2 \times \frac{1 \text{ mol Ca(NO}_3)_2}{164 \text{ g Ca(NO}_3)_2} \times \frac{6.02 \times 10^{23} \text{ Ca(NO}_3)_2 \text{ units}}{1 \text{ mol Ca(NO}_3)_2} = 2.41 \times 10^{22} \text{ units Ca(NO}_3)_2$

(f) $0.184 \text{ mol } CS_2 \times \frac{6.02 \times 10^{23}\ CS_2 \text{ molecules}}{1 \text{ mol } CS_2} = 1.11 \times 10^{23}\ CS_2 \text{ molecules}$

(21) (a) $1.84 \times 10^{22} \text{ Ar atoms} \times \frac{1 \text{ mol Ar}}{6.02 \times 10^{23} \text{ Ar atoms}} = 0.0306 \text{ mol Ar}$

(b) $9.24 \times 10^{24} \text{ units KOH} \times \frac{1 \text{ mol KOH}}{6.02 \times 10^{23} \text{ units KOH}} = 15.3 \text{ mol KOH}$

(22) (a) $4.11 \times 10^{22}\ N_2O \text{ molecules} \times \frac{1 \text{ mol } N_2O}{6.02 \times 10^{23}\ N_2O \text{ molecules}} \times \frac{44.0 \text{ g } N_2O}{1 \text{ mol } N_2O} = 3.00 \text{ g } N_2O$

(b) $1.03 \times 10^{24} \text{ K atoms} \times \frac{1 \text{ mol K}}{6.02 \times 10^{23} \text{ K atoms}} \times \frac{39.1 \text{ g K}}{1 \text{ mol K}} = 66.9 \text{ g K}$

(23) $0.500 \text{ carat} \times \frac{200 \text{ mg C}}{1 \text{ carat}} \times \frac{1 \text{ g}}{1000 \text{ mg}} \times \frac{1 \text{ mol C}}{12.0 \text{ g C}} \times \frac{6.02 \times 10^{23} \text{ C atoms}}{1 \text{ mol C}} = 5.02 \times 10^{21} \text{ C atoms}$

(24) $495 \text{ g } H_2O \times \frac{1 \text{ mol } H_2O}{18.0 \text{ g } H_2O} \times \frac{6.02 \times 10^{23}\ H_2O \text{ molecules}}{1 \text{ mol } H_2O} = 1.66 \times 10^{25}\ H_2O \text{ molecules}$

(25) (a & b) $3.61 \text{ g } F_2 \times \frac{1 \text{ mol } F_2}{38.0 \text{ g } F_2} \times \frac{6.02 \times 10^{23}\ F_2 \text{ molecules}}{1 \text{ mol } F_2} \times \frac{2 \text{ F atoms}}{1\ F_2 \text{ molecule}} = 1.14 \times 10^{23}\ F_2 \text{ atoms (b)}$

$= 5.72 \times 10^{22}\ F_2 \text{ molecules (a)}$

(c) $3.61 \text{ g F} \times \frac{1 \text{ mol F}}{19.0 \text{ g F}} \times \frac{6.02 \times 10^{23} \text{ F atoms}}{1 \text{ mol F}} = 1.14 \times 10^{23} \text{ F atoms}$

(d) $3.61 \times 10^{23} \text{ F atoms} \times \frac{1 \text{ mol F}}{6.02 \times 10^{23} \text{ F atoms}} \times \frac{19.0 \text{ g F}}{1 \text{ mol F}} = 11.4 \text{ g F}$

(e) $3.61 \times 10^{23}\ F_2 \text{ molecules} \times \frac{1 \text{ mol } F_2}{6.02 \times 10^{23}\ F_2 \text{ molecules}} \times \frac{38.0 \text{ g } F_2}{1 \text{ mol } F_2} = 22.8 \text{ g } F_2$

(26) 6 and 10 are both divisible by 2, so the simplest formula of C_6H_{10} is C_3H_5. C_7H_{10} cannot be "reduced," and is therefore a simplest formula. It also happens to be a molecular formula.

	Element	Grams	Moles	Mole ratio	Formula ratio	Simplest formula	
(27)	C	52.2	4.35	2.00	2		
	H	13.0	12.9	5.92	6		
	O	34.8	2.18	1.00	1	C_2H_6O	
(28)	Fe	11.89	0.213	1.00	2		
	O	5.10	0.319	1.50	3	Fe_2O_3	g O = 16.99 − 11.89 = 5.10
(29)	C	17.2	1.43	1.00	1		
	H	1.44	1.43	1.00	1		
	F	81.4	4.28	3.00	3	CHF_3	

(30)	C	38.7	3.23	1.00	1		
	H	9.7	9.6	2.97	3		$\frac{62.0}{31.0} = 2$
	O	51.6	3.23	1.00	1	CH_3O	$C_2H_6O_2$
(31)	C	24.8	2.07	1.00	1		$\frac{97}{48.5} = 2$
	H	2.1	2.1	1.0	1		
	Cl	73.1	2.06	1.00	1	CHCl	$C_2H_2Cl_2$

(64) True: c; false, a, b, d, e, f.

(65) Hardly—about 2.3 pounds

$$10^{25}\text{ Cu atoms} \times \frac{1\text{ mol Cu}}{6.02 \times 10^{23}\text{ Cu atoms}} \times \frac{63.6\text{ g Cu}}{1\text{ mol Cu}} = 1.06 \times 10^3\text{ g} = 1.06\text{ kg}$$

(66) $$85.0\text{ g }P_4 \times \frac{1\text{ mol }P_4}{124\text{ g }P_4} \times \frac{6.02 \times 10^{23}\ P_4\text{ molecules}}{1\text{ mol }P_4} \times \frac{4\text{ P atoms}}{1\ P_4\text{ molecule}} = 1.65 \times 10^{24}\text{ P atoms}$$

$$= 4.13 \times 10^{23}\ P_4\text{ molecules}$$

(67) $$2.95 \times 10^{22}\text{ air ``molecules''} \times \frac{1\text{ mol ``air''}}{6.02 \times 10^{23}\text{ air ``molecules''}} \times \frac{29\text{ g air}}{1\text{ mol ``air''}} = 1.42\text{ g air}$$

(68) $$5.62 \times 10^{23}\ C_8H_{18}\text{ molecules} \times \frac{1\text{ mol }C_8H_{18}}{6.02 \times 10^{23}\ C_8H_{18}\text{ molecules}} \times \frac{114\text{ g }C_8H_{18}}{1\text{ mol }C_8H_{18}} = 106\text{ g }C_8H_{18}$$

(69) $$86.9\text{ g }CO_2 \times \frac{12.0\text{ g C}}{44.0\text{ g }CO_2} = 23.7\text{ g C} \qquad 35.5\text{ g }H_2O \times \frac{2.02\text{ g H}}{18.0\text{ g }H_2O} = 3.98\text{ g H}$$

	Element	Grams	Moles	Mole ratio	Formula ratio	Simplest formula	
	C	23.7	1.98	1	1		
	H	3.98	3.94	1.99	2	CH_2	
(70) (a)	Co	42.4	0.720	1.00	1		
	S	23.0	0.717	1.00	1		
	O	34.6	2.16	3.02	3	$CoSO_3$	
(b)	$CoSO_3$	26.1	0.188	1.00	1		43.0 g hydrate
	H_2O	16.9	0.939	4.99	5	$CoSO_3 \cdot$ 5 H_2O	<u>26.1</u> g $CoSO_3$ 16.9 g H_2O
(71)	C	76.4	6.37	7.03	7		$\frac{330}{110} = 3$
	H	9.2	9.11	10.1	10		
	O	14.5	0.906	1.00	1	$C_7H_{10}O$	$C_{21}H_{30}O_3$

CHAPTER 8

Answers to Equation-Balancing Exercises

(1) $4\ Na + O_2 \rightarrow 2\ Na_2O$

(2) $H_2 + Cl_2 \rightarrow 2\ HCl$

(3) $4\ P + 3\ O_2 \rightarrow 2\ P_2O_3$

(4) $KClO_4 \rightarrow KCl + 2\ O_2$

(5) $Sb_2S_3 + 6\ HCl \rightarrow 2\ SbCl_3 + 3\ H_2S$

(6) $2\ NH_3 + H_2SO_4 \rightarrow (NH_4)_2SO_4$

(7) $CuO + 2\ HCl \rightarrow CuCl_2 + H_2O$

(8) $Zn + Pb(NO_3)_2 \rightarrow Zn(NO_3)_2 + Pb$

(9) $2\ AgNO_3 + H_2S \rightarrow Ag_2S + 2\ HNO_3$

(10) $2\ Cu + S \rightarrow Cu_2S$

(11) $2\ Al + 2\ H_3PO_4 \rightarrow 3\ H_2 + 2\ AlPO_4$

(12) $2\ NaNO_3 \rightarrow 2\ NaNO_2 + O_2$

(13) $Mg(ClO_3)_2 \rightarrow MgCl_2 + 3\ O_2$

(14) $2\ H_2O_2 \rightarrow 2\ H_2O + O_2$

(15) $2\ BaO_2 \rightarrow 2\ BaO + O_2$
(16) $H_2CO_3 \rightarrow H_2O + CO_2$
(17) $Pb(NO_3)_2 + 2\ KCl \rightarrow PbCl_2 + 2\ KNO_3$
(18) $2\ Al + 3\ Cl_2 \rightarrow 2\ AlCl_3$
(19) $4\ P + 5\ O_2 \rightarrow 2\ P_2O_5$
(20) $NH_4NO_2 \rightarrow N_2 + 2\ H_2O$
(21) $3\ H_2 + N_2 \rightarrow 2\ NH_3$
(22) $Cl_2 + 2\ KBr \rightarrow Br_2 + 2\ KCl$
(23) $BaCl_2 + (NH_4)_2CO_3 \rightarrow BaCO_3 + 2\ NH_4Cl$
(24) $MgCO_3 + 2\ HCl \rightarrow MgCl_2 + CO_2 + H_2O$
(25) $2\ P + 3\ I_2 \rightarrow 2\ PI_3$
(26) $2\ PbO_2 \rightarrow 2\ PbO + O_2$
(27) $2\ Al + 6\ HCl \rightarrow 2\ AlCl_3 + 3\ H_2$
(28) $Fe_2(SO_4)_3 + 3\ Ba(OH)_2 \rightarrow 3\ BaSO_4 + 2\ Fe(OH)_3$
(29) $2\ Al + 3\ CuSO_4 \rightarrow Al_2(SO_4)_3 + 3\ Cu$
(30) $2\ KClO_3 \rightarrow 2\ KCl + 3\ O_2$
(31) $3\ Mg + N_2 \rightarrow Mg_3N_2$
(32) $2\ C_6H_{14} + 19\ O_2 \rightarrow 12\ CO_2 + 14\ H_2O$
(33) $3\ FeCl_2 + 2\ Na_3PO_4 \rightarrow Fe_3(PO_4)_2 + 6\ NaCl$
(34) $Li_2O + HOH \rightarrow 2\ LiOH$
(35) $2\ HgO \rightarrow 2\ Hg + O_2$
(36) $CaSO_4 \cdot 2\ H_2O \rightarrow CaSO_4 + 2\ H_2O$
(37) $2\ C_3H_7CHO + 11\ O_2 \rightarrow 8\ CO_2 + 8\ H_2O$
(38) $NaHCO_3 + HCl \rightarrow NaCl + H_2O + CO_2$
(39) $Bi(NO_3)_3 + 3\ NaOH \rightarrow Bi(OH)_3 + 3\ NaNO_3$
(40) $FeS + 2\ HBr \rightarrow FeBr_2 + H_2S$
(41) $Zn(OH)_2 + H_2SO_4 \rightarrow ZnSO_4 + 2\ HOH$
(42) $P_4O_{10} + 6\ H_2O \rightarrow 4\ H_3PO_4$
(43) $C_4H_9OH + 6\ O_2 \rightarrow 4\ CO_2 + 5\ H_2O$
(44) $CaC_2 + 2\ H_2O \rightarrow C_2H_2 + Ca(OH)_2$
(45) $3\ CaCO_3 + 2\ H_3PO_4 \rightarrow Ca_3(PO_4)_2 + 3\ CO_2 + 3\ H_2O$
(46) $PCl_5 + 4\ H_2O \rightarrow H_3PO_4 + 5\ HCl$
(47) $CaI_2 + H_2SO_4 \rightarrow 2\ HI + CaSO_4$
(48) $C_3H_7COOH + 5\ O_2 \rightarrow 4\ CO_2 + 4\ H_2O$
(49) $Mg(CN)_2 + 2\ HCl \rightarrow 2\ HCN + MgCl_2$
(50) $(NH_4)_2S + HgBr_2 \rightarrow 2\ NH_4Br + HgS$

Questions and Problems

(1) Two molecules of benzene react with fifteen molecules of oxygen to produce twelve molecules of carbon dioxide and six molecules of water.
(2) $2\ Ca(s) + O_2(g) \rightarrow 2\ CaO(s)$
(3) $4\ P(s) + 5\ O_2(g) \rightarrow P_4O_{10}(s)$; also $P_4(s) + 5\ O_2(g) \rightarrow P_4O_{10}(s)$
(4) $2\ K(s) + F_2(g) \rightarrow 2\ KF(s)$
(5) $Si(s) + 2\ Cl_2(g) \rightarrow SiCl_4(s)$
(6) $O_2(g) + 2\ F_2(g) \rightarrow 2\ OF_2(g)$
(7) $3\ Mg(s) + N_2(g) \rightarrow Mg_3N_2(s)$

(8) $2\ NaCl(s) \rightarrow 2\ Na(s) + Cl_2(g)$

(9) $2\ HgO(s) \rightarrow 2\ Hg(\ell) + O_2(g)$

(10) $H_2CO_3(aq) \rightarrow H_2O(\ell) + CO_2(g)$

(11) $2\ H_2O_2(\ell) \rightarrow 2\ H_2O(\ell) + O_2(g)$

(12) $C_3H_8(g) + 5\ O_2(g) \rightarrow 3\ CO_2(g) + 4\ H_2O(\ell)$

(13) $2\ C_2H_2(g) + 5\ O_2(g) \rightarrow 4\ CO_2(g) + 2\ H_2O(\ell)$

(14) $2\ CH_3CHO(\ell) + 5\ O_2(g) \rightarrow 4\ CO_2(g) + 4\ H_2O(\ell)$

(15) $C_{12}H_{22}O_{11}(s) + 12\ O_2(g) \rightarrow 12\ CO_2(g) + 11\ H_2O(\ell)$

(16) $Mg(s) + H_2SO_4(aq) \rightarrow MgSO_4(aq) + H_2(g)$ Redox

(17) $Br_2(\ell) + 2\ NaI(aq) \rightarrow I_2(aq) + 2\ NaBr(aq)$ Redox

(18) $MgCl_2(aq) + 2\ NaF(aq) \rightarrow 2\ NaCl(aq) + MgF_2(s)$ Precipitation

(19) $KOH(aq) + HNO_3(aq) \rightarrow KNO_3(aq) + HOH(\ell)$ Neutralization

(20) $Mg(OH)_2(s) + 2\ HCl(aq) \rightarrow MgCl_2(aq) + 2\ HOH(\ell)$ Neutralization

(21) $BaCl_2(aq) + Na_2SO_4(aq) \rightarrow BaSO_4(s) + 2\ NaCl(aq)$ Precipitation

(22) $Ba(s) + 2\ HOH(\ell) \rightarrow Ba(OH)_2(aq) + H_2(g)$ Redox

(23) $AgNO_3(aq) + KBr(aq) \rightarrow AgBr(s) + KNO_3(aq)$ Precipitation

(24) $2\ Li(s) + MnCl_2(aq) \rightarrow Mn(s) + 2\ LiCl(aq)$ Redox

(25) $Na_2S(aq) + 2\ AgNO_3(aq) \rightarrow Ag_2S(s) + 2\ NaNO_3(aq)$ Precipitation

(26) $2\ NaOH(aq) + H_2C_2O_4(aq) \rightarrow Na_2C_2O_4(aq) + 2\ HOH(\ell)$ Neutralization

(27) $2\ NaIO_3(aq) + CuSO_4(aq) \rightarrow Cu(IO_3)_2(s) + Na_2SO_4(aq)$ Precipitation

(28) $Mg(s) + NiCl_2(aq) \rightarrow MgCl_2(aq) + Ni(s)$ Redox

(29) $AgNo_3(aq) + KI(aq) \rightarrow AgI(s) + KNO_3(aq)$ Precipitation

(30) $Pb(NO_3)_2(aq) + CuSO_4(aq) \rightarrow PbSO_4(s) + Cu(NO_3)_2(aq)$ Precipitation

(31) $Zn(s) + H_2O(g) \rightarrow ZnO(s) + H_2(g)$

(32) $BaO(s) + H_2O(\ell) \rightarrow Ba(OH)_2(aq)$

(33) $3\ FeO(s) + 2\ Al(s) \rightarrow 3\ Fe(s) + Al_2O_3(s)$

(34) $Fe_2O_3(s) + 3\ CO(g) \rightarrow 2\ Fe(s) + 3\ CO_2(g)$

(70) True: a, b, c, d, e; false: f, g, h. Note: While c and d are "the truth," they are not "the whole truth." Compounds can also be reactants in combination reactions, and products in decomposition reactions.

(71) (a) Redox: $Pb + Cu(NO_3)_2 \rightarrow Cu + Pb(NO_3)_2$
(b) Neutralization: $Mg(OH)_2 + 2\ HBr \rightarrow MgBr_2 + 2\ HOH$
(c) Burning: $C_5H_{10}O + 7\ O_2 \rightarrow 5\ CO_2 + 5\ H_2O$
(d) Precipitation: $Na_2CO_3 + CaSO_4 \rightarrow CaCO_3 + Na_2SO_4$
(e) Decomposition: $2\ LiBr \rightarrow 2\ Li + Br_2$
(f) Precipitation: $NH_4Cl + AgNO_3 \rightarrow AgCl + NH_4NO_3$
(g) Combination: $Ca + Cl_2 \rightarrow CaCl_2$
(h) Redox: $F_2 + 2\ NaI \rightarrow I_2 + 2\ NaF$
(i) Precipitation: $Zn(NO_3)_2 + Ba(OH)_2 \rightarrow Zn(OH)_2 + Ba(NO_3)_2$
(j) Redox: $Cu + NiCl_2 \rightarrow Ni + CuCl_2$

(72) Never.

(73) (1) Combination: $S + O_2 \rightarrow SO_2$. (2) Combination: $2\ SO_2 + O_2 \rightarrow 2\ SO_3$. (3) Combination: $SO_3 + H_2O \rightarrow H_2SO_4$.

(74) $H_2SO_4 + CaCO_3 \rightarrow CaSO_4 + H_2CO_3$. $H_2SO_4 + CaCO_3 \rightarrow CaSO_4 + CO_2 + H_2O$

(75) $Ag + 2\ H_2S + O_2 \rightarrow AgS_2 + 2\ H_2O$

(76) $S + 3\ F_2 \rightarrow SF_6$. $S + Cl_2 \rightarrow SCl_2$. $2\ S + Br_2 \rightarrow S_2Br_2$

(77) Redox: $3\ H_2 + WO_3 \rightarrow W + 3\ H_2O$

CHAPTER 9

(1) $3.40 \text{ mol } C_4H_{10} \times \frac{13 \text{ mol } O_2}{2 \text{ mol } C_4H_{10}} = 22.1 \text{ mol } O_2$

(2) $4.68 \text{ mol } C_4H_{10} \times \frac{8 \text{ mol } CO_2}{2 \text{ mol } C_4H_{10}} = 18.7 \text{ mol } CO_2$

(3) $0.568 \text{ mol } CO_2 \times \frac{10 \text{ mol } H_2O}{8 \text{ mol } CO_2} = 0.710 \text{ mol } H_2O$

(4) $1.42 \text{ mol } O_2 \times \frac{2 \text{ mol } C_4H_{10}}{13 \text{ mol } O_2} \times \frac{58.1 \text{ g } C_4H_{10}}{1 \text{ mol } C_4H_{10}} = 12.7 \text{ g } C_4H_{10}$

(5) $9.43 \text{ g } O_2 \times \frac{1 \text{ mol } O_2}{32.0 \text{ g } O_2} \times \frac{10 \text{ mol } H_2O}{13 \text{ mol } O_2} = 0.227 \text{ mol } H_2O$

(6) $78.4 \text{ g } C_4H_{10} \times \frac{1 \text{ mol } C_4H_{10}}{58.1 \text{ g } C_4H_{10}} \times \frac{8 \text{ mol } CO_2}{2 \text{ mol } C_4H_{10}} \times \frac{44.0 \text{ g } CO_2}{1 \text{ mol } CO_2} = 237 \text{ g } CO_2$

(7) $43.8 \text{ g } H_2O \times \frac{1 \text{ mol } H_2O}{18.0 \text{ g } H_2O} \times \frac{13 \text{ mol } O_2}{10 \text{ mol } H_2O} \times \frac{32.0 \text{ g } O_2}{1 \text{ mol } O_2} = 101 \text{ g } O_2$

(8) $47.1 \text{ g Al} \times \frac{1 \text{ mol Al}}{27.0 \text{ g Al}} \times \frac{1 \text{ mol } Fe_2O_3}{2 \text{ mol Al}} \times \frac{160 \text{ g } Fe_2O_3}{1 \text{ mol } Fe_2O_3} = 1.40 \times 10^2 \text{ g } Fe_2O_3$

(9) $100 \text{ g } C_{600}H_{1000}O_{500} \times \frac{1 \text{ mol } C_{600}H_{1000}O_{500}}{1.62 \times 10^4 \text{ g } C_{600}H_{1000}O_{500}} \times \frac{50 \text{ mol } C_{12}H_{22}O_{11}}{1 \text{ mol } C_{600}H_{1000}O_{500}}$

$\times \frac{342 \text{ g } C_{12}H_{22}O_{11}}{1 \text{ mol } C_{12}H_{22}O_{11}} = 106 \text{ g } C_{12}H_{22}O_{11}$

(10) $600 \text{ g } NaHCO_3 \times \frac{1 \text{ mol } NaHCO_3}{84.0 \text{ g } NaHCO_3} \times \frac{1 \text{ mol } H_2SO_4}{2 \text{ mol } NaHCO_3} \times \frac{98.1 \text{ g } H_2SO_4}{1 \text{ mol } H_2SO_4} = 350 \text{ g } H_2SO_4$

(11) $5.00 \text{ kg } Na_2SO_4 \times \frac{1 \text{ kmol } Na_2SO_4}{142 \text{ kg } Na_2SO_4} \times \frac{4 \text{ kmol NaCl}}{2 \text{ kmol } Na_2SO_4} \times \frac{58.5 \text{ kg NaCl}}{1 \text{ kmol NaCl}} = 4.12 \text{ kg NaCl}$

(12) (a) $1.90 \text{ kg } C_7H_8 \times \frac{1 \text{ kmol } C_7H_8}{92.1 \text{ kg } C_7H_8} \times \frac{3 \text{ kmol } HNO_3}{1 \text{ kmol } C_7H_8} \times \frac{63.0 \text{ kg } HNO_3}{1 \text{ kmol } HNO_3} = 3.90 \text{ kg } HNO_3$

(b) $1.90 \text{ kg } C_7H_8 \times \frac{1 \text{ kmol } C_7H_8}{92.1 \text{ kg } C_7H_8} \times \frac{1 \text{ kmol } C_7H_5N_3O_6}{1 \text{ kmol } C_7H_8} \times \frac{227 \text{ kg } C_7H_5N_3O_6}{1 \text{ kmol } C_7H_5N_3O_6} = 4.86 \text{ kg } C_7H_5N_3O_6$

(13) $778 \text{ kg } Fe_2O_3 \times \frac{1 \text{ kmol } Fe_2O_3}{160 \text{ kg } Fe_2O_3} \times \frac{2 \text{ kmol Fe}}{1 \text{ kmol } Fe_2O_3} \times \frac{55.9 \text{ kg Fe}}{1 \text{ kmol Fe}} = 544 \text{ kg Fe}$

$778 \text{ kg } Fe_2O_3 \times \frac{2(55.9) \text{ kg Fe}}{160 \text{ kg } Fe_2O_3} = 544 \text{ kg Fe}$

(14) $81.2 \text{ g } NH_3 \times \frac{1 \text{ mol } NH_3}{17.0 \text{ g } NH_3} \times \frac{1 \text{ mol } NH_4HCO_3}{1 \text{ mol } NH_3} \times \frac{79.1 \text{ g } NH_4HCO_3}{1 \text{ mol } NH_4HCO_3} = 378 \text{ g } NH_4HCO_3$

(15) $448 \text{ g } Na_2CO_3 \times \frac{1 \text{ mol } Na_2CO_3}{106 \text{ g } Na_2CO_3} \times \frac{2 \text{ mol } NaHCO_3}{1 \text{ mol } Na_2CO_3} \times \frac{84.0 \text{ g } NaHCO_3}{1 \text{ mol } NaHCO_3} = 710 \text{ g } NaHCO_3$

(16) $0.500 \text{ ton } Ca(H_2PO_4)_2 \times \frac{2000 \text{ lb}}{1 \text{ ton}} \times \frac{1 \text{ kg}}{2.20 \text{ lb}} \times \frac{1 \text{ kmol } Ca(H_2PO_4)_2}{234 \text{ kg } Ca(H_2PO_4)_2} \times \frac{1 \text{ kmol } Ca_3(PO_4)_2}{1 \text{ kmol } Ca(H_2PO_4)_2} \times \frac{310 \text{ kg } Ca_3PO_4}{1 \text{ kmol } Ca_3PO_4}$

$\times \frac{100 \text{ kg rock}}{79.4 \text{ kg } Ca_3PO_4} = 758 \text{ kg rock}$

(17) $40.1 \text{ kg sludge} \times \frac{23.1 \text{ kg AgCl}}{100 \text{ kg sludge}} \times \frac{1 \text{ kmol AgCl}}{144 \text{ kg AgCl}} \times \frac{4 \text{ kmol NaCN}}{2 \text{ kmol AgCl}} \times \frac{49.0 \text{ kg NaCN}}{1 \text{ kmol NaCN}} = 6.30 \text{ kg NaCN}$

(18) $41.9 \text{ g } Cu_2S \times \frac{1 \text{ mol } Cu_2S}{159 \text{ g } Cu_2S} \times \frac{2 \text{ mol Cu}}{1 \text{ mol } Cu_2S} \times \frac{63.6 \text{ g Cu}}{1 \text{ mol Cu}} = 33.5 \text{ g Cu (theo)}$

$\frac{29.2 \text{ g Cu (actual)}}{33.5 \text{ g Cu (theo)}} \times 100 = 87.2\% \text{ yield}$

(19) $557 \text{ kg NaCl} \times \frac{1 \text{ kmol NaCl}}{58.5 \text{ kg NaCl}} \times \frac{2 \text{ kmol HCl}}{2 \text{ kmol NaCl}} \times \frac{36.5 \text{ kg HCl (theo)}}{1 \text{ kmol HCl}} \times \frac{82.6 \text{ kg HCl (actual)}}{100 \text{ kg HCl (theo)}}$

$= 287 \text{ kg HCl (actual)}$

(20) $38.5\text{ g CCl}_4\text{ (actual)} \times \frac{100\text{ g CCl}_4\text{ (theo)}}{85\text{ g CCl}_4\text{ (actual)}} \times \frac{1\text{ mol CCl}_4}{154\text{ g CCl}_4} \times \frac{2\text{ mol S}_2\text{Cl}_2}{1\text{ mol CCl}_4} \times \frac{135\text{ g S}_2\text{Cl}_2}{1\text{ mol S}_2\text{Cl}_2} = 79\text{ g S}_2\text{Cl}_2$

(21) $5.95\text{ g MnO}_2 \times \frac{1\text{ mol MnO}_2}{86.9\text{ g MnO}_2} \times \frac{1\text{ mol Cl}_2}{1\text{ mol MnO}_2} \times \frac{71.0\text{ g Cl}_2}{1\text{ mol Cl}_2} = 4.86\text{ g Cl}_2\text{ (theo)}$

$\frac{4.22\text{ g Cl}_2\text{ (actual)}}{4.86\text{ g Cl}_2\text{ (theo)}} \times 100 = 86.8\%\text{ yield}$

(22) $397\text{ kg CH}_3\text{OH} \times \frac{1\text{ mol CH}_3\text{OH}}{32.0\text{ g CH}_3\text{OH}} \times \frac{2\text{ mol HCHO}}{2\text{ mol CH}_3\text{OH}} \times \frac{30.0\text{ g HCHO (theo)}}{1\text{ mol HCHO}} \times \frac{84.9\text{ g HCHO (actual)}}{100\text{ g HCHO (theo)}}$
$= 316\text{ g HCHO (actual)}$

(23) $105\text{ kg Cl}_2\text{ (actual)} \times \frac{100\text{ kg Cl}_2\text{ (theo)}}{61\text{ kg Cl}_2\text{ (actual)}} \times \frac{1\text{ kmol Cl}_2}{71.0\text{ kg Cl}_2} \times \frac{2\text{ kmol NaCl}}{1\text{ kmol Cl}_2} \times \frac{58.5\text{ kg NaCl}}{1\text{ kmol NaCl}} \times \frac{100\text{ kg water}}{9.6\text{ kg NaCl}}$
$= 3.0 \times 10^3\text{ kg water}$

(24) $74.4\text{ g NH}_3 \times \frac{1\text{ mol NH}_3}{17.0\text{ g NH}_3} \times \frac{1\text{ mol NH}_4\text{NO}_3}{1\text{ mol NH}_3} \times \frac{80.0\text{ g NH}_4\text{NO}_3}{1\text{ mol NH}_4\text{NO}_3} = 350\text{ g NH}_4\text{NO}_3$

$159\text{ g NHO}_3 \times \frac{1\text{ mol HNO}_3}{63.0\text{ g HNO}_3} \times \frac{1\text{ mol NH}_4\text{NO}_3}{1\text{ mol HNO}_3} \times \frac{80.0\text{ g NH}_4\text{NO}_3}{1\text{ mol NH}_4\text{NO}_3} = 202\text{ g NH}_4\text{NO}_3$

202 g NH_4NO_3 will form from the limiting reagent, HNO_3.

$159\text{ g HNO}_3 \times \frac{1\text{ mol HNO}_3}{63.0\text{ g HNO}_3} \times \frac{1\text{ mol NH}_3}{1\text{ mol HNO}_3} \times \frac{17.0\text{ g NH}_3}{1\text{ mol NH}_3} = 42.9\text{ g NH}_3\text{ react}$

$74.4\text{ g NH}_3\text{ (start)} - 42.9\text{ g NH}_3\text{ (react)} = 31.5\text{ g NH}_3\text{ remain}$

(25) $3.19\text{ g CCl}_3\text{CHO} \times \frac{1\text{ mol CCl}_3\text{CHO}}{148\text{ g CCl}_3\text{CHO}} \times \frac{1\text{ mol (ClC}_6\text{H}_4)_2\text{CHCCl}_3}{1\text{ mol CCl}_3\text{CHO}} \times \frac{355\text{ g (ClC}_6\text{H}_4)_2\text{CHCCl}_3}{1\text{ mol (ClC}_6\text{H}_4)_2\text{CHCCl}_3}$
$= 7.65\text{ g (ClC}_6\text{H}_4)_2\text{CHCCl}_3$

$4.54\text{ g C}_6\text{H}_5\text{Cl} \times \frac{1\text{ mol C}_6\text{H}_5\text{Cl}}{113\text{ g C}_6\text{H}_5\text{Cl}} \times \frac{1\text{ mol (ClC}_6\text{H}_4)_2\text{CHCCl}_3}{2\text{ mol C}_6\text{H}_5\text{Cl}} \times \frac{355\text{ g (ClC}_6\text{H}_4)_2\text{CHCCl}_3}{1\text{ mol (ClC}_6\text{H}_4)_2\text{CHCCl}_3}$
$= 7.13\text{ g (ClC}_6\text{H}_4)_2\text{CHCCl}_3 = \text{theoretical yield}$

$4.54\text{ g C}_6\text{H}_5\text{Cl} \times \frac{1\text{ mol C}_6\text{H}_5\text{Cl}}{113\text{ g C}_6\text{H}_5\text{Cl}} \times \frac{1\text{ mol CCl}_3\text{CHO}}{2\text{ mol C}_6\text{H}_5\text{Cl}} \times \frac{148\text{ g CCl}_3\text{CHO}}{1\text{ mol CCl}_3\text{CHO}} = 2.97\text{ g CCl}_3\text{CHO}$

3.19 g (start) − 2.97 g (used) = 0.22 g CCl_3CHO remain unreacted.

(26) $135\text{ g Na}_2\text{CO}_3 \times \frac{1\text{ mol Na}_2\text{CO}_3}{106\text{ g Na}_2\text{CO}_3} = \frac{2\text{ mol HNO}_3}{1\text{ mol Na}_2\text{CO}_3} \times \frac{63.0\text{ g HNO}_3}{1\text{ mol HNO}_3} = 160\text{ g HNO}_3$

135 g Na_2CO_3 will neutralize only 160 g HNO_3.

$135\text{ g Na}_2\text{CO}_3 \times \frac{1\text{ mol Na}_2\text{CO}_3}{106\text{ g Na}_2\text{CO}_3} \times \frac{1\text{ mol CO}_2}{1\text{ mol Na}_2\text{CO}_3} \times \frac{44.0\text{ g CO}_2}{1\text{ mol CO}_2} = 56.0\text{ g CO}_2$

(27) $239\text{ mg Ca(OH)}_2 \times \frac{1\text{ mmol Ca(OH)}_2}{74.1\text{ mg Ca(OH)}_2} \times \frac{1\text{ mmol SnF}_2}{1\text{ mmol Ca(OH)}_2} \times \frac{157\text{ mg SnF}_2}{1\text{ mmol SnF}_2} = 506\text{ mg SnF}_2\text{ required}$

The dentist is short by 506 mg (required) − 305 mg (used) = 201 mg SnF_2.

(28) $28.9\text{ g HgO} \times \frac{1\text{ mol HgO}}{217\text{ g HgO}} \times \frac{1\text{ mol O}_2}{2\text{ mol HgO}} \times \frac{22.4\text{ L O}_2}{1\text{ mol O}_2} = 1.49\text{ L O}_2$

(29) $C_7H_{16} + 11\ O_2 \rightarrow 7\ CO_2 + 8\ H_2O$

$$72.0\text{ g } C_7H_{16} \times \frac{1\text{ mol } C_7H_{16}}{100\text{ g } C_7H_{16}} \times \frac{11\text{ mol } O_2}{1\text{ mol } C_7H_{16}} \times \frac{22.4\text{ L } O_2}{1\text{ mol } O_2} \times \frac{100\text{ L air}}{21\text{ L } O_2} = 8.4 \times 10^2\text{ L air}$$

(30) $$0.345\text{ L NO} \times \frac{1\text{ mol NO}}{22.4\text{ L NO}} \times \frac{2\text{ mol } NO_2}{2\text{ mol NO}} \times \frac{22.4\text{ L } NO_2}{1\text{ mol } NO_2} = 0.345\text{ L } NO_2$$

For gases measured at the same temperature and pressure, volume is proportional to moles. The first and third steps of the stoichiometry pattern therefore cancel and the unit path becomes L given → L wanted. Coefficients from the equation become the conversion factor:

$$0.345\text{ L NO} \times \frac{2\text{ L } NO_2}{2\text{ L NO}} = 0.345\text{ L } NO_2$$

More about this appears in Chapter 13.

(31) $C_3H_8(g) + 5\ O_2(g) \rightarrow 3\ CO_2(g) + 4\ H_2O(\ell) + 2220\text{ kJ}$
$C_3H_8(g) + 5\ O_2(g) \rightarrow 3\ CO_2(g) + 4\ H_2O(\ell) \quad \Delta H = -2220\text{ kJ}$

(32) $CaO(s) + H_2O(\ell) \rightarrow Ca(OH)_2(s) + 65.3\text{ kJ}$
$CaO(s) + H_2O(\ell) \rightarrow Ca(OH)_2(s) \quad \Delta H = -65.3\text{ kJ}$

(33) $2\ Al_2O_3(s) + 3\ C(s) + 2160\text{ kJ} \rightarrow 3\ CO_2(g) + 4\ Al(s)$
$2\ Al_2O_3(s) + 3\ C(s) \rightarrow 3\ CO_2(g) + 4\ Al(s) \quad \Delta H = +2160\text{ kJ}$

(34) $$356\text{ kJ} \times \frac{2\text{ mol } H_2O}{572\text{ kJ}} \times \frac{18.0\text{ g } H_2O}{1\text{ mol } H_2O} \times \frac{1\text{ mL } H_2O}{1.00\text{ g } H_2O} = 22.4\text{ mL } H_2O$$

(35) $$454\text{ g } C_6H_{12}O_6 \times \frac{1\text{ mol } C_6H_{12}O_6}{180\text{ g } C_6H_{12}O_6} \times \frac{2.82 \times 10^3\text{ kJ}}{1\text{ mol } C_6H_{12}O_6} = 7.11 \times 10^3\text{ kJ}$$

(36) $$1.50 \times 10^3\text{ g } C_4H_{10} \times \frac{1\text{ mol } C_4H_{10}}{58.1\text{ g } C_4H_{10}} \times \frac{5.77 \times 10^3\text{ kJ}}{2\text{ mol } C_4H_{10}} = 7.45 \times 10^4\text{ kJ}$$

(37) $$454\text{ g Al} \times \frac{1\text{ mol Al}}{27.0\text{ g Al}} \times \frac{1.97 \times 10^3\text{ kJ}}{4\text{ mol Al}} \times \frac{1\text{ kw-hr}}{3.60 \times 10^3\text{ kJ}} = 2.30\text{ kw-hr}$$

(76) b and c false; others true.

(77) (a) $$35\text{ g } N_2 \times \frac{1\text{ mol } N_2}{28.0\text{ g } N_2} \times \frac{1\text{ mol } Na_2CO_3}{1\text{ mol } N_2} \times \frac{106\text{ g } Na_2CO_3}{1\text{ mol } Na_2CO_3} = 133\text{ g } Na_2CO_3$$

(b) $$35\text{ g } N_2 \times \frac{1\text{ mol } N_2}{28.0\text{ g } N_2} \times \frac{3\text{ mol CO}}{1\text{ mol } N_2} \times \frac{22.4\text{ L CO}}{1\text{ mol CO}} = 84\text{ L CO}$$

(78) (a) $$68.7\text{ L } NH_3 \times \frac{3\text{ L } O_2}{4\text{ L } NH_3} = 51.5\text{ L } O_2$$ See answer to Problem 30.

(b) $$68.7\text{ L } NH_3 \times \frac{1\text{ mol } NH_3}{22.4\text{ L } NH_3} \times \frac{2\text{ mol } N_2}{4\text{ mol } NH_3} \times \frac{28.0\text{ g } N_2}{1\text{ mol } N_2} = 42.9\text{ g } N_2$$

(79) $$42.2\text{ g } K_2O \times \frac{1\text{ mol } K_2O}{94.2\text{ g } K_2O} \times \frac{3\text{ mol } O_2}{4\text{ mol } K_2O} \times \frac{22.4\text{ L } O_2\text{ (theo)}}{1\text{ mol } O_2} \times \frac{76.2\text{ L } O_2\text{ (actual)}}{100\text{ L } O_2\text{ (theo)}} = 5.73\text{ L } O_2$$

(80) $$578\text{ g } CH_3OH\text{ (actual)} \times \frac{100\text{ g } CH_3OH\text{ (theo)}}{78.6\text{ g } CH_3OH\text{ (actual)}} \times \frac{1\text{ mol } CH_3OH}{32.0\text{ g } CH_3OH} \times \frac{1\text{ mol CO}}{1\text{ mol } CH_3OH} \times \frac{22.4\text{ L CO}}{1\text{ mol CO}} = 515\text{ L CO}$$

(81) $NaCl + AgNO_3 \rightarrow AgCl + NaNO_3$

$$2.056\text{ g AgCl} \times \frac{1\text{ mol AgCl}}{143.4\text{ g AgCl}} \times \frac{1\text{ mol NaCl}}{1\text{ mol AgCl}} \times \frac{58.44\text{ g NaCl}}{1\text{ mol NaCl}} = 0.8379\text{ g NaCl}$$

1.6240 g mix − 0.83 g NaCl = 0.7861 g $NaNO_3$

$$\frac{0.8379\text{ g NaCl}}{1.6240\text{ g sample}} \times 100 = 51.59\%\text{ NaCl} \qquad \frac{0.7861\text{ g } NaNO_3}{1.6240\text{ g sample}} \times 100 = 48.41\%\text{ } NaNO_3$$

(82) $FeS_2 + 2\ O_2 \rightarrow 2\ SO_2 + Fe$

$$0.65\ L\ SO_2 \times \frac{1\ mol\ SO_2}{22.4\ L\ SO_2} \times \frac{1\ mol\ FeS_2}{2\ mol\ SO_2} \times \frac{55.9\ g\ Fe}{1\ mol\ FeS_2} = 0.81\ g\ Fe$$

$$\frac{0.81\ g\ Fe}{19.86\ g\ sample} \times 100 = 4.1\%\ Fe$$

(83) $3\ Cu \rightarrow 3\ Cu(NO_3)_2 \rightarrow 3\ Cu(OH)_2 \rightarrow 3\ CuO \rightarrow 3\ CuCl_2 \rightarrow Cu_3(PO_4)_2$

$$2.637\ g\ Cu_3(PO_4)_2 \times \frac{1\ mol\ Cu_3(PO_4)_2}{380.6\ g\ Cu_3(PO_4)_2} \times \frac{3\ mol\ Cu}{1\ mol\ Cu_3(PO_4)_2} \times \frac{63.55\ g\ Cu}{1\ mol\ Cu} = 1.321\ g\ Cu$$

$$\frac{1.321\ g\ Cu}{1.382\ g\ sample} \times 100 = 95.59\%\ Cu$$

(84) $$50.0\ mL \times \frac{1.19\ g\ soln}{1\ mL} \times \frac{17.0\ g\ NaOH}{100\ g\ soln} \times \frac{1\ mol\ NaOH}{40.0\ g\ NaOH} \times \frac{1\ mol\ Zn(NO_3)_2}{2\ mol\ NaOH} \times \frac{189\ g\ Zn(NO_3)_2}{1\ mol\ Zn(NO_3)_2} = 23.9\ g\ Zn(NO_3)_2$$

CHAPTER 10

(1) Na^+ Mg^{2+} Al^{3+} P^{3-} S^{2-} Cl^-.

(2) O^{2-} and F^-; less common, N^{3-}.

(3) S^{2-}, K^+, and Ca^{2+}; less common, P^{3-} and Sc^{3+}.

(4) (a) Krypton, Z = 36; (b) selenide ion, Se^{2-}; (c) Rb^+.

(5) S^{2-}.

(6) Ca^{2+}, K^+, Cl^-, Br^-.

(8) $2\ K\cdot + \cdot\ddot{\underset{..}{S}}: \rightarrow 2\ K^+ \left[:\ddot{\underset{..}{S}}:\right]^{2-}$ K_2S

(9) Ions are formed when neutral atoms lose or gain electrons. The electron(s) that is(are) lost by one atom is(are) gained by another, effectively a transfer of electrons from one atom to another. The attraction between the ions produced makes up the "ionic bond." Covalent bonds are formed when a pair of electrons is shared by the two bonded atoms. Effectively the electrons belong to both atoms, spending some time near each nucleus.

(10) K—Cl bond is ionic, formed by "transferring" an electron from a potassium atom to a chlorine atom. Cl—Cl bond is covalent, formed by two chlorine atoms sharing a pair of electrons.

(11) $:\ddot{\underset{..}{Cl}}\cdot + \cdot\ddot{\underset{..}{I}}: \rightarrow :\ddot{\underset{..}{Cl}}:\ddot{\underset{..}{I}}:$

(12) Nonmetal atoms are usually one, two, or possibly three or four short of an "octet" of electrons, and achieve that octet most easily by gaining the missing electrons. When two nonmetal atoms combine, the easiest way for both atoms to reach the octet is to gain each other's electrons, or share them, forming a covalent bond. If the second atom is a metal, however, it has one, two, or possibly three electrons more than an octet. It reaches the octet by giving its electrons to the nonmetal, becoming a positive ion itself, and making the nonmetal atom a negative ion. The two atoms form an ionic bond.

(15) Cl—Cl; Br—Cl; I—Cl; F—Cl. The Cl—Cl bond is nonpolar; the other bonds are polar.

(16) In Br—Cl and I—Cl, chlorine is the more electronegative; in F—Cl, fluorine is the negative pole.

(17) "Electronegativity is a measure of the relative ability of two atoms to attract the pair of electrons forming a single covalent bond between them." Because noble gases do not normally form bonds, electronegativity numbers are not assigned to them.

(38) True: c, d, e, f, g, h, j; false: a, b, i.

(39) H^+ has no electron configuration because it has no electrons.

(40) Decreasing. The ions are isoelectronic, and in order of increasing nuclear charge.

(41) c.

(42) e. Cesium (Z = 55) has the lowest electronegativity, and fluorine (Z = 9) has the highest. The electronegativity difference is therefore greater than any other pair, so the bond is the most ionic.

(43) Al^{3+} Ar Cl^- K.

(44) 4p from bromine and 2p from oxygen.

(45) Potassium atoms are larger because of the 4s electron that is lost when the atom forms a K^+ ion.

(46) Ionic bonds do not appear in molecular compounds, but covalent bonds exist in polyatomic ions that are present in many ionic compounds.

(47) The electronegativity of A is higher than the electronegativity of B, because electronegativities are highest at the top and right side of the periodic table and lowest at the bottom left. Because X is in a lower period than Y, the electronegativity of X should be larger than that of Y. But because Y is farther to the right, the electronegativity of X should be smaller than that of Y. Therefore no prediction can be made for X and Y.

(48) A bond between identical atoms is completely nonpolar. No "transfer" of electrons from one atom to another is ever absolutely complete, so there is a small amount of covalency in the most ionic bond.

CHAPTER 11

(1)

```
      ..          ..           ..
H—Br:      H—S:      H—P—H
      ..          |           |
                  H           H
  HBr            H2S          PH3
```

(2)

```
                                      [     ..     ]2-
                                      [    :O:     ]
 ..   ..                              [ ..  |   .. ]
:F—O:          :C≡O:                  [:O—S—O:     ]
 ..   |                               [ ..  |   .. ]
     :F:                              [    :O:     ]
      ..                              [     ..     ]
  OF2           CO                         SO4 2-
```

(3)

```
                [      ..     ]-           ..
                [     :O:     ]           :O:
  ..  ..        [  ..  |   .. ]       ..   |   ..
[:I—O:]-        [ :O—Br—O:    ]     H—O—S—O—H
  ..  ..        [  ..  |   .. ]       ..   |   ..
                [     :O:     ]           :O:
                [      ..     ]            ..
 ClO-               BrO4-                H2SO4
```

(4) C_4H_{10}:

```
   H  H  H  H
   |  |  |  |
H—C—C—C—C—H
   |  |  |  |
   H  H  H  H
```

C_4H_8:

```
      H  H  H                   H  H  H  H
      |  |  |                   |  |  |  |
H—C=C—C—C—H        or       H—C—C=C—C—H
   |     |  |                   |        |
   H     H  H                   H        H
```

C_4H_6:

```
          H  H               H           H
          |  |               |           |
H—C≡C—C—C—H   or   H—C—C≡C—C—H   or
          |  |               |           |
          H  H               H           H

H     H  H      H         H          H  H
 \    |  |     /           \         |  |
  C=C—C=C          or       C=C=C—C—H
 /          \              /         |
H            H            H          H
```

(5)

```
     ..            ..             ..
    :F:           :F:            :F:
     |             |         ..   |   ..
 H—C—H        H—C—H        :F—C—F:
     |             |         ..   |   ..
     H            :F:            :F:
                   ..             ..
   CH3F          CH2F2           CF4
```

(6) See page 497.

(7) C_5H_{10}:

```
   H  H  H  H  H              H  H  H  H  H
   |  |  |  |  |              |  |  |  |  |
H—C=C—C—C—C—H    or    H—C—C=C—C—C—H
         |  |  |              |        |  |
         H  H  H              H        H  H
```

(8) C_3H_6O:

```
   H  ˙Ö˙ H                 H  H
   |  ||  |                 |  |    Ö:
H—C—C—C—H     or     H—C—C—C//
   |      |                 |  |   \
   H      H                 H  H    H
```

(9) HCOOH:

```
       Ö:
      //
H—C
      \
       Ö—H
```

(10) $[H—\ddot{S}:]^-$ (HS⁻) H—S—H (H_2S) H—S—S—H (H_2S_2) H—S—S—S—H (H_2S_3)

H—S—S—S—S—H (H_2S_4) H—S—S—S—S—S—H (H_2S_5)

(12) The Lewis diagram for AsI_3 is at the right. The formula AsI_5 suggests that the central arsenic atom forms five bonds involving five electron pairs, or ten electrons. This is two more electrons than the eight in an octet.

```
:Ï—Äs—Ï:
     |
    :Ï:
```

(13) Substance	Electron-pair geometry	Molecular geometry
(13) BeH_2	Linear	Linear
CF_4	Tetrahedral	Tetrahedral
OF_2	Tetrahedral	Bent
(14) IO_4^-	Tetrahedral	Tetrahedral
ClO_2^-	Tetrahedral	Bent
CO_3^{2-}	Trigonal planar	Trigonal planar
(15) C in C_2H_5OH	Tetrahedral	Tetrahedral
(16) N in CH_3NH_2	Tetrahedral	Trigonal pyramid
(17) C in C_2H_4	Trigonal planar	Trigonal planar
(18) C in HCHO	Trigonal planar	Trigonal planar

(20) HCl, with an electronegativity difference of 0.9, is more polar than HI, with an electronegativity difference of 0.4. The halogen end is more negative in both molecules.

(21) H_2O is more polar than H_2S. Both bond and molecular polarity decrease for the hydrides of column 6A from oxygen to tellurium. H_2Te has zero electronegativity difference in its bonds, and the molecule is essentially nonpolar.

(22)

```
  ˙O˙              ˙O˙    H
 ↗   ↖            ↗   ↖  /
H     H          H     C
                      / \
                     H   H
 Water             Methanol
```

Bond angles around oxygen atoms are approximately equal, according to electron repulsion principle. H—O bond is much more polar than C—O bond (electronegativity differences 1.4 vs 0.4). Bonding electrons are therefore displaced more toward oxygen in the water molecule, which is the more polar of the two.

(46) True: a, c, d, g, i, j; false: b, e, f, j, h, k.

(47)

```
          H           H               H
          |             \           /
H—C≡C—C—H              C=C=C
          |             /           \
          H           H               H
```

(48)

```
    ..              ..
H—O—C—C—O—H
    ..   ||  ||    ..
        .O.. O.
```

(49)

```
        H     H
         \   /
    H      C      H
     \   /   \   /
  H—C         C—H
      \       /
     H—C—C—H
       /     \
      H       H
```

(50) See Table 12.2.

(51) Except for the elements, the diagrams are the same. All are 5-atom species, and all have 32 electrons. SiO_4^{4-} and CI_4 are also 5-atom species with 32 electrons, so they also have the same Lewis diagrams.

CHAPTER 12

(1) Copper; krypton; manganese; nitrogen.

(2) Na; H or H_2; Si; Pb.

(3) Calcium ion; chromium(III) ion; zinc ion; phosphide ion; bromide ion.

(4) Li^+; NH_4^+; N^{3-}; F^-; Hg^{2+}.

(5) Nitric acid; sulfurous acid; $HClO_4$; H_2SeO_4.

(6) SO_4^{2-}; ClO_2^-; iodate ion; hypobromite ion.

(7) HCO_3^-; $H_2PO_4^-$; hydrogen sulfate ion.

(8) K_2S; $Cu(NO_3)_2$; $NaHCO_3$.

(9) Magnesium sulfite; aluminum fluoride; lead carbonate [or lead(II) carbonate].

(10) Sulfur dioxide; dinitrogen oxide; PBr_3; HI.

(11) HSO_3^-; KNO_3; manganese(II) sulfate; sulfur trioxide.

(12) Bromate ion; nickel hydroxide; AgCl; SiF_6.

(13) TeO_4^{2-}; $FePO_4$; sodium acetate; hydrogen sulfide (or dihydrogen sulfide).

(14) Monohydrogen phosphate ion (or hydrogen phosphate ion); copper(II) oxide; $Na_2C_2O_4$; NH_3.

(15) HClO; $CrBr_2$; potassium hydrogen carbonate; sodium dichromate.

(16) Cobalt(III) oxide; sodium sulfite; HgI_2; $Al(OH)_3$.

(17) $Ca(H_2PO_4)_2$; $KMnO_4$; ammonium iodate; selenic acid.

(18) Mercury(I) chloride; periodic acid; $CoSO_4$; $Pb(NO_3)_2$.

(19) UF_3; BaO_2; manganese(II) chloride; sodium chlorite.

(20) Potassium tellurate; zinc carbonate; $CrCl_2$; $HC_2H_3O_2$.

(21) $BaCrO_4$; $CaSO_3$; copper(I) chloride; silver nitrate.

(22) Sodium peroxide; nickel carbonate; FeO; $H_2S(aq)$.

(23) Zn_3P_2; $CsNO_3$; ammonium cyanide; disulfur decafluoride.

(24) Dinitrogen trioxide; lithium permanganate; In_2Se_3; $Hg_2(SCN)_2$.

CHAPTER 13

(3) Because gas molecules are widely spaced, as much air as desired may be put into a tire. This compressibility produces a softer, more comfortable ride than a liquid-filled or solid tire would. The open spacing also gives air a lower density, contributing less to the weight of the automobile. The constant motion of gas molecules causes them to fill the tire and exert pressure uniformly. Because molecular collisions are without loss of energy, tire pressure remains constant, except for leaks and variations due to temperature changes.

(4) This is evidence that gas particles are moving.

(5) Because gas molecules are more widely spaced than liquid molecules, the gas is less dense and therefore rises through the liquid.

(6) Gas particles are widely separated compared to the same number of particles close together in the liquid state.

(7) When gas particles are pushed close to each other, the intermolecular attractions become significant. The molecules are no longer independent. The ideal gas model is violated, so the gas does not behave ideally.

(8) Particles move in straight lines until they hit something—eventually the walls of the container, thereby filling it.

(11)

atm	PSI	inches Hg	cm Hg	mm Hg	torr	Pa	kPa
1.84	27.0	55.1	1.40×10^2	1.40×10^3	1.40×10^3	1.86×10^5	186
0.946	13.9	28.3	71.9	719	719	9.59×10^4	95.9
0.959	14.1	28.7	72.9	729	729	9.72×10^4	97.2
0.984	14.5	29.4	74.8	748	748	9.97×10^4	99.7
1.03	15.2	30.9	78.5	785	785	1.05×10^5	105
0.163	2.40	4.88	12.4	124	124	1.65×10^4	16.5
1.16	17.1	34.8	88.5	885	885	1.18×10^5	118
0.902	13.3	27.0	68.6	686	686	9.14×10^4	91.4

(12) 752 torr + 284 torr = 1036 torr

(13) Reducing volume increases pressure and breaks balloon.

(14) $5.83\text{ L} \times \frac{2.18\text{ atm}}{5.03\text{ atm}} = 2.53\text{ L}$

(15) $3.19\text{ L} \times \frac{664\text{ torr}}{529\text{ torr}} = 4.00\text{ L}$

(16) $959\text{ torr} \times \frac{1.91\text{ L}}{(1.91 + 2.45)\text{L}} = 420\text{ torr}$

(18) Yes—water freezes at 0°C. 273 − 8 = 265 K

(19) 801 + 273 = 1074 K

(20) 246 − 273 = −27°C. You can't swim in ice.

(21) $1.20\text{ L} \times \frac{313\text{ K}}{288\text{ K}} = 1.30\text{ L}$

(22) $14.2\text{ m}^3 \times \frac{291\text{ K}}{315\text{ K}} = 13.1\text{ m}^3$

(23) $4.26\text{ atm} \times \frac{315\text{ K}}{292\text{ K}} = 4.60\text{ atm}$

(24) $(355 + 14.7)\text{ psi} \times \frac{296\text{ K}}{255\text{ K}} = 429\text{ psi abs.}$ $429 - 14.7 = 414\text{ psi gauge}$

(26) $8.42\text{ L} \times \frac{725\text{ torr}}{760\text{ torr}} \times \frac{273\text{ K}}{308\text{ K}} = 7.12\text{ L}$

(27) $6.29\text{ L} \times \frac{1\text{ atm}}{1.86\text{ atm}} \times \frac{238\text{ K}}{273\text{ K}} = 2.95\text{ L}$

(28) $1.00\text{ L} \times \frac{844\text{ torr}}{748\text{ torr}} \times \frac{287\text{ K}}{408\text{ K}} = 0.794\text{ L}$

(29) $0.140\text{ m}^3 \times \frac{(125 + 14.7)\text{psi}}{751\text{ torr}} \times \frac{760\text{ torr}}{14.7\text{ psi}} \times \frac{286\text{ K}}{306\text{ K}} = 1.26\text{ m}^3$

(30) $P = \frac{nRT}{V} = \frac{23.5\text{ mol}}{9.81\text{ L}} \times \frac{0.0821\text{ L} \cdot \text{atm}}{\text{mol} \cdot \text{K}} \times 296\text{ K} = 58.2\text{ atm}$

(31) $V = \frac{gRT}{(MM)P} = \frac{28.6\ g\ SO_2}{850\ torr} \times \frac{62.4\ L \cdot torr}{mol \cdot K} \times \frac{1\ mol\ SO_2}{64.1\ g\ SO_2} \times 313\ K = 10.3\ L\ SO_2$

(32) $n = \frac{PV}{RT} = \frac{1.62\ atm}{290\ K} \times \frac{mol \cdot K}{0.0821\ L \cdot atm} \times 5.24\ L = 0.357\ mol$

(33) $T = \frac{(MM)PV}{gR} = \frac{40.0\ g\ Ar}{1\ mol\ Ar} \times \frac{1\ L}{10.3\ g\ Ar} \times \frac{mol \cdot K}{0.0821\ L \cdot atm} \times 6.43\ atm = 304\ K = 31°C$

(34) $g = \frac{(MM)PV}{RT} = \frac{17.0\ g\ NH_3}{1\ mol} \times \frac{4.76\ atm}{298\ K} \times \frac{mol \cdot K}{0.0821\ L \cdot atm} \times 6.64\ L = 22.0\ g\ NH_3$

(35) $MM = \frac{gRT}{PV} = \frac{1.31\ g}{1\ L} \times \frac{293\ K}{749\ torr} \times \frac{62.4\ L \cdot torr}{mol \cdot K} = 32.0\ g/mol$

(36) $\frac{g}{V} = \frac{(MM)P}{RT} = \frac{29\ g}{mol} \times \frac{1\ atm}{298\ K} \times \frac{mol \cdot K}{0.0821\ L \cdot atm} = 1.2\ g/L$

(37) $MM = \frac{gRT}{PV} = \frac{0.625\ g}{L} \times \frac{1373\ K}{1.10\ atm} \times \frac{0.0821\ L \cdot atm}{mol \cdot K} = 64.0\ g/mol$

$n = 64.0/32.1 = 2$ Molecular formula: S_2

(38)

Element	Grams	Moles	Mole ratio	Formula ratio	Simplest formula
C	85.7	7.14	1	1	
H	14.3	14.2	1.98	2	CH_2

$MM = \frac{gRT}{PV} = \frac{29.4\ g}{2.84\ atm} \times \frac{533\ K}{3.60\ L} \times \frac{0.0821\ L \cdot atm}{mol \cdot K} = 126\ g/mol$

$n = 126/14 = 9 \qquad (CH_2)_9 = C_9H_{18}$

(40) $28.4\ g\ C_3H_8 \times \frac{22.4\ L\ C_3H_8}{44.0\ g\ C_3H_8} = 14.5\ L\ C_3H_8$

(41) $1.50\ L\ N_2 \times \frac{28.0\ g\ N_2}{22.4\ L\ N_2} = 1.88\ L\ N_2$

(42) $\frac{20.2\ g\ Ne}{22.4\ L\ Ne} = 0.902\ g/L$

(43) $\frac{1.63\ g}{1\ L} \times \frac{22.4\ L}{1\ mol} = 36.5\ g/mol$

(44) $\frac{2.63\ g}{2.10\ L} \times \frac{22.4\ L}{1\ mol} = 28.1\ g/mol$

(45) $8.55\ g\ NaHCO_3 \times \frac{1\ mol\ NaHCO_3}{84.0\ g\ NaHCO_3} \times \frac{1\ mol\ CO_2}{1\ mol\ NaHCO_3} \times \frac{0.0821\ L \cdot atm}{mol \cdot K} \times \frac{598\ K}{0.949\ atm} = 5.27\ L\ CO_2$

(46) $CH_4 + 2\ O_2 \rightarrow CO_2 + 2\ H_2O$

$35.0\ L\ CH_4 \times \frac{mol \cdot K}{62.4\ L \cdot torr} \times \frac{749\ torr}{295\ K} \times \frac{1\ mol\ CO_2}{1\ mol\ CH_4} \times \frac{44.0\ g\ CO_2}{1\ mol\ CO_2} = 62.7\ g\ CO_2$

(47) $1.00 \times 10^3\ g\ CaCO_3 \cdot MgCO_3 \times \frac{1\ mol\ CaCO_3 \cdot MgCO_3}{184\ g\ CaCO_3 \cdot MgCO_3} \times \frac{2\ mol\ CO_2}{1\ mol\ CaCO_3 \cdot MgCO_3} \times \frac{62.4\ L \cdot torr}{mol \cdot K} \times \frac{498\ K}{825\ torr} = 409\ L\ CO_2$

(48) $1.39\ L\ NO_2 \times \frac{mol \cdot K}{62.4\ L \cdot torr} \times \frac{751\ torr}{296\ K} \times \frac{1\ mol\ Sn}{4\ mol\ NO_2} \times \frac{119\ g\ Sn}{1\ mol\ Sn} = 1.68\ g\ Sn$

$\frac{1.68}{3.54} \times 100 = 47.5\%\ Sn \qquad 52.5\%\ Pb$

(50) $155\ L\ O_2 \times \frac{2\ L\ NO_2}{1\ L\ O_2} = 3.10 \times 10^3\ L\ NO_2$

(51) $525 \text{ L CO} \times \frac{440 \text{ torr}}{160 \text{ torr}} \times \frac{296 \text{ K}}{1973 \text{ K}} \times \frac{1 \text{ L } O_2}{2 \text{ L CO}} = 108 \text{ L } O_2$

(53) $0.319 + 0.605 + 0.456 = 1.38$ atm

(54) $762 - 25 = 737$ torr

(109) True: a, b, c, f, h, j, k; false: d, e, g, i, l, m, n.

(110) Volume at start, 350 cm^3. Volume at end, $350 - 309 = 41$ cm^3. Compression ratio = 350/41 = 8.5.

(111) (a) 1.0 atm (b) $1.0 \text{ atm} + 60 \text{ psi} \times \frac{1 \text{ atm}}{15 \text{ psi}} = 5.0 \text{ atm}$

(c) $n = \frac{PV}{RT} = \frac{1.0 \text{ atm}}{295 \text{ K}} \times \frac{\text{mol} \cdot \text{K}}{0.0821 \text{ L} \cdot \text{atm}} \times 0.39 \text{ L} = 0.016 \text{ mol}$

(d) $n = \frac{PV}{RT} = \frac{1.0 \text{ atm}}{295 \text{ K}} \times \frac{\text{mol} \cdot \text{K}}{0.0821 \text{ L} \cdot \text{atm}} \times 1.5 \text{ L} = 0.062 \text{ mol}$

(e) $n = \frac{PV}{RT} = \frac{5.0 \text{ atm}}{295 \text{ K}} \times \frac{\text{mol} \cdot \text{K}}{0.0821 \text{ L} \cdot \text{atm}} \times 1.5 \text{ L} = 0.31 \text{ mol}$

(f) 0.31 mol full − 0.062 mol empty = 0.25 mol to be added $0.25 \text{ mol} \times \frac{1 \text{ stroke}}{0.016 \text{ mol}} = 16 \text{ strokes}$

(g) With each stroke you are pumping against a higher pressure.

(112) Pressure when full: $1.0 \text{ atm} + 30 \text{ psi} \times \frac{1 \text{ atm}}{15 \text{ psi}} = 3.0 \text{ atm}$

$n(\text{full}) = \frac{PV}{RT} = \frac{3.0 \text{ atm}}{295 \text{ K}} \times \frac{\text{mol} \cdot \text{K}}{0.0821 \text{ L} \cdot \text{atm}} \times 41 \text{ L} = 5.1 \text{ mol}$

$n(\text{empty}) = \frac{PV}{RT} = \frac{1.0 \text{ atm}}{295 \text{ K}} = \frac{\text{mol} \cdot \text{K}}{0.0821 \text{ L} \cdot \text{atm}} \times 41 \text{ L} = 1.7 \text{ mol}$

$5.1 - 1.7 = 3.4$ mol added $3.4 \text{ mol} \times \frac{1 \text{ stroke}}{0.016 \text{ mol}} = 213 \text{ strokes}$

CHAPTER 14

(3) Substances with strong intermolecular attractions tend to have low vapor pressures. The strong attractions make the evaporation rate slow at a given temperature. A relatively low vapor concentration—and thus a low equilibrium vapor pressure—is therefore sufficient to make the condensation rate equal to the evaporation rate.

(5) Motor oil is more viscous than water. From this it may be predicted that intermolecular attractions are stronger in motor oil, as strong attractions lead to internal resistance to liquid flow, which is the property called viscosity.

(6) Surface tension is greater in mercury than in water, indicating stronger intermolecular attractions.

(7) Soap reduces intermolecular attractions so the soapy water is able to penetrate fabrics and cleanse them throughout.

(8) Gas.

(9) Of the three compounds, only N_2O is a liquid at $-90°C$, and therefore only N_2O possesses the property of viscosity. If a solid is considered more "viscous" than a liquid, NO_2 is the most viscous.

(11) HBr and NF_3: dipole; C_2H_2: dispersion; C_2H_5OH: hydrogen bonds.

(12) The high melting points of ionic compounds suggest correctly that ionic bonds are stronger than dipole forces. Both are electrical in character, ionic forces arising from a nearly complete transfer of electrons, and dipole forces from nonsymmetrical distribution of electrical charge within molecules.

(13) CCl_4, because it is larger, as suggested by its higher molecular weight.

(14) H_2S, because it is slightly more polar.

(15) Fluorine, oxygen, and nitrogen. The electronegativity difference between hydrogen and these elements is large enough to shift the bonding electron pair away from the hydrogen atom. If the molecule is polar, a hydrogen bond develops between the hydrogen atom of one molecule and the fluorine, oxygen, or nitrogen atom of another.

(16) (a) Dispersion forces; (b) dipole forces.

(17) (a), (c), (d).

(18) C_6H_{14}, a larger molecule with higher molecular weight than C_3H_8, will have stronger intermolecular attractions and therefore higher melting and boiling points.

(19) SO_2 molecules are larger than CO_2 molecules, and would therefore be expected to have stronger intermolecular attractions. The stronger the intermolecular attractions, the lower the vapor pressure. The prediction therefore is that CO_2 has the higher vapor pressure.

(21) Evaporation at constant rate begins immediately when liquid is introduced. At that time condensation rate is zero. Net rate of increase in vapor concentration is a maximum, so rate of vapor pressure increase is a maximum at start. Later condensation rate is more than zero, but less than evaporation rate. Net rate of increase in vapor concentration is less than initially, so rate of vapor pressure increase is less than initially. At equilibrium, evaporation and condensation rates are equal. Vapor concentration and therefore vapor pressure remain constant.

(22) All of the liquid evaporated before the vapor concentration was high enough to yield a condensation rate equal to the evaporation rate. At lower than equilibrium vapor concentration, the vapor pressure is lower than the equilibrium vapor pressure.

(23) Concentration is measured in moles/liter, n/V. Solving ideal gas equation for concentration, $\frac{n}{V} = \frac{p}{RT} = \left(\frac{1}{RT}\right)$ p. At constant temperature 1/RT is a constant, so pressure and concentration are proportional.

(24) (a) Second and third boxes have greatest pressure—the equilibrium pressure. (b) First box probably has least, having all evaporated before reaching equilibrium. Only possible exception is if vapor pressure just reached equilibrium pressure as last molecule evaporated in first box.

(25) 93 torr, the vapor pressure of the water at which equilibrium is reached.

(27) Gas, because vapor pressure is greater than surrounding pressure.

(28) High boiling liquids have strong intermolecular attractions, and therefore require high energy to escape from the liquid to form a gas. Evaporation rate is therefore slow, quickly equaled by condensation rate at low vapor concentration, or vapor pressure.

(29) More energy is required to vaporize X, so it would have the higher boiling point and lower vapor pressure.

(31) A, molecular; B, metallic.

(32) 29.3 kJ/6.04 g = 4.85 kJ/g

(33) $16 \text{ g Cu} \times \frac{4.81 \text{ kJ}}{\text{g}} = 77 \text{ kJ}$

(34) $18.3 \text{ kJ} \times \frac{1 \text{ g}}{0.371 \text{ kJ}} = 49.3 \text{ g}$

(35) $744 \text{ g } CCl_2F_2 \times \frac{1 \text{ mol } CCl_2F_2}{121 \text{ g } CCl_2F_2} \times \frac{35 \text{ kJ}}{1 \text{ mol } CCl_2F_2} = 2.2 \times 10^2 \text{ kJ}$

(36) $35.4 \text{ g Au} \times \frac{64.0 \text{ J}}{\text{g}} = 2.27 \times 10^3 \text{ J} = 2.27 \text{ kJ}$

(37) $\frac{7.08 \times 10^3 \text{ J}}{46.9 \text{ g}} = 151 \text{ J/g}$

(38) $11.3 \times 10^3 \text{ J} \times \frac{1 \text{ g}}{105 \text{ J}} = 108 \text{ g}$

(39) Horizontally, energy, or heat; vertically, temperature.

(40) D E F.

(41) G.

(42) The solid substance melts completely at constant temperature K.

(43) Energy = O − N.

(44) $Q_1 = 127\text{ g} \times \dfrac{2.1\text{ J}}{\text{g}\cdot{}^\circ\text{C}} \times (0 - 11)^\circ\text{C} = 2.9 \times 10^3\text{ J} \quad = 2.9\text{ kJ}$

$Q_2 = 127\text{ g} \times \dfrac{335\text{ J}}{\text{g}} = 4.25 \times 10^4\text{ J} = \quad 42.5\text{ kJ}$

$Q_3 = 127\text{ g} \times \dfrac{4.18\text{ J}}{\text{g}\cdot{}^\circ\text{C}} \times (21 - 0)^\circ\text{C} = 1.1 \times 10^4\text{ J} = \underline{11\text{ kJ}}$

$Q = Q_1 + Q_2 + Q_3 = \quad 56\text{ kJ}$

(45) $Q_1 = 25.1\text{ kg} \times \dfrac{0.452\text{ kJ}}{\text{kg}\cdot{}^\circ\text{C}} \times (1535 - 1645)^\circ\text{C} = -1.25 \times 10^3\text{ kJ}$

$Q_2 = 25.1\text{ kg} \times \dfrac{-267\text{ kJ}}{1\text{ kg}} = -6.70 \times 10^3\text{ kJ}$

$Q_3 = 25.1\text{ kg} \times \dfrac{0.444\text{ kJ}}{\text{kg}\cdot{}^\circ\text{C}} \times (33 - 1535) = -16.7 \times 10^3\text{ kJ}$

$Q = Q_1 + Q_2 + Q_3 = -24.7 \times 10^3\text{ kJ}$

(92) True: a, c, f, h, i, o; false: b, d, e, g, j, k, l, m, n, p, q.

(93) Heat lost by lemonade = Heat gained by ice. Let M = mass of ice.

$175\text{ g} \times \dfrac{4.18\text{ J}}{\text{g}\cdot{}^\circ\text{C}} \times (23 - 5)^\circ\text{C} = \text{Mg} \times \dfrac{2.1\text{ J}}{\text{g}\cdot{}^\circ\text{C}} \times 8^\circ\text{C} + \text{Mg} \times \dfrac{335\text{ J}}{\text{g}} + \text{Mg} \times \dfrac{4.18\text{ J}}{\text{g}\cdot{}^\circ\text{C}} \times 5^\circ\text{C}$

$M = 35\text{ g}$

(94) Dissolve the compounds and check for electrical conductivity. The ionic potassium sulfate solute will conduct, while the molecular sugar solute will not.

(95) Both molecules have dispersion and dipole–dipole forces. CH_3OH has hydrogen bonding and CH_3F does not. The molecules are about the same size. It is reasonable to predict stronger intermolecular forces in CH_3OH, and therefore higher boiling point and lower equilibrium vapor pressure.

(96) Without a regular and uniform structure in an amorphous solid, some intermolecular bonds are stronger than others. The weak bonds break at a lower temperature than the strong bonds.

(97) Large molecules having strong dispersion forces may have stronger intermolecular attractions than small molecules with hydrogen bonding, and therefore exhibit greater viscosity.

(98) As temperature drops, the equilibrium vapor pressure drops below the atmospheric vapor pressure. The air becomes first saturated, then supersaturated, and condensation (dew) begins to form.

CHAPTER 15

(4) If solute A is very soluble, its 10 grams per 100 grams of solvent concentration may be quite *dilute* compared to the possible concentration. If solute B is only slightly soluble, its 5 grams per 100 grams of solvent may be close to its maximum solubility, and therefore *concentrated*.

(5) Drop a small amount of solute into the solution. If the solution is unsaturated, the solute will dissolve; if saturated, it will simply settle to the bottom; if supersaturated, it will promote additional crystallization.

(6) Any quantity units of solute over any quantity units of solvent may be used to express solubility.

(9) Dissolving and crystallization must occur at equal rates for an equilibrium to exist. This can happen only if solute is in contact with a saturated solution. The solution could be poured off, however, leaving a saturated solution that is not in contact with solute.

(11) Carbon tetrachloride because both solute and solvent are nonpolar.

(12) Water is polar and has hydrogen bonding. The same is true of methanol, so methanol probably would not work. A nonpolar solvent is more apt to dissolve what a polar solvent does not. Cyclopentane is more promising.

(13) The statement is true only if the air contains carbon dioxide. The solubility of a gas in a liquid depends on the partial pressure of that gas over the liquid. It is independent of the partial pressures of other gases or of the total pressure.

(14) (18.5 g salt/135 g soln)100 = 13.7% salt.

(15) $65.0 \text{ g soln} \times \frac{13.0 \text{ g solute}}{100 \text{ g soln}} = 8.45 \text{ g solute.}$

(16) $\frac{20.0 \text{ g } C_{12}H_{22}O_{11}}{0.100 \text{ kg } H_2O} \times \frac{1 \text{ mol } C_{12}H_{22}O_{11}}{342 \text{ g } C_{12}H_{22}O_{11}} = 0.585 \text{ m}$

(17) $0.0800 \text{ kg } H_2O \times \frac{4.00 \text{ mol } CO(NH_2)_2}{1 \text{ kg } H_2O} \times \frac{60.0 \text{ g } CO(NH_2)_2}{1 \text{ mol } CO(NH_2)_2} = 19.2 \text{ g } CO(NH_2)_2$

(18) $90.9 \text{ g } HC_2H_3O_2 \times \frac{1 \text{ mol } HC_2H_3O_2}{60.0 \text{ g } HC_2H_3O_2} \times \frac{1000 \text{ mL } H_2O}{1.40 \text{ mol } HC_2H_3O_2} = 1.08 \times 10^3 \text{ mL}$

(19) $\frac{23.5 \text{ g } Na_2SO_4}{0.600 \text{ L}} \times \frac{1 \text{ mol } Na_2SO_4}{142 \text{ g } Na_2SO_4} = 0.276 \text{ M } Na_2SO_4$

(20) $\frac{1.2 \times 10^2 \text{ g } Na_2S_2O_3 \cdot 5 \; H_2O}{1.250 \text{ L}} \times \frac{1 \text{ mol } Na_2S_2O_3}{248 \text{ g } Na_2S_2O_3 \cdot 5 \; H_2O} = 0.387 \text{ M } Na_2S_2O_3$

(21) $0.400 \text{ L} \times \frac{0.800 \text{ mol } Na_2CO_3}{1 \text{ L}} \times \frac{106 \text{ g } Na_2CO_3}{1 \text{ mol } Na_2CO_3} = 33.9 \text{ g } Na_2CO_3$

(22) $0.750 \text{ L} \times \frac{0.600 \text{ mol } HC_2H_3O_2}{1 \text{ L}} \times \frac{60.0 \text{ g } HC_2H_3O_2}{1 \text{ mol } HC_2H_3O_2} = 27.0 \text{ g } HC_2H_3O_2$

(23) $0.0150 \text{ mol HCl} \times \frac{1 \text{ L}}{0.850 \text{ mol HCl}} = 0.0176 \text{ L} = 17.6 \text{ mL}$

(24) $75.0 \text{ g } NH_3 \times \frac{1 \text{ mol } NH_3}{17.0 \text{ g } NH_3} \times \frac{1 \text{ L}}{15 \text{ mol } NH_3} = 0.29 \text{ L}$

(25) $0.0650 \text{ L} \times \frac{2.20 \text{ mol NaOH}}{1 \text{ L}} = 0.143 \text{ mol NaOH}$

(26) $0.0293 \text{ L} \times \frac{0.482 \text{ mol } H_2SO_4}{1 \text{ L}} = 0.0141 \text{ mol } H_2SO_4$

(27) $\frac{0.100 \text{ L conc}}{2.00 \text{ L dil}} \times \frac{12 \text{ mol HCl}}{1 \text{ L conc}} = 0.60 \text{ M HCl}$

(28) $0.500 \text{ L dil} \times \frac{6.0 \text{ mol } NH_3}{\text{L dil}} \times \frac{1 \text{ L conc}}{15 \text{ mol } NH_3} = 0.20 \text{ L conc}$

(29) $\frac{18.0 \text{ g HCl}}{100 \text{ g soln}} \times \frac{1.09 \text{ g soln}}{0.001 \text{ L}} \times \frac{1 \text{ mol HCl}}{36.5 \text{ g HCl}} = 5.38 \text{ M HCl}$

(31) 1 eq/mol HF; 2 eq/mol $H_2C_2O_4$

(32) 2 eq/mol $Zn(OH)_2$; 1 eq/mol RbOH

(33) 40.0 g HF/eq; 90.0 g $H_2C_2O_4$/2 eq = 45.0 g $H_2C_2O_4$/eq

(34) 99.4 g $Zn(OH)_2$/2 eq = 49.7 g $Zn(OH)_2$/eq; 103 g RbOH/eq

(35) $\frac{17.2 \text{ g } HC_2H_3O_2}{0.300 \text{ L}} \times \frac{1 \text{ eq } HC_2H_3O_2}{60.0 \text{ g } HC_2H_3O_2} = 0.956 \text{ N } HC_2H_3O_2$

(36) $\frac{9.79 \text{ g } NaHCO_3}{0.500 \text{ L}} \times \frac{1 \text{ eq } NaHCO_3}{84.0 \text{ g } NaHCO_3} = 0.233 \text{ N } NaHCO_3$

(37) $0.600 \text{ L} \times \frac{2.00 \text{ eq KOH}}{1 \text{ L}} \times \frac{56.1 \text{ g KOH}}{1 \text{ eq KOH}} = 67.3 \text{ g KOH}$

(38) $0.250\text{ L} \times \frac{0.500\text{ eq }H_2C_2O_4}{1\text{L}} \times \frac{126\text{ g }H_2C_2O_4 \cdot 2\,H_2O}{2\text{ eq}} = 7.88\text{ g }H_2C_2O_4 \cdot 2\,H_2O$

(39) (a) 0.423 N HCl; (b) 0.826 N H_2SO_4

(40) $2.0\text{ L dil} \times \frac{0.5\text{ eq HCl}}{1\text{ L dil}} \times \frac{1\text{ L conc}}{12\text{ eq HCl}} = 0.083\text{ L} = 83\text{ mL}$

(41) $\frac{0.025\text{ L conc}}{0.400\text{ L dil}} \times \frac{15\text{ eq }HNO_3}{1\text{ L conc}} = 0.94\text{ N }HNO_3$

(42) $2.25\text{ L} \times \frac{0.871\text{ eq }H_2SO_4}{1\text{ L}} = 1.96\text{ eq }H_2SO_4$

(43) $0.0385\text{ eq HCl} \times \frac{1\text{ L}}{0.371\text{ eq HCl}} = 0.104\text{ L} = 104\text{ mL}$

(44) $AgNO_3 + NaCl \rightarrow AgCl + NaNO_3$

$$0.0500\text{ L} \times \frac{0.855\text{ mol }AgNO_3}{1\text{ L}} \times \frac{1\text{ mol AgCl}}{1\text{ mol }AgNO_3} \times \frac{144\text{ g AgCl}}{1\text{ mol AgCl}} = 6.16\text{ g AgCl}$$

(45) $Ba(NO_3)_2 + 2\,NaF \rightarrow BaF_2 + 2\,NaNO_3$

$$0.0400\text{ L} \times \frac{0.436\text{ mol NaF}}{1\text{ L}} \times \frac{1\text{ mol }BaF_2}{2\text{ mol NaF}} \times \frac{175\text{ g }BaF_2}{1\text{ mol }BaF_2} = 1.53$$

(46) $2\,NaOH + CuSO_4 \rightarrow Na_2SO_4 + Cu(OH)_2$

$$0.450\text{ L} \times \frac{0.125\text{ mol }CuSO_4}{1\text{ L}} \times \frac{1\text{ mol }Cu(OH)_2}{1\text{ mol }CuSO_4} \times \frac{97.6\text{ g }Cu(OH)_2}{1\text{ mol }Cu(OH)_2} = 0.549\text{ g }Cu(OH)_2$$

$$0.250\text{ L} \times \frac{0.350\text{ mol NaOH}}{1\text{ L}} \times \frac{1\text{ mol }Cu(OH)_2}{2\text{ mol NaOH}} \times \frac{97.6\text{ g }Cu(OH)_2}{1\text{ mol }Cu(OH)_2} = 0.427\text{ g }Cu(OH)_2\text{ precipitate}$$

(47) $0.0500\text{ L} \times \frac{1.20\text{ mol HCl}}{1\text{ L}} \times \frac{1\text{ mol }Cl_2}{4\text{ mol HCl}} \times \frac{22.4\text{ L }Cl_2}{1\text{ mol }Cl_2} = 0.336\text{ L }Cl_2$

(48) $Na_2CO_3 + 2\,HCl \rightarrow 2\,NaCl + H_2O + CO_2$

$$1.24\text{ g }Na_2CO_3 \times \frac{1\text{ mol }Na_2CO_3}{106\text{ g }Na_2CO_3} \times \frac{2\text{ mol HCl}}{1\text{ mol }Na_2CO_3} \times \frac{1\text{ L}}{0.715\text{ mol HCl}} = 0.0327\text{ L} = 32.7\text{ mL}$$

(49) $1.359\text{ g }KH(IO_3)_2 \times \frac{1\text{ mol }KH(IO_3)_2}{389.9\text{ g }KH(IO_3)_2} \times \frac{1\text{ mol KOH}}{1\text{ mol }KH(IO_3)_2} \times \frac{1}{0.03214\text{ L}} = 0.1084\text{ M KOH}$

(50) $H_2C_2O_4 + 2\,NaOH \rightarrow Na_2C_2O_4 + 2\,HOH$

(a) $\frac{3.290\text{ g }H_2C_2O_4 \cdot 2\,H_2O}{0.5000\text{ L}} \times \frac{1\text{ mol }H_2C_2O_4}{126.0\text{ g }H_2C_2O_4 \cdot 2\,H_2O} = 0.05222\text{ M }H_2C_2O_4$

(b) $0.02500\text{ L} \times \frac{0.05222\text{ mol }H_2C_2O_4}{1\text{ L}} \times \frac{2\text{ mol NaOH}}{1\text{ mol }H_2C_2O_4} \times \frac{1}{0.3010\text{ L}} = 0.08674\text{ M NaOH}$

(51) $0.0150\text{ L} \times \frac{0.100\text{ mol NaOH}}{1\text{ L}} \times \frac{1\text{ mol }H_2C_2O_4}{2\text{ mol NaOH}} \times \frac{126.0\text{ g }H_2C_2O_4 \cdot 2\,H_2O}{1\text{ mol }H_2C_2O_4} = 0.0945\text{ g }H_2C_2O_4 \cdot 2\,H_2O$

(52) $0.01680\text{ L} \times \frac{0.629\text{ mol }AgNO_3}{1\text{ L}} \times \frac{1\text{ mol }Cl^-}{1\text{ mol }AgNO_3} \times \frac{1}{0.02500\text{ L}} = 0.423\text{ M }Cl^-$

(53) $Na_2CO_3 + 2\,HCl \rightarrow 2\,NaCl + H_2O + CO_2$

$$41.24\text{ mL} \times \frac{0.244\text{ mmol HCl}}{1\text{ mL}} \times \frac{1\text{ mmol }Na_2CO_3}{2\text{ mmol HCl}} \times \frac{106\text{ mg }Na_2CO_3}{1\text{ mmol }Na_2CO_3} = 533\text{ mg }Na_2CO_3$$

$$\frac{533\text{ mg }Na_2CO_3}{694\text{ mg sample}} \times 100 = 76.8\%\ Na_2CO_3$$

(54) At 1 eq/mol, 0.1084 M KOH = 0.1084 N KOH

(55) (a) $\frac{3.290 \text{ g } H_2C_2O_4 \cdot 2\ H_2O}{0.5000 \text{ L}} \times \frac{2 \text{ eq } H_2C_2O_4}{126.0 \text{ g } H_2C_2O_4 \cdot 2\ H_2O} = 0.1044 \text{ N } H_2C_2O_4$

(b) $\frac{25.00 \times 0.1044}{30.10} = 0.08671$ N NaOH

(56) $N = \frac{15.0 \times 0.882}{12.8} = 1.03$ N acid

(57) $N = \frac{28.4 \times 0.424}{25.0} = 0.482 \text{ N } NiCl_2$

(58) $N = \frac{32.6 \times 0.208}{20.0} = 0.339 \text{ N } H_3PO_4$

(59) Using Equation 15.16, VN = eq, $\frac{0.631 \text{ g}}{0.0156 \times 0.562 \text{ eq}} = 72.0$ g/eq

(60) A colligative property of a solution is independent of the identity of the solute. Specific gravity, with opposite effects from different solutes, is not a colligative property.

(61) $\frac{50.0 \text{ g } C_6H_{12}O_6}{0.100 \text{ kg } H_2O} \times \frac{1 \text{ mol } C_6H_{12}O_6}{180 \text{ g } C_6H_{12}O_6} = 2.78$ m

$\Delta T_b = 0.52 \times 2.78 = 1.4°C \qquad T_b = 101.4°C$
$\Delta T_f = -1.86 \times 2.78 = -5.17 \qquad T_f = -5.17°C$

(62) $\frac{4.34 \text{ g } C_6H_4Cl_2}{0.0650 \text{ kg } C_{10}H_8} \times \frac{1 \text{ mol } C_6H_4Cl_2}{147 \text{ g } C_6H_4Cl_2} = 0.454$ m

$\Delta T_f = -6.9 \times 0.454 = -3.1°C \qquad T_f = 80.2 - 3.1 = 77.1°C$

(63) $m = \frac{\Delta T_b}{K_b} = \frac{0.84}{0.52} = 1.6$ m

(64) $\frac{26.0 \text{ g solute}}{0.380 \text{ kg}} = 68.4 \text{ g solute/kg } H_2O$; $m = \frac{-1.18}{-1.86} = 0.634 \qquad \frac{68.4 \text{ g}}{0.634 \text{ mol}} = 108$ g/mol

(65) $\frac{12.0 \text{ g solute}}{0.0800 \text{ kg } C_{10}H_8} = 150$ g/kg; $m = \frac{71.3 - 80.2}{-6.9} = 1.29\ (1.3) \qquad \frac{150}{1.29} = 116 = 1.2 \times 10^2$ g/mol

(66) $\frac{1.40 \text{ g } NH_2CONH_2}{0.0163 \text{ kg solvent}} \times \frac{1 \text{ mol } NH_2CONH_2}{60.0 \text{ g } NH_2CONH_2} = 1.43 \text{ m } NH_2CONH_2 \qquad K_b = \frac{3.92}{1.43} = 2.74$

(134) True: b, e, i (see Question 13), k, n; false: a, c, d, f, g, h, j, l, m.

(135) The graph will be like Figure 14.8 with dissolving rate in place of evaporation rate, and crystallization rate in place of condensation rate. (a) At equilibrium. (b) It is constant at constant temperature. (c) Time zero. (d) Assuming rates are per unit of surface area: (1) curve would stretch to right because more time would be needed to reach equilibrium; (2) curve would be compressed to left because less time would be needed to reach equilibrium; (3) curve would be flattened because both rates would become lower, with an unpredictable effect on time needed to reach equilibrium. (e) No effect. (f) Crystallization rate drops until concentration is built up to saturation as before. (g) Crystallization rate drops, all solute dissolves before reaching saturation, and both operations cease at time X.

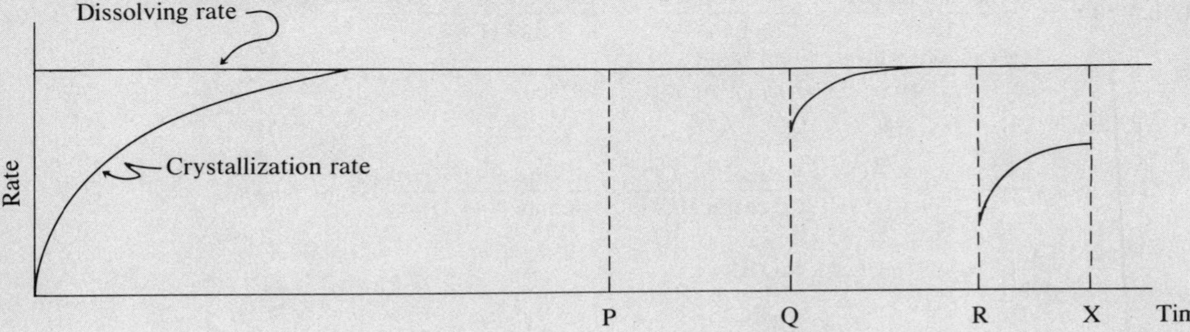

(136) The bubbles are dissolved air (nitrogen, oxygen) that become less soluble at higher temperatures.

(137) There is a boiling point elevation.

(138) $2\ KI + Pb(NO_3)_2 \rightarrow PbI_2 + 2\ KNO_3$

(a) $0.0200\ L \times \frac{0.530\ mol\ Pb(NO_3)_2}{1\ L} \times \frac{1\ mol\ PbI_2}{1\ mol\ Pb(NO_3)_2} \times \frac{461\ g\ PbI_2}{1\ mol\ PbI_2} = 4.88\ g\ PbI_2$

$0.0600\ L \times \frac{0.322\ mol\ KI}{1\ L} \times \frac{1\ mol\ PbI_2}{2\ mol\ KI} \times \frac{461\ g\ PbI_2}{1\ mol\ PbI_2} = 4.45\ g\ PbI_2$ precipitate

(b) Total volume $= 0.0600\ L + 0.0200\ L = 0.0800\ L$

$\frac{0.0600\ L\ KI(aq)}{0.0800\ L\ final} \times \frac{0.322\ mol\ KI}{1\ L\ KI(aq)} \times \frac{1\ mol\ K^+}{1\ mol\ KI} = 0.242\ M\ K^+$

(c) $0.0600\ L \times \frac{0.322\ mol\ KI}{1\ L} \times \frac{1\ mol\ PbI_2}{2\ mol\ KI} \times \frac{1\ mol\ Pb^{2+}}{1\ mol\ PbI_2} = 0.00966\ mol\ Pb^{2+}$ in ppt

$0.0200\ L \times \frac{0.530\ mol\ Pb(NO_3)_2}{1\ L} \times \frac{1\ mol\ Pb^{2+}}{1\ mol\ Pb(NO_3)_2} = 0.0106\ mol\ Pb^{2+}$ total

$\frac{0.0106 - 0.00966\ mol\ Pb^{2+}\ in\ solution}{0.0800\ L} = 0.01\ M\ Pb^{2+}$

(139) A small sample of pure air is a homogeneous mixture, and therefore a solution. The "atmosphere," even if it was pure air, is a very tall sample that becomes less dense at higher elevations. The atmosphere is therefore not homogeneous, and consequently it is not a solution.

(140) No.

(141) The density of a solution must be known to convert concentrations based on mass only (percentage, molality) to those based on volume (molarity, normality).

CHAPTER 16

(3) LiF: $Li^+(aq) + F^-(aq)$ $Mg(NO_3)_2$: $Mg^{2+}(aq) + 2\ NO_3^-(aq)$ $FeCl_3$: $Fe^{3+}(aq) + 3\ Cl^-(aq)$

(4) NH_4NO_3: $NH_4^+(aq) + NO_3^-(aq)$ $Ca(OH)_2$: $Ca^{2+}(aq) + 2\ OH^-(aq)$ Li_2SO_3: $2\ Li^+(aq) + SO_3^{2-}(aq)$

(5) $HCHO_2$: $HCHO_2(aq)$ HCl: $H^+(aq) + Cl^-(aq)$ $HC_4H_4O_6$: $HC_4H_4O_6(aq)$

(6) HI: $H^+(aq) + I^-(aq)$ H_2SO_4: $2\ H^+(aq) + SO_4^{2-}(aq)$ $H_3C_2H_5O_7$: $H_3C_2H_5O_7(aq)$

(7) $Ba(s) + Zn^{2+}(aq) \rightarrow Ba^{2+}(aq) + Zn(s)$

(8) No reaction.

(9) $3\ Mg(s) + 2\ Al^{3+}(aq) \rightarrow 3\ Mg^{2+}(aq) + 2\ Al(s)$

(10) $Ba^{2+}(aq) + CO_3^{2-}(aq) \rightarrow BaCO_3(s)$

(11) $Co^{2+}(aq) + 2\ OH^-(aq) \rightarrow Co(OH)_2(s)$

(12) $Fe^{2+}(aq) + S^{2-}(aq) \rightarrow FeS(s)$

(13) No reaction.

(14) $Mg^{2+}(aq) + 2\ F^-(aq) \rightarrow MgF_2(s)$

$3\ Zn^{2+}(aq) + 2\ PO_4^{3-}(aq) \rightarrow Zn_3(PO_4)_2(s)$

(15) $H^+(aq) + C_6H_5O_7^-(aq) \rightarrow HC_6H_5O_7(aq)$

(16) $H^+(aq) + OH^-(aq) \rightarrow HOH(\ell)$

(17) $H^+(aq) + C_4H_4O_6^-(aq) \rightarrow HC_4H_4O_6(aq)$

(18) $2\ H^+(aq) + CO_3^{2-}(aq) \rightarrow H_2O(\ell) + CO_2(g)$

(19) $NH_4^+(aq) + OH^-(aq) \rightarrow NH_3(aq) + HOH(\ell)$

(20) $2\ Ag^+(aq) + CO_3^{2-}(aq) \rightarrow Ag_2CO_3(s)$

(21) $Al^{3+}(aq) + PO_4^{3-}(aq) \rightarrow AlPO_4(s)$

(22) $Ca(s) + 2\ Ag^+(aq) \rightarrow Ca^{2+}(aq) + 2\ Ag(s)$

(23) $2\ H^+(aq) + SO_3^{2-}(aq) \rightarrow H_2O(\ell) + SO_2(aq)$

(24) $H^+(aq) + OH^-(aq) \rightarrow HOH(\ell)$

(25) $AgCl(s) + I^-(aq) \rightarrow AgI(s) + Cl^-(aq)$

(26) $2\ Li(s) + Cu^{2+}(aq) \rightarrow 2\ Li^+(aq) + Cu(s)$

(27) $H^+(aq) + C_7H_5O_2^-(aq) \rightarrow HC_7H_5O_2(aq)$

(28) $NH_4^+(aq) + OH^-(aq) \rightarrow NH_3(aq) + HOH(\ell)$

(29) $Ca^{2+}(aq) + 2\ F^-(aq) \rightarrow CaF_2(s)$

(30) $2\ K(s) + Zn^{2+}(aq) \rightarrow 2\ K^{+}(aq) + Zn(s)$

(31) $2\ H^{+}(aq) + Ba(OH)_2(s) \rightarrow 2\ HOH(\ell) + Ba^{2+}(aq)$

(32) $2\ H^{+}(aq) + Cu(OH)_2(s) \rightarrow 2\ HOH(\ell) + Cu^{2+}(aq)$

(33) $2\ Al(s) + 6\ OH^{-}(aq) \rightarrow 3\ H_2(g) + 2\ AlO_3^{3-}(aq)$

(34) $H_3PO_4(aq) + OH^{-}(aq) \rightarrow HOH(\ell) + H_2PO_4^{-}(aq)$
$H_3PO_4(aq) + 2\ OH^{-}(aq) \rightarrow 2\ HOH(\ell) + HPO_4^{2-}(aq)$

(70) True: b, c, d, i, j, k, l; false: a, e, f, g, h, m, n.

CHAPTER 17

(2) See Section 17.1 summary. The concepts are in agreement regarding acids, but not bases.

(4) As electron pair acceptors and donors, the hydrogen and hydroxide ions may be classified as a Lewis acid and a Lewis base, respectively.

(5)

```
       :C̈l:                          :C̈l:
        |                             |
  :C̈l—Al    +  [ :C̈l: ]⁻   →   [ :C̈l—Al—C̈l: ]⁻
        |                             |
       :C̈l:                          :C̈l:
```

$AlCl_3$ + Cl^- → $AlCl_4^-$

Aluminum in aluminum chloride is able to accept an electron pair, so it qualifies as a Lewis acid. The chloride ion can contribute the electron pair to the bond, so it is a Lewis base.

(6) HOH and H_2CO_3; H_2O and CO_3^{2-}.

(7) Forward: acid, HNO_2; base, CN^-. Reverse: acid, HCN; base NO_2^-.

(8) HNO_2 and NO_2^-; CN^- and HCN.

(9) HSO_4^- and SO_4^{2-}; $C_2O_4^{2-}$ and $HC_2O_4^-$.

(10) $H_2PO_4^-$ and HPO_4^{2-}; HCO_3^- and H_2CO_3.

(12) CO_3^{2-}; $H_2PO_4^-$; SO_4^{2-}; Br^-.

(13) HI; $H_2C_2O_4$; HSO_3^-; NH_4^+; H_2O.

(14) $HC_7H_5O_2(aq) + SO_4^{2-} \rightleftharpoons C_7H_5O_2^{-}(aq) + HSO_4^{-}(aq)$; reverse.

(15) $H_2C_2O_4(aq) + NH_3(aq) \rightleftharpoons HC_2O_4^{-}(aq) + NH_4^{+}(aq)$; forward.

(16) $H_3PO_4(aq) + CN^{-}(aq) \rightleftharpoons H_2PO_4^{-}(aq) + HCN(aq)$; forward.

(17) $H_2BO_3^{-}(aq) + NH_4^{+}(aq) \rightarrow H_3BO_3(aq) + NH_3(aq)$; reverse.

(18) $HPO_4^{2+}(aq) + HC_2H_3O_2(aq) \rightarrow H_2PO_4^{-}(aq) + C_2H_3O_2^{-}(aq)$; forward.

(19) Water ionizes very slightly and does not produce enough ions to light an ordinary conductivity device. With a sufficiently sensitive detector, water displays a very weak conductivity.

(22) $pH = 8$; $[H^+] = 10^{-8}$; $[OH^-] = 10^{-6}$; weakly basic.

(23) $pH = 1$; $pOH = 13$; $[OH^-] = 10^{-13}$; strongly acidic.

(24) $pOH = 2$; $pH = 12$; $[H^+] = 10^{-12}$; strongly basic.

(25) $pOH = 10$; $[OH^-] = 10^{-10}$; $[H^+] = 10^{-4}$; weakly acidic.

(26) $[H^+] = 2.4 \times 10^{-7}$; $[OH^-] = 4.2 \times 10^{-8}$; $pOH = 7.38$.

(27) $pOH = 10.96$; $pH = 3.04$; $[H^+] = 9.1 \times 10^{-4}$.

(28) $[OH^-] = 2.9 \times 10^{-6}$; $[H^+] = 3.5 \times 10^{-9}$; $pH = 8.46$.

(29) $pH = 1.14$; $pOH = 12.86$; $[OH^-] = 1.4 \times 10^{-13}$.

(60) True: a, b, c, e, f, g, h; false: d, i, j, k, l.

(61) $OH^- + NH_3 \rightarrow HOH + NH_2^-$

(63) $pCl = 7.126$

(64) When a proton is removed from an H_2X species, the single positive charge is being pulled away from a particle with a single minus charge, HX^-. When a proton is removed from a HX^-, the single positive charge is being pulled away from a particle with a double minus charge, X^{2-}. The loss of the second proton is more difficult, so the HX^- is a weaker acid than H_2X.

(65) There can be no proton transfer reaction without a proton—an H^+ ion.

(66) Carbonate ion is a proton acceptor: $H^+ + CO_3^{2-} \rightarrow HCO_3^-$.

(67) Arrhenius and B–L acids are limited to hydrogen ions, whereas anything with an empty valence orbital is a potential acid by the Lewis theory. The Lewis theory is the broadest of the three.

CHAPTER 18

(2) (a), (b), and (c), oxidation; (d), reduction.

(3) Oxidation.

(4) Reduction.

(5)

$$\begin{array}{r} Ni^{2} + 2e^- \rightarrow Ni \\ Mg \rightarrow Mg^{2+} + 2e^- \\ \hline Ni^{2+} + Mg \rightarrow Ni + Mg^{2+} \end{array}$$

(6)

$$\begin{array}{r} PbO_2 + SO_4^{2-} + 4H^+ + 2e^- \rightarrow PbSO_4 + 2\ H_2O \\ Pb + SO_4^{2-} \rightarrow PbSO_4 + 2e^- \\ \hline PbO_2 + 2\ SO_4^{2-} + 4\ H^+ + Pb \rightarrow 2\ PbSO_4 + 2\ H_2O \end{array}$$

(7) +2, −1, +1, +5.

(8) +5, −3, +7, +3.

(9) (a) Copper reduced from +2 to 0; (b) cobalt reduced from +3 to +2.

(10) (a) Sulfur oxidized from +4 to +6; (b) P oxidized from −3 to 0.

(11) (a) F oxidized from −1 to 0; (b) Mn reduced from +6 to +4.

(12) Hydrogen is the reducing agent, and copper oxide is the oxidizing agent.

(13) BrO_3^- is the oxidizing agent, HNO_2 the reducing agent.

(15) Ag^+ is a stronger oxidizer than H^+, based on their relative positions in Table 18.1. This means that Ag^+ has a stronger attraction for electrons than does H^+.

(16) Al, H_2, Fe^{2+}, Cl^-.

(17) $Ni + Zn^{2+} \rightleftharpoons Ni^{2+} + Zn$; reverse.

(18) $2\ Fe^{3+} + Co \rightleftharpoons 2\ Fe^{2+} + Co^{2+}$; forward.

(19) $\frac{1}{2}\ O_2 + 2\ H^+ + Ca \rightleftharpoons H_2O + Ca^{2+}$; forward.

(20)

$$\begin{array}{r} SO_4^{2-} + 4\ H^+ + 2e^- \rightarrow SO_2 + 2\ H_2O \\ (Ag \rightarrow Ag^+ + e^-)2 \\ \hline SO_4^{2-} + 4\ H^+ + 2\ Ag \rightarrow SO_2 + 2\ H_2O + 2\ Ag^+ \end{array}$$

(21)

$$\begin{array}{r} NO_3^- + 10\ H^+ + 8e^- \rightarrow NH_4^+ + 3\ H_2O \\ (Zn \rightarrow Zn^{2+} + 2e^-)4 \\ \hline NO_3^- + 10\ H^+ + 4\ Zn \rightarrow NH_4^+ + 3\ H_2O + 4\ Zn^{2+} \end{array}$$

(22)

$$\begin{array}{r} Cr_2O_7^{2-} + 14\ H^+ + 6e^- \rightarrow 2\ Cr^{3+} + 7\ H_2O \\ (Fe^{2+} \rightarrow Fe^{3+} + e^-)6 \\ \hline Cr_2O_7^{2-} + 14\ H^+ + 6\ Fe^{2+} \rightarrow 2\ Cr^{3+} + 7\ H_2O + 6\ Fe^{3+} \end{array}$$

(23)

$$\begin{array}{r} (MnO_4^- + 4\ H^+ + 3e^- \rightarrow MnO_2 + 2\ H_2O)2 \\ (2\ I^- \rightarrow I_2 + 2e^-)3 \\ \hline 2\ MnO_4^- + 8\ H^+ + 6\ I^- \rightarrow 2\ MnO_2 + 4\ H_2O + 3\ I_2 \end{array}$$

(24)

$$\begin{array}{r} 2\ BrO_3^- + 12\ H^+ + 10e^- \rightarrow Br_2 + 6\ H_2O \\ (2\ Br^- \rightarrow Br_2 + 2e^-)5 \\ \hline 2\ BrO_3^- + 12\ H^+ + 10\ Br^- \rightarrow 6\ Br_2 + 6\ H_2O \\ BrO_3^- + 6\ H^+ + 5\ Br^- \rightarrow 3\ Br_2 + 3\ H_2O \end{array}$$

(50) True: a, c, g; false: b, d, e, f.

(51) This "property of an acid" is more correctly described as the property of an acid (hydrogen ion) acting as an oxidizing agent. The H^+ ion reacts with only those metals whose ions are weaker oxidizing agents, located above hydrogen in Table 18.1.

(52) Water is available in large amounts in any aqueous solution, as is H^+ in an acidic solution.

(53) Figure 16.2 is an electrolytic cell, in which the "electromotive force" that moves the charges through the circuit is *outside* the cell. Figure 18.1 is a voltaic cell that is the *source* of the electromotive force that moves the charged particles.

(54) Copper is the cathode, and zinc is the anode. The positively charged ion, the *ca*tion moves toward the negatively charged *ca*thode, and the *an*ion moves toward the *an*ode. In all electrolytic systems, oxidation occurs at the anode, and reduction at the cathode.

(55) In Figure 16.2 the electrons are flowing toward the positively charged pole of the dry cell. It is the removal of electrons from the anode of the electrolytic cell by the dry cell that gives the anode its positive charge.

(56) In simple element ↔ monatomic ion redox reaction, the statement is correct. The *element* oxidized or reduced can always be identified by a change in oxidation number. The oxidizing or reducing agent, however, is a *species,* which may be an element, a monatomic ion, or a polyatomic ion, such as MnO_4^-.

CHAPTER 19

(3) Only a small portion of the fluoride ion present will be converted to HF, or most of the HF will be changed to F^-.

(5) ΔE is negative; the reaction is exothermic. $\Delta E = c - b$. Activation energy $= a - b$.

(6) The activation energy is greater for the reverse reaction: $a - c$.

(7) See text for general discussion of activation energy. All other things being equal, the reaction with the lower activation energy will be faster because a larger fraction of reacting particles will be able to engage in reaction-producing collisions.

(10) Rate varies directly as reactant concentration. As A concentration increases, rate increases; as B concentration decreases, rate decreases.

(11) At the beginning, when both reactants are at highest concentration.

(12) Nitrogen and hydrogen concentrations decrease, ammonia increases.

(13) Reverse shift to use up some of the added reactant in reverse direction.

(14) Forward direction to replenish some of the product that was removed.

(15) Reverse shift to use up some of the added reactant in reverse direction.

(16) Reverse shift to reduce total number of gaseous molecules and thereby reduce the pressure that had been created by reduced volume.

(17) Forward direction to increase total number of gaseous molecules and thereby increase the pressure that had been reduced by larger volume.

(18) Reverse direction to use up some of the added energy.

(19) Heat the system to produce a shift in forward direction.

(20) (a) and (d), forward; (b) and (c), reverse.

(21) $\dfrac{[SO_3]^2}{[SO_2]^2[O_2]}$

(22) $\dfrac{[CH_4][H_2S]^2}{[H_2]^4[CS_2]}$

(23) $[Cd^{2+}][OH^-]^2$

(24) $\dfrac{[H^+][NO_2^-]}{[HNO_2]}$

(25) $\dfrac{[Ag^+][CN^-]^2}{[Ag(CN)_2^-]}$

(27) The equilibrium constant is very small, so the reaction is favored in the reverse direction.

(28) $HC_2H_3O_2 \rightleftharpoons H^+ + C_2H_3O_2^-$. The equilibrium constant, $\dfrac{[H^+][C_2H_3O_2^-]}{[HC_2H_3O_2]}$, will be small. A weak acid ionizes only slightly, producing very small concentrations of the numerator species compared to the almost unchanged concentration of the denominator species.

(29) (a) Forward, with large K. (b) Reverse, with small K.

(30) $[Cd^{2+}] = [S^{2-}] = 8.8 \times 10^{-14}$ $\quad K_{sp} = (8.8 \times 10^{-14})^2 = 7.7 \times 10^{-27}$

(31) $\dfrac{1.0 \times 10^{-3}\text{ g CuBr}}{0.100\text{ L}} \times \dfrac{1\text{ mol CuBr}}{144\text{ g CuBr}} = 6.9 \times 10^{-5}\text{ M CuBr}$

$[Cu^+] = [Br^-] = 6.9 \times 10^{-5}$ $\quad K_{sp} = (6.9 \times 10^{-5})^2 = 4.8 \times 10^{-9}$

(32) $[Ca^{2+}] = [CO_3^{2-}] = s \qquad s^2 = 8.7 \times 10^{-9} \qquad s = 9.3 \times 10^{-5}$

$$0.100\ L \times \frac{9.3 \times 10^{-5}\ mol\ CaCO_3}{L} \times \frac{100\ g\ CaCO_3}{1\ mol\ CaCO_3} = \frac{9.3 \times 10^{-4}\ g\ CaCO_3}{100\ mL}$$

(33) $BaF_2 \rightleftharpoons Ba^{2+} + 2\ F^- \qquad K_{sp} = [Ba^{2+}][F^-]^2 = 1.7 \times 10^{-6}$
solubility $= s = [Ba^{2+}] \qquad (s)(2s)^2 = 4\ s^3 = 1.7 \times 10^{-6}$
$[F^-] = 2s \qquad s = 7.5 \times 10^{-3}$

(34) $K_{sp} = [Ba^{2+}][CO_3^{2-}] = 0.10[CO_3^{2-}] = 8.1 \times 10^{-9}$
solubility $= [CO_3^{2-}] = 8.1 \times 10^{-8}$

(35) $[H^+] = 10^{-1.93} = 0.012 = [C_3H_5O_3^-]$
$K_a = [H^+][C_3H_5O_3^-]/[HC_3H_5O_3] = (0.012)^2/1.0 = 1.4 \times 10^{-4}$
$(0.012/1.0) \times 100 = 1.2\%$ ionized

(36) $[H^+] = \sqrt{K_a[HA]} = \sqrt{(4.6 \times 10^{-4})(0.1)} = 6.8 \times 10^{-3}$
pH = 2.2 (answer rounded off to 1 sig. fig. to match 0.1)

(37) $[H^+] = K_a \times \frac{[HNO_2]}{[NO_2^-]} = 4.6 \times 10^{-4} \times \frac{0.75}{0.25} = 1.38 \times 10^{-3} \qquad pH = 2.86$

(38) $\frac{[HC_7H_5O_2]}{[C_7H_5O_2^-]} = \frac{[H^+]}{K_a} = \frac{10^{-4.80}}{6.5 \times 10^{-5}} = 0.24$

(39) 0.058 mol Cl_2 at start
<u>0.031</u> mol Cl_2 at end
0.027 mol Cl_2 used = 0.027 mol PCl_2 used
= 0.027 mol PCl_5 formed

0.069 mol PCl_3 at start
<u>0.027</u> mol PCl_3 used
0.042 mol PCl_3 at end

$$K = \frac{[PCl_5]}{[PCl_3][Cl_2]} = \frac{0.027}{(0.031)(0.042)} = 21$$

(80) True: b, d, f, g, i, j, l, m, n, o; false: a, c, e, h, k, p.

(81) Kinetic energies are greater at Time 1 because at Time 2 some of that energy has been converted to potential energy of activated complex.

(82) Increase [AB], heat the system, and introduce catalyst.

(83) (a) High pressure to force reaction to the smaller number of gaseous product molecules. (b) High temperature, at which all reaction rates are faster.

(84) A manufacturer cannot use an equilibrium, which is a closed system from which no product can be removed.

(85) The higher temperature is used to speed the reaction rate to an acceptable level. Lower pressure is dictated by limits of mechanical design and safety.

(86) $$\frac{1.6 \times 10^{-52}\ mol\ HgS}{1\ L} \times \frac{233\ g\ HgS}{1\ mol\ HgS} \times \frac{1\ lb}{454\ g} \times \frac{1\ L}{1000\ cm^3} \times \frac{2.54^3\ cm^3}{1\ in^3} \times \frac{12^3\ in^3}{1\ ft^3} \times \frac{5280^3\ ft^3}{1\ mile^3} = 2.7 \times 10^{-14}\ lb\ HgS$$

(87) (a) $Ca(OH)_2 \rightleftharpoons Ca^{2+} + 2\ OH^-$.
(b) (1) Adding a strong base or soluble calcium compound would increase $[OH^-]$ and $[Ca^{2+}]$, respectively, causing a shift in the reverse direction and reducing solubility of $Ca(OH)_2$. (2) Adding an acid to reduce $[OH^-]$ by forming water; adding a cation that will reduce $[OH^-]$ by precipitation; or adding an anion whose calcium salt is less soluble than $Ca(OH)_2$ would cause a shift in the forward direction, increasing the solubility of $Ca(OH)_2$.
(c) (1) Any anion whose calcium salt is less soluble than $Ca(OH)_2$ will cause a forward shift, increasing $[OH^-]$. (2) An acid that will form water with OH^- or a cation that will precipitate OH^- will reduce $[OH^-]$.

(88) Add NO_2: R–I–I–D–I. Reduce temperature: F–D–D–I–I.
Add N_2: None. Remove NH_3: R–D–I–D–D. Add catalyst: None.

CHAPTER 20

(7) $^{212}_{82}Pb \rightarrow {}^{212}_{83}Bi + {}^{0}_{-1}e$; $^{231}_{90}Th \rightarrow {}^{231}_{91}Pa + {}^{0}_{-1}e$

(8) $^{228}_{90}Th \rightarrow {}^{224}_{88}Ra + {}^{4}_{2}He$; $^{222}_{86}Rn \rightarrow {}^{218}_{84}Po + {}^{4}_{2}He$

(9) The half-life of a radioactive substance is the time required for half of the sample to decay. The fraction of a sample remaining after the passage of six half-lives is $(1/2)^6$, or $1/64$.

(10) (a) $n = 1/5.2 = 0.192$ half-lives. $0.5^{0.192} = 0.875 = 88\%$ remains. $100 - 88 = 12\%$ lost.
(b) $R = X \times 0.5^n = 125 \times 0.5^{18/5.2} = 11$ g remain.

(11) $138/227 = 0.608$ remains after 0.71 half-lives (from graph). 20.0 yr/0.71 HL = 28 years per half-life.

(12) $1.9 \times 10^4/7.1 \times 10^4 = 0.268$ remains after 1.90 half-lives (from graph) $1.90\ \cancel{\text{HL}} \times \dfrac{3.8\ \text{days}}{1\ \cancel{\text{HL}}} = 7.2$ days

(14) UCl_4. Because of the lower molar weight of UCl_4, there are more moles of uranium in 100 grams of UCl_4 than in 100 grams of UBr_4. Only the radioactive element, uranium, contributes to radioactivity. The radioactivity of 0.10 mole of UCl_4 will be the same as that of 0.10 mole of UBr_4, inasmuch as both samples contain the same number of moles of uranium.

(15) Lead is the stable end product of natural radioactive decay series. It is constantly being produced in natural radioactivity.

(16) Nuclear bombardment involves directing a nuclear particle to strike another nucleus, producing a nuclear reaction.

(19) $^{99}_{43}Tc$ **(20)** $^{239}_{94}Pu$ **(21)** $^{59}_{27}Co$

(52) True: a, d, f, g, j, k; false: b, c, e, h, i, l, n. The answer to (m) is left to you.

(54) The mass values used in the calculations must be masses of the radioactive isotope only, or some masses directly proportional to them. If the radioactive isotope is in a compound, there will be a changing mass of the radioactive isotope, a fixed mass of stable isotopes of the same element, and a fixed mass of other elements in the compound, as well as the mass of the decay products in all mass measurements. The amount of radioactive substance can be measured directly with a Geiger counter in the form of disintegrations per second, or some such quantity, as indicated in Problem 12.

(56) Presumably it takes an infinite time for all of a sample of radioactive matter to decay.

(57) $$1\ \text{lb} \times \frac{454\ \text{g}}{1\ \text{lb}} \times \frac{1\ \text{mol U}}{238\ \text{g U}} \times \frac{2.0 \times 10^{10}\ \text{kJ}}{1\ \text{mol U}} \times \frac{1\ \text{ton coal}}{2.5 \times 10^{7}\ \text{kJ}} = 1.5 \times 10^{3}\ \text{tons coal}$$

CHAPTER 21

(5)

```
     H  H  H  H  H  H  H  H  H
     |  |  |  |  |  |  |  |  |
  H—C—C—C—C—C—C—C—C—C—H
     |  |  |  |  |  |  |  |  |
     H  H  H  H  H  H  H  H  H
```

$CH_3CH_2CH_2CH_2CH_2CH_2CH_2CH_2CH_3$ or $CH_3(CH_2)_7CH_3$

(6) Of the compounds listed (a) and (e) are isomers, both having the molecular formula C_9H_{20}.

(8)

```
              |2  |1
             —C—C—
              |   |
  |   |   |4  |3
 —C—C—C—C—
  |   |   |   |
          |5  |6  |7  |8
         —C—C—C—C—
          |   |   |   |
         —C—
          |
```

Octane. It is possible to count out an eight-carbon chain.

```
            |
           —C—
     |  |  |  |  |  |  |
(9) —C—C—C—C—C—C—C—
     |  |  |  |  |  |  |
```

```
            |    |
           —C—  —C—
      |  |  |  |  |  |  |  |
(10) —C—C—C—C—C—C—C—C—
      |  |  |  |  |  |  |  |
```

(11) 3-ethylpentane. **(12)** 3-chloroheptane. **(13)** 1-bromo-4,5-dichlorohexane.

```
      Br
      |  |
(14) —C—C—
      |  |
      Br
```

```
      Br       Br Br
      |  |  |  |  |
(15) —C—C—C—C—C—
      |  |  |  |  |
      Br
```

```
      Br Br
      |  |  |  |
(16) —C—C—C—C—
      |  |  |  |
      Cl Br
```

(18) H—C≡C—H; acetylene, or ethyne.

```
          H              H  H           H         H
          |              |  |           |         |
H—C≡C—C—H     H—C≡C—C—C—H     H—C—C≡C—C—H
          |              |  |           |         |
          H              H  H           H         H
   Propyne          1-butyne               2-butyne
```

(19) The double bond is between the first and second carbons in 1-hexene, the second and third in 2-hexene, and the third and fourth in 3-hexene.

(21) *cis*-2-pentene.

```
        H  H  H  H  H  H
        |  |  |  |  |  |
(22) H—C—C—C—C—C=C   1-hexene
        |  |  |  |     |
        H  H  H  H     H
```

```
        H  H  H  H  H
        |  |  |  |  |
(23) H—C—C—C—C=C   1-pentene
        |  |  |     |
        H  H  H     H
```

(24) A double or triple bond must be present in order for an addition reaction to occur.

(25) $C_3H_6 + H_2 \rightarrow C_3H_8$

```
     H  Cl    H  Cl    H  Cl      H  Cl H  Cl H  Cl
     |  |     |  |     |  |       |  |  |  |  |  |
(27) C=C  +  C=C  +  C=C  →  —C—C—C—C—C—C—
     |  |     |  |     |  |       |  |  |  |  |  |
     H  H     H  H     H  H       H  H  H  H  H  H
```

(28) Aromatic compounds have benzene ring structures; aliphatic compounds have open chain structures.

(29) (a) and (c), *m*-dichlorobenzene or 1,3-dichlorobenzene; (b) *p*-dichlorobenzene, or 1,4-dichlorobenzene.

```
        H  H  H  H  H
        |  |  |  |  |
(31) H—C—C—C—C—C—OH
        |  |  |  |  |
        H  H  H  H  H
```

(32) In a secondary alcohol, the carbon to which the hydroxyl group is attached is bonded to two other carbon atoms, giving a minimum of three carbon atoms in the molecule.

```
   H  H  H  H  H              H  H  H  H  H
   |  |  |  |  |              |  |  |  |  |
H—C—C—C—C—C—H          H—C—C—C—C—C—H
   |  |  |  |  |              |  |  |  |  |
   H  O  H  H  H              H  H  O  H  H
      |                             |
      H                             H
```

```
       O           O           O
      / \         / \         / \
(33) H   H       R   H       R   R'
     Water       Alcohol      Ether
```

```
          O
         / \
(34) CH₃   C₃H₇
```

CH_3–O–C_3H_7

(35) $$\mathrm{R{-}\overset{\displaystyle H}{\underset{\displaystyle H}{\overset{|}{\underset{|}{C}}}}{-}OH} + \tfrac{1}{2}\,O_2 \rightarrow \mathrm{R{-}\underset{\displaystyle H}{\underset{|}{C}}{=}O} + H_2O$$

(38) Carboxyl groups are polar, and hydrogen bonding is present. This leads to relatively strong intermolecular attractions and therefore high boiling points.

(39) $RCOOH \rightarrow RCOO^- + H^+$

(41) Methylpropylamine.

(86) True: b, c, d, g, h, j, k, n, o, p; false: a, e, f, i, l, m.

(87) The carbon atoms form a continuous chain, but at the tetrahedral angle, the chain is not straight.

(88) C_8H_{16}—C_9H_{18}—alkenes; C_5H_{12}—C_6H_{14}—alkanes.

(89) (a) planar; (b) linear; (c) irregular zig-zag.

(90) The molecule is not planar because of the tetrahedral arrangement around the —CH_3 carbon.

(91) Alcohols are most apt to be soluble in water because their structures are more like that of water. The molecules are polar, and hydrogen bonding is present.

(92) The H—O bond is stronger in CH_3COOH, a weak acid, than in H_2SO_4, a strong acid.

(93) $$\begin{array}{c} \mathrm{H} \\ | \\ \mathrm{H{-}C{-}O{-}H} \\ | \\ \mathrm{H} \end{array} + \begin{array}{c} \mathrm{\quad\;\; H\;\;\; H\;\;\; H} \\ \mathrm{\quad\;\; |\;\;\;\; |\;\;\;\; |} \\ \mathrm{H{-}O{-}C{-}C{-}C{-}H} \\ \mathrm{\quad\;\; |\;\;\;\; |\;\;\;\; |} \\ \mathrm{\quad\;\; H\;\;\; H\;\;\; H} \end{array} \rightarrow \begin{array}{c} \mathrm{\;\; H\quad\;\;\; H\;\;\; H\;\;\; H} \\ \mathrm{\;\; |\quad\;\;\;\; |\;\;\;\; |\;\;\;\; |} \\ \mathrm{H{-}C{-}O{-}C{-}C{-}C{-}H} \\ \mathrm{\;\; |\quad\;\;\;\; |\;\;\;\; |\;\;\;\; |} \\ \mathrm{\;\; H\quad\;\;\; H\;\;\; H\;\;\; H} \end{array} + \mathrm{HOH}$$

(94) $$\mathrm{CH_3{-}\overset{\displaystyle H}{\underset{\displaystyle OH}{\overset{|}{\underset{|}{C}}}}{-}C_2H_5}$$

2-butanol, a secondary alcohol.

(95) $$\mathrm{CH_3{-}\overset{\displaystyle H}{\underset{\displaystyle NH_2}{\overset{|}{\underset{|}{C}}}}{-}\overset{\displaystyle O}{\overset{\|}{C}}{-}\underset{\displaystyle H}{\underset{|}{N}}{-}\overset{\displaystyle H}{\underset{\displaystyle H}{\overset{|}{\underset{|}{C}}}}{-}COOH}$$

Glossary

absolute zero—the temperature predicted by extrapolation of experimental data where translational kinetic energy theoretically becomes zero; the zero of the absolute temperature scale, which is equivalent to $-273.15°C$.

acid—a substance that yields hydrogen (hydronium) ions in aqueous solution (Arrhenius definition); a substance that donates protons in chemical reaction (Brönsted–Lowry definition); a substance that forms covalent bonds by accepting a pair of electrons (Lewis definition).

acidic solution—an aqueous solution in which the hydrogen-ion concentration is greater than the hydroxide-ion concentration; a solution in which the pH is less than 7.

actinides—elements 90 (Th) through 103 (Lr).

activated complex—an intermediate molecular species presumed to be formed during the interaction (collision) of reacting molecules in a chemical change.

activation energy—the energy barrier that must be overcome to start a chemical reaction.

alcohol—an organic compound consisting of an alkyl group and at least one hydroxyl group, having the general formula ROH.

aldehyde—a compound consisting of a carbonyl group bonded to a hydrogen on one side, and a hydrogen, alkyl, or aryl group on the other, having the general formula RCHO.

aliphatic hydrocarbon—an alkane, alkene or alkyne.

alkaline—basic; having pH greater than 7.

alkaline earth metal—a metal from Group 2A of the periodic table.

alkali metal—a metal from Group 1A of the periodic table.

alkane—a saturated hydrocarbon containing only single bonds, in which each carbon atom is bonded to four other atoms.

alkene—an unsaturated hydrocarbon containing a double bond, and each carbon atom that is double bonded is bonded to a maximum of three atoms.

alkyl group—an alkane hydrocarbon group lacking one hydrogen atom, having the general formula C_nH_{2n+1}, and frequently symbolized by the letter R.

alkyne—an unsaturated hydrocarbon containing a triple bond, in which each carbon atom that is triple-bonded is bonded to a total of two atoms.

alpha (α) particle—the nucleus of a helium atom, often emitted in nuclear disintegration.

amide—a derivative of a carboxylic acid in which the hydroxyl group is replaced by a $—NH_2$ group, and having the general formula $RCONH_2$.

amine—an ammonia derivative in which one or more hydrogens are replaced by an alkyl group.

amorphous—a substance that is without definite structure or shape.

amphiprotic, amphoteric—a substance that can act as an acid or a base.

angstrom—a length unit equal to 10^{-10} m.

anhydride (anhydrous)—a substance that is without water, or from which water has been removed.

anion—a negatively charged ion.

anode—the electrode at which oxidation occurs in an electrochemical cell.

aqueous—pertaining to water.

aromatic hydrocarbon—a hydrocarbon containing a benzene ring.

atmosphere (pressure unit)—a unit of pressure based on atmospheric pressure at sea level, and capable of supporting a mercury column 760 mm high.

atom—the smallest particle of an element that can combine with atoms of other elements in forming chemical compounds.

atomic mass—*see atomic weight.*

atomic mass unit (amu)—a unit of mass that is exactly $1/12$ of the mass of an atom of carbon-12.

atomic number (Z)—the number of protons in an atom of an element.

atomic weight—the number that expresses that average mass of the atoms of an element compared to the mass of an atom of carbon-12 at a value of exactly

12; the average mass of the atoms of an element expressed in atomic mass units.

Avogadro's number—the number of carbon atoms in exactly 12 grams of carbon-12; the number of units in 1 mole (6.02×10^{23}).

barometer—laboratory device for measuring atmospheric pressure.

base—a substance that yields hydroxide ions in aqueous solution (Arrhenius definition); a substance that accepts protons in chemical reaction (Brönsted–Lowry definition); a substance that forms covalent bonds by donating a pair of electrons (Lewis definition).

basic solution—an aqueous solution in which the hydroxide ion concentration is greater than the hydrogen ion concentration; a solution in which the pH is greater than 7.

beta (β) particle—a high energy electron, often emitted in nuclear disintegration.

binary compound—a compound consisting of two elements.

boiling point—the temperature at which vapor pressure becomes equal to the pressure above a liquid; the temperature at which vapor bubbles form spontaneously anyplace within a liquid.

boiling point elevation—the difference between the boiling point of a solution and the boiling point of the pure solvent.

bombardment (nuclear)—the striking of a target nucleus by an atomic particle, causing a nuclear change.

bond—*see chemical bond.*

bond angle—the angle formed by the bonds between two atoms that are bonded to a common central atom.

bonding electrons—the electrons transferred or shared in forming chemical bonds; valence electrons.

buffer—a solution that resists a change in pH.

calorie—a unit of heat equal to 4.184 joules.

calorimeter—laboratory device for measuring heat flow.

carbonyl group—an organic functional group, $>C{=}O$, characteristic of aldehydes and ketones.

carboxyl group—an organic functional group, $-C(=O)-O-H$, characteristic of carboxylic acids.

carboxylic acid—an organic acid containing the carboxyl group, having the general formula RCOOH.

catalyst—a substance that increases the rate of a chemical reaction by lowering activation energy. The catalyst is either a nonparticipant in the reaction, or it is regenerated. *See inhibitor.*

cathode—the negative electrode in a cathode ray tube; the electrode at which reduction occurs in an electrochemical cell.

cation—a positively charged ion.

cell, electrolytic—a cell in which electrolysis occurs as a result of an externally applied electrical potential.

cell, galvanic—*see cell, voltaic.*

cell, voltaic—a cell in which an electrical potential is developed by a spontaneous chemical change. Also called a *galvanic cell.*

chain reaction—a reaction that has, as a product, one of its own reactants; that product becomes a reactant, thereby allowing the original reaction to continue.

charge cloud—*see electron cloud.*

chemical bond—a general term that sometimes includes all of the electrostatic attractions among atoms, molecules, and ions, but more often refers to covalent and ionic bonds. *See covalent bond, ionic bond.*

chemical change—a change in which one or more substances disappear and one or more new substances are formed.

chemical family—a group of elements having similar chemical properties because of similar valence electron configuration, appearing in the same column of the periodic table.

chemical properties—the types of chemical change a substance is able to experience.

cloud chamber—a device in which condensation tracks form behind radioactive emissions as they travel through a supersaturated vapor.

colligative properties—physical properties of mixtures that depend upon concentration of particles irrespective of their identity.

colloid—a nonsettling dispersion of aggregated ions or molecules intermediate in size between the particles in a true solution and those in a suspension.

combustion—the process of burning.

compound—a pure substance that can be broken down into two or more other pure substances by a chemical change.

concentrated—adjective for a solution with a relatively large amount of solute per given quantity of solvent or solution.

condense—to change from a vapor to a liquid or solid.

condensation—the act of condensing.

conjugate acid–base pair—a Brönsted–Lowry acid and the base derived from it when it loses a proton; or a Brönsted–Lowry base and the acid developed from it when it accepts a proton.

coordinate covalent bond—in which both bonding electrons are furnished by only one of the bonded atoms.

coulomb—a unit of electrical charge.

covalent bond—the chemical bond between two atoms that share a pair of electrons.

crystalline solid—a solid in which the ions and/or molecules are arranged in a definite geometric pattern.

decompose—to change chemically into simpler substances.

density—the mass of a substance per unit volume.

diatomic—that which has two atoms.

dilute—adjective for a solution with a relatively small amount of solute per given quantity of solvent or solution.

dipole—a polar molecule.

diprotic acid—an acid capable of yielding two protons per molecule in complete ionization.

dispersion forces—weak electrical attractions between molecules, temporarily produced by the shifting of electrons within molecules.

dissolve—to pass into solution.

distillation—the process of separating components of a mixture by boiling off and condensing the more volatile component.

dynamic equilibrium—a state in which opposing changes occur at equal rates, resulting in zero net change over a period of time.

electrode—conductor by which electric charge enters or leaves an electrolyte.

electrolysis—passage of electric charge through an electrolyte.

electrolyte—a substance which, when dissolved, yields a solution that conducts electricity; a solution or other medium that conducts electricity by ionic movement.

electrolytic cell—*see cell, electrolytic.*

electron—subatomic particle carrying a unit negative charge and having a mass of 9.1×10^{-28} gram, or 1/1837 of the mass of a hydrogen nucleus, found outside the nucleus of the atom.

electron (charge) cloud—region of space around or between atoms that is occupied by electrons.

electron configuration—the orbital arrangement of electrons in ions or atoms.

electron-dot diagram (structure)—*see Lewis diagram.*

electronegativity—a scale of the relative ability of an atom of one element to attract the electron pair that forms a single covalent bond with an atom of another element.

electron orbit—the circular or elliptical path supposedly followed by an electron around an atomic nucleus, according to the Bohr theory of the atom.

electron orbital—a mathematically described region in space within an atom in which there is a high probability that an electron will be found.

electron-pair geometry—a description of the distribution of bonding and unshared electron pairs around a bonded atom.

electron-pair repulsion—the principle that electron-pair geometry is the result of repulsion between electron pairs around a bonded atom, causing them to be as far apart as possible.

electrostatic force—force of attraction or repulsion between electrically charged objects.

element—a pure substance that cannot be decomposed into other pure substances by ordinary chemical means.

empirical formula—a formula that represents the lowest integral ratio of atoms of the elements in a compound.

endothermic—a change that absorbs energy from the surroundings, having a positive ΔH, an increase in enthalpy.

energy—the ability to do work.

enthalpy—the heat content of a chemical system.

enthalpy of reaction—*see heat of reaction.*

equilibrium—*see dynamic equilibrium.*

equilibrium constant—with reference to an equilibrium equation, the ratio in which the numerator is the product of concentrations of the species on the right side of the equation, each raised to a power corresponding to its coefficient in the equation, and the denominator is the corresponding product of the species on the left side of the equation; symbol: K, K_c, or K_{eq}.

equilibrium vapor pressure—*see vapor pressure.*

equivalent—that quantity of an acid (or base) that yields or reacts with one mole of H^+ (or OH^-) in a chemical reaction; that quantity of a substance that gains or loses one mole of electrons in a redox reaction.

ester—an organic compound formed by the reaction between a carboxylic acid and an alcohol, having the general formula R—CO—OR′.

ether—an organic comound in which two alkyl groups are bonded to the same oxygen, having the general formula R—O—R′.

excited state—the state of an atom in which one or more electrons have absorbed energy—become "excited''—to raise them to energy levels above ground state.

exothermic reaction—a reaction that gives off energy to its surroundings.

family—*see chemical family.*

fission—a nuclear reaction in which a large nucleus splits into two smaller nuclei.

formula, chemical—a combination of chemical symbols and subscript numbers that represents the elements in a pure substance and the ratio in which the atoms of the different elements appear.

formula unit—a real (molecular) or hypothetical (ionic) unit particle represented by a chemical formula.

formula weight—the weight in amu of one formula unit of a substance; the molar weight of formula units of a substance.

fractional distillation—separation of a mixture into fractions whose components boil over a given temperature range.

freezing point depression—the difference between the freezing point of a solution and the freezing point of the pure solvent.

fusion—the process of melting; also, a nuclear reaction in which two small nuclei combine to form a larger nucleus.

galvanic cell—*see cell, voltaic.*

gamma (γ) ray—a high-energy photon emission in radioactive disintegration.

Geiger counter—an electrical device for detecting and measuring the intensity of radioactive emission.

ground state—the state of an atom in which all electrons occupy the lowest possible energy levels.

group (periodic table)—the elements comprising a vertical column in the periodic table.

half-life ($t_{1/2}$)—the time required for the disintegration of one half of the radioactive atoms in a sample.

half-reaction—the oxidation or reduction half of an oxidation–reduction reaction.

halide ion—F^-, Cl^-, Br^-, or I^-.

halogen—the name of the chemical family consisting of fluorine, chlorine, bromine, and iodine; any member of the halogen family.

heat of fusion (solidification)—the heat flow when one gram of a substance changes between a solid and a liquid at constant pressure and temperature. *See also molar heat of fusion (solidification).*

heat of reaction—change of enthalpy in a chemical reaction.

heat of vaporization (condensation)—the heat flow when one gram of a substance changes between a liquid and a vapor at constant pressure and temperature.

heterogeneous matter—matter having a nonuniform composition, usually with visibly different parts or phases.

homogeneous matter—matter having a uniform appearance and uniform properties throughout.

homologous series—a series of compounds in which each member differs from the one next to it by the same structural unit.

hydrate—a crystalline solid that contains water of hydration.

hydrocarbon—an organic compound consisting of carbon and hydrogen.

hydrogen bond—an intermolecular bond (attraction) between a hydrogen atom in one molecule and a highly electronegative atom (fluorine, oxygen, or nitrogen) of another polar molecule; the polar molecule may be of the same subtance containing the hydrogen, or a different substance.

hydronium ion—a hydrated hydrogen ion, H_3O^+.

hydroxyl group—an organic functional group, —OH, characteristic of alcohols.

ideal gas—a hypothetical gas that behaves according to the ideal gas model over all ranges of temperature and pressure.

ideal gas equation—the equation $PV = nRT$ that relates quantitatively the pressure, volume, quantity, and temperature of an ideal gas.

immiscible—insoluble (usually used only in reference to liquids).

indicator—a substance that changes from one color to another, used to signal the end of a titration.

inhibitor—a substance added to a chemical reaction to retard its rate; sometimes called a negative catalyst.

ion—an atom or group of covalently bonded atoms that is electrically charged because of an excess or deficiency of electrons.

ion-combination reaction—when two solutions are combined, the formation of a precipitate or molecular compound by a cation from one solution and an anion from the second solution.

ionic bond—the chemical bond arising from the attraction forces between oppositely charged ions in an ionic compound.

ionic compound—a compound in which ions are held by ionic bonds.

ionic equation—a chemical equation in which dissociated ionic compounds are shown in ionic form.

ionization—the formation of an ion from a molecule or atom.

ionization energy—the energy required to remove an electron from an atom or ion.

isoelectronic—having the same electron configuration.

isomers—two compounds having the same molecular formulas, but different structural formulas and different physical and chemical properties.

isotopes—two or more atoms of the same element that have different atomic masses because of different numbers of neutrons.

IUPAC—International Union of Pure and Applied Chemistry.

joule—the SI energy unit, defined as a force of one newton applied over a distance of one meter; 1 joule = 0.239 calorie.

K—the symbol for the kelvin, the absolute temperature unit; the symbol for an equilibrium constant. K_a is the constant for the ionization of a weak acid; K_{sp} is the constant for the equilibrium between a slightly soluble ionic compound and a saturated solution of its ions; K_w is the constant for the ionization of water.

Kelvin temperature scale—an absolute temperature scale on which the degrees are the same size as Celsius degrees, with 0 K at absolute zero, or −273.15°C.

ketone—a compound consisting of a carbonyl group bonded on each side to an alkyl group, having the general formula R—CO—R′.

kinetic energy—energy of motion; translational kinetic energy is equal to ½ mass × (velocity)^2.

kinetic molecular theory—the general theory that all matter consists of particles in constant motion, with different degrees of freedom distinguishing among solids, liquids, and gases.

kinetic theory of gases—that portion of the kinetic molecular theory that describes gases and from which the model of an ideal gas is developed.

lanthanides—elements 58 (Ce) through 71 (Lu).

Le Chatelier's Principle—if an equilibrium system is subjected to a change, processes occur that tend to counteract partially the initial change, thereby bringing the system to a new position of equilibrium.

Lewis diagram, structure, or **symbol**—a diagram representing the valence electrons and covalent bonds in an atomic or molecular species.

limiting reagent—the reactant first totally consumed in a reaction, thereby determining the maximum yield possible.

line spectrum—the spectral lines that appear when light emitted from a sample is analyzed in a spectroscope.

macromolecular crystal—a crystal made up of a large but indefinite number of atoms covalently bonded to each other to form a huge molecule.

manometer—a laboratory device for measuring gas pressure.

mass—a property reflecting the quantity of matter in a sample.

mass number—the total number of protons plus neutrons in the nucleus of an atom.

mass spectroscope—a laboratory device whereby a flow of gaseous ions may be analyzed in regard to their charge and/or mass.

matter—that which occupies space and has mass.

metal—a substance that possesses metallic properties, such as luster, ductility, malleability, good conductivity of heat, and electricity; an element that loses electrons to form monatomic cations.

miscible—soluble (usually used only in reference to liquids).

mixture—a sample of matter containing two or more pure substances.

molality—solution concentration expressed in moles of solute per kilogram of solvent.

molar heat of fusion (solidification)—the heat flow when one mole of a substance changes between a solid and a liquid at constant temperature and pressure.

molar heat of vaporization (condensation)—the heat flow when one mole of a substance changes between a liquid and a vapor at constant temperature and pressure.

molarity—solution concentration expressed in moles of solute per liter of solution.

molar volume—the volume occupied by one mole, usually of a gas.

molar weight—the mass of one mole of any substance.

mole—that quantity of any species that contains the same number of units as the number of atoms in exactly 17 grams of carbon-12.

molecular compound—a compound whose fundamental particles are molecules rather than ions.

molecular crystal—a molecular solid in which the molecules are arranged according to a definite geometric pattern.

molecular geometry—a description of the shape of a molecule.

molecular weight (mass)—the number that expresses the average mass of the molecules of a compound compared to the mass of an atom of carbon-12 at a value of exactly 12; the average mass of the molecules of a compound expressed in atomic mass units.

molecule—the smallest unit particle of a pure substance

that can exist independently and possess the identity of the substance.

monatomic—that which has only one atom.

monomer—the individual chemical structural unit from which a polymer may be developed.

monoprotic acid—an acid capable of yielding one proton per molecule in complete ionization.

negative catalyst—*see inhibitor.*

net ionic equation—an ionic equation from which all spectators have been removed.

neutralization—the reaction between an acid and a base to form a salt and water; any reaction between an acid and a base.

neutron—an electrically neutral subatomic particle having a mass of 1.7×10^{-24} gram, approximately equal to the mass of a proton, or 1 atomic mass unit, found in the nucleus of the atom.

newton—SI unit of force, equal to $1 \text{ kg} \cdot \text{m}^2/\text{sec}^2$.

noble gas—the name of the chemical family of relatively unreactive elemental gases appearing in Group 0 of the periodic table.

nonelectrolyte—a substance which, when dissolved, yields a solution that is a nonconductor of electricity; a solution or other fluid that does not conduct electricity by ionic movement.

nonpolar—pertaining to a bond or molecule having a symmetrical distribution of electric charge.

normal boiling point—the temperature at which a substance boils in an open vessel at one atmosphere pressure.

normality—solution concentration in equivalents per liter.

nucleus—the extremely dense central portion of the atom that contains the neutrons and protons which constitute nearly all the mass of the atom and all of the positive charge.

octet rule—the general rule that atoms tend to form stable bonds by sharing or transferring electrons until the atom is surrounded by a total of eight electrons.

orbit—*see electron orbit.*

orbital—*see electron orbital.*

organic chemistry—the chemistry of carbon compounds other than carbonates, cyanides, carbon monoxide, and carbon dioxide.

oxidation—chemical reaction with oxygen; a chemical change in which the oxidation number (state) of an element is increased; also, the loss of electrons in a redox reaction.

oxidation number—a number assigned to each element in a compound, ion, or elemental species by an arbitrary set of rules. Its two main functions are to organize and simplify the study of oxidation–reduction reactions and to serve as a base for one branch of chemical nomenclature.

oxidation state—*see oxidation number.*

oxidizer, oxidizing agent—the substance that takes electrons from another species, thereby oxidizing it.

oxyacid—an acid that contains oxygen.

oxyanion—an anion that contains oxygen.

partial pressure—the pressure one component of a mixture of gases would exert if it alone occupied the same volume as the mixture at the same temperature.

Pauli exclusion principle—the principle that says, in effect, that no more than two electrons can occupy the same orbital.

period (periodic table)—a horizontal row of the periodic table.

pH—a way of expressing hydrogen-ion concentration; the negative of the logarithm of the hydrogen-ion concentration.

phase—a visibly distinct part of a heterogeneous sample of matter.

physical change—a change in the physical form of a substance without changing its chemical identity.

physical properties—properties of a substance that can be observed and measured without changing the substance chemically.

pOH—a way of expressing hydroxide ion concentration; the negative logarithm of the hydroxide ion concentration.

polar—pertaining to a bond or molecule having an unsymmetrical distribution of electric charge.

polyatomic—pertaining to a species consisting of two or more atoms; usually said of polyatomic ions.

polymer—a chemical compound formed by bonding two or more monomers; frequently, in plastics, a huge macromolecule.

polymerization—the reaction in which monomers combine to form polymers.

polyprotic acid—an acid capable of yielding more than one proton per molecule on complete ionization.

potential energy—energy possessed by a body by virtue of its position in an attractive or repulsive force field.

precipitate—a solid that forms when two solutions are mixed.

pressure—force per unit area.

principal energy level(s)—the main energy levels within the electron arrangement in an atom. They are quantized by a set of integers beginning at n = 1 for the lowest level, n = 2 for the next, and so forth; also called the principal quantum number.

proton—a subatomic particle carrying a unit positive charge and having a mass of 1.7×10^{-24} gram, almost the same as the mass of a neutron, found in the nucleus of the atom.

pure substance—a sample consisting of only one kind of matter, either compound or element.

quantization of energy—the existence of certain discrete electron energy levels within an atom such that electrons may have any one of these energies, but no energy between two such levels.

quantum mechanical model of the atom—an atomic concept that recognizes four quantum numbers by which electron energy levels may be described.

R—a symbol used to designate any alkyl group; the ideal gas constant, having a value of 0.0821 L atm/mol K.

radioactivity—spontaneous emission of rays and/or particles from an atomic nucleus.

redox—a term coined from REDuction–OXidation to refer to oxidation–reduction reactions.

reducer, reducing agent—the substance that loses electrons to another species, thereby reducing it.

reduction—a chemical change in which the oxidation number (state) of an element is reduced; also, the gain of electrons in a redox reaction.

reversible reaction—a chemical reaction in which the products may react to re-form the original reactants.

salt—the product of a neutralization reaction other than water; an ionic compound containing neither the hydrogen ion, H^+, oxide ion, O^{2-}, nor hydroxide ion, OH^-.

saturated hydrocarbon—a hydrocarbon that contains only single bonds, in which each carbon atom is bonded to four other atoms.

saturated solution—a solution of such concentration that it is or would be in a state of equilibrium with excess solute present.

SI unit—a unit associated with the International System of Units.

significant figures—the digits in a measurement that are known to be accurate plus one doubtful digit.

soluble—a substance that will dissolve in a suitable solvent.

solubility—the quantity of solute that will dissolve in a given quantity of solvent, or in a given quantity of solution, at a specified temperature, to establish an equilibrium between the solution and excess solute; frequently expressed in grams of solute per 100 grams of solvent.

solubility product constant—*under K, see* K_{sp}.

solute—the substance dissolved in the solvent; sometimes not clearly distinguishable from the solvent (see below), but usually the lesser of the two.

solution—a homogeneous mixture of two or more substances of molecular or ionic particle size, the concentration of which may be varied, usually within certain limits.

solution inventory—a precise identification of the chemical species present in a solution, in contrast with the solute from which they may have come; i.e., sodium ions and chloride ions, rather than sodium chloride.

solvent—the medium in which the solute is dissolved; *see solute.*

specific gravity—the ratio of the density of a substance to the density of some standard, usually water at 4°C.

specific heat—the quantity of heat required to raise the temperature of one gram of a substance one degree Celsius.

spectator (ion)—a species present at the scene of a reaction but not a participant in it.

spectroscope—a laboratory instrument used to analyze spectra.

spectrum (*plural:* **spectra**)—the result of a dispersion of a beam of light into its component colors; also the result of a dispersion of a beam of gaseous ions into its component particles, distinguished by mass and electric charge.

spontaneous—a change that appears to take place by itself, without outside influence.

stable—that which does not change spontaneously.

standard temperature and pressure (STP)—arbitrarily defined conditions of temperature (0°C) and pressure (1 atmosphere) at which gas volumes and quantities are frequently measured and/or compared.

stoichiometry—the quantitative relationships between the substances involved in a chemical reaction, established by the equation for the reaction.

STP—abbreviation for standard temperature and pressure (see above).

strong acid—an acid that ionizes almost completely in aqueous solution; an acid that loses its protons readily.

strong base—a base that dissociates almost completely in aqueous solution; a base that has a strong attraction for protons.

strong electrolyte—a substance which, when dissolved,

yields a solution that is a good conductor of electricity because of nearly complete ionization or dissociation.

strong oxidizer (oxidizing agent)—an oxidizer that has a strong attraction for electrons.

strong reducer (reducing agent)—a reducer that releases electrons readily.

sublevel—the levels into which the principal energy levels are divided according to the quantum mechanical model of the atom; usually specified s, p, d, and f.

supersaturated—a state of solution concentration which is greater than the equilibrium concentration (solubility) at a given temperature and/or pressure.

suspension—a mixture that gradually separates by settling.

tetrahedral—related to a tetrahedron; usually used in reference to the orientation of four covalent bonds radiating from a central atom toward the vertices of a tetrahedron, or to the 109°28′ angle formed by any two corners of the tetrahedron and the central atom as its vertex.

tetrahedron—a regular four-sided solid, having congruent equilateral triangles as its four faces.

thermal—having to do with heat.

thermochemical equation—a chemical equation which includes an energy term, or for which ΔH is indicated.

thermochemical stoichiometry—stoichiometry expanded to include the energy involved in a chemical reaction, as defined by the thermochemical equation.

titration—the controlled and measured addition of one solution into another.

torr—a unit of pressure equal to the pressure unit millimeter of mercury.

transition element; transition metal—an element from one of the B groups or Group 8 of the periodic table.

transmutation—conversion of an atom from one element to another by means of a nuclear change.

transuranium elements—man-made elements whose atomic numbers are greater than 92.

triprotic acid—an acid capable of yielding three protons in complete ionization.

unsaturated hydrocarbon—a hydrocarbon that contains one or more multiple bonds.

valence electrons—the highest energy s and p electrons in an atom that determine the bonding characteristics of an element.

van der Waals forces—a general term for all kinds of weak intermolecular attractions.

vapor—a gas.

vaporize, vaporization—changing from a solid or liquid to a gas.

vapor pressure—the partial pressure exerted by a specific vapor component in a mixture of gases. Frequently refers to the partial pressure of a vapor that is in equilibrium with its liquid state at a given temperature.

voltaic cell—*see cell, voltaic.*

volatile—that which vaporizes easily.

water of crystallization, water of hydration—water molecules that are included as structural parts of crystals formed from aqueous solutions.

weak acid—an acid that ionizes only slightly in aqueous solution; an acid that does not donate protons readily.

weak base—a base that dissociates only slightly in aqueous solution; a base that has a weak attraction for protons.

weak electrolyte—a substance which, when dissolved, yields a solution that is a poor conductor of electricity because of limited ionization or dissociation.

weak oxidizer (oxidizing agent)—an oxidizer that has a weak attraction for electrons.

weak reducer (reducing agent)—a reducer that does not release electrons readily.

weight—a measure of the force of gravitational attraction of a body to the earth.

yield—the amount of product from a chemical reaction.

Z—atomic number.

Index